Optical Fiber Communication Systems

For a complete listing of the *Artech House Optoelectronics Library*,
turn to the back of this book.

Optical Fiber Communication Systems

Leonid Kazovsky
Sergio Benedetto
Alan Willner

Artech House
Boston • London

Library of Congress Cataloguing-in-Publication Data
Kazovsky, Leonid G.
Optical fiber communication systems/Leonid Kazovsky, Sergio Benedetto, Alan Willner.
p. cm.
Includes bibliographical references and index.
ISBN 0-89006-756-2 (alk. paper)
1. Optical communications. 2. Fiber optics. 3. Digital communications. I. Benedetto, Sergio. II. Willner, Alan E. III. Title.
TK5103.59.K39 1996
621.382'75—dc20 96-27860
CIP

British Library Cataloguing in Publication Data
Kazovsky, Leonid
Optical fiber communication systems
1. Optical fibers 2. Fiber optics 3. Optical communications I. Title II. Benedetto, Sergio III. Willner, Alan
621.3'8275

ISBN 0-89006-756-2

Cover design by Jennifer Makower

685 Canton Street
Norwood, MA 02062

International Standard Book Number: 0-89006-756-2
Library of Congress Catalog Card Number: 96-27860

10 9 8 7 6 5 4 3 2

To the memory of my mother, Frida Kazovsky, and my father, Grigori Kazovsky.

L.K.

To the memory of my dear father, Carlo Benedetto.

S.B.

To my loving wife and parents and to the memory of my grandparents.

A.E.W.

Contents

Introduction

In the 1960s, telephone network engineers in the United States and other developed countries faced the problem of the increasing traffic between central offices. The interoffice, or trunk, traffic was at that time frequently transmitted using the so-called T-carrier approach, which was introduced in 1962 [1]. The T-carrier approach used digitized voice channels having the bit rate of 64 Kbps per voice channel. Twenty-four digitized voice channels were time-division multiplexed to yield a composite 1.544-Mbps digital stream known as DS-1. DS-1 signals were carried by wire pair circuits installed in ducts under the streets; every 2 km, these signals had to be cleaned and reshaped (*repeated*).

The increasing voice traffic exhausted the capacity of the wire pair circuits; in many cases, there was no room for additional wires in the ducts. Thus, a new higher-capacity transmission medium was needed. Many alternatives were considered and investigated; several have been almost completely developed. One technology—fiber optics—dominated because of the superior qualities of the glass fiber (very low attenuation and dispersion); others, like coaxial cable and microwave, are also used today. Many other contenders have been abandoned altogether.

Thus, the optical fiber communications technology was developed to address a fairly narrow but well-defined problem of congestion in trunk traffic running at some 1.5 Mbps over a modest distance of 2 km. From that modest beginning, a powerful new optical fiber communication technology has evolved with capabilities far exceeding the original application. Today optical fiber systems carry many gigabits per second of traffic over many thousands of kilometers in trans-Atlantic and trans-Pacific applications. They do so without electronic repeaters, using instead optical amplifiers. Optical fiber systems are also used to distribute television programming, interconnect computers, and perform numerous other tasks.

The enormous progress of the optical fiber communication technology is unparalleled and is rivaled by only one field—computers. One can only speculate about the reasons underlying this success. It may be attributed to the basic physics (the very low intrinsic attenuation of glass, less than 0.2 dB/km) or to the extraordinary talent of the people who developed this technology, an outstanding group of physicists and communication engineers.

Once optical fiber communications became a widespread and important technology (in fact, the technology of choice in many applications), it became necessary to offer courses in that field in most universities. A number of good books have been written to support such courses, including [1–7]. Most current textbooks follow the (by now) traditional path, essentially repeating the history of technological developments in the field of optical fiber communications. A typical book deals in depth with optical fibers (perhaps some four chapters), lasers and LEDs (perhaps some three to five chapters), photodetectors and receivers (perhaps some four chapters) and adds a chapter or two to deal with systems. Newer textbooks (and newer editions of old textbooks) include more recent issues, like optical amplification, coherent detection, and soliton transmission. The resulting distribution of material tilts textbooks and their associated courses toward device- and physics-oriented students. At the same time, there is a great need to educate a much larger number of communication systems-oriented students and to provide them a working knowledge of optical fiber communications, including advanced technologies, without converting them into device scientists.

That need, coupled with the current trend toward more systems-oriented research and development, led to the writing of this textbook. Our textbook is designed for electrical engineering students and places much less emphasis on devices than earlier textbooks. Our textbook focuses on systems, especially on newer technologies: coherent detection, optical amplifiers, solitons, and multichannel systems and networks. The roots of this book go back to 1990, when Leonid Kazovsky left Bellcore and joined Stanford University as a professor of electrical engineering. He developed two new courses at Stanford, "Introduction to Optical Fiber Communications" and "Advanced Optical Fiber Communications." The former is a mezanine course for senior undergraduate students and graduate students; it is typically attended by some 50 students per year. The latter is an advanced course for graduate students; it is typically attended by some 15 students per year. Notes developed by Leonid Kazovsky for those two courses formed the foundation for this textbook.

This book was written by all three authors, and it takes its material from notes developed by each of the authors for teaching optical fiber communications courses at their respective schools. The resulting textbook is designed for electrical engineering students who want to specialize in communication systems engineering. This textbook may be used for a one-semester, two-semester, or two-quarter sequence in optical fiber communications.

The introductory course may cover all of Chapter 1 ("Basic Optical Fiber Communications Components"); all of Chapter 3 ("Basic Intensity Modulation/Direct Detection Systems"), and several sections from Chapter 4 ("Coherent Detection"), Chapter 5 ("Optical Amplifiers"), Chapter 6 ("Solitons"), and Chapter 7 ("Multichannel Systems"). We recommend the following sections of Chapters 4–7 for the introductory course: the first six sections of Chapter 4; Sections 5.1, 5.2, and 5.4; Sections 6.1 and 6.2; and Sections 7.1 and 7.2. Chapter 2 does not have to be covered in class if the students have a basic knowledge of random variables and random processes. If they do not, the introductory class should cover Sections 2.1, 2.2, 2.3, 2.4, and 2.8 before beginning Chapter 3.

The introductory course is a prerequisite for the advanced course. The advanced course may focus on coherent systems, optical amplifiers, solitons, and multichannel systems. We recommend the following sections of Chapters 4–7 for the advanced course: Sections 4.7–4.13; Sections 5.2, 5.3, 5.5, and 5.6; Sections 6.3–6.16; and Sections 7.2–7.8. Chapter 2 does not have to be covered in class if the students have a basic knowledge of random variables and random processes. If they do not, the advanced course should cover Sections 2.5, 2.6, and 2.7.

In a school that uses a quarter (rather than semester) system, the foregoing suggestions have to be modified to create shorter courses. Depending on his or her interests, the instructor may drop one or even two of the advanced chapters (Chapters 4–7).

Each chapter contains problems that can be used for homework assignments and exams. No solutions manual is available for the problems at this time.

This first edition undoubtedly contains bugs. The authors will be grateful to readers pointing out any bugs. Comments should be addressed to the publisher.

While the authors alone are responsible for the bugs in this book, they cannot take credit for its material. It is a textbook, not a monograph; a vast majority of its developments, figures, and other material were first published in various papers, conference talks, and other books. We tried to acknowledge the source of each item whenever it was known. In some cases, the material is "classic," that is, difficult to attribute to any single source. In such cases, we tried to indicate what other textbooks contain a similar discussion or figure. Any possible omissions are unintentional and, if discovered, will be corrected in future editions.

It would be impossible to create this book without the work of many talented scientists and engineers (unfortunately, too numerous to be listed here) and without interaction with and help of numerous friends, colleagues, and graduate students.

Leonid Kazovsky is particularly grateful to his Bellcore and Stanford friends and colleagues, especially Peter Kaiser, Nim Cheung, T. P. Lee, and Jim Gimlett of Bellcore and Joe Goodman and Bob Byer of Stanford. His work would be impossible without the generous help of U.S. government sponsors: Air Force, ONR, ARPA, NSF, BMDO, and SDC. The help of and collaboration with Brian Hendrickson (Air Force and ARPA), Bert Hui and Bob Leheny (ARPA), Lou Lome (BMDO), Max McCurry and K. Pathak (SDC), and Ken Davis and R. Madan (ONR) were extremely important. Linn Mollenauer of AT&T Bell Labs looked through an early draft of Chapter 6, provided useful comments, and supplied several reprints and preprints of his papers.

The interaction with and help of Stanford graduate students also were important, especially those of Thomas Fong, Michael Hickey, Robert Kalman, Pierluigi Poggiolini, Frank Yang, Allen Lu, Tad Hofmeister, Derek Mayweather, Sanjay Agrawal, Silviu Savin, and Steven Gemelos. Marli Williams, Jennifer Beltran, and Libbi Hendelsman worked hard to prepare the manuscript for publication. Last but not least, Leonid Kazovsky is grateful to his wife, Ilana, and his daughter, Galit, whose love and support made it possible for him to work on this book along with the numerous other responsibilities of a Stanford professor, principal investigator of a number of research projects, teacher, consultant, associate editor of several journals, author, and homeowner.

Sergio Benedetto wishes to express his appreciation to Pierluigi Poggiolini, a researcher at the Politecnico di Torino whose thesis in 1989 was the starting point for the foundation of a new group of Optical Communications in the Dipartimento di Elettronica. Fruitful interactions with graduate students have been very important in the genesis of the book. Particularly those with Andrea Carena, Vittorio Curri, and Roberto Gaudino, who also provided some numerical examples and drew the related figures. Emilio Casaccia carefully read a preliminary version of some chapters and contributed to their improvement with useful comments. He also helped to draw most of the figures in Chapters 2 and 4. Finally, Luciano Brino contributed with his drawing of the remaining figures. Sergio Benedetto wants to express his deep gratitude to his wife Fernanda, his daughters Mariachiara and Cecilia, and his son Giovanni, whose support, love, and understanding have greatly contributed to the realization of this project.

As is stated in *Ethics of our Fathers*, "Who is a wise person? He who learns from everyone." It is in this spirit that Alan Willner has many individuals to thank for their help and inspiration. Several of his Ph.D. students and one post-doctoral associate were instrumental in teaching him much insight about optical communications and for help with this manuscript: Imran Hayee, Dr. Syang-Myau Hwang, Dr. James Leight, Dr. David Norte, Dr. Eugene Park, Dr. William Shieh, and Dr. Xingyu Zou. Moreover, the many graduate students who have taken his course have helped refine his method of conveying the material. The key inspirational figures for this book are, of course, Alan's mentors: Prof. Richard Osgood, Jr., Dr. Ivan Kaminow, and Dr. Tingye Li. Maintaining an active research program while writing a book would be impossible if not for generous funding support from the NSF-sponsored Presidential Faculty Fellows and Young Investigator Award (Drs. Crawford, Harvey, and Ortega), the David and Lucile Packard Foundation Fellowship in Science and Engineering, ARPA (Drs. Hui, Leheny, and Yang), and BMDO and AFOSR (Drs. Craig and Lome). The environment at USC was quite supportive, and Alan is grateful to his colleagues and to the staff at USC (particularly Milly Montenegro). He would especially like to thank Joshua Davis for his tireless help with this manuscript. Alan hopes this book brings joy and pride to his parents, Gerald and Sondra Willner; may they live a long (i.e., 120 years) and happy life. Finally and most importantly, he would like to express his deep appreciation to his wife, Dr. Michelle Frida Green Willner, for her support, love, strength, insight, and incredible understanding during the preparation of this book.

Leonid Kazovsky, Stanford University, Stanford, CA
Sergio Benedetto, Politecnico di Torino, Torino, Italy
Alan Willner, University of Southern California, Los Angeles, CA

October 1996

References

[1] Personick, S. D., *Optical Transmission Systems*, New York: Plenum Press, 1981.
[2] Jones, W. B., *Introduction to Optical Fiber Communication Systems*, New York: Holt, Rinehart & Winston, 1988.
[3] Keiser, G., *Optical Fiber Communications*, 2d Ed., New York: McGraw-Hill, 1991.
[4] Palais, J. C., *Fiber Optic Communications*, 2d Ed., Englewood Cliffs, N. J.: Prentice-Hall, 1992.
[5] Gowar, J., *Optical Communication Systems*, 2d Ed., London: Prentice-Hall International, 1991.
[6] Senior, J., *Optical Fiber Communications: Principles and Practice*, 2d Ed., London: Prentice-Hall International, 1992.
[7] Agrawal, G. P., *Fiber-Optic Communications Systems*, New York: John Wiley & Sons, 1992.

Chapter 1

Basic Optical Fiber Communications Components

This chapter deals with the basic components of optical fiber communication systems: fibers, lasers, and photodiodes. The structure and the principle of operation of intensity-modulated direct-detection systems are also briefly discussed. The material in Chapters 1 and 3 is also available in a number of optical communications textbooks [1–8]. Our coverage is close to that of [1–8] but is condensed to free students' time for other topics.

1.1 INTRODUCTION

Basic optical fiber communication systems use *intensity modulation* (IM) and *direct detection* (DD). Virtually all currently deployed systems are IM-DD.

Figure 1.1 shows the basic IM-DD system. It consists of a light source, either a laser or a *light emitting diode* (LED); a fiber; and an optical receiver based on either PIN photodiode or an *avalanche photodiode* (APD). Most current optical systems are digital. A high-performance digital system might transmit at bit rates of many gigabits per second over long unrepeated distances. Such a system would typically use single-mode single-frequency lasers, single-mode fibers, and PIN based receivers.

When the distances and the bit rates involved are more modest, the system might employ LEDs and multimode fibers. While systems of that type are no longer installed in telecommunications networks, they are useful for computer interconnects, intracar communications, and other similar short-distance applications.

Still another class of IM/DD systems evolved in the late 1980s and early 1990s: analog systems. Such systems carry analog signals for cable television and antenna remoting.

Digital and analog applications need somewhat different component and system designs. The goals of digital system designers are a high bit rate, R_b, and a low *bit-error*

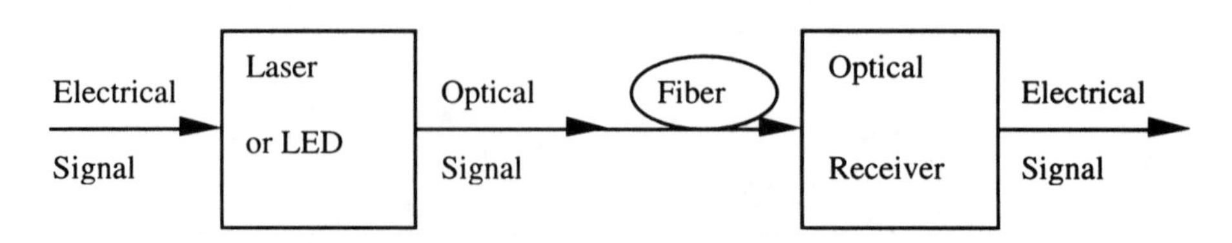

Figure 1.1 The basic IM/DD communication system.

ratio (BER). The goals of analog system designers are large bandwidth, B; low distortions[1]; large *signal-to-noise ratio* (SNR); and large dynamic range.

In this book, we focus our attention on digital systems. It takes a wide variety of specialists to build an optical communication system. As shown in Figure 1.2,

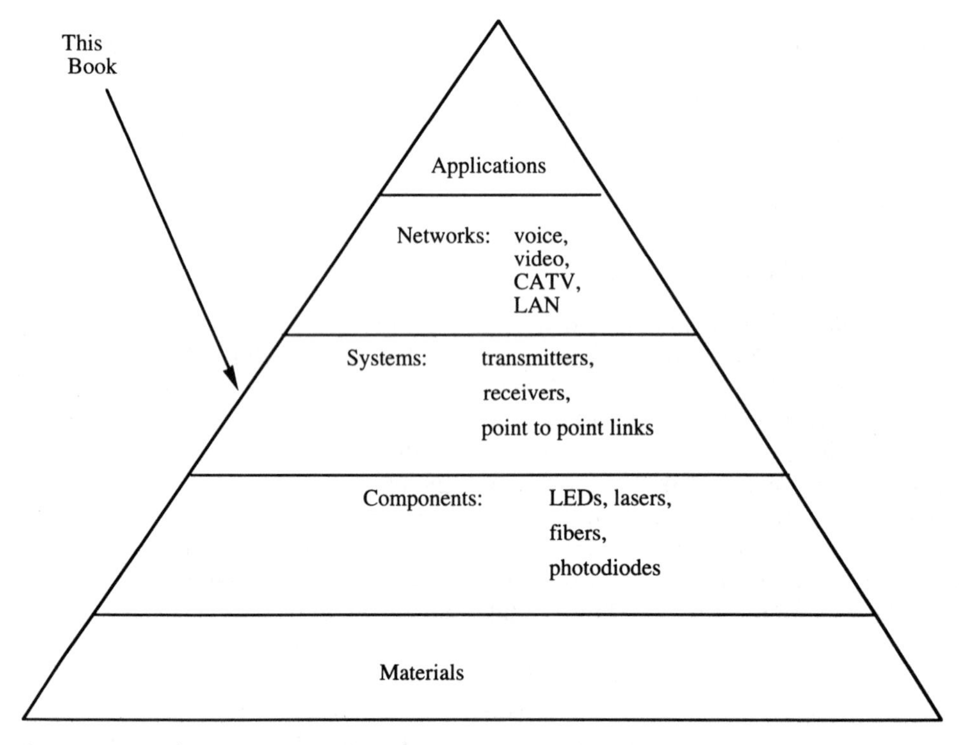

Figure 1.2 Optical communications pyramid.

1. The level of nonlinear distortions is quantified by the following parameters: composite second-order (CSO) distortions, composite triple-beat (CTB) distortions, and spurious-free dynamic range (SFDR).

material scientists make new artificial materials; using those materials, device physicists make optoelectronic devices, such as lasers and photodiodes; system designers make subsystems and systems, such as transmitters and receivers, based on optoelectronic devices; networking specialists make networks using those systems; and, finally, actual applications are designed (such as voice and video conferencing) utilizing the networks.

This book focuses on systems. However, some understanding of optoelectronic components is needed to facilitate system design. The goal of this chapter is to provide such understanding. The chapter does *not* aim to educate a device physicist—there are plenty of specialized books devoted to fibers [4], active devices [9,10], and so on. Rather, the goal is to provide enough knowledge to a system engineer to allow him or her to design optical communication systems.

The three main parts of this chapter are devoted to fibers, light sources, and photodiodes. Each part begins with a brief reminder of the fundamental physics involved and proceeds to the device structure, properties, and input/output relationships.

1.2 FIBERS

1.2.1 The Refractive Index and the Laws of Reflection and Refraction

A plane electromagnetic wave, such as a beam of light, travels in vacuum at the speed $c = 3 \cdot 10^8$ m/s. Different light frequencies correspond to different colors. For example, the frequency of $f = 500$ THz (1 THz (terahertz) = 10^{12} Hz) corresponds to the red color. The two most important frequencies for optical fiber communications are in the infrared region: f = 231 THz, corresponding to the vacuum wavelength of $\lambda = c/f$ = 1.3 μm; and f = 194 THz, corresponding to the vacuum wavelength of $\lambda = c/f$ = 1.55 μm. The first frequency corresponds to the minimum dispersion in regular single-mode telecommunications fiber; the second corresponds to the minimum attenuation.

The frequency of light is set by the source; for example, a given laser may emit at 231 THz. Both the speed and the wavelength of light depend on its frequency and on the propagation medium. For example, a light beam having the frequency of 231 THz has the wavelength of 1.3 μm and speed of $3 \cdot 10^8$ m/s in vacuum. In a typical fiber, light from the same source will have the same frequency (231 THz) but will propagate with a different speed (approximately $2 \cdot 10^8$ m/s) and have a different wavelength (less than 1 μm).

The ratio of the speed of light in vacuum, c, to the speed of light in a medium, v, is called the refractive index:

$$n = \frac{c}{v} \tag{1.1}$$

The refractive index may (and generally does) depend on light frequency. This phenomenon is known as *chromatic dispersion* and is a fundamental limitation of optical fiber communication systems.

Here are a few examples. The refractive index of glass (telecommunications fibers are made from glass) is 1.520 in the near infrared region; the refractive index of water is 1.333; the refractive index of air is 1.000.

The wavelength is, by definition, the distance traveled by light during one light period. Hence, the relationships between the vacuum wavelength, λ; the medium wavelength, λ_m; and the light frequency and velocity look as follows:

$$\lambda = \frac{c}{f} \tag{1.2}$$

$$\lambda_m = \frac{v}{f} \tag{1.3}$$

$$\lambda_m = \frac{\lambda}{n} \tag{1.4}$$

Figure 1.3 shows what happens when a beam of light travels in one medium toward an interface with another.

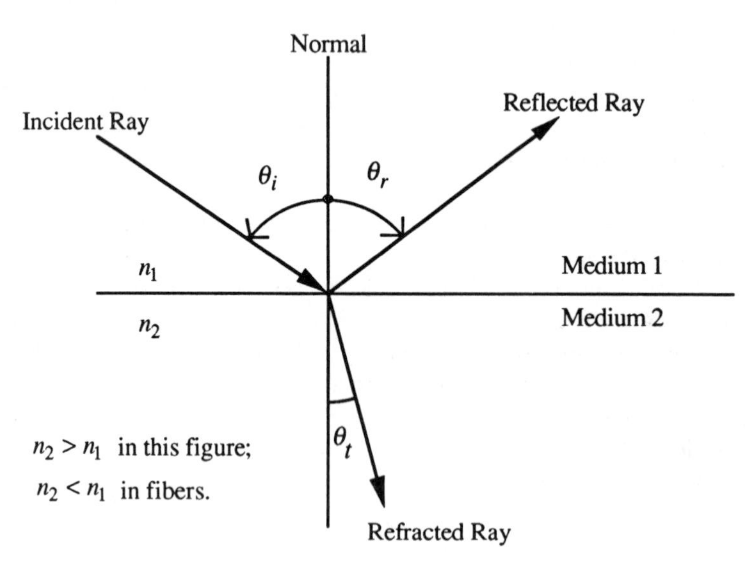

Figure 1.3 Laws of reflection and refraction.

In general, two new beams may (but do not always) emerge when the incident ray impinges on the interface: a reflected ray that propagates back into medium 1 and a refracted ray that propagates into medium 2. The angle between the incident ray and the normal to the interface is denoted by θ_i; the angle between the normal and the reflected ray is denoted by θ_r; and the angle between the normal and the refracted ray is denoted by θ_t (t stands for transmission). The three quantities—θ_i, θ_r, and θ_t—are related as follows:

$$\theta_r = \theta_i \tag{1.5}$$

$$\frac{\sin \theta_t}{\sin \theta_i} = \frac{n_1}{n_2} \tag{1.6}$$

where n_1 and n_2 are the refractive indices of medium 1 and medium 2, respectively. Equation (1.6) is known as Snell's law. It has been observed experimentally and can, therefore, be taken as an experimental fact. However, it also can be derived directly from the more fundamental Maxwell equations.

1.2.2 Total Internal Reflection

Figure 1.3 illustrates the case when $n_2 > n_1$. In that case, the refracted beam travels closer to the normal than the incident beam [see (1.6)].

In fibers and slab waveguides, $n_2 < n_1$. As a result, the refracted beam travels farther from the normal and closer to the interface, as per (1.6). At some value of θ_i, the refracted beam has $\theta_t = 90°$, that is, it travels along the interface. The corresponding value of the incident angle is known as the critical angle, θ_c. When the incident angle is larger than θ_c, there is no refracted beam—all the power of the incident beam is reflected back into medium 1. This phenomenon is known as total internal reflection.

Let us find the value of the critical angle. It follows from Snell's law (1.6) that

$$\sin \theta_t = \frac{n_1}{n_2} \cdot \sin \theta_i \tag{1.7}$$

When $\theta_i = \theta_c$, $\theta_t = 90°$, $\sin \theta_t = 1$, and (1.7) yields the following expression for the critical angle:

$$\sin \theta_c = \frac{n_2}{n_1} \tag{1.8}$$

To summarize, when the incident angle θ_i is smaller than θ_c, both the refracted ray and the reflected ray are generated at the interface; when $\theta_i > \theta_c$, there is no refracted ray.

1.2.3 Step-Index Fibers and Slab Waveguides: Ray Theory

Both step-index fibers and slab waveguides utilize total internal reflection to guide light (or in general any electromagnetic wave). Figures 1.4 and 1.5 illustrate a step-index fiber and a slab waveguide, respectively.

The central region of the fiber is called the core, and the central region of the slab waveguide is called the slab; the peripheral region is called the cladding in both cases. The refractive index of the central region is denoted by n_1, and the refractive region of the cladding is denoted by n_2. A useful fiber/waveguide must have $n_1 > n_2$. Then light is guided by the total internal reflection between the central and peripheral regions.

While the x-y cross sections of the fiber and the slab waveguide are different from each other (Fig. 1.6), their x-z cross-sections are identical (Fig. 1.7). Hence, one can hope that their waveguiding properties are, on a fundamental level, similar.

For both fiber and slab waveguide, the dependence of the refractive index on the x-coordinate is called the refractive index profile. In the simplest case, the refractive index profile has a single step, as shown in Figure 1.8. As we will see later in the chapter, more sophisticated profiles may be beneficial.

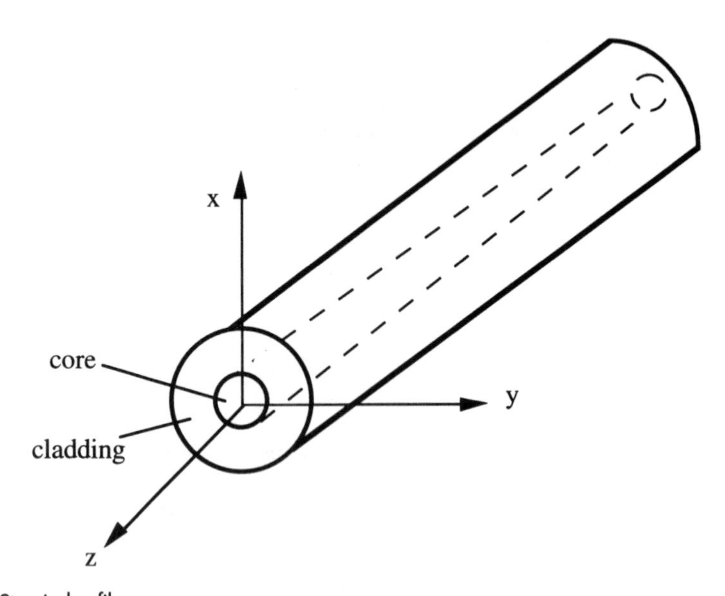

Figure 1.4 Step-index fiber.

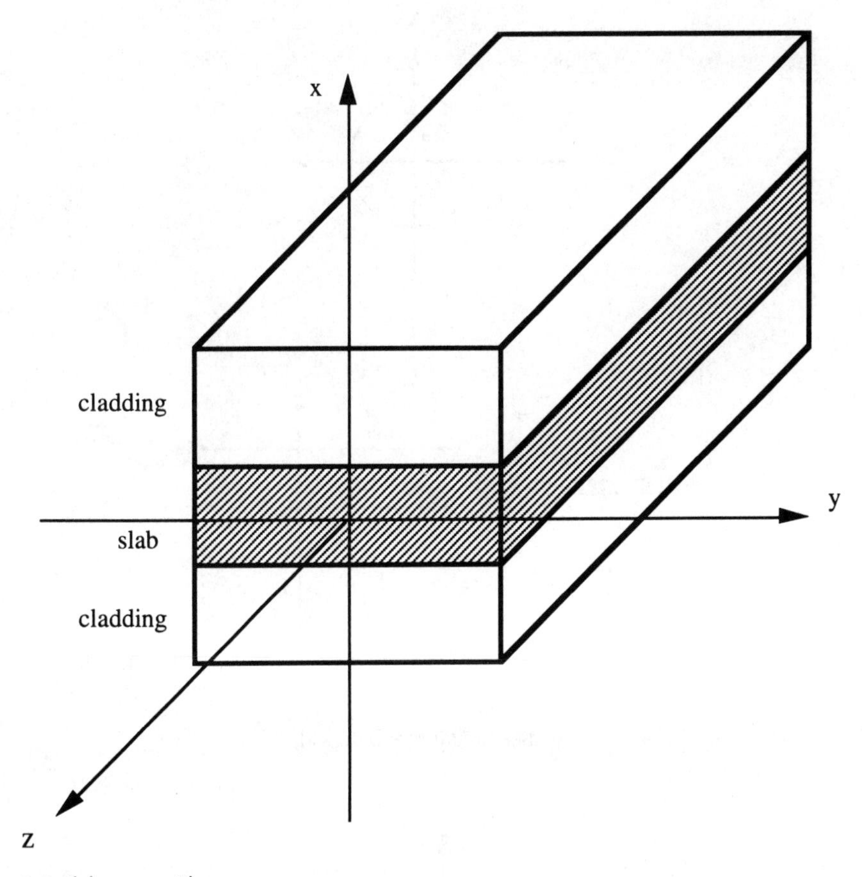

Figure 1.5 Slab waveguide.

Let us investigate what happens when a light beam is launched into the fiber or slab waveguide, as shown in Figure 1.9.

First, the incoming light impinges on the interface between the waveguide[2] and the outside world. Part of the incoming power is lost due to reflection (not shown in Figure 1.9), and part propagates into the waveguide as a refracted beam.

Next, the refracted beam impinges on the core-cladding interface. If the angle of incidence at that interface is smaller than the critical angle, some of the power will be lost. Multiple reflections at the core-cladding interface will soon attenuate the beam to a negligible level, and it will not propagate through any appreciable distance.

However, if the angle of incidence at the core-cladding interface is sufficiently

2. In the rest of this chapter, the term *waveguide* refers to both fibers and slab waveguides.

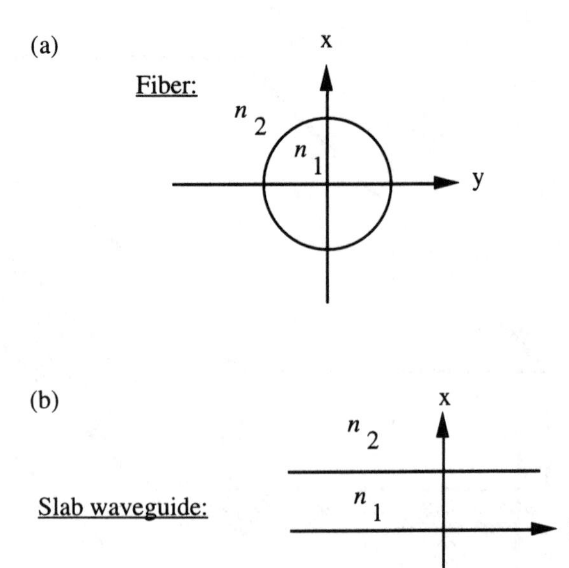

Figure 1.6 The x-y cross-sections of (a) fiber and (b) slab waveguide.

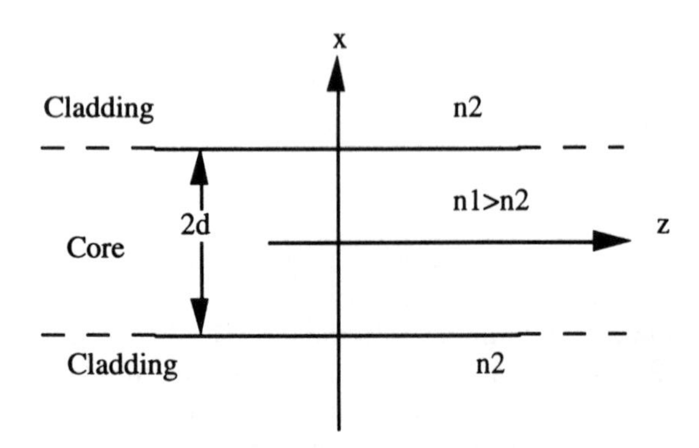

Figure 1.7 The x-z cross-section of fiber and slab waveguide.

large, the entire power will be reflected back into the core via total internal reflection. In that case, the light beam will propagate through multiple reflections at the core-cladding interface over extremely long distances (hundreds of kilometers) without any need for amplification or regeneration.

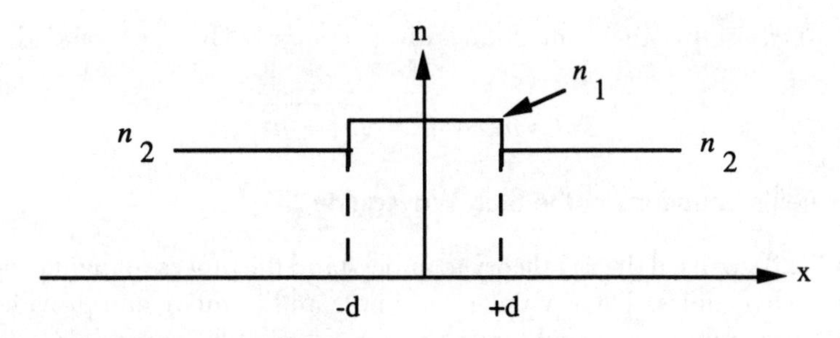

Figure 1.8 The refractive index profile for step-index fibers and slab waveguides.

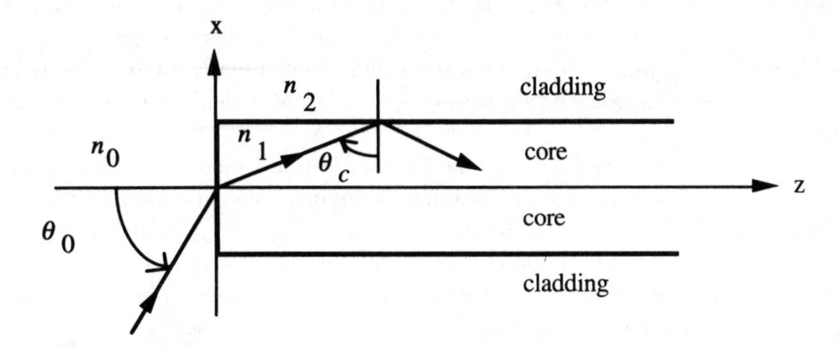

Figure 1.9 Light beam launched into the fiber or slab waveguide: numerical aperture.

It can be shown that the total internal reflection at the core-cladding interface takes place if and only if the angle of incidence at the external waveguide interface θ_0 is sufficiently small:

$$n_0 \sin \theta_0 \leq n_1 \cos \theta_c \tag{1.9}$$

where n_0 is the refractive index of the outside medium.

Exercise 1.1

Derive (1.9).

The maximum value of $n_0 \sin \theta_0$ is known as the numerical aperture NA:

$$NA \equiv n_0 \max [\sin \theta_0] \tag{1.10}$$

In most cases, the outside medium is air, and $n_0 = 1$. Then (1.9) and (1.10) yield

$$NA = n_1 \cos \theta_c = \sqrt{n_1^2 - n_2^2} \tag{1.11}$$

1.2.4 Maxwell's Equations in the Slab Waveguide

In Section 1.2.3, we used the ray theory to understand the process of light propagation in dielectric waveguides. The ray theory is simple and intuitive and provides a clear physical picture of multimode light propagation. Essentially, each mode is associated with a ray propagating with a certain angle with respect to the z-axis; different modes correspond to different angles.

Unfortunately, the ray theory is inadequate for single-mode fibers that are used universally in most telecommunications applications. One has to use a more powerful tool—Maxwell's equations—to understand single-mode fibers and, in general, single-mode waveguides. The goal of this section is to apply Maxwell's equations to the slab waveguide.

We will write Maxwell's equations for the slab and for the cladding separately (see Fig. 1.5). Both are assumed to be linear, homogeneous, and isotropic dielectrics. Further, we will be interested in the monochromatic light, so that the dependence of both electrical field E and magnetic field H on time t is given by $\exp(j2\pi ft)$, where f is the light frequency in hertz. Under the foregoing conditions, Maxwell's equations yield the so-called wave equations:

$$\nabla^2 E + \omega^2 \mu\varepsilon E = 0 \tag{1.12}$$

$$\nabla^2 H + \omega^2 \mu\varepsilon H = 0 \tag{1.13}$$

where ω is the light frequency in radians per second, and μ and ε are the magnetic constant and the dielectric constant, respectively. The light frequency in radians per second (ω) is related to the light frequency in hertz (f) as follows: $\omega = 2\pi f$.

In systems theory, f is used more frequently than ω (this keeps the proper input/output relationships using Fourier transforms). Hence, we will use f when dealing with system issues (such as in Chapter 3). In fiber and waveguide theory, ω is used more frequently than f (this simplifies propagation and dispersion equations). Hence, we will use ω in this chapter.

The values of μ and ε in the slab are generally different from those in the cladding. When it will be necessary to distinguish between the values of μ and ε in the slab and in the cladding, we will use the following notation:

In the slab and in the fiber core:

$$\varepsilon = \varepsilon_1; \ \mu = \mu \tag{1.14}$$

In the cladding:

$$\varepsilon = \varepsilon_2; \; \mu = \mu \tag{1.15}$$

The slab and the cladding are assumed to have different dielectric constants but identical magnetic constants.

Returning now to (1.12) and (1.13), ∇^2 is the Laplacian operator defined by

$$\nabla^2 \equiv \frac{\partial^2}{\partial x^2} + \frac{\partial^2}{\partial y^2} + \frac{\partial^2}{\partial z^2} \tag{1.16}$$

Exercise 1.2

Derive (1.12) and (1.13) from Maxwell's equations.

Slab waveguides can support two groups of modes: *transversal electric* (TE) and *transversal magnetic* (TM). Referring to Figure 1.5, TE modes have their electric field parallel to the y-axis, while TM modes have their magnetic field parallel to the y-axis.

We are interested now in TE modes. Hence,

$$|E_y| > 0; \; E_x = 0; \; E_z = 0 \tag{1.17}$$

Using (1.17) in (1.12), we obtain

$$\frac{\partial^2 E_y}{\partial x^2} + \frac{\partial^2 E_y}{\partial y^2} + \frac{\partial^2 E_y}{\partial z^2} = -\omega^2 \mu \varepsilon E_y \tag{1.18}$$

Exercise 1.3

Derive (1.18).

Slab modes are uniform in the y-direction:

$$\frac{\partial E}{\partial y} = 0 \tag{1.19}$$

Hence,

$$\frac{\partial E_y}{\partial y} = 0 \tag{1.20}$$

and

$$\frac{\partial^2 E_y}{\partial y^2} = 0 \tag{1.21}$$

We are interested in modes propagating in the z-direction. Thus, the dependence of E and E_y on z is given by $\exp(-j\beta z)$, where β is the propagation constant in inverse meters (m^{-1}). Hence,

$$\frac{\partial E_y}{\partial z} = -j\beta E_y \tag{1.22}$$

and

$$\frac{\partial^2 E_y}{\partial z^2} = -\beta^2 E_y \tag{1.23}$$

Substituting (1.21) and (1.23) into (1.18), we obtain

$$\frac{d^2 E_y}{dx^2} = (\beta^2 - \omega^2 \mu\varepsilon) E_y \tag{1.24}$$

Solutions to (1.24) can be expressed either in the form of exponents or in the form of sines/cosines. Within the slab, sine/cosine solutions are more convenient:

Cosine, or even, solution:

$$E_y(x) = A \cos kx, \ |x| \le d \tag{1.25}$$

Sine, or odd, solution:

$$E_y(x) = A \sin kx, \ |x| \le d \tag{1.26}$$

where A is the electrical field amplitude in the slab, and k describes the rate of field variation along the x-axis; k has the units of inverse meters. Equation (1.25) describes even modes, while (1.26) describes odd modes. Different values of k give rise to different transversal modes.

Equations (1.25) and (1.26) don't seem to impose any restrictions on the value of k. However, as we shall see shortly, k can take on discrete values only. Each value of k corresponds to a different mode, and the number of permissible values of k determines the number of modes the waveguide can support.

Substituting (1.25) and/or (1.26) into (1.24), we obtain

$$k^2 = \omega^2\mu\varepsilon_1 - \beta^2 = k_1^2 - \beta^2 \tag{1.27}$$

where ε_1 is the slab dielectric constant, and k_1 (m^{-1}) is the propagation constant of a plane wave in the slab material:

$$k_1 = \omega^2\mu\varepsilon_1 \tag{1.28}$$

In the cladding (see Fig. 1.5 or Fig. 1.6), it is more convenient to express solutions of (1.24) in the exponential form:[3]

$$E_y(x) = C\exp(-\alpha|x|), \qquad |x| \geq d \tag{1.29}$$

where C is the field amplitude in the cladding, and α is the constant describing the rate of field attenuation in the cladding.

Substituting (1.29) into (1.24), we obtain

$$\alpha^2 = \beta^2 - k_2^2 \tag{1.30}$$

and

$$\beta^2 = \alpha^2 + k_2^2 \tag{1.31}$$

where k_2 is the propagation constant of a plane wave in the cladding material:

$$k_2^2 = \omega^2\mu\varepsilon_2 \tag{1.32}$$

It follows from (1.27) and (1.31) that the value of the waveguide propagation constant β lies between the values of propagation constants of the slab and of the cladding:

$$k_2 < \beta < k_1 \tag{1.33}$$

—an intuitively plausible result.

3. Strictly speaking, (1.29) applies to even modes only. For odd modes, it has to be modified as will be discussed in Section 1.2.6.

1.2.5 Even Propagation Modes

This section deals with even propagation modes stemming from and partially described by (1.25). We derive the eigenvalue equations from the boundary conditions and discuss the field distribution.

The boundary condition for the tangential electrical field states that the field must be continuous and, therefore, identical on both sides of the slab-cladding interface:

$$E_{y'}(d-0) = E_{y'}(d+0) \tag{1.34}$$

where the left side of (1.34) is determined by (1.25), while the right side of (1.34) is determined by (1.29). Substitution of (1.25) and (1.29) into (1.34) yields

$$A \cos kd = C \exp(-\alpha d) \tag{1.35}$$

Hence,

$$C = A \cos kd \cdot \exp(\alpha d) \tag{1.36}$$

Substituting (1.36) into (1.29), we obtain

$$E_{y'}(x) = A \cos kd \exp[-\alpha\,(|x| - d)], \qquad |x| \geq d \tag{1.37}$$

Now, let us write the boundary condition for the tangential magnetic field, H_z. H_z can be found from the following Maxwell's equation:

$$\frac{dE_{y'}}{dx} = -j\omega\mu H_z \tag{1.38}$$

It follows from (1.38) that

$$H_z = \frac{j}{\omega\mu} \cdot \frac{dE_{y'}}{dx} \tag{1.39}$$

In the slab ($|x| \leq d$), (1.25) describes E_y; substituting (1.25) into (1.39), we obtain

$$H_z(x) = \frac{-jAk}{\omega\mu} \sin kx, \qquad |x| \leq d \tag{1.40}$$

In the cladding, (1.37) describes E_y; substituting (1.37) into (1.39), we obtain

$$H_z(x) = \frac{\mp jA\alpha}{\omega\mu}\cos(kd)\exp[\alpha(d - |x|)], \qquad |x| \geq d \tag{1.41}$$

where minus should be taken for $x > d$ and plus should be taken for $x < -d$.

The boundary condition for the magnetic field states that the tangential magnetic field H_z must be continuous and, therefore, identical on both sides of the slab-cladding interface:

$$H_z(d - 0) = H_z(d + 0) \text{ and } H_z(-d + 0) = H_z(-d - 0) \tag{1.42}$$

where the left side is determined by (1.40), while the right side is determined by (1.41). Substituting (1.40) and (1.41) into (1.42), we obtain

$$\frac{-jAk}{\omega\mu}\sin kd = \frac{-jA\alpha}{\omega\mu}\cos kd \tag{1.43}$$

After obvious simplifications, (1.43) yields

$$\tan kd = \frac{\alpha}{k} \tag{1.44}$$

Equation (1.44) is one of the two equations needed to find the two unknown variables, k and α. To derive the second equation, we combine (1.27) and (1.30); the result is

$$\beta^2 = k_1^2 - k^2 = k_2^2 + \alpha^2 \tag{1.45}$$

It follows from (1.45) that

$$\alpha^2 = k_1^2 - k_2^2 - k^2 \tag{1.46}$$

Substituting (1.28) and (1.32) into (1.46), we obtain

$$\alpha^2 = \omega^2(\mu\varepsilon_1 - \mu\varepsilon_2) - k^2 \tag{1.47}$$

Note that the phase velocity of light in the slab (v_1) and in the cladding (v_2) are given by

$$v_1^2 = \frac{1}{\mu\varepsilon_1} \tag{1.48}$$

$$v_2^2 = \frac{1}{\mu\varepsilon_2} \tag{1.49}$$

Substituting (1.48) and (1.49) into (1.47), we obtain

$$\alpha^2 = k_0^2(n_1^2 - n_2^2) - k^2 \tag{1.50}$$

where k_0 is the light propagation constant in a vacuum, and n_1 and n_2 are the refractive indices of the slab and of the cladding, respectively:

$$k_0 = \frac{\omega}{c} = \frac{2\pi}{\lambda} \tag{1.51}$$

$$n_1 = \frac{c}{v_1} \tag{1.52}$$

$$n_2 = \frac{c}{v_2} \tag{1.53}$$

Equation (1.50) is the second needed equation for the two unknowns, α and k. Taken together, (1.44) and (1.50) give the system of eigenvalue equations:

$$\tan kd = \frac{\alpha}{k} \tag{1.54}$$

$$\alpha^2 = k_0^2(n_1^2 - n_2^2) - k^2 \tag{1.55}$$

The values of α and k satisfying (1.54) and (1.55) are called the eigenvalues. Equations (1.54) and (1.55) can be combined into a single equation with respect to kd:

$$\tan kd = \sqrt{\frac{V^2}{(kd)^2} - 1} \tag{1.56}$$

where V is known as the normalized cut-off frequency (we will see why shortly) and is defined as

$$V \equiv dk_0 \sqrt{n_1^2 - n_2^2} = dk_0 \sqrt{2n\Delta n} \tag{1.57}$$

where n is the average refractive index, and Δn is the refractive index difference between the slab and the cladding:

$$n \equiv \frac{n_1 + n_2}{2} \tag{1.58}$$

$$\Delta n \equiv n_1 - n_2 \tag{1.59}$$

Since $k_0 = \frac{2\pi}{\lambda} = \frac{2\pi f}{c}$, (1.57) for the normalized cut-off frequency can be rewritten as follows:

$$V = \sqrt{2}\pi \frac{2dn}{\lambda} \sqrt{\frac{\Delta n}{n}} \tag{1.60}$$

or

$$V = \frac{2\sqrt{2}\pi}{c} dfn \sqrt{\frac{\Delta n}{n}} \tag{1.61}$$

Equation (1.60) shows that V depends on just two key waveguide parameters: $2dn/\lambda$, the ratio of slab thickness $2d$ to the wavelength in the medium λ/n, and $\Delta n/n$, the relative difference of the refractive indices.

Equation (1.61) explains why V is called the normalized *frequency*: it is directly proportional to the light frequency f. To understand why V is called the (normalized) *cut-off* frequency, note that (1.56) has roots only if

$$kd < V \tag{1.62}$$

In other words, V is indeed a "cut-off"—there are no roots if kd is larger than V.

One way to solve (1.56) is to plot its left side and its right side versus kd, as shown in Figure 1.10 for a particular case of $V = 5$. Intersection points between the left and right sides of (1.56) give the eigenvalues: the x-axis coordinates give the eigenvalues of kd, and the values of tan kd (that can be read from the y-axis) give the eigenvalues of α/k [see (1.54)].

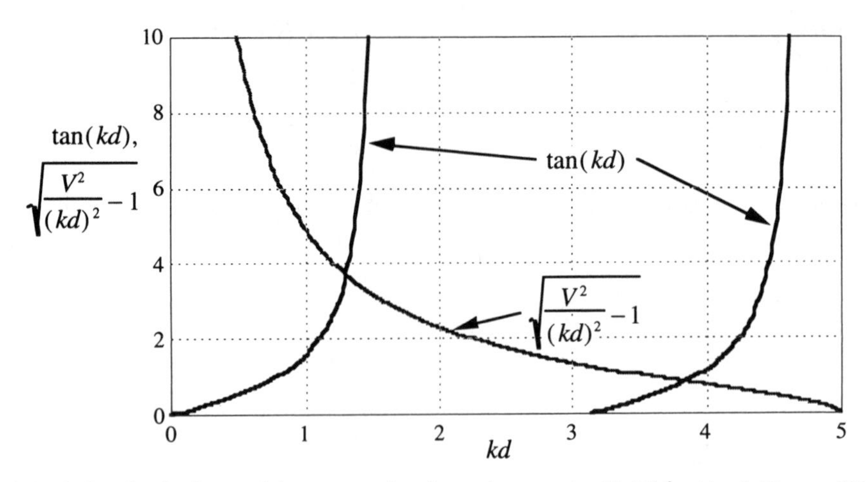

Figure 1.10 Graphical solution of the even-modes eigenvalue equation (1.56) for $V = 5$. (*Source:* [20].)

How many even modes can a given waveguide support? To answer this question, note that each π interval of kd generates one even mode, up to $kd = V$. Hence,

$$\textit{number of even modes} \cong V/\pi \tag{1.63}$$

The reason we use $\cong$ and not = in (1.63) is that the number of modes must be an integer, while V/π in general can be any real number.

The first even mode corresponds to the first branch of the tangent in Figure 1.10; it is called the *fundamental mode.* The fundamental mode can be always excited, irrespective of the wavelength and the waveguide thickness. However, subsequent (so-called higher order) modes may or may not be excited, depending on the value of V.

Consider a numerical example: $\lambda = 1.55$ µm; $n = 1.5$; $\Delta n = 0.1\%$; and $2d = 45$ µm. These parameters have been selected to yield $V = 5$, as in Figure 1.10. The two resulting eigenvalues are listed in Table 1.1.

Figures 1.11 and 1.12 show field distributions $E(x)$ for both even modes. Both

Table 1.1
Eigenvalues for Even Modes; $V = 5$

Mode	*kd*	α/k	k, m^{-1}	α, m^{-1}
First mode (fundamental)	1.31	3.75	$58.2 \cdot 10^3$	$218 \cdot 10^3$
Second even mode	3.84	0.840	$171 \cdot 10^3$	$143 \cdot 10^3$

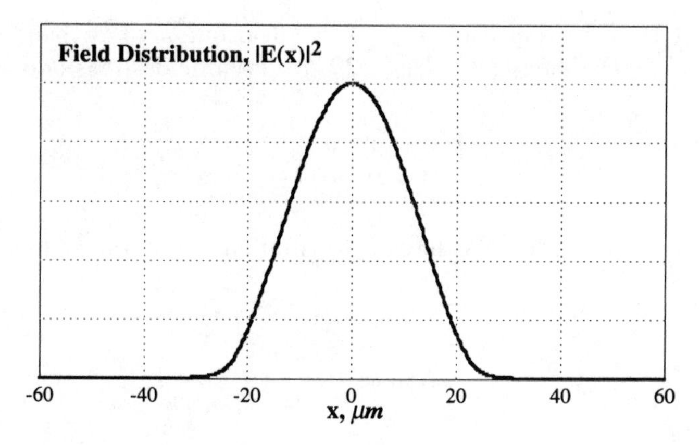

Figure 1.11 Field distribution for the fundamental mode. (*Source:* [11].)

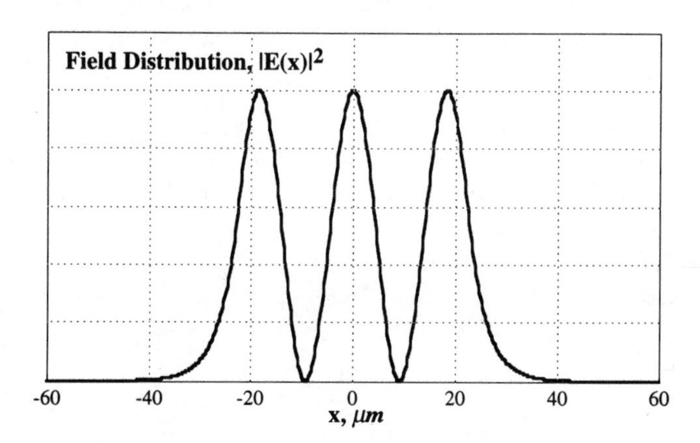

Figure 1.12 Field distribution for the second even mode. (*Source:* [11].)

have peaks at the center of the waveguide, at $x = 0$. However, the fundamental mode has only one peak, while the second even mode has three peaks. Both modes attenuate in the cladding.

If the wavelength is decreased, V will increase [see (1.60)], and the number of modes will increase, too. If the wavelength is increased, V will decrease, and the number of modes will decrease, too.

1.2.6 Odd Propagation Modes

The procedure for obtaining odd modes is almost identical to the procedure for obtaining even modes outlined in Section 1.2.5. Two details are different: (1) the sine distribution (1.26) is used instead of the cosine distribution (1.25); and (2) the field distribution in the cladding [given by (1.29) for even modes] has odd symmetry:

$$E_y(x) = \begin{cases} C \cdot \exp(-\alpha x), & x \geq d \\ -C \cdot \exp(\alpha x), & x \leq -d \end{cases}$$

Following the procedure for the sine distribution, we obtain the following eigenvalue equation:

$$\tan kd = -\frac{k}{\alpha} \tag{1.64}$$

This equation needs to be solved in conjunction with the second eigenvalue equation, (1.55):

$$\alpha^2 = k_0^2(n_1^2 - n_2^2) - k^2 \tag{1.65}$$

Equation (1.55), which is repeated above as (1.65), is valid for both odd and even modes. Combining (1.64) and (1.65), we obtain a single eigenvalue equation with respect to *kd:*

$$\tan kd = -\frac{1}{\sqrt{\frac{V^2}{(kd)^2} - 1}} \tag{1.66}$$

where V is the normalized cut-off frequency defined by (1.57).

Similar to the case of even modes, (1.66) can be solved graphically. Figure 1.13 illustrates the solution for $V = 5$.

How many odd modes can a given waveguide support? To answer this question, note that each π interval of kd generates one odd mode, up to $kd = V$. Hence,

$$\textit{number of odd nodes} \cong V/\pi \tag{1.67}$$

Consider the same numerical example we used in Section 1.2.5: $\lambda = 1.55$ µm; $n = 1.5$; $\Delta n = 0.1\%$; and $2d = 45$ *µm*, so that $V = 5$. The two resulting eigenvalues are listed in Table 1.2.

Figures 1.14 and 1.15 show field distributions $E(x)$ for both modes. Both have zeros at the center of the waveguide, at $x = 0$; compare this with even modes, Figures 1.11 and 1.12 (even modes have their maxima at the center).

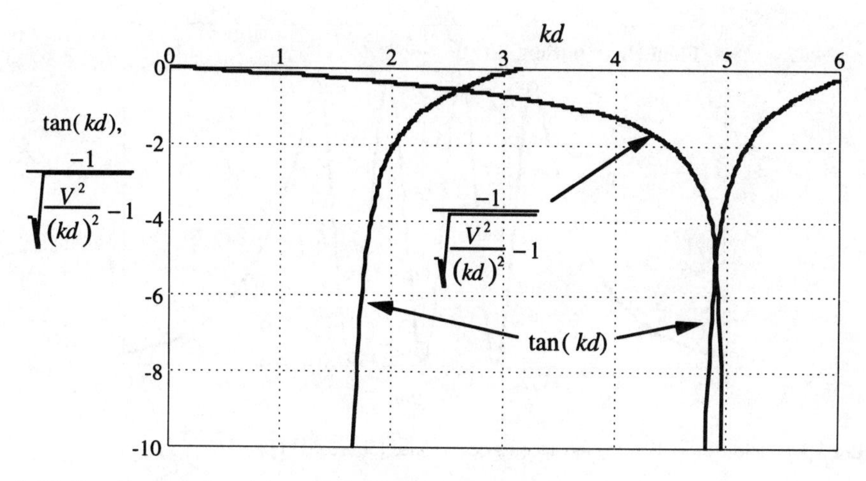

Figure 1.13 Graphical solution of the odd-modes eigenvalue equation (1.66) for $V = 5$. (*Source:* [11].)

Table 1.2
Eigenvalues for Odd Modes; $V = 5$

Mode	*kd*	α/k	*k, l/m*	α*, l/m*
First odd mode	2.60	1.66	$116 \cdot 10^3$	$192 \cdot 10^3$
Second odd mode	4.91	0.2	$218 \cdot 10^3$	$43.6 \cdot 10^3$

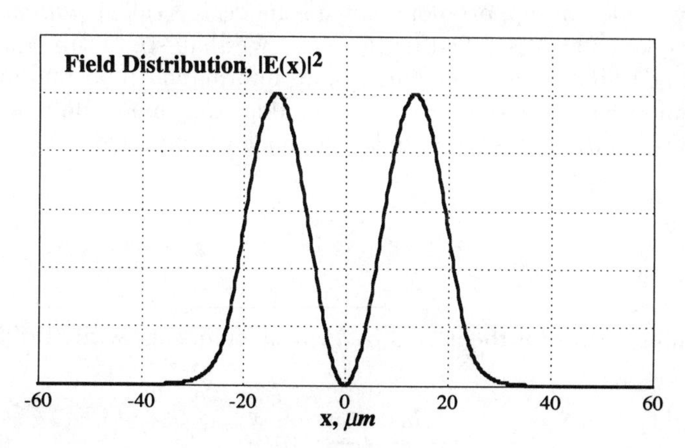

Figure 1.14 Field distribution for the first odd mode. (*Source:* [11].)

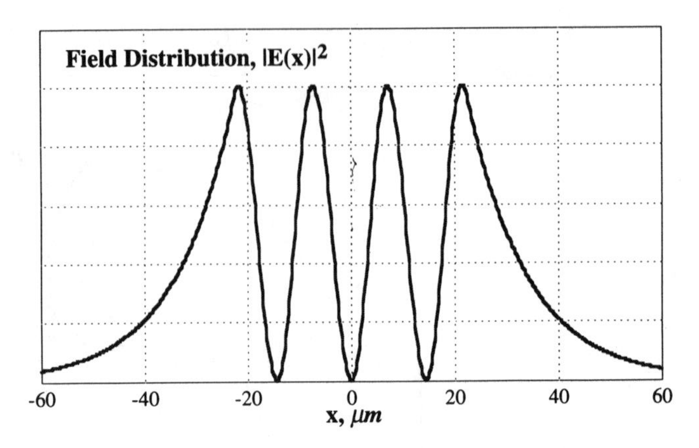

Figure 1.15 Field distribution for the second odd mode. (*Source:* [11].)

1.2.7. Number of Modes and Single-Mode Fibers

We saw in Sections 1.2.4 and 1.2.5 that a waveguide can, in general, support both even and odd modes. The number of even modes was given by (1.63), and the number of odd modes was given by (1.67). Combining those two expressions, we find the total number of modes a waveguide can support:

$$\textit{total number of modes} \cong 2V/\pi \tag{1.68}$$

The reader is reminded that the reason ≅ and not = is used in (1.68) is that the number of modes must be an integer, while $2V/\pi$ in general can be any real number.

If a waveguide can support more than one mode, it is called *multimode*. If it can support only one mode, it is called *single mode*. We shall see in subsequent sections that single-mode fibers can carry much more information than multimode fibers. Hence, it is important to develop a waveguide that can support only one mode.

Equation (1.68) suggests that only one mode is supported if

$$V \leq \frac{\pi}{2} \cong 1.57 \tag{1.69}$$

Combining (1.60) for the normalized cut-off frequency with (1.69), we obtain

$$\frac{2dn}{\lambda}\sqrt{\frac{\Delta n}{n}} \leq 0.35 \tag{1.70}$$

Equation (1.70) shows that one can manipulate just two physical quantities to make a waveguide single mode:

- The optical size of the waveguide $2dn/\lambda$, that is, the number of wavelengths across the waveguide;
- The relative refractive index difference, $\Delta n/n$.

If the product of these two quantities is small (< 0.35), the waveguide is single mode; if not, it is multimode.

Equation (1.70) has been derived for slab waveguides rather than for fibers. The analysis for fibers can be found in many textbooks [1–4]. The mathematical details are more complicated and tedious, but the basic procedure is the same:

1. Write out the Maxwell's equations for the core and for the cladding and specify boundary conditions.
2. Find general solutions of the Maxwell's equations and combine them with boundary conditions to yield eigenvalue equations.
3. Solve eigenvalue equations to find the mode distribution and the numerical value of the propagation constant, β, for each mode.

The results of that procedure are fundamentally similar to the results we have obtained for the slab waveguide: The number of modes is finite and can be made one with a proper fiber design. However, the field distribution generally is complicated. For that reason, the field distribution of the fundamental mode is frequently approximated by the Gaussian function. This approximation is highly accurate for weakly guiding fibers having

$$\frac{\Delta n}{n} << 1 \tag{1.71}$$

Because (1.71) is satisfied in all telecommunication fibers, Gaussian field approximation can be safely used for the fundamental mode of such fibers.

Table 1.3 summarizes the Gaussian approximation for weakly guiding circular fibers; the parameters used are defined in Table 1.4.

Like slab waveguides, fibers are single mode if the normalized cut-off frequency is smaller than a critical value. Similar to (1.69), the single-mode condition is ([4], p. 320)

$$V \leq 2.405 \tag{1.72(a)}$$

where V is defined by (1.57), with d replaced by the fiber radius. Since single-mode fibers and waveguides can carry much more information than their multimode

Table 1.3
Gaussian Approximation for Circular Fibers

Even HE_{11} mode (*x-polarized*)	Odd HE_{11} mode (*y-polarized*)
$E_x = \exp\left\{-\frac{1}{2}\frac{r^2}{r_0^2}\right\} \exp(i\beta z)$	$E_y = \exp\left\{-\frac{1}{2}\frac{r^2}{r_0^2}\right\} \exp(i\beta z)$
$H_y = \left(\frac{\varepsilon_0}{\mu_0}\right)^{1/2} n_{co} E_x$	$H_x = -\left(\frac{\varepsilon_0}{\mu_0}\right)^{1/2} n_{co} E_y$

$$n^2(R) = n_{co}^2\{1 - 2\Delta f(R)\}; \quad V = kpn_{co}(2\Delta)^{\frac{1}{2}}; \quad R = \frac{r}{p}; \quad R_o = \frac{r_0}{p}$$

Variational equation for the propagation constant, β

$$U^2 = \frac{V^2}{2\Delta} - p^2\beta^2 = \frac{1}{R_0^2} + V^2\left\{f(0) + \int_0^\infty \frac{df(R)}{dR} \exp\left\{-\frac{R^2}{R_0^2}\right\} dR\right\}$$

Equation for spot size, r_0

$$\frac{1}{V^2} = \int_0^\infty \frac{df(R)}{dR} R^2 \exp\left\{-\frac{R^2}{R_0^2}\right\} dR; \qquad r_0 = pR_0$$

Source: [4]. © 1983 Chapman and Hall.

counterparts, we focus almost exclusively on single-mode fibers and waveguides in the rest of this book.

1.2.8 Phase Velocity

A monochromatic wave traveling in the z-direction, such as a mode in a fiber or in a waveguide, is given by

$$E(t,z) = A \exp[j(\omega t - \beta z)] \tag{1.72(b)}$$

where ω is the wave frequency in radians per second, and β is the propagation constant in inverse meters. The phase velocity is defined as the velocity an observer must maintain to observe the field with a constant phase. In other words, to find the value of the phase velocity, one must trace a point of constant phase:

$$\omega t - \beta z = const \tag{1.73}$$

Table 1.4
Waveguide or Fiber Parameters

Parameter	*Expression*
Refractive-index profile	$n(x,y)$ or $n(r)$
Profile representation	$n^2(x,y) = n_{co}^2\,[1 - 2\Delta f(x,y)]$
$(f \geq 0)$	$n^2(r) = n_{co}^2\,[1 - 2\Delta f(r)]$
Group index	$n_g = n - \lambda \dfrac{\partial n}{\partial \lambda}$
Profile height parameter	$\Delta = \dfrac{n_{co}^2 - n_{cl}^2}{2n_{co}^2} = \dfrac{\sin^2\theta_c}{2}$
Complement of the critical angle	$\theta_c = \sin^{-1}\left\{1 - \dfrac{n_{cl}^2}{n_{co}^2}\right\}^{\frac{1}{2}} = \cos^{-1}\left(\dfrac{n_{cl}}{n_{co}}\right)$
Maximum core index	n_{cos} $(f = 0)$
Uniform cladding index	$n_{cl} = n_{co}(1 - 2\Delta)^{\frac{1}{2}}$ $(f = 1)$
Profile shape	$S = 1 - f$
Profile volume	$\Omega = \int_{A\infty} (1 - f)dA = \int_{A\infty} SdA$
Infinite cross section	A_∞
Core cross section	A_{CO}
Core radius or half-width	p
Free-space wavelength	λ
Free-space wave number	$k = \dfrac{2\pi}{\lambda} = \dfrac{V}{pn_{co}(2\Delta)^{1/2}}$
Waveguide or fiber normalized cut-off frequency	$V = kp(n_{co}^2 - n_{cl}^2)^{1/2} = kpn_{co}(2\Delta)^{1/2}$ $= kpn_{co}\sin\theta_c$
Weak-guidance or paraxial approximation	$n_{co} \cong n_{cl}$ *or* $\Delta << 1$; $\Delta \cong \dfrac{n_{co} - n_{cl}}{n_{co}}$; $\theta_c \cong (2\Delta)^{1/2} \cong \left(1 - \dfrac{n_{cl}^2}{n_{co}^2}\right)^{1/2} \cong \cos^{-1}\left(\dfrac{n_{cl}}{n_{co}}\right)$

Source: [4].

The phase velocity, v, must give $z = vt$ such that (1.73) holds true. Substituting $z = vt$ into (1.73), we obtain

$$\omega t - \beta vt = const \text{ for } \forall t \tag{1.74}$$

Equation (1.74) can be valid only if

$$v = \frac{\omega}{\beta} \tag{1.75}$$

Equation (1.75) specifies the relationship between the phase velocity, frequency, and propagation constant of an electromagnetic wave. The ratio of the speed of light to the phase velocity is known as the refractive index (or as the phase refractive index):

$$n \equiv \frac{c}{v} \tag{1.76}$$

Equation (1.76) is similar to (1.1) but is more precise since we specify now that the velocity in question is the phase velocity.

In free space (but not in a waveguide), $v = c$ and $n = 1$ *if the wave propagates along the z-axis.* If the wave does not propagate along the z-axis, the phase velocity can be larger than the speed of light, c. Figure 1.16 illustrates such a case: a plane wave propagates at an angle α with respect to the z-axis. Its "natural" (i.e., measured along the direction of propagation) wavelength is λ. However, its wavelength along the z-axis is larger than λ and is equal to

$$\lambda' = \frac{\lambda}{\cos \alpha} > \lambda \tag{1.77}$$

The field along the z-axis is given by $A \cdot \exp[j(\omega t - \beta z)]$, with β being given by

$$\beta = \frac{2\pi}{\lambda'} = \frac{2\pi}{\lambda} \cdot \cos \alpha < k \tag{1.78}$$

Thus, the propagation constant along the z-axis in Figure 1.16 is smaller than the vacuum propagation constant.

It follows from (1.75) and (1.78) that the phase velocity along the z-axis is

$$v = \frac{\omega}{\beta} = \frac{\omega\lambda}{2\pi \cos \alpha} = \lambda f \cdot \frac{1}{\cos \alpha} = \frac{c}{\cos \alpha} > c \tag{1.79}$$

Thus, the phase velocity *can* be larger than the speed of light. This result does not contradict Einstein's relativity theory and does not mean that information can travel faster than the speed of light, since the phase velocity characterizes only the rate of phase change in space, not the rate of power or envelope propagation (the latter is characterized by the group velocity).

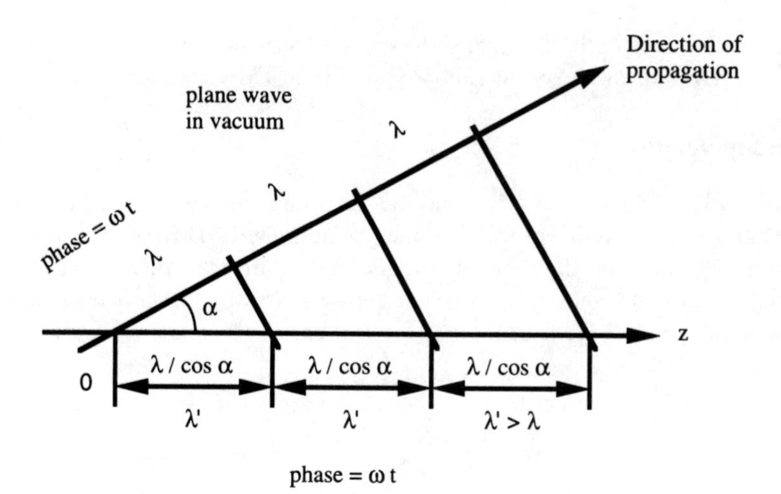

Figure 1.16 A plane wave propagating at an angle α with respect to the z-axis. Its wavelength along the z-axis is λ′, and λ′ is larger than λ.

We showed in (1.33) that in the waveguide aligned with the z-axis,

$$k_2 < \beta < k_1 \tag{1.80}$$

It follows from (1.75) and (1.80) that the phase velocity, v, of a mode in the waveguide/fiber is bounded by the speeds of light in the slab/core and in the cladding:

$$v_1 < v < v_2 < c \tag{1.81}$$

In conclusion, we list the relationships between the mode wavelength, λ_{mode}; the mode propagation constant, β; and the mode (phase) refractive index, n:

$$\lambda_{mode} = \frac{2\pi}{\beta} \tag{1.82}$$

$$\beta = \frac{2\pi}{\lambda_{mode}} \tag{1.83}$$

$$\lambda_{mode} = \frac{\lambda}{n} \tag{1.84}$$

Both the mode wavelength, λ_{mode}, and the mode phase velocity, v, depend, in general, on the light frequency (ω), on the mode type (TE or TM), and on the mode number.

1.2.9 Group Velocity

A modulated lightwave is, strictly speaking, not monochromatic: Fourier theory indicates that it must contain several frequency components. Different frequency components travel through the fiber with generally different phase velocities and, therefore, accumulate generally different phase shifts. In other words, fiber may introduce a *phase distortion.* In many practical cases, the relative bandwidth of the optical signal is small:

$$\frac{B}{f} << 1 \tag{1.85}$$

where B is the optical signal bandwidth, and f is the optical frequency. In such cases, the impact of the fiber-induced phase distortion can be characterized by the *group velocity* and by the group velocity dispersion.

Group velocity is, by definition, the velocity of the envelope (amplitude modulation) of the optical signal and is, in general, different from the phase velocity.

The easiest way to derive an expression for the group velocity and to understand the difference between the group velocity and the phase velocity is to consider a simple test signal and to investigate its propagation through a fiber. Thus, let us consider an amplitude-modulated (AM) optical signal:

$$e_{AM}(t, z = 0) = E(1 + m \cos \omega_1 t) \cos \omega_c t \tag{1.86}$$

where $z = 0$ means that we are launching the wave at the entrance to a fiber, E is the field amplitude, m is the modulation depth, ω_1 is the modulation frequency (we assume cosine modulation), and ω_c is the light frequency. In most cases, $\omega_1 << \omega_c$. Note that the cosine modulation of the field assumed in (1.86) does *not* correspond to the cosine modulation of the power or intensity since the light power is proportional to the field squared.

Using trigonometry and Euler's formula, we can rewrite (1.86) as follows:

$$e_{AM}(t, z = 0) = E\,Re\left\{\exp\,(j\omega_c t) + \frac{m}{2}\exp\,[j(\omega_c - \omega_1)t] + \frac{m}{2}\exp\,[j(\omega_c + \omega_1)t]\right\} \tag{1.87}$$

Equation (1.87) indicates that our signal, $e_{AM}(t, z = 0)$, contains three different frequency components having the frequencies of ω_c, $\omega_c - \omega_1$, and $\omega_c + \omega_1$. Each of these components will travel at its own phase velocity and accumulate its own phase shift.

In general, we must know β for each value of ω to find the output field. It is convenient to expand β into a Taylor power series:

$$\beta(\omega) = \beta_c + \dot{\beta}\Delta\omega + \frac{1}{2}\ddot{\beta}(\Delta\omega)^2 + \frac{1}{6}\dddot{\beta}(\Delta\omega)^3 + \ldots \tag{1.88}$$

where β_c is the value of β at the "carrier" frequency, ω_c; $\Delta\omega$ is the difference between ω and ω_c; and $\dot{\beta}$, $\ddot{\beta}$, and $\dddot{\beta}$ are the derivatives of β with respect to ω evaluated at $\omega = \omega_c$:

$$\Delta\omega \equiv \omega - \omega_c \tag{1.89}$$

$$\dot{\beta} = \left.\frac{\partial\beta}{\partial\omega}\right|_{\omega = \omega_c} \tag{1.90}$$

$$\ddot{\beta} = \left.\frac{\partial^2\beta}{\partial\omega^2}\right|_{\omega = \omega_c} \tag{1.91}$$

$$\dddot{\beta} = \left.\frac{\partial^3\beta}{\partial\omega^3}\right|_{\omega = \omega_c} \tag{1.92}$$

As we will see shortly, $\dot{\beta}$ does not lead to envelope distortion, while $\ddot{\beta}$ and higher-order terms do. For that reason, $\ddot{\beta}$ is known as the first-order dispersion coefficient, $\dddot{\beta}$ is known as the second-order dispersion coefficient, and so on.

In this section, we neglect the dispersion (it will be dealt with in Section 1.2.10); hence, (1.88) yields

$$\beta(\omega) \cong \beta_c + \dot{\beta}\Delta\omega \tag{1.93}$$

Thus, the propagation constants of the three frequency components of the AM signal (1.87) are given by:

$$\text{at } \omega_c - \omega_1\text{: } \beta = \beta_c - \Delta\beta \tag{1.94}$$

$$\text{at } \omega_c\text{: } \beta = \beta_c \tag{1.95}$$

$$\text{at } \omega_c + \omega_1\text{: } \beta = \beta_c + \Delta\beta \tag{1.96}$$

where

$$\Delta\beta \equiv \dot{\beta}\omega_1 \tag{1.97}$$

Because the propagation constants of the three frequency components are now known, we can write the expression for the field at the output of a z-meters-long fiber (we neglect attenuation):

$$\begin{aligned} e_{AM}(t, z) = E\, Re\{&\exp[j(\omega_c t - \beta_c z)]\} \\ &+ \frac{m}{2}\exp\{j[(\omega_c - \omega_1)t - (\beta_c - \Delta\beta)z]\} \\ &+ \frac{m}{2}\exp\{j[(\omega_c + \omega_1)t - (\beta_c + \Delta\beta)z]\} \end{aligned} \tag{1.98}$$

Equation (1.98) is obtained by adding an appropriate phase shift, $\beta(\omega) \cdot z$, to each of the three components of (1.87).

Equation (1.98) can be manipulated into the following form:

$$e_{AM}(t, z) = E[1 + m\cos(\omega_1 t - \Delta\beta z)]\cos(\omega_c t - \beta_c z) \tag{1.99}$$

Equation (1.99) shows that the phase shift accumulated by the carrier (i.e., light) is equal to $\beta_c z$, while the phase shift accumulated by the modulation is equal to $\Delta\beta z$.

The group velocity, v_g, is defined as the velocity one must maintain to observe a constant phase *of the envelope*. In other words, when $z = v_g t$, $\omega_1 t - \Delta\beta z$ must be constant:

$$\omega_1 t - \Delta\beta z = const \text{ for } \forall t \tag{1.100}$$

Substituting $z = v_g t$ and (1.97) into (1.100), we obtain

$$\omega_1 t - \dot{\beta}\omega_1 v_g t = const \text{ for } \forall t \tag{1.101}$$

Equation (1.101) has a solution if and only if $const = 0$, in which case

$$v_g = \frac{1}{\dot{\beta}} = \frac{\partial\omega}{\partial\beta} \tag{1.102}$$

Comparison of (1.102) with (1.75) indicates that while both the phase velocity, v, and the group velocity, v_g, are determined by the shape of the β-versus-ω curve, they are, in general, different. Even though ω is an independent variable and β depends

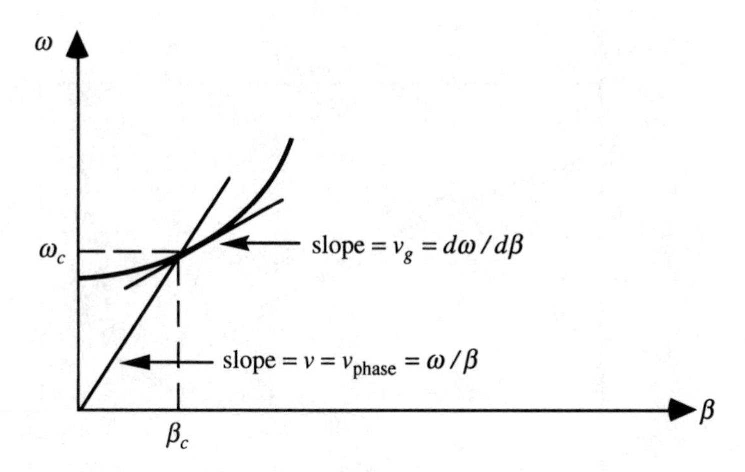

Figure 1.17 The shape of the ω-versus-β curve determines both the group velocity, v_g, and the phase velocity, v. However, in general, $v \neq v_g$.

on ω, it is customary to illustrate the difference between v and v_g on the ω-versus-β plane, as shown in Figure 1.17.

What happens if the optical signals (1.86) and (1.99) are detected by two separate photodiodes? Since photodiodes respond to optical signal power only (and not to signal phase), the photocurrents produced in both cases are *identical*, except for a phase difference of $\Delta\beta z$. Thus, the term $\dot{\beta}$ in (1.88) does *not* lead to the detected signal distortion, even though the optical signal is distorted [compare (1.86) with (1.99)].

We conclude by rewriting (1.99) once more to emphasize the difference between the phase velocity and the group velocity:

$$e_{AM}(t, z) = e = E\left[1 + m \cdot \cos \omega_1\left(t - \frac{z}{v_g}\right)\right] \cdot \cos \omega_c\left(t - \frac{z}{v}\right). \qquad (1.103)$$

1.2.10 Attenuation and Dispersion

This section deals with two main fiber phenomena that limit the performance of optical fiber communication systems: attenuation and dispersion. At high power levels, fiber nonlinearities also affect system performance; the impact of fiber nonlinearities is out of the scope of this section.

1.2.10.1 Attenuation

Because telecommunication fibers are made from silica, their attenuation is determined by silica's attenuation. Figure 1.18 shows silica attenuation versus wavelength.

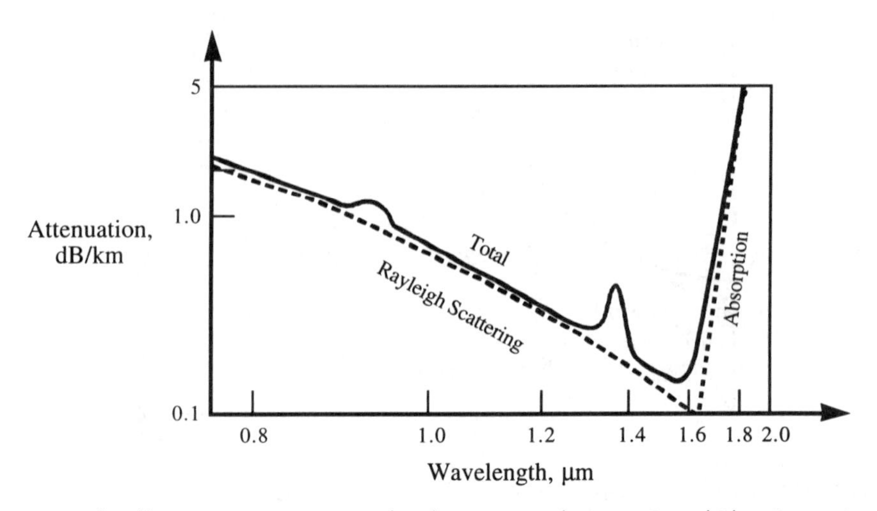

Figure 1.18 Silica fiber attenuation versus wavelength. *From: Introduction to Optical Fiber Communication Systems* by William B. Jones, Jr. © 1988 by Oxford University Press, Inc. Reprinted by permission.

The two broken lines correspond to the theoretical limits due to scattering and absorption of light by silica molecules. Inspection of Figure 1.18 reveals that the silica fiber attenuation is now close to the theoretical limit; the two parasitic peaks are due to residual contaminants. The two low-loss regions are (1) around 1.3 μm and (2) around 1.55 μm. The 1.55-μm region has the lowest attenuation. Both low-loss regions, or windows, are used for communications. In some short-distance applications, such as computer interconnects, the wavelength of 0.8 μm and other wavelengths can be used, too.

Systems running at 1.3 μm were developed before those running at 1.5 μm, but many new systems are designed to operate at 1.55 μm due to the lower fiber loss and the availability of 1.55-μm erbium-doped fiber amplifiers (see Chap. 5). The exact value of fiber attenuation at 1.55 μm depends on the fiber chemical composition (i.e., on the silica purity and on the dopants added to the base silica); 0.22 dB/km is fairly common, with some fibers having the attenuation as low as 0.17 dB/km.

Given fiber attenuation A in dB/km (Fig. 1.18) and transmitter power P_T, one can find the optical power at the end of the link (at the receiver). The received power, P_R, can be found as follows:

$$P_R = P_T \cdot 10^{-AL/10} \tag{1.104}$$

where L is the link length in kilometers. Equation (1.104) assumes that power levels are sufficiently low to neglect fiber nonlinearities.

A receiver needs a certain minimum amount of optical power to function; that amount is called *receiver sensitivity* and will be discussed in depth in Chapter 3. It follows from (1.104) that the maximum transmission distance for an *unamplified* link is given by

$$L_{max} = \frac{10}{A} \log_{10} \frac{transmitter\ power}{receiver\ sensitivity} \tag{1.105}$$

Note that fiber attenuation has a much stronger impact on the transmission distance than either transmitter power or receiver sensitivity, because the latter parameters affect L_{max} logarithmically.

1.2.10.2 Dispersion

Different light components (different modes and/or different frequencies, or colors) travel through the fiber at generally different velocities. This phenomenon is known as *dispersion*. When a composite optical signal travels through the fiber, it can be distorted due to fiber dispersion.

In particular, pulses used in digital systems may become longer due to fiber dispersion by a certain amount we will denote by $\Delta\tau$. Obviously, it is desirable to keep $\Delta\tau$ within a reasonably small fraction of the bit duration, T, and the following rule of thumb is frequently used:

$$\Delta\tau \leq \frac{1}{4} T \tag{1.106}$$

$$R_b \leq \frac{1}{4\Delta\tau} \tag{1.107}$$

where $R_b \equiv 1/T$ is the information transmission rate in bits per second. There is nothing magical about the factor 4 in (1.106) and (1.107); one can use factor 3 or 5 just as well. Factor 3 will result in a larger dispersion-induced penalty, while factor 5 will result in a smaller penalty.

If a fiber supports several modes, the dominant dispersion mechanism is the *modal* or *intermodal* dispersion (i.e., different modes traveling with different velocities). If a fiber supports only one mode, the dominant dispersion mechanism is the *chromatic* dispersion (i.e., different light frequencies traveling with different velocities).

Modal Dispersion in Multimode Fibers

Different modes travel at different speeds through multimode fibers; the result is the modal dispersion that can be evaluated using Figure 1.19.

The fastest mode in Figure 1.19 corresponds to ray 1, which travels directly to the destination along the fiber axis. The travel time for that mode is

$$\tau_{fast} = \frac{Ln_1}{c} \tag{1.108}$$

where L is the fiber length, and n_1 is the core refractive index.

The slowest mode corresponds to ray 2, which travels with the maximum angle θ with respect to the fiber axis; that mode will travel the longest distance. The travel time for the slowest mode is

$$\tau_{slow} = \frac{Ln_1}{c \cos \theta} \tag{1.109}$$

where θ is the angle between ray 2 and the fiber axis.

According to Snell's law, when $\theta = \theta_{max}$,

$$\cos \theta_{max} = \frac{n_1}{n_2} \tag{1.110}$$

Substituting (1.110) into (1.109) and subtracting (1.108), we obtain a simple estimate of modal dispersion:

$$\Delta\tau = \frac{Ln}{c} \cdot \frac{\Delta n}{n} \tag{1.111}$$

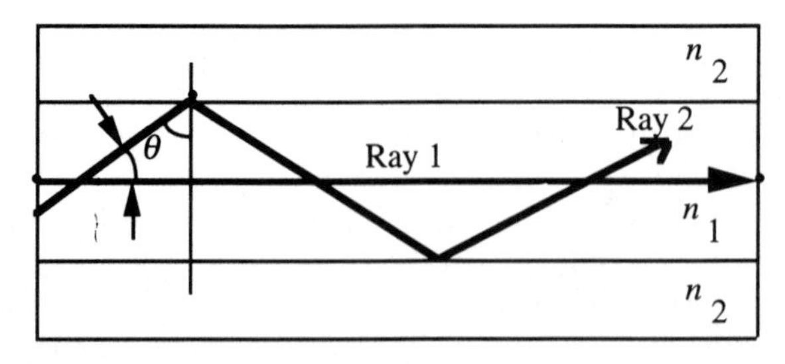

Figure 1.19 A simple estimate of modal dispersion.

Finally, substitution of (1.111) into (1.107) yields an estimate of the maximum bit rate one can achieve with multimode fibers:

$$R_b \leq \frac{1}{4\Delta\tau} = \frac{1}{4} \cdot \frac{c}{Ln} \cdot \frac{n}{\Delta n} \tag{1.112}$$

In a typical multimode fiber, $\Delta n/n = 0.01$ yielding the bit rate of only 10 Mbps·km. This number is well below the current telecommunication needs, so the use of multimode fiber is limited to short-distance communications, such as interconnection of computers and/or computer components and communications between automobile subsystems.

To improve the bandwidth of multimode fibers, a graded-index profile of Figure 1.20 can be (and is) used. The resulting fiber is known as the graded-index multimode fiber.

Figure 1.20 indicates that the refractive index varies gradually as a function of the distance from the center of the fiber. The following profile is widely used:

$$n(r) = \begin{cases} n_1\sqrt{1 - 2\Delta\left(\dfrac{r}{a}\right)^{\alpha}}, & r \leq a \\ n_1\sqrt{1 - 2\Delta} = n_2\,, & r \geq a \end{cases} \tag{1.113}$$

where Δ is the maximum relative refractive index difference between the core and the cladding, and α is a numerical parameter determining the shape of the refractive index profile. A particular value of $\alpha = 2$ can be shown to be optimum and is used in practical fibers.

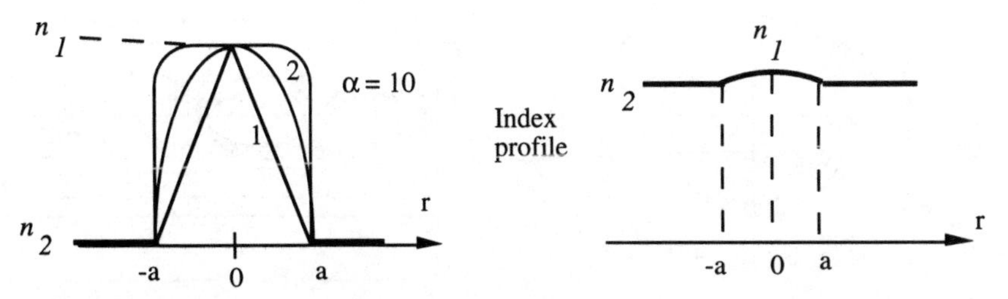

Figure 1.20 Refractive index profile of graded-index fibers.

Figure 1.21 shows the overall cross section of a graded-index fiber along with mode trajectories.

Inspection of Figure 1.21 reveals that modes corresponding to larger values of θ spend more time in a region of the core where the refractive index is lower. Therefore, they have a larger velocity; this factor partially compensates for their longer propagation path. In other words, graded-index fiber design partially compensates for modal dispersion. It can be shown that the propagation delay difference between the fastest and the slowest mode in graded-index fibers is given by

$$\Delta\tau = \frac{L\,N_1\,\Delta^2}{8c} \tag{1.114}$$

where N_1 is the core group refractive index corresponding to n_1.

Comparison of (1.114) with the corresponding result for step-index fibers, (1.111), reveals that $\Delta\tau$ is quadratic in $n_1 - n_2$ for graded-index fibers, while it is linear in $n_1 - n_2$ for step-index fibers. Since $n_1 - n_2$ is very small, graded-index fibers greatly outperform step-index fibers.

Combining (1.114) with (1.107), we obtain the following estimate of the graded-index fiber bandwidth:

$$R_b \le \frac{2c}{NL\Delta^2} \tag{1.115}$$

For typical parameter values, (1.115) yields about 4 Gbps·km; in practice, some 2 Gbps·km is achievable. While this value is much larger than the 10 Mbps·km limit we obtained for step-index fibers, it is still well below the current telecommunications

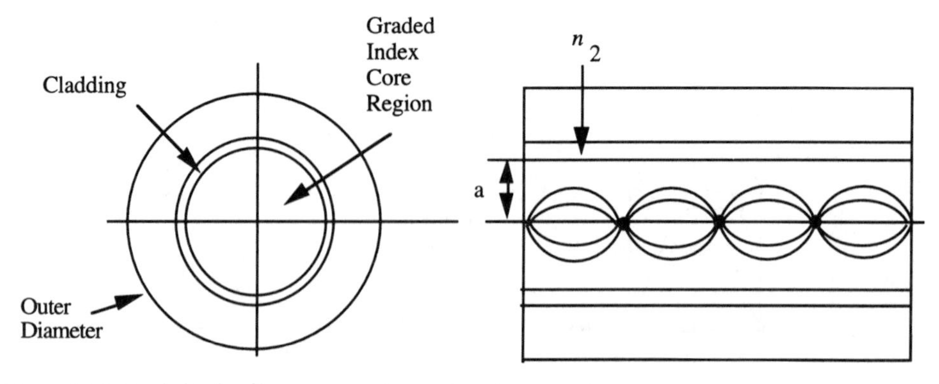

Figure 1.21 Graded-index fiber: cross section and mode trajectories.

needs. Thus, even with a graded-index profile, multimode fibers are confined to short-distance applications.

Chromatic Dispersion in Single-Mode Fibers

We saw in Section 1.2.7 that single-mode fibers support only one mode; they are, therefore, not subject to modal dispersion. The dominant dispersion mechanism in single-mode fibers is the so-called chromatic dispersion: different light colors propagate through fiber with different velocities.

Mathematically, the chromatic dispersion can be characterized either by $\ddot{\beta}$, the second derivative of the β-versus-ω characteristic [see (1.88)] or by the so-called *group velocity dispersion* (GVD) coefficient normally denoted by D. Of course, the two coefficients, $\ddot{\beta}$ and D, are related since they describe the same physical phenomenon. To understand their relationship, let us first find the pulse spread due to chromatic dispersion in terms of $\ddot{\beta}[\text{sec}^2/(\text{m rad}^2)]$.

Let τ be the propagation delay at a particular value of ω:

$$\tau = \frac{L}{v_g} \tag{1.116}$$

where L is the fiber length, and v_g is the group velocity corresponding to ω. Since v_g is in general frequency-dependent, it follows from (1.116) and (1.102) that

$$\frac{\partial \tau}{\partial \omega} = L\frac{\partial}{\partial \omega}\frac{1}{v_g} = L\frac{\partial^2 \beta}{\partial \omega^2} = L\ddot{\beta} \tag{1.117}$$

If a signal has a frequency spectral width of $\Delta\omega$, then the difference in propagation time of parts of this signal at the opposite extremes of the spectrum will be

$$\Delta\tau = \left|\frac{\partial \tau}{\partial \omega}\right|\Delta\omega = \left|\frac{\partial^2 \beta}{\partial \omega^2}\right| \cdot L \cdot \Delta\omega = \left|\ddot{\beta}\right| L\Delta\omega \tag{1.118}$$

Thus, the pulse spread is proportional to $\ddot{\beta}$, fiber length L, and signal spectrum width $\Delta\omega$.

Now, let us evaluate the pulse spread in terms of the GVD coefficient D [ns/(km nm)], defined as

$$D = \frac{1}{L}\frac{\partial \tau}{\partial \lambda} \tag{1.119}$$

It follows from (1.119) that

$$D = \frac{1}{L}\frac{\partial \tau}{\partial \lambda} = \frac{1}{L}\frac{\partial \tau}{\partial \omega} \cdot \frac{\partial \omega}{\partial \lambda} \tag{1.120}$$

It follows from (1.2) that

$$\frac{\partial \omega}{\partial \lambda} = \frac{\partial}{\partial \lambda}\left(\frac{2\pi c}{\lambda}\right) = -\frac{2\pi c}{\lambda^2} \tag{1.121}$$

Substituting (1.118) and (1.121) into (1.120), we obtain

$$D = \frac{1}{L} \cdot L \cdot \ddot{\beta} \cdot \left(-\frac{2\pi c}{\lambda^2}\right) = -\frac{2\pi c}{\lambda^2}\ddot{\beta} \tag{1.122}$$

Equation (1.122) specifies the relationship between the two dispersion coefficients, $\ddot{\beta}$ and D. Note that the nominal units of $\ddot{\beta}$ are $sec^2/(m\ rad^2)$, while the nominal units of D are sec/m^2. Typically, D is expressed in ps/(nm·km).

The expression for the pulse spread in terms of D can be obtained either by direct integration of (1.120) or by substituting (1.122) into (1.118); the result is, of course, the same in both cases:

$$\Delta\tau = |D| \cdot \Delta\lambda \cdot L \tag{1.123}$$

where $\Delta\lambda$ is the optical signal spectral width expressed in wavelength units (typically, nm).

The values of $\ddot{\beta}$ and of the GVD coefficient D depend on two chromatic dispersion phenomena:

- Material dispersion. The silica dielectric constant, ε, and, therefore, the refractive index, n, depend on the light frequency, ω.
- Waveguide dispersion. The propagation constant, β, depends on the light frequency, ω, in a nonlinear fashion even if ε were independent of ω.

Figure 1.22 shows the material dispersion, the waveguide dispersion, and the total fiber dispersion, D, versus wavelength for two fiber designs: conventional single-mode fiber and dispersion-shifted fiber, which will be discussed in Section 1.2.11.

Inspection of Figure 1.22 reveals several interesting conclusions:

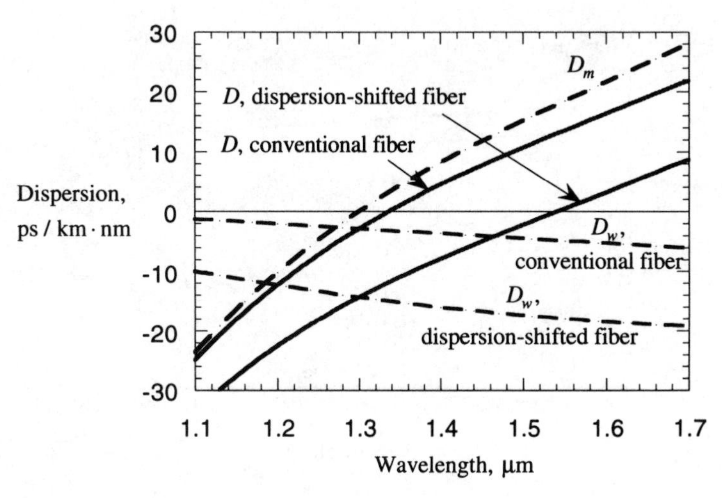

Figure 1.22 Total dispersion, *D*, and contributions of material dispersion, D_m, and waveguide dispersion, D_w, for conventional single-mode fibers and dispersion-shifted fibers. The zero-dispersion wavelength shifts to a higher value because of the waveguide contribution. (*After*: [8,14]. © 1986 IEEE.)

- For $\lambda > 1.3$ μm, the material dispersion is positive ($D_m > 0$), while the waveguide dispersion is negative ($D_w < 0$).
- At a certain wavelength, the waveguide dispersion cancels the material dispersion, giving the total dispersion of zero: $D = D_m + D_w = 0$; that wavelength is known as the zero-dispersion wavelength, λ_{ZD}.
- The zero-dispersion wavelength depends on fiber type; it is approximately equal to 1.3 μm for conventional single-mode fibers and 1.55 μm for dispersion-shifted fibers.

Thus, with conventional, that is, non-dispersion-shifted (see Section 1.2.11) fibers, it is desirable to operate close to the wavelength of 1.3 μm if the impact of dispersion is to be minimized. For example, if one is able to maintain the laser wavelength within, say, 10 nm from the zero-dispersion wavelength, the resulting dispersion is, from Figure 1.22, equal to 1 ps/(nm · km). Substituting this value into (1.123) and (1.123) in (1.107), we obtain

$$R_b \leq \frac{1}{4 \cdot 10^{-3} \cdot \Delta\lambda \cdot L} \text{ Gbps} \tag{1.124}$$

For example, if the laser spectral width is 1 nm, (1.124) yields 250 Gbps·km—a much larger value than that for multimode fibers.

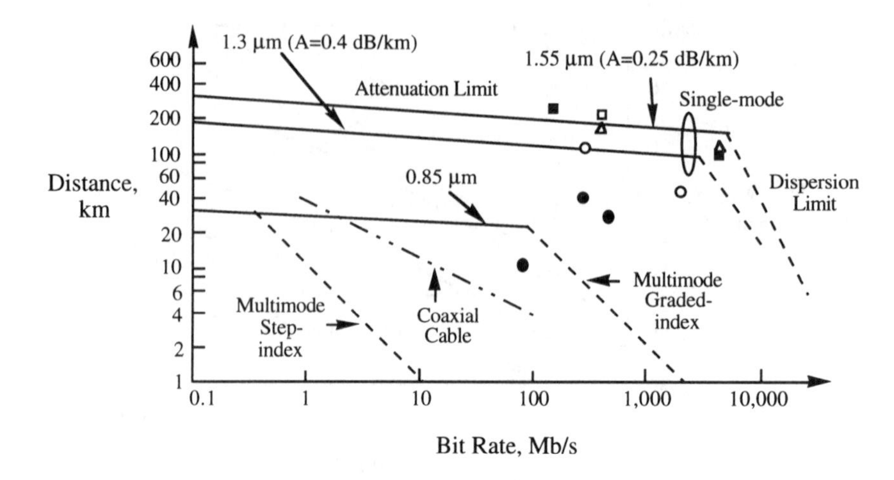

Figure 1.23 Maximum transmission distance versus bit rate for several types of optical fiber communication systems. The limits shown are *not* fundamental and are easily broken. (*Source:* [15]. © 1985 IEEE.)

The dispersion limits (1.112), (1.115), and (1.124) can be combined with the attenuation limit (1.105) into a single plot like the one shown in Figure 1.23. However, keep in mind that these limits refer to particular system designs and are not fundamental. For example, we can increase the attenuation limit (it is nearly parallel to the horizontal axis) by using optical amplifiers. We can increase the dispersion limit using narrow-spectrum transmitters and various dispersion-management techniques, such as keeping a tighter control of laser wavelength, using dispersion shifted or dispersion-flattened fibers, and using dispersion compensators. The latter option is discussed next.

Compensation for Chromatic Dispersion

Because chromatic dispersion is a linear phenomenon, the resulting distortion can be compensated for (i.e., corrected). Several dispersion compensation techniques have been developed and successfully demonstrated. For example, one can use an additional piece of fiber having dispersion opposite to that of the transmission fiber. Such fiber is known as dispersion-compensating fiber (DCF). Dispersion compensation is a subject of intensive current research and has been used in several impressive system experiments.

For example, Onaka et al. used dispersion compensation to transmit 55 wavelength-division-multiplexed channels, each running at 20 Gbps, for a total of 1.1 Tbps [16]. The transmission medium was 150 km of conventional single-mode fiber with the zero-dispersion wavelength of 1.3 μm. Since the transmission wavelength was 1.55 μm to allow the use of erbium-doped fiber amplifiers, the resulting dispersion was +15.2 ps/nm/km (see Fig. 1.22), and the dispersion slope was +0.064 ps/nm^2/km. To compensate for the transmission fiber dispersion, three segments of DCF were used (they were inserted into the transmission fiber every 50 km and colocated with erbium-doped amplifiers). The dispersion of the three DCF segments at 1545 nm was −800, −700, and −650 ps/nm, respectively. Each DCF had a large negative dispersion of −103 ps/nm/km and a large negative dispersion slope of −0.18 ps/nm^2/km. As a result, both $\ddot{\beta}$ and $\dddot{\beta}$ (in other words, both D and $\dot{D}$) have been compensated, leading to very small total dispersion for the entire 150-km transmission link: just 191 ps/nm.

While the foregoing experimental results look impressive at the time of writing of this book, they most probably will be surpassed by even more impressive results by the time this book is published.

The potential of dispersion-compensating techniques has now been realized by equipment vendors. It has been reported that some high-speed commercial systems already use dispersion compensation.

1.2.11 Dispersion-Shifted and Dispersion-Flattened Fibers

A conventional single-mode fiber has a fairly large mode field diameter. A typical commercial fiber of that type has the core diameter of 8.3 μm, refractive index delta of 0.37%, and mode field diameter of 8.7 μm. As a result, its zero-dispersion wavelength is about 1.3 μm (see Fig. 1.22), while its minimum-attenuation wavelength is 1.55 μm (see Fig. 1.18).

For optimum design of long-distance systems, it is highly desirable to shift the zero-dispersion wavelength of the fiber to the minimum-attenuation wavelength of 1.55 μm (the latter is fixed by the silica composition of the fiber). The resulting fiber is known as the dispersion-shifted fiber.

Dispersion-shifted fiber design involves a smaller mode field diameter leading to a larger waveguide dispersion and therefore shifting the fiber zero-dispersion wavelength to the right, toward longer wavelengths (see Fig. 1.22).

Dispersion-shifted fiber has excellent dispersion and attenuation properties at 1.5 μm but has high dispersion at 1.3 μm. To achieve small dispersion at *both* 1.3 μm and 1.5 μm, a more sophisticated index profile has to be used (Fig. 1.24). The resulting fiber is known as dispersion-flattened fiber.

Figure 1.25 compares the dispersion of conventional single-mode fiber with that of dispersion-shifted and dispersion-flattened fibers. Inspection of Figure 1.25 reveals that the dispersion-flattened fiber has excellent low dispersion over a huge range of wavelengths from 1.3 μm up to 1.6 μm corresponding to the bandwidth of 43.3 THz.

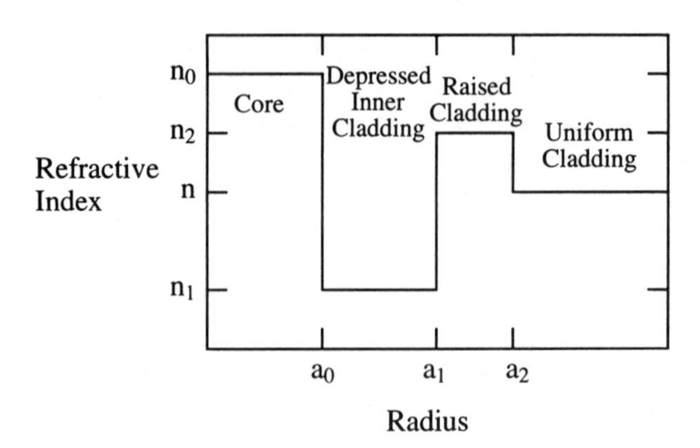

Figure 1.24 Refractive index profile of triple-clad fiber. (*Source:* [17]. © 1993 IEEE.)

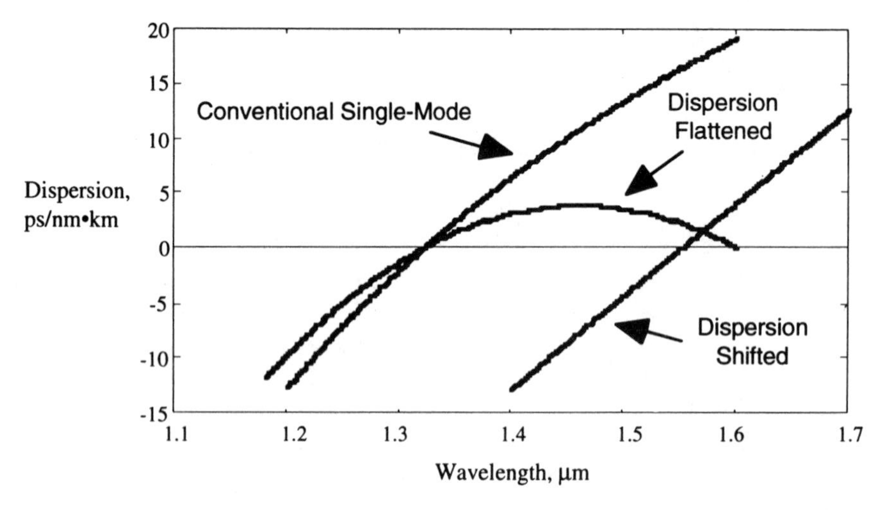

Figure 1.25 Dispersion versus wavelength for conventional single-mode fiber, dispersion-shifted fiber, and dispersion-flattened fiber. (*Source:* [8]. © 1992 John Wiley & Sons, Inc. Reprinted by permission of John Wiley & Sons, Inc.)

1.2.12 Polarization-Maintaining and Single-Polarization Fibers

Single-mode fibers discussed in previous subsections are not really single-mode: they support *two* modes with orthogonal states of polarization, say, E_x and E_y. In the ideal fiber, the difference between the two modes of "single-mode" fiber is immaterial, since both have theoretically identical propagation constants. However, a practical fiber can have some core eccentricity and different pressure along the x- and y-axes. As a result, the two modes are not completely degenerate (see more about this issue in Chapter 6).

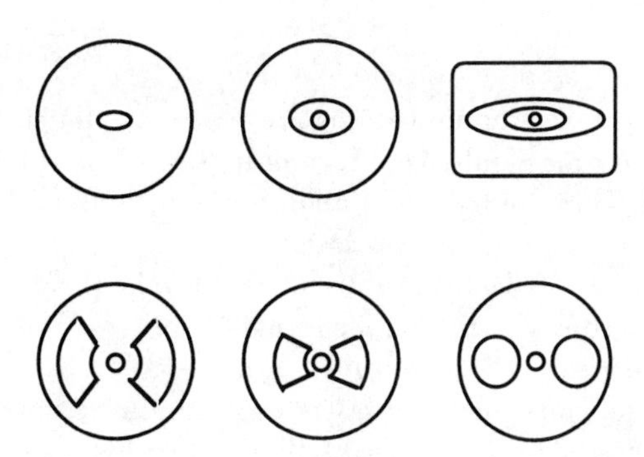

Figure 1.26 Representative designs of polarization-sensitive fibers. (*Source:* [18]. © 1988 Academic Press.)

Furthermore, some systems, such as coherent systems and certain filters, are polarization-sensitive and require light with a definite *state of polarization* (SOP) to function properly.

Polarization-maintaining and single-polarization fibers are designed to address these applications. Ideally, polarization-maintaining fibers maintain the SOP once the light is launched into the fiber. A different input SOP results in a different output SOP, but a given input SOP always results in the same output SOP.

Single-polarization fibers ideally transmit the light in one SOP only; all other SOPs are attenuated (ideally, to zero) while the one transmitted always results in the same output SOP.

Practical fibers for polarization-sensitive applications are based on asymmetric cores or stress members, as shown in Figure 1.26. They have two eigen-axes, say, x and y. Light with the SOP aligned along an eigen-axis emerges at the other end of the fiber with the same SOP. However, light with the SOP not aligned along one of the eigen-axes, will experience a change of its SOP as it propagates along the fiber; its output SOP is, therefore, unpredictable.

Depending on the material used for stress members, the attenuation for the two eigen-SOPs can be different from each other. If one of them is much larger than the other, the fiber behaves similarly to the single-polarization fiber. If not, the fiber behaves similarly to the polarization-maintaining fiber if the light is launched along one of the two eigen-axes.

1.3 LIGHT SOURCES

This section discusses the two most important light sources for optical communications: light-emitting diodes (LEDs) and semiconductor lasers.

1.3.1 LEDs

Most light sources for telecommunications are based on the electron-hole recombination in semiconductor materials. That recombination results in the release of energy. If the energy is released in the form of a photon, the recombination process is said to be *radiative;* if not, it is said to be nonradiative.

The recombination of electrons with holes can take place either spontaneously or as a result of an external stimulus, in the form of another photon. In the former case, the resulting emission is said to be *spontaneous;* in the latter, it is said to be *stimulated.*

LEDs use spontaneous emission; lasers use stimulated emission (they must still rely on spontaneous emission to start the oscillation process). We begin our discussion with LEDs and then proceed to lasers.

1.3.1.1 The P-N Junction

Figure 1.27 illustrates the energy level diagrams for a semiconductor p-n junction.

When the p-type and n-type materials are isolated, as in Figure 1.27(a), their Fermi levels are quite different. The Fermi level in the n-type material is close to the conduction band heavily populated by electrons. The Fermi level in the p-type material is close to the valence band heavily populated by holes.

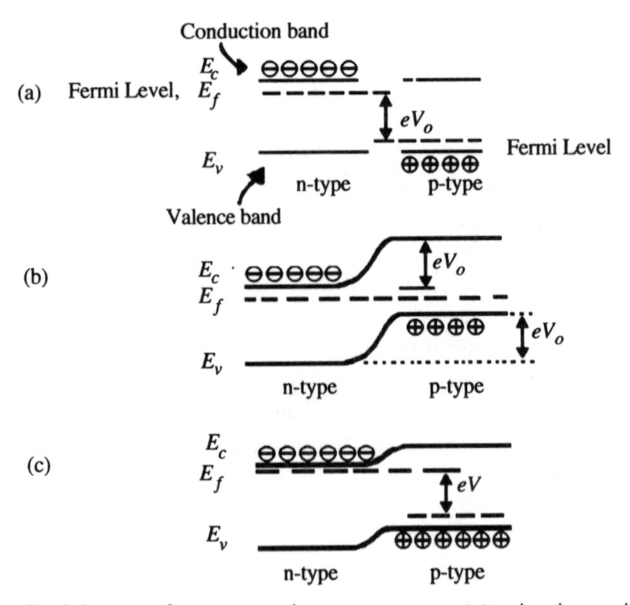

Figure 1.27 Energy level diagrams for a semiconductor p-n junction: (a) isolated p- and n-type materials, (b) the p-n junction in equilibrium, and (c) the p-n junction with an applied forward-bias voltage V.

When the two materials are brought together, a junction is formed, as shown in Figure 1.27(b). The Fermi level is now the same across the junction. As a result, a potential barrier is formed, and neither electrons nor holes can move across the junction; therefore, electron-hole recombinations are, to a first approximation, impossible.

When a forward-bias voltage, V, is applied to the junction, the potential barrier is reduced by eV, see Figure 1.27(c). As a result, electrons and holes are now free to move into the junction area. There, they recombine and produce spontaneous emission. A fraction of spontaneous emissions can be intercepted and used productively for communications, display, or other applications.

Surface-emitting diodes (Figure 1.28) collect the light in the direction perpendicular to the junction; edge-emitting diodes (Figure 1.29) collect the light in the direction parallel to the junction.

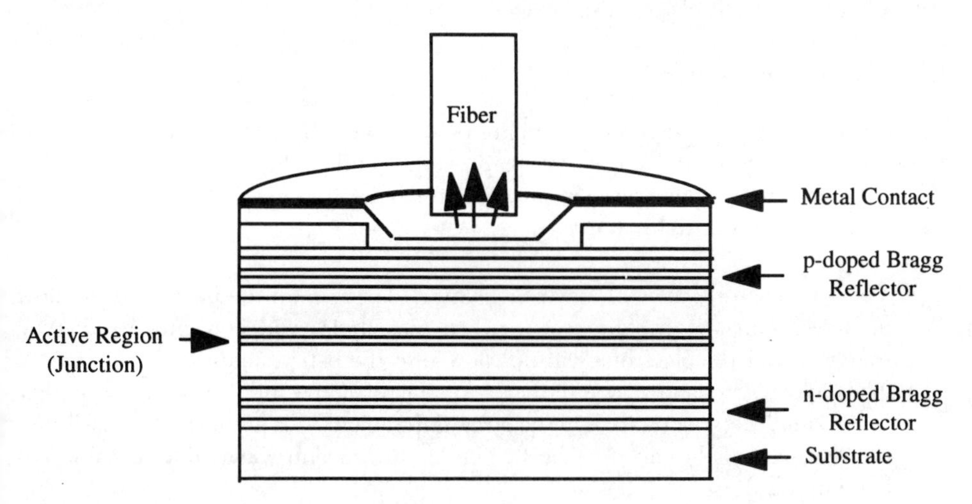

Figure 1.28 A surface-emitting diode. (*Source:* [19]. © 1995 IEEE.)

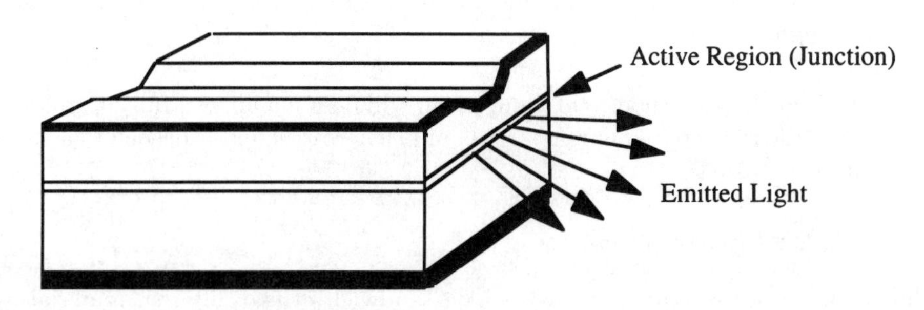

Figure 1.29 Edge-emitting diode. (*Source:* [20].)

Both surface-emitting and edge-emitting structures can be used to construct both LEDs and lasers. Surface-emitting lasers, like the one shown in Figure 1.28, incorporate gratings to provide optical feedback (without gratings, the device will function as an LED). Such devices are known as vertical-cavity surface-emitting lasers (VCSELs). A vast majority of currently deployed lasers are of the edge-emitting design shown in Figure 1.29.

The electron-hole recombinations creating photons take place in a narrow region around the junction. The width of the recombination region is an important device parameter. Yet in the case of the simple p-n junction in Figure 1.27, that width cannot be easily controlled by the device designer. Rather, it depends on the physics of the diffusion process of the electrons and holes. *Heterostructures* provide a full control over the width of the combination region and are used in practically all modern telecommunications lasers and LEDs. The next two sections explore the structure and the principle of operation of heterostructures.

1.3.1.2 Single Heterostructure

Figure 1.30 shows a *p-n* heterojunction made from two different materials: a narrow-gap n-type semiconductor and a wide-gap *p*-type semiconductor.

A heterostructure has three important properties that make it extremely useful for the construction of light sources:

- The barrier to the flow of electrons is much larger than the barrier to the flow of holes. Thus, with a modest forward bias, holes will flow into the n-type material, but the electrons will not flow into the p-type material.
- The dielectric constant, ε (and, therefore, the refractive index, $n = \sqrt{\mu\varepsilon}$), is higher in the small-gap region than in the large-gap region. This points to the possibility of creating a dielectric waveguide, similar to the slab waveguide discussed in Section 1.2.3.
- Light generated in material 1 has the photon energy Eg_1, which is smaller than the energy gap in material 2, Eg_2. Hence, light generated in material 1 is not absorbed in material 2; that points to the possibility of highly efficient surface-emitting light sources.

The single heterostructure in Figure 1.30 still does not allow a full control over the width of the recombination region. A double heterostructure is needed to achieve that goal.

1.3.1.3 Double Heterostructure

While a single heterostructure consists of a sandwich of two different materials, a double heterostructure consists of a sandwich of three different materials:

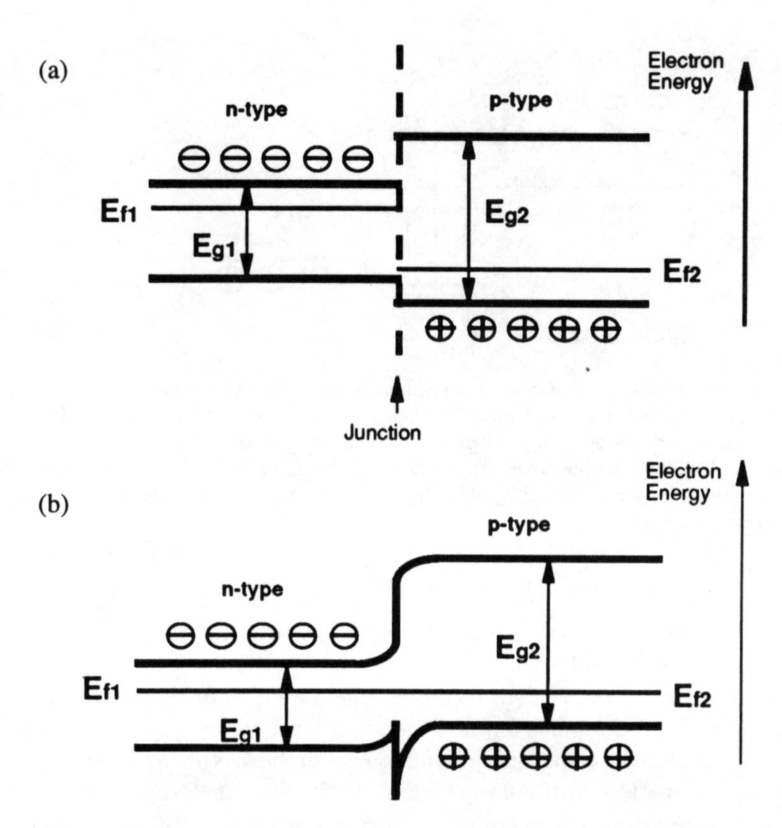

Figure 1.30 An n-p heterojunction made from a narrow-gap n-type semiconductor and a wide-gap p-type semiconductor. (a) Energy level diagram for the isolated semiconductors. (b) Energy level diagram for the n-p heterojunction. Note that the barrier to the flow of electrons is greater than in the corresponding n-p junction with equal-band-gap n and p materials, while the barrier to the flow of holes is smaller. (*Source:* [5]. © 1992. Reprinted by permission of Prentice-Hall, Inc., Upper Saddle River, NJ.)

- A heavily-doped N-type material having a large energy gap;
- A very lightly doped n-type material having a smaller energy gap;
- A heavily-doped P-type material having a large energy gap.

Figure 1.31 shows the structure of the resulting N-n-P double heterojunction.

When no voltage is applied to the double heterojunction in Figure 1.31, neither electrons (from the N-type material) nor holes (from the P-type material) can enter the central n-type material. Hence, recombinations are impossible, and no light is produced.

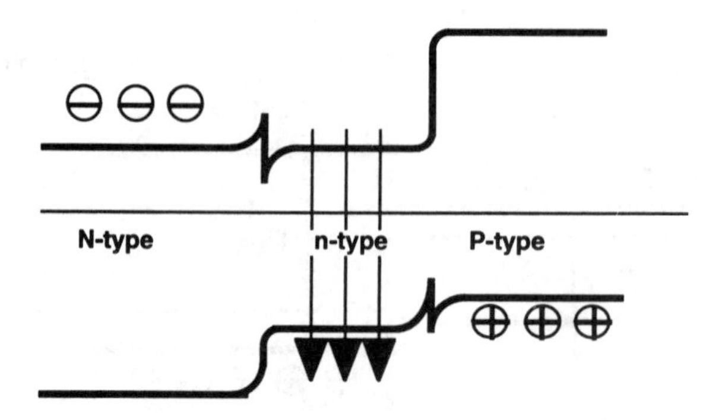

Figure 1.31 An N-n-P double heterojunction. The enhanced barriers to the flow of electrons in one direction and holes in the other are at different heterojunctions. The concentration of minority carriers in the center, narrow-band-gap, *n* region is substantially greater than in the homojunction diode. Recombination radiation takes place in a well-defined region that has dimensions selected by the device designer. (*Source:* [5]. © 1992. Reprinted by permission of Prentice-Hall, Inc., Upper Saddle River, NJ.)

However, when a forward bias is applied to the double heterojunction in Figure 1.31, both potential barriers (at the N-n and n-P interfaces) are reduced. As a result, electrons flow from left (from the N-type material) to right (into the n-type material). At the same time, holes flow from right (from the P-type material) to left (into the n-type material). Electrons and holes meet in the n-type material, recombine, and produce photons. The width of the recombination region is equal to the thickness of the n-type material and is established during the manufacturing process. It is, therefore, under full control of the device designer.

In addition, the N-n-P sandwich forms a dielectric waveguide consisting of a high-refractive-index core (n-type material) and low-refractive-index cladding layers (N-type and P-type materials). Photons generated in the core are, therefore, guided to device facets (provided, of course, they satisfy the usual propagation conditions; see Sections 1.2.3–1.2.7). This feature is used to create efficient edge-emitting lasers and LEDs.

1.3.1.4 LED Physical Structure

The LED physical structure consists, to a first approximation, of a double heterostructure buried in a thin layer grown on a semiconductor substrate (Figure 1.32).

The result is known as the buried double-heterostructure diode. Specific materials used depend on the wavelength desired; one specific example (a surface-emitting InGaAsP/InP LED) is shown in Figure 1.33.

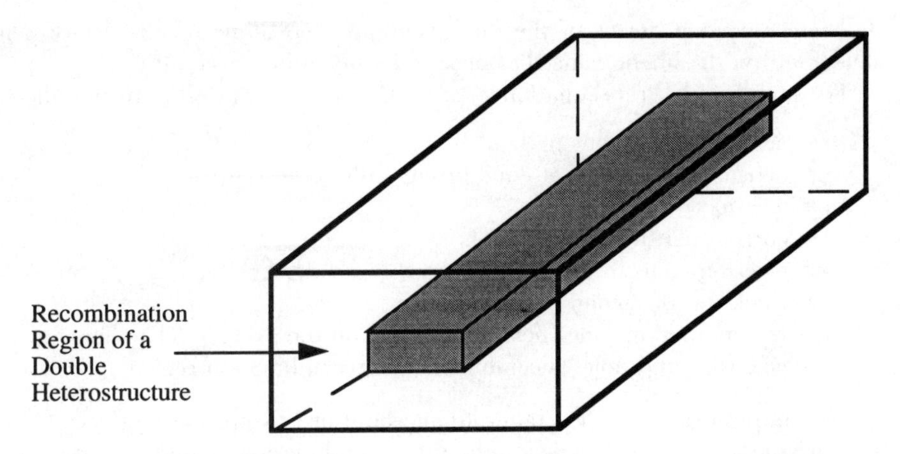

Figure 1.32 The double heterojunction is buried in the body of the device (by means of material regrowth after the junction is formed) to isolate it from potentially detrimental effects of the environment.

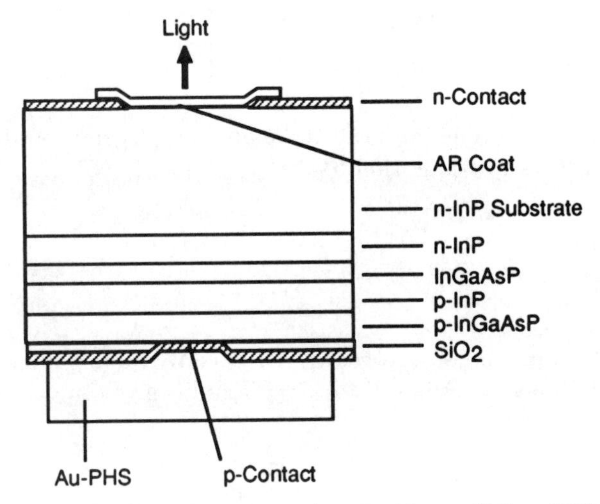

Figure 1.33 Cross-sectional diagram of an InGaAsP/InP LED. (*Source:* [21]. © 1991 Artech House, Inc.)

1.3.1.5 The LED Rate Equation

The physical processes in a semiconductor light source can be nicely modeled and understood using the so-called *rate equations,* which relate the device drive current and its physical dimensions to the density of photons and electrons in the recombination region.

There are many forms of the rate equations, depending on the device being modeled and on the phenomena that need to be taken into account.

The simplest LED rate equation is specified in terms of the following parameters:

$n \equiv$ density of electrons in $1/m^3$;
$n_0 \equiv$ density of electrons at equilibrium with no current;
$\Delta n \equiv n - n_0$;
$J \equiv$ current density in A/m^2;
$q \equiv$ electron charge;
$d \equiv$ thickness of recombination region;
$\tau_r \equiv$ recombination time (how much time, on the average, elapses before electrons and holes recombine in the recombination region).

The simplest rate equation takes into account just two physical effects: (1) the drive current brings new carriers into the recombination region at the rate of J/qd carriers/second; and (2) the recombination process "kills" carriers (and creates photons) at the rate $\Delta n/\tau_r$. Thus, the simplest rate equation looks as follows:

$$\frac{d(\Delta n)}{dt} = \frac{J}{qd} - \frac{\Delta n}{\tau_r} \quad \left[\frac{1}{m^3 s}\right] \tag{1.125}$$

The recombination time (also called the lifetime) depends on the material and on the density of holes in equilibrium P_0:

$$\tau_r = \frac{1}{B_r P_0} \tag{1.126}$$

where B_r is the recombination coefficient. The steady-state equilibrium solution to (1.125) is easily found by setting $d(\Delta n)/dt = 0$ and is given by

$$\Delta n = \frac{J\tau_r}{dq} \tag{1.127}$$

The corresponding rate of recombinations is

$$\textit{rate of recombinations} = \frac{\Delta n}{\tau_r} = \frac{J}{dq} \tag{1.128}$$

Using (1.128), we can find the LED output power as follows:

$$P_{out} = [rate\ of\ recombinations] \cdot \eta \cdot h\upsilon \cdot [volume]$$

$$= \frac{J}{dq} \cdot \eta \cdot h\upsilon \cdot [d \cdot Area]$$

$$= \frac{\eta h\upsilon}{q} \cdot I = \frac{\eta hc}{q\lambda} \cdot I \tag{1.129}$$

where υ is the light frequency, I is the LED drive current, and η is the quantum efficiency, that is, the fraction of recombinations that produces "useful" photons. Equation (1.129) implies that the LED output power is directly proportional to its drive current.

Depending on the definition of "useful" photons, the quantum efficiency, η, may mean quite different things. For example, we can define the "internal" quantum efficiency as

$$\eta_{int} = \frac{radiative\ recombinations}{total\ recombinations} \tag{1.130}$$

and the "external" quantum efficiency as

$$\eta_{ext} = \frac{power\ in\ fiber}{power\ emitted\ from\ the\ active\ region} \tag{1.131}$$

Figure 1.34 illustrates some of the mechanisms keeping the quantum efficiency below unity.

Equation (1.127) is the steady-state solution of (1.125). The transient solution to a step-like current density increase from J_1 to J_2 also can be easily obtained, because (1.125) is a first-order differential equation:

$$\Delta n = \frac{J_2 \tau_r}{dq} - \frac{(J_2 - J_1)}{dq} \exp\left(-\frac{t}{\tau_r}\right) \tag{1.132}$$

Inspection of (1.132) reveals that the LED time constant is equal to τ_r. Since τ_r is of the order of 1 nsec, LEDs are essentially low-speed devices, with the maximum modulation rates below 1 GHz. This limits their performance. Additional factors limiting LEDs' performance are their broad emission spectrum leading to large dispersion and their low output power.

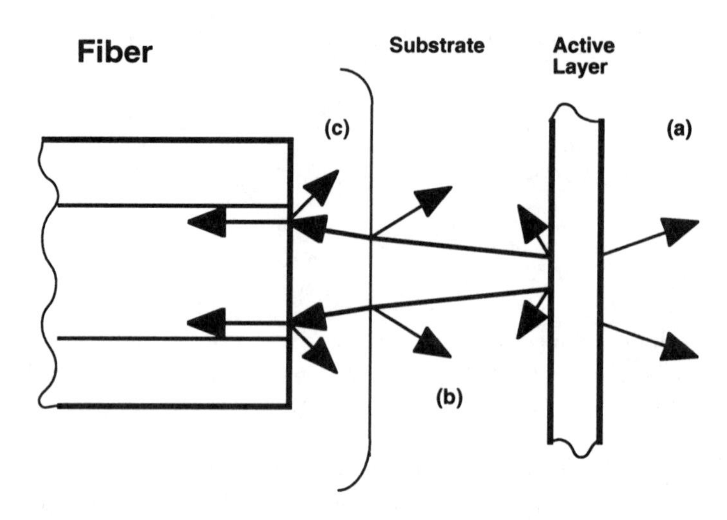

Figure 1.34 Quantum efficiency. Useful output, that is, light power propagating in the fiber, is less than the light power produced in the active region of the LED because of (a) radiation in directions other than toward the fiber, (b) partial reflection at the substrate-air and air-fiber interfaces, and (c) absorption in the substrate and other semiconductor materials between the active region and the output interface. (*From Introduction to Optical Fiber Communication Systems* by William B. Jones, Jr. © 1988 by Oxford University Press, Inc. Reprinted by permission.)

1.3.1.6 LED Output Spectrum

An LED generates a photon when a free electron in the conduction band, with the energy E_c, recombines with a hole in the valence band, with the energy E_v. The energy of the resulting photon, E_{ph}, is equal to the difference between E_c and E_v:

$$E_{ph} = E_c - E_v \tag{1.133}$$

Since both the conduction band and the valence band have a finite width, ΔE, the photon energy, E_{ph}, falls into the interval, $2\Delta E$. Fundamental physics predicts (and experiments confirm) that

$$\Delta E_{ph} = 2\Delta E \simeq 3.3kT \tag{1.134}$$

where k is Boltzmann's constant, and T is the absolute temperature.[4] Hence, an LED will emit a range of optical frequencies given by

4. Equation (1.134) gives a first-order approximation. A more accurate analysis requires evaluation of the density of states and occupation density.

$$\Delta\upsilon = \frac{3.3kT}{h} \tag{1.135}$$

where h is Planck's constant. We recall now that the frequency range, $\Delta\upsilon$, and the wavelength range, $\Delta\lambda$, are related by

$$\Delta\lambda = c\frac{\Delta\upsilon}{\upsilon^2} = \frac{\lambda^2}{c}\Delta\upsilon \tag{1.136}$$

Combining (1.135) with (1.136), we obtain the LED emission wavelength range:

$$\Delta\lambda = \frac{\lambda^2}{c}\Delta\upsilon = \frac{\lambda^2}{c}\cdot\frac{3.3kT}{h} \tag{1.137}$$

Note that $\Delta\lambda$ depends both on the emission wavelength and on the absolute temperature. This fact gives a convenient way to verify (1.137) experimentally. Figure 1.35 shows the emission spectrum of an LED at four different temperatures [22]. Inspection of Figure 1.35 reveals that $\Delta\lambda$ is indeed increasing with T, as predicted by (1.137).

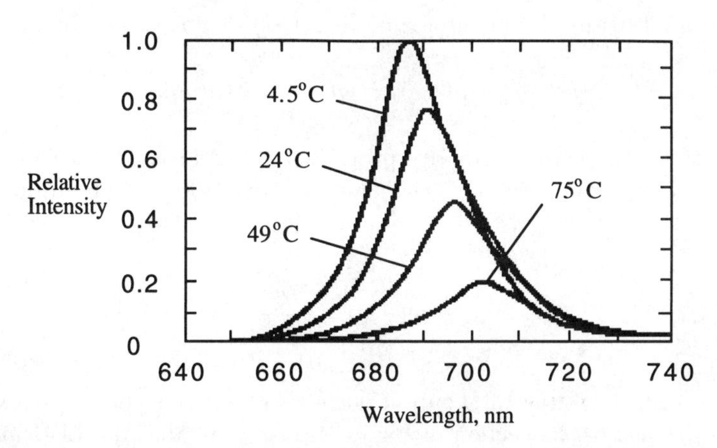

Figure 1.35 Emission spectrum of an LED at four different temperatures. (*Source:* [22]. © 1994 IEEE.)

1.3.1.7 LED Modulation Response

In this section, we investigate what happens when an LED is driven by a DC bias current *and* an AC modulation current, so that the current density J is given by

$$J = J_0[1 + m_j \exp(j\omega t)] \tag{1.138}$$

where J_0 reflects the bias, m_j is the current (and current density) modulation depth, and ω is the modulation frequency. Because the differential equation (1.125) describing the LED is linear, the resulting Δn has the same functional form as J:

$$\Delta n = N_0 \{1 + m_N \exp[j(\omega t - \theta)]\} \tag{1.139}$$

where m_N is the modulation depth of Δn, θ is the phase shift in Δn with respect to J, and N_0 is the steady-state value of Δn given [see (1.127)] by

$$N_0 = \frac{J_0 \tau_r}{dq} \tag{1.140}$$

To find m_N and θ, we substitute (1.138) and (1.139) into (1.125), the differential equation describing the LED; omitting the DC terms from the result, we obtain:

$$j\omega m_N N_0 \exp[j(\omega t - \theta)] = \frac{J_0 m_j}{dq} \exp(j\omega t) - \frac{N_0 m_N}{\tau_r} \exp[j(\omega t - \theta)] \tag{1.141}$$

Substituting (1.140) in (1.141) and canceling $1/dq \exp(j\omega t)$, we obtain

$$j\omega\tau_r m_N \exp(-j\theta) = mj - m_N \exp(-j\theta) \tag{1.142}$$

Because the LED output power is proportional to Δn [see (1.128) and (1.129)], the output power modulation index is equal to $m_N \exp(-j\theta)$, which, in turn, is equal to [see (1.142)]

$$m_N \exp(-j\theta) = \frac{mj}{1 + j\omega\tau_r} \tag{1.143}$$

Figure 1.36 shows the LED output power modulation response, m_N, versus the modulation frequency. Inspection of Figure 1.34 reveals that the LED modulation response is identical to that of a first-order low-pass filter with the 3-dB modulation bandwidth:

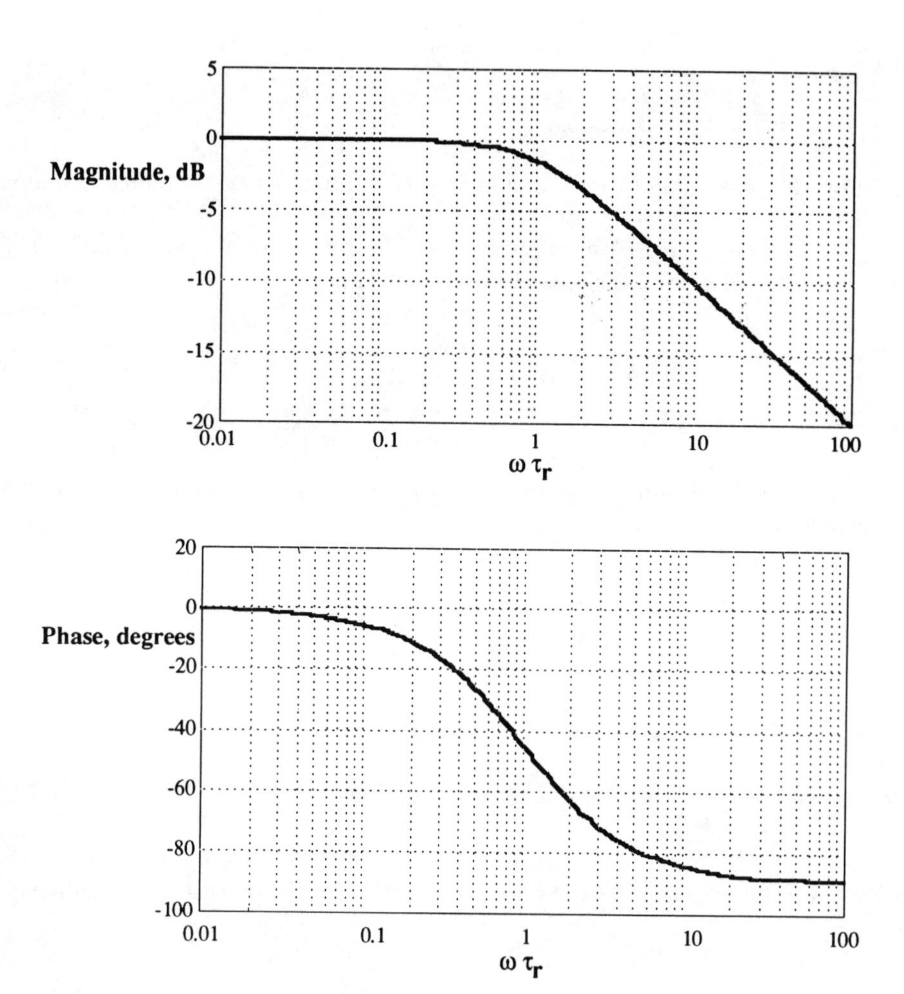

Figure 1.36 LED modulation response m_N versus modulation frequency. (*Source:* [11].)

$$\omega_{3dB} = \frac{1}{\tau_r} \tag{1.144}$$

Equation (1.144) shows that the LED modulation bandwidth depends only on the recombination lifetime, τ_r. Once the LED material has been selected, the LED designer has no control over the LED ultimate bandwidth. As a result, LEDs are essentially slow devices with the bandwidth limited to $1/\tau_r$, or below 1 GHz for practical semiconductor materials.

1.3.2 Lasers

1.3.2.1 The Fabry-Perot Resonator

The Fabry-Perot resonator shown in Figure 1.37 is fundamental to an understanding of lasers.

Consider an optical plane wave traveling from left to right within the Fabry-Perot resonator in Figure 1.37. Its electric field is given by

$$E(t, x) = A \exp\left(-\frac{\alpha_s x}{2}\right) \exp\left[j(\omega t - \beta x)\right] \tag{1.145}$$

where A is the field amplitude at the left mirror, α_s is the intensity attenuation coefficient, x is the distance from the left mirror, ω is the light frequency in radians per second, and β is the propagation constant in inverse meters. The complex amplitude of the field E at point x is given by

$$A^*(x) \equiv A \exp\left(-\frac{\alpha_s x}{2}\right) \exp(-j\beta x) \tag{1.146}$$

Hence, (1.145) can be rewritten as

$$E(t,x) = A^* \exp(j\omega t) \tag{1.147}$$

In lasers, the active medium provides a gain characterized by the intensity gain coefficient g, in inverse meters. As a result, (1.146) for lasers has to be modified as follows:

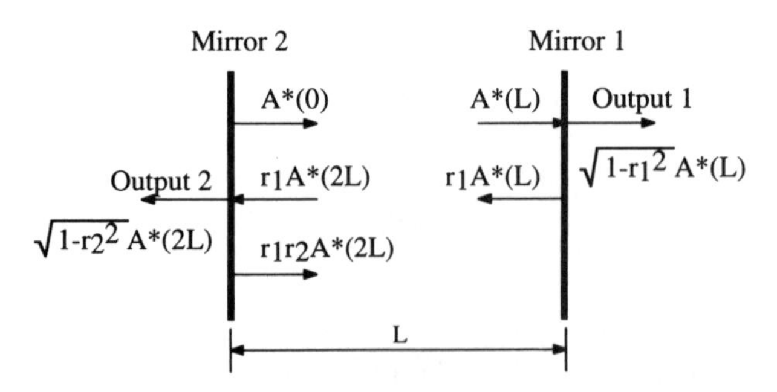

Figure 1.37 A Fabry-Perot resonator consisting of two plane parallel mirrors, with reflection coefficients r_1 and r_2, separated by distance L. (*Source:* [12]. © 1993. Reprinted by permission of Prentice-Hall, Inc., Upper Saddle River, NJ.)

$$A^*(x) \equiv A \exp\left(-\frac{(\alpha_s - g)x}{2}\right) \exp(-j\beta x) \tag{1.148}$$

Let a wave with the complex amplitude (1.148) propagate to the right mirror, be reflected, propagate back to the left mirror, and be reflected again. The result is the following field at the left mirror (just after reflection):

$$E(t,0) = r_1 r_2 A \exp[(g - \alpha_S)L]\exp[j(\omega t - 2\beta L)] = B^* \exp(j\omega t) \tag{1.149}$$

where r_1 is the reflection coefficient at Mirror 1, r_2 is the reflection coefficient at Mirror 2, L is the length of the Fabry-Perot resonator, and B^* is the field amplitude at the left mirror after one round trip given by

$$B^* \equiv r_1 r_2 A \exp[(g - \alpha_S)L]\exp(-j2\beta L) \tag{1.150}$$

For steady-state oscillations, B^* (the field amplitude after a round trip around the cavity) must be equal to $A^*(0)$ (the initial field amplitude):

$$B^* = A^*(0) \tag{1.151}$$

Substituting (1.148) and (1.150) into (1.151), we obtain the following two conditions:

$$r_1 r_2 A \exp[(g - \alpha_s)L] = A \tag{1.152}$$

and

$$\exp(-j2\beta L) = 1 \tag{1.153}$$

Equations (1.152) and (1.153) are known as the *amplitude condition* and the *phase condition*, respectively. The amplitude condition (1.152) can be satisfied only if the gain is sufficiently large:

$$g \geq g_t \equiv \alpha_s + \frac{1}{L}\ln\frac{1}{r_1 r_2} = \alpha_s + \frac{1}{2L}\ln\frac{1}{R_1 R_2} \tag{1.154}$$

where g_t is the threshold gain, and R_1 and R_2 are the power reflection coefficients of the two mirrors:

$$R_1 = |r_1|^2 \tag{1.155}$$

$$R_2 = |r_2|^2 \tag{1.156}$$

The phase condition (1.153) can be satisfied only if

$$2\beta L = 2\pi m \tag{1.157}$$

where m is an arbitrary integer. Recall that

$$\beta = \frac{2\pi n}{\lambda} \tag{1.158}$$

where n is the refractive index, and λ is the vacuum wavelength. Substituting (1.158) into (1.157), we obtain

$$m\lambda = 2Ln \tag{1.159}$$

or

$$\lambda = \frac{2Ln}{m} \tag{1.160}$$

The values of λ satisfying (1.160) give the resonant wavelengths, or modes, of the Fabry-Perot resonator. Equation (1.160) shows that there is an infinite number of modes. For some of those modes, the amplitude condition (1.154) may be satisfied, too, giving *modes of oscillation*, or *lasing modes.*

Let us find the wavelength and the frequency spacing between adjacent modes. From (1.160), for the mth mode,

$$m = \frac{2Ln}{\lambda_m} = \frac{2Ln}{c} f_m \tag{1.161}$$

where λ_m and f_m are the wavelength and the frequency of the mth mode, respectively. Similarly, for the mode number $(m - 1)$,

$$m - 1 = \frac{2Ln}{\lambda_{m-1}} = \frac{2Ln}{c} f_{m-1} \tag{1.162}$$

Subtracting (1.162) from (1.161), we obtain

$$1 = \frac{2Ln}{c}(f_m - f_{m-1}) = \frac{2Ln}{c}\Delta f \tag{1.163}$$

where Δf is the frequency spacing between adjacent modes. Hence, the mode frequency spacing, Δf, and the mode wavelength spacing, $\Delta\lambda$, are given by

$$\Delta f = \frac{c}{2Ln} \tag{1.164}$$

$$\Delta\lambda = \frac{\lambda^2}{2Ln} \tag{1.165}$$

Equations (1.164) and (1.165) predict the spacing between longitudinal modes; they do *not* take into account transversal modes. They also indicate that the mode spacing (in hertz) depends only on the resonator length and material.

1.3.2.2 Physical Structure of a Semiconductor Laser

Fundamentally, a semiconductor laser can have the following components:

- An optical waveguide to guide the generated light. To support a single transversal mode, both the thickness and the width of the waveguide must be sufficiently small [see (1.70)].
- A gain, or active, section. Gain is provided by the stimulated emission taking place when an optical wave traveling through an optical waveguide interacts with the electrons and holes in the recombination region of a double heterostructure.
- An optical feedback to facilitate oscillations. In the simplest optical oscillator—a Fabry-Perot laser—the feedback is provided by light reflections at the two ends of the laser stemming from the refractive index discontinuity between the laser semiconductor material and the surrounding air.
- A frequency-selective element (usually grating) that can be used to select just one mode of the Fabry-Perot cavity.

A laser *must* have (1) an optical waveguide, (2) a gain (or active) section, and (3) an optical feedback. A frequency-selective component may or may not be included, depending on whether the presence of several modes is tolerable.

The optical waveguide and the gain section can overlap: a single heterostructure can provide both light guiding and gain.

The physical structure of a Fabry-Perot laser can be very similar or even identical to that of an edge-emitting LED (see Figure 1.32). In fact, some factories are rumored to sell Fabry-Perot lasers that have been manufactured but failed to lase as LEDs. Further, *any* laser acts as an LED below "threshold," defined as the minimum current needed to make the laser oscillate (Fig. 1.38).

Above threshold, the gain provided by the active section (through stimulated emission) offsets device losses stemming from the losses within the device and from the optical power escaping through the two end facets of the laser. The device is then said to have a positive net gain. The positive net gain, in conjunction with the feedback (provided either by end reflections or by gratings; see Sections 1.2.3–1.2.5), make the device lase at one or more modes defined by (1.160).

The modes defined by (1.160) are known as longitudinal, or Fabry-Perot,

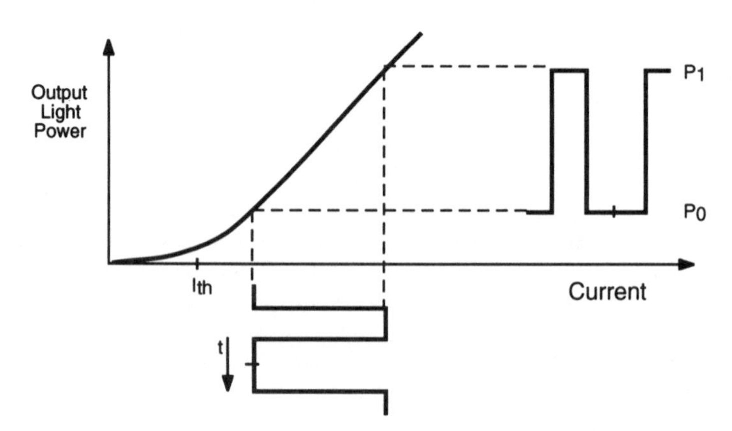

Figure 1.38 Light versus drive current characteristic of a semiconductor laser. Below threshold, the device acts as an LED producing spontaneous emission. Above threshold, the device oscillates (lases) using the gain provided by the stimulated emission. (*Source:* [12]. © 1993. Reprinted by permission of Prentice-Hall, Inc., Upper Saddle River, NJ.)

modes. If the optical waveguide is insufficiently narrow or insufficiently thin to support just one transversal mode [see (1.70)], several transversal modes can oscillate simultaneously. To prevent this from happening, both the thickness and the width of the optical waveguide must be carefully controlled. Semiconductor laser physicists developed elaborate techniques to control both very accurately.

The waveguide thickness is controlled by the epitaxial growth of a double heterostructure, as illustrated in Figure 1.39. The same double heterostructure can be

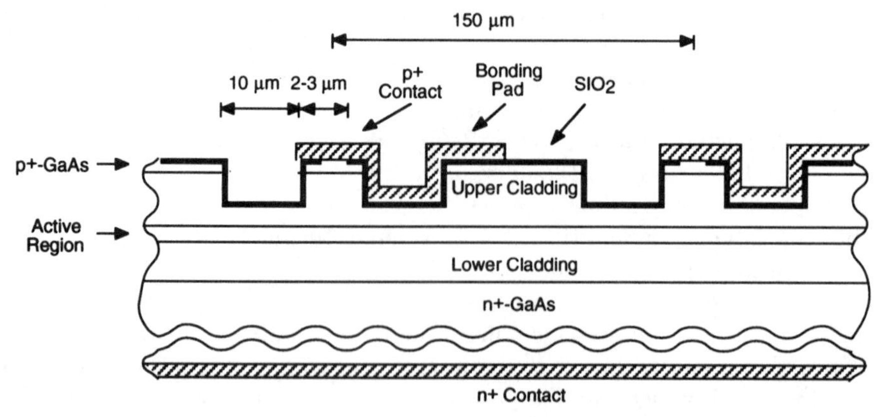

Figure 1.39 Cross-sectional schematic diagrams of a ridge-waveguide laser array structure. (*Source:* [23]. © 1995 IEEE.)

(but is not always) used both as a waveguide and as an active layer. Since the thickness of an epitaxial layer is easily controlled and can be made less than one wavelength, it is fairly easy to satisfy (1.70) in the vertical direction. A semiconductor laser physicist would say that "vertical confinement is an easy problem."

Horizontal confinement is the control of the width of the optical waveguide and is a more difficult problem. In early lasers, horizontal confinement was achieved by using a narrow conductive strip as the top contact supplying the drive current to the laser, as shown in Figure 1.40(a). The idea was that the electrons would then flow through the device in a fairly narrow formation, and the width of that formation would determine the active width of the waveguide. This principle became known as gain guiding and is still used in some devices.

Gain guiding is a simple technique. Unfortunately, electrons do tend to spread in gain-guiding lasers, leading to fairly unstable waveguide width.

Figure 1.40(b) shows an improved solution that constrains the width of the electron flow (and, therefore, the useful width of the active section) with insulating regions; Figure 1.41 shows a laser that utilizes that approach.

In many modern lasers, so-called *index guiding* is used. It is based on the creation of a rectangular optical waveguide limited on all four sides by the regions of lower refractive index, as shown in Figure 1.40(c). The lower and the upper regions are created by a usual double heterostructure. After the double heterostructure has been grown, the regions lying to the left and to the right of the waveguide are removed by etching, and a new insulating material with a lower refractive index is grown instead. This creates the horizontal confinement. Finally, a few additional layers are grown on top of the laser to provide an ohmic contact. The resulting structure is known as the *buried heterostructure* (BH) and is used in many modern lasers. (See Figure 1.48 in Section 1.3.2.11, the cross section of a commercial BH laser.)

1.3.2.3 Laser Output Spectrum: Spectral Width and Linewidth

If a laser waveguide supports only one transversal mode, the laser is frequently said to be single mode. A single-mode (in the foregoing sense) laser can, however, oscillate in several longitudinal modes [defined by (1.160)] simultaneously. Such a laser can be (and sometimes is) called a single-transversal-mode, multilongitudinal-mode laser.

To simplify the terminology and avoid the confusion, we will call a laser single mode if its waveguide supports a single transversal mode, whether it oscillates in a single longitudinal mode or in multiple longitudinal modes. We will call a laser single frequency if it supports a single transversal mode *and* a single longitudinal mode. We will call a laser multifrequency if it oscillates at several light frequencies simultaneously, be they longitudinal or transversal modes. Figure 1.42 illustrates the output spectrum of a multifrequency laser.

A laser oscillates at frequencies (modes) at which the gain is sufficiently large

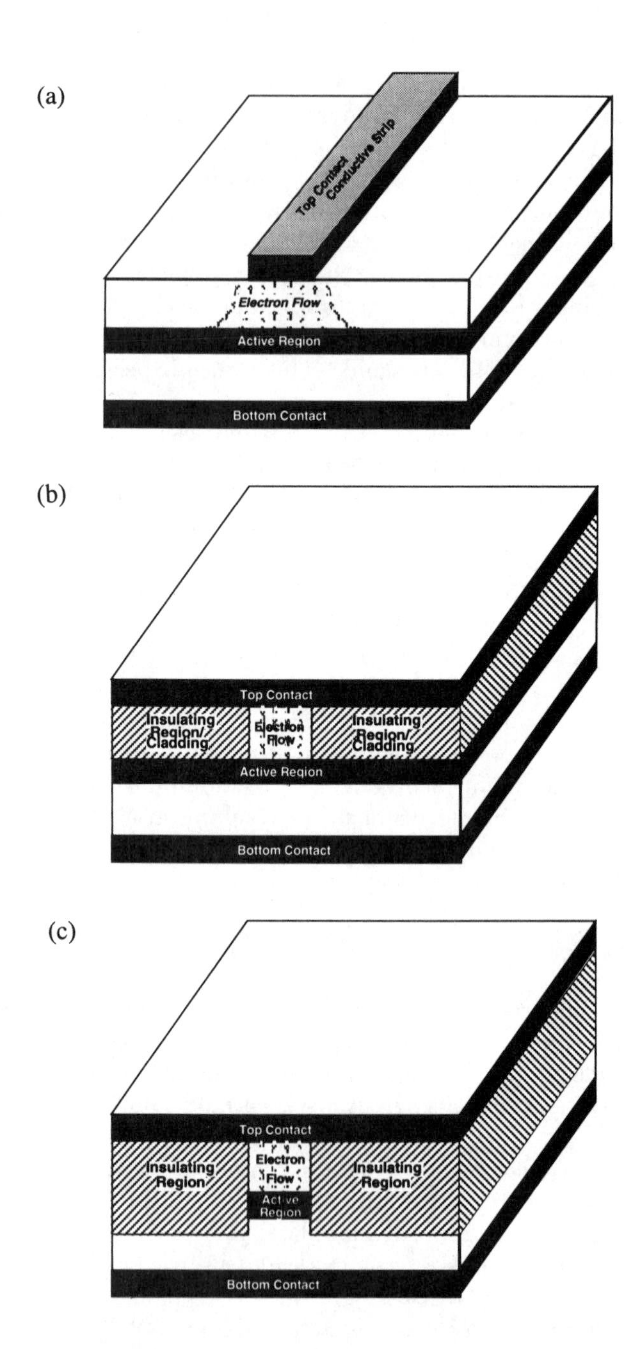

Figure 1.40 Three techniques of horizontal confinement: (a) gain guiding, (b) insulating regions cladding, and (c) index guiding. (*Source:* [20].)

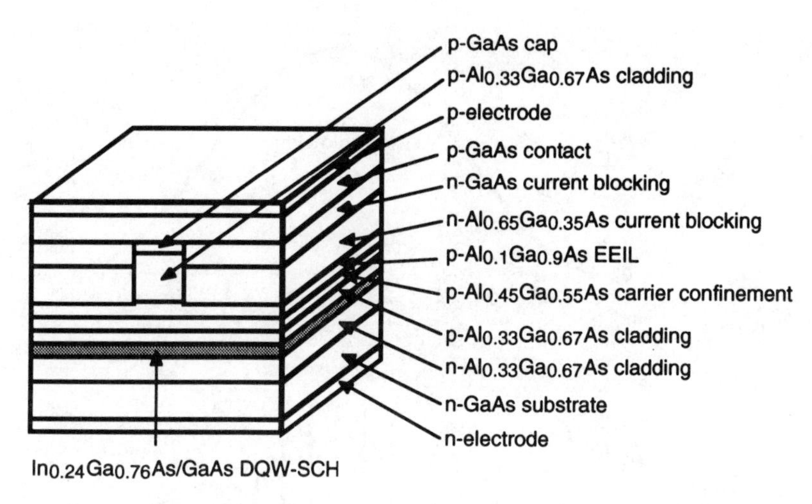

Figure 1.41 Schematic drawing of a 0.98-μm InGaAs-AlGaAs strained quantum-well heterostructure laser. (*Source:* [24]. © 1995 IEEE.)

so that the amplitude condition (1.154)—and the phase condition (1.160)—are satisfied. The width of the gain curve is determined by the semiconductor material the laser is built from, while the spacing between the longitudinal modes of a Fabry-Perot laser is determined by the laser length. Because the former is much larger than the latter, Fabry-Perot lasers can run at several longitudinal modes at once. The exact number of modes depends on the drive current. Typically, as we increase the drive current and exceed the threshold, one mode appears in the output spectrum (the one having the highest net gain). Then, as we increase the drive current further, additional modes (so-called side modes) may appear. The power of side modes is smaller than that of the first (main) mode, because side modes lie further from the peak of the gain curve and, therefore, have smaller gain.

The total width of the emission spectrum, including all oscillating modes, is known as the laser spectral width. The width of an individual mode is known as the linewidth. For a single-frequency laser, the spectral width is equal to the linewidth. For a multifrequency laser, the spectral width is *not* equal to the linewidth (see Fig. 1.42). The linewidth of single-frequency lasers is determined by the frequency (or phase) noise. This subject is further discussed in Section 1.3.2.10.

To eliminate side modes, we can use gratings etched into the laser body. The resulting devices are known as *distributed feedback* (DFB) and *distributed Bragg reflection* (DBR) lasers. DFB and DBR lasers are discussed in Section 1.3.2.5. Since both DFB and DBR lasers are based on Bragg reflections, a brief discussion of Bragg reflections is provided next.

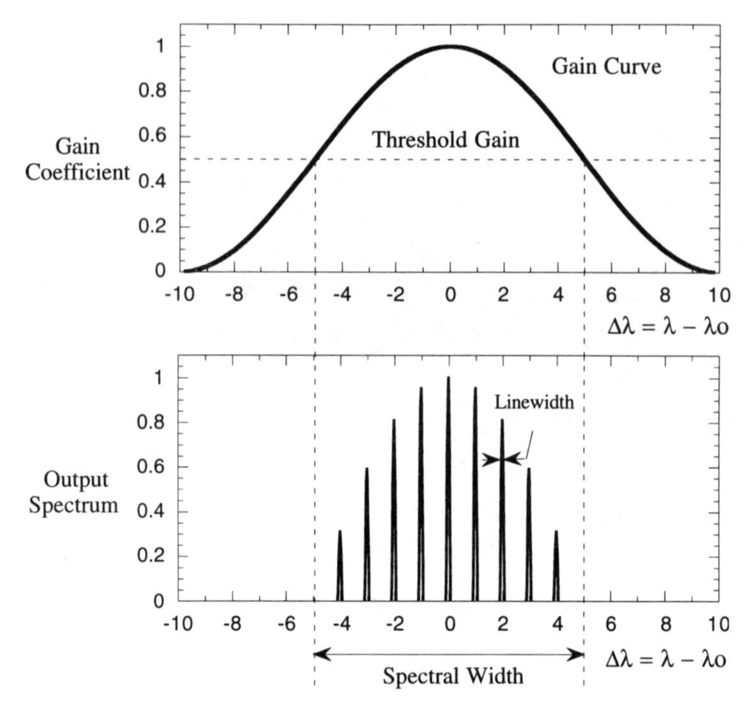

Figure 1.42 Gain curve and output spectrum of a multifrequency laser. (*Source:* [20].)

1.3.2.4 Bragg Reflections

Consider the experiment shown in Figure 1.43. A plane optical wave, E_{in}, travels through several weakly reflecting mirrors, $M_1, M_2, \ldots, M_N$. Each mirror transmits most of the incident power and reflects a small fraction of the incident power. Thus, $|t| \approx 1$ and $|r| << 1$, where r and t are the transmission coefficient and the reflection coefficient, respectively. Since at each mirror a fraction of the light energy is reflected back, the following reflected waves, E_{ri}, and transmitted waves, E_{ti}, are created at each mirror:

$$E_{r1}(0) = E_{in}(0) \cdot r; \quad E_{t1}(0) = E_{in}(0) \cdot t \tag{1.166}$$

$$E_{r2}(\Lambda) = E_{t1}(0) \cdot r \cdot e^{-j\beta\Lambda}; \quad E_{t2}(\Lambda) = E_{t1}(0) \cdot t \cdot e^{-j\beta\Lambda} \tag{1.167}$$

$$E_{r3}(2\Lambda) = E_{t2}(\Lambda) \cdot r \cdot e^{-j\beta\Lambda}; \quad E_{t3}(2\Lambda) = E_{t2}(\Lambda) \cdot t \cdot e^{-j\beta\Lambda} \tag{1.168}$$

$$\vdots$$

$$\begin{aligned} E_{rN}[(N-1)\Lambda] &= E_{t(N-1)}[(N-2)\Lambda] \cdot r \cdot e^{-j\beta\Lambda}; \\ E_{tN}[(N-1)\Lambda] &= E_{t(N-1)}[(N-2)\Lambda] \cdot t \cdot e^{-j\beta\Lambda} \end{aligned} \tag{1.169}$$

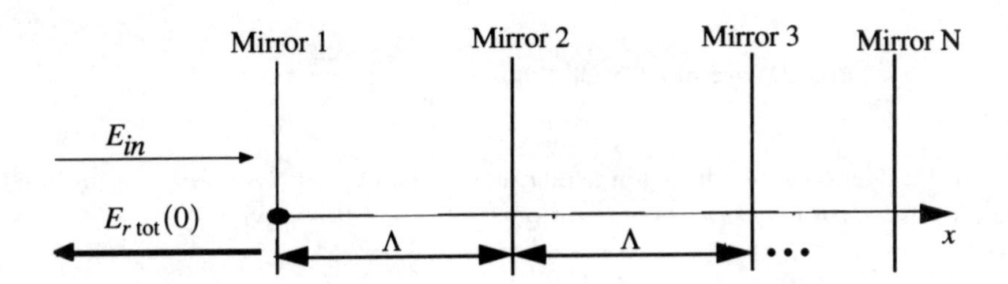

Figure 1.43 The incident wave E_{in} travels through N partially (and weakly) reflective mirrors. At each mirror, the wave is partially transmitted and partially reflected.

where $E_{rk}[(k-1)\Lambda]$ is the reflected wave generated by the kth mirror at its location, $(k-1)\Lambda$; $E_{ik}[(k-1)\Lambda]$ is the transmitted wave generated by the kth mirror at its location, $(k-1)\Lambda$; Λ is the structure period (i.e., the spacing between adjacent mirrors); β is the light propagation constant; r is the mirrors' reflection coefficient; and t is the mirrors' transmission coefficient.

To find the total reflected wave at $x = 0^-$, we need to add reflections from all mirrors while taking into account the additional phase delay due to light propagation between the mirrors and the mirrors' transmission coefficient, t:

$$E_{rtot}(0) = E_{r1}(0) + E_{r2}(\Lambda) \cdot e^{-j\beta\Lambda} \cdot t + E_{r3}(2\Lambda) \cdot e^{-2j\beta\Lambda} \cdot t^2 + \ldots + E_{rN}[(N-1)\Lambda] \cdot e^{(N-1)j\beta\Lambda} \cdot t^{N-1} \tag{1.170}$$

We neglected multiple reflections between mirrors in (1.170).

Substituting (1.166) to (1.169) in (1.170), we obtain

$$E_{rtot}(0) = rE_{in}(0)[1 + t^2 e^{-2j\beta\Lambda} + t^4 e^{-4j\beta\Lambda} + t^{2(N-1)} e^{-2(N-1)j\beta\Lambda}] \tag{1.171}$$

Since the expression in the brackets is a geometric series, (1.171) can be rewritten as

$$E_{rtot}(0) = rE_{in}(0) \cdot \frac{1 - M^N}{1 - M} \tag{1.172}$$

where

$$M \equiv t^2 \cdot e^{-2j\beta\Lambda} \tag{1.173}$$

Equations (1.172) and (1.173) indicate that, with a large number of mirrors, a strong total reflection may be obtained even if the reflection from each individual mirror is weak ($|t| << 1$). The condition for strong reflections is, from (1.172) and (1.173), given by

$$\arg(M) = 2\arg t - 2\beta\Lambda = 2\arg t - 2\cdot\frac{2\pi}{\lambda}\Lambda = n\cdot 2\pi \tag{1.174}$$

where $\arg(x)$ denotes the phase angle of x, and n is an arbitrary integer. Assuming for simplicity that $\arg t = 0$ (a generalization for $\arg t \neq 0$ is obvious), (1.174) can be rewritten as

$$2\Lambda = m\lambda \tag{1.175}$$

where $m = -n$ is an arbitrary integer. The condition for strong reflections (1.175) is known as the Bragg condition.

More general forms of the Bragg condition (1.175) describe more sophisticated cases when (1) the light travels at an angle to the mirror's array, and (2) the array moves, as in the case of interaction of light with sound. Descriptions of these more general cases can be found elsewhere, for example, in [9].

In our case, (1.175) is quite sufficient. It describes not only the case of partially reflective mirrors, but a general case of any periodic perturbation, including grating.

Let us summarize the physical meaning of (1.175). It states that a grating acts as a strongly reflective mirror if its period is equal to one-half the light wavelength (in which case, the grating is said to be of the first order) or two halves of the light wavelength (in which case, the grating is said to be of the second order), and so on. This principle is used in both DFB and DBR lasers.

1.3.2.5 Distributed Feedback and Distributed Bragg Reflection Lasers

Both DFB and DBR lasers are designed to oscillate in just one longitudinal (and transversal) mode, that is, to produce single-frequency (monochromatic) light. Further, both rely on grating as distributed reflectors (see Section 1.3.2.4). However, their physical structures are different. Figure 1.44 compares the cross sections of a DFB and a DBR laser.

In modern DFB lasers, a grating is etched into the entire length of the laser except for a narrow region in the center whose length is equal to one-quarter the wavelength, $\lambda/4$. These lasers are thus known as quarter-wavelength-shifted DFB lasers. One way to understand the physics of this structure is to consider it as a short $\lambda/4$ resonator surrounded by two distributed reflectors. The active region, the optical waveguide, and the grating cover the entire length of the DFB laser (except for the $\lambda/4$ break in the grating).

In DBR lasers, the optical waveguide also runs through the entire length of the device. However, the active region and the distributed reflectors do not overlap and cover only parts of the device length. In some DBR lasers, the distributed reflector provides only one "mirror"; the other "mirror" is provided by the device-air interface,

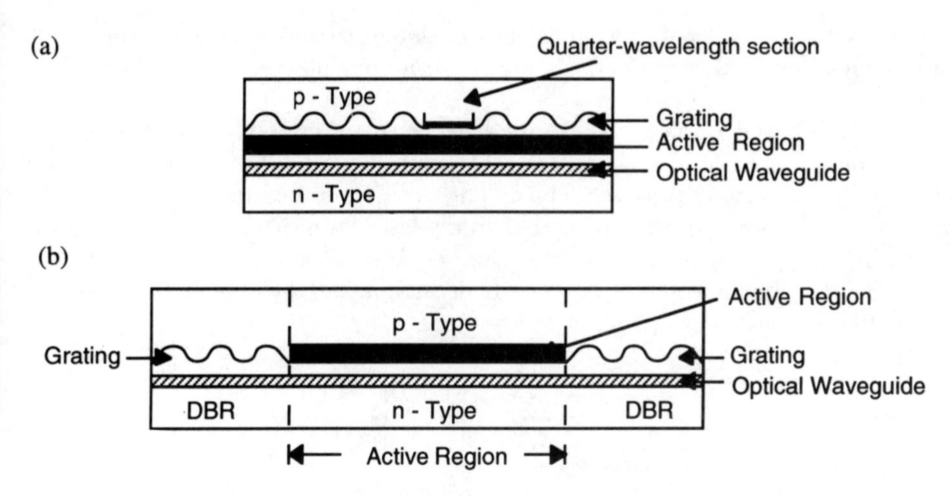

Figure 1.44 Schematic illustration of distributed feedback (DFB) and distributed Bragg reflector (DBR) semiconductor lasers: (a) DFB laser and (b) DBR laser. (*After:* [25]. © 1993 Chapman and Hall.)

just as in Fabry-Perot lasers. In other DBR lasers, like the one shown in Figure 1.44(b), two gratings are used, one on each side of the active section, to provide both "mirrors."

Since the physical structures of DFB and DBR lasers are different from each other, their optical and electrical properties are also different. DFB lasers tend to have a broader modulation bandwidth, smaller frequency and intensity noise, better side-mode suppression, and smaller linewidth. However, DFB lasers have overlapping gain and grating sections and therefore suffer from chirp: when the drive current changes, both output power and frequency change simultaneously. DBR lasers have the distinct advantage of having the distributed reflector separate from the gain section. Thus, we can control the DBR laser output power (by changing the current through the active section) and the laser frequency/wavelength (by changing the current through the grating section) independently.

Recently, more sophisticated lasers have been developed that combine a DFB laser with an electroabsorption modulator. Such devices keep the DFB laser current fixed and use a separate modulator section for information transmission. As a result, the chirp is greatly reduced.

DFB lasers have been available commercially since the mid-1980s; a few companies are beginning to offer DBR lasers commercially at the time of the writing of this book.

1.3.2.6 Laser Rate Equations

Rate equations are powerful analytical tools that provide an understanding of many important properties of semiconductor lasers. There are many different versions of rate equations. Depending on the phenomena we are interested in studying, we can include in rate equations a variety of effects ranging from modulation response to noise to chirp.

We will consider here only the simplest version of the rate equations. That version is valid on the average, that is, it does not account for variations of the photon and electron densities along the laser. Thus, "our" version of the rate equations is valid when the times of interest are smaller than the laser "transit time" (of the order of 4 ps). In other words, the bandwidth of interest is smaller than some 100 GHz.

The rate equations relate the electron density, n, and the photon density, ϕ (the units of both n and ϕ are $1/\mathrm{m}^3$) as follows:

$$\frac{dn}{dt} = \frac{J}{qd} - \frac{n}{\tau_{sp}} - Cn\phi \qquad \left[\frac{1}{m^3\ sec}\right] \tag{1.176}$$

$$\frac{dn}{dt} = Cn\phi + \delta\frac{n}{\tau_{sp}} - \frac{\phi}{\tau_{ph}} \qquad \left[\frac{1}{m^3\ sec}\right] \tag{1.177}$$

where J is the current density in A/m^2; q is the electron charge; d is the thickness of recombination region; τ_{sp} is the spontaneous emission lifetime; C is a constant accounting for the strength of stimulated emission; τ_{ph} is the photon lifetime in the laser cavity (it accounts for the loss of photons due to cavity losses and emission of light through laser facets); and δ is the fraction of spontaneous photons that adds to the lasing mode; δ is small and usually can be neglected.

The photon lifetime, τ_{ph}, can be related to the total loss coefficient, $\delta_t[1/m]$ as follows:

$$loss\ after\ t_{[sec]} = loss\ after\ x_{[meters]} \tag{1.178}$$

Hence,

$$\exp(-t/\tau_{ph}) = \exp(-\alpha_t x) \tag{1.179}$$

The time, t, and the distance, x, are related by

$$x = vt = \frac{ct}{n_{ri}} \tag{1.180}$$

where v is the light velocity in the laser, and n_{ri} is the refractive index. Substituting (1.180) into (1.179), we obtain, after simple transformations, the following result:

$$\tau_{ph} = \frac{n_{ri}}{\alpha_t c} \tag{1.181}$$

Rate equations are based on the following physical model. The first equation, (1.176), tells that the number of active electrons in the recombination region is increased by the pump current (the first term on the right side) and is decreased by spontaneous recombinations (the second term on the right side) and by the stimulated emission (the third term on the right side). The second rate equation, (1.177), tells that the number of photons in the lasing mode(s) is increased by the stimulated emission (the first term on the right side) and by the spontaneous emission (the second term on the right side); it is decreased by the cavity losses due to waveguide attenuation and light emission through laser facets (the third term on the right side).

1.3.2.7 The Steady-State Solution of the Laser Rate Equations

In the steady-state,

$$\frac{dn}{dt} = 0; \qquad \frac{d\phi}{dt} = 0 \tag{1.182}$$

For lasing, $d\phi/dt$ must be positive for small ϕ. Hence, (1.177) yields (neglecting δ):

$$Cn - \frac{1}{\tau_{ph}} \geq 0 \tag{1.183}$$

At threshold, the left side of (1.183) is equal to zero. Hence, denoting the threshold carrier density by n_{th}, we obtain

$$Cn_{th} - \frac{1}{\tau_{ph}} = 0 \tag{1.184}$$

It follows from (1.184) that the threshold carrier density, n_{th}, is uniquely determined by the photon lifetime:

$$n_{th} = \frac{1}{C\tau_{ph}} \tag{1.185}$$

Further, it follows from (1.177) that the steady-state value of the carrier density, n, above threshold is equal to n_{th} [we can verify it by substituting (1.185) into (1.177)]. This phenomenon is known as *clamping*.

We can now find the threshold current density J_{th}. By definition of the threshold, when $J = J_{th}$, the carrier density $n = n_{th}$ and the photon density $\phi = 0$. Hence, from (1.176),

$$\frac{J_{th}}{qd} - \frac{n_{th}}{\tau_{sp}} = 0 \tag{1.186}$$

It follows from (1.186) that the threshold current density is given by

$$J_{th} = \frac{n_{th} qd}{\tau_{sp}} \tag{1.187}$$

We can rewrite (1.187) as

$$\frac{n_{th}}{\tau_{sp}} = \frac{J_{th}}{qd} \tag{1.188}$$

To find ϕ_s, the steady-state photon density, we substitute (1.188) into (1.176):

$$0 = \frac{J}{qd} - \frac{J_{th}}{qd} - Cn_{th}\,\phi_s \tag{1.189}$$

It follows from (1.189) that the steady-state photon density, ϕ_s, is given by

$$\phi_s = \frac{1}{Cn_{th}} \cdot \frac{J - J_{th}}{qd} \tag{1.190}$$

Finally, substituting (1.185) into (1.190), we obtain

$$\phi_s = \frac{\tau_{ph}}{qd}(J - J_{th}),\; J > J_{th} \tag{1.191}$$

Equation (1.191) shows that the steady-state photon density (and, therefore, laser output power) is directly proportional to the drive current above the threshold. This relationship is illustrated by the plot in Figure 1.38.

1.3.2.8 Laser Modulation: Step Response

In digital transmission systems, the information is transmitted in the form of binary pulses. The largest bit rate we can achieve is, therefore, limited by

$$R_b \le \frac{1}{\Delta t_{01} + \Delta t_{10}} \tag{1.192}$$

where Δt_{01} is the time required to switch the laser from binary 0 to binary 1, and Δt_{10} is the time required to switch the laser from binary 1 to binary 0. To evaluate Δt_{10} and Δt_{01}, we have to evaluate the so-called step response, defined as the time evolution of the laser output when its drive current is switched from binary 0 to binary 1 or from binary 1 to binary 0.

To develop an understanding of the physics involved, this section evaluates the laser step response for binary 0 to binary 1 transmissions. It is assumed that binary 0s drive the laser below the threshold, while binary 1s drive the laser above the threshold, as shown in Figure 1.45.

The resulting analysis is fairly general and covers the case when both J_0 and J_1 are above the threshold as a special case. We note that when both J_0 and J_1 are above the threshold, the transient time is smaller. Thus, one can achieve a higher transmission rate by keeping both J_0 and J_1 above the threshold.

We begin our analysis by noting that, since $J_0 < J_{th}$, at the switching moment $t = 0$,

$$n = n_1 < n_{th} \tag{1.193}$$

and

$$\phi = 0 \tag{1.194}$$

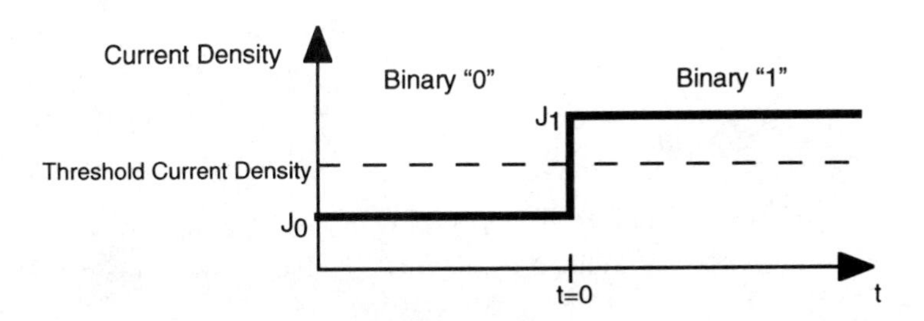

Figure 1.45 A step drive current. (*From Optical Fiber Communications Systems* by William B. Jones, Jr., © 1988 by Oxford University Press, Inc. Reprinted by permission.)

As long as the laser is below the threshold, $n < n_{th}$, and (1.176) yields

$$\frac{dn}{dt} = \frac{J}{qd} - \frac{n}{\tau_{sp}} \tag{1.195}$$

The solution of (1.195) is

$$n - n_1 = \frac{(J_2 - J_1)\tau_{sp}}{qd}\left[1 - \exp\left(-\frac{t}{\tau_{sp}}\right)\right] \tag{1.196}$$

From (1.196), the delay to reach the threshold $n = n_{th}$ is

$$t_d = \tau_{sp} \ln\left(\frac{J_2 - J_1}{J_2 - J_{th}}\right) \tag{1.197}$$

After the carrier density, n, reaches the threshold value, n_{th}, the photon density is no longer zero, and the simplified rate equation, (1.195), is no longer valid. Thus, for $n > n_{th}$, we must return to the full rate equations, (1.176) and (1.177).

To facilitate their solution, we express the carrier density, n, and the photon density, ϕ, in terms of their equilibrium values, n_{th} and ϕ_s:

$$n = n_{th} + \Delta n \tag{1.198}$$

$$\phi = \phi_s + \Delta\phi \tag{1.199}$$

where Δn and $\Delta\phi$ are the deviations of n and ϕ from their steady-state values.

Next, we substitute (1.198) and (1.199) into (1.176) and (1.177), and simplify the result by dropping the product $\Delta\phi \cdot \Delta n$ (for simplicity, it is assumed to be small). After transformations, the result is

$$\frac{d(\Delta n)}{dt} = -\left(C\phi_s + \frac{1}{\tau_{sp}}\right)\Delta n - Cn_{th}\Delta\phi \tag{1.200}$$

$$\frac{d(\Delta\phi)}{dt} = C\phi_s\Delta n \tag{1.201}$$

To reduce the system of (1.200) and (1.201) to a single equation, we differentiate (1.200), isolate $d(\Delta\phi)/dt$, and substitute it into (1.201); after transformations, the result is

$$\frac{d^2(\Delta n)}{dt^2} + 2\alpha\frac{d(\Delta n)}{dt} + \omega_0^2\,\Delta n = 0 \tag{1.202}$$

where α is the laser damping constant, and ω_0 is the laser resonant frequency (also called the relaxation oscillation frequency) given by

$$2\alpha \equiv C\phi_s + \frac{1}{\tau_{sp}} = \frac{\phi_s}{n_{th}\tau_{ph}} + \frac{1}{\tau_{sp}} \tag{1.203}$$

and

$$\omega_0^2 \equiv C^2\, n_{th}\phi_s = \frac{\phi_s}{n_{th}\tau_{ph}^2} \tag{1.204}$$

The solution of (1.202) is

$$\Delta n = D \exp(-\alpha t)\sin \omega t \tag{1.205}$$

where the amplitude, D, is to be determined from initial conditions, and ω is the oscillations' frequency given by

$$\omega \equiv \sqrt{\omega_0^2 - \alpha^2} \tag{1.206}$$

To find $\Delta\phi$, we substitute (2.205) into (2.201) and integrate the result:

$$\Delta\phi = \frac{C\phi_s D}{\omega} \exp(-\alpha t) \cos \omega t \tag{1.207}$$

where $\exp(-\alpha t)$ has been assumed to be slowly varying with respect to $\cos \omega t$.

Now, we find the amplitude, D, from the initial condition (at $t = 0$, $n = n_{th}$, and $\phi = 0$) and substitute it into (1.205) and (1.207). The result is

$$\Delta n = \frac{\omega}{C} \exp(-\alpha t) \sin \omega t \tag{1.208}$$

and

$$\Delta\phi = -\phi_s \exp(-\alpha t)\cos \omega t \tag{1.209}$$

If the damping constant is much smaller than the laser resonant frequency ($\alpha << \omega_0$), then $\omega \approx \omega_0$, and (1.204) yields

$$\frac{\omega}{C} \cong \frac{\omega_0}{C} = \frac{C\sqrt{n_{th}\phi_s}}{C} = \sqrt{n_{th}\,\phi_s} \tag{1.210}$$

Substituting (1.210) into (1.208), we obtain the following approximate expression for Δn:

$$\Delta n = \sqrt{n_{th}\phi_s}\exp(-\alpha\omega t)\sin\omega t \tag{1.211}$$

Taken together, (1.196), (1.198), and (1.211) specify the time evolution of the carrier density, n, both before and after the laser reaches the threshold. Similarly, (1.199) and (1.207), taken together, specify the time evolution of the photon density, ϕ, after the laser reaches the threshold (below the threshold, $\phi = 0$). Figure 1.46 shows the time evolution of both n and ϕ as predicted by those expressions for the following parameter values: $d = 0.2\mu$m, $\tau_{sp} = 2$ nsec, $n_{ri} = 3.5$, $J_{th} = 0.75$ A/nm^2, and $\tau_{ph} = 9$ psec.

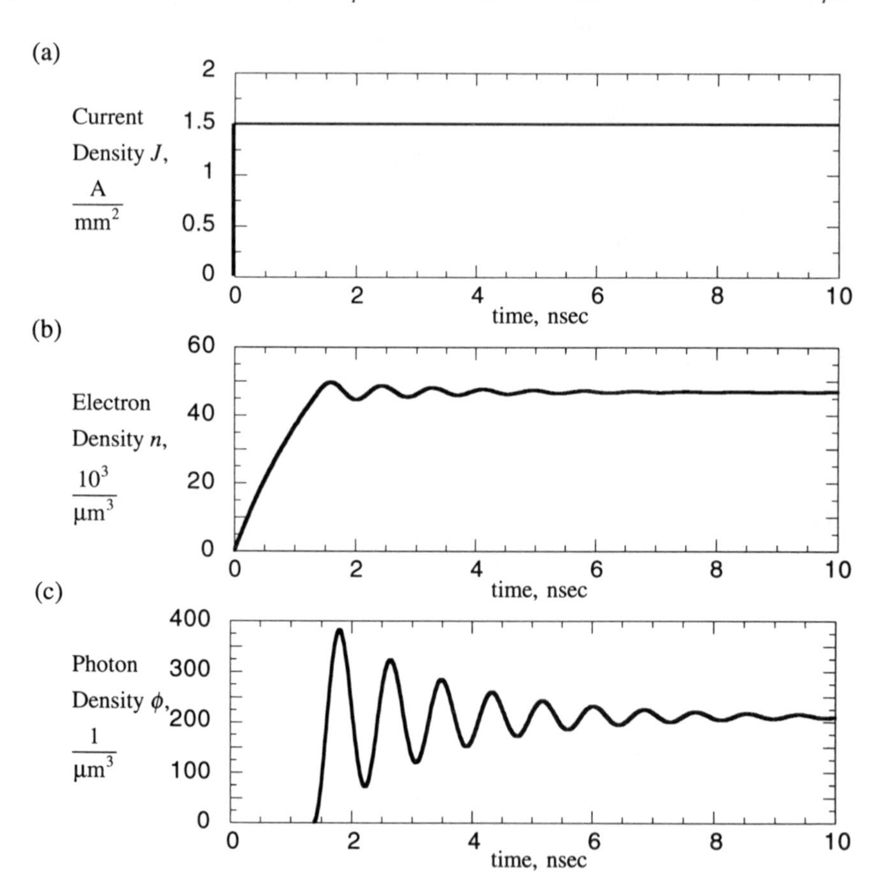

Figure 1.46 A step-function input current (a) is applied to a laser at $t = 0$. As a result, the carrier density, n begins to grow immediately as shown in (b). However, the photon density, ϕ, begins to grow only after the carrier density, n, exceeds the threshold value n_{th}, as shown in (c). This figure has been calculated using expressions (2.186), (2.191), (2.197), (2.198), (2.199), (2.209), and (2.211) for the following parameter values: $d = 0.2\ \mu$m, $\tau_{sp} = 2$ nsec, $n_n = 3.5$, $J_{th} = 0.75$ A/mm^2, and $\tau_{ph} = 9$ psec. The resulting values of ω_0 and α are $7.45 \cdot 10^9$ rad/sec and $5.0 \cdot 10^8$ sec^{-1}, respectively. (*Source:* [20].)

Inspection of Figure 1.46 reveals that the laser settling time consists of (1) the delay time, t_d, defined by (1.197) (the laser produces no light for $t < t_d$) and (2) the time needed for damped oscillations to settle. For example, in Figure 1.46, the delay time, t_d, is ~ 1.5 nsec, the oscillations' settling time is 6.5 nsec, and the total transition time Δt_{01} is ~ 8 nsec. We can eliminate the first item (but not the second one) by keeping the laser above the threshold for both binary 0s and binary 1s.

1.3.2.9 Laser Modulation: Sinusoidal Frequency Response

While the bulk of information in modern networks is digital, analog transmission systems are important, too. Analog links are used for TV transmission and distribution, antenna remoting, and other applications.

Analog links are characterized by transmitting one or two *continuous-wave* (CW) sinusoidal test signals. In this section, we evaluate the laser response to a CW sinusoidal excitation of the form

$$J = J_0 \left[1 + m_j \exp(j\omega_m t)\right] \tag{1.212}$$

where J_0 is the bias current density (the laser is assumed to be biased above the threshold), m_j is the current modulation depth (it is also called modulation coefficient), and ω_m is the modulation frequency.

We are interested in $\Delta\phi$, the deviation of ϕ from the steady-state condition. The low-frequency solution, $\Delta\phi$ ($\omega_m = 0$), is identical to the steady-state solution (1.190):

$$\Delta\phi(\omega_m = 0) = \frac{1}{Cn_{th}} \cdot \frac{m_j J_0}{qd} \tag{1.213}$$

where $m_j J_0$ is the current density swing due to modulation.

To find a general solution for an arbitrary ω_m, we proceed as with the step-function case (see Section 1.3.2.8):

1. Substitute (1.198), (1.199), (1.212), and (1.213) into the rate equations (1.176) and (1.177).
2. Simplify the result by dropping the product $\Delta\phi \cdot \Delta n$ (which is more justified in this section since analog systems frequently operate with a fairly small modulation depth).
3. Differentiate the first rate equation, isolate $d(\Delta\phi)/dt$, and substitute it into the second rate equation.

The result of this procedure is the following differential equation:

$$\frac{d^2(\Delta n)}{dt^2} + 2\alpha \frac{d(\Delta n)}{dt} + \omega_0^2 \, \Delta n = \frac{j\omega_m m_j J_0}{qd} \exp\,(j\omega_m t) \tag{1.214}$$

where α is the damping constant given by (1.203), and ω_0 is the laser resonant frequency given by (1.204).

Equation (1.214) is linear, and the excitation (1.212) is sinusoidal. Hence, the resulting Δn is sinusoidal, too:

$$\Delta n = m_n \exp(j\omega_m t) \tag{1.215}$$

where m_n is the modulation index of Δn. Note that m_n is generally complex, reflecting both amplitude and phase change. To evaluate m_n, we substitute (1.215) into (1.214): The result is

$$[-\,\omega_m^2 + j2\alpha\omega_m + \omega_0^2]\; m_n n_{th} = \frac{j\omega_m m_j J_0}{qd} \tag{1.216}$$

Isolating m_n, we obtain

$$m_n = \frac{j\omega_m J_0 m_j}{n_{th} qd} \cdot \frac{1}{(\omega_0^2 - \omega_m^2) + j2\alpha\omega_m} \tag{1.217}$$

To find the photon density variation, $\Delta\phi$, we substitute (1.217) into (1.215), (1.215) into (1.198), and finally (1.198) into (1.201). Integration of the result yields

$$\Delta\phi = \frac{C\phi_s n_{th} m_n}{j\omega_m} \exp\,(j\omega_m t)$$

$$= \frac{C\phi_s J_0 m_j}{\omega_0^2 qd} \cdot \frac{1}{\left(1 - \dfrac{\omega_m^2}{\omega_0^2}\right) + \dfrac{j2\alpha\omega_m}{\omega_0^2}} \exp\,(j\omega_m t) \tag{1.218}$$

Equation (1.218) for $\Delta\phi$ can be broken into three parts:

- The low-frequency (or DC) value, $\Delta\phi$ $(\omega_m = 0)$, given by (1.213);
- The frequency-dependent term $M(\omega_m)$, defined by

$$M(\omega_m) \equiv \frac{1}{1 - \left(\dfrac{\omega_m}{\omega_0}\right)^2 + j\,\dfrac{2\alpha\omega_m}{\omega_0^2}} \tag{1.219}$$

- The term $\exp(j\omega_m t)$, reflecting the modulation imposed.

Thus,

$$\Delta\phi(\omega_m) = \Delta\phi(\omega_m = 0)M(\omega_m)\exp(j\omega_m t) \tag{1.220}$$

Note that the frequency response of a laser diode is fully defined by $M(\omega_m)$.

Figure 1.47 shows the laser modulation response, $M(\omega_m)$, as per (1.219). Note that the laser modulation bandwidth is limited by the laser resonant frequency, ω_0. The value of ω_0, in turn, depends on the laser threshold (i.e., on n_{th}) and on the laser

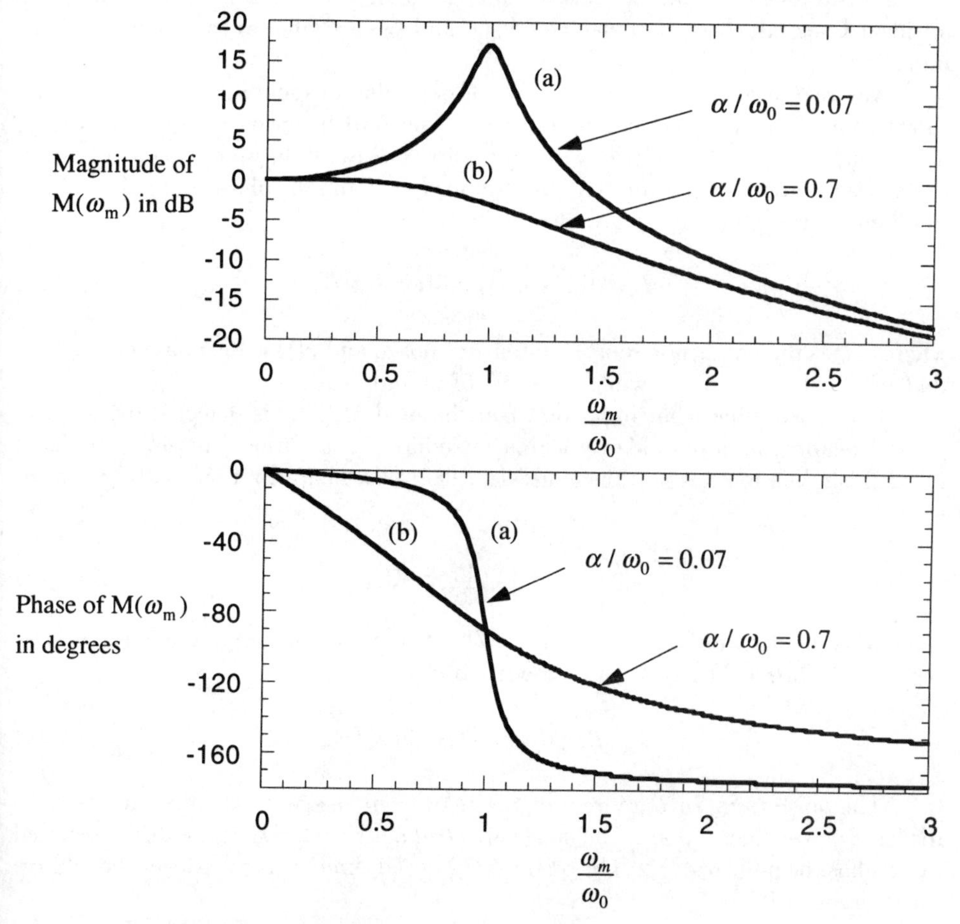

Figure 1.47 Magnitude and phase plot of the laser modulation response versus modulation frequency for (a) $\alpha/\omega_0 = 0.07$ and (b) $\alpha/\omega_0 = 0.7$. (*Source:* [20].)

output power (i.e., on ϕ_s); see (1.204). Thus, a laser with lower losses and, therefore, lower threshold is also likely to be faster. Also, a laser becomes faster (i.e., has a larger modulation bandwidth) if it is driven hard (much like human beings).

1.3.2.10 Relative Intensity Noise, Phase and Frequency Noise, Chirp

This section deals with several important parasitic effects in semiconductor laser diodes. All of them share the same starting point, the equation for the ideal laser output field:

$$E = A \cos(\omega t + \phi) \tag{1.221}$$

where A is the field amplitude, ω is the laser frequency, and ϕ is the laser phase. For an ideal laser, all three parameters—A, ω, and ϕ—are time-independent constant numbers.

Any real laser has a certain amount of noise due to spontaneous emission. The spontaneous emission is accounted for by the deterministic terms $-n/\tau_{sp}$ and $+\delta n/\tau_{sp}$ in (1.176) and (1.177), respectively. In reality, however, the spontaneous emission process is random. As a result, both the amplitude, A, and the phase, ϕ, in (1.221) are randomly varying in time in any real laser:

$$E = A[1 + n(t)] \cos[\omega t + \phi(t)] \tag{1.222}$$

where $n(t)$ is the amplitude (not the intensity) noise, and $\phi(t)$ is the phase noise. Note that $n(t)$ is dimensionless, while the units of $\phi(t)$ are radians.

Let us examine what happens when the field (1.222) is detected by an ideal photodetector (i.e., a photodetector that introduces no additional noise). If the field (1.222) is taken to represent the equivalent field (see Chap. 2), the resulting current is

$$I = R \cdot A^2 \cdot [1 + n(t)]^2 \tag{1.223}$$

where R is the detector's responsivity in A/W. Typically, the amplitude noise, $n(t)$, is very small, so that (1.223) can be rewritten as

$$I \simeq R \cdot A^2 \, [1 + 2n(t)] \tag{1.224}$$

The noise term $2n(t)$ represents the *relative intensity noise* (RIN); it describes the laser power fluctuations. The *power spectral density* (PSD) of the RIN is denoted by ζ; it has the units of 1/Hz. The quantity 10 log ζ is known as the RIN or the relative excess noise (REN):

$$REN = RIN \equiv 10 \log \zeta, \text{ dB/Hz} \tag{1.225}$$

RIN can be as low as −165 dB/Hz for a good laser. Note that RIN in dB/Hz does *not* scale linearly with system bandwidth, while the resulting noise power does. For example, if the system bandwidth is 1 GHz and the RIN is −160dB/Hz, the resulting RIN-induced noise power is

$$bandwidth \cdot \zeta = bandwidth \cdot 10^{RIN/10} = 10^9 \cdot 10^{-16} = 10^{-7} = -70 \text{ dB} \qquad (1.226)$$

Further, the actual PSD of the RIN varies with frequency in a manner similar to the laser modulation response (see Fig. 1.47): the RIN is more or less constant at "low" frequencies, peaks at the laser resonant frequency (frequently, several gigahertz), and then rolls off. The value of the RIN normally quoted in laser data sheets corresponds to low frequencies.

The phase noise, $\phi(t)$ (in radians), is related to the frequency noise, $\dot{\phi}(t)$ (in radians per second), as follows:

$$\phi(t) = \int_{-\infty}^{t} \dot{\phi}(t_1)dt_1 \qquad \text{in radians} \qquad (1.227)$$

and

$$\dot{\phi}(t) = \frac{d\phi(t)}{dt} \qquad \text{radians per second} \qquad (1.228)$$

The PSD of the frequency noise, $\dot{\phi}(t)$, has the units of (rad^2/sec^2)/Hz = rad^2/sec = rad^2 · Hz.

Sometimes, the frequency noise is defined differently:

$$f(t) = \frac{1}{2\pi}\frac{d\phi(t)}{dt} = \frac{1}{2\pi}\dot{\phi}(t) \qquad \text{in cycles per second (hertz)} \qquad (1.229)$$

This definition is used to yield the frequency noise in hertz. If the frequency noise is defined by (1.229), the phase noise is related to the frequency noise by

$$\phi(t) = 2\pi \int_{-\infty}^{t} f(t_1)dt_1 \qquad \text{in radians} \qquad (1.230)$$

The PSD of the frequency noise, $f(t)$, has the units of Hz^2/Hz = Hz.

The PSD of the frequency noise is uniquely related to the laser linewidth, $\Delta\upsilon$ (see Section 1.3.2.3) by the following equations [26]:

$$\text{PSD of } \dot{\phi} = 4\pi\Delta\upsilon \tag{1.231}$$

$$\text{PSD of } f = \frac{\Delta\upsilon}{\pi} \tag{1.232}$$

where $\Delta\upsilon$ is the *full-width half-maximum* (FWHM) linewidth. Equations (1.231) and (1.232) are valid only if the PSD of the frequency noise is frequency independent. The corresponding frequency noise is known as the white frequency noise, and the resulting laser output spectrum is known as the Lorentzian spectrum.

So far, we have looked at the laser amplitude and phase taking into account noise but assuming that the laser is not modulated (i.e., it is driven by a constant bias current only). When the laser is modulated (i.e., its current is changing), both laser power and laser frequency change. The latter phenomenon is known as *chirp;* the magnitude of chirp is some 100 MHz–1 GHz for each 1-mA current change in a typical laser.

Chirp broadens the laser output spectrum. Thus, some engineers distinguish between the *static* laser spectrum and linewidth and the *dynamic* laser spectrum and linewidth. The word *static* means "measured with no modulation applied," while the word *dynamic* means "measured with modulation applied." The two sets of parameters (static and dynamic) are very different from each other and yet are sometimes confused in the literature: static parameters are used instead of dynamic parameters, and vice versa.

In intensity-modulated systems (and all currently deployed systems are intensity modulated), chirp is generally detrimental: chirp broadens laser spectrum and, therefore, distorts transmission system output via fiber dispersion (see Section 1.2.10). However, chirp is beneficial in frequency-modulated systems (see Chapter 4); in fact, frequency-modulated systems are based on laser chirp.

1.3.2.11 Laser Package

The most fundamental component of a laser package is, of course, the laser chip. We have examined its structure and properties in the preceding sections. Figure 1.48

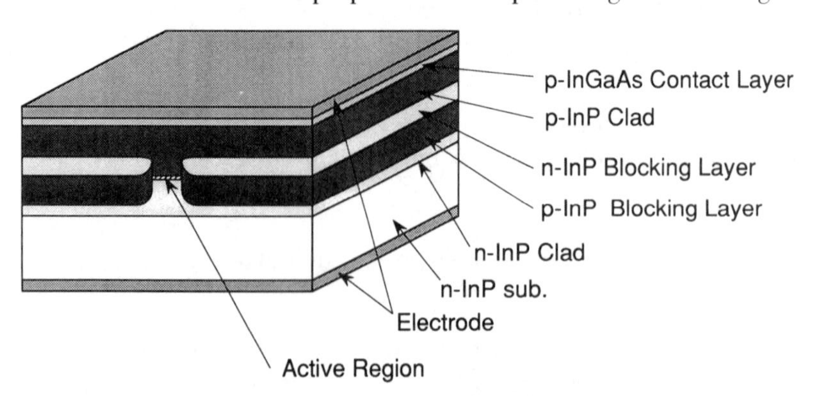

Figure 1.48 The cross section of a commercial buried heterostructure DFB laser. (Courtesy of Furukawa, Inc.)

shows a cross section of a commercial DFB laser. Figure 1.49 shows how a laser chip is mounted on a heat sink.

Even though the laser chip is the most sophisticated component residing in a laser package, the size, the weight, and the cost of the laser package are dominated by the auxiliary components needed to run the laser (Fig. 1.50): the substrate; the thermoelectric cooler needed to keep the laser temperature constant; the thermistor used as a temperature sensor; the photodiode sensing the laser power; the lens collimating the laser output field; the optical isolator isolating the laser from reflections; the fiber coupling lens; and the fiber mount.

Figures 1.51 and 1.52 show the internal and external views of a laser package.

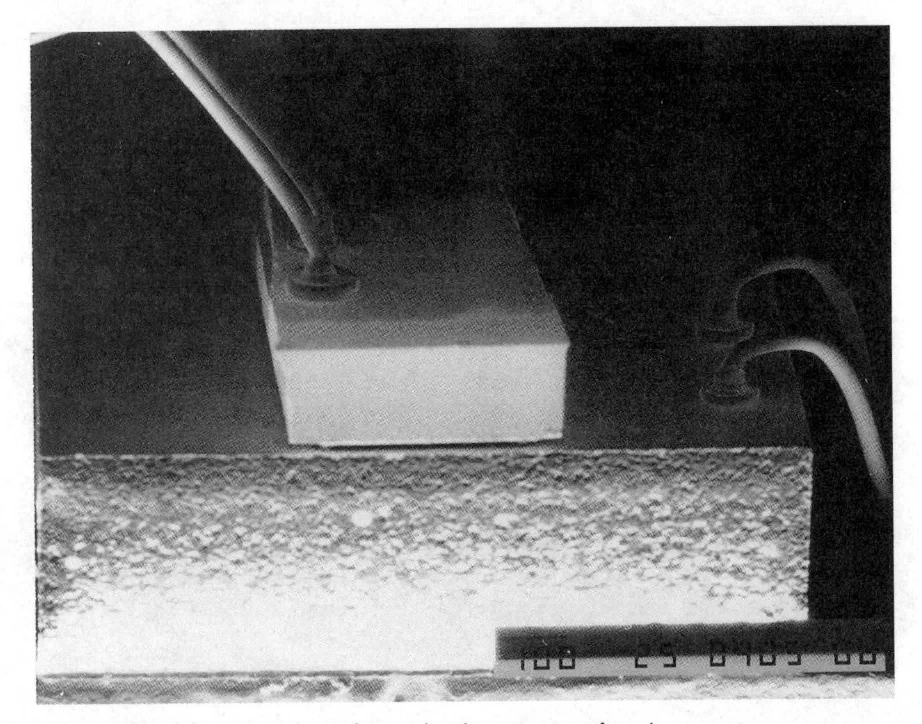

Figure 1.49 A laser chip mounted on a heat sink. (Photo courtesy of Furukawa, Inc.)

Photodiode
Laser
Submount
Thermistor
TE Cooler
Substrate
Fiber

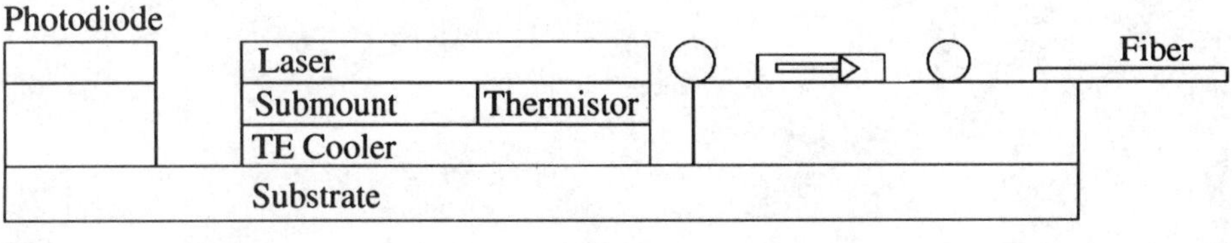

Figure 1.50 Schematic of a laser package. (Courtesy of Furukawa, Inc.)

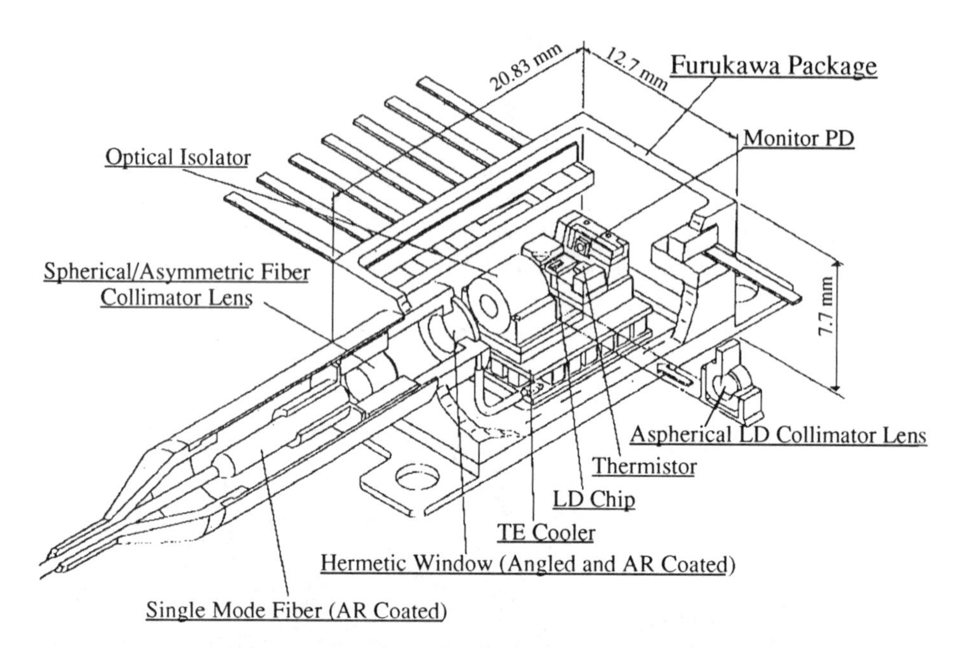

Figure 1.51 Internal view of a laser package. (Courtesy of Furukawa, Inc.)

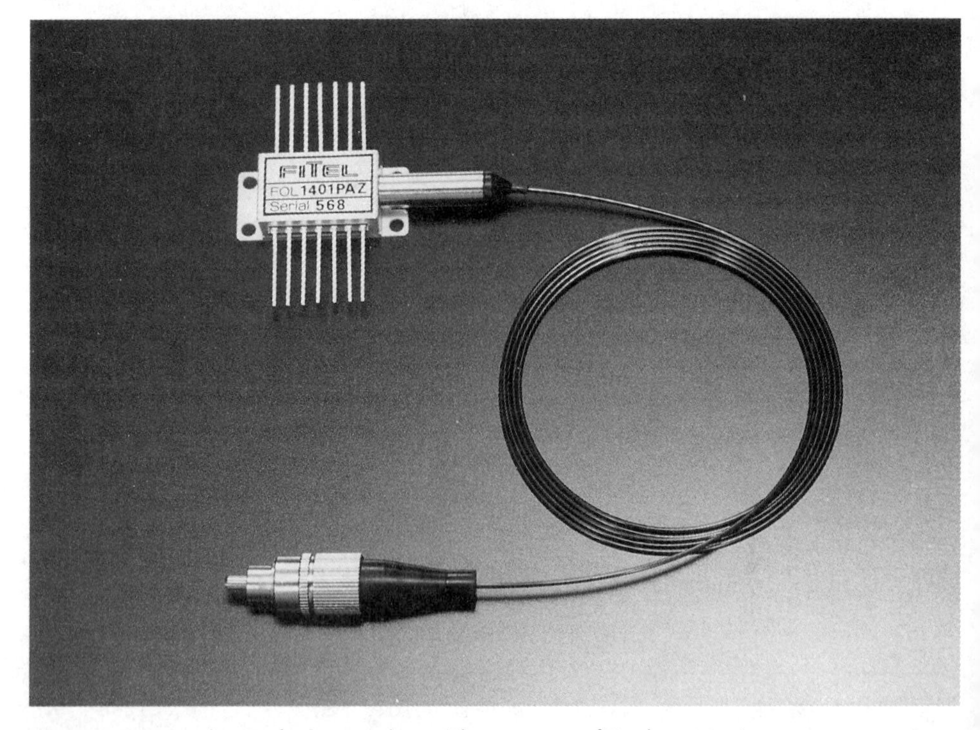

Figure 1.52 External view of a laser package. (Photo courtesy of Furukawa, Inc.)

1.4 PHOTODETECTORS

1.4.1 The PIN Photodiode

The most common photodetector is the PIN photodiode, which consists of a very lightly doped "intrinsic" region sandwiched between the p- and n-type regions, as shown in Figure 1.53.

A reverse-biased PIN photodiode is a nearly ideal photodetector, with essentially infinite internal impedance. Because a photodiode's output current is proportional to its input optical power, *responsivity* normally is used to characterize a photodiode's efficiency:

$$\textit{responsivity, or } R \equiv \frac{\textit{output current, in A}}{\textit{input optical power, in W}} \frac{A}{W} \tag{1.233}$$

While engineers usually use responsivity, physicists frequently prefer *quantum efficiency*, defined as

$$\textit{quantum efficiency, or } \eta \equiv \frac{\textit{number of output electrons}}{\textit{number of input photons}} \tag{1.234}$$

The quantum efficiency is dimensionless and is always smaller than unity in PIN photodiodes.

The responsivity, R, and the quantum efficiency, η, are, of course, related. It is easy to show that

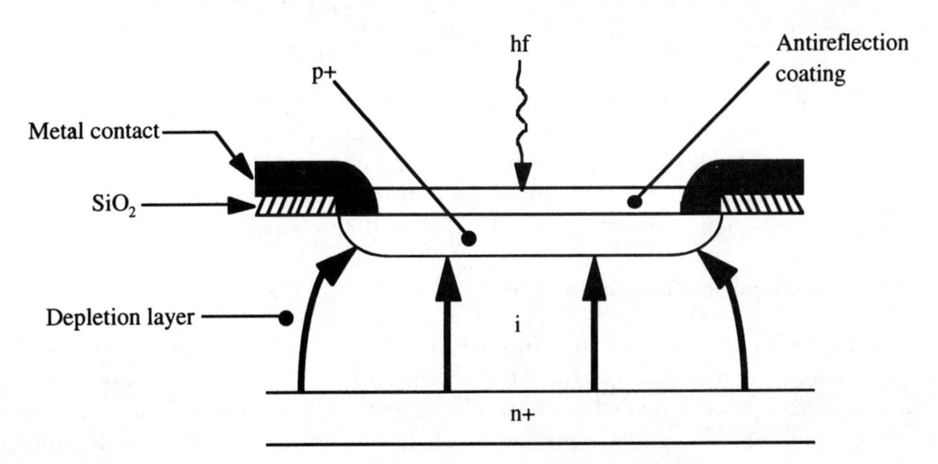

Figure 1.53 Schematic diagram of a front illuminated silicon PIN photodiode. (*Source:* [1]. © 1992. Reprinted by permission of Prentice-Hall, Inc., Upper Saddle River, NJ.)

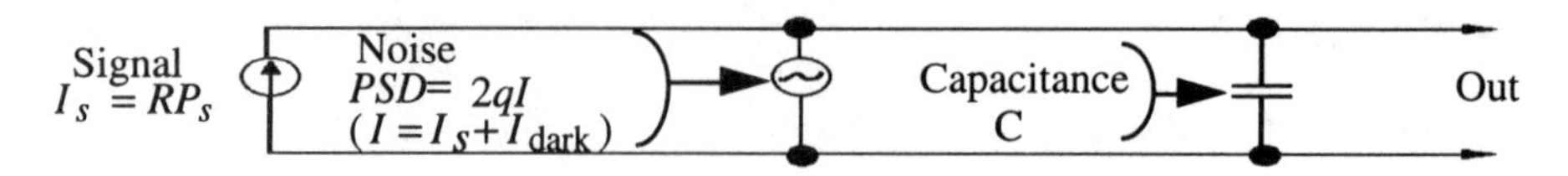

Figure 1.54 An equivalent circuit of a PIN photodiode. P_s is the instantaneous optical power, and I_{dark} is the dark current.

$$R = \frac{q\eta}{h\nu} \tag{1.235}$$

where q is the electron charge, and ν is the light frequency.

The capacitance of the reverse-biased PIN photodiode is an important parameter that ultimately limits the performance of optical receivers. Figure 1.54 shows a simple equivalent circuit of a reverse-biased PIN photodiode.

The equivalent circuit of Figure 1.54 has a reasonable accuracy up to very high frequencies, when parasitic inductances have to be taken into account. Essentially, the equivalent circuit reflects the three main features of a PIN photodiode:

- Conversion of the instantaneous optical signal power, P_s, into electrical current:

$$I_s = R \cdot P_s \tag{1.236}$$

- Generation of the shot noise during the conversion process, with the power spectral density

$$Shot\ noise\ PSD = 2qI = 2q(I_s + I_{dark}) \tag{1.237}$$

where $I \equiv I_s + I_{dark}$ is the total current flowing through the PIN photodiode, I_s is the signal current, and I_{dark} is the dark current;

- Shunting of the photodetector output by its own parasitic capacitance.

When a photodiode is connected to external circuitry, the circuitry noise contaminates the signal current. The impact of that effect can be somewhat reduced by amplifying the signal current before it leaves the photodiode.

1.4.2 The Avalanche Photodiode (APD)

The APD amplifies the photocurrent by the so-called *avalanche multiplication* process. A large bias voltage applied to the APD accelerates the electrons born through the photodetection process (called primary electrons in APDs). The primary electrons travel through the APD and accumulate a substantial energy due to the high applied voltage. The high energy of the primary electrons allows them to create secondary electrons (and holes) through ionization. The process repeats itself many times, so that the APD output current is much stronger than the primary current generated by the photons-to-electrons conversion. Figure 1.55 illustrates the APD avalanche multiplication process.

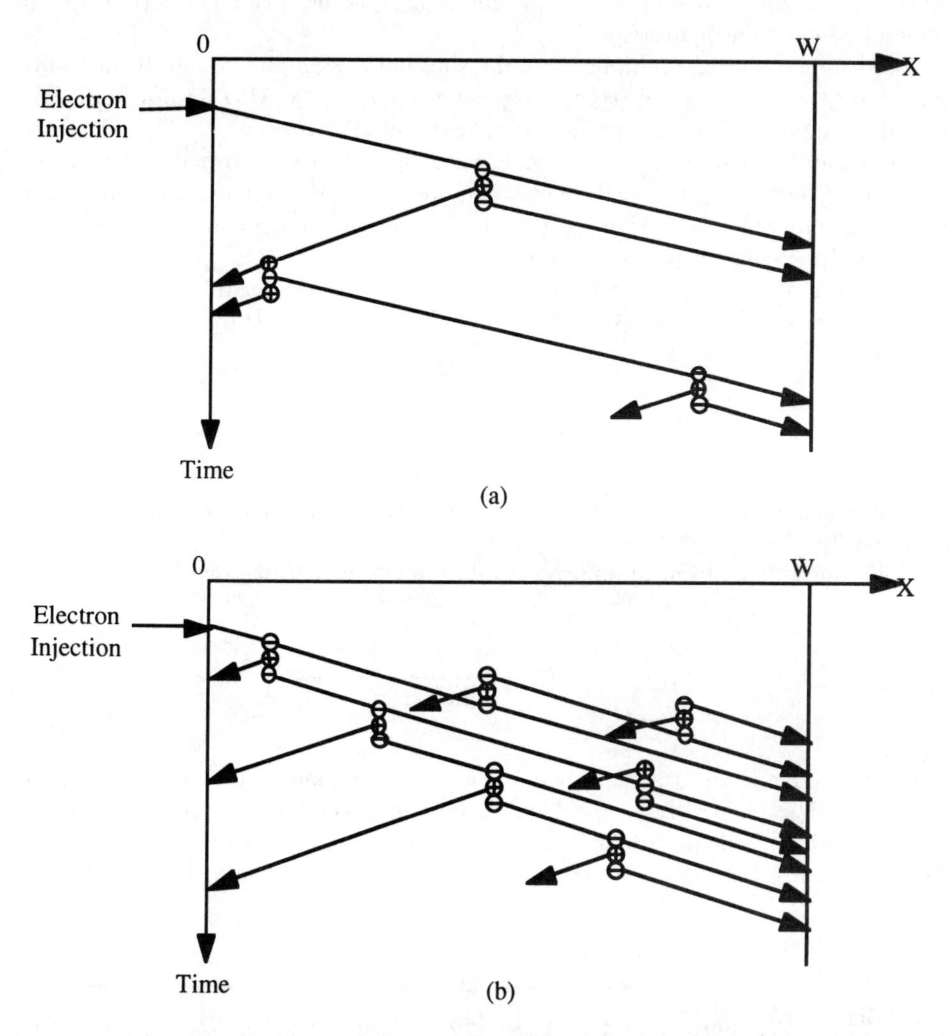

Figure 1.55 Schematic time evolution of the avalanche process in (a) materials where the electron (α') and hole (β') ionization coefficients are roughly equal and (b) where $\alpha' >> \beta'$. Here W indicates the depletion region width. The top diagram is typical of avalanche occurring in Ge and GaAs, whereas the bottom corresponds to the case of Si. In InP, $\beta' = 2\alpha'$. (*Source:* [27]. © 1977 Academic Press.)

The internal gain of the APD is denoted by M and is given by

$$M = I_{APD}/I_{primary}. \tag{1.238}$$

where I_{APD} is the APD output current, and $I_{primary}$ is the (primary) current due to photons-to-electrons conversion.

Of course, during multiplication, the shot noise is amplified, too. If the multiplication process were ideal (completely deterministic), the APD output noise *PSD* would be equal to M^2 times the shot noise PSD [equal to $2qI$; see (1.237)]. However, the actual multiplication process is random: a given primary electron does not always produce M output electrons. Rather, sometimes it produces more than M output electrons, sometimes less. Thus, M reflects the average number of output electrons produced by a single primary electron.

As a result, the actual noise PSD at APD's output is larger than $M^2 \cdot 2qI$, by a factor that is known as the APD noise factor F. Thus, the APD output noise PSD is given by

$$APD\ noise\ PSD = 2qIM^2F \tag{1.239}$$

The foregoing discussion leads to the APD equivalent circuit shown in Figure 1.56.

Because the multiplication process takes time, the APD bandwidth is limited to

$$B_{APD} \approx \frac{1}{2\pi M \tau_1} \tag{1.240}$$

where τ_1 is the effective transit time of the avalanche region. Note that increasing the APD gain, M, increases the output signal current and, therefore, suppresses the impact of noise added by the external circuit; at the same time, it increases the noise added by the APD itself [see (1.239)] and decreases the APD bandwidth [see (1.240)].

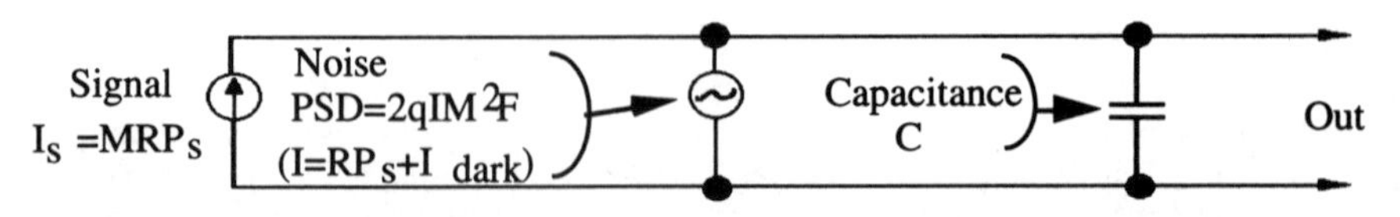

Figure 1.56 An equivalent circuit of an APD photodiode. P_s is the instantaneous optical power, and I_{dark} is the primary dark current.

Problems

Some of the problems below were created by teaching assistants at Stanford University.[5] Others are modified versions of problems published in other textbooks. Still others are inspired by current technical papers.

Fibers

Problem 1.1

A uniform plane wave with (vacuum) wavelength $\lambda = 1.55\ \mu m$ travels along the x-axis in an unbounded medium, with the propagation constant $k_x = 5.2 \cdot 10^6 (m^{-1})$.

a. What is the phase velocity along x? How does it compare to c?
b. What is the index of refraction of the medium?

Problem 1.2

A step-index fiber has $n_1 = 1.52$, $n_2 = 1.49$.

a. The fiber is immersed in water ($n_0 = 1.33$). What is the maximum acceptance angle for light incident from the water onto the input facet?
b. The fiber is immersed in air. What is the numerical aperture?

Problem 1.3

Consider a symmetric dielectric slab waveguide.

a. Prove that $k_{1x}d \leq V$ for a guided mode.
b. When $k_{1x}d = V$, the mode reaches cut-off. What happens to the transverse propagation constant and field in the cladding at cut-off?
c. Prove that the lowest-order mode is never cut off, no matter how small d/λ is.

Problem 1.4

A slab waveguide has $n_1 = 1.52$, $n_2 = 1.50$, $d = 3\ \mu m$; light with $\lambda = 0.85\ \mu m$ is used.

a. How many TE modes can propagate in the waveguide?
b. Find a good approximation for the largest eigenvalue, $k_{1x}d$ (three significant digits).

5. Dr. Thomas Fong contributed many of them when he was a teaching and research assistant in Prof. Kazovsky's group at Stanford University. Dr. Fong is now with Bell Laboratories.

Problem 1.5

A communication link has the following parameters: fiber length = 100 km, $\lambda = 1.55\ \mu m$, D = 17 ps/(nm·km). If we assume that the source bandwidth is due to the modulation only, what is the maximum information transmission rate?

Problem 1.6

A Burrus-type LED with radius r_s is butt-coupled to a fiber of core radius $a > r_s$. The LED has the modified Lambertian radiance pattern shown in Figure 1.57.

$$B(\theta) = B_0(\cos\theta)^m \quad (m \geq 1),$$

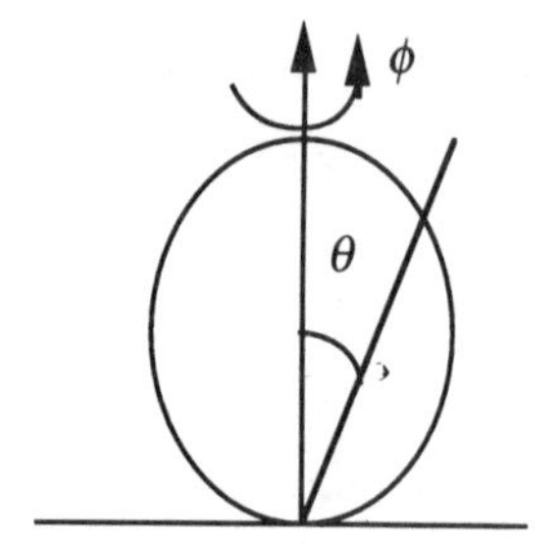

Figure 1.57 Lambertian radiation pattern.

where $B(\theta)$ is the power emitted per area element per unit solid angle in the direction of a ray making the angle θ with respect to the fiber axis.

a. Neglecting possible reflection losses between the LED and fiber, calculate the fraction η of the total LED output captured by the fiber. Express the result in terms of the fiber acceptance angle. *Note:* You need to integrate over the solid angle ($d\Omega = \sin\theta d\theta d\phi$) as well as over area.
b. Show that for $m = 1$ (Lambertian source) the coupling efficiency η can simply be expressed in terms of the fiber numerical aperture (NA).

Problem 1.7

A step-index fiber has the core index of refraction $n_1 = 1.5$, the cladding index of refraction $n_2 = 1.478$, and the core radius $r = 25\ \mu m$.

a. What is the NA and maximum acceptance angle (θ_{max}) of this fiber?
b. Consider a guided ray traveling at the steepest angle with respect to the fiber axis. How many reflections are there per meter for this ray?

Problem 1.8

Assume that a narrow light beam propagates exactly along the axis of a bent glass cylinder, as shown in Figure 1.58. What is the smallest radius of the curvature R such

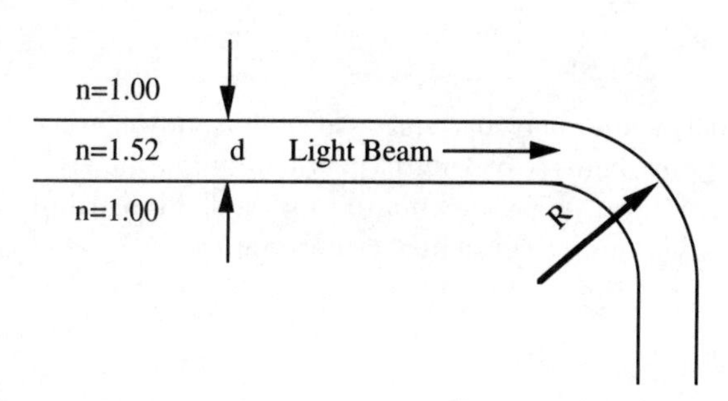

Figure 1.58 Fiber bond.

that there is no leakage of light outside the cylinder? Derive an expression for R and evaluate it for $d = 10\ \mu\text{m}$. *Hint:* Consider the first two reflection points. The second reflection can, in principle, occur either on the outside wall or on the inside wall. Consider both cases.

Problem 1.9

A dielectric slab waveguide, as shown in Figure 1.59, has $n_1 = 1.43$, $n_2 = 1.42$, and wavelength $\lambda = 1550$ nm. What is the maximum slab thickness, $2d$, resulting in single-mode propagation?

Problem 1.10

Sketch a plot of β versus ω for (a) odd and (b) even TE modes in the dielectric slab waveguide shown in Figure 1.59 when $n_1 = 1.521$, $n_2 = 1.520$, $d = 10\ \mu\text{m}$, and $0 < f < 750$ THz, where f is the light frequency. Indicate the frequencies at which new modes appear. What happens as ω goes to infinity?

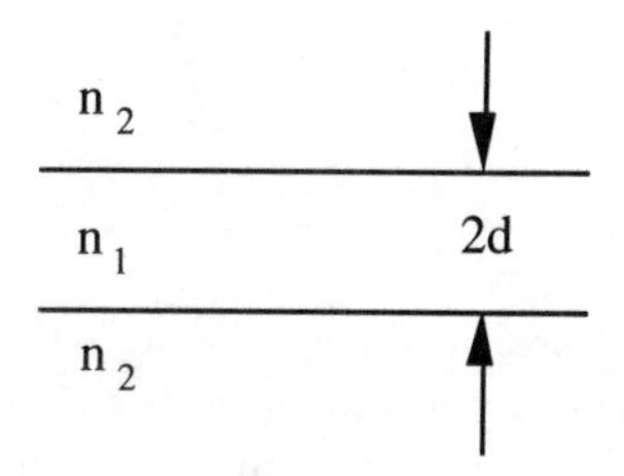

Figure 1.59 Waveguide cross-section.

Problem 1.11

Consider TE modes of the dielectric slab waveguide shown in Figure 1.59, with the following parameters: light wavelength in vacuum, $\lambda_0 = 1.55\ \mu m$; $n_1 = 1.521$; $n_2 = 1.520$; and $d = 8\ \mu m$. Write down and plot the field distributions ($|E_y|$ versus x) for all modes. Include numerical values of parameters.

Problem 1.12

Numerical aperture of an optical fiber:

a. Consider a step index fiber excited by a surface-emitting LED. Assume that the LED is a Lambertian source with the intensity pattern $I(\theta) = B\cos\theta$. Find the expressions for:

1. The total output power of the LED:

$$P_{LED} = \int_{r=0}^{a}\int_{\theta=0}^{\pi/2}\int_{\phi=0}^{2\pi}\int_{\varphi=0}^{2\pi} I(\theta)\cdot r\sin\theta\cdot dr d\theta d\phi d\varphi$$

2. The power coupled to the fiber:

$$P_{fiber} = \int_{r=0}^{a}\int_{\theta=0}^{\theta_{max}}\int_{\phi=0}^{2\pi}\int_{\varphi=0}^{2\pi} I(\theta)\cdot r\sin\theta\cdot dr d\theta d\phi d\varphi$$

where θ_{max} is the maximum acceptance angle of the fiber.

3. The coupling efficiency η_c of the LED into the fiber:

$$\eta_c = \frac{P_{fiber}}{P_{LED}}$$

b. Using the result of part a(3), prove that η_c is proportional to the square of the NA of the fiber.

c. Calculate the coupled power and the coupling efficiency in part a for the following fiber and LED specifications: core radius, $a = 25\ \mu m$; refractive index of the core, $n_1 = 1.5$; refractive index of the cladding, $n_2 = 1.478$; and $B = 200\ W/cm^2/Sr$.

Problem 1.13

Bandwidth measurement for optical fiber:

A very narrow pulse (ideally a unit impulse function) of optical power $x(t)$ is transmitted into a fiber and produces an output waveform. The output pulse, $y(t)$, corresponds to the impulse response of the fiber. Assume that the output pulse is Gaussian:

$$y(t) = \frac{1}{\sigma\sqrt{2\pi}} \exp\left(-t^2/2\sigma^2\right)$$

where 2σ is the rms pulse width. Show that the FWHM (full-width half-maximum) bandwidth is

$$BW = \frac{\sqrt{\ln 2}}{\sqrt{2}\pi\sigma}$$

The FWHM bandwidth is the frequency at which the frequency response is equal to one-half its maximum value.

Problem 1.14

For propagation in a slab waveguide, such as that given in Figure 1.60, the following condition is satisfied:

$$k_2 \le \beta \le k_1$$

Using this inequality, find the range of the angle θ that the propagation vector makes with the side of the waveguide. Express your answer in terms of the critical angle, θ_c, for total internal reflection.

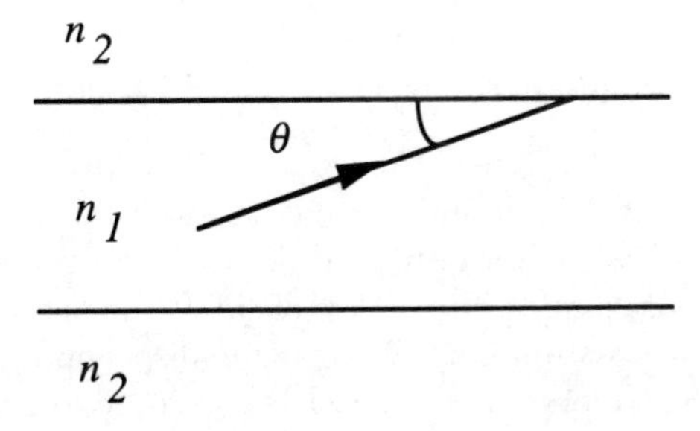

Figure 1.60 Propagation in a slab waveguide.

Problem 1.15 Determining Fiber Dimensions

A step-index fiber has a core with the refractive index $n_1 = 1.4258$ and cladding with the refractive index $n_2 = 1.4205$. This fiber is to be used for both 1.3-μm and 1.55-μm operations. What is the maximum core diameter for which this fiber can be used as a single-mode fiber?

Problem 1.16 Determining Fiber Parameters

Light traveling in air strikes on the core area of the fiber end surface at an angle $\theta = 33$ degrees, where θ is measured between the incoming ray and the surface of the fiber end. Upon striking the fiber, part of the beam is reflected and part is refracted. The reflected and the refracted beams make an angle of 90 degrees with each other.

a. What is the refractive index of the core of this fiber?
b. If the cladding index of refraction $n_2 = 1.50$, what is the NA and the maximum acceptance angle (θ_{max}) of this fiber?

Problem 1.17 Fiber Dispersion

Consider a 10-km-long multimode step-index fiber with NA = 0.30. If the refractive index of the core is 1.450, estimate the bandwidth of the fiber.

LEDs and Lasers

Problem 1.18

A silicon LED has a modulation bandwidth of 100 MHz. The current density in the LED is $J(t) = J_0\, u(t)$ where $u(t)$ is the step function. What is the 10–90% rise time for the optical output power?

Problem 1.19

A GaAs LED operating at room temperature is used as the transmitter for a binary system. It is modulated with a current pulse of 50 mA for a binary 1 and 0 mA for a binary 0. The quantum efficiency of the LED is $\eta = 0.1$. Assume that $\Delta E_{ph} = 3.5$ kT.

a. What are the wavelength and the spectral width of the LED?
b. What is the optical power output of the LED when the current is on?
c. The LED light is coupled into 50 km of single-mode fiber with the dispersion coefficient of 30 ps/(nm·km). What is the dispersion of the link?
d. If the operating temperature of the LED is reduced to 100K, what effects does that have on the link dispersion?

Problem 1.20

Derive an equation for the disturbance, $\Delta n(t)$, of an LED when the diode current is

a. a step pulse:

$$J(t) = \begin{cases} 0 & \text{for } t < 0 \\ J_1 & \text{for } t > 0 \end{cases}$$

b. an exponential decay function:

$$J(t) = \begin{cases} 0 & \text{for } t < 0 \\ J_1 + J_2 \exp(-t/\tau) & \text{for } t > 0 \end{cases}$$

Problem 1.21

A GaAs LED has a recombination lifetime of 5 ns. It is forward biased with a DC current of 20 mA and sinusoidally modulated at 100 MHz by an AC current with an amplitude of 1 mA. Calculate the resulting modulation index, m_N, and phase angle, θ, of Δn.

Problem 1.22

A GaAs laser has the following characteristics: refractive index = 3.4; attenuation constant = 900 m^{-1}; cavity length = 0.6 mm. One mirror has 0.95 reflectivity; the reflectivity of the other mirror is not yet specified. The gain has a Lorentzian profile given by:

$$g(\lambda) = \frac{2000}{1 + \left(\frac{\lambda - \lambda_0}{\Delta\lambda}\right)^2} \qquad (m^{-1})$$

where $\lambda_0 = 0.94$ µm, and $\Delta\lambda = 25$ nm.

a. Calculate the mode spacing between lasing modes (in nm).
b. What minimum value of the reflectivity of the second mirror is necessary to achieve lasing?
c. Assume for the rest of the problem that the reflectivity of the second mirror is 0.5. Estimate how many modes oscillate simultaneously.
d. What is the photon lifetime in this laser?

Problem 1.23

A GaAs laser has the following parameters: bandgap = 1.4 eV; length = 400 µm; thickness and width of the active region = 0.6 µm and 1.2 µm, respectively. The refractive index is 3.4, and the photon lifetime is 4 ps. The threshold current density is 250 A/cm^2; the laser is pumped with 700 A/cm^2.

a. What is the photon density in the cavity?
b. How much optical power travels in one direction in the optical cavity?
c. What is the laser output power if its facet reflectivity = 0.5?

Problem 1.24

Consider the modulation transfer function, $M(\omega_m)$, of a laser, given by (1.219). The constraint $|M(\omega_0)| = 1$ is imposed to achieve a relatively flat response.

a. For what value of α/ω_0 is this constraint satisfied?
b. Under these circumstances, $|M(\omega_m)|$ passes through an extremum between 0 and ω_m. Find ω_m for that extremum and the corresponding value of $|M|$.
c. Show that if τ_{sp} is very large, α/ω_0 can be expressed in terms of ϕ_s and n_{th}. (There is no constraint on α/ω_0 here.)

Problem 1.25

Short-wavelength LEDs can be made from the material $Ga_{1-x}A\ell_xAs$, where x denotes the alloy fraction. The bandgap energy for such a material is E_g(in eV) = $1.424 + 1.266x + 0.266x^2$. Given that x must satisfy $0 \le x \le 0.37$, what is the wavelength range that can be covered with such LEDs?

Problem 1.26

Consider an LED with $E_g = 0.9$ eV, T = 320K, $\tau_r = 0.5$ ns driven by 70 mA of DC current. The internal quantum efficiency is 45%. The LED output is coupled into a graded-index fiber with $n_1 = 1.48$, $n_2 = 1.47$, $\alpha = 0.3dB/km$, and $L = 30km$. The coupling efficiency into the fiber is 1%.

a. Calculate the fiber output power.
b. Calculate the LED radiation bandwidth.
c. The LED is now amplitude modulated at a frequency f. Calculate the maximum f that can be used on this communication link.

Problem 1.27

A 0.85-μm GaAs laser has a cavity length of 0.35 mm, an effective absorption coefficient of 1200 m^{-1}, and facet reflectivities of 0.5.

a. What is the minimum gain for which lasing can occur?
b. Assume that the shape of the gain versus wavelength (λ) is Gaussian:

$$g(\lambda) = g_0 \exp\left[-\frac{(\lambda - \lambda_0)^2}{2\sigma^2}\right]$$

The center wavelength, λ_0, is 0.85 μm; with $g_0 = 5000\ \text{m}^{-1}$; and the width of the gain spectrum σ, is 32 nm. Estimate the number of modes that will be active in the laser.

c. The current in the laser is $J = 1.5\, J_{th}\, u(t)$, where $J_{th} = 3 \cdot 10^6\ \text{Am}^{-2}$ and $u(t)$ is the step function. Assume that $d = 0.7$ μm. What are the photon lifetime and the steady-state photon concentration for $t \to \infty$?

Problem 1.28

A semiconductor laser is pulse modulated for use in an optical fiber system. It has been shown that

$$\frac{d^2(\Delta n)}{dt^2} + 2\alpha \frac{d(\Delta n)}{dt} + \omega_0^2 (\Delta n) = 0$$

Show that the resonant frequency and damping constant of the laser can be expressed as

$$\omega_0^2 = \frac{1}{\tau_{sp}\, \tau_{ph}} \left(\frac{J}{J_{th}} - 1 \right) \qquad 2\alpha = \frac{1}{\tau_{sp}} \frac{J}{J_{th}}$$

Problem 1.29

a. A current pulse with the density amplitude J_p ($J_p > J_{th}$) is applied to an unbiased laser diode. Show that the time needed for the onset of the stimulated emission is

$$t_d = \tau_{sp} \ln \left(\frac{J_p}{J_p - J_{th}} \right)$$

where τ_{sp} is the mean lifetime, and J_{th} is the threshold current density.

b. A laser is prebiased to current density J_B ($J_B < J_{th}$), and the current density in the active region during a current pulse is $J = J_B + J_p$. Show that in this case

$$t_d = \tau_{sp} \ln \left(\frac{J_p}{J_p + J_B - J_{th}} \right)$$

Problem 1.30

Consider the optical cavity resonator in Figure 1.61.

Assume that propagation through the cavity goes in only one direction. The power reflectivities of the mirrors, $M1$, $M2$, and $M3$ are R_1, R_2, and R_3, respectively.

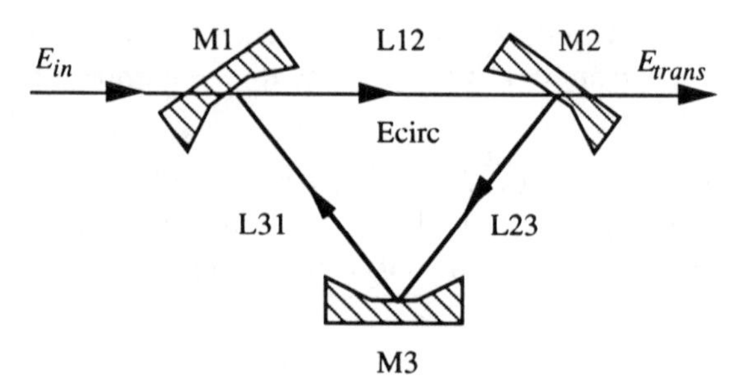

Figure 1.61 Ring optical cavity resonator.

The perimeter of the cavity is $p = L12 + L23 + L31$. The cavity material has an intensity attenuation constant of α_s (in m^{-1}), a gain of g (in m^{-1}), and an index of refraction n.

a. For this resonator, derive the minimum gain of the cavity required to achieve oscillation, in terms of the system parameters.
b. Derive the longitudinal mode spacing (in hertz) in terms of the system parameters.
c. Due to the temperature changes, the cavity may undergo a change in length. How much will the mode spacing change if the perimeter increases by $\Delta\%$?
d. Assume that it is possible to construct a semiconductor laser with the structure shown in Figure 1.61. Given that $p = 350\ \mu m$, R_1 and $R_3 = 0.98$, $R_2 = 0.8$, the attenuation constant $\alpha_s = 300\ m^{-1}$, and the index of refraction $n = 3.3$, calculate the photon lifetime, τ_{ph}.

Problem 1.31

Consider a 100-Mbps optical fiber link employing pulse modulation on a semiconductor laser. The laser is set for a peak current density, J_1, of $3 \cdot 10^6\ Am^{-2}$ and the threshold current density, J_{th}, is $2 \cdot 10^6\ Am^{-2}$. The time constant for spontaneous decay, τ_{sp}, is 2 ns.

a. If the system permits a maximum turn-on delay $t_\alpha = 20\%$ of the bit duration, determine the maximum achievable value of the ratio of the current density for binary 1 and binary 0 (J_1/J_0) (ignore the effect of fiber dispersion).
b. For the J_0 obtained in part (a), determine the time, including the turn-on delay, needed for the optical power to settle down to within 5% of its steady state value for a transition from binary 0 to binary 1.

Problem 1.32

In practical systems, there is usually a rise time associated with a step pulse. Assume that the current density $J(t)$ in Figure 1.62 is applied to an unbiased semiconductor laser.

Assuming that $\tau_{sp} << t_d < t_r$, derive an approximate expression for the time needed for the onset of the stimulated emission, t_d, in terms of the rise time, t_r, of the pulse shape and the laser parameters.

Problem 1.33

A 1.3-μm InGaAsP laser has the following dimensions of the active region: length $L = 300$ μm, width $w = 1.3$ μm, and thickness $d = 0.7$ μm. The loss coefficient is $\alpha_s = 500\ \text{m}^{-1}$, $\tau_{sp} = 5$ ns, the index of refraction in the active region is 3.6, and the facet reflectivities are $R_1 = R_2 = 0.32$.

a. Calculate the photon lifetime, τ_{ph}.
b. If the threshold current density, $J_{th} = 1.4 \times 10^6\ \text{A/m}^2$ and the excitation current density $J = 2.1 \times 10^6\ \text{A/m}^2$, calculate the steady-state photon concentration, ϕ_s.
c. Suppose the current density in this semiconductor diode laser is $J = 0$ for $t < 0$ and $J = 1.5J_{th}$ for $t > 0$. Calculate the delay to reach threshold t_d. Calculate α, ω_0 and ω.
d. For the conditions of part (c), roughly sketch graphs of the electron density, n, versus time and the photon density, ϕ, versus time. Show the important features.

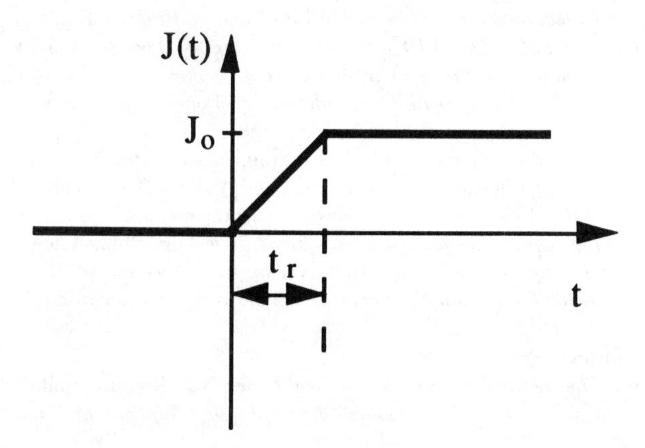

Figure 1.62 Current pulse with linear rise.

Detectors

Problem 1.34

A GaAs PIN photodiode on average generates one electron-hole pair per three incident photons. Assume all the electrons are collected.

a. Calculate the quantum efficiency of the device.
b. Calculate the mean output photocurrent when the received optical power is 10^{-7}W at a wavelength of 0.8 μm.
c. Determine the wavelength above which this photodiode will cease to operate, that is, the long wavelength cut-off point, λ_c.

Problem 1.35

An APD operates at 1.55 μm and has quantum efficiency = 0.3, gain = 10^3, transit time = 10 ps, k = 0.1; the avalanches are initiated by electrons.

a. Calculate the 3-dB bandwidth of the detector.
b. Calculate the output photocurrent if 2 mW of optical power is received.
c. Under the condition of (b), calculate the total root-mean-square noise current in 10 MHz of bandwidth.

$$i_{rms} = (i_n^2)^{\frac{1}{2}}$$

References

[1] Senior, J., *Optical Fiber Communications: Principles and Practice*, 2d Ed., Englewood Cliffs, NJ: Prentice-Hall International, 1992.
[2] Gowar, J., *Optical Communication Systems*, 2d Ed., London: Prentice-Hall International, 1991.
[3] Jones, W.B., *Introduction to Optical Fiber Communication Systems*, New York: Holt, Rinehart & Winston, 1988. Distributed by Oxford University Press.
[4] Snyder, A. W., and J. D. Love, *Optical Waveguide Theory*, London and New York: Chapman & Hall, 1983.
[5] Palais, J. C., *Fiber Optic Communications*, 3d Ed., Englewood Cliffs, NJ: Prentice-Hall, 1992.
[6] Keiser, G., *Optical Fiber Communications*, 2d Ed., New York: McGraw-Hill, 1991.
[7] Personick, S. D., *Optical Transmission Systems*, New York: Plenum Press, 1981.
[8] Agrawal, G. P., *Fiber-Optic Communications Systems*, New York: John Wiley & Sons, 1992.
[9] Yariv, A., *Optical Electronics*, New York: Holt, Rinehart & Winston, 1985.
[10] Tsang, W. T., *Semiconductor and Semimetals*, Lightwave Communications Technology, Vol. 22, Orlando, FL: Academic Press, 1985.
[11] Fong, T., unpublished.
[12] Green, P. E., Jr., *Fiber Optic Networks*, Englewood Cliffs, N.J.: Prentice-Hall, 1993.
[13] Jeunhomme, L. B., *Single-Mode Fiber Optics—Principles and Applications*, New York: Marcel Dekker, 1983.
[14] Ainslie, B. J., and C. R. Day, "A Review of Single-Mode Fibers With Modified Dispersion Characteristics," *IEEE Journal of Lightwave Technology*, LT-4: 967, 1986.

[15] Henry, P. S., "Lightwave Primer." *IEEE Journal of Quantum Electronics*, QE-21: 1862–1879, Dec. 1985.
[16] Onaka, H., et al., "1.1 Tb/s WDM Transmission Over a 150 km 1.3 μm Zero-Dispersion Single-Mode Fiber," OFC '96, San Jose, CA, February 1996. Paper PD-19. *Digest of Postdeadline Papers, Part B*, pp. PD19-1–PD19-5.
[17] Li, Y. W., C. D. Hussey, and T. A. Birks, "Triple-Clad Single-Mode Fibers for Dispersion Shifting," *IEEE Journal of Lightwave Technology*, 11(11):1812–1819, 1993.
[18] Miller, S. E., and I. P. Kaminow, *Optical Fiber Telecommunications II*, London: Academic Press, 1988.
[19] Reiner G., et al., "Optimization of Planar Be-Doped InGaAs VCSEL's With Two-Sided Output," *IEEE Photonics Technology Letters*, 7(7):730–732, 1995.
[20] Gemelos, S. M., T. Hofmeister, C.-L. Lu, D. Mayweather, S. Savin, and F. Yang, unpublished.
[21] Fukuda, M., *Reliability and Degradation of Semiconductor Lasers and LEDs*, Norwood: Artech House Inc., 1991.
[22] Tanaka, Y., T. Toyama, and R. Tohmon, "A Novel Temperature-Stable Light-Emitting Diode," *IEEE Transaction on Electron Devices*, 41(7):1125–1127, 1994.
[23] Woodhouse, J. D., et al., "Uniform Linear Arrays of Strained-Layer InGas-AlGaAs Quantum-Well Ridge-Waveguide Diode Lasers Fabricated by ECR-IBAE," *IEEE Journal of Quantum Electronics*, 31(8):1357–1363, 1995.
[24] Hamamoto, K., H. Chida, T. Miyazaki, and S. Ishikawa, "High-Power 0.98-μm Strained Quantum-Well Lasers Fabricated Using In Situ Monitored Reactive Ion Beam Etching," *IEEE Photonics Technology Letters*, 7(6):602–603, 1995.
[25] Agrawal, G. P., and N. K. Dutta, *Semiconductor Lasers*, New York: Chapman and Hall, 1993.
[26] Kazovsky, L. G., "Impart of Laser Phase Noise on Optical Heterodyne Communication Systems," *Journal of Optical Communications*, 7:66–78, 1986.
[27] Stillman, G. E., and C. M. Wolfe, "Avalanche Photodiodes." Chap. 5 in Semiconductors and Semimetals, Vol. 12, edited by R. K. Willardson and A. C. Beer, N.Y.: Academic Press, 1977.

Selected Bibliography

Kressel, H., "Semiconductor Devices for Optical Communication," *Topics in Applied Physics*, Vol. 39, 2d Ed., Heidelberg: Springer-Verlag, 1982.
Miller, S. E., and A. G. Chynoweth, *Optical Fiber Telecommunications*, London: Academic Press, 1979.
Panish and Hayashi, *Applied Solid State Science*, Vol. 4, London: Academic Press, 1974.
Wada, O., et al., "Performance and Reliability of High-Radiance InGaAsP/InP DH LED's operating in the 1.15–1.5 μm Wavelength Region." *IEEE Journal of Quantum Electronics*, QE-18: 368–373, March 1982.

Chapter 2

The Optical Communication Toolbox

2.1 INTRODUCTION

Signal and system theory, probability, random processes, and detection theory are essential mathematical tools to understand, analyze, and design communications systems in general and optical communication systems in particular.

Extensive and consolidated bibliographies exist on these subjects, and we assume that the reader has some familiarity with them. For that reason, the way those subjects are treated in this chapter is aimed mainly at consolidating a common background of notions and notations that will be used throughout the book. As a rule, proofs will be omitted (although some are referred to the problems at the end of the chapter), and lengthy analytical details are skipped. Focus is rather on a few definitions and concepts whose choice has been dictated by the *optical* characterization of communication systems we are interested in.

The line of thought in this chapter follows that of Chapter 2 in Benedetto, Biglieri, and Castellani [1].

2.2 PROBABILITY AND RANDOM VARIABLES

The goal of a communication system is to transfer information from the source to the end user. This process, for the end user, consists in the removal of some uncertainty about the message sent. Usually, the end user knows the set of possible messages from which the source can choose and must decide which message was actually transmitted on the basis of the received message, which is different from the transmitted one since it has been modified in its travel from the source to the end user.

This process is inherently probabilistic, in the sense that at least part of the transformations on the transmitted signal are random, and that, in practice, there will always be a nonzero probability that the message decided by the end user will be different from the one the source transmitted. This is the *error probability* of the system. We need thus to state the basic notions of the theory of probability and random variables.

2.2.1 Probability

A *random experiment*, like the transmission of information, is an experiment with a number of possible outcomes. Its sample space, S, is the set of all possible outcomes (called sample points). When the experiment consists in rolling a die, $S = 1, 2, 3, 4, 5, 6$. Event A is a subset of the sample space, that is, something that can—or cannot—happen when the experiment is performed. In the die-rolling example, the event A = "the outcome is an odd number" is the subset $A = 1, 3, 5$. The complement of event A, denoted by $\overline{A}$, consists of all points of S that are not in A. For example, if A is the previously defined event, $\overline{A} = 2, 4, 6$. A and $\overline{A}$ are *mutually exclusive* events, meaning that the occurrence of A prevents the occurrence of $\overline{A}$. In general, two events are mutually exclusive if they have no outcomes in common.

The union of two events A and B, denoted by $A \cup B$, is the subset of S that contains all the points in A and B. Of course, $A \cup \overline{A} = S$. The intersection of two events A and B, denoted by $A \cap B$, is the subset of S that contains all the points common to A and B. Of course, $A \cap \overline{A} = \phi$, ϕ being the null (impossible) event.

To each event A we associate a probability, $P(A)$. It is a real number that satisfies the three axioms of probability:

- $P(A) \leq 1$;
- $P(S) = 1$;
- $P(A \cup B) = P(A) + P(B)$ if A and B are mutually exclusive.

If the sample space has n outcomes, and if they are equally likely, then to each point s_i, $i = 1, \ldots, n$, the probability $P(s_i) = 1/n$ is associated.

Consider two experiments, with sample spaces S_1 and S_2. The combined experiment, to which a pair of possible outcomes (s_i, s_j), the first belonging to S_1 and the second to S_2, is associated, has a sample space formed by the Cartesian product of the sample spaces S_1 and S_2. To each point of the combined experiment there corresponds a *joint* probability $P(s_i, s_j) \overset{\text{def}}{=} P[s_i \cap s_j]$ satisfying the probability axioms. For example, assume that the two experiments consist in flipping a coin. The joint sample space will be $S = (TT, TH, HT, HH)$.

Consider now the two events A and B from the first and second experiment and form a joint event (A, B) of the combined experiment with joint probability $P(A, B)$. We say that A and B are statistically independent if and only if $P(A, B) = P(A)P(B)$.

Suppose now that we want to compute the probability of A knowing that B has occurred. This is called the conditional probability of the event A given B and is defined as

$$P(A|B) = \frac{P(A, B)}{P(B)} \tag{2.1}$$

provided that $P(B) > 0$. Similarly, we define

$$P(B|A) = \frac{P(A, B)}{P(A)} \tag{2.2}$$

so that the joint probability can be expressed as

$$P(A,B) = P(A)P(A|B) = P(B)P(B|A) \tag{2.3}$$

Intuitively, when two events are statistically independent, knowing that one has occurred does not provide useful information on the occurrence of the other. In fact, using previous results, we have that $P(A|B) = P(A)$ if A and B are independent.

Suppose now that we want to compute the joint probability of m events $A_1, \ldots, A_m$. If they are mutually exclusive, for the third axiom of the probability, we have

$$P\,[\bigcup_i^m A_i] = \sum_i^m P(A_i) \tag{2.4}$$

If they are not mutually exclusive, (2.4) cannot be applied. The following useful inequality always holds (*union bound*):

$$P\,[\bigcup_i^m A_i] \leq \sum_{i=1}^m P(A_i) \tag{2.5}$$

A second useful relationship in communication theory is Bayes theorem, which concerns m events A_i forming a partition of the sample space, that is, such that $A_i \cap A_j = \phi,\ \forall i, j\ j \neq i$, and $\cup_i^m A_i = S$, and an arbitrary event B with associated nonzero probability $P(B)$. The theorem states that

$$P(A_i|B) = \frac{P(A_i, B)}{P(B)} = \frac{P(B|A_i)}{\sum_{j=1}^m P(B|A_j)P(A_j)} \tag{2.6}$$

Bayes theorem is applied in communications in a situation where the events A_i represent source messages and B a received message. It allows computation of the *a posteriori* probabilities $P(A_i|B)$, conditioned on the received message, from the knowledge of the *a priori* probabilities $P(B|A_i)$, which depend on the communication channel and are usually known or easy to compute.

2.2.2 Random Variables

Let S be the sample space of an experiment. A *random variable* (RV) X is a real-valued function $X(s)$ of the points of S. Through the correspondence between sample points and real numbers defined by the function $X(s)$, the probabilities of the sample points are transferred to the real values assumed by the RV X. As an example, consider the experiment of rolling a fair die and let $X(s) = s$. In this case, the RV X will assume the six values (1, 2, 3, 4, 5, 6) with equal probabilities 1/6.

Up to now, we have considered experiments with discrete sample spaces. Dealing with communications experiments, we will also encounter continuous sample spaces. A typical case is represented by the thermal noise generated in electronics circuits, which can be represented by a continuous RV assuming values in the whole real axis.

Dealing with RVs requires the ability to compute the probability of events related to them. In particular, since an RV assumes real values, we need to compute the probability for it to lie within a given interval (a, b), that is, $P[a < X \le b]$.[1] This problem can be solved through the knowledge of the *probability distribution function* (PDF) $F_X(x)$ of the RV X, defined as

$$F_X(x) \overset{\text{def}}{=} P[X \le x], \qquad -\infty < x < \infty \tag{2.7}$$

The function $F_X(x)$, owing to its definition, possesses the following properties:

- It is a nondecreasing function within its limited range (0,1).
- $F_X(-\infty) = 0$.
- $F_X(\infty) = 1$.

As an example, for the previous case of die rolling, the PDF is shown in Figure 2.1. It is a staircase function, with discontinuities located in correspondence of the values assumed by the RV and jump sizes equal to the probability of assuming those particular values (1/6 for each jump in this example). This structure is typical to all discrete PDFs, whose analytical form is

$$F_X(x) = \sum_{i=1}^{n} P(x_i)\mathrm{u}(x - x_i) \tag{2.8}$$

1. We use the notation $P[\cdot]$ to denote the probability of an event as a concept, and $P(\cdot)$ to denote its numerical value.

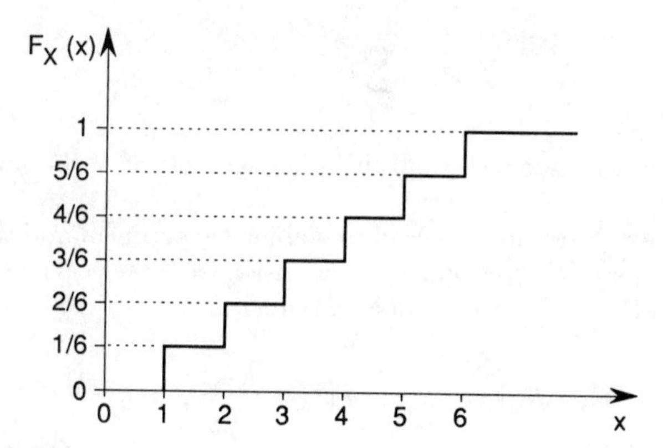

Figure 2.1 Probability distribution function for the die-rolling experiment.

where n is the number of values assumed by X, $\{x_i\}$ the set of those values, $P(x_i)$ the probability for X to assume the value x_i, and u(x) the unitary step function.

For continuous RVs, instead, the PDF is a continuous function.

The knowledge of the PDF of an RV X allows us to compute the probability of X to lie within a real interval as

$$P[a < X \leq b] = F_X(b) - F_X(a) \tag{2.9}$$

(See Problem 2.2.)

A second function that contains all the information needed to work with an RV X like the PDF is its *probability density function* (pdf) $f_X(x)$, defined as

$$f_X(x) \stackrel{\text{def}}{=} \frac{dF_X(x)}{dx} \tag{2.10}$$

From its definition, the pdf has the following properties:

- It is a nonnegative function.
- $F_X(x) = \int_{-\infty}^{x} f_X(s) ds$.
- $P[a < X \leq b] = \int_a^b f_X(s) ds$.

For discrete RVs, the pdf contains a number of delta functions [2,p.72] located in correspondence of the jumps of the PDF, so that its analytical form is in general

$$f_X(x) = \sum_{i=1}^{n} P(x_i)\delta(x - x_i) \tag{2.11}$$

where $\delta(x)$ is the delta function, and the other symbols have the same meaning as in (2.8).

Consider now the sample space of a combined experiment and two RVs, X_1 and X_2, defined on it (joint RVs). As a natural extension from the previous definitions, we define the joint PDF of the two random variables as

$$F_{X_1X_2}(x_1, x_2) \stackrel{\text{def}}{=} P[X_1 \leq x_1, X_2 \leq x_2] \tag{2.12}$$

and the joint pdf as

$$f_{X_1X_2}(x_1, x_2) \stackrel{\text{def}}{=} \frac{\partial^2}{\partial x_1 \partial x_2} F_{X_1X_2}(x_1, x_2) \tag{2.13}$$

The knowledge of the joint pdf permits us to obtain the two *marginal* pdf's through

$$f_{X_1}(x_1) = \int_{-\infty}^{\infty} f_{X_1X_2}(x_1, x_2)dx_2 \tag{2.14}$$

and

$$f_{X_2}(x_2) = \int_{-\infty}^{\infty} f_{X_1X_2}(x_1, x_2)dx_1 \tag{2.15}$$

(See Problem 2.2.)

In the same way, we can define multiple (more than two) RVs.

Useful expressions concern the conditional PDF and pdf, obtained (see Problem 2.3) as

$$F_{X_1 \mid X_2}(x_1, x_2) \stackrel{\text{def}}{=} P[X_1 \leq x_1 \mid X_2 = x_2] = \frac{\int_{-\infty}^{\infty} f_{X_1X_2}(v, x_2)dv}{f_{X_2}(x_2)} \tag{2.16}$$

and

$$f_{X_1 \mid X_2}(x_1, x_2) = \frac{f_{X_1X_2}(x_1, x_2)}{f_{X_2}(x_2)} \tag{2.17}$$

Two (or more) joint RVs are statistically independent if and only their joint PDF and pdf factor into the product of the corresponding marginal distributions

$$F_{X_1X_2}(x_1, x_2) = F_{X_1}(x_1)F_{X_2}(x_2) \tag{2.18}$$

and

$$f_{X_1X_2}(x_1, x_2) = f_{X_1}(x_1)f_{X_2}(x_2) \tag{2.19}$$

2.2.2.1 Functions of Random Variables

Communication systems deal with quantities that can be represented by RVs and modify those quantities in various ways during the transmission and reception processes. Thus, it is important to solve the following problem: given an RV, X, and a real-valued function, $y = g(x)$, which defines a new RV, $Y = g(X)$, in the sense that each value x assumed by X is transformed by the function $g(\cdot)$ into a value assumed by Y, find the PDF (or pdf) of Y.

Example 2.1

Let the new RV Y be

$$Y = aX + b$$

with $a > 0$ and b constant. By definition, we can write

$$F_Y(y) = P[Y \leq y] = P[aX + b \leq y] = P\left[X \leq \frac{y-b}{a}\right] = F_X\left(\frac{y-b}{a}\right)$$

Differentiating, we also obtain

$$f_Y(y) = \frac{1}{a} f_X\left(\frac{y-b}{a}\right)$$

As an example, if X is a uniform RV in the interval (0,1), and $a = 2$, $b = 3$, the pdf of Y is depicted in Figure 2.2.

The preceding example concerned a monotonic transformation. In general, if the equation $y = g(x)$ defined by the transformation has m real roots $\{x_i\}$, the pdf of Y is expressed as

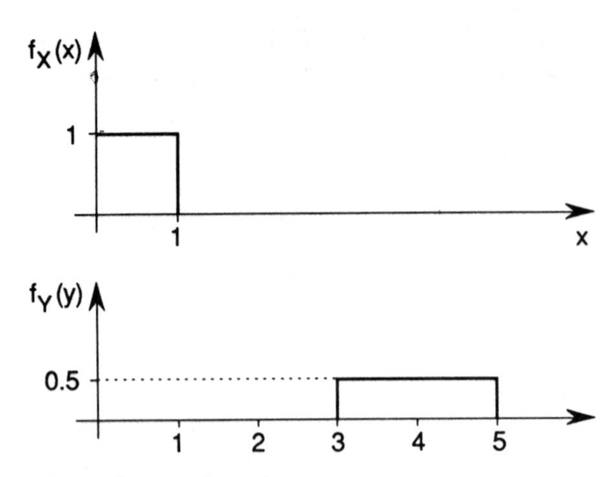

Figure 2.2 Probability density function of a uniform RV.

$$f_Y(y) = \sum_{i=1}^{m} \frac{f_X[x_i(y)]}{|g'[x_i(y)]|} \tag{2.20}$$

where $g'(x)$ denotes the derivative of $g(x)$.

2.2.2.2 Averages, Moments, Characteristic Function

Sometimes it is difficult to completely characterize the RVs through their joint or marginal pdf's; in other cases, for the applications at hand, the required effort may not be justified. Partial knowledge through appropriately defined quantities is then sought. Most used parameters are the *moments* of an RV. The definition of the kth moment is

$$m_X^{(k)} \stackrel{\text{def}}{=} \mathrm{E}(X^k) = \int_{-\infty}^{\infty} x^k f_X(x)dx \tag{2.21}$$

where E denotes expectation or statistical average. Of great importance are the first moment, obtained for $k = 1$, also called the mean value of X, and denoted simply as m_X and the second moment, for $k = 2$, related to physical quantities like power.

It is sometimes important to characterize the form of the pdf of an RV in its behavior around its mean value. A suitable parameter to this end is the variance of X

$$\sigma_x^2 \stackrel{\text{def}}{=} \mathrm{E}[(X - m_X)^2] = \int_{-\infty}^{\infty} (x - m_X)^2 f_X(x)dx \tag{2.22}$$

The variance of X represents physically the moment of inertia of a mass distribution, $f_X(x)$, and measures the dispersion of the values of X around its mean value.

Mean value, second moment, and variance are related by

$$\sigma_X^2 = \mathrm{E}(X^2) - m_X^2$$

Moments can be defined over two or more joint RVs. Definitions are straightforward. We report here only the correlation between two RVs

$$\mathrm{E}(X_i X_j) = \int_{-\infty}^{\infty}\int_{-\infty}^{\infty} x_i x_j f_{X_i X_j} dx_i dx_j$$

and the *covariance*

$$\sigma_{X_i X_j} = \mathrm{E}[(X_i - m_{X_i})(X_j - m_{X_j})] = \mathrm{E}(X_i X_j) - m_{X_i} m_{X_j}$$

Two RVs are said to be *uncorrelated* if their covariance is zero.

The complete set of moments of an RV, X, can be obtained from the knowledge of its characteristic function, $C_X(s)$, defined as

$$C_X(s) \overset{\text{def}}{=} \mathrm{E}(e^{jsX}) = \int_{-\infty}^{\infty} e^{jsx} f_X(x)dx \tag{2.23}$$

From its definition, we see that the characteristic function of X is strictly related to the inverse Fourier transform of its pdf. It can be easily shown that (see Problem 2.4) the moments of X are related to its characteristic function by

$$m_X^{(k)} = (-j)^k \left. \frac{d^k C_X(s)}{ds^k} \right|_{s=0} \tag{2.24}$$

An important application of the characteristic function occurs when we want to statistically characterize an RV, Y, obtained as the sum of n independent RVs, X_i. In fact, by definition, we have

$$C_Y(s) = E(e^{jsY}) = \mathrm{E}\left[\exp\left(js\sum_{i=1}^{n} X_i\right)\right] = \mathrm{E}\left[\prod_{i=1}^{n} e^{jsX_i}\right] = \prod_{i=1}^{n} \mathrm{E}(e^{jsX_i}) = \prod_{i=1}^{n} C_{X_i}(s)$$

where the last two steps are made possible by the independence of the RV X_is.

2.3 SOME IMPORTANT PROBABILITY DISTRIBUTIONS

We introduce here some discrete and continuous distributions concerning RVs that will be used in the rest of the chapter.

2.3.1 The Binomial Distribution

Consider a discrete RV, X, assuming two values, 1 and 0, with probabilities p and $1 - p$, respectively, and let Y be the sum of n independent RVs, X_i, with the same distribution of X, that is,

$$Y = \sum_{i=1}^{n} X_i$$

It is easy to verify that Y assumes $n + 1$ integer values in the range $(0, n)$ and that the probability $P(Y = k)$ of assuming the integer value k in that range is given by

$$P[Y = k] = \binom{n}{k} p^k (1 - p)^{n-k} \tag{2.25}$$

obtained as the product of the probability of a particular combination of k 1s and $n - k$ 0s by the number of such combinations. The resulting pdf becomes thus

$$f_Y(y) = \sum_{k=0}^{n} \binom{n}{k} p^k (1 - p)^{n-k} \delta(y - k) \tag{2.26}$$

Figure 2.3 shows the pdf for $n = 9$ and $p = 0.5$.

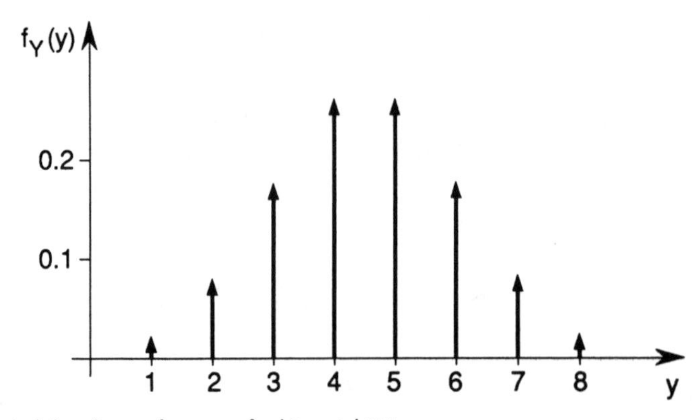

Figure 2.3 Probability density function of a binomial RV.

The mean and the variance of a binomial RV are np and $np(1-p)$, respectively (see Problem 2.5).

2.3.2 The Poisson Distribution

The Poisson distribution, which plays an important role in optical communication, describes the phenomenon of electrical carrier generation in a photodetector. It can be obtained from the binomial distribution as a limit approximation for n very large and p very small, so that the product np is significant, that is, n and $1/p$ are comparable.

The pdf of a Poisson RV is

$$f_X(x) = \sum_{k=0}^{\infty} \frac{\lambda^k}{k!} e^{-\lambda} \delta(x-k), \qquad \lambda > 0 \tag{2.27}$$

The parameter λ represents both the mean value and the variance of X (see Problem 2.5).

2.3.3 The Gaussian Distribution

Gaussian RVs play a fundamental role in probability theory and, as a consequence, also in optical communication theory. Under mild assumptions, an RV obtained as the sum of a large number of independent and identically distributed RVs approaches a Gaussian distribution (central limit theorem) [2,Chap. 8]. Thus, Gaussian distributions accurately model physical phenomena whose behavior depends on a large number of independent contributions, like thermal noise generation in electronic circuits.

The Gaussian pdf and PDF are given by

$$f_X(x) = \frac{1}{\sqrt{2\pi}\sigma} e^{-\left[(x-\mu)^2/2\sigma^2\right]} \tag{2.28}$$

and

$$F_X(x) = \frac{1}{2}\left[1 + \operatorname{erf}\left(\frac{(x-\mu)}{\sqrt{2}\sigma}\right)\right] \tag{2.29}$$

where

$$\operatorname{erf}(z) \stackrel{\text{def}}{=} \frac{2}{\sqrt{\pi}} \int_0^z e^{-x^2}\, dx$$

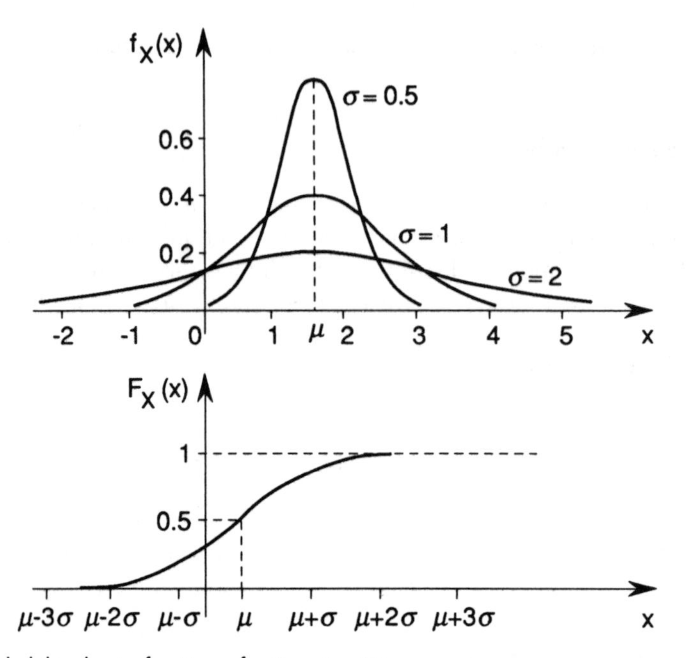

Figure 2.4 Probability density functions of a Gaussian RV.

Equation (2.28) represents a family of curves defined by the two parameters μ and σ^2. The first one is the mean value of X, and the vertical line $x = \mu$ represents a symmetry axis of the set of curves. The second parameter σ is the so-called standard deviation of X (its square is the variance of X), and the curves (Fig. 2.4) become more and more spread around their symmetry axis as long as the value of σ increases.

2.3.4 The χ-Square Distribution

As we shall see in Chapter 4, some of the more important practical receivers for coherent optical communications perform quadratic transformations on the received signal. Because the received signal can often be modeled as a Gaussian RV, it is important to find the distribution of the RV obtained as the square of a Gaussian RV with mean μ and variance σ^2.

Easy computations lead to the following noncentral χ-square pdf:

$$f_X(x) = \frac{1}{\sqrt{2\pi x}\sigma} e^{-\left[(x+\mu^2)/2\sigma^2\right]} \cosh\left(\frac{\sqrt{x}\mu}{\sigma^2}\right) \tag{2.30}$$

The parameter μ^2 is the noncentrality parameter. The particular case $\mu = 0$ leads to the central χ-square distribution.

In some cases, the decision variable in the receiver is obtained at the end of a diversity process, in which we combine more than one received signal. In that case, we need to generalize the previous distribution to the case of an RV, Y, obtained as the sum of the squares of n independent Gaussian RVs, X_i, with mean values μ_i and equal variance σ^2. Obtaining the pdf of Y is made easier by evaluating first the characteristic function and then inverse-Fourier transforming (see Problem 2.6). Denoting by $m^2 = \Sigma_{i=1}^{n} \mu_i^2$, the resulting pdf is

$$f_Y(y) = \frac{1}{2\sigma^2}\left(\frac{y}{m^2}\right)^{(n-2)/4} e^{-[(y+m^2)/2\sigma^2]} \cdot I_{\frac{n}{2}-1}\left(\frac{\sqrt{y}m}{\sigma^2}\right), \qquad y \geq 0 \tag{2.31}$$

where $I_a(\cdot)$ is the ath-order modified Bessel function of the first kind. The pdf (2.31) is the noncentral χ-square with n degrees of freedom and noncentrality parameter m^2. For $m = 0$, it particularizes to the central χ-square with n degrees of freedom

$$f_Y(y) = \frac{1}{\sigma^n 2^{n/2}\,\Gamma\,(n/2)} \cdot y^{(n/2-1)}\, e^{-(y/2\sigma^2)}, \qquad y \geq 0 \tag{2.32}$$

where $\Gamma(x)$ is the gamma function, defined as

$$\Gamma(x) = (x-1)!, \qquad x \text{ integer} > 0 \tag{2.33}$$

$$\Gamma(x) = \int_0^\infty t^{x-1}\, e^{-t}\, dt, \qquad x > 0 \tag{2.34}$$

$$\Gamma(1/2) = \sqrt{\pi},\ \Gamma(3/2) = \frac{\sqrt{\pi}}{2} \tag{2.35}$$

In Figure 2.5, the pdf (2.32) is plotted for some values of n.

2.3.5 The Rayleigh and Rice Distributions

Closely related to χ-square distributions are Rayleigh and Rice RVs. The first is obtained as the square root of a central χ-square with 2 degrees of freedom, which gives (see Problem 2.7)

$$f_X(x) = \frac{x}{\sigma^2}\, e^{-(x^2/2\sigma^2)} \qquad x \geq 0 \tag{2.36}$$

The Rayleigh RV characterizes the envelope of a narrow-band Gaussian noise. For different values of σ, Rayleigh pdf's are plotted in Figure 2.6.

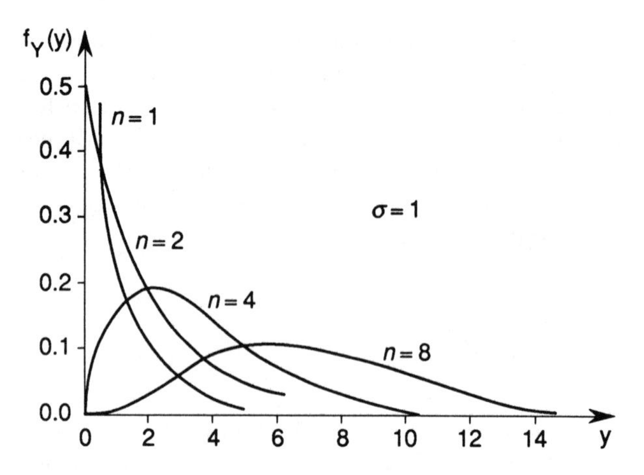

Figure 2.5 Probability density functions of a χ-square RV.

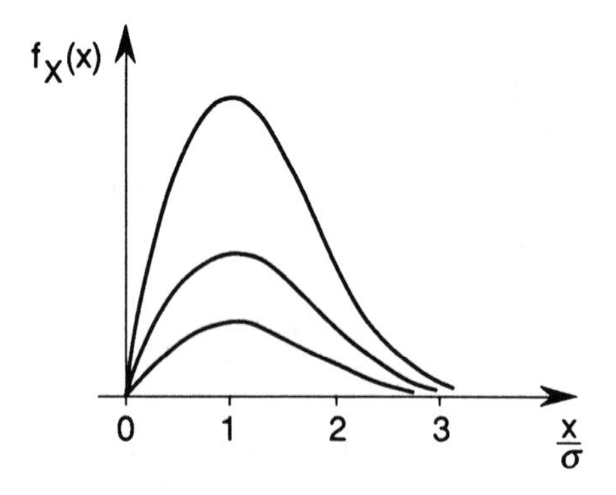

Figure 2.6 Probability density functions of a Rayleigh RV.

The second pdf, the Rice distribution, refers to an RV obtained as the square root of a noncentral χ-square with 2 degrees of freedom and is given by (see Problem 2.7)

$$f_X(x) = \frac{x}{\sigma^2} \mathrm{e}^{-\left[(x+m^2)/2\sigma^2\right]} I_0 \left(\frac{\sqrt{x}m}{\sigma^2} \right), \qquad x \geq 0 \tag{2.37}$$

where $I_0(\cdot)$ is the modified Bessel function of the first kind and zeroth order, with integral representation

$$I_0(z) = \frac{1}{2\pi} \int_0^{2\pi} e^{z \cos \theta} \, d\theta$$

Integrating (2.37), we obtain the PDF of a Rice RV:

$$F_X(x) = 1 - Q\left(\frac{m}{\sigma}, \frac{x}{\sigma}\right) \tag{2.38}$$

where the function $Q(\cdot, \cdot)$ is the Marcum Q-function [1,App. A], defined as

$$Q(a, b) \overset{\text{def}}{=} \int_b^{\infty} e^{-\left[(a^2+z^2)/2\right]} I_0(az) z dz \tag{2.39}$$

The Rice RV characterizes the envelope of a narrowband signal perturbed by additive Gaussian noise. For different values of m and $\sigma = 1$, Rice pdf's are plotted in Figure 2.7.

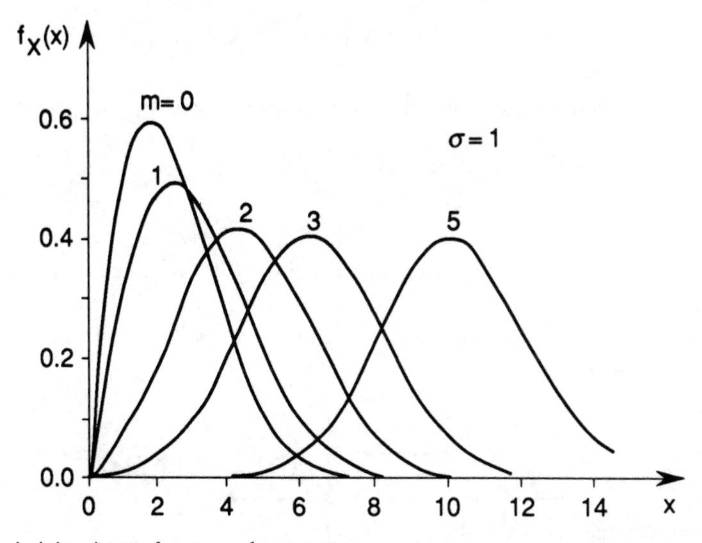

Figure 2.7 Probability density functions of a Rice RV.

2.4 SIGNALS AND SYSTEMS

In this section, we summarize the basic concepts of the linear system theory, in both time and frequency domains.

2.4.1 Continuous-Time Signals and Systems

2.4.1.1 Time Domain

A *continuous-time signal* is a real or complex function, $x(t)$, of the real variable t (the time), assumed to range from $-\infty$ to $+\infty$. A *continuous-time system* is a mapping of a signal, $x(t)$, the system input, into a signal, $y(t)$, that represents the system output or system response:

$$y(t) = S[x(t)] \tag{2.40}$$

A continuous system is linear if for any pair of input signals, $x'(t)$ and $x''(t)$, and for any complex numbers a' and $''$ the following holds:

$$S[a'x'(t) + a''x''(t)] = a'S[x'(t)] + a''S[x''(t)] \tag{2.41}$$

A continuous system is *time invariant* if (2.40) implies that

$$S[x(t - \tau)] = y(t - \tau) \tag{2.42}$$

for all τ. Let $\delta(t)$ denote the delta function, characterized by the property

$$\int_{-\infty}^{\infty} \delta(t)\phi(t)dt = \phi(0) \tag{2.43}$$

valid for all functions $\phi(t)$ continuous at the origin. The response of a time-invariant linear continuous system to the input $\delta(t)$ is called the *impulse response*, $h(t)$, of the system. For a generic input, $x(t)$, the output (Fig. 2.8) is obtained through the *convolution integral*:

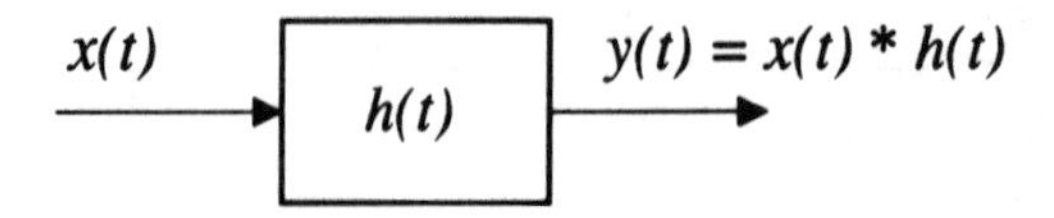

Figure 2.8 A continuous-time linear system.

$$y(t) = x(t) * h(t) \stackrel{\text{def}}{=} \int_{-\infty}^{\infty} x(\tau)h(t-\tau)d\tau \tag{2.44}$$

If $h(t) = 0$ for all $t < 0$, the system is called *causal,* from the fact that the output at time t depends only on the values of the input for times less than or equal to t. If $h(t) = A\delta(t - t_0)$ the system is called "memoryless," because the output at time t depends only on the input value at the same time.

2.4.1.2 Frequency Domain

Given a signal, $x(t)$, we denote its Fourier transform as $X(f)$:

$$X(f) = \int_{-\infty}^{\infty} x(t)e^{-j2\pi ft}dt \tag{2.45}$$

The inverse Fourier transform permits the recovery of $x(t)$ from $X(f)$:

$$x(t) = \int_{-\infty}^{\infty} X(f)e^{-j2\pi ft}dt \tag{2.46}$$

$X(f)$ is called the *amplitude spectrum* of $x(t)$, because $X(f)df$ represents the contribution of the spectral component at the frequency f to the signal construction in (2.46).

A fundamental property of linear systems, stemming directly from the theory of the Fourier transform, is that the amplitude spectrum of the output of a linear system is related to the input and to the transfer function of the system through the simple relation

$$Y(f) = X(f)H(f) \tag{2.47}$$

where $H(f)$ is the Fourier transform of the impulse response $h(t)$ of the system, also called the *transfer function* of the system.

Example 2.2

Consider a real signal, $x(t)$. Its amplitude spectrum is characterized from the fact that its real part is an even function of f, whereas its imaginary part is an odd function of f. In fact, from (2.45), we have

$$\begin{aligned} \Re\{X(f)\} &= \int_{-\infty}^{\infty} x(t)\cos(2\pi ft)dt \\ \Im\{X(f)\} &= -\int_{-\infty}^{\infty} x(t)\sin(2\pi ft)dt \end{aligned} \tag{2.48}$$

Analogously, the magnitude, $|X(f)|$, is an even function, and the phase, $\arg[X(f)]$, is an odd function.

2.4.2 Discrete-Time Signals and Systems

2.4.2.1 Time Domain

A *discrete-time signal* is a sequence of real or complex numbers, denoted by (x_n), where n represents the *discrete time* and assumes in general all positive and negative integer values. A *discrete-time system* is a mapping of the sequence (x_n), the system input, into the sequence (y_n), the system output or system response:

$$y_n = S[(x_n)] \tag{2.49}$$

A discrete system is linear if for any pair of input signals, (x_n') and (x_n''), and for any complex numbers, a' and a'', the following holds:

$$S[(a'x_n' + a''\,x_n'')] = a'S[(x_n')] + a''S[(x_n'')] \tag{2.50}$$

A discrete system is *time invariant* if (2.49) implies that

$$S[(x_{n-k})] = y_{n-k} \tag{2.51}$$

for all integers k. Let (δ_n) denote the sequence

$$\delta_n = \begin{cases} 1, & n = 0 \\ 0, & n \neq 0 \end{cases}$$

The *discrete impulse response*, h_n, of a discrete system is its response to the input, δ_n. When the system is linear, its output can be obtained through the discrete convolution (Figure 2.9)

$$y_n = \sum_{k=-\infty}^{\infty} x_k h_{n-k} \tag{2.52}$$

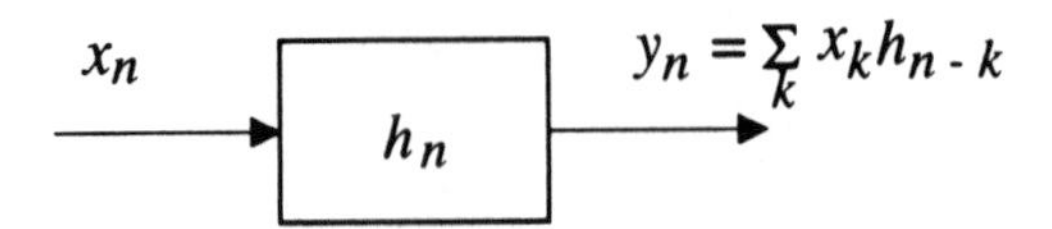

Figure 2.9 A discrete-time linear system.

By extension from the continuous case, we call the system causal if the output, y_n, depends only on $(x_k)^n_{-\infty}$, and memoryless if y_n depends only on x_n.

2.4.2.2 Frequency Domain

Given a discrete signal, x_n, we denote by $X(f)$ its Fourier transform:

$$X(f) = \sum_{n=-\infty}^{\infty} x_n \, e^{-jn2\pi f} \tag{2.53}$$

$X(f)$ is the amplitude spectrum of x_n. Because it is a periodic function of f with period 1, we can consider it limited to the interval $-1/2 \leq f \leq 1/2$. The inverse Fourier transform permits the recovery of x_n from $X(f)$

$$x_n = \int_{-1/2}^{1/2} X(f) e^{jn2\pi f} df \tag{2.54}$$

The Fourier transform $H(f)$ of the impulse response (h_n) of a discrete linear system is called the transfer function of the system. As for continuous signals and systems, the amplitude spectrum of the output of a linear system is related to the input one and to the transfer function of the system through the relation in (2.47).

2.5 RANDOM PROCESSES

In this section, we will extend the notions of Sections 2.2 and 2.4 to continuous and discrete random processes.

2.5.1 Continuous-Time Random Processes

A *continuous-time random process* $x(t)$ is a family of real or complex signals defined on some probability space.[2] For a given t, $x(t)$ represents an RV with the pdf $f(x,t)$. The statistical characterization of a random process can be done by giving the joint pdf of the n-tuple of RVs $x(t_1), \ldots, x(t_n)$ for every n and for every n-tuple of time instants. A random process is called stationary if $x(t)$ has the same statistical properties as $x(t - \tau)$ for every real τ, and cyclostationary with period T if $x(t)$ has the same

2. Every book on statistical communication or on stochastic processes uses a notation for random processes that is different from the one used for deterministic signals. This differentiation permits one to distinguish easily between the process and its individual realizations. However, it collides invariably with the limited number of distinct ways of writing the same letters and ends up representing two different things with the same notation. Here, we have to deal with scalar and vector deterministic signals, their amplitude spectra, scalar and vector stochastic processes, and their power spectra. Thus, we choose simplicity and appeal to the reader's intuition. The same notation, $x(t)$, will then represent a scalar deterministic signal or a stochastic process according to the context. Its Fourier transform will be denoted by the same capital letter $X(f)$ and its power spectrum by $G_x(f)$.

statistical properties as $x(t+kT)$ for all k integers. The *mean value* of the process $x(t)$ is the deterministic signal

$$m(t) \overset{\text{def}}{=} \mathrm{E}[x(t)] \tag{2.55}$$

where E means statistical expectation. The autocorrelation function of $x(t)$ is the function

$$R_x(t_1, t_2) \overset{\text{def}}{=} \mathrm{E}[x(t_1)x^*(t_2)] \tag{2.56}$$

For stationary processes, $m(t)$ does not depend on time, and the autocorrelation function depends only on the difference $t_1 - t_2$. Consequently, we can write

$$R_x(t_1 - t_2) = \mathrm{E}[x(t_1)x^*(t_2)] \tag{2.57}$$

For cyclostationary processes, $m(t)$ is a periodic function with period T, and the autocorrelation function has the property

$$R_x(t + \tau, t) = R_x(t + \tau + kT, t + kT) \tag{2.58}$$

When only the mean and autocorrelation properties are satisfied, the process is called wide-sense stationary (or wide-sense cyclostationary). Most of the random processes encountered in digital communications are wide-sense cyclostationary.

2.5.1.1 Gaussian Random Processes

A real random process $x(t)$ is called Gaussian if, for every n and for any n-tuple of time instants, the n RVs $x(t_1), \ldots, x(t_n)$ have a jointly Gaussian pdf. For notation compactness, we associate to the n RVs a row vector

$$\mathbf{x} \overset{\text{def}}{=} [x(t_1), \ldots, x(t_n)] \tag{2.59}$$

so that the joint pdf can be written as

$$f(\mathbf{x}) = \frac{1}{(2\pi)^{n/2}(\det\boldsymbol{\Lambda})^{1/2}} \exp\left[-\frac{1}{2}(\mathbf{x} - \boldsymbol{\mu})\boldsymbol{\Lambda}^{-1}(\mathbf{x} - \boldsymbol{\mu})'\right] \tag{2.60}$$

where

$$\boldsymbol{\mu} \overset{\text{def}}{=} \mathrm{E}[\mathbf{x}]$$
$$\boldsymbol{\Lambda} \overset{\text{def}}{=} \mathrm{E}[(\mathbf{x} - \boldsymbol{\mu})'(\mathbf{x} - \boldsymbol{\mu})] \tag{2.61}$$

are the mean vector and the covariance matrix of the process, respectively. The joint pdf (2.60) generalizes the pdf of a single Gaussian RV already introduced in (2.28).

Consider now a complex random process

$$x(t) = x_P(t) + jx_Q(t)$$

We say that this process is Gaussian if the joint pdf of the $2n$ RVs for every n and for any n-tuple of time instants

$$x_P(t_1), x_P(t_2), \ldots, x_P(t_n), x_Q(t_1), x_Q(t_2), \ldots, x_Q(t_n)$$

is Gaussian, that is, it has the form given in (2.60).

Gaussian random processes are important in statistical communication theory, because they model quite accurately the thermal and shot noise, disturbances always present in communication systems. Gaussian processes have the following properties:

- The output of a linear system whose input is a Gaussian random process is still a Gaussian random process.
- If $x(t)$ is a real wide-sense stationary Gaussian process, then it is also stationary.
- If $x(t)$ is a complex wide-sense stationary Gaussian process, then it is also stationary if and only if the average $E[x(t_1)x(t_2)]$ is a function only of the time difference $t_1 - t_2$.

2.5.1.2 Poisson Random Processes

Consider the time axis and a sequence of points, t_n, randomly chosen on it. The points are called *homogeneous Poisson points* if they satisfy the following properties:

- The RV $N(t_1, t_2)$ representing the number of points in the time interval (t_1, t_2) of length $t = t_2 - t_1$ is a Poisson RV with parameter λt, that is,

$$P[N(t_1, t_2) = k] = \frac{e^{-\lambda t}(\lambda t)^k}{k!} \tag{2.62}$$

  The Poisson distribution has mean and variance λt, where λ is the average number of points in unitary time (*density*).

- If the intervals (t_1, t_2) and (t_3, t_4) are nonoverlapping, the RVs $N(t_1, t_2)$ and $N(t_3, t_4)$ are independent.

Using the sequence t_n, we form the Poisson random process

$$x(t) \stackrel{\text{def}}{=} N(0, t) \tag{2.63}$$

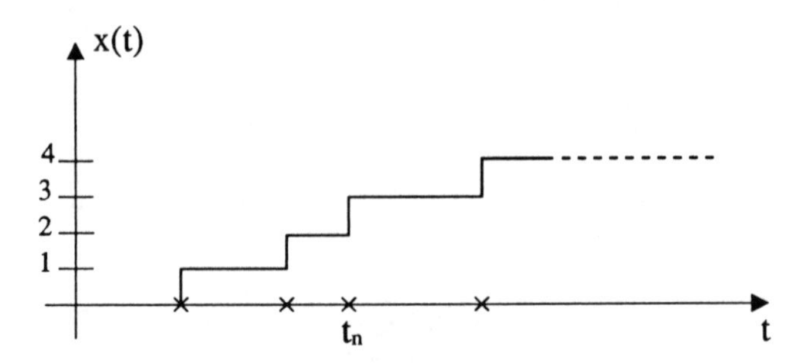

Figure 2.10 A realization of a Poisson random process.

It is a family of staircase functions, with discontinuities at times t_n, as shown in Figure 2.10. Its autocorrelation function equals (see Problem 2.8)

$$R_x(t_1, t_2) = \begin{cases} \lambda t_2 + \lambda^2 t_1 t_2, & t_1 \geq t_2 \\ \lambda t_1 + \lambda^2 t_1 t_2, & t_1 \leq t_2 \end{cases} \tag{2.64}$$

so that the process is not stationary.

When the points t_n have a nonuniform density, $\lambda(t)$ (inhomogeneous Poisson points), all the preceding results hold, provided that the parameter λt is replaced by the integral $\int_{t_1}^{t_2} \lambda(\tau)d\tau$, $t_2 - t_1 = t$.

2.5.1.3 Filtered Poisson Processes

The Poisson random process $x(t)$ previously defined can be also represented as

$$x(t) = \sum_n \mathrm{u}(t - t_n)$$

where $\mathrm{u}(t)$ is the unit step function, defined as

$$\mathrm{u}(t) \stackrel{\text{def}}{=} \begin{cases} 1, & t \geq 0 \\ 0, & t < 0 \end{cases}$$

It can be obtained as the output of a linear system with impulse response $\mathrm{u}(t)$ to an input $\Sigma_n \, \delta(t - t_n)$. In general, we call *filtered Poisson process* $s(t)$ a random process obtained as the response of a linear system with impulse response $h(t)$ to the input $\Sigma_n \, \delta(t - t_n)$ (Fig. 2.11)

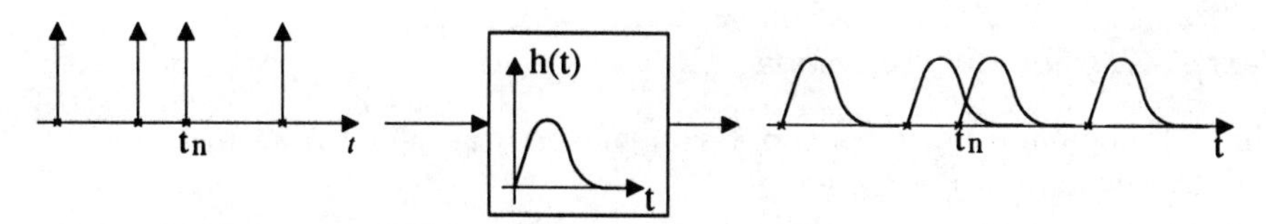

Figure 2.11 A realization of a filtered Poisson process.

$$s(t) = \sum_n h(t - t_n)$$

The mean $m_s(t)$ and the variance $\sigma_s^2(t)$ of $s(t)$ are given by the generalized Campbell's theorem [2,p. 632]:

$$m_s(t) = \int_{-\infty}^{\infty} \lambda(\alpha) h(t - \alpha) d\alpha, \qquad \sigma_s^2(t) = \int_{-\infty}^{\infty} \lambda(\alpha) h^2(t - \alpha) d\alpha \tag{2.65}$$

We will also need the autocorrelation function of the process, which is given by

$$R_s(t_1, t_2) = \int_{-\infty}^{\infty} \lambda(\alpha) h(t_1 - \alpha) h(t_2 - \alpha) d\alpha + \int_{-\infty}^{\infty} \lambda(\alpha) h(t_1 - \alpha) d\alpha \int_{-\infty}^{\infty} \lambda(\alpha) h(t_2 - \alpha) d\alpha \tag{2.66}$$

When the density, λ, of a homogeneous filtered Poisson process is large, compared with the inverse of the time constant of $h(t)$, the instantaneous value of the process $s(t)$ is determined by the contribution of many individual events, so that its first-order statistics tend to be Gaussian [3,p.179].

In actual system applications, the density $\lambda(t)$ of the process is itself a random process, because it contains the information to be transmitted. In that case, the inhomogeneous Poisson process is called a *doubly stochastic* Poisson process [3,Chap. 6]. Previous results still apply, provided they are considered as results conditioned to a given information message. A successive average with respect to the statistics of the information message will yield the final result.

When the underlying Poisson point process is homogeneous, (2.65) and (2.66) become

$$m = \lambda \int_{-\infty}^{\infty} h(t) dt = \lambda H(0), \qquad \sigma^2 = \lambda \int_{-\infty}^{\infty} h^2(t) dt \tag{2.67}$$

$$R_s(t_1, t_2) = \lambda \int_{-\infty}^{\infty} h(t_1 - t_2 + \alpha) h(\alpha) d\alpha + \lambda \left[\int_{-\infty}^{\infty} h(t) dt \right]^2 = R_s(t_1 - t_2)$$

so that the process is wide-sense stationary.

Example 2.3: The PIN Photodiode

The PIN photodiode (see Section 1.4.1) converts the signal from the optical to the electrical domain in the receiver.

For system purposes, a PIN diode can be characterized by the equivalent circuit shown in Figure 2.12, where the resistance-capacitance (RC) part of the circuit accounts for the frequency response of the device, and the current $i(t)$ of the current generator is a random process that depends on the statistics of the incoming photons.

We denote as (t_k) the sequence of time instants in which a pair of electron holes is created. The statistics of the carrier pair generation is Poisson, that is, the number, N_t, of pairs generated in t seconds is an RV with the pdf

$$P[N_t = n] = \frac{(\lambda t)^n e^{-\lambda t}}{n!} \tag{2.68}$$

where λ is the intensity of the Poisson process.[3] If we neglect the dark current (see Section 1.4.1), λ is related to the incident optical power P by

$$\lambda = \eta \frac{P}{h\nu} \tag{2.69}$$

where η, a positive number lower than 1, is the quantum efficiency of the photodiode, h is Planck's constant, and ν is the frequency of the incoming light supposed to be monochromatic.

When a carrier pair is created by a photon, the photodiode produces the current pulse $h(t - t_k)$ with duration τ and shape depending on the photodiode response time

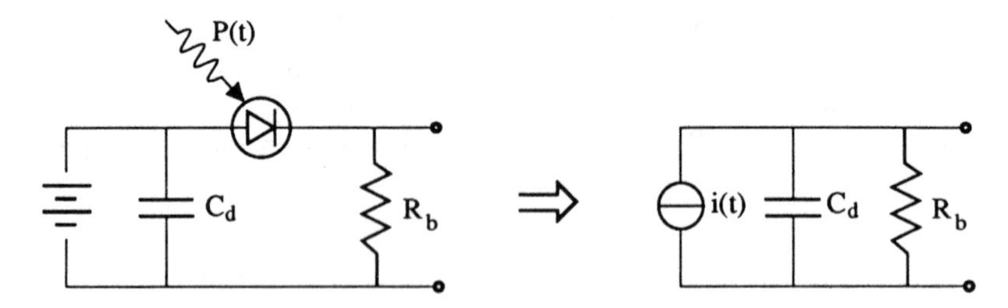

Figure 2.12 System characterization of a PIN photodiode.

3. We consider only the case of homogeneous Poisson processes, having the intensity λ constant with time. The most general case can be dealt with using the generalized Campbell's theorem and the filtered Poisson processes with variable intensity.

and whose integral equals the electron charge, q. Thus, the current emitted by the photodiode is the sum of such pulses

$$i(t) = \sum_k h(t - t_k) \tag{2.70}$$

We will now find the statistical characteristics of the current $i(t)$. From the previously introduced Campbell's theorem we can write

$$E[i(t)] = \lambda \int_{-\infty}^{\infty} h(z)dz = \lambda q \tag{2.71}$$

$$E[i^2(t)] = \lambda \int_{-\infty}^{\infty} h^2(z)dz. \tag{2.72}$$

The spectral density[4] of $i(t)$, derived in Example 2.6, is

$$G_i(f) = \lambda^2 q^2 \delta(f) + \lambda |H(f)|^2 \tag{2.73}$$

where $H(f)$ is the Fourier transform of $h(t)$.

From (2.73) the photodiode current can be decomposed into two components as

$$i(t) = i_S(t) + i_n(t) = q\lambda + i_n(t) = q\frac{\eta}{h\nu}P + i_n(t) = RP + i_n(t) \tag{2.74}$$

having introduced the *responsivity* $R \overset{\text{def}}{=} q\eta/h\nu$ of the photodiode. The first component represents the DC signal current, whereas the second is the *shot noise* contribution, with spectral density $\lambda|H(f)|^2$. When the transit time of the carriers through the depletion layer (time constant) of the photodiode is small enough with respect to the inverse of the bandwidth of interest, we can approximate $H(f)$ by $H(0)$, so that the spectral density of $i_n(t)$ becomes equal to λq^2, and $i_n(t)$ is a zero-mean white-noise process. To determine its actual pdf, we suppose the intensity λ to be so large[5] that the average time between carrier creation, $1/\lambda$, is much smaller than the time constant of the photodiode. In this case, the instantaneous value of the current is the contribution of many current pulses, and we can invoke the central limit theorem to say that $i_n(t)$ tends to be Gaussian.

4. The meaning of $G_i(f)$ is the spectral density of the mean square value of the current produced by the photodiode.
5. This hypothesis is easily satisfied in the case of coherent optical communication systems, as we will see in Chapter 4, and for optically amplified direct-detection systems (Chap. 5).

As an example, consider a "good" photodiode with a response time of 100 ps, to be applied in a coherent communication system working at a data rate of 1 Gbps and at a wavelength near 1.5 μm. The white-noise hypothesis is verified, as 100 ps < $[1\ \text{Gbps}]^{-1}$ = 1 ns. Moreover, the Gaussian hypothesis is based on the inequality 100 ps >> $1/\lambda = h\nu/\eta P$, which is satisfied, for reasonable values of η, by optical powers larger than −50 dBm, which is largely overcome in coherent and optically amplified systems.

Marked and Filtered Poisson Processes

A generalization of filtered Poisson processes is needed when the events that happen at the Poisson time instants are in some sense different. That is the case of avalanche photodiodes (APDs), described in Example 2.4, which generate several electric carriers in connection with every incident photon. The number of generated carriers is random and varies with time. We call these processes marked (and filtered) Poisson processes. They are described as

$$s(t) = \sum_n g_n h(t - t_n)$$

where (g_n) is a sequence of RVs assumed to be independent and with the same mean m_g and variance σ_g^2. The mean, the variance, and the autocorrelation function of $s(t)$ are in this case [3,Chap. 4]

$$m(t) = m_g \int_{-\infty}^{\infty} \lambda(\alpha)h(t - \alpha)d\alpha$$

$$\sigma^2(t) = (m_g^2 + \sigma_g^2) \int_{-\infty}^{\infty} \lambda(\alpha)h^2(t - \alpha)d\alpha$$

$$R_s(t_1, t_2) = (m_g^2 + \sigma_g^2) \int_{-\infty}^{\infty} \lambda(\alpha)h(t_1 - \alpha)h(t_2 - \alpha)d\alpha + m_g^2 \int_{-\infty}^{\infty} \lambda(\alpha)h(t_1 - \alpha)d\alpha \int_{-\infty}^{\infty} \lambda(\alpha)h(t_2 - \alpha)d\alpha \quad (2.75)$$

Example 2.4: The Avalanche Photodiode

APDs differ from PIN photodiodes in that they present a *gain* in the carrier pairs generation, that is, each primary carrier accelerated by the strong electric field that reverse-biases the junction produces several secondary carriers (avalanche effect). The previous system model still holds, with the addition of the random *avalanche multiplication gain*, g_k, as a factor in (2.70), which thus becomes

$$i(t) = \sum_k g_k h(t - t_k) \tag{2.76}$$

The mean and the mean square of the produced current are easily obtainable as (see Problem 2.9)

$$E[i(t)] = \mathrm{E}[g_k]\lambda \int_{-\infty}^{\infty} h(z)dz = G\lambda q \tag{2.77}$$

$$E[i^2(t)] = \mathrm{E}[g_k^2]\lambda \int_{-\infty}^{\infty} h^2(z)dz = G_2\lambda \int_{-\infty}^{\infty} h^2(z)dz \tag{2.78}$$

where we have defined as G and G_2 the average and the mean square of the RV g_k supposed to be constant with k.

The ratio

$$\frac{G_2}{G^2} > 1$$

is called *excess noise* and represents the price to be paid in terms of the *signal-to-noise ratio* (SNR) for getting the avalanche multiplication. Typical values of the excess noise are around 10. That translates directly in a shot noise to SNR penalty of 10 dB. However, in the overall SNR budget, which also takes into account the circuit noise, the presence of the avalanche gain has a beneficial effect, as we will see in Chapter 3.

2.5.2 Discrete-Time Random Processes

A *discrete-time random process*, or *random sequence* (x_n) is a sequence of real or complex RVs defined on some probability space. For a given n, x_n represents an RV with pdf $f(x,n)$. The statistical characterization of a random sequence can be done by giving the joint pdf of the n-tuple of RVs $x_{i1}, \ldots, x_{i_n}$ for every n and for every n-tuple of integers, i_n. A random sequence is called stationary if (x_n) has the same statistical properties as (x_{n+k}) for every integer k. The *mean value* of the random sequence (x_n) is the sequence (m_n) of numbers

$$m_n \overset{\text{def}}{=} \mathrm{E}[x_n] \tag{2.79}$$

The *autocorrelation* of (x_n) is the two-index sequence $r_{n,m}$ such that

$$r_{n,m} \overset{\text{def}}{=} \mathrm{E}[x_n x_m^*]. \tag{2.80}$$

For stationary sequences, m_n does not depend on n, and the autocorrelation depends only on the difference $n - m$. Consequently, we can write

$$r_{n-m} = \mathrm{E}[x_n x_m^*] \tag{2.81}$$

Also here, we will call *wide-sense stationary* a random sequence for which only the mean and autocorrelation stationarity properties are satisfied. Most of the messages emitted by a source of information are wide-sense stationary sequences.

2.6 SPECTRAL ANALYSIS

We have already seen in (2.47) that, when passing through a linear system, the frequency components $X(f)$ of the input signal are weighted separately by the transfer function $H(f)$. A similar relationship holds for the signal power spectrum.

2.6.1 Deterministic Signals

Given a signal, $x(t)$, its power, P_x, is defined as[6]

$$P_x \stackrel{\text{def}}{=} \lim_{a\to\infty} \frac{1}{a} \int_{-a/2}^{a/2} |x(t)|^2 dt, \tag{2.82}$$

provided that the limit exists. For periodic signals with period T, the definition, with $a = T$, does not need the limit. The autocorrelation function of $x(t)$ is defined as the time average:

$$R_x(\tau) \stackrel{\text{def}}{=} \lim_{a\to\infty} \frac{1}{a} \int_{-a/2}^{a/2} x(t)x^*(t+\tau)dt \tag{2.83}$$

From (2.83), it is clear that $P_x = R_x(0)$. The power spectrum $G_x(f)$, a nonnegative function of the frequency f such that

$$P_x = \int_{-\infty}^{\infty} G_x(f)df \tag{2.84}$$

is the Fourier transform of the autocorrelation function

6. In this definition we do not assign to the signal a particular physical meaning. When the context determines it as a current, as voltage, or as an electric field, it will be implicitly assumed that the relationship between the squared signal and its power contains a physical parameter (a resistance, etc.) with unitary value.

$$G_x(f) \stackrel{\text{def}}{=} \int_{-\infty}^{\infty} R_x(\tau) e^{-i2\pi f \tau} d\tau \tag{2.85}$$

When passing through the linear system $H(f)$, each frequency component of the power spectrum of a random process is independently weighted by $|H(f)|^2$, so that the power spectrum of the output random process will be given by

$$G_y(f) = G_x(f)|H(f)|^2 \tag{2.86}$$

2.6.2 Stationary Random Processes

The *average* power of a wide-sense stationary continuous random process $x(t)$ is defined as

$$P_x \stackrel{\text{def}}{=} \mathrm{E}[\,|\,x^2(t)\,|\,] \tag{2.87}$$

The *autocorrelation function* of $x(t)$ is defined as the statistical average

$$R_x(\tau) \stackrel{\text{def}}{=} \mathrm{E}[x(t)x^*(t+\tau)] \tag{2.88}$$

The (average) power spectrum $G_x(f)$ is still the Fourier transform of the autocorrelation function

$$G_x(f) = \int_{-\infty}^{\infty} R_x(\tau) e^{-i2\pi f \tau} d\tau \tag{2.89}$$

Also in this case, when passing through the linear system $H(f)$, the power spectrum of the output random process is given by (2.86).

Example 2.5: White Noise

A process, $n(t)$, with autocorrelation function

$$R_n(\tau) = \frac{N_0}{2} \delta(\tau) \tag{2.90}$$

and, consequently, power spectrum

$$G_n(f) = \frac{N_0}{2}, \qquad -\infty < f < \infty \tag{2.91}$$

is called a *white* random process, or *white noise*. It is an abstraction without physical meaning, because its power is clearly infinite. However, it is useful in all cases where the actual process has an approximately constant spectrum over a frequency range wider than the system bandwidth. We will see in the rest of the chapter that this is a reasonable approximation for both shot and thermal noises.

At the output of a linear system with transfer function $H(f)$, the average noise power is given by

$$P_n = \frac{N_0}{2} \int_{-\infty}^{\infty} |H(f)|^2 df$$

In such cases, the equivalent noise bandwidth of the system is defined as

$$B_{eq} \stackrel{\text{def}}{=} \frac{1}{2} \int_{-\infty}^{\infty} |H(f)|^2 df \tag{2.92}$$

The presence of the factor 1/2 in (2.92) means that we define the bandwidth only for positive frequencies, so that, for example, the telephone bandwidth will be equal to 3,000 Hz, although the amplitude spectrum is defined from −3,000 to 3,000 Hz.

With definition (2.92), the noise power at the system output is written simply as

$$P_n = (N_0/2)(2B_{eq}) = N_0 B_{eq}$$

Example 2.6: Power Spectrum of Poisson Processes

We want to compute the power spectrum of the homogeneous filtered and marked Poisson process. From the autocorrelation function of (2.75), with $\lambda(t) = \lambda$ constant for the homogeneity of the process, we get

$$R_s(t_1, t_2) = (m_g^2 + \sigma_g^2)\lambda \int_{-\infty}^{\infty} h(t_1 - \alpha)h(t_2 - \alpha)d\alpha$$
$$+ m_g^2\lambda^2 \int_{-\infty}^{\infty} h(t_1 - \alpha)d\alpha \int_{-\infty}^{\infty} h(t_2 - \alpha)d\alpha \tag{2.93}$$

From (2.93) we see that the autocorrelation depends only on the difference $t_1 - t_2$, so that the process is wide-sense stationary. Taking the Fourier transform of the autocorrelation function, we thus obtain (see Problem 2.10)

$$G_s(f) = (m_g^2 + \sigma_g^2)\lambda\, |H(f)|^2 + m_g^2\lambda^2 H^2(0)\delta(f) \tag{2.94}$$

When the time constant of $h(t)$ is small enough compared to the inverse of the system bandwidth (like the case in Example 2.3), we can approximate $H(f)$ by its value at the origin $H(0)$, so that the continuous part of the power spectrum becomes constant and the process is approximately white.

2.6.3 Nonstationary Random Processes

For nonstationary random processes, the (2.87) would lead to a power that depends on time. To overcome that problem, the average power for a nonstationary random process is defined as

$$P_x \stackrel{\text{def}}{=} \lim_{a\to\infty} \frac{1}{a} \int_{-1/a}^{1/a} \mathrm{E}[\,|\,x^2(t)\,|\,]dt \tag{2.95}$$

With this definition, a power spectral density can still be defined, provided that the process belongs to the class of the *harmonizable process* [4,p.474]. The power spectrum of a harmonizable random process can be obtained as follows. Define first the function

$$\Gamma_x(f_1, f_2) \stackrel{\text{def}}{=} \mathrm{E}[X(f_1)X^*(f_2)] \tag{2.96}$$

where

$$X(f_1) = \int_{-\infty}^{\infty} x(t)e^{-j2\pi f_1 t}dt \tag{2.97}$$

is a new random process in the variable f_1. The function $\Gamma_x(f_1, f_2)$ can also be obtained as the two-dimensional Fourier transform of the autocorrelation function $R_x(t_1, t_2)$.

If $\Gamma_x(f_1, f_2)$ can be written as

$$\Gamma_x(f_1, f_2) = G_x(f_1)\delta(f_1 - f_2) + \Delta_x(f_1, f_2) \tag{2.98}$$

where $G_x(f_1)$ represents a line mass located on the bisector $f_1 = f_2$, as in Figure 2.13, and the function $\Delta_x(f_1, f_2)$ has no line masses located on the same bisector, then the line mass $G_x(f_1)$ is the power spectrum of $x(t)$. A few examples will clarify the above definitions.

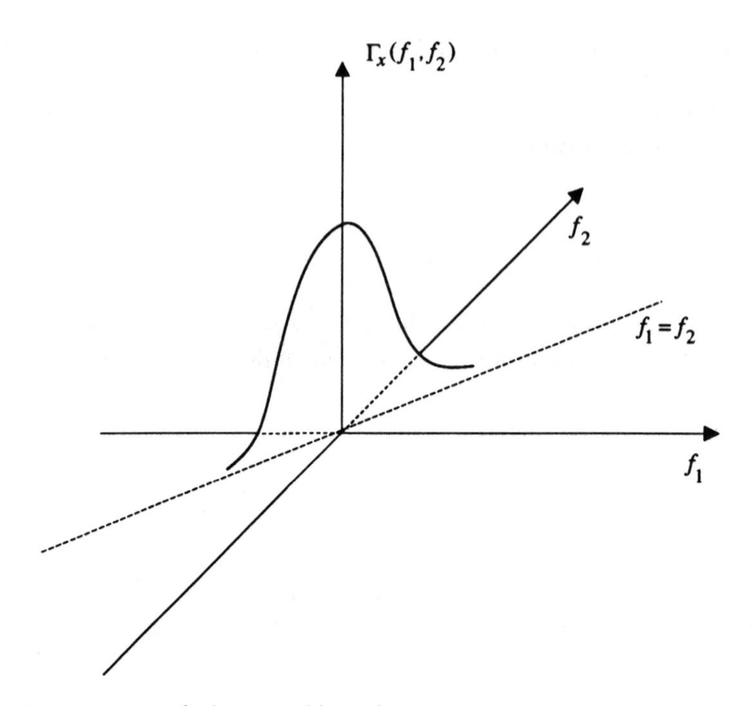

Figure 2.13 Power spectrum of a harmonizable random process.

Example 2.7: Stationary Processes

This example shows how the general definition in (2.98) reduces to the one already given when the random process is stationary. Let $x(t)$ be a wide-sense stationary random process, so that its autocorrelation function depends only on $t_1 - t_2$. Taking the two-dimensional Fourier transform of it, we obtain

$$\begin{aligned}\Gamma_x(f_1, f_2) &= \int_{-\infty}^{\infty}\int_{-\infty}^{\infty} R_x(t_1 - t_2)e^{-j2\pi(f_1 t_1 - f_2 t_2)}dt_1 dt_2 \\ &= \int_{-\infty}^{\infty}\int_{-\infty}^{\infty} R_x(t_1 - t_2)e^{-j2\pi[f_1(t_1-t_2)+t_2(f_1-f_2)]}dt_1 dt_2 \\ &= \int_{-\infty}^{\infty} R_x(\tau)e^{-j2\pi f_1 \tau}d\tau \cdot \delta(f_1 - f_2)\end{aligned} \tag{2.99}$$

which is the same as (2.89).

Example 2.8: Cyclostationary Processes

Let $x(t)$ be a cyclostationary process with period T. Using the property (2.58), we can expand its autocorrelation function in Fourier series as

$$R_x(t+\tau,t) = \sum_{n=-\infty}^{\infty} r_n(\tau)e^{jn2\pi t/T} \tag{2.100}$$

with

$$r_n(\tau) = \frac{1}{T}\int_{-T/2}^{T/2} R_x(t+\tau,t)e^{-jn2\pi t/T}dt$$

Taking the two-dimensional Fourier transform of the autocorrelation function yields

$$\begin{aligned}\Gamma_x(f_1,f_2) &= \int_{-\infty}^{\infty}\int_{-\infty}^{\infty} R_x(t_1,t_2)e^{-j2\pi(f_1t_1-f_2t_2)}dt_1dt_2 \\ &= \sum_{n=-\infty}^{\infty}\int_{-\infty}^{\infty}\int_{-\infty}^{\infty} r_n(\tau)e^{jn2\pi t/T}e^{-j2\pi[f_1\tau+(f_1-f_2)t]}dtd\tau \\ &= \sum_{n=-\infty}^{\infty} G_n(f_1)\delta(f_1-f_2-n/T)\end{aligned} \tag{2.101}$$

where $G_n(f)$ is the Fourier transform of $g_n(t)$. This result shows that $\Gamma_x(f_1,f_2)$ consists of line masses located on the parallel lines $f_1 = f_2 + n/T$, n integer. The power spectrum is then

$$G_x(f) = G_0(f)$$

Consider now a linearly modulated digital signal, which can be written in the form

$$x(t) = \sum_{n=-\infty}^{\infty} a_n s(t-nT) \tag{2.102}$$

where (a_n) is a sequence of real or complex RVs representing the information to be transmitted that are statistically characterized by

$$E[a_n] = \mu,\ E[a_n a_m^*] = \sigma^2\rho_{n-m} + |\mu|^2$$

with $\rho_0 = 1$ and $\rho_\infty = 0$. The signal in (2.102) is cyclostationary, so that, applying the method just described (see Problem 2.11), we can compute its power spectrum, obtaining

$$G_x(f) = G_x^c(f) + G_x^d(f)$$

where $G_x^c(f)$ and $G_x^d(f)$ are the continuous and discrete part of the spectrum, respectively, given by

$$G_x^c(f) = \frac{\sigma^2}{T} \mid S(f) \mid^2 \left[\sum_{m=-\infty}^{\infty} \rho_m e^{-j 2\pi f m T} \right]$$

$$G_x^d(f) = \frac{\mid \mu \mid^2}{T^2} \mid S(f) \mid^2 \sum_{m=-\infty}^{\infty} \delta(f - m/T) \qquad (2.103)$$

where $S(f)$ is the Fourier transform of the shaping function $s(t)$. As a consequence of (2.103), $\mu = 0$ is a sufficient condition for the spectrum to have no discrete lines. When (a_n) is a sequence of zero-mean, uncorrelated RVs, the power spectrum simplifies to

$$G_x(f) = G_x^c(f) = \frac{\sigma^2}{T} \mid S(f) \mid^2$$

In general, we see from (2.103) that the form of the power spectrum can be modeled acting on two separate contributions, namely, the shaping function $s(t)$ and the autocorrelation of the sequence (a_n). The latter possibility is studied in connection with the so-called line codes [1,Chap. 9].

2.7 NARROWBAND SIGNALS AND SYSTEMS

We have already seen that the amplitude spectrum of a signal (or the transfer function of a system) exhibits a certain degree of symmetry around the zero frequency. In particular, for a real signal $x(t)$, the real part of $X(f)$ is even and the imaginary part is odd, so that we can reconstruct completely $x(t)$ through an inverse Fourier transform even if we only know $X(f)$ for $f > 0$. This is equivalent to saying that if we pass our signal through a linear system with transfer function[7]

$$H(f) = 2\mathrm{u}(f)$$

7. The choice of the value 2 for the coefficient of $\mathrm{u}(t)$ is completely arbitrary, and any other value will do. This value, however, simplifies the subsequent relationships.

we can still reconstruct the original signal from the output of the linear system. The linear system has an impulse response equal to

$$h(t) = \left[\delta(t) + j\frac{1}{\pi t}\right]$$

so that the output, called the *analytic signal* associated to $x(t)$, will be

$$\mathring{x}(t) \overset{\text{def}}{=} x(t) + j\hat{x}(t)$$

where

$$\hat{x}(t) \overset{\text{def}}{=} \frac{1}{\pi}\int_{-\infty}^{\infty} \frac{x(\tau)}{t - \tau}\, d\tau$$

is the *Hilbert transform* of $x(t)$. Knowing the analytic signal, it is possible to obtain the real signal through

$$x(t) = \Re\mathring{x}(t)$$

Example 2.9

When $x(t) = \cos 2\pi f_0 t$, its Hilbert transform becomes $\hat{x}(t) = \sin 2\pi f_0 t$, so that the analytic signal is $\mathring{x}(t) = e^{j2\pi f_0 t}$. This is the standard rotating-vector representation of the electrotechnics.

Example 2.10

Consider now two real signals, $x(t)$ and $z(t)$, whose amplitude spectra are nonoverlapping, as shown in Figure 2.14. Their product, $y(t) = x(t)z(t)$, from the definition of an analytic signal, is

$$\mathring{y}(t) = x(t)\mathring{z}(t)$$

In particular, when $z(t) = \cos 2\pi f_0 t$, we obtain

$$\mathring{y}(t) = x(t)e^{j2\pi f_0 t}$$

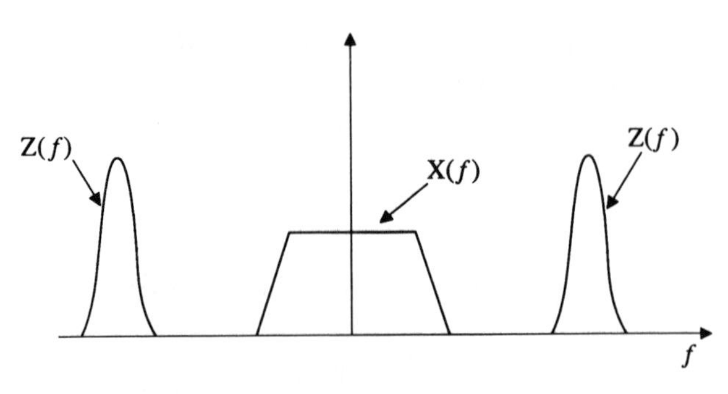

Figure 2.14 Nonoverlapping amplitude spectra of two real signals.

2.7.1 Narrowband Signals and Complex Envelopes

A *narrowband* signal (or system) is characterized by the fact that its spectrum is concentrated around some nonzero frequency; more formally, we define as narrowband a signal whose amplitude (or power) spectrum is zero outside the interval (f_1, f_2), which does not include the origin (Fig. 2.15). A signal whose spectrum is concentrated around the origin is called *baseband.* We can think of a narrowband signal $x(t)$ as the product of a baseband signal and a sinusoid with a frequency belonging to the interval (f_1, f_2), so that

$$\mathring{x}(t) = \tilde{x}(t)e^{j2\pi f_0 t} \tag{2.104}$$

where $\tilde{x}(t)$ is a signal, generally complex, whose spectrum lies around the origin and is obtained by down-shifting the positive-frequency spectrum of $x(t)$ by f_0 (see Fig. 2.15). The signal $\tilde{x}(t)$ is called the *complex envelope* of $x(t)$. Defining

$$\tilde{x}(t) \stackrel{\text{def}}{=} x_c(t) + jx_s(t)$$

we have the following representation for the narrowband signal $x(t)$

$$x(t) = \Re\mathring{x}(t) = x_c(t) \cos 2\pi f_0 t - x_s(t) \sin 2\pi f_0 t \tag{2.105}$$

and for the real and imaginary parts of $\tilde{x}(t)$

$$\begin{aligned} x_c(t) &= \Re\,\tilde{x}(t) = \Re[\mathring{x}(t)e^{-j2\pi f_0 t}] \\ &= x(t) \cos 2\pi f_0 t + \hat{x}(t) \sin 2\pi f_0 t \\ x_s(t) &= \Im\tilde{x}(t) = \Im[\mathring{x}(t)e^{-j2\pi f_0 t}] \\ &= -x(t) \sin 2\pi f_0 t + \hat{x}(t) \cos 2\pi f_0 t \end{aligned} \tag{2.106}$$

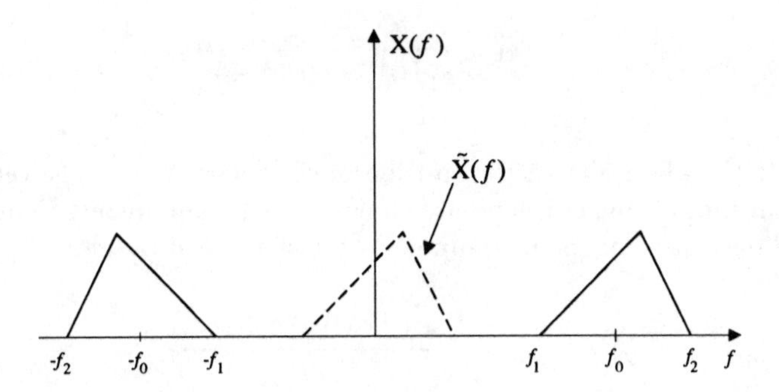

Figure 2.15 Amplitude spectrum of a narrowband signal.

From (2.105) it is clear that we can obtain the real and imaginary parts of the complex envelope of the narrowband signal $x(t)$, as shown in Figure 2.16, where the filters are ideal lowpass filters.

From (2.104) we can think of the analytic signal associated to a narrowband signal as a vector rotating with angular frequency $2\pi f_0$ with *instantaneous envelope*

$$A_x(t) \stackrel{\text{def}}{=} |\, \tilde{x}(t)\, | = \sqrt{x_c^2(t) + x_s^2(t)}$$

and the *instantaneous phase* depurated from the carrier phase $2\pi f_0 t$:

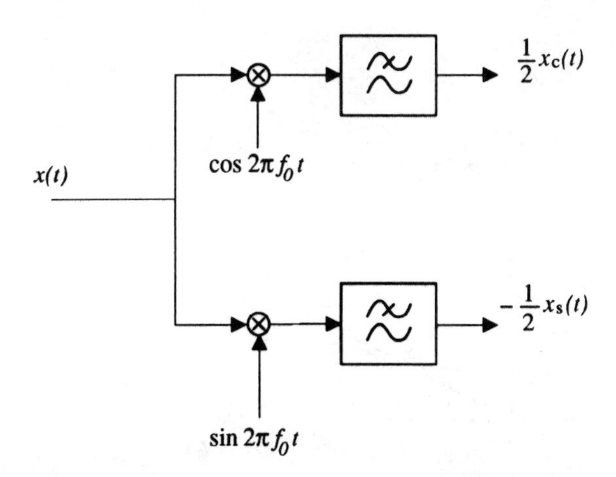

Figure 2.16 Linear transformations to obtain the real and imaginary parts of the complex envelope of a narrowband signal.

$$\phi_x(t) \stackrel{\text{def}}{=} \arg\left[\tilde{x}(t)\right] = \tan^{-1}\frac{x_s(t)}{x_c(t)}$$

as in Figure 2.17, where the vector and the block diagram of a receiver capable of deriving the instantaneous envelope are shown. The instantaneous frequency is obtained as the derivative of the instantaneous phase divided by 2π:

$$f_x(t) \stackrel{\text{def}}{=} f_0 + \frac{1}{2\pi}\frac{\dot{x}_s(t)x_c(t) - x_s(t)\dot{x}_c(t)}{x_c^2(t) + x_s^2(t)}$$

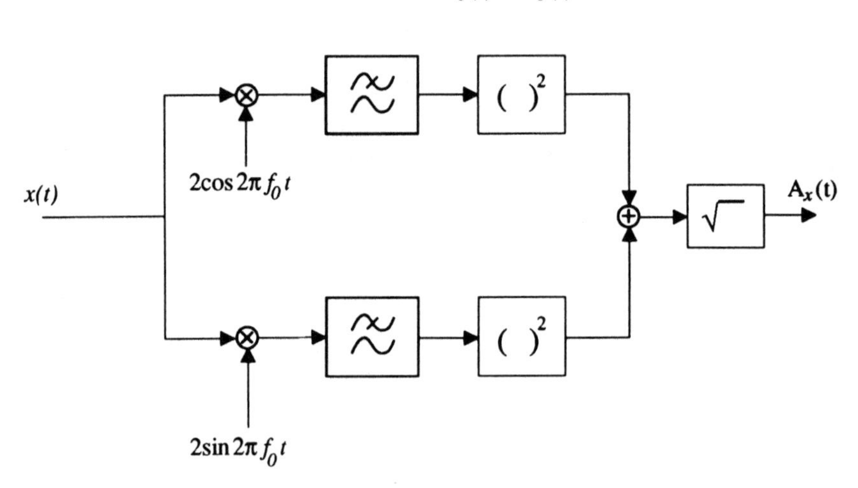

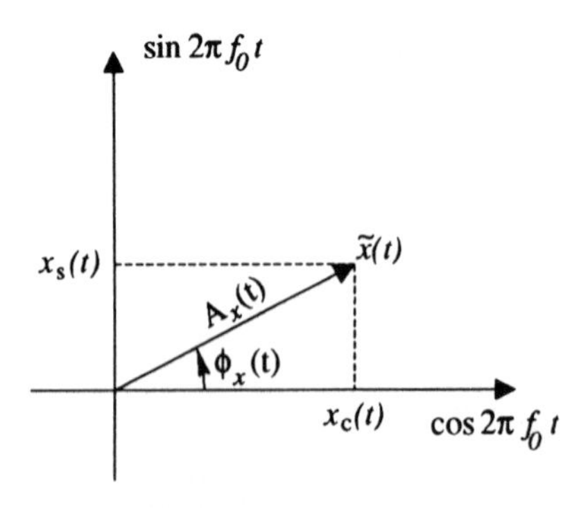

Figure 2.17 Transformations to obtain the instantaneous envelope of narrowband signal.

Using the instantaneous envelope and phase, we can represent a narrowband signal as

$$x(t) = A_x(t)\cos[2\pi f_0 t + \phi_x(t)]$$

2.7.1.1 Narrowband Random Processes

Consider now a real, narrowband, wide-sense stationary random process $n(t)$ with power spectrum $G_n(f)$. If we pass the random process through a linear system with transfer function $H(f) = 2\mathrm{u}(f)$, we obtain its analytic representation

$$\mathring{n}(t) \stackrel{\text{def}}{=} n(t) + j\hat{n}(t) \tag{2.107}$$

with the possible representations of $n(t)$ as

$$n(t) = \Re\mathring{n}(t) = \Re[\tilde{n}(t)e^{j2\pi f_0 t}] = n_c(t)\cos 2\pi f_0 t - n_s(t)\sin 2\pi f_0 t \tag{2.108}$$

where we have used the complex envelope

$$\tilde{n}(t) = n_c(t) + jn_s(t)$$

As for the power spectrum of $\mathring{n}(t)$, we can use (2.86) to obtain

$$G_{\mathring{n}}(f) = 4G_n(f)\mathrm{u}^2(f) = 4G_n(f)\mathrm{u}(f)$$

The power spectrum of the complex envelope is easily derived as

$$G_{\tilde{n}}(f) = G_{\mathring{n}}(f+f_0) = 4G_n(f+f_0)\mathrm{u}(f+f_0) \tag{2.109}$$

which shows that the spectrum of the complex envelope is the spectrum of the analytic signal translated around the origin. Finally, considering the real and imaginary components of the complex envelope, it can be shown (see Problem 2.12) that the following hold:

$$\begin{aligned} R_{n_c}(\tau) &= R_{n_s}(\tau) \\ R_{\tilde{n}}(\tau) &= 2[R_{n_c}(\tau) + j\,R_{n_s n_c}(\tau)] \end{aligned} \tag{2.110}$$

where

$$R_{n_s n_c}(\tau) \stackrel{\text{def}}{=} \mathrm{E}[n_s(t+\tau)n_c(t)]$$

From the previous equalities, we can draw the following conclusions:

- Because $R_{\tilde{n}}(0) = \mathrm{E}|\tilde{n}(t)|^2$ must be a real quantity, we have

$$\mathrm{E}[n_s(t)n_c(t)] = 0$$

  That is, for any given t, $n_c(t)$ and $n_s(t)$ are uncorrelated RVs. As a special case, when $n(t)$ is Gaussian, $n_c(t)$ and $n_s(t)$ are independent RVs.

- From (2.110)

$$\mathrm{E}|\tilde{n}(t)|^2 = 2\mathrm{E}|n_c(t)|^2 = 2\mathrm{E}|n_s(t)|^2$$

  so that, from (2.109)

$$\mathrm{E}|n(t)|^2 = \mathrm{E}|n_c(t)|^2 = \mathrm{E}|n_s(t)|^2$$

- If the power spectrum of the process $n(t)$ is symmetric around f_0, the spectrum of $\tilde{n}(t)$ is an even function, and thus $R_{\tilde{n}}(\tau)$ is a real function, and

$$\mathrm{E}[n_s(t+\tau)n_c(t)] = 0 \text{ for all } \tau$$

  which means that the two components of $\tilde{n}(t)$ are uncorrelated (or independent when $n(t)$ is Gaussian) random processes. In this situation,

$$G_{\tilde{n}}(f) = 2G_{n_c}(f) = 2G_{n_s}(f)$$

2.7.1.2 Narrowband White Noise

We will see in the following chapters that the common form of the signal entering the *intermediate-frequency* (IF) section of a receiver for optical communication is a composite signal made by the addition of the information-bearing signal and a Gaussian noise process whose spectrum is approximately constant over a frequency range much wider than the signal bandwidth. To limit the noise power entering the receiver section where decisions are taken, we limit the spectrum of this composite signal with a linear system (the IF filter) whose transfer function is approximately constant within the signal bandwidth and approximately zero outside it. Thus, at the output of the IF filter, we have the sum of the information signal unaltered plus a narrowband noise with power spectrum constant over the signal bandwidth. The latter item is called *narrowband white noise* and is thus a zero-mean, stationary random process whose power spectrum is constant over a frequency range not including the origin. Figure 2.18 presents an example of the power spectra of the narrowband white noise and its components.

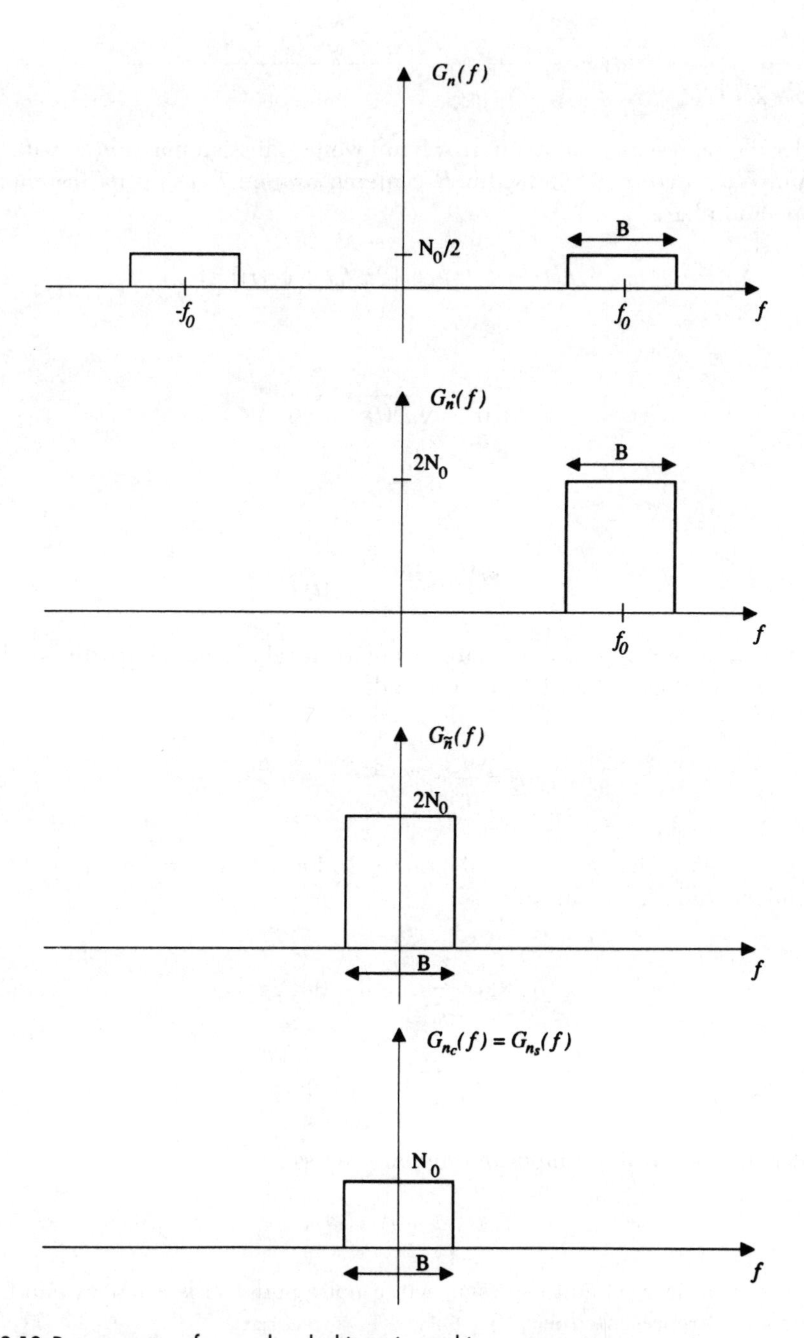

Figure 2.18 Power spectra of narrowband white noise and its components.

Example 2.11

Consider the representation of a narrowband white Gaussian noise, $n(t)$, with power spectrum $N_0/2$ over the bandwidth B centered around f_0 using its instantaneous envelope and phase:

$$n(t) = A_n(t) \cos[2\pi f_0 t + \phi_n(t)]$$

where

$$A_n(t) = \sqrt{n_c^2(t) + n_s^2(t)}$$

and

$$\phi_n(t) = \tan^{-1} \frac{n_s(t)}{n_c(t)}$$

For a given t, the RV $A_n(t)$ is the square root of a central χ-square RV with two degrees of freedom and thus a Rayleigh RV with pdf

$$f_{A_n}(a) = \frac{a}{\sigma^2} e^{-(a^2/2\sigma^2)}, \qquad a > 0 \tag{2.111}$$

where $\sigma^2 = N_0 B$ is the variance of the noise. As for $\phi(t)$, it turns out [5,p.22] to be independent from $A_n(t)$ with pdf

$$f_{\phi_n}(\theta) = \frac{1}{2\pi}, \qquad 0 < \theta < 2\pi \tag{2.112}$$

Example 2.12

Consider the case of the composite random process

$$y(t) = x(t) + n(t)$$

where $n(t)$ is a narrowband Gaussian white noise and $x(t)$ is a narrowband deterministic signal represented as

$$x(t) = A_x \cos(2\pi f_0 t + \phi_x)$$

We are interested in the statistics of the instantaneous envelope of $y(t)$:

$$A_y(t) = | \mathring{n}(t) + \mathring{x}(t) | = \sqrt{(A_x \cos \phi + n_c)^2 + (A_x \sin \phi + n_s)^2}$$

For a given t, the RV $A_y(t)$ is the square root of a noncentral χ-square RV with two degrees of freedom and the noncentrality parameter A_x^2; thus, it is a Rice RV with the pdf

$$f_{A_y} = \frac{a}{\sigma^2} e^{-[(a^2+A_x^2)/2\sigma^2]} I_0\left(\frac{aA_x}{\sigma^2}\right) \tag{2.113}$$

2.7.2 Bandpass Systems and Lowpass Representation

Paralleling the treatment devoted to narrowband signals, we can define a bandpass (linear) system as one having a transfer function $H(f)$ that is zero outside the interval (f_1, f_2) not including the origin. The analytic representation of $h(t)$ is

$$\mathring{h}(t) = \tilde{h}(t)e^{j2\pi f_0 t} \tag{2.114}$$

where $\tilde{h}(t)$ is the inverse Fourier transform of the function lying around the origin and obtained by down-shifting the positive-frequency transfer function $H(f)$ by f_0. The generally complex function $\tilde{h}(t)$ is called the *lowpass (equivalent) representation* of $h(t)$.

Consider now a narrowband signal, $x(t)$, passing through the bandpass linear system with impulse response $h(t)$. Considering the analytic signal representations of both signal and system, we can easily obtain the analytic signal associated to the output $y(t)$ as

$$\mathring{y}(t) = \frac{1}{2}\int_{-\infty}^{\infty} \mathring{h}(\tau)\mathring{x}(t-\tau)d\tau \tag{2.115}$$

Substituting the complex envelope representation, we obtain

$$\mathring{y}(t) = \frac{1}{2} e^{j2\pi f_0 t}\int_{-\infty}^{\infty} \tilde{h}(\tau)\tilde{x}(t-\tau)d\tau \tag{2.116}$$

which shows that the complex envelope of the output is obtained by the complex envelope of the input and lowpass representation of the system as

$$\tilde{y}(t) = \frac{1}{2}\int_{-\infty}^{\infty} \tilde{h}(\tau)\tilde{x}(t-\tau)d\tau \tag{2.117}$$

The conclusion is that, when considering narrowband signals and linear bandpass systems, we can perform all the computations using the complex envelope representations, provided we use the same frequency, f_0, to represent signals and systems. This property is of crucial importance in the simulation of communication systems.

2.8 ELEMENTS OF DETECTION THEORY

We address here the problem of detecting, at best, one (generally complex) signal, $s_i(t)$, from a finite set, $\{s_i(t)\}_{i=1}^{M}$, based on the knowledge of a composite signal, $x(t)$, that is the sum of a (possibly modified) version of $s_i(t)$ and of a white Gaussian random process $n(t)$ independent of the signal. Using the complex envelope description for signals and omitting the tilde for notational simplicity, we base our decision on the received signal:

$$x(t) = s_i(t)e^{j\phi(t)} + n(t) \tag{2.118}$$

where $\phi(t)$ is an unknown random phase and $n(t)$ the complex envelope of a white Gaussian random process with two-sided power spectral density $N_0/2$ independent of $s_i(t)$.

2.8.1 One Signal in Additive Noise

Consider first the case $M = 1$, where the detection process consists in deciding whether the received signal $x(t)$ contains the only possible transmitted signal $s(t)$ or not. We will suppose first that the receiver is able to determine exactly the phase ϕ, so that we can assume $\phi = 0$. This means that the receiver has a complete knowledge of the transmitted signal. Then we will relax this hypothesis.

2.8.1.1 One Known Signal in Additive Noise

The received signal in the symbol interval $(0, T)$ is given by

$$x(t) = s(t) + n(t) \tag{2.119}$$

where, with obvious notation,

$$x(t) = x_c(t) + jx_s(t),\ s(t) = s_c(t) + js_s(t),\ n(t) = n_c(t) + jn_s(t) \tag{2.120}$$

Formally, the receiver must choose between the two hypotheses

$$\begin{aligned} H_0: &\quad x(t) = n(t) \\ H_1: &\quad x(t) = s(t) + n(t) \end{aligned} \tag{2.121}$$

We want to reduce a continuous-time problem to a discrete one. To that end, we consider an orthonormal sequence, $(\psi_1(t), \ldots, \psi_n(t), \ldots)$, complete for any complex signals in the interval $(0, T)$, and choose

$$\psi_1(t) = \frac{1}{\sqrt{E_s}} s(t)$$

where E_s is the energy of $s(t)$. Then we expand the received signal in an orthonormal series and use for detection the coefficients x_i, $i = 1, \ldots, \infty$ expressed as

$$x_i \stackrel{\text{def}}{=} \int_0^T x(t)\psi_i^*(t)dt \tag{2.122}$$

Because of the particular choice of the orthonormal basis, the coefficients x_i will have the form

$$x_1 = \sqrt{E_s} + n_1,\ x_i = n_i,\ i = 2, 3, \ldots, \infty \text{ under hypothesis } H_1 \tag{2.123}$$

where

$$n_i \stackrel{\text{def}}{=} \int_0^T n(t)\psi_i^*(t)dt$$

and

$$x_i = n_i,\ i = 1, 2, \ldots, \infty \text{ under hypothesis } H_0 \tag{2.124}$$

We see that, under both hypotheses, the components x_i, $i > 1$ contain only noise, are independent from each other and from x_1, and, as such, do not add any useful information to be exploited in the detection process. As a consequence, they can be discarded and the optimum detection based on the *sufficient statistics*, x_1. The decision will be made on the value assumed by x_1 according to

$$\begin{aligned} &\text{Choose } H_0 \text{ if } x_1 \in \chi_0 \\ &\text{Choose } H_1 \text{ if } x_1 \in \chi_1 \end{aligned}$$

where χ_0 and χ_1 are real intervals defining the decision regions.

The detection strategy is optimized by choosing a decision rule that minimizes the *average error probability:*

$$\begin{aligned} P_e &= \int_{\chi_0} p_1 f(x \mid H_1) dx + \int_{\chi_1} p_0 f(x \mid H_0) dx \\ &= \int_{\Re} p_1 f(x \mid H_1) dx - \int_{\chi_1} p_1 f(x \mid H_1) dx + \int_{\chi_1} p_0 f(x \mid H_0) dx \\ &= p_1 - \int_{\chi_1} [p_1 f(x \mid H_1) - p_0 f(x \mid H_0)] dx \end{aligned} \tag{2.125}$$

where $p_0 \overset{\text{def}}{=} P[H_0], p_1 \overset{\text{def}}{=} P[H_1]$, $\Re$ is the real line, and $f(\cdot|H_i)$ is a conditional pdf. To minimize P_e, we must maximize the contribution to the integral of the term in brackets in the last expression of (2.125), which in turn implies that we include in χ_1 the values of x_1 such that $p_1 f(x|H_1) > p_0 f(x|H_0)$ and in χ_0 the remaining values. If we define the log-likelihood ratio among hypotheses H_0 and H_1 as

$$\lambda(x) \overset{\text{def}}{=} \log \frac{f(x \mid H_1)}{f(x \mid H_0)} \tag{2.126}$$

the optimum decision rule becomes

Choose H_0 if $\lambda(x_1) \leq \log(p_0/p_1)$
Choose H_1 if $\lambda(x_1) > \log(p_0/p_1)$

This rule is called, for evident reasons, *maximum-a-posteriori* decision strategy. In the important case where $p_0 = p_1 = 1/2$ (always assumed to be true in the remainder of this book), the decision strategy becomes

Choose H_0 if $\lambda(x_1) \leq 0$
Choose H_1 if $\lambda(x_1) > 0$

is called *maximum-likelihood* (ML) decision strategy. Since x_1 is a complex-valued Gaussian RV with variance $N_0/2$, zero-mean under the hypothesis H_0, and mean $\sqrt{E_s}$ under H_1, the log-likelihood ratio is easily seen to be (see Problem 2.13)

$$\lambda(x_1) = \frac{2}{N_0} \Re \left\{ \int_0^T x(t) s^*(t) dt \right\} - \frac{1}{N_0} \int_0^T |s(t)|^2 dt \tag{2.127}$$

which directly translates into an optimum receiver whose block diagram is represented in Figure 2.19.

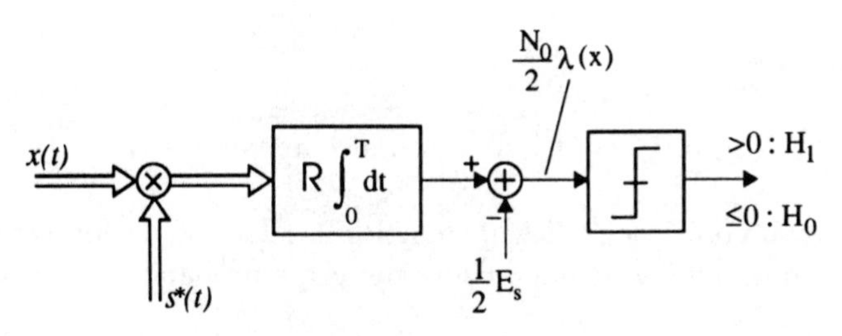

Figure 2.19 Block diagram of the receiver for the ML detection of one known signal in additive noise.

Example 2.13: Binary Unipolar AM Modulation

Binary unipolar AM modulation consists in transmitting the signal $s(t)$ in correspondence to, for example, a 1 binary symbol and nothing in correspondence to a 0 symbol. We want to evaluate the error probability, P_e, for the ML decision strategy. The log-likelihood ratio (2.127), after multiplication for $N_0/2$, as in Figure 2.19, is easily seen to be a Gaussian RV with mean $E_s/2$ and variance $(E_s N_0)/2$. Thus, the error probabilities conditioned by the hypotheses H_0 and H_1 are the same, so that the average error probability becomes, with obvious notations,

$$P_e = \frac{1}{2}\left[P(e \mid H_0) + P(e \mid H_1)\right] = \frac{1}{2}\operatorname{erfc}\left(\frac{1}{2}\sqrt{\frac{E_s}{N_0}}\right) \tag{2.128}$$

where we have introduced the function $\operatorname{erfc}(\cdot) = 1 - \operatorname{erf}(\cdot)$.

2.8.1.2 One Signal With Unknown Phase in Additive Noise

We consider now the case where the receiver is not able to (or does not want to) estimate the phase ϕ, which is assumed to be an RV uniformly distributed in the interval $(0, 2\pi)$. Since the receiver has no information about the values of the phase ϕ, it will base its decisions on a version of the log-likelihood ratio, which is obtained by averaging over ϕ the ratio considered in Section 2.8.1.1.

We proceed as before by first expanding the received signal $x(t)$ in series of orthonormal functions, obtaining the coefficients x_i in the form

$$x_1 = \sqrt{E_s}e^{j\phi} + n_1,\ x_i = n_i,\ i = 2,3, \ldots \infty \text{ under hypothesis } H_1 \tag{2.129}$$

and

$$x_i = n_i,\ i = 1,2, \ldots \infty \text{ under hypothesis } H_0 \tag{2.130}$$

As before, we retain only the sufficient statistics, x_1. The whole optimization process proceeds as before, aiming at minimizing the error probability, P_e, which is now defined as

$$P_e = \mathrm{E}_\phi P_{e|\phi} \tag{2.131}$$

where $P_{e|\phi}$ is the error probability conditioned over a given value of the RV ϕ and has the same expression as before, and E_ϕ is the average operator over ϕ. As a consequence, the optimum ML decision strategy turns out to be the following:

$$\begin{array}{l}\text{Choose } H_0 \text{ if } \mathrm{E}_\phi \lambda(x_1, \phi) \leq 0 \\ \text{Choose } H_1 \text{ if } \mathrm{E}_\phi \lambda(x_1, \phi) > 0\end{array}$$

where $\lambda(x_1, \phi)$ is the log-likelihood ratio conditioned to a value of the RV ϕ. The log-likelihood ratio $\lambda(x_1)$, obtained after averaging over ϕ, becomes (see Problem 2.14)

$$\lambda(x_1) = \log\left[I_0\left(\frac{2}{N_0}\left|\int_0^T x(t)s^*(t)dt\right|\right)\right] - \frac{1}{N_0}\int_0^T |\, s(t)\,|^2 dt \tag{2.132}$$

Because both the log and I_0 are monotonically increasing functions and as such can be inverted, the optimum decision strategy requires the comparison of the envelope $|\int_0^T x(t)s^*(t)dt|$ with a suitable threshold, a_{th} (chosen to minimize the error probability), as shown in the block diagram of Figure 2.20.

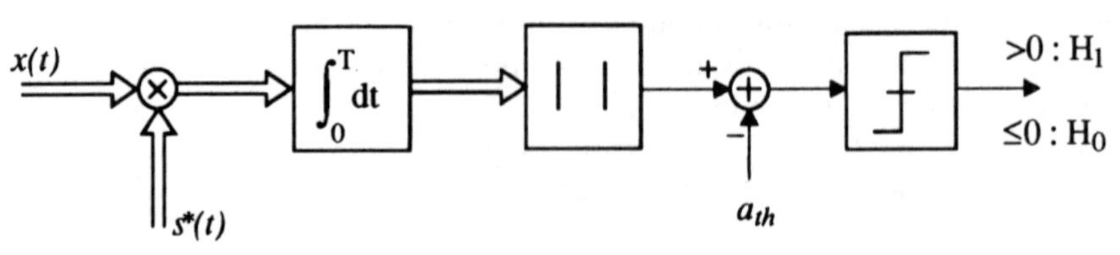

Figure 2.20 Block diagram of the receiver for the ML detection of one signal with unknown phase in additive noise.

Example 2.14: On-Off Modulation

In the case of on-off modulation, which is the most used scheme for digital optical transmission, where it is known as IM-DD (intensity modulation, direct detection), the transmitted signal is based on unipolar AM modulation. The detection process on the received signal is incoherent, so that decision at the receiver side is based on the envelope of the received signal checked against a suitable threshold. Contrary to Example 2.13, the statistics of the received signal are different in the two hypotheses, H_1 and H_0. In fact, in the first case, the envelope $|\int_0^T x(t)s^*(t)dt|$ becomes equal to $Y_1 \overset{\text{def}}{=} |E_s + n_1|$, n_1 being a complex Gaussian RV with zero mean and variance $(E_s N_0)/2$, so that Y_0 is a Rice RV. In the second case, the envelope becomes $Y_0 \overset{\text{def}}{=} |n_0|$, where n_0 is a complex Gaussian RV with the same characteristics of n_1, so that Y_0 is a Rayleigh RV.

Denoting by A_{th} the normalized decision threshold, $a_{th}/\sqrt{E_S}$ (see Fig. 2.20), the average error probability becomes

$$P_e \overset{\text{def}}{=} \frac{1}{2}\left[P(e \mid H_0) + P(e \mid H_1)\right] = \frac{1}{2}\left[e^{-(A_{th}^2/N_0)} + 1 - Q\left(\sqrt{\frac{2E_s}{N_0}}, \sqrt{\frac{2A_{th}^2}{N_0}}\right)\right] \tag{2.133}$$

2.8.2 M Signals in Additive Noise

We will extend the results of Section 2.8.1 to the more general case where the detection process consists in deciding which signal was transmitted, among the possible M, on the basis of the received signal (2.118). As before, we will consider first the case where the receiver has a complete knowledge of the signal, so that we can assume $\phi = 0$, and then treat the incoherent situation, where ϕ is unknown.

2.8.2.1 M Known Signals in Additive Noise

The received signal in the symbol interval $(0, T)$ is given by

$$x(t) = s_i(t) + n(t) \tag{2.134}$$

where $s_i(t)$ is the complex envelope of the transmitted information signal. Formally, the receiver must choose between the M hypotheses:

$$H_j : x(t) = s_j(t) + n(t) \tag{2.135}$$

As before, to reduce a continuous-time problem to a discrete one, we consider first an orthonormal sequence, $(\psi_1(t), \ldots, \psi_n(t), \ldots)$, complete for any complex signals in the interval $(0, T)$, whose first L orthonormal signals, $L \leq M$, form a basis for the representation of the M information signals. We can thus expand the received signal using the series and discard all the components x_l, $l > L$ so that finally the decision will be made among the hypotheses

$$H_j: \mathbf{x} = \mathbf{s}_j + \mathbf{n} \tag{2.136}$$

where $\mathbf{x}$ is the L-vector with components

$$x_i \stackrel{\text{def}}{=} \int_0^T x(t)\psi_i^*(t)dt$$

and similarly $\mathbf{s_j}$ is the L-vector with components

$$s_{ji} \stackrel{\text{def}}{=} \int_0^T s_j(t)\psi_i^*(t)dt$$

and $\mathbf{n}$ has components

$$n_i \stackrel{\text{def}}{=} \int_0^T n(t)\psi_i^*(t)dt$$

Proceeding the same way as before (see Problem 2.16), we finally obtain the ML decision rule

$$\text{Choose } H_j \text{ if } \lambda_j(x) = \max_i \lambda_i(x)$$

where

$$\lambda_i(x) = \frac{2}{N_0} \Re\left\{\int_0^T x(t)s_i^*(t)dt\right\} - \frac{1}{N_0}\int_0^T |s_i(t)|^2 dt, \qquad i = 1, \ldots, M \tag{2.137}$$

The optimum receiver structure is represented in Figure 2.21.

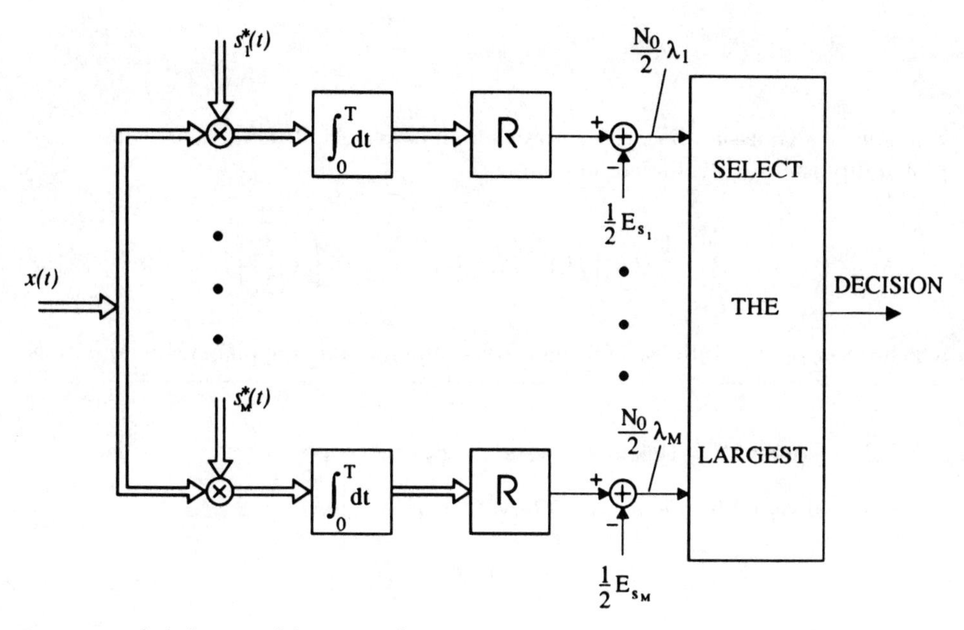

Figure 2.21 Block diagram of the receiver for the ML detection of *M* known signals in additive noise.

Example 2.15: Binary PSK Modulation

A widely used modulation scheme consists of two signals with the same frequency and phases, $\theta_1 = 0$, $\theta_2 = \pi$, so that the complex envelope associated with them is

$$s_1(t) = \sqrt{\frac{E_s}{T}}u_T(t),\ s_2(t) = -\sqrt{\frac{E_s}{T}}u_T(t)$$

$u_T(t)$ being the unit step function, equal to 1 in $(0, T)$ and 0 elsewhere. It can be shown [1] that a basis for representing the two signals consists of the signal

$$\psi_1(t) = \sqrt{\frac{1}{T}}u_T(t)$$

Applying the ML decision rule based on the log-likelihood ratios (2.137), and dropping the biasing terms that account for the signals' power, equal in our case for both signals, we obtain, under the hypothesis of $s_1(t)$ transmitted,

$$\lambda_1 = \frac{2}{N_0} E_s + n_1, \; \lambda_2 = -\frac{2}{N_0} E_s + n_1$$

where n_1 is a Gaussian RV with zero mean and variance $2E_S/N_0$. The average error probability is easily evaluated, obtaining

$$P_e = \frac{1}{2}\left[P(e \mid H_1) + P(e \mid H_2)\right] = \frac{1}{2}\,\mathrm{erfc}\left(\sqrt{\frac{E_s}{N_0}}\right) \tag{2.138}$$

which improves by 3 dB in SNR the results obtained for unipolar AM modulation.

2.8.2.2 M Signals With Unknown Phase in Additive Noise

The received signal in the symbol interval $(0, T)$ is given by

$$x(t) = s_i(t)e^{j\phi_i} + n(t) \tag{2.139}$$

where $s_i(t)$ is the complex envelope of the transmitted information signal, and $\{\phi_i\}$ are the carrier or channel phase shifts for the M signals, assumed to be independent RVs with the same uniformly distributed pdf in the interval $(0, 2\pi)$.

Formally, the receiver must choose between the M hypotheses

$$H_j : x(t) = s_j(t)e^{j\phi j} + n(t) \tag{2.140}$$

As before, to reduce a continuous-time problem to a discrete one, we consider first an orthonormal sequence $(\psi_1(t), \ldots, \psi_n(t), \ldots)$ complete for any complex signals in the interval $(0, T)$, whose first L orthonormal signals, $L \leq M$, form a basis for the representation of the M information signals. We can thus expand the received signal using the series and discard all the components x_l, $l > L$ so that finally the decision will be made among the hypotheses

$$H_j : \mathbf{x} = \mathbf{s}_j e^{j\phi_j} + \mathbf{n} \tag{2.141}$$

where $\mathbf{x}$ is the L-vector with components

$$x_i \overset{\mathrm{def}}{=} \int_0^T x(t)\psi_i^*(t)dt$$

and similarly $\mathbf{s}_j$ is the L-vector with components

$$s_{ji} \stackrel{\text{def}}{=} \int_0^T s_j(t)\psi_i^*(t)dt$$

and the noise vector **n** has components

$$n_i \stackrel{\text{def}}{=} \int_0^T n(t)\psi_i^*(t)dt$$

Proceeding the same way as in the case of one signal with unknown phase, that is, evaluating first the conditional likelihood ratios and then averaging with respect to the unknown phases, we end up with the following ML decision rule (see Problem 2.17):

$$\text{Choose } H_j \text{ if } \lambda_j(x) = \max_i \lambda_i(x)$$

where

$$\lambda_i(x) = \log\left[I_0\left(\frac{2}{N_0}\left|\int_0^T x(t)s_i^*(t)dt\right|\right)\right] - \frac{1}{N_0}\int_0^T |s_i(t)|^2 dt, \qquad i = 1, \ldots, M \quad (2.142)$$

The optimum receiver structure is represented in Figure 2.22. Note that, in the case of signals with the same energy, the second term on the right side of (2.142) can be dropped; furthermore, being that both the log and I_0 are strictly increasing functions,

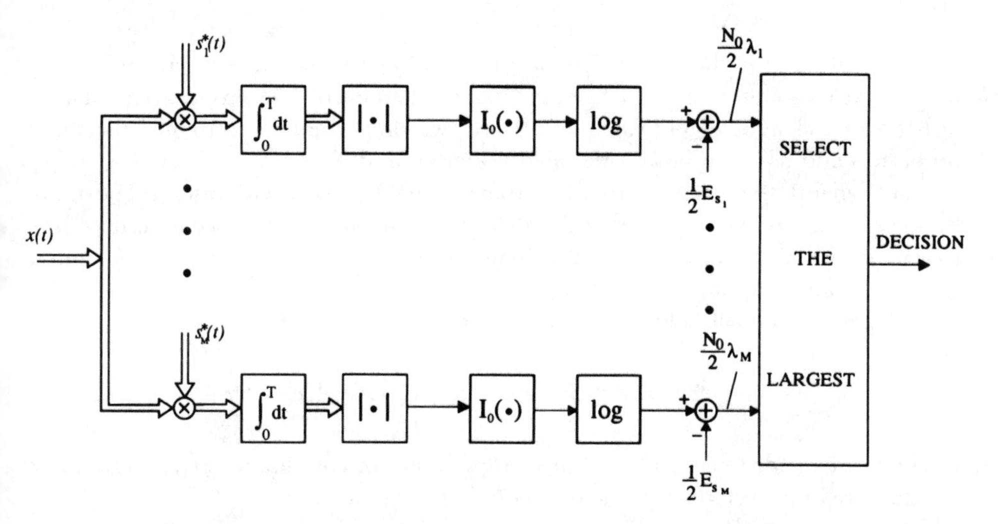

Figure 2.22 Block diagram of the receiver for the ML detection of M signals with unknown phase in additive noise.

maximization of the $\lambda_i s$ can be performed comparing the simplified log-likelihood ratios

$$\lambda_i(x) = \left| \int_0^T x(t) s_i^*(t) dt \right| \tag{2.143}$$

2.9 FROM LIGHT TO SIGNALS

We have considered the main properties of signals as they are encountered and used in communications theory. In this section, we will relate the abstract signals analyzed so far to the physical quantities characterizing light propagation.

Light is an electromagnetic wave described by two interdependent field vectors: the electric field **e** and the magnetic field **h.** The two field vectors obey Maxwell's equations, which describe the mutual interactions between **e** and **h.**

For vacuum propagation, Maxwell's equations can be manipulated into the form of two concise vector expressions [6,Chap. 3]:

$$\nabla^2 \mathbf{e} = \epsilon_0 \mu_0 \frac{\partial^2 \mathbf{e}}{\partial t^2}$$

$$\nabla^2 \mathbf{h} = \epsilon_0 \mu_0 \frac{\partial^2 \mathbf{h}}{\partial t^2}$$

where ϵ_0 and μ_0 are the vacuum *permittivity* and *permeability*, respectively.

Maxwell's equations were used in Chapter 1 to derive the propagation characteristics of the light in optical fibers. Here, we only want to introduce the main properties and assumptions on the field vectors **e** and **h.**

In general, these vectors are three-dimensional and vary with time and position. Because we are interested in light characteristics that can be expressed in terms of the electric field **e,** we will focus on that field and refer to it as the *optical field.* A similar treatment can be applied to the magnetic field.[8]

A generic optical field can be written as

$$\mathbf{e}(x,y,z,t) = \mathbf{e}(\mathbf{r},t) = e_x(\mathbf{r},t)\mathbf{x} + e_y(\mathbf{r},t)\mathbf{y} + e_z(\mathbf{r},t)\mathbf{z} \tag{2.144}$$

where **r** is a vector joining the origin with the point of coordinates (x,y,z) and **x,y,z** are the three unit vectors that define the reference system.

8. We denote the actual signals and vectors with lowercase letters and the complex envelopes of signals and vectors, as defined in Section 2.7, with uppercase letters.

We assume hereafter that the lightwave is quasi-monochromatic, that is, we can write each component of the vector in (2.144) as:

$$e_x(\mathbf{r},t) = E_{0x}(\mathbf{r},t) \cos\left[2\pi f_0 t + \phi_x(\mathbf{r},t)\right] \tag{2.145}$$

Under the quasi-monochromatic assumption, we can use the complex analytic signal representation introduced in Section 2.7 for the components in (2.144):

$$e_x(\mathbf{r},t) = \Re\left[E_x(\mathbf{r},t)e^{j2\pi f_0 t}\right] \tag{2.146}$$

where $E_x(\mathbf{r},t) \stackrel{\text{def}}{=} E_{0x}(\mathbf{r},t)e^{j\phi_x(\mathbf{r},t)}$ *is the complex envelope (see Section 2.7) of* $e_x(\mathbf{r},t)$ supposed to exhibit a *slow* variation with time or, equivalently, to have a bandwidth much lower than f_0. In the language of communication theory, this is also called the *narrow-bandwidth* assumption.

We saw in Section 2.7 that the complex envelope can be used to replace the actual signal for all linear operations performed on it, in the sense that the actual signal after the linear operations can be derived from its transformed complex envelope through the relation (2.146).

Suppose now, for simplicity, that $\mathbf{e}(\mathbf{r},t)$ represents a plane wave propagating along the direction of the z-axis and consider it for $z = 0$. In that case, the optical field lies in the transversal (x,y) plane, and its complex envelope is

$$\mathbf{E}(t) = E_{0x}(t)e^{j\phi_x(t)}\,\mathbf{x} + E_{0y}e^{j\phi_y(t)}\,\mathbf{y} \tag{2.147}$$

As we will see, the *polarization* of the lightwave, which can be defined as the time course of the direction of the optical field vector $\mathbf{e}(\mathbf{r},t)$, plays a crucial role in coherent optical communications. The state of polarization of the light depends on four parameters, E_{0x}, E_{0y}, ϕ_x, and ϕ_y, as seen in Example 2.16.

Example 2.16: Polarization of a Monochromatic Wave

In the case of a monochromatic plane wave traveling in the z direction, the x and y components of the field are periodic functions of time oscillating at frequency f_0, that is,

$$e_x(t) = E_{0x}\cos(2\pi f_0 t + \phi_x)$$

$$e_y(t) = E_{0y}\cos(2\pi f_0 t + \phi_y) \tag{2.148}$$

Previous equations are the parametric equations of the ellipse

$$\frac{x^2}{|E_{0x}^2|} + \frac{y^2}{|E_{0y}^2|} - 2\cos\phi\,\frac{xy}{|E_x|\,|E_y|} = \sin^2\phi \tag{2.149}$$

where $\phi = \phi_x - \phi_y$ is the phase difference between the two field components. At a fixed value of z, the tip of the optical field vector rotates periodically with time in the (x,y) plane tracing out an ellipse. At a fixed time, t, the locus of the tip of the optical field vector follows a helical trajectory in space lying on the surface of an elliptical cylinder.

The state of polarization of the lightwave is determined by the shape of the ellipse, that is, the direction of the major axis and the ratio of the minor axis to the major axis of the ellipse, which in turn are completely specified by the ratio of the magnitudes E_{0y}/E_{0x} and the phase difference $\phi = \phi_x - \phi_y$.

Some particular cases deserve a brief presentation.

- *Linear Polarization.* The lightwave is linearly polarized when one of the components vanishes, or when the phase difference ϕ is 0 or π. In the former case, the light is polarized in the direction of the nonvanishing component, whereas in the latter, it is polarized along a straight line of slope $\pm E_{0y}/E_{0x}$. The elliptical cylinder collapses into a tilted plane on which the locus of the tip of the optical field vector follows a sinusoidal trajectory, as shown in Figure 2.23.

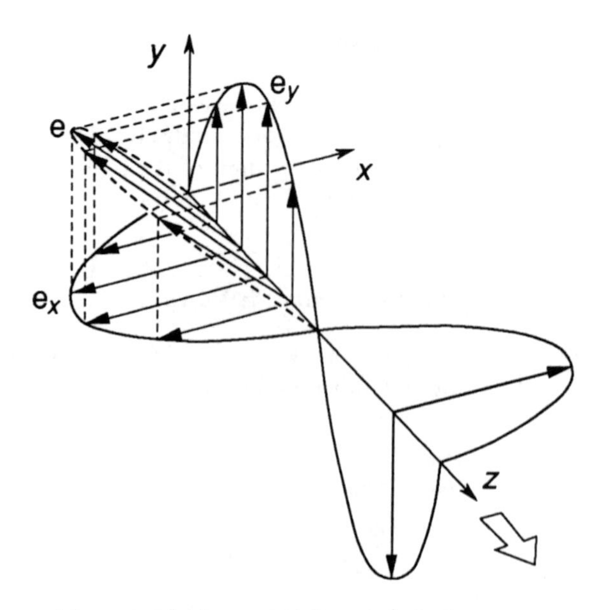

Figure 2.23 Trajectory of the optical field vector with linear polarization.

- *Circular Polarization.* When $E_{0x} = E_{0y}$ and $\phi = \pm\pi/2$, (2.149) becomes

$$x^2 + y^2 = E_{0x}^2$$

which is the equation of a circle. In that case, the light is said to be circularly polarized, right or left according to the sign of ϕ. The elliptical cylinder becomes a circular one, on which the locus of the tip of the optical field vector follows a helical trajectory, as shown in Figure 2.24.

When the light is quasi-monochromatic, the four parameters that define the state of polarization depend on time. In such a case [7, Chap. 10], if we watch the lightwave during a time interval much longer than the inverse of f_0, the amplitude and the phase of the two complex envelopes, E_x and E_y, will vary somehow, either independently or in some correlated fashion. If the variations are completely uncorrelated, the ellipse describing the polarization state may change shape, orientation, and handedness in such a way that, practically speaking, no existing detector could discern any particular state. Thus, we conclude that the lightwave is *unpolarized*.

On the other hand, if the ratio E_{0y}/E_{0x} were constant even though both terms varied, and the phase difference $\phi = \phi_x - \phi_y$ were constant as well, the wave would be completely polarized, exactly as it was in the monochromatic case.

Between these two extremes of completely polarized and unpolarized light is the condition of partial polarization. In particular, it can be shown that any quasi-monochromatic wave can be represented as the sum of a polarized and an unpolarized wave, where the two are independent and either can be zero.

A second fundamental property of electromagnetic waves is that they transport energy. Defining as **S** the so-called *Poynting vector*, whose magnitude represents the

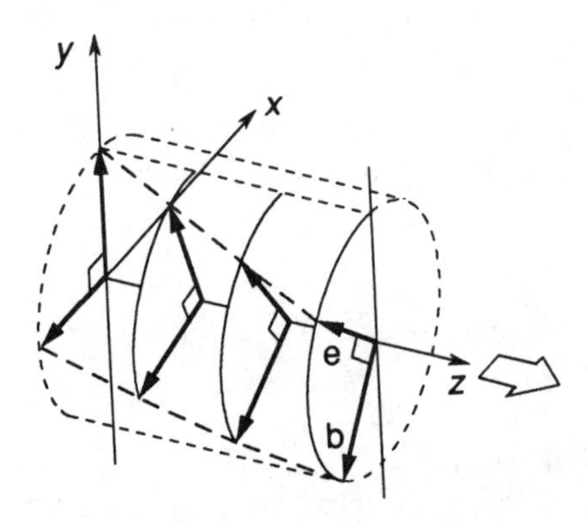

Figure 2.24 Trajectory of the optical field vector with circular polarization.

energy per unit time (the instantaneous power) flowing across a unit area, measured in watts per square meter, it can be shown that

$$\mathbf{S}(t) = \frac{1}{\mu_0}\mathbf{E}(t) \times \mathbf{H}(t) \tag{2.150}$$

where the symbol × means vector product. For isotropic media, the optical fields lie in a plane that is orthogonal to the direction of propagation of the wave. Thus, from (2.150), we argue that energy flows in the direction of the propagation of the wave, with instantaneous intensity, known as *irradiance*, equal to

$$I(t) \stackrel{\text{def}}{=} |\mathbf{S}(t)| = \frac{1}{2Z}|\mathbf{E}(t)|^2 \tag{2.151}$$

where the factor 1/2 takes into account that we are interested in the power of the actual signal rather than the complex envelope, and $Z = \sqrt{\mu_0/\epsilon_0}$ is the vacuum wave impedance.

In optical communication receivers, the incoming lightwave is converted to the electronic domain by means of a photodetector, a device that produces an output electric current proportional to the light power incident on its surface. In this case, denoting by $\mathbf{n}_p$ the unit vector orthogonal to the photodiode surface S, we can compute the instantaneous power "collected" by a completely matched photodiode:

$$P(t) = \int_s I(t)\mathbf{z} \cdot \mathbf{n}_p ds = \frac{1}{2Z}\int_s |\mathbf{E}(t)|^2 \mathbf{z} \cdot \mathbf{n}_p ds, \tag{2.152}$$

Example 2.17

For a monochromatic plane wave traveling in the z direction and incident on a photodiode with surface parallel to the (x,y) plane, we have that the collected power is given by

$$P = \frac{S}{2Z}(E_{0x}^2 + E_{0y}^2) \tag{2.153}$$

where S is the photodiode surface.

Under the plane wave hypothesis and assuming that $(S/2Z) = 1$, the relationship between the incident power and the complex envelope of the optical field becomes:[9]

$$P(t) = E_{0x}^2(t) + E_{0y}^2(t) \tag{2.154}$$

In the following chapters, the analytic signal associated with the actual signal will be denoted as $E(t)e^{j2\pi f_0 t}$, where $E(t)$ is the complex envelope. The associated instantaneous optical power will be computed as $P(t) = |E(t)|^2$.

The *sensitivity* of an optical communication system is defined as the received optical power permitting achievement of a given value of the bit error probability, usually 10^{-9}. Instead of measuring the received optical power in watts, it is customary to use the *number of photons per bit* N_T, which represents the average number of photons an ideal photon counter sitting at the receiver input would count in a time duration of T seconds. For a monochromatic wave at frequency f_0, the two quantities are related through

$$N_T = \frac{PT}{h f_0} \tag{2.155}$$

where T is the signaling period, that is, the time duration of a transmitted symbol (which equals the bit duration for binary modulations), and $h = 6.624 \cdot 10^{-34}$J/Hz is Planck's constant.

Problems

Problem 2.1

Prove that the knowledge of the probability distribution function permits the evaluation of the probability for an RV to lie in the interval (a, b) as

$$P[a < X \leq b] = F_X(b) - F_X(a)$$

Problem 2.2

Prove (2.14) between joint and marginal pdf's.

9. A similar simple relationship was assumed in Section 2.6, where the power was expressed as the squared magnitude of a generic signal, which could in turn represent a voltage or a current, thus implying the presence of a unitary impedence somewhere in the relationship.

Problem 2.3

Derive (2.16) between conditional PDF and pdf.

Problem 2.4

Derive the expression (2.24) of the moments of an RV as a function of its characteristic function.

Problem 2.5

Evaluate the mean and the variance of a binomial RV with the pdf in (2.26) and of a Poisson RV with the pdf in (2.27).

Problem 2.6

Derive (2.31) of the pdf of a noncentral χ-square RV. (*Hint:* It is convenient first to compute the characteristic function and then perform an inverse Fourier transform.)

Problem 2.7

Derive the pdfs in (2.36) and (2.37) of Rayleigh and Rice RVs.

Problem 2.8

Derive the autocorrelation function in (2.64) of a Poisson random process.

Problem 2.9

Evaluate the mean and the variance of the output current (2.76) of an APD.

Problem 2.10

Derive the power spectrum (2.94) of a homogeneous Poisson random process.

Problem 2.11

Derive the continuous and discrete parts (2.103) of the spectrum of a linearly modulated digital signal.

Problem 2.12

Derive the expressions (2.110) for the correlation functions of a complex narrowband random process.

Problem 2.13

Derive (2.127) of the log-likelihood ratio for the optimum detection of one known signal in additive Gaussian noise.

Problem 2.14

Derive (2.132) of the log-likelihood ratio for the optimum detection of one signal with unknown phase in additive Gaussian noise.

Problem 2.15

Derive the average error probability in (2.133) for the on-off modulation.

Problem 2.16

Derive the ML decision rule and (2.137) of the log-likelihood ratio for the optimum detection of M known signals in additive Gaussian noise.

Problem 2.17

Derive the ML decision rule and (2.142) of the log-likelihood ratio for the optimum detection of M signals with unknown phases in additive Gaussian noise.

References

[1] Benedetto, S., E. Biglieri, and V. Castellani, *Digital Transmission Theory*, Englewood Cliff, NJ: Prentice-Hall, 1987.
[2] Papoulis, A., *Probability, Random Variables and Stochastic Processes*, New York: McGraw-Hill, 1991.
[3] Snyder, D., *Random Point Processes*, New York: John Wiley & Sons, 1975.
[4] Loeve, M., *Probability Theory*, New York: Van Nostrand Reinhold, 1963.
[5] Schwartz, T., W. R. Bennett, and S. Stein. *Communication Systems and Techniques*, New York: McGraw-Hill, 1966.
[6] Hecht, E., *Optics*, Reading, MA: Addison-Wesley, 1987.
[7] Born, M., and E. Wolf, *Principles of Optics*, London: Pergamon Press, 1964.

Chapter 3

The Basic Binary Optical Communication System

3.1 INTRODUCTION

We described in Chapter 1 the physical characteristics of the main components of an optical communications link, namely LEDs and lasers, fibers, and photodetectors.

In this chapter, we will explain how these components are used to build an optical digital communication system based on the on-off modulation of the light source and direct detection of the received lightwave impinging on a photodetector. This is a very simple (and rudimentary, if compared to the sophisticated modulation formats designed and implemented for systems operating on telephone lines or radio-relay links) system, yet it is capable of reaching transmission speeds of tens of gigabits per second.

Applied in the first generation of digital optical links back in 1978, the IM-DD (intensity modulation, direct detection) system is still the most widely used for optical communications. The vast majority of bits transmitted around the world employ this modulation scheme. The invention and coming into service of *erbium-doped fiber amplifiers* (EDFAs) has recently reinforced IM-DD, allowing performance close to the quantum limit. With a limited, yet significant exception, we will limit ourselves in this chapter to the case of systems that do not use optical amplification (optically amplified systems are treated in Chap. 5). System impairments due to nonlinear phenomena in the fiber are not considered.

The topics treated in this chapter are covered by most existing books devoted to optical communication systems, like [1–12]. The level of treatment varies from reference books, like [4–6], to a more class-oriented style, as in [3,7,8,12]. Quite often, the approach to the analysis and design is heuristic, and the simplicity of the analysis leads to a lack of mathematical rigor.

This chapter takes inspiration from several sources, like [2,7,12,13] while attempting to pursue an original path. Most of the figures and results reflect newly derived (or rederived) results. The analytical approach should permit the analysis of systems in their generality, with a few simplifying hypotheses clearly stated. After the most general results are obtained, specific examples are given referring to simpler and practical cases.

An optical communication system (like all communication systems) consists of a *transmitter*, a *communication channel*, and a *receiver* connected as shown in Figure 3.1. The transmitter converts the electronic signal to be transmitted into an optical signal. This operation can be performed either through direct modulation of the light source or via an external modulator (the latter case is shown in Fig. 3.1). Direct modulation is simpler, but it induces a spectrum broadening of the emitted lightwave, known as frequency chirping (described in Chap. 1). Frequency chirp interacts with the group-velocity dispersion of the fiber and severely affects the pulse-shape degrading system performance.

External modulation, which minimizes or avoids frequency chirping, may be preferable in cases of long unrepeated distances and high information speed. The most common external modulators make use of electro-optic effect [5,Chap. 10].

The communication channel is the optical fiber. We will not consider in this book the case of unguided optical communication systems, in which the optical beam propagates in the free space. The two main parameters of the fiber are *attenuation* and *dispersion.* As seen in Chapter 1, both have the effect of limiting the length of the unrepeated link, that is, the maximum distance at which we can reliably transmit a light signal without inserting a repeater/regenerator.

Finally, the receiver includes a photodetector, typically, a PIN or an APD (avalanche photodiode), which converts the incoming light into an electronic signal. Because the electronic signal can be small and distorted, it needs amplification and equalization before entering the decision device. The decision device samples the signal and restores the transmitted bits by comparing the samples with a suitable threshold.

In this chapter, we will first review briefly the main characteristics of the key system components: transmitter, fiber, and receiver. Then we will consider in more detail the physical limitations induced by the fiber (the analysis builds on and expands the treatment in Chap. 1). The third part of the chapter will be devoted to the description and analysis of the receiver. We will characterize the system noises and evaluate the system performance in terms of *bit error probability* while focusing on *shot* and *thermal* noises. Finally, we will briefly examine other causes of system impairment: extinction ratio, relative intensity noise, timing jitter, modal noise, and modal partition noise.

3.2 MAIN SYSTEM COMPONENTS

3.2.1 Transmitter

The information source is assumed to emit a stationary sequence of binary symbols (bits) at a *bit rate* of $R_b = 1/T$ bit/s, where T is the time interval between consecutive emissions, also known as the *symbol period* or *symbol interval.*

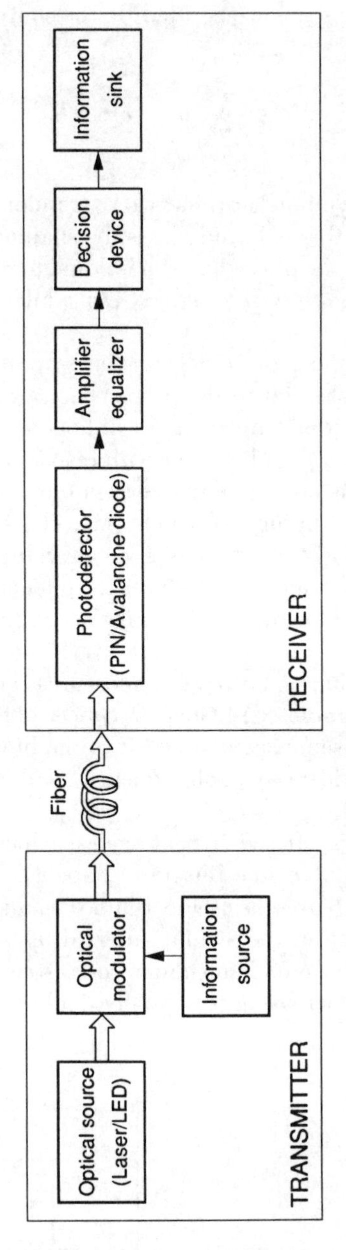

Figure 3.1 Block diagram of the basic IM-DD optical binary communication system.

The electronic binary sequence emitted by the source can be written as

$$z(t) = \sum_{n=-\infty}^{\infty} a_n g(t - nT) \tag{3.1}$$

where a_n (the information symbols) are binary RVs (random variables) assuming, with probability 1/2, the values 0 and 1, and $g(t)$ is the shaping pulse, normally a square pulse of duration T'. When the pulse duration is equal to the symbol period ($T' = T$), the shaping pulse is called *nonreturn to zero* (NRZ), while for $T' < T$ it is called *return to zero* (RZ).

The choice of the shaping pulse depends on several system concerns, such as symbol synchronization (essential to the proper choice of the sampling instants in which decisions on transmitted information symbols are made) and effects of AC coupling. NRZ pulses show a poor behavior with respect to both. A long sequence of 1 (or 0) bits generates a constant signal that does not present variations at the symbol rate $1/T$, thus making the timing extraction (i.e., synchronization) very difficult. Moreover, a sequence of NRZ pulses has a mean value that depends on the bit sequence, and that, after AC coupling, produces the phenomenon of *baseline wander,* which hampers the ability of the receiver circuitry to distinguish between the presence and the absence of pulses.

To reduce these drawbacks, two countermeasures can be used: data scrambling and line coding. Binary scramblers [14,Chap. 9] consist of feedback shift registers that produce a pseudo-random sequence and add it to the binary information sequence. This operation creates pseudo-random bit sequences, with a good 1 and 0 balance and an almost "white" spectral density.

A different solution is offered by line codes, which encode blocks of m information bits into blocks of n line bits [14,Chap. 9]. The simplest solution has $m = n = 1$, and consists in representing the 1 and 0 as an RZ pulse with $T' = 0.5\,T$ and a different position within the symbol interval, as shown in Figure 3.2. This solution, known as split-phase or Manchester code, solves the problem of baseline wander, at the expense of increasing (essentially doubling) the signal bandwidth.

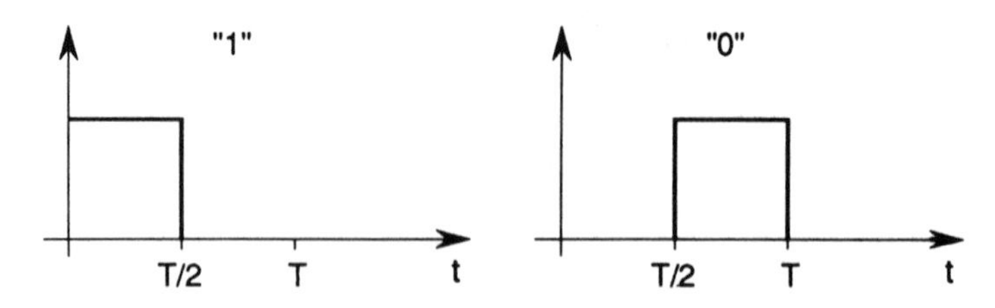

Figure 3.2 Pulses of the Manchester code.

The modulation bandwidth of the light source or of the external modulator must be large enough to match the bit rate of the source. The parameter controlling this ability is the *rise time* of the transmitter, defined as the time during which the response to an input step function increases from 10% to 90% of the final steady-state value. As an example, consider the simple case in which the transmitter transfer function can be approximated as that of the first-order system in Figure 3.3. The system will respond to a step function of amplitude V, with a voltage across the capacitance of the form

$$v_C(t) = V\,(1 - e^{-t/RC}) \tag{3.2}$$

The rise time, T_R, can be easily found from (3.2):

$$T_R \simeq 2.2RC \tag{3.3}$$

The 3-dB system bandwidth is $B_S = 1/(2\pi RC)$. If we select $B_S = R_b = 1/T$ for NRZ pulses and $B_S = 2R_B = 2/T$ for RZ pulses, where R_b is the bit rate, we obtain the following relations between the transmitter rise time and bit rate:

$$T_R \leq \begin{cases} 0.35/R_b \text{ for NRZ digital format} \\ 0.175/R_b \text{ for Manchester encoded digital format} \end{cases}$$

The light source responds to the driving sequence of amplitude-modulated electronic pulses with a sequence of light pulses that form a bandpass signal, $x(t)$. To the actual real signal $x(t)$ we can associate, as seen in Section 2.7, an *analytic* signal,

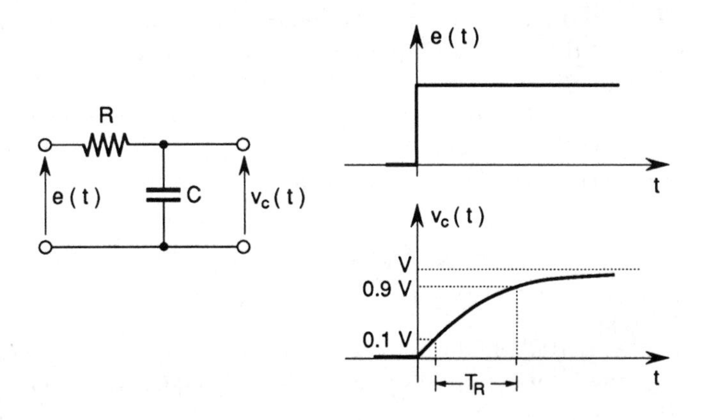

Figure 3.3 Rise time of the transmitter assumed to behave as a simple RC first-order system.

$\mathring{x}(t)$, whose real part is $x(t)$. In the following equations, instead of the analytic signal $\mathring{x}(t)$, we will use its complex envelope, $\tilde{x}(t)$

$$\tilde{x}(t) = \mathring{x}(t)e^{-j2\pi f_0 t} \tag{3.4}$$

where f_0 is the central optical (carrier) frequency. The complex envelope of bandpass signals was introduced in Section 2.7; here, for simplicity of notation, we will denote it as $x(t)$. Use of the signal complex envelope permits us to neglect the optical carrier term in the equations.

Then the transmitted signal, $x(t)$, consisting of a sequence of modulated light pulses, can be written as

$$x(t) = \sum_{n=-\infty}^{\infty} a_n s_T(t - nT) \tag{3.5}$$

The form of the individual pulse, $s_T(t)$, depends on the physical characteristics of the LED or laser. In the case of direct modulation of the light source, the frequency chirping must be included in the model (the same may or may not be true for external modulation). For simplicity, we will consider a Gaussian pulse shape, $s_T(t)$, of the pulses:

$$s_T(t) = A_T \exp\left(-\frac{(1 - jc)t^2}{2\tau_T^2}\right), \qquad -\infty < t < \infty \tag{3.6}$$

In (3.6) the peak amplitude, A_T, is related to the power of $x(t)$, whereas τ_T (T denotes that we are considering the transmitted pulse) is the *pulse width*, that is, the value of t at which the pulse amplitude with $c = 0$ becomes equal to $A_T/\sqrt{e}$ (see Fig. 3.4, where the pulse (3.6) is plotted in the case $c = 0$). The pulse width is measured in picoseconds. The rise time, T_R, of the Gaussian pulse (3.6) is $T_R = 1.55\ \tau_T$.

Figure 3.5 plots three pulses $s_T(t)$ from (3.6) with $c = 0$, equal energy, and different values of τ_T. In Figure 3.6 we report the ratio between the energy, $E_S(t)$, contained in the interval $(-t,t)$ and the total energy, E_S, versus the normalized time, t/τ_T. It appears clearly that satisfying the inequality,

$$\tau_T \leq \frac{T}{4} \tag{3.7}$$

guarantees that almost 100% of the pulse energy is contained in a duration T. In other words, the transmitted pulse duration is approximately equal to $4\tau_T$ for the selected pulse shape.

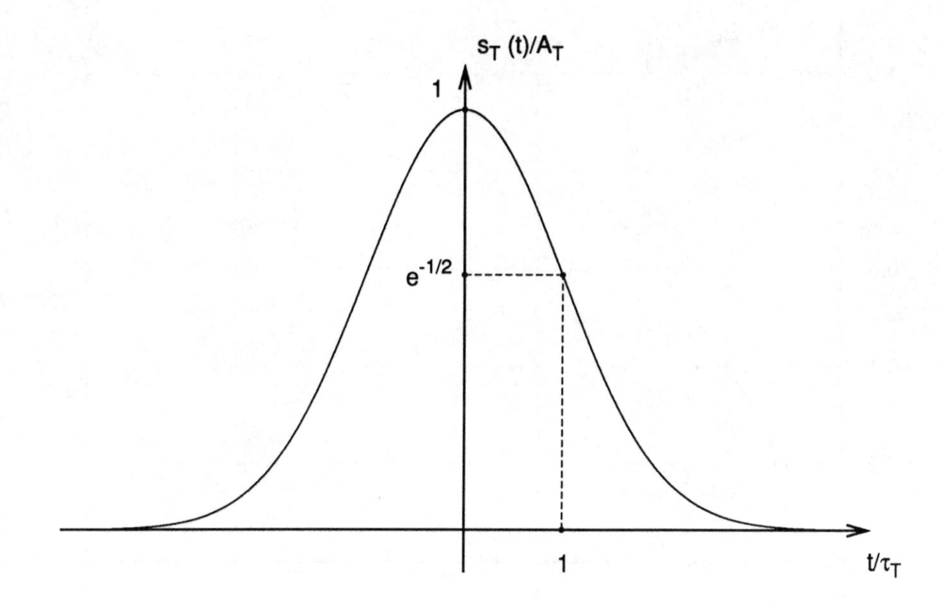

Figure 3.4 Plot of a Gaussian-shaped pulse versus normalized time variable.

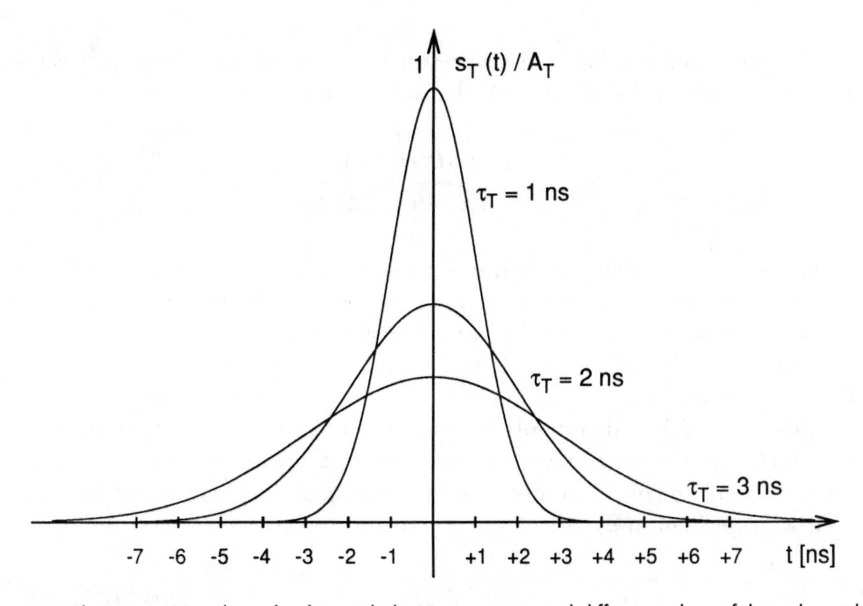

Figure 3.5 Three Gaussian-shaped pulses with the same energy and different values of the pulse width τ_T.

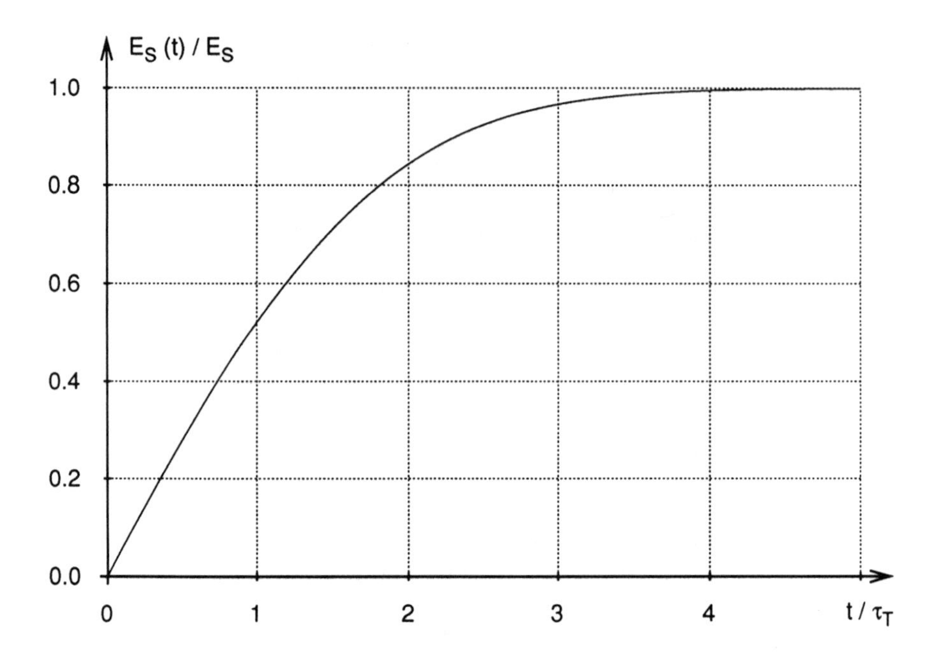

Figure 3.6 Fractional energy contained in the interval $(-t,t)$ for Gaussian-shaped pulses versus normalized time.

The dimensionless parameter c accounts for the frequency chirp. The frequency variation is given by the derivative of the pulse phase:

$$f(t) \overset{\text{def}}{=} \frac{1}{2\pi}\frac{d\phi_s(t)}{dt} = \frac{1}{2\pi}\frac{ct}{\tau_T^2} \tag{3.8}$$

Equation (3.8) shows that the frequency variation is linear with time through a coefficient proportional to c. We have chosen a linear model for the frequency chirp, because in this case the shape of the pulse does not change during propagation, thus allowing simple analytical formulations. Agrawal discusses the validity of this model [12,Chap. 5], while Agrawal and Potasek describe a more refined model [15]. Physically, in the case of directly modulated semiconductor lasers, c is often referred to as the *linewidth enhancement factor* and amounts to a few negative units.

The effect of chirping is a spectrum broadening of the pulse, as can be seen from the Fourier transform of $s_T(t)$:

$$S_T(f) = A_T\tau_T\sqrt[4]{\frac{4\pi^2}{1+c^2}}\exp\left(-\frac{2\pi^2 f^2\tau_T^2}{1+c^2}\right)\cdot\exp\left[\frac{1}{2}\tan^{-1}c - j\left(\frac{2\pi^2 f^2\,\tau_T^2 c}{1+c^2}\right)\right] \tag{3.9}$$

The pulse spectrum (3.9) exhibits a broadening factor:

$$\Delta f = \frac{1}{2\pi}\sqrt{\frac{1+c^2}{\tau_T^2}} \tag{3.10}$$

The normalized energy spectrum, $|S_T(f)|^2$, of $s_T(t)$ is plotted in Figure 3.7 to show the spectral broadening induced by c.

The transmitted *power spectrum* can be evaluated using the technique explained in Section 2.6 (or, more directly, the results of Example 2.8):

$$G_x(f) = \frac{1}{4T^2}|S_T(f)|^2 \sum_{n=-\infty}^{\infty} \delta\left(f - \frac{n}{T}\right) + \frac{1}{2T}|S_T(f)|^2 \tag{3.11}$$

where the first term on the right side is the discrete part of the spectrum due to the nonzero mean value of the RVs a_n. Considering only the continuous part of the spectrum (which carries the transmitted information), we can evaluate the average AC power of $x(t)$ as

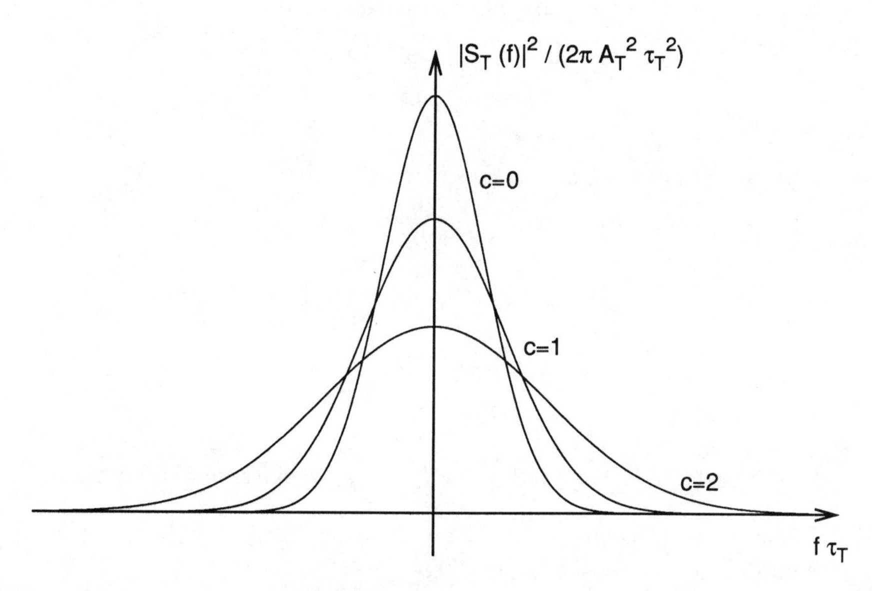

Figure 3.7 Plot of the normalized amplitude spectrum of a Gaussian-shaped pulse for three values of the chirp parameter, c.

$$\overline{P}_x = \frac{1}{2T}\int_{-\infty}^{\infty}|S_T(f)|^2 df = \frac{1}{2T}\int_{-\infty}^{\infty} s_T^2(t)dt \tag{3.12}$$

Substituting (3.6) into (3.12), we can relate the power, $\overline{P}_x$, to the pulse parameters as

$$\overline{P}_x = \frac{A_T^2 \tau_T \sqrt{\pi}}{2T} \tag{3.13}$$

The power is normally measured in decibels above one milliwatt according to the definition

$$P\,(\text{dBm}) \stackrel{\text{def}}{=} 10\log_{10}\frac{P}{1mW} \tag{3.14}$$

3.2.2 Fiber

When the optical power is low, the optical fiber is a linear, time-invariant, dispersive medium with respect to the optical fields. Assume an input field to the fiber of the form (3.5) with $s_T(t)$ being the Gaussian pulse of (3.6) and apply the basic propagation equations in Chapter 1 (see Problem 3.1). The expression of the output field after propagating a distance z along the fiber turns out to be

$$y(t,z) = n\sum_{-\infty}^{\infty} a_n s_R(t - nT, z) \tag{3.15}$$

where the received pulse, $s_R(t,z)$, is given by

$$s_R(t,z) = A_R(z)\, M_R(t,Z)e^{j\phi_R(t,Z)} \tag{3.16}$$

The terms $A_R(z)$ and $M_R(t,Z)$, $\phi_R(t,Z)$ account for the following physical phenomena:

- Fiber attenuation, accounted by

$$A_R(z) = A_T e^{-\alpha z} \tag{3.17}$$

where α is the fiber attenuation coefficient parameter, measured in km^{-1}.

- Fiber dispersion, accounted by the two terms $M_R(t,Z)$ and $\phi_R(t,Z)$, reflecting the magnitude and phase distortions:

$$M_R(t,Z) = \frac{1}{\sqrt[4]{(1+cZ)^2+Z^2}} \exp\left\{-\frac{(t-t_D)^2}{2\tau_T^2[(1+cZ)^2+Z^2]}\right\} \tag{3.18}$$

$$\phi_R(t,Z) = -\frac{1}{2}\tan^{-1}\left(\frac{Z}{1+cZ}\right) - \frac{t^2[(1+cZ)c - Z]}{2\tau_T^2[(1+cZ)^2+Z^2]} \tag{3.19}$$

In previous relationships, we used the fiber delay

$$t_D \stackrel{\text{def}}{=} \frac{z}{v_g}, \qquad [\text{s}] \tag{3.20}$$

where v_g is the group velocity, and the normalized propagation distance

$$Z \stackrel{\text{def}}{=} \frac{z}{L_D} \tag{3.21}$$

where L_D is the *dispersion length* of the fiber defined as

$$L_D \stackrel{\text{def}}{=} \frac{\tau_T^2}{\ddot{\beta}}, \qquad [\text{km}] \tag{3.22}$$

in which $\ddot{\beta}$ is the second derivative of the propagation constant β (see Section 1.2.10), called the dispersion parameter.

Recall that a Gaussian signal passing through a linear system with Gaussian impulse response remains Gaussian. Hence, the fiber behaves as a linear system with lowpass equivalent impulse response:

$$h_F(t,z) = \frac{1}{\sqrt{2\pi\ddot{\beta}z}} \exp\left(\frac{t^2}{2\ddot{\beta}z} + \frac{\pi}{4} + n\pi\right) \tag{3.23}$$

Thus, the relation between input and output fields is $s_R(t,z) = s_T(t) * h_F(t,z)$, where $*$ means convolution. In other words, we can say that input Gaussian pulses propagate into a single-mode fiber keeping their shape, with modifications that can be accounted for as if the pulses were filtered by a linear system with a Gaussian impulse response whose variance increases with the propagation distance z.

Figure 3.8 shows the magnitude, $M_R(t,Z)$, of the received pulse as a function of the normalized time, $(t - t_D)/\tau_T$. The first curve for $Z = 0$ corresponds to the transmitted pulse, whereas the other two show the dispersed pulse after traveling a

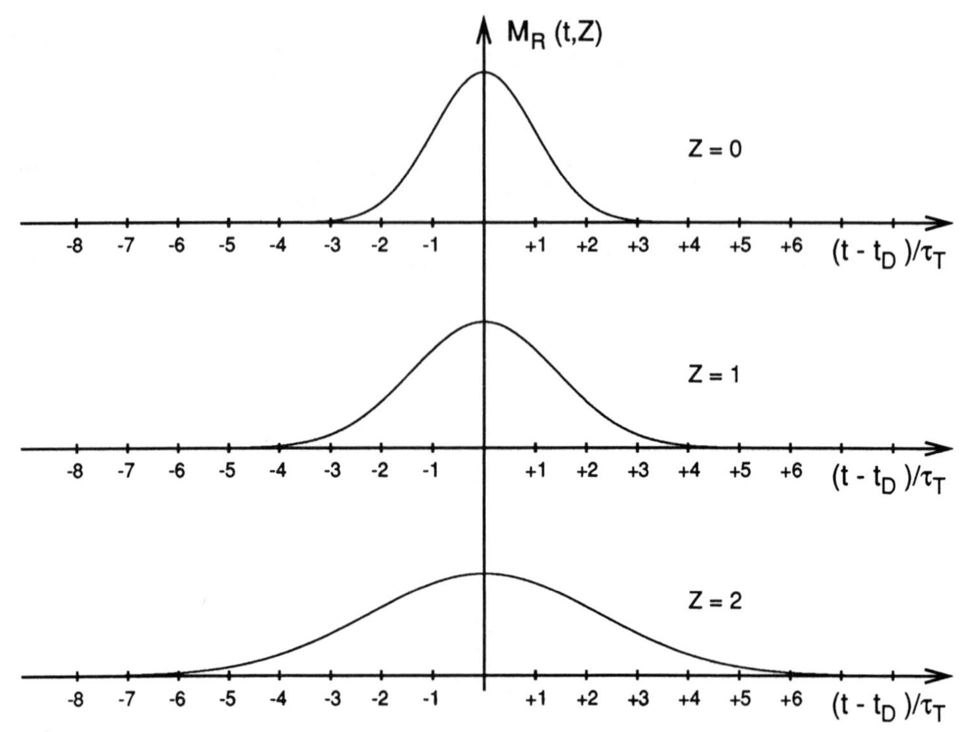

Figure 3.8 Magnitude of received Gaussian-shaped pulses for three different values of the normalized traveled distance, *Z*.

distance equal to one and two dispersion lengths. Example 3.1 clarifies the practical implications of Figure 3.8.

Example 3.1

Consider a light source emitting at a wavelength of 1,550 nm and driven by an information signal at the bit rate of 2.5 Gbps, which launches a Gaussian pulse, $s_T(t)$, as in (3.6), with pulse width τ_T = 100 ps. The fiber group-velocity dispersion coefficient is $D = 16$ ps/(nm · km), and the laser chirp parameter is $c = 0$. Recall the relation between D and $\ddot{\beta}$ (see Section 1.2.10):

$$D = -\frac{2\pi c\ddot{\beta}}{\lambda^2} \tag{3.24}$$

Using (3.22) and (3.21), a dispersion length, $L_D = 500$ km, and a normalized propagation distance, $Z = 2$, are obtained, meaning a propagation distance of $z = 1{,}000$ km. From Figure 3.8, we deduce that after propagating 1,000 km each pulse is spread significantly over more than three symbol periods and thus interferes with two adjacent pulses, degrading the system performance.

Fiber dispersion broadens individual pulses forming the received signal $y(t,z)$ of (3.15) over several adjacent symbol intervals and gives rise to the phenomenon of *intersymbol interference.* To visualize the effect of intersymbol interference on the received signal, the so-called *eye diagram* is used. It is obtained by slicing the received signal into segments of T seconds and superimposing the slices. Two examples are shown in Figures 3.9 and 3.10. They correspond to the system in Example 3.1 for a propagation distance of 0 and 750 km, respectively, and depict the eye diagram of the electrical signal after photodetection and integrate-and-dump filtering.

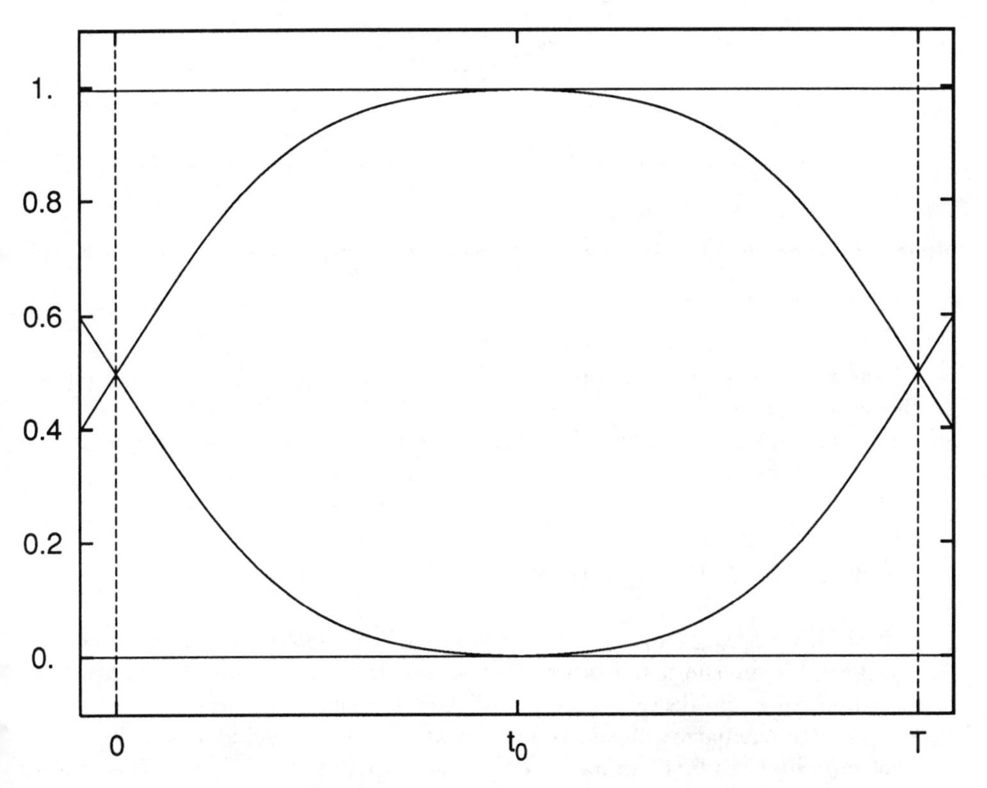

Figure 3.9 Eye diagram of system described in Example 3.1 without the effects of fiber dispersion.

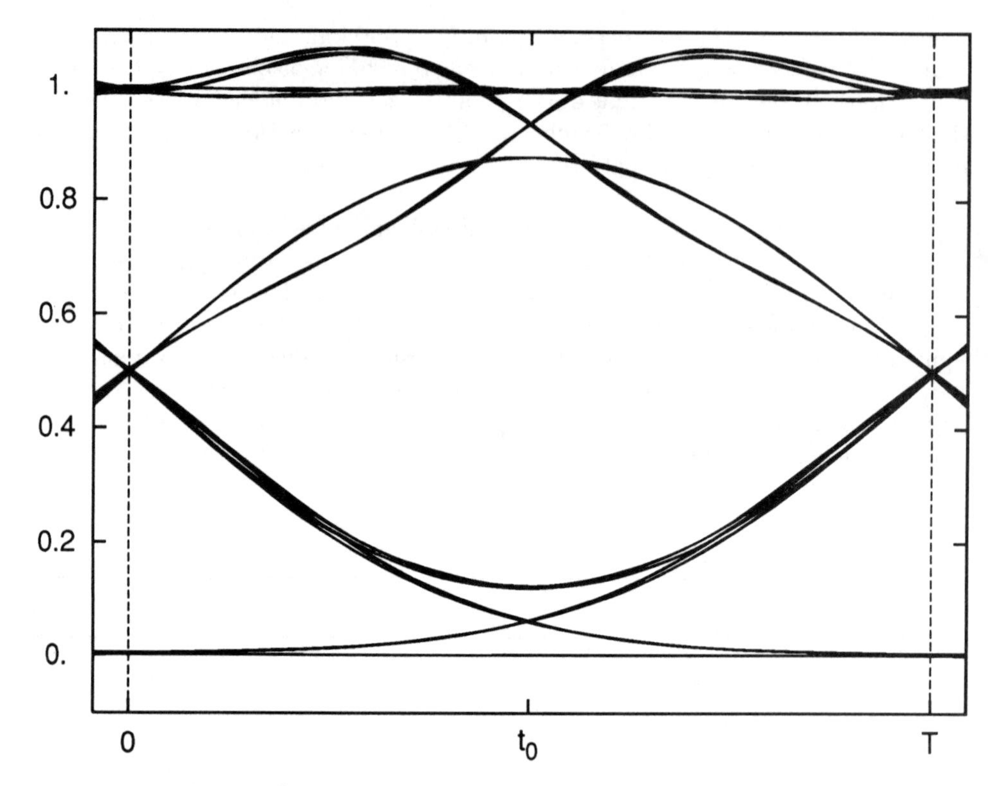

Figure 3.10 Eye diagram of system described in Example 3.1 corresponding to a propagation distance of 750 km.

The form of the eye diagram provides useful information on receiver performance. For example, consider the eye diagrams in Figures 3.9 and 3.10. To make a decision on the transmitted symbols, the received signal is sampled at a particular time instant, t_0, within each interval of duration T; then the obtained sample is compared to a threshold. The sampling instant, t_0, is chosen to minimize the intersymbol interference or, equivalently, to maximize the eye diagram opening.

Intersymbol interference has two detrimental effects:

- First, it reduces the *eye opening*, that is, the useful amplitude to be compared against the threshold; this brings the two waveforms (corresponding to the 0 and 1 bits) close and decreases the guard interval against the noise.
- Second, it can narrow the horizontal *eye width*, thus reducing the possible range of sampling instants and increasing the sensitivity to their fluctuations, known as timing jitter.

The first effect can be seen in a comparison of the two eye diagrams in Figures 3.9 and 3.10. The second effect is not visible in the figures because of the particular choice of the pulses.

Experimentally, eye diagrams are obtained using an oscilloscope whose time axis is synchronized with the symbol period.

3.2.3 Receiver

The receiver processes the received optical signal to recover the transmitted information bits. The reliability of its decisions depends on the received power through a complex mechanism involving various types of noise present in practical optical receivers. Receiver analysis is the subject of Section 3.5. In particular, we will relate the bit error probability of the system to the received average optical power[1], $\overline{P}_R$, or, equivalently, to the *average received number of photons per bit*, $\overline{N}_R$. The two quantities are related by

$$\overline{P}_R = \frac{\overline{E}_R}{T} = \frac{\overline{N}_R h\nu}{T} = \overline{N}_R R_b h\nu \tag{3.25}$$

where $\overline{E}_R$ is the average received energy per bit, $h = 6.62559 \cdot 10^{-34}$ J (Planck's constant), ν is the light frequency, and $h\nu$ is the energy quantum of a photon.

We define receiver *sensitivity* as the *minimum average number of photons per bit* that guarantees a particular value of the bit error probability. Alternatively, one could also define the sensitivity in terms of received power. The particular value of the bit error probability depends on the application; a common value is 10^{-9}.

3.3. PHYSICAL LIMITATIONS IMPOSED BY FIBER

Both fiber attenuation and dispersion impose a limit on the link length, L. By link length, we mean either the distance between transmitter and receiver, for single-hop systems, or the distance between two adjacent regenerators, in the case of multihop systems. For multihop systems that employ optical amplification (which will be treated briefly in Section 3.4.5), L will denote the total distance between transmitter and receiver.

3.3.1 Attenuation Limits

The fiber attenuation parameter, α, introduced in (3.17), is usually expressed in decibels per kilometer as

1. By average power, we mean average with respect to the symbol period, T, and to the information symbol. In the following, we will use the notation $\overline{P}$ (and $\overline{E}$, $\overline{N}$) for the average power (and energy, number of photons per bit), and P for the peak power (and E, N for energy and number of photons per bit).

$$\alpha_{dB} \overset{\text{def}}{=} -\frac{10}{L}\log_{10}\frac{\overline{P}_R}{\overline{P}_T} \tag{3.26}$$

where $\overline{P}_T$ and $\overline{P}_R$ are the transmitted and received powers, respectively, L is the link length, and α_{dB} is related to α as

$$\alpha_{dB} = \alpha 10 \log_{10} e \tag{3.27}$$

For a given system sensitivity expressed in terms of average received number of photons per bit, $\overline{N}_R$, substitution of (3.25) into (3.26) leads to the following relationship between the link length, L, and the bit rate, R_b:

$$L = \frac{10}{\alpha_{dB}}\log_{10}\frac{\overline{P}_T}{\overline{N}_R R_b h\nu} \tag{3.28}$$

As an example, consider a laser diode operating at a wavelength of 1,550 nm that couples to the fiber a power, $\overline{P}_T$, of 0 dBm. At this wavelength, a single-mode fiber has a typical attenuation α_{dB} = 0.2 dB/km. Assume a receiver sensitivity equal to $\overline{N}_R = 1{,}000$ photons per bit; the photon energy is equal to $h\nu = 1.28 \times 10^{-19}$ J, so that (3.28) becomes

$$L = 644.6 - 50 \log_{10} R_b \tag{3.29}$$

Equation (3.29) is plotted in Figure 3.11 together with the similar curves referring to the case of light sources emitting at wavelengths of 800 and 1,300 nm. The following assumptions are used: Transmitted power is the same in all cases, fiber attenuation is 3 and 0.5 dB/km, respectively, and receiver sensitivity is 100 and 200 received photons per bit, respectively. It is evident that it is advantageous to use the wavelength of 1,550 nm, where, as seen in Chapter 1, the fiber presents a minimum of attenuation.

Equation (3.28) shows that, to increase the link length for a given bit rate and operating wavelength, we can either increase the transmitted power, $\overline{P}_T$, or decrease the receiver sensitivity, $\overline{N}_R$. The first solution requires the use of a higher-power laser or an optical amplifier to boost the signal, whereas the latter solution is made possible by coherent reception and/or, again, optical amplification. Coherent receivers are discussed in Chapter 4, optical amplifiers in Chapter 5.

In the design of practical systems, one must include in the power budget other causes of power loss, like connectors and splices that join adjacent fiber segments. On top of that, it is common practice to add a system margin to account for unforeseen power penalties that might occur during the system lifetime. Typical connector and

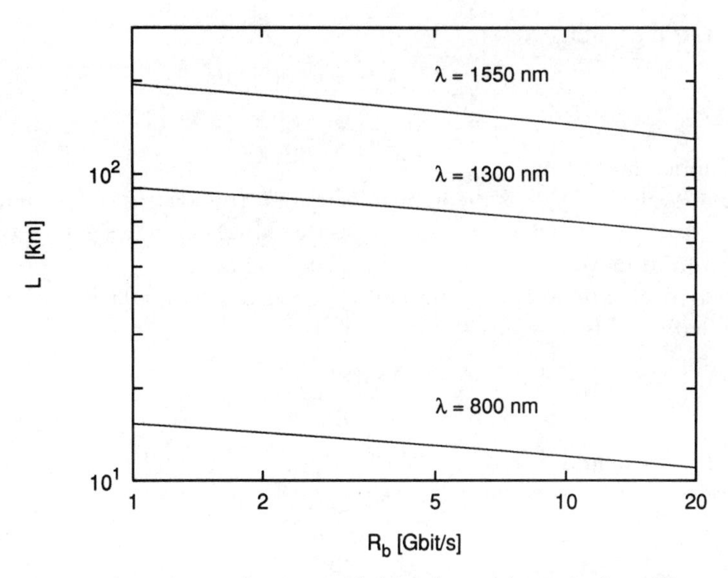

Figure 3.11 Attenuation limits showing the maximum link length versus bit rate for three different wavelengths. System parameters are as follows: $\overline{P}_T = 0$ dBm, $\overline{N}_R = 1{,}000$, 200, 100 photons/bit, $\alpha_{dB} = 0.2$, 0.5, 3 dB/km at wavelengths of 1,550, 1,300, and 850 nm, respectively.

splice losses are on the order of 0.3 dB per connector and 0.1 dB per splice, whereas the system margin typically ranges from 3 to 6 dB.

3.3.2 Heuristic Dispersion Limits

We will first recall the heuristic dispersion relations obtained in Section 1.2.10 between the link length, L, and the information bit rate, R_b. Those relations require that the pulse broadening, $\Delta\tau$, be less than or equal to $T/4$, where T is the symbol period. Then we will discuss more thoroughly the case of single-mode fibers and narrow-linewidth light sources.

3.3.2.1 Multimode Fibers

Dispersion in multimode fibers is due mainly to *modal dispersion* and leads (see Section 1.2.10) to the following inequalities:

$$R_b \leq \frac{c}{4n} \cdot \frac{n}{\Delta n} \cdot \frac{1}{L} \tag{3.30}$$

for step-index fibers and

$$R_b \leq \frac{2c}{n_1} \cdot \frac{n^2}{(\Delta n)^2} \cdot \frac{1}{L} \tag{3.31}$$

for graded-index fibers.

In (3.30) and (3.31), c is the light velocity, $n = (n_1 + n_2)/2$ is the average (core and cladding) refractive index, $\Delta n = n_1 - n_2$ is the difference between the two indices, n_1 is the core refractive index, and L is the link length.

Using typical parameters for multimode fibers, like $\Delta n/n = 0.01$, $n = 1.4925$, and $n_1 = 1.5$, we obtain the numerical relationships

$$R_b \leq \frac{4.98 \cdot 10^{-3}}{L} \tag{3.32}$$

for step-index fibers and

$$R_b \leq \frac{4}{L} \tag{3.33}$$

for graded-index fibers. In both inequalities (plotted in Figure 3.12), R_b is measured in gigabits per second and L in kilometers.

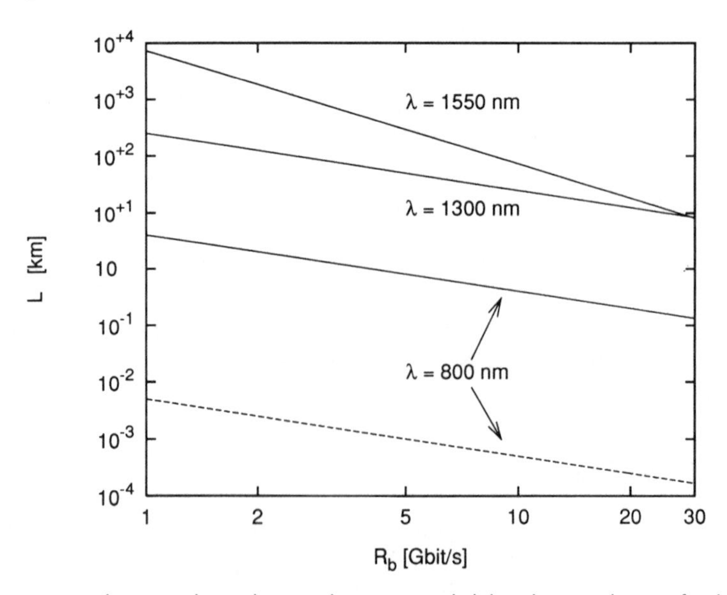

Figure 3.12 Heuristic dispersion limits showing the maximum link length versus bit rate for three different wavelengths. The dashed curve corresponds to step-index fibers. The curve referring to 1,300 nm assumes a source with a linewidth of 1 nm and a dispersion $D = 1$ ps/(nm·km). The curve referring to 1,550 nm assumes a narrow linewidth source and a dispersion $D = 16$ ps/(nm · km).

3.3.2.2 Single-Mode Fibers

Single-mode fibers support only one mode, thus avoiding modal dispersion. Dispersion is due to *chromatic dispersion* and can be quantitatively ascribed to one of two parameters: D (defined as group velocity dispersion coefficient in Section 1.2.10) or $\ddot{\beta}$, the second derivative of the propagation constant β with respect to the angular light frequency ω in radians per second. They are related through (3.24). At the wavelengths used for optical communications, D and $\ddot{\beta}$ assume values of the same order of magnitude and opposite signs. Assuming that the fiber is linear, the dispersion effects on the propagating pulse [see (3.18)] depend only on the magnitude of $\ddot{\beta}$ (or D) for zero chirp. When chirp is not zero, the sign of the product $c \cdot \ddot{\beta}$, where c is the chirp parameter introduced in (3.6), also influences the dispersion effects.

Nonlinear phenomena in the fiber, on the other hand, produce different effects depending on the sign of dispersion. A typical example is provided by modulation instability, as described, for example, in [16]. For that reason, it is important to distinguish between systems that use fibers with positive or negative dispersion. It is common practice to say that a system is operating in the normal dispersion regime when the fiber presents a negative value of D (or, equivalently, a positive value of $\ddot{\beta}$) and in the anomalous dispersion regime when D is positive (or $\ddot{\beta}$ is negative).

Imposing the rule-of-thumb $\Delta\tau \leq (T/4)$ leads to the following inequality, derived in Section 1.11:

$$R_b \leq \frac{1}{4|D|\Delta\lambda} \cdot \frac{1}{L} \tag{3.34}$$

where $\Delta\lambda$ is the signal bandwidth expressed in terms of the occupied wavelength interval.

For systems that employ lasers with several longitudinal modes, $\Delta\lambda$ is essentially determined by the laser spectral width. On the other hand, for single-longitudinal-mode lasers with small bandwidth, $\Delta\lambda$ depends on the modulation bandwidth and, consequently, on the information rate through

$$\Delta\lambda \simeq \frac{\lambda^2}{c} R_b \tag{3.35}$$

where λ is the central light wavelength; expression (3.35) assumes that the chirp is negligible. Substituting (3.35) into (3.34) yields

$$R_b^2 \leq \frac{c}{4|D|\lambda^2} \cdot \frac{1}{L} \tag{3.36}$$

Notice that in (3.36) R_b^2 is proportional to $1/L$, instead of R_b being proportional to $1/L$, as encountered so far.

As seen in Section 1.2.10 (see Fig. 1.22), for conventional single-mode fibers, the parameter D is small in the 1,300-nm window and increases up to 16–20 ps/(nm·km) in the 1,550-nm window.

Consider first a system operating in the 1,300-nm window, with $D = 1$ ps/(nm·km) and $\Delta\lambda = 1$ nm. From (3.34), we obtain

$$R_b \leq \frac{250}{L} \tag{3.37}$$

where R_b is expressed in gigabits per second and L in kilometers.

In the case of a system operating in the 1,550-nm range, we assume that $D = 16$ ps/(nm·km) and a single-longitudinal-mode laser. Using (3.36), we obtain

$$R_b^2 \leq \frac{2.1 \times 10^3}{L} \tag{3.38}$$

where R_b is expressed in gigabits per second and L in kilometers. As an example, for $L = 1$ km we have a bit rate of 45.8 Gbps.

Relations (3.37) and (3.38) are plotted in Figure 3.12.

3.3.3 Power Penalty Due to Fiber Dispersion

We have seen that the fiber dispersion gives rise to intersymbol interference, which, in turn, reduces the noise immunity and makes more critical the symbol synchronization. In this section, we will derive a limit to the product (link length) × (information rate) in terms of *peak distortion*, defined as the largest value of intersymbol interference, assuming single-mode fibers.

Consider the optical received field, $y(t, z)$, given by (3.15) and the received pulse given by (3.16); assume an ideal noiseless receiver formed by a photodetector and an integrate-and-dump filter. The detected photocurrent is equal to the squared magnitude of the optical field; the filter integrates the detected signal over the symbol period, T.

We define the peak distortion, d_p, as the eye closure normalized with respect to the received energy. In formulae, it satisfies the bound

$$d_p \leq 1 - \frac{Y_1 - Y_0}{E_R} \tag{3.39}$$

where Y_1 is the minimum value of the sample at the output of the integrate-and-dump filter when the transmitted symbol in the considered period is binary 1, Y_0 the maximum value when the transmitted symbol is binary 0, and E_R the received peak energy. We have normalized the eye opening with respect to the received energy so as to consider only the effects of fiber dispersion and not those of fiber attenuation.

To obtain the system penalty, we must evaluate Y_1 and Y_0. In case of binary 1 transmitted, the minimum amplitude at the output of the integrate-and-dump filter corresponds to a burst of binary 1s preceding and following the symbol at hand, assuming that all interfering pulses are added to the useful sample with a phase difference of π. With these hypotheses, Y_1 turns out to be (see Problem 3.2)

$$Y_1 \geq \int_{-T/2}^{T/2} |s_R(t,L)|^2\, dt - 2\int_{-T/2}^{T/2} |s_R(t,L)| \sum_{n\neq 0} |s_R(t-nT,L)|dt + \int_{-T/2}^{T/2} \left(\sum_{n\neq 0} |s_R(t-nT,L)| \right)^2 dt \tag{3.40}$$

where $s_R(t, L)$ is the received pulse (3.16) for $z = L$.

Similarly, in the case of binary 0 transmitted, the expression for Y_0 is (see Problem 3.2)

$$Y_0 \leq \int_{-T/2}^{T/2} \left(\sum_{n\neq 0} |s_R(t,L)| \right)^2 dt \tag{3.41}$$

Substituting (3.40) and (3.41) into (3.39), using expression (3.16) for $s_R(t, z)$ and performing the integrations, we obtain the following upper bound on the peak distortion (see Problem 3.2):

$$d_p \leq \operatorname{erfc}(\xi) + 2\sum_{n=1}^{+\infty} \exp(-n^2\xi^2)\,\{\operatorname{erf}[(n+1)\xi] - \operatorname{erf}[(n-1)\xi]\} \tag{3.42}$$

where

$$\xi = \frac{T}{2\tau_T\sqrt{(1+cZ)^2 + Z^2}} \tag{3.43}$$

In (3.43), c is the chirp parameter defined in (3.6), and Z is the normalized distance (already defined in (3.21)):

$$Z = \frac{z}{L_D} \tag{3.44}$$

Assume now that the detected current is corrupted by additive Gaussian noise, $n(t)$, independent from the signal for simplicity, with zero mean and variance σ_n^2 (this assumption will be discussed in detail in Section 3.4, and corresponds to a receiver in which thermal noise is dominant on shot noise). In the absence of intersymbol interference, assuming perfect synchronization and a decision threshold equal to $E_R/2$, the bit error probability is easily evaluated using the results of Section 2.8; the result is

$$P_e = \frac{1}{2}\,\mathrm{erfc}\left(\frac{E_R}{2\sigma_n}\right) \tag{3.45}$$

where $\mathrm{erfc}(z)$ is the complementary error function, defined as

$$\mathrm{erfc}(z) \overset{\mathrm{def}}{=} \frac{2}{\sqrt{\pi}} \int_z^{\infty} e^{-x^2} dx$$

Removing the assumption of no intersymbol interference and using the peak distortion defined in (3.39) to evaluate the eye closure, we obtain a simple upper bound to the bit error probability in the presence of intersymbol interference:

$$P_e \leq \frac{1}{2}\,\mathrm{erfc}\left[\frac{E_R(1 - d_p)}{2\sigma_n}\right] \tag{3.46}$$

Finally, we can relate the peak distortion (and, consequently, the fiber length, L) to the SNR (signal-to-noise ratio) (or, equivalently, sensitivity) penalty, δ_{SNR}, in decibels:

$$\delta_{SNR} = -20 \log_{10}(1 - d_p) \tag{3.47}$$

Through (3.42), the penalty, δ_{SNR}, depends on the normalized variable, ξ, defined in (3.43), that includes all system parameters that influence the performance: link length z, dispersion length (and, through it, fiber dispersion), information rate $R_b = 1/T$, transmitted pulse width τ_T, and chirp parameter c. Thus, for a given system, we first compute ξ; then, from it, d_p; and, finally, the SNR penalty. To facilitate those steps, we have plotted in Figure 3.13 the function $\delta_{SNR}(\xi)$ implicitly defined through (3.47) and (3.42).

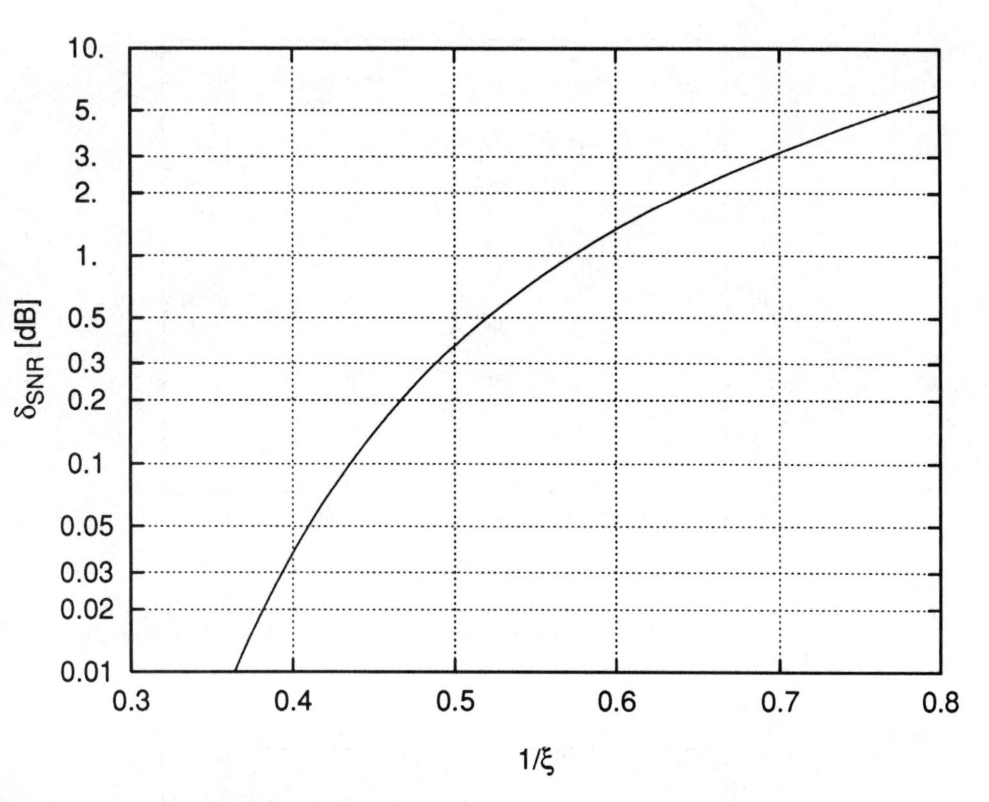

Figure 3.13 SNR penalty due to fiber dispersion versus the normalized parameter $1/\xi$.

Example 3.2

Consider a system that operates at a wavelength of 1,550 nm, uses a conventional single-mode fiber with $\beta = -20$ ps^2/km, and transmits at a rate of 5 Gbps. The transmitter employs a DFB (distributed feedback) laser and an external modulator, so we can assume $c = 0$. The pulse width, τ_1, is 50 ps, so that almost all pulse energy is contained within T. Let us evaluate the SNR penalty due to dispersion after a propagation length of 100 km. Using the system parameters in (3.43), we obtain $\xi \simeq 1.56$ and $1/\xi = 0.64$. Using this value as abscissa in the curve of Figure 3.13, we obtain a penalty $\delta_{SNR} \simeq 1.5$ dB.

This analysis permits us to evaluate the influence of various system parameters on the maximum length of the link for a given dispersion penalty. Figures 3.14 and 3.15 show the dependence of the link length on the width of the transmitted pulse

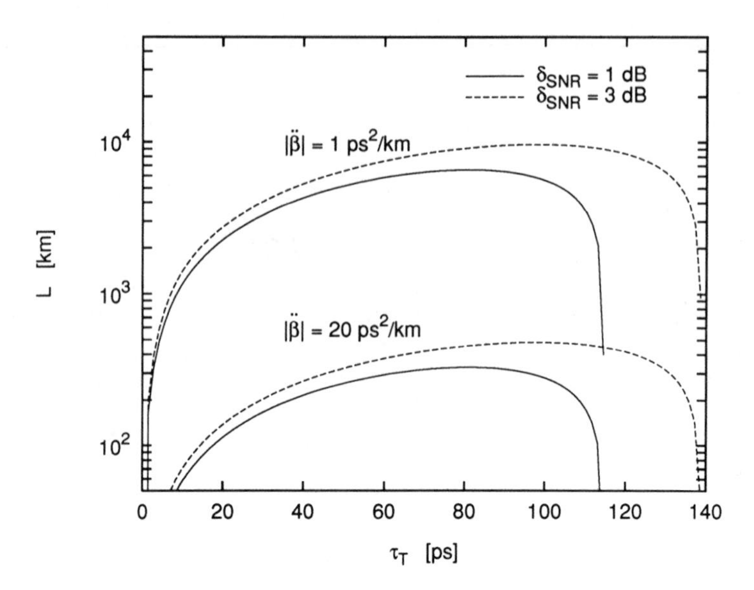

Figure 3.14 Maximum link length, L, versus transmitted Gaussian-shaped pulse width for a bit rate $R_b = 2.5$ Gbps, two values of dispersion parameter $|\ddot{\beta}|$ and two values of the SNR penalty, δ_{SNR}.

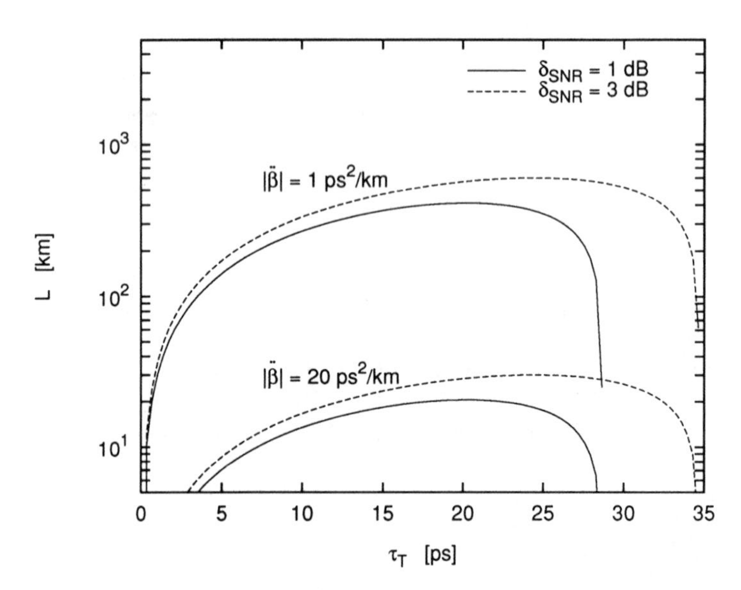

Figure 3.15 Maximum link length, L, versus transmitted Gaussian-shaped pulse width for a bit rate $R_b = 10$ Gbps, two values of dispersion parameter magnitude $|\ddot{\beta}|$, and two values of the SNR penalty, δ_{SNR}.

(corresponding to a given percentage of energy contained in the symbol interval), for two bit rates of 2.5 and 10 Gbps, respectively. As the figures show, there is an optimum transmitted pulse width for each value of $\ddot{\beta}$. We have chosen here two numerical values of $|\ddot{\beta}|$ typical for a conventional (20 ps^2/km) and dispersion-shifted (1 ps^2/km) single-mode fiber at 1,550 nm. A physical explanation of the optimum value is as follows: When the transmitted input pulse is too narrow, the impact of broadening is stronger. On the other hand, if the input pulse is too long, even a small propagation broadening is enough to induce strong intersymbol interference. In the following discussion, we always optimize the transmitted pulse width.

Figure 3.16 shows the dependence of the link length, L, on the magnitude of the dispersion parameter, $|\ddot{\beta}|$, for the two bit rates, 2.5 and 10 Gbps. As intuition suggests, increasing the dispersion decreases the maximum link length.

To obtain curves comparable to the heuristic ones derived in Section 3.3.2, Figure 3.17 plots the maximum link length as a function of the system bit rate.

To see the effect of chirp on performance, in Figure 3.18 we plot the maximum link length for a 1-dB penalty versus the bit rate for three values of the chirp parameter, $c = -1, 0, 1$. The curves show that $c \neq 0$ of the same sign as $\ddot{\beta}$ is detrimental, whereas a c opposite in sign can improve the system performance. Unfortunately, c is negative for semiconductor lasers, and $\ddot{\beta}$ is also negative for standard single-mode fiber at 1,550 nm.

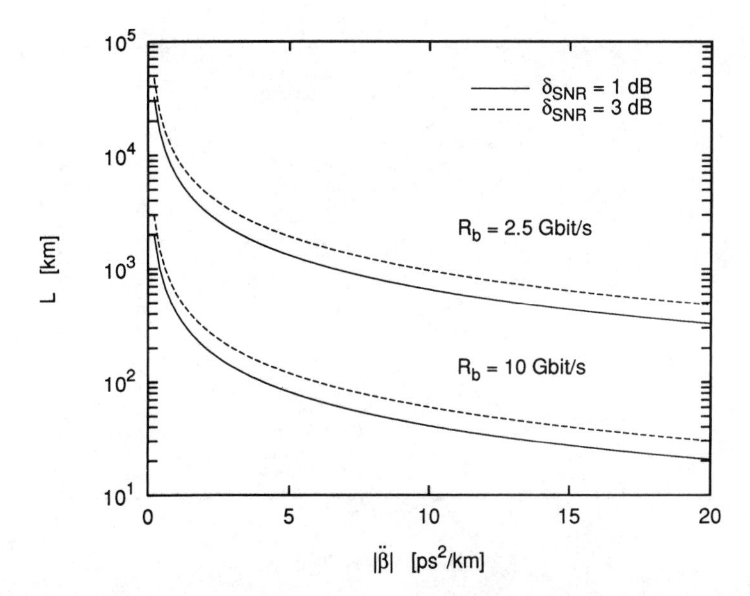

Figure 3.16 Maximum link length, L, versus dispersion parameter magnitude, $|\ddot{\beta}|$, for two bit rates and two values of the SNR penalty, δ_{SNR}.

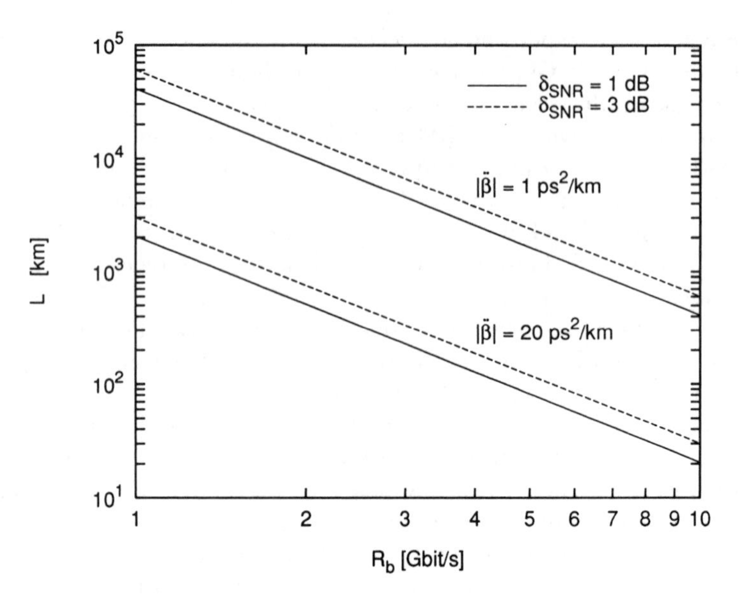

Figure 3.17 Maximum link length, L, versus bit rate, R_b, for transmitted Gaussian-shaped pulses with optimized width for two values of dispersion parameter magnitude $|\ddot{\beta}|$ and two values of the SNR penalty, δ_{SNR}.

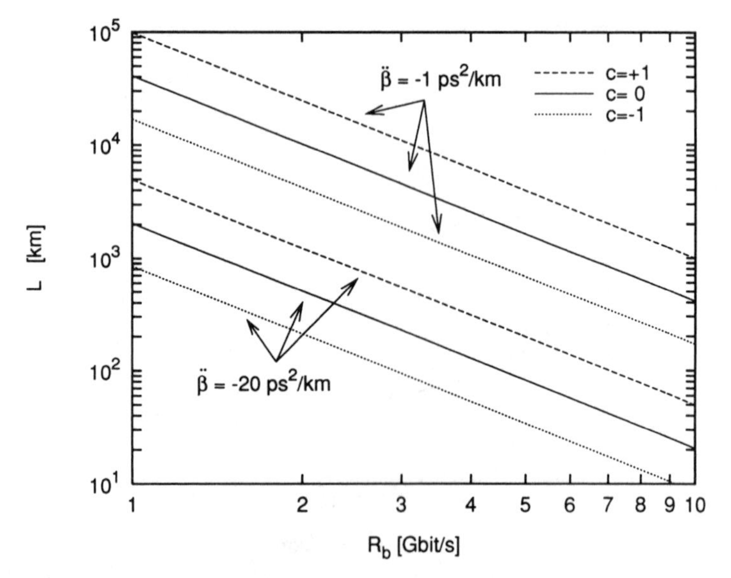

Figure 3.18 Maximum link length, L, versus bit rate, R_b, for two values of dispersion parameter, $\ddot{\beta}$, and three values of the chirp parameter, c. The SNR penalty is δ_{SNR} = 1 dB.

Finally, comparison of the curves of Figure 3.17 with the heuristic curves of Figure 3.12 for the case of 1,550 nm shows that the latter were somewhat optimistic. For example, for R_b = 10 Gbps, we obtain from Figure 3.17 a distance of 20 km, whereas Figure 3.12 yields over 50 km.

3.3.4 Combined Attenuation and Dispersion Limitations

3.3.4.1 Heuristic Limitations

We will now plot the maximum link length as a function of the bit rate, taking into account both fiber attenuation and dispersion. That will permit us to distinguish the two regimes of a system: the attenuation-limited regime and the dispersion-limited regime.

In Figure 3.19, we have merged the attenuation limits of Figure 3.11 and the heuristic dispersion limits of Figure 3.12. Let us start following the abscissas of Figure 3.19 from the left, that is, from the lower bit rates. All systems are first limited by attenuation. Then, as the bit rate increases, the curves come to a point where the limits imposed by the attenuation and those due to the dispersion equal each other (this can be spotted in the figure as the points where curves change slope). From that point

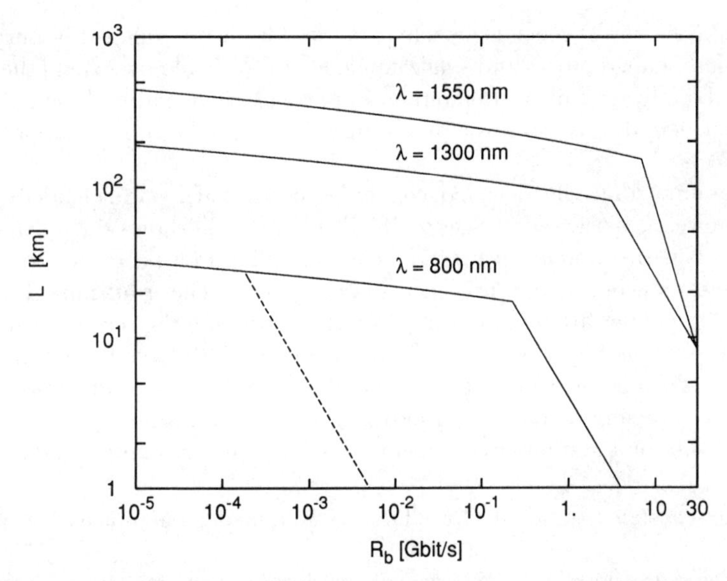

Figure 3.19 Merge of attenuation and heuristic dispersion limits showing the maximum link length, L, versus the bit rate, R_b, for three different wavelengths. The parameter values are the same as in Figure 3.11 and Figure 3.12.

on, the systems become dispersion limited. Figure 3.19 reflects three generations of optical communication systems:

- The first generation used multimode fibers and GaAs LEDs emitting in the first window (850 nm). These systems become dispersion limited at $R_b \cdot L$ products of a few megabits per second per kilometer for step-index fibers and a few gigabits per second per kilometer for graded-index fibers.
- The second generation used single-mode fibers and multi-longitudinal-mode lasers in the second window (1,300 nm), which become dispersion limited at $R_b \cdot L$ products of hundreds gigabits per second per kilometer.
- The third generation, finally, employs single-mode fibers and single-longitudinal-mode lasers in the third window (1,550 nm). These systems become dispersion limited at $R_b \cdot L$ products of a few terabits per second per kilometer.

Dispersion limits in Figure 3.19 do not take into account the effects of laser chirp, which imposes more severe system limitations.

The attenuation limit can be further increased in third-generation systems employing dispersion-shifted fibers (see Section 1.2.10), in which the zero-dispersion point is moved to 1,550 nm.

3.3.4.2 Error Probability Limitations

In Section 3.3.3, we analyzed dispersion penalties for narrow-linewidth sources emitting Gaussian-shaped pulses and single-mode fibers. We will now extend the analysis to include the effects of fiber attenuation. In particular, we will find the relationship between link length and bit error probability, including both attenuation and dispersion effects.

We assume that the detected current is corrupted by independent additive Gaussian noise, as discussed in Section 3.3.3. Figure 3.20 shows the maximum link length L, versus the system bit rate, R_b, corresponding to a bit error probability of 10^{-9}. The following assumptions have been used: The transmitted power is $\overline{P}_T = 0$ dBm, the fiber attenuation is $\alpha_{dB} = 0.2$ dB/km, and the receiver sensitivity is $\overline{N}_R = 1{,}000$ photons/bit. The results shown in Figure 3.20 have been obtained (see Problem 3.3) using the worst case bound to the error probability based on the minimum eye opening defined by Y_1 and Y_0 in (3.40) and (3.41).

Three different system environments are illustrated in Figure 3.20:

- The first system uses an idealized fiber without dispersion; it is used as reference only.
- The second system employs a dispersion-shifted fiber with $|\ddot{\beta}| = 1$ ps^2/km. We notice from the results that the dispersion effects become noticeable for bit rates greater than 15 Gbps.

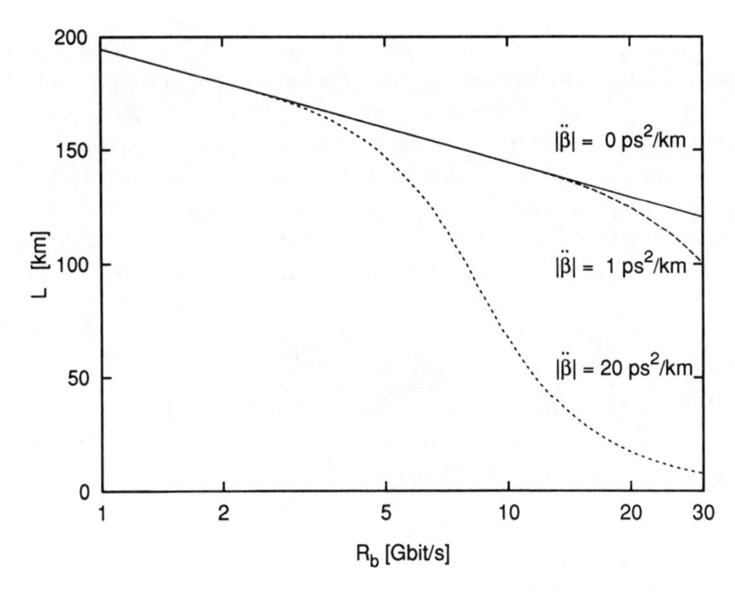

Figure 3.20 Maximum link length, L, versus bit rate, R_b, yielding a bit error probability of 10^{-9}. The curves take into account both attenuation and dispersion limitations for three values of dispersion parameter magnitude, $|\ddot{\beta}|$. Fiber attenuation is 0.2 dB/km, transmitted power 0 dBm, and receiver sensitivity 1,000 photons/bit. The source has narrow linewidth, and Gaussian-shaped pulses are assumed.

- The third system uses a conventional fiber with $|\ddot{\beta}| = 20$ ps^2/km. In this case, dispersion becomes the dominant system limitation for bit rates greater than 5 Gbps.

The results shown in Figure 3.20 suggest several important system level conclusions. Achieving long distances at high bit rates is not possible using conventional fibers at 1,550 nm, unless dispersion management is used (see Chapter 1). In that case, in fact, the system is dispersion limited, and optical amplification and coherent detection are of little help. On the other hand, systems that use dispersion-shifted fibers (and external modulators to avoid chirp) are attenuation limited up to almost 20 Gbps; in such systems, optical amplifiers and coherent detection can be used to extend the transmission distance.

These topics will be treated further in Chapters 4 and 5.

3.3.5 Attenuation and Dispersion Limits in Optically Amplified Links

We will consider now the effects of fiber attenuation and dispersion in optically amplified links; the *amplifier spontaneous emission* (ASE) noise due to optical amplifier(s) is assumed to be the dominant noise source. We neglect all other sources of impairment, such as fiber nonlinearities and limited amplifier bandwidth.

The receiver model is shown in Figure 3.21; $y(t, z)$ is the received optical field after propagating a fiber length z, and $n(t)$ is a complex white Gaussian noise with two-sided spectral density, $G_n/2$, due to ASE. The value of $G_n/2$ will be specified according to the link type. The receiver includes an optical filter with bandwidth B_{opt}, an ideal photodiode, and an integrate-and-dump electrical filter followed by the decision circuit that compares the integrated signal with a threshold and makes decisions accordingly.

We make the following assumptions [17]: (1) Both optical and electric filters are integrating filters over T_{opt} and $T_{el} = T$ seconds, respectively, and (2) the ratio $b_r \stackrel{\text{def}}{=} T_{el}/T_{opt}$ is sufficiently large (it is always true in practice).

The signal d_0 at the input of the decision circuit is a χ^2 RV (see Section 2.3) with mean and variance given by (see Problem 3.4)

$$E[d_0] = 2G_n b_r + \int_{t_0-T/2}^{t_0+T/2} |y(t, L)|^2 dt$$

$$\sigma_{d_0}^2 = (2G_n)^2 b_r + 4G_n \int_{t_0-T/2}^{t_0+T/2} |y(t, L)|^2 dt \tag{3.48}$$

In (3.48), the first two terms on the right side (which depend on b_r) account for the ASE · ASE product induced by the quadratic photodetection process, and the second term in the variance expression is due to the signal · ASE product.

To simplify the analysis, we assume that d_0 is a Gaussian random variable. Let us define an RV, Y, accounting for dispersion:

$$Y \stackrel{\text{def}}{=} \int_{t_0-T/2}^{t_0+T/2} |y(t, L)|^2 dt \tag{3.49}$$

where $y(t, L)$ is the received signal, including intersymbol interference due to fiber dispersion.

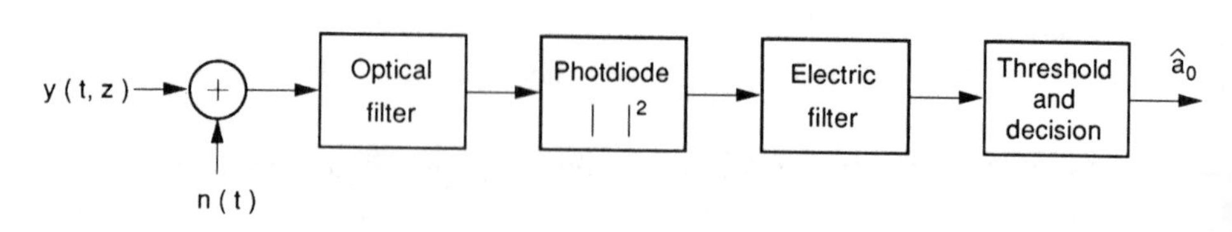

Figure 3.21 Block diagram of the receiver for optically amplified IM-DD links.

The error probability is then given by

$$P(e) = \min_{V_{th}} \mathrm{E}_Y[P(e, V_{th}| Y)] \tag{3.50}$$

where V_{th} is the decision threshold, and E_Y means average with respect to the RV Y. The exact evaluation of (3.50) requires repeated use of Gauss quadrature rules (see Appendix B) to perform the averaging with respect to the intersymbol interference variable Y, and optimization of V_{th}.

Instead, we derive a simpler upper bound to $P(e)$ based on the worst-case value of Y. This is the so-called *worst-case bound* (see [14, Chap. 6]). Recalling from Section 3.3.3 the definitions (3.41) and (3.40) of Y_0 and Y_1, corresponding to the levels of maximum closure of the eye diagram, we obtain

$$\min_{\mathbf{a}}(Y|a_0 = 1) = Y_1, \qquad \max_{\mathbf{a}}(Y|a_0 = 0) = Y_0$$

where $\mathbf{a}$ denotes the sequence of information symbols, a_n.

Let us choose the threshold value, $V_{th,s}$, that makes equal the two conditional error probabilities:

$$P(e, V_{th,s}|a_0 = 1) = P(e, V_{th,s}|a_0 = 0) \tag{3.51}$$

This yields the following bit error probability (see Problem 3.4):

$$P_b(e) \leq \frac{1}{2}\,\mathrm{erfc}\left(\frac{Y_1 - Y_0}{2\sqrt{2}(\sqrt{G_n^2 b_r + G_n Y_1} + \sqrt{G_n^2 b_r + G_n Y_0})}\right) \tag{3.52}$$

We will use now (3.52) to find the maximum link length, L, as a function of the bit rate, under the constraint that $P(e) \leq 10^{-9}$. To compare the results with the previous ones obtained for nonamplified links, we consider the same three values of the dispersion parameter $|\ddot{\beta}| = 0, 1, 20$ ps^2/km as in Fig. 3.20. Other system parameters are as follows:

- $\lambda = 1{,}550$ nm, fiber attenuation = 0.2 dB/km;
- Gaussian pulses with transmitted width τ_T optimized to achieve the lowest error probability;
- Optical amplifiers' noise figure: $F = 4$ dB. As will be discussed in Chapter 5, the noise figure is related to the spontaneous emission factor, n_{sp}, through

$$F = 2n_{sp}$$

All other noise sources are assumed to be negligible.

- Receiver optical filter bandwidth: $B_{opt} = 1/T_{opt} = 125$ GHz, corresponding to 1 nm.

Both single-hop and multihop links will be considered.

Dispersion-shifted fibers enhance the nonlinear phenomena. For that reason, the low dispersion of long-haul optically amplified links is normally obtained with a dispersion map that employs in each hop a longer span of dispersion-shifted fiber at a wavelength ensuring a slightly positive $\ddot{\beta}$, cascaded with a shorter span of standard fiber with a highly positive $\ddot{\beta}$ [18]. Thus, $\ddot{\beta}(z)$ is not constant along the link. The following results, however, assume that the power density in the fiber is low enough not to produce nonlinear effects, and thus they still hold for nonconstant $\ddot{\beta}$, provided that

$$\ddot{\beta} = \frac{1}{L}\int_0^L \ddot{\beta}(z)\, dz \tag{3.53}$$

3.3.5.1 Single-Hop Link

The first system we will consider is a single-hop link with an optical booster amplifier in the transmitter and an optical preamplifier in the receiver. The launched power is 7 dBm (just below the threshold of nonlinear phenomena). For a single-hop link, neglecting the noise due to booster amplifier with respect to that of the receiver preamplifier, ASE noise spectral density is given by (see Chapter 5)

$$\frac{G_n}{2} = n_{sp} h\nu$$

The maximum link length yielding an error probability of 10^{-9} is plotted in Figure 3.22 as a function of the bit rate. The curves show that for $|\ddot{\beta}| = 20$ ps^2/km the dispersion becomes the dominant impairment for bit rates higher than 2.5 Gbps. Large improvements are obtained by using a dispersion shifted fiber (or an appropriate dispersion map). For $|\ddot{\beta}| = 1$ ps^2/km, the dispersion effects become relevant only for bit rates greater than 15 Gbps.

Comparison of the curves in Figure 3.22 with the analogous curves of Figure 3.20 (Figure 3.20 shows the case of no amplification) reveals the great transmission distance improvement due to optical amplifiers. The improvement is due to the higher transmitted power and higher receiver sensitivity (lower value of N_R).

3.3.5.2 Multihop Links

When the actual distance between the transmitter and the receiver is beyond the one achievable with a single-hop link, one can regenerate the signal through optical-to-

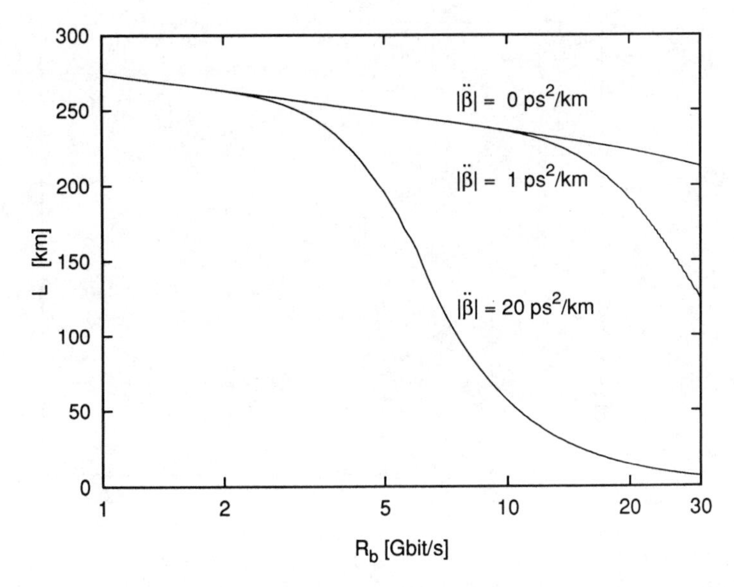

Figure 3.22 Maximum link length, L, versus bit rate, R_b, yielding a bit error probability of 10^{-9} for single-hop optically amplified systems at 1,550 nm. The curves take into account both attenuation and dispersion limitations for three values of dispersion parameter magnitude, $|\ddot{\beta}|$. Fiber attenuation is 0.2 dB/km and transmitted power 7 dBm. The preamplifier has a noise figure of 4 dB, and the receiver uses an optical filter with 1-nm bandwidth.

electronic conversion, demodulation, and electronic-to-optical conversion, after the signal travels a maximum distance dictated by previous results. Alternatively, one can insert properly spaced optical amplifiers.

Long multihop amplified links without regeneration, like the latest generation of trans-Atlantic and trans-Pacific optical systems, pose serious problems to the designer, because the dispersion acts on the whole multihop link length (which can amount to many thousands of kilometers).

Below, we analyze a multihop system that makes use of in-line optical amplifiers, with a constant spacing of 50 km. It is designed to keep the signal power at the input of each amplifier at −13 dBm, and the amplifier gains are set to exactly compensate the fiber attenuation. For an N_h-hop link (i.e., a link using $N_h + 1$ optical amplifiers), ASE noise spectral density at the receiver photodiode is equal to

$$\frac{G_n}{2} = (N_h + 1)n_{sp}h\nu$$

In Figure 3.23, the overall maximum link length is plotted as a function of the

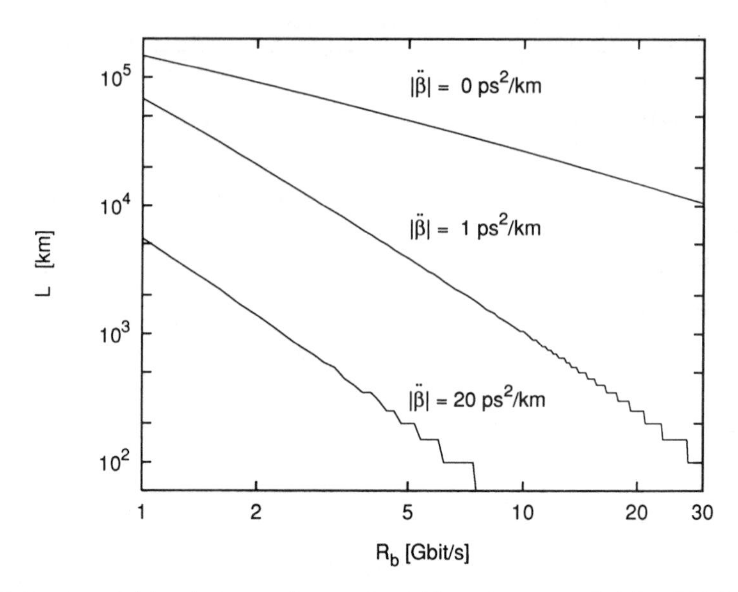

Figure 3.23 Maximum link length, L, versus bit rate, R_b, yielding a bit error probability of 10^{-9} for multihop optically amplified systems at 1,550 nm. The curves take into account both attenuation and dispersion limitations for three values of dispersion parameter magnitude, $|\ddot{\beta}|$. Fiber attenuation is 0.2 dB/km, the hop length is equal to 50 km, the amplifier gain compensates fiber attenuation, and the signal power at amplifiers input is -13 dBm.

bit rate on a log-log scale. Due to the long distances involved, dispersion effects become extremely important, as one may anticipate. For bit rates above 5 Gbps, conventional fibers cannot be used, because the signal cannot go beyond the first hop. With the dispersion of 1 ps/km, on the other hand, one can reach 6,000 km at 5 Gbps. To go further, with both conventional and dispersion-shifted fibers, dispersion-compensation techniques must be used.

3.4 RECEIVER CHARACTERISTICS

The goal of the receiver is to extract with high reliability the transmitted information from the received optical signal. The main receiver elements are shown in the block diagram of Figure 3.24. The incident optical power is converted to electronic domain by either a PIN or an APD. The electronic signal generated by the photodiode enters a low-noise preamplifier followed by a power amplifier that includes an automatic gain control. The equalizer-shaping filter is used to eliminate or at least reduce intersymbol interference. Then the signal is sampled every T seconds and compared with a threshold to make decisions on the transmitted bits. Sampling is driven by a timing

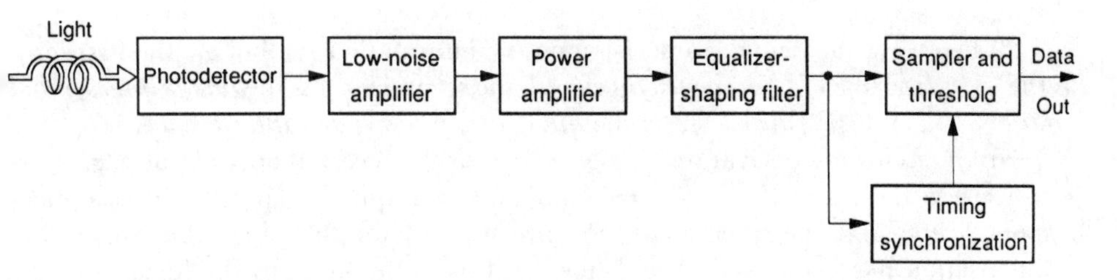

Figure 3.24 Block diagram of the receiver of an IM-DD optical binary communication system.

synchronization circuit that provides the clock frequency $1/T$ and a suitable timing phase.

We will examine each receiver block with the aim of characterizing it from a system point of view. We start with a statistical noise characterization of the photodetector, which will permit us to derive the fundamental limit to the receiver performance known as the *quantum limit.* Then we procede to the equivalent circuit of the linear part of the receiver, composed of the low-noise amplifier, the power amplifier, and the equalizer-shaping filter. The noise generated by the low-noise amplifier is shown to influence significantly system performance. Finally, we derive an expression for the bit error probability as a function of the detected signal and system noises and use that expression to obtain the receiver sensitivity.

The received optical field has the form

$$y(t) = \sum_{n=-\infty}^{\infty} a_n s_R(t - nT) \tag{3.54}$$

where a_n are the information symbols, binary RVs assuming with equal probability the values 0 and 1, and $s_R(t)$ is the received optical pulse. The detected current (see Section 1.2.2) is proportional to the incident optical power, $P_R(t)$, which is proportional to the magnitude squared of the received field (3.54). Thus, omitting the proportionality constant, we have

$$\begin{aligned} P_R(t) = |y(t)|^2 &= \sum_{n=-\infty}^{\infty} \sum_{m=-\infty}^{\infty} a_n a_m \, s_R(t - nT) s_R^*(t - mT) \\ &= a_0 |s_R(t)|^2 + \sum_{n\neq 0, n=-\infty}^{\infty} a_n |s_R^2(t - nT)| + \sum_{n=-\infty}^{\infty} \sum_{m\neq n} a_n a_m s_R(t - nT) s_R^*(t - mT) \end{aligned} \tag{3.55}$$

where the symbol * means *complex conjugate.*

When we want to make a decision on the information symbol a_0, the first term in the right side of (3.55) is the *useful signal*, the second term is the *linear intersymbol interference*, and the third term is the *quadratic intersymbol interference.*

In the following derivations, we assume that the received optical pulse, $s_R(t)$, is zero outside the interval $(0, T)$. This assumption requires that the source optical response time and the fiber dispersion are low or, for the fiber, that dispersion compensation has been used [19]. Under that assumption, both the linear and the quadratic intersymbol interferences disappear, because the instantaneous received power in the interval $(0, T)$ is simply $P_R(t) = a_0|s_R(t)|^2$. The general analysis, including intersymbol interference, is quite cumbersome. If needed, it can be done by using (3.55) as a starting point. We will see that the receiver performance depends on, besides thermal noise, filtered and sampled versions of $P_R(t)$, which are easily derived from (3.55).

Details of the general analysis can be found in [1,20–22]. References [1,20] make the Gaussian hypothesis on the overall receiver noise, whereas [21] is completely general. All four references, however, assume that the fiber behaves linearly with respect to the launched power, that is that the output power can be obtained through a convolution of the input power with the fiber impulse response. As a consequence, only the linear intersymbol interference is taken into account. The hypothesis does not hold true, especially for narrow-linewidth light sources and single-mode fibers.

Thus, we assume a received optical power of the form

$$P_R(t) = 2\overline{P}_R \sum_{n=-\infty}^{\infty} a_n p(t - nT) \tag{3.56}$$

where the power pulse, $p(t)$, is

$$p(t) \stackrel{\text{def}}{=} \frac{1}{2\overline{P}_R} |s_R(t)|^2$$

Without loss of generality, we also assume that

$$\frac{1}{T}\int_0^T p(t) = 1 \tag{3.57}$$

so that $\overline{P}_R$ in (3.56) represents the *average received power* in each symbol interval.

3.4.1 Quantum Limit to Receiver Sensitivity

In this section, we will show that a receiver acting on an intensity-modulated light has a performance limit set by the granular nature of the light.[2]

Let us assume an ideal receiver consisting of a device able to count the received photons, sensitive to the point that even a single received photon is detected. Consider now the interval $(0,T)$ and the reliability of the decision $\hat{a}_0$ on the transmitted RV a_0.

According to (3.56), when $a_0 = 0$, the received power is zero and no photons are detected. Thus, the conditional bit error probability $P_e(0) \overset{\text{def}}{=} P[\hat{a} = 1 | a_0 = 0]$ is zero, and the average bit error probability $P(e)$ is equal to $P_e(1)/2$.

To evaluate $P_e(1)$, we recall the statistical properties of the incoming light. As seen in Section 2.5, the photon arrival is governed by a doubly stochastic Poisson process with intensity

$$\lambda_R(t) = \frac{P_R(t)}{h\nu} \tag{3.58}$$

where $h = 6.62559 \cdot 10^{-34}$ J (Planck's constant) and ν is the light frequency. The probability of receiving n photons in the interval $(0,T)$ is given by the Poisson distribution

$$P[N_R(T) = n] = \frac{\left[\int_0^T \lambda(t)dt\right]^n}{n!} \cdot e^{-\int_0^T \lambda(t)dt}, \qquad n = 0,1, \ldots \tag{3.59}$$

Now, substitute (3.58) into (3.59), take into account (3.56), and define $\overline{N}_R$ as the average number of received photons in the interval $(0,T)$

$$\overline{N}_R = \frac{\overline{P}_R T}{h\nu} \tag{3.60}$$

Then (3.59) yields

$$P[N_R(T) = n] = \frac{(2a_0\overline{N}_R)^n}{n!} e^{-2a_0\overline{N}_R}, \qquad n = 0,1, \ldots \tag{3.61}$$

2. Certain nonclassical states of light, called *squeezed states*, could be used in principle to achieve better performance. Unfortunately, these particular states degenerate in the presence of optical losses; for that and other reasons, their practical use seems unlikely [23].

The receiver makes an error when a binary 1 is transmitted ($a_0 = 1$), and the number of received photons is zero, that is,

$$P(e) = \frac{1}{2} P[N_R(T) = 0|a_0 = 1] = \frac{1}{2} e^{-2\overline{N}_R} \tag{3.62}$$

The sensitivity of the ideal receiver, or *quantum limit* sensitivity, depends on the required value of $P(e)$; for $P(e) = 10^{-9}$, the quantum limit sensitivity is

$$\overline{N}_{Rm} \simeq 10 \text{ photons/bit}$$

We will see later in this chapter that the behavior of practical receivers can be significantly worse than that predicted by the quantum limit. But first we need to characterize receiver noise sources.

3.4.2 Detected Signal and Shot Noise

As can be seen from Figure 3.24, the received optical power impinges on the photodetector, a PIN or APD inversely biased, as shown in Figure 3.25. In the figure, we also show the equivalent circuit of the photodiode, including the linear part of the receiver. The photodiode is represented by a current generator, and the overall receiver transfer function is generally a *transimpedance* that transforms the input current in an output voltage. As a consequence, the transfer function has the dimension of Ω. Because we prefer to deal with dimensionless transfer functions and impulse responses, we split the transfer function into the cascade of a resistance, R_t, transforming the input current into a voltage, $v_i(t)$, and a dimensionless transfer function, $H_L(f)$. The resistance, R_t, includes the bias resistance, R_{bias}, the photodiode differential resistance, and the input resistance of the first amplifier, whereas the photodiode differential capacitance has been included in the transfer function, $H_L(f)$ (L standing for *linear*), which also accounts for the linear transfer function of the low-noise amplifier, power amplifier, and equalizer-shaping filter in Figure 3.24.

The photodiode gives rise to three currents: the first, $i_s(t)$, represents the signal component; the second is the *dark* current, i_d; and the third is the shot noise, $i_{sh}(t)$.

The photodetection mechanism can be better understood using Figure 3.26, where the optoelectrical conversion has been split into three parts. The first part accounts for the photodiode quantum efficiency, η, defined as the percentage of optical incident power that effectively produces the photocurrent. The second part is the photodetection, in which the impinging field produces primary electron-hole pairs.

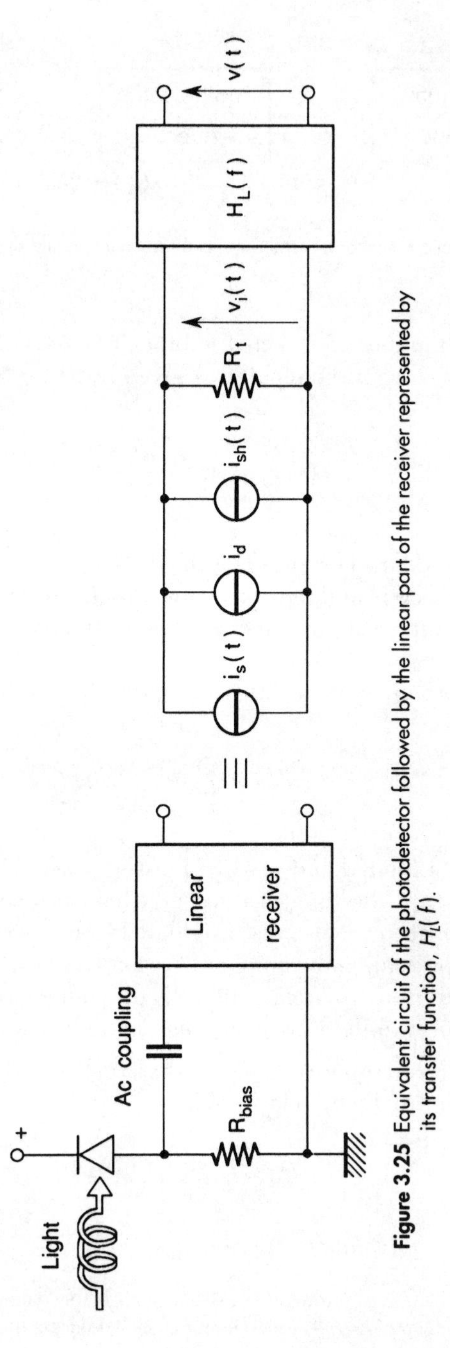

Figure 3.25 Equivalent circuit of the photodetector followed by the linear part of the receiver represented by its transfer function, $H_L(f)$.

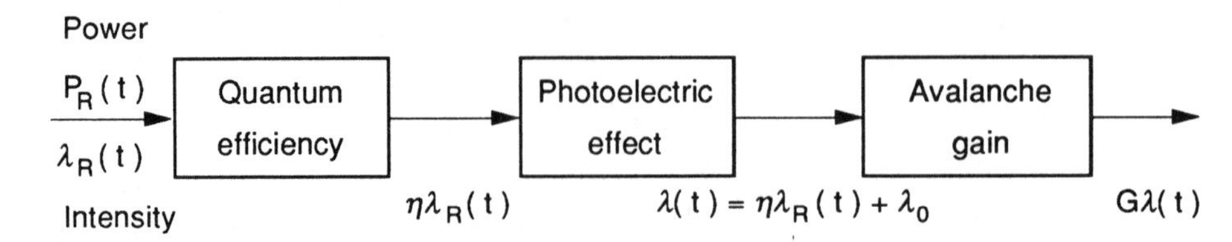

Figure 3.26 Photodetection mechanism split into its three conceptually distinct steps.

Using (3.58), the number of detected photons per second (equal to the number of generated primary electron-hole pairs) is given by

$$\lambda(t) = \eta\lambda_R(t) = \frac{\eta P_R(t)}{h\nu} + \lambda_0 \tag{3.63}$$

where λ_0 represents dark current. The third part of Figure 3.26 is present only in avalanche photodetectors and represents the amplification due to the avalanche multiplication. It is a statistical phenomenon: Each primary pair produces on average G secondary pairs.

The photocurrent, as shown in Examples 2.3 and 2.4, has the form

$$i(t) = \sum_k g_k i_0(t - t_k) \tag{3.64}$$

The current $i(t)$ is a filtered and marked Poisson process (described in Section 2.5). The moments $\{t_k\}$ are the time instants at which primary electron-hole pairs are formed; $i_0(t - t_k)$ is the current pulse originated from the electron-hole pair at $t = t_k$; and g_k is an RV representing the number of secondary pairs induced by the avalanche mechanism at t_k (the particular case of a PIN photodiode corresponds to $g_k = 1$). The sequence of $\{g_k\}$ forms a discrete wide-sense stationary process. The (time-independent) average and mean square of g_k are $G \stackrel{\text{def}}{=} \mathrm{E}[g_k]$ and $G_2 \stackrel{\text{def}}{=} \mathrm{E}[g_k^2]$; they are related by the following approximate relationship:

$$G_2 = G^{2+x}, \qquad x > 0 \tag{3.65}$$

where x is the *excess avalanche noise factor*.[3]

3. The reader should be aware of the fact that we changed here the notations with respect to Section 1.2.2, where G was named M and $G^x = F$. The need for the change resides in the fact that F is used in this chapter, as a standard notation, to denote the noise figure.

Let us assume that the current pulse, $i_0(t)$, has a duration much smaller than that of the power pulse, $p(t)$, so that

$$i_0(t) = q\delta(t) \tag{3.66}$$

where $q = 1.60210^{-19}$ C is the electron charge. Using the generalized Campbell theorem, introduced in Section 2.5, we can obtain the characteristics of the three currents produced by the photodetector. The signal current is

$$i_s(t) = G[\lambda(t) - \lambda_0] * i_0(t) = GRP_R(t) = 2GR\overline{P}_R \sum_{n=-\infty}^{\infty} a_n p(t - nT), \qquad [\text{A}] \tag{3.67}$$

where $R = (\eta q)/(h\nu)$ is the *responsivity* of the photodetector.

The dark current is

$$i_d = G\lambda_0 * i_0(t) = G\lambda_0 q, \qquad [\text{A}] \tag{3.68}$$

The shot noise current is a random process whose autocorrelation function can be found as shown in Section 2.5; it is given by

$$R_{i_{sh}}(t_1,t_2) \stackrel{\text{def}}{=} \text{E}[i_s(t_1)i_s(t_2)] = G_2 q^2 \lambda(t_1)\delta(t_1 - t_2) \tag{3.69}$$

Equation (3.69) depends on both t_1 and t_2 and not only on the difference $t_1 - t_2$; thus, $i_s(t)$ is, in general, a nonstationary random process. This contrasts with the model for shot noise adopted in most textbooks, where it is assumed to be a white random process.

We now show the hypotheses on which this model is based. Assuming a constant received power[4] $P_R(t) = 2\overline{P}_R$, the intensity of the Poisson process underlying the received power is

$$\lambda = 2R\overline{P}_R/q + \lambda_0 \tag{3.70}$$

The autocorrelation function, in turn, is given by

$$R_{i_{sh}}(t_1,t_2) = G_2 q(2R\overline{P}_R + q\lambda_0)\delta(t_1 - t_2) \tag{3.71}$$

4. The received power is constant, for a given transmitted symbol, only in the absence of intersymbol interference and for a constant waveform, $p(t)$.

and corresponds to a stationary white random process with two-sided (we always consider two-sided spectral densities) spectral density:

$$G_{i_{sh}}(f) = G_2 q(2R\overline{P}_R + q\lambda_0), \qquad -\infty < f < \infty \tag{3.72}$$

Consider the voltage, $v(t)$, at the output of the linear receiver in Figure 3.25. It can be decomposed into three components. The first two are the signal, $v_s(t)$, and the dark voltage, v_d; they are derived by filtering the corresponding currents. Let us define the equivalent impulse response, $r(t)$, as the convolution of the power pulse, $p(t)$, and the dimensionless impulse response, $h_L(t)$:

$$r(t) = p(t) * h_L(t) \tag{3.73}$$

Then the voltages due to the signal and to the dark current are given by

$$v_s(t) = 2GRR_t\overline{P}_R \sum_{n=-\infty}^{\infty} a_n r(t - nT) \tag{3.74}$$

$$v_d = GR_t q\lambda_0 \int_{-\infty}^{\infty} h_L(t)dt = R_t i_d H_L(0)$$

The shot noise voltage, $v_{sh}(t)$, is a zero-mean nonstationary random process. Its variance is easily evaluated using the generalized Campbell theorem:

$$\sigma_{sh}^2(t) = G_2 R_t^2 q^2 \lambda(t) * h_L^2(t) = 2G_2 R_t^2 Rq\overline{P}_R \sum_{n=-\infty}^{\infty} a_n w(t - nT) + \sigma_d^2 \tag{3.75}$$

where we have defined the noise variance pulse, $w(t)$, and the voltage variance due to the dark current, σ_d^2, as follows:

$$w(t) \stackrel{\text{def}}{=} p(t) * h_L^2(t)$$

$$\sigma_d^2 \stackrel{\text{def}}{=} G_2 R_t^2 q^2 \lambda_0 \int_{-\infty}^{\infty} h_L^2(t)dt \tag{3.76}$$

Figure 3.27 shows the transformations of the received signal yielding the useful received signal and the shot noise variance as per (3.74) and (3.75).

As is apparent from Figure 3.27, the shot noise has two components. One is due to the dark current (it is independent from the received signal), and the other (normally

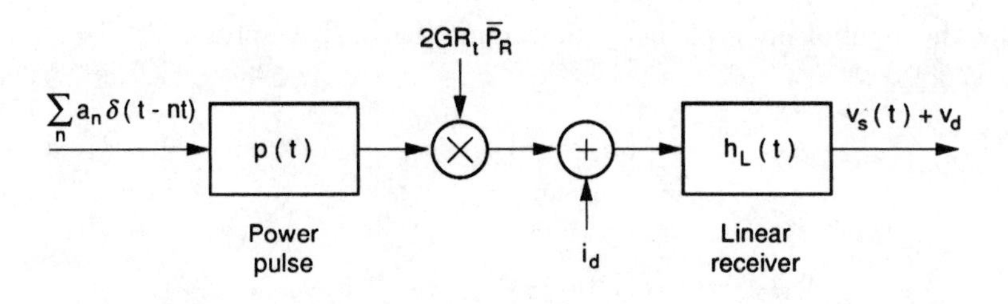

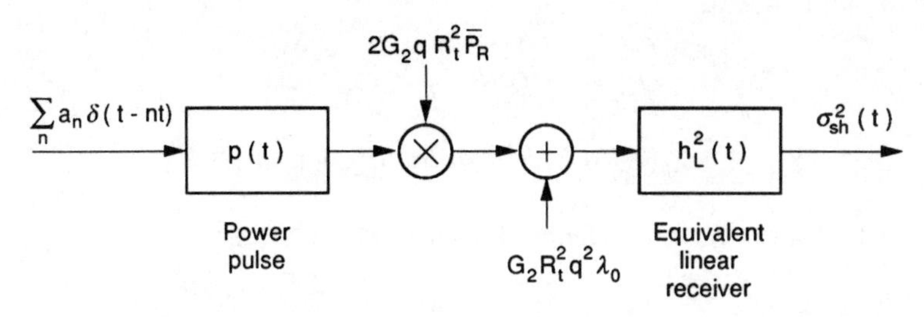

Figure 3.27 Equivalent linear system representing the transformations operated on the received optical power to obtain the output electrical voltage signal and shot noise variance.

the most significant one) is produced by the signal. As a consequence, it affects the binary 1 and the binary 0 differently.

Example 3.3

Consider an integrate-and-dump receiver, with impulse response

$$h_L(t) = \frac{1}{T} \mathrm{u}_T(t)$$

where $\mathrm{u}_T(t)$ is the rectangular pulse of unitary amplitude and duration T. According to (3.73) and (3.76), we obtain

$$r(t) = \frac{1}{T} \int_0^t p(t')dt'$$

$$w(t) = \frac{1}{T^2} \int_0^t p(t')dt'$$

Consider the symbol interval $(0,T)$. Sampling at $t = T$, we obtain

$$r(T) = \frac{1}{T}\int_0^T p(t)dt = 1$$

$$w(T) = \frac{1}{T}$$

From (3.74) and (3.75),

$$v_s(T) = 2a_0 GRR_t \overline{P}_R$$

$$\sigma_{sh}^2(T) = G_2 R_t^2 R_b q(2a_0 R\overline{P}_R + q\lambda_0) \quad (3.77)$$

where $R_b = 1/T$ is the bit rate. Define the shot noise SNR as

$$\left.\frac{S}{N}\right|_{sh} \overset{\text{def}}{=} \frac{v_s^2(T)}{\sigma_{sh}^2(T)}$$

Substituting (3.77) into that definition, we obtain

$$\left.\frac{S}{N}\right|_{sh} = \frac{(2a_0 GR\overline{P}_R)^2}{G_2 q\, R_b(2a_0 R\overline{P}_R + q\lambda_0)}$$

For $a_0 = 1$ and large received power, the last result simplifies to

$$\left.\frac{S}{N}\right|_{sh} = \frac{2R\overline{P}_R}{G^x qR_b} = \frac{2\eta\overline{N}_R}{G^x} \quad (3.78)$$

Equation (3.78) is the shot-noise SNR limit for integrate-and-dump receivers using avalanche photodetectors. When the signal-dependent shot noise is dominant, a PIN photodiode yields better performance (this corresponds to $x = 0$ in (3.78)). This is precisely what happens in optically amplified receivers, where thermal noise is negligible with respect to amplifier spontaneous emission noise (see Chap. 5).

We have already evaluated the quantum limit to receiver sensitivity, based on the abstract receiver capable of detecting even a single received photon. To evaluate the limit of practical receivers, we need to find the output voltage induced by a single received photon, in the absence of dark current and avalanche gain, and for a unit quantum efficiency, η. Considering that the voltage, $v_s(T)$, in (3.77) is produced by the power $2\overline{P}_R = 2\overline{N}_R h\nu/T$, we immediately obtain the following expression for the voltage v_{ph} due to a single photon:

$$v_{ph} = \frac{v_s(T)|_{a_0=1, G=1, \eta=1}}{2\overline{N}_R} = \frac{R_t q}{T} \tag{3.79}$$

where the presence of the symbol interval T in the denominator is due to the fact that the voltage is obtained by integrating the photocurrent induced by the photon over the interval $(0,T)$. For an information bit rate of 1 Gbps and a resistance of 50 Ω, the voltage then becomes $v_{ph} = 8.01 \cdot 10^{-9}$ V.

3.4.2.1 Intersymbol Interference and Equalization

The signal, $v_s(t)$, in (3.74) represents the useful component of the output voltage, $v(t)$. The voltage due to the dark current is a DC term that can be blocked with a capacitor or accounted for by a proper setting of the threshold value. The decisions on the transmitted symbols, a_n, are made according to samples of $v_s(t)$ taken every T seconds. To make these decisions more reliable, that is, to increase the noise immunity, we should reduce as much as possible the intersymbol interference. Here, we may have intersymbol interference at the electronic level, even if the received optical pulse, $p(t)$, is entirely contained in the symbol interval $(0,T)$. In fact, the convolution in (3.73) between $p(t)$ and the impulse response, $h_L(t)$, of the overall linear part of the receiver can broaden the pulse, $r(t)$, beyond a T-second duration.

We treated the phenomenon of intersymbol interference in Section 3.2.2. The reader is referred there for a description of the eye diagram and its significance.

The equalizer-shaping filter can be designed to compensate the distortions of the preceding receiver blocks. An in-depth treatment of intersymbol interference can be found in [14, Chaps. 6 and 7]. Here, we only state the Nyquist criterion, ensuring that the samples $v_{sk} \overset{\text{def}}{=} v_s(t_0 + kT)$ of the useful signals are unaffected by intersymbol interference. The Nyquist criterion states that the intersymbol interference can be eliminated if the overall transfer function, $R(f)$, satisfies the following condition:

$$R_{eq}(f) \overset{\text{def}}{=} \sum_{k=-\infty}^{\infty} R\left(f + \frac{k}{T}\right) = \text{constant} \tag{3.80}$$

where the equivalent transfer function, $R_{eq}(f)$, is obtained by cutting the transfer function, $R(f)$, in slices of width $1/T$ and piling them up in the Nyquist interval $[-1/(2T), 1/(2T)]$.

The Nyquist criterion (3.80) shows that the intersymbol interference does not depend directly on the actual transfer function of the system but rather on the transfer function of the system impulse response sampled every T seconds, which is equal to $R_{eq}(f)$. This stems from the fact that intersymbol interference is a phenomenon that acts after sampling the received signal with a sampling frequency $1/T$.

An important class of transfer functions satisfying the Nyquist criterion is the raised cosine transfer function, shown in Figure 3.28 and described by the following expression:

$$R(f) = \begin{cases} T, & |f| \le \dfrac{1-\alpha}{2T} \\ \dfrac{T}{2}\left\{1 - \sin\left[\dfrac{\pi T}{\alpha}\left(|f| - \dfrac{1}{2T}\right)\right]\right\}, & \dfrac{1-\alpha}{2T} \le |f| \le \dfrac{1+\alpha}{2T} \\ 0, & |f| \ge \dfrac{1+\alpha}{2T} \end{cases} \tag{3.81}$$

where the parameter α is the *roll-off* factor, ranging in the interval (0,1). For $\alpha = 0$, the transfer function, $R(f)$, occupies the minimum bandwidth, $1/(2T) = R_b/2$ Hz; for $\alpha = 1$, the bandwidth occupation doubles. The minimum bandwidth transfer function

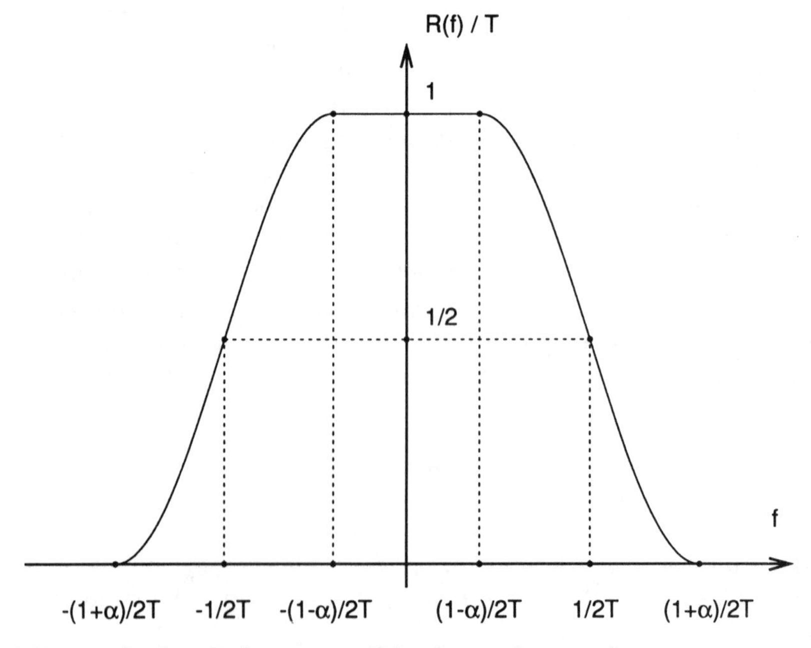

Figure 3.28 Normalized amplitude spectrum, $R(f)/T$, of a raised-cosine pulse.

is not physically realizable, and, moreover, it does not tolerate errors in timing synchronization. That means that even a very small timing error produces an infinite peak distortion. The sensitivity to timing errors decreases with increasing values of α. Typical values of the roll-off range in the interval (0.2,0.8).

The solution suggested by the Nyquist criterion requires an accurate knowledge of the impulse responses $p(t)$ and $h_L(t)$. When $p(t)$ and $h_L(t)$ are not known, we can resort to adaptive equalization, which involves a filter whose impulse response can be modified iteratively to minimize intersymbol interference [14].

Apart from unwanted system effects coming from the physical behavior of the light source (such as extinction ratio, RIN, modal noise), to be discussed at the end of this chapter, the model derived here for the signal term and the variance of shot noise is quite general, the only assumption being the absence of intersymbol interference in the received signal at the optical level. When the assumption on the $(0,T)$ duration of the received optical pulse is not true, the previous analysis needs to be extended by using the general expression of $P_R(t)$ given by (3.55) into (3.63) and proceeding consequently to obtain the three output voltage terms.

To be applied to a particular receiver configuration, the described model requires the knowledge of the received optical pulse, $p(t)$, and of the impulse response, $h_L(t)$, of the linear part of the receiver. As seen from (3.74) and (3.75), both the useful signal and the shot noise variance depend on time. Thus, to find the system bit error probability, the sampling instant must be specified.

3.4.3 Receiver Thermal Noise

Beyond the shot noise due to signal and dark current, the thermal noise due to the photodiode load resistor and receiver amplifiers affects the system performance. Thermal (or Johnson) noise is due to the random thermal motion of electrons in circuit elements that manifests itself as a fluctuating current even in the absence of an applied voltage. We will assume, as it frequently happens in practice, that the receiver noise is dominated by the front-end amplifier of the linear channel in Figure 3.24. This can be explained by recalling that, in a cascade of several circuits, the output noise contribution of each circuit in terms of noise figure is divided by the product of the gains of the preceding circuits. As a consequence, when the first amplifier gain is high enough, the noise contributions of the power amplifier, and, a fortiori, of the equalizer-shaping filter, can be neglected.

The noise figure of an amplifier is defined as the ratio of the SNR at the amplifier input to that at the amplifier output. Thus, it represents the deterioration in the SNR due to amplifier noise. Analytically, as we will see, the noise figure can be accounted for by a multiplicative factor of the spectral density of the thermal noise current of the amplifier input impedance. The noise figure can be evaluated from the physical circuit of the amplifier [24].

We will deal with both current and voltage noises, and we will need spectral densities and variances. Given a noise current, $i(t)$, its two-sided spectral density, measured in A^2/Hz, will be denoted by $G_i(f)$, and its variance by σ_i^2, measured in A^2. The two quantities are related by

$$\sigma_i^2 = \int_{-\infty}^{\infty} G_i(f)df \tag{3.82}$$

The same relationship holds between the spectral density of a voltage noise, $G_v(f)$, measured in V^2/Hz and its variance, σ_v^2, measured in V^2. When a noise signal (current or voltage) is passed through a linear system with transfer function $H(f)$, its spectral density is modified through multiplication by $|H(f)|^2$.

A resistance, R, at absolute temperature, T_0, is represented by an equivalent circuit consisting of a noiseless resistance, R, in parallel with a noise current generator, $i(t)$, with spectral density $G_i = 2kT_0/R$, where $k = 1.38 \cdot 10^{-23}$ JK^{-1} is the Boltzmann constant.

The noise characteristics of an amplifier depend on the active electronic devices employed and the way they are biased and embedded in the amplifier circuit. The general model of Figure 3.29 can normally be adopted. The noisy amplifier is represented by an ideal noiseless amplifier with the same gain and transfer function; the noise sources have been extracted and lumped into two input noise generators, current and voltage, denoted by i_A and e_A [24]. The two noise sources are assumed to be independent, zero-mean, white, and Gaussian, with spectral densities G_{i_A} and G_{e_A}, respectively. Their contribution to the output noise variance depends on the particular receiver structure.

Consider now an amplifier with an input capacitive impedance $Z_t = R_t/(1 + j2\pi f R_t C_t)$, as shown in Figure 3.30; we also included the current noise, $i_t(t)$, due to R_t with spectral density $G_{i_t} = 2kT_0/R_t$. To use dimensionless transfer functions, we split the impedance $Z_t(f)$ into a cascade of the resistance, R_t, and a dimensionless transfer function, $H_t(f)$:

$$Z_t(f) = R_t H_t(f) = R_t \cdot \frac{1}{1 + j2\pi f R_t C_t} \tag{3.83}$$

as shown in the second part of Figure 3.30.

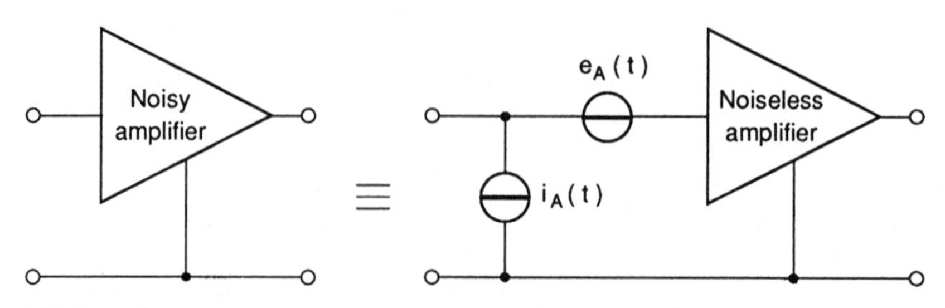

Figure 3.29 Equivalent input noise sources representation of a noisy amplifier.

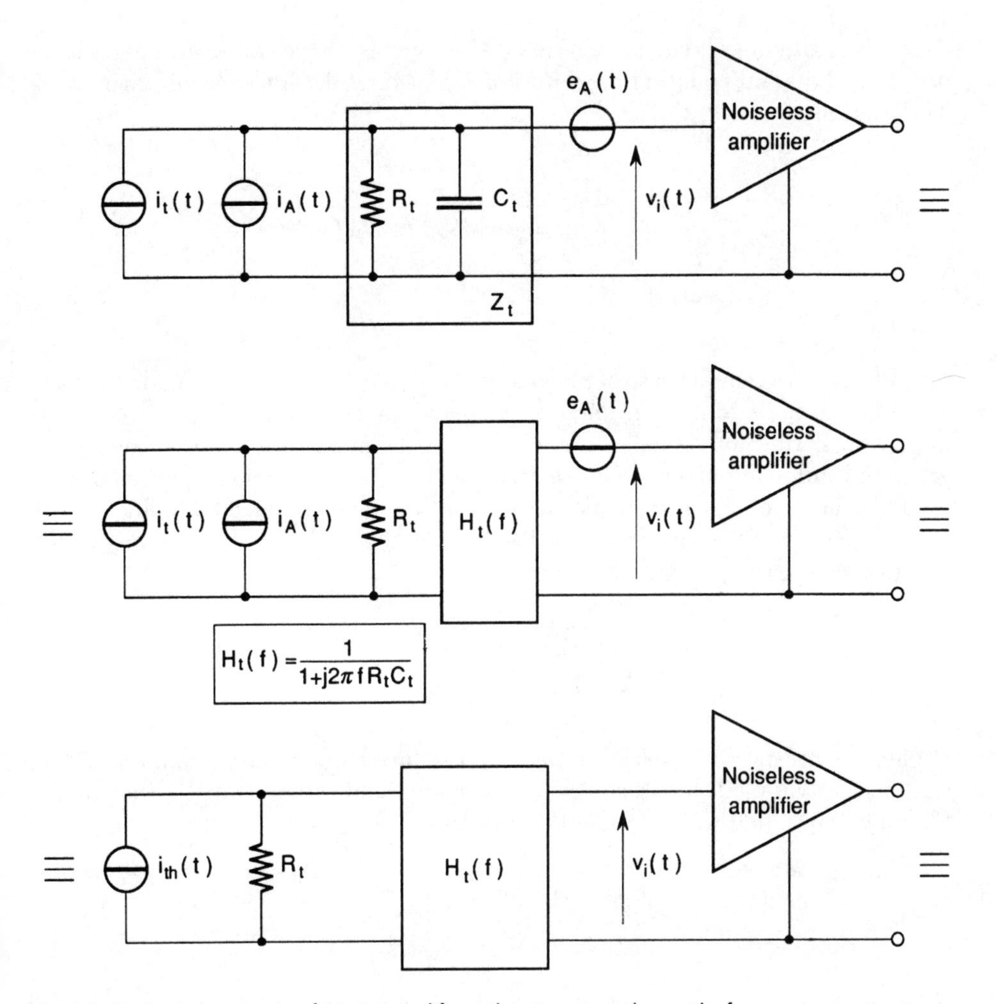

Figure 3.30 Equivalent circuits of a noisy amplifier with its input impedance. The first two use two input noise sources, whereas the third uses the noise figure.

An equivalent representation of the amplifier can be obtained through the amplifier noise figure, $F(f)$, as shown in the third part of Figure 3.30. In this representation, the effect of the amplifier noise has been combined with the current spectral density of its input resistance as follows:

$$G_{i_{th}}(f) \stackrel{\text{def}}{=} G_{i_t} F(f) = \frac{2kT_0}{R_t} F(f) \tag{3.84}$$

The noise figure, $F(f)$, can be determined by comparing the two equivalent schemes of Figure 3.30 and equating the output noise spectral densities (see Problem 3.21). The result is

$$F(f) = 1 + \frac{R_t}{2kT_0}\left[G_{i_A} + \frac{G_{e_A}}{R_t^2|H_t(f)|^2}\right] \tag{3.85}$$

Example 3.4

Consider now the simple case of a real impedance, $Z_t(f) = R_t = 50\Omega$. With this value, the amplifier noise sources, i_A and e_A, can be neglected because their contribution is in general significantly smaller. Assuming that the amplifier-equalizer has an integrate-and-dump impulse response, $h_e(t)$, equal to that of Example 3.3, we want to find the RMS noise voltage at the equalizer output. It can be obtained by multiplying the spectral density of the voltage across the resistance by the integral of the square of the impulse response of the amplifier-equalizer.

$$\sigma_{th} = \sqrt{\frac{2kT_0 R_t}{T^2}\int_0^T u_T^2(t)dt} = \sqrt{\frac{2kT_0 R_t}{T}} \tag{3.86}$$

Let us compare it with the output voltage, v_{ph}, induced by a single photon. This voltage has been computed for the case of integrate-and-dump transfer function in Example 3.3. Using (3.79), we obtain the ratio (for T_0 = 290K)

$$\frac{\sigma_{th}}{v_{ph}} = \frac{1}{q}\sqrt{\frac{2kT_0 T}{R_t}} = 2{,}497 \tag{3.87}$$

Equation (3.87) shows that the thermal noise is much stronger than the voltage produced by a single photon. In this case, the single hole-electron pair produced by a single photon would not be observable.

3.4.4 A Comprehensive Receiver Model

To obtain the complete receiver model, we combine the models of the photodiode (Fig. 3.25) and of the amplifier (Fig. 3.30) and add the transfer function, $H_e(f)$, of the amplifiers and equalizer-shaping filter. The result is a complete model of the receiver shown in Figure 3.31.

The resistance, R_t, accounts for the photodiode bias resistance, R_{bias}, the

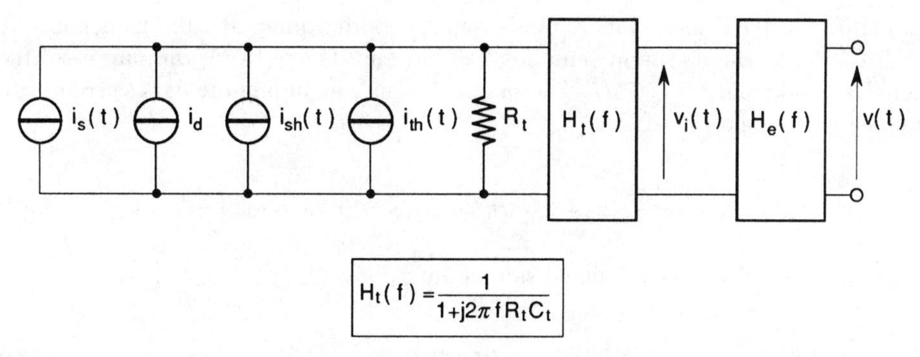

Figure 3.31 Complete equivalent circuit of the receiver with the current sources representing the useful signal, the dark current, the shot noise, and the overall load resistance and amplifier thermal noises.

photodiode resistance, and the amplifier input resistance, whereas the capacitance, C_t, in the transfer function, $H_t(f)$, includes the photodiode differential capacitance, C_d, the parasitic capacitance, and the amplifier input capacitance.

In the model, the gain block A representing the overall gain of the amplifiers is omitted, since it affects equally both signal and noise and leaves unchanged the SNR. That, by the way, is an advantage of using the equivalent noise sources at the amplifier input.

In Figure 3.31, the current generator, $i_s(t)$, represents the useful signal current given by (3.67), i_d represents the dark current of (3.68), $i_{sh}(t)$ represents the shot noise current, and, finally, $i_{th}(t)$ represents the thermal and amplifier noises with spectral density (3.84).

The same physical interpretation applies to the four components of the output voltage:

$$v(t) = v_s(t) + v_d + v_{sh}(t) + v_{th}(t) \tag{3.88}$$

To evaluate the bit error probability, we need to know the output signal voltage corresponding to the transmission of a 1 information symbol and the variances of noise voltages corresponding to two possible values (0 and 1) of the information symbol. Of the four $v(t)$ components in (3.88), one (v_d), defined in (3.68), is constant; another ($v_{th}(t)$) is a stationary random process, being a filtered white noise; the remaining two ($v_s(t)$ and $v_{sh}(t)$) depend on the transmitted information sequence. Assuming that the sequence $\mathbf{a}$ of a_n is a discrete stationary random process, (3.74) and (3.75) show that the signal component, $v_s(t)$, and the variance, $\sigma_{sh}^2(t)$, are cyclostationary random processes (see Section 2.6) with period T. Thus, $v(t)$ is a cyclostationary random process with period equal to the symbol period T, and, consequently, the performance of the receiver is the same in all symbol intervals. To simplify the notation, we assume that the received data bit is processed over the interval $(0,T)$, denote by t_0 $(0 \le t_0 \le T)$ the optimum sampling instant; and denote by x_n the sample of the signal $x(t)$ taken at $t = t_0 + nT$. The sampled signal, including the dark current, can thus be written as

$$v_s(t_0) + v_d \stackrel{\text{def}}{=} v_{s0} + v_d = 2GRR_t\overline{P}_R a_0 r_0 + X_s + v_d \stackrel{\text{def}}{=} a_0 z_0 + X_s + v_d \qquad (3.89)$$

where we defined the useful signal sample by z_0:

$$z_0 \stackrel{\text{def}}{=} 2GRR_t\overline{P}_R r_0 \qquad (3.90)$$

In (3.89), we separated the useful signal component (the first term on the right side) from the intersymbol interference (the second term on the right side)

$$X_s \stackrel{\text{def}}{=} 2GRR_t\overline{P}_R \sum_{n\neq 0} a_n r_n \qquad (3.91)$$

The thermal noise is a Gaussian random process, and thus its sample v_{th0} is a Gaussian random variable. As to shot noise, we assume that the received light power is sufficiently large (see discussion in Example 2.3); then, the shot noise also can be treated as a Gaussian process.[5] The total noise is given by

$$v_{n0} \stackrel{\text{def}}{=} v_{sh0} + v_{th0} \qquad (3.92)$$

Since the thermal noise and the shot noise are independent, the total noise, v_{n0}, is a Gaussian random variable with zero-mean and variance

$$\sigma_{n0}^2(a_0, X_\sigma) = \sigma_{sh0}^2(a_0, X_\sigma) + \sigma_{th}^2 \qquad (3.93)$$

where we have used 0 in the subscript of the shot noise variance to make explicit the fact that it depends on the sampling instant. The thermal noise variance can be evaluated from the spectral density (3.84) as follows:

$$\sigma_{th}^2 = \int_{-\infty}^{\infty} G_{th}(f)df = 2kT_0R_t \int_{-\infty}^{\infty} F(f)|H_t(f)H_e(f)|^2 df \qquad (3.94)$$

5. This is likely to be accurate for a transmitted information symbol $a_0 = 1$. When $a_0 = 0$, shot noise reduces to the dark current contribution. In that case, however, the thermal noise becomes dominant, and the Gaussian approximation still makes sense.

The shot noise variance, $\sigma^2_{sh0}(a_0, X_\sigma)$ is derived from (3.75) with $w(t) = p(t) * h_t(t) * h_e(t)$; it is given by

$$\sigma^2_{sh0}(a_0, X_\sigma) = 2G_2 R R_t^2 q \overline{P}_R a_0 w_0 + X_\sigma + \sigma_d^2 \tag{3.95}$$

where we have kept separate the contributions of the useful symbol (the first term on the right side) from the contribution X_σ of the intersymbol interference (the second term on the right side) and from the contribution of the dark voltage (the third term on the right side). The intersymbol interference contribution is given by

$$X_\sigma \stackrel{\text{def}}{=} 2G_2 R R_t^2 q \overline{P}_R \sum_{n \neq 0} a_n w_n \tag{3.96}$$

Let us now define the SNR $\Lambda_1(X_\sigma)$ as the ratio between the squared useful part of the signal when $a_0 = 1$ and the overall noise variance when $a_0 = 1$. It is given by

$$\Lambda_1(X_\sigma) \stackrel{\text{def}}{=} \frac{z_0^2}{\sigma^2_{n0}(1, X_\sigma)} = \frac{(2GRR_t \overline{P}_R r_0)^2}{2G_2 RR_t^2 q \overline{P}_R w_0 + X_\sigma + \sigma_d^2 + \sigma_{th}^2} \tag{3.97}$$

Let us also define the SNR $\Lambda_0(X_\sigma)$ as the ratio between the squared useful part of the signal when $a_0 = 1$ and the overall noise variance when $a_0 = 0$. It is given by

$$\Lambda_0(X_\sigma) \stackrel{\text{def}}{=} \frac{z_0^2}{\sigma^2_{n0}(0, X_\sigma)} = \frac{(2GRR_t \overline{P}_R r_0)^2}{X_\sigma + \sigma_d^2 + \sigma_{th}^2} \tag{3.98}$$

The SNR Λ_1 has a precise physical meaning; it is the SNR in the decision sample when transmitting a binary 1. However, Λ_0 has no clear physical meaning; its signal part corresponds to a binary 1 transmitted, while its noise part corresponds to the binary 0. The ratio between $\Lambda_1(X_\sigma)$ and $\Lambda_0(X_\sigma)$ is an important system parameter, because it represents the ratio between the noise variances corresponding to 0 and 1, and gives a quantitative information on the relative importance of shot and receiver thermal noises. More important, Λ_0 makes nice, symmetric bit error probability formulae (see Section 3.4.5).

As (3.97) and (3.98) make clear, the SNRs Λ_1 and Λ_0 depend in general on the whole transmitted information sequence through the shot noise variance, and this makes the performance evaluation rather complicated. Imposing the Nyquist criterion on $r(t)$ [and thus cancelling X_s from (3.89)] does not completely solve the problem, because X_σ may still be different from zero.

3.4.5 Case of No Intersymbol Interference

We will consider a special case of no intersymbol interference:

$$r_n = 0, \, w_n = 0, \, \forall n \neq 0 \tag{3.99}$$

Then the two random variables, X_s and X_σ, are identically equal to zero, so that the sampled signal becomes

$$v_{s0} = 2GRR_t \overline{P}_R a_0 r_0 \tag{3.100}$$

Then (3.97) and (3.98), by using (3.76) for σ_d^2, yield

$$\Lambda_1 \overset{\text{def}}{=} \frac{z_0^2}{\sigma_n^2(1)} = \frac{(2GRR_t \overline{P}_R r_0)^2}{2G_2RR_t^2 q \overline{P}_R w_0 + G_2 R_t^2 q^2 \lambda_0 \int_{-\infty}^{\infty} h_L^2(t)dt + \sigma_{th}^2} \tag{3.101}$$

and

$$\Lambda_0 \overset{\text{def}}{=} \frac{z_0^2}{\sigma_n^2(0)} = \frac{(2GRR_t \overline{P}_R r_0)^2}{G_2 R_t^2 q^2 \lambda_0 \int_{-\infty}^{\infty} h_L^2(t)dt + \sigma_{th}^2} \tag{3.102}$$

where $h_D(t) = h_t(t) * h_e(t)$, is the impulse response of the cascade of the amplifier and equalizer. We now introduce the amplifier figure of merit and express Λ_1 and Λ_0 in terms of that figure.

3.4.5.1 Amplifier Figure of Merit

The amplifier figure of merit [1] can be used to characterize the receiver quality from the noise point of view. The figure of merit states how far the receiver is from the quantum limit performance, or, in other words, how much the thermal noise overwhelms the minimum possible output signal.

The figure of merit of the amplifier [6] is denoted by ρ_A and is defined as

$$\rho_A \overset{\text{def}}{=} \frac{\text{RMS output voltage due to thermal noise}}{\text{output voltage due to a single received photon}} = \frac{\sigma_{th}}{v_{ph}} \tag{3.103}$$

Using (3.100) for the sampled signal, (3.79) for the definition of v_{ph}, and (3.94) for

6. An amplifier is better when its figure of merit is lesser, so, in a sense, ρ is a figure of *demerit*, much like the noise figure.

the thermal noise variance, σ_{th}^2, we obtain the following expression for the figure of merit:

$$\rho_A = \frac{\sigma_{th}}{v_{ph}} = \frac{T}{qr_0}\sqrt{\frac{2kT_0}{R_t}\int_{-\infty}^{\infty} F(f)|H_t(f)H_e(f)|^2 df} \tag{3.104}$$

Equation (3.104) shows that ρ_A depends on the receiver noise through the noise figure, $F(f)$, and on its linear transfer function. The parameter ρ_A can be easily measured (see [1]). The SNRs Λ_1 and Λ_0 can be expressed in terms of the figure of merit as follows (see derivation in Appendix A):

$$\Lambda_1 = \frac{(2\eta G\overline{N}_R)^2}{G_2(2\eta\overline{N}_R A_0 + N_0 A_1) + \rho_A^2} \tag{3.105}$$

and

$$\Lambda_0 = \frac{(2\eta G\overline{N}_R)^2}{G_2 N_0 A_1 + \rho_A^2} \tag{3.106}$$

where N_0 is the nonstimulated number of hole-electron pairs generated in a bit period (the counterpart of the stimulated pairs $\eta\overline{N}_R$) due to the dark current:

$$N_0 \stackrel{\text{def}}{=} \lambda_0 T \tag{3.107}$$

The SNRs (3.105) and (3.106) depend on two parameters, A_0 and A_1, that can be expressed in terms of the sampled impulse responses of the system or, equivalently, in terms of the transfer functions (see derivation in Appendix A):

Time-Domain Expressions

$$A_0 = \frac{w_0 T}{r_0^2} \tag{3.108}$$

$$A_1 = \frac{TH_{L2}(0)}{r_0^2} \tag{3.109}$$

where

$$H_{L2}(f) \overset{\text{def}}{=} \mathcal{F}[h_L^2(t)] = H_L(f) * H_L(f) \tag{3.110}$$

and $\mathcal{F}$ denotes Fourier transform.

Frequency-Domain Expressions

$$A_0 = \frac{B_{RL2} B_L}{B_{RL}^2} \tag{3.111}$$

$$A_1 = \frac{B_L}{B_{RL}^2 T} \tag{3.112}$$

with

$$B_L \overset{\text{def}}{=} \frac{\int_{-\infty}^{\infty} |H_L(f)|^2 df}{H_L^2(0)} \tag{3.113}$$

$$B_{RL} \overset{\text{def}}{=} \frac{\int_{-\infty}^{\infty} P(f)\, H_L\, (f) e^{-j2\pi f t_0} df}{P(0)\, H_L(0)} = \frac{\int_{-\infty}^{\infty} R(f) e^{-j2\pi f t_0} df}{R(0)} \tag{3.114}$$

$$B_{RL2} \overset{\text{def}}{=} \frac{\int_{-\infty}^{\infty} P(f)\, H_{L2}\, e^{-j2\pi f t_0} df}{P(0)\, H_{L2}(0)} = \frac{\int_{-\infty}^{\infty} W(f) e^{-j2\pi f t_0} df}{W(0)} \tag{3.115}$$

where t_0 is the sampling instant.

To get a physical insight into frequency-domain expressions (3.114) and (3.115), note that B_{RL} and B_{RL2} have the dimension of hertz. They are equal to twice the bandwidth of the linear systems, $R(f) = P(f) H_L(f) e^{-j2\pi f t_0}$ and $S(f) = P(f) H_{L2}(f) e^{-j2\pi f t_0}$, respectively. The parameter B_L has the dimension of hertz, too, and is equal to twice the normalized equivalent noise bandwidth of the linear receiver.

Example 3.5

Consider the integrate-and-dump receiver presented in Example 3.3. In that receiver, $h_L(t) = (1/T) u_T(t)$. Let us apply the frequency-domain expressions. Recall from (3.57) that $P(0) = \int_0^T p(t) dt = T$. Hence,

$$B_L = B_{RL} = B_{RL2} = \frac{1}{T}$$

Further, $A_0 = A_1 = 1$. Therefore,

$$\Lambda_1 = \frac{(2\eta G\overline{N}_R)^2}{G_2(2\eta\overline{N}_R + N_0) + \rho_A^2} \tag{3.116}$$

$$\Lambda_0 = \frac{(2\eta G\overline{N}_R)^2}{G_2 N_0 + \rho_A^2} \tag{3.117}$$

3.5 PERFORMANCE EVALUATION

3.5.1 Bit Error Probability

Evaluating the performance of a digital optical communication system is different from the case of radio signals transmitted over additive Gaussian noise channels, like satellite links, radio relay links, or conventional electric transmission lines (coaxial cables or twisted-pair telephone channels). One reason is that part of the noise (the shot noise) depends on the signal.

We saw in Section 3.4 that the sampled signal at the output of the amplifier-equalizer in the receiver can be written as

$$v_0 = a_0 z_0 + X_s + v_d + v_{sh0} + v_{th0} \tag{3.118}$$

where $z_0 \stackrel{\text{def}}{=} 2GRR_t\overline{P}_R r_0$. The first term on the right side is the useful signal, the second term is the intersymbol interference defined by (3.91), the third term is the voltage due to the dark current, the fourth term is the shot noise with variance $\sigma_{sh0}^2\,(a_0, X_\sigma)$ defined in (3.95), and the fifth term is the thermal noise with variance σ_{th}^2 defined in (3.94).

The decision circuit (the "sample and threshold" block in Fig. 3.24) makes its decision by comparing v_0 with a threshold, V_{th}: The decision will be in favor of a transmitted 1 if $v_0 > V_{th}$ and of a transmitted 0 if $v_0 \leq V_{th}$. The signal part of v_0, in the absence of intersymbol interference, ranges in the interval $(v_d, z_0 + v_d)$. Hence, we set the threshold at $V_{th} = v_d + \beta z_0$, $0 \leq \beta \leq 1$.

Since both the signal and the shot noise variance depend on the information symbol being transmitted, we have two different expressions for the bit error probability corresponding to $a_0 = 1$ and $a_0 = 0$. Assuming that the RV a_0 assumes the two values 1 and 0 with probability 0.5, the average bit error probability is equal to

$$P(e) = \frac{1}{2}[P_e(1) + P_e(0)] \tag{3.119}$$

where $P_e(1)$ is the conditional error probability when transmitting a binary 1, and $P_e(0)$ is the conditional error probability when transmitting a binary 0.

Let us evaluate separately the two conditional error probabilities. Each conditional error probability, $P_e(0)$ and $P_e(1)$, is evaluated in two steps. First, the bit error probability conditioned to a particular value assumed by the RVs X_s [the intersymbol interference on the signal defined in (3.91)] and X_σ [the contribution of the intersymbol interference to the shot noise variance as for (3.96)] is evaluated. Then the result is averaged with respect to X_s and X_σ. This procedure yields the following expression for $P_e(1)$:

$$\begin{aligned}P_e(1) &= \mathrm{E}_{X_s,X_\sigma}\left[P_e(1\,|X_s = x_s, X_\sigma = x_\sigma)\right]\\ &= \mathrm{E}_{X_s,X_\sigma}\left[P(z_0 + x_s + v_{sh0} + v_{th0} \le \beta z_0)\right]\\ &= \frac{1}{2}\,\mathrm{E}_{X_s,X_\sigma}\left\{\mathrm{erfc}\left[\frac{(1-\beta)\,z_0 + x_s}{\sqrt{2[\sigma_{sh0}^2(1, x_\sigma) + \sigma_{th}^2]}}\right]\right\}\end{aligned} \tag{3.120}$$

where in the last step we have exploited the fact that the sum $v_{sh0} + v_{th0}$ is a zero-mean, Gaussian RV with variance $\sigma_{n0}^2\,(1, x_\sigma) \overset{\text{def}}{=} \sigma_{sh0}^2\,(1, x_\sigma) + \sigma_{th}^2$.

The same procedure applied to $P_e(0)$ yields

$$P_e(0) = \frac{1}{2}\,\mathrm{E}_{X_s,X_\sigma}\{P_e(0|X_s = x_s, X_\sigma = x_\sigma)\} = \frac{1}{2}\,\mathrm{E}_{X_s,X_\sigma}\left\{\mathrm{erfc}\left[\frac{\beta z_0 - x_s}{\sqrt{2[\sigma_{sh0}^2(0, x_\sigma) + \sigma_{th}^2]}}\right]\right\} \tag{3.121}$$

Recalling now the definitions in (3.97) and (3.98) of the SNRs $\Lambda_1(x_\sigma)$ and $\Lambda_0(x_\sigma)$ and using them in (3.120) and (3.121), we obtain

$$P_e(1) = \frac{1}{2}\,\mathrm{E}_{X_s,X_\sigma}\{P_e(1|x_s, x_\sigma)\} = \frac{1}{2}\,\mathrm{E}_{X_s,X_\sigma}\left\{\mathrm{erfc}\left[\frac{(1-\beta)^2\Lambda_1(x_\sigma)}{2} + \frac{x_s^2\Lambda_1(x_\sigma)}{2z_0^2} + \frac{x_s(1-\beta)\Lambda_1(x_\sigma)}{z_0}\right]^{1/2}\right\} \tag{3.122}$$

and

$$P_e(0) = \frac{1}{2}\,\mathrm{E}_{X_s,X_\sigma}\{P_e(0|x_s, x_\sigma)\} = \frac{1}{2}\,\mathrm{E}_{X_s,X_\sigma}\left\{\mathrm{erfc}\left[\frac{\beta^2\Lambda_0(x_\sigma)}{2} + \frac{x_s^2\Lambda_0(x_\sigma)}{2z_0^2} - \frac{x_s\beta}{z_0}\right]^{1/2}\right\} \tag{3.123}$$

Results (3.122) and (3.123) are very general. A critical step in their evaluation consists of the evaluation of averages with respect to the two RVs X_s and X_σ. When only a few interfering samples, r_n and w_n [recall (3.91) and (3.96)] are different from zero, the two quantities X_s and X_σ are discrete RVs that assume only a limited number of values. Then averages can be easily evaluated using a double summation with respect to the discrete values assumed by the RVs. When the intersymbol interference is large, the averages can be performed using Gaussian quadrature rules, as explained in Appendix C. Notice that the two RVs X_s and X_σ are *not* independent, because their randomness derives from the same transmitted information sequence.

To summarize, evaluation of the two conditional error probabilities $P_e(1)$ and $P_e(0)$ consists of the following steps. We specify them for $P_e(1)$, as an example.

1. Set $a_0 = 1$.
2. Set the dummy variables $A = 0$ and $n = 0$.
3. Using the received pulse waveform $p(t)$ and the overall impulse response $h_L(t)$ of the linear part of the receiver, choose the optimum sampling instant t_0 (e.g., based on the maximization of the eye opening) and evaluate the significant samples r_n (i.e., those whose ratio to the useful one, r_0, lies below a certain threshold, say 0.01), and w_n, of the impulse responses $r(t)$ and $w(t)$.
4. For a given information sequence **a**, with the constraint $a_0 = 1$, evaluate a pair of values x_{si} and $x_{\sigma j}$ assumed by the RVs X_s and X_σ.
5. Using r_0, w_0, the average received power $\overline{P}_R$ the photodetector characteristics (R, G, and G_2), and $x_{\sigma j}$, evaluate the useful signal, z_0, and the shot noise variance, $\sigma_{sh0}^2\ (1, x_{\sigma j})$.
6. Evaluate the receiver noise figure, $F(f)$, as per (3.85) and the thermal noise variance as per (3.94).
7. Evaluate the SNR, Λ_1, as per (3.97).
8. Evaluate the bit error probability, $P_e(1|x_{si}, x_{\sigma j})$ conditioned on x_{si} and $x_{\sigma j}$ as per (3.122); let A be its value.
9. Evaluate the accumulated conditional error probabilities through $A = A + P_e(1|x_{si}, x_{\sigma j})$ and set $n = n + 1$.
10. Change the information sequence **a**, always with the constraint $a_0 = 1$, and return to step 4. When all possible information sequences have been exhausted, evaluate $P_e(1)$ by dividing the final value of A by n (the number of examined information sequences).

3.5.2 A Simpler Case: Bit Error Probability Without Intersymbol Interference

Previous results (3.122) and (3.123) are quite general. Let us now consider a simple special case in which the receiver is free of intersymbol interference, so that (3.99) applies.

Then the two SNRs assume the form (3.101) and (3.102), and the conditional bit error probabilities (3.122) and (3.123) simplify to

$$P_e(1) = \frac{1}{2}\,\mathrm{erfc}[(1 - \beta)\sqrt{\Lambda_1/2}] \tag{3.124}$$

and

$$P_e(0) = \frac{1}{2}\,\mathrm{erfc}[\beta\sqrt{\Lambda_0/2}] \tag{3.125}$$

From (3.124) and (3.125), we observe that, in general, the two conditional error probabilities are different. This is due to the fact that the probability density functions of the voltage sample v_0 depend on a_0. A qualitative plot of $f_v(x|a_0 = 1)$ and $f_v(x|a_0 = 0)$ is shown in Figure 3.32. It is clear that the variance is larger when $a_0 = 1$ is transmitted. To make equal the conditional error probabilities, the threshold should be set to $V_{th}^* = v_d + \beta_S z_0$, where β_S is equal to (see Problem 3.27)

$$\beta_S = \frac{\sigma_n(0)}{\sigma_n(0) + \sigma_n(1)} \tag{3.126}$$

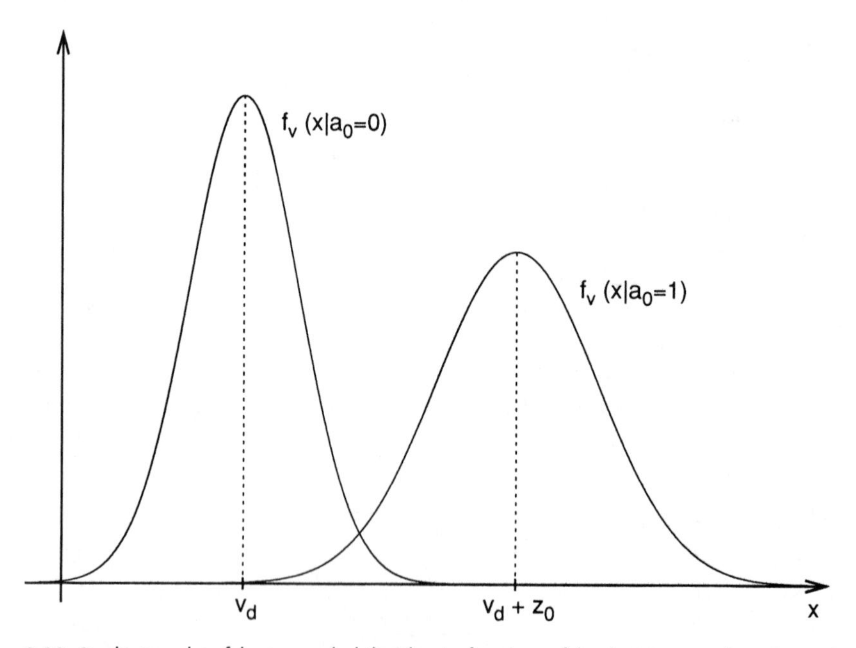

Figure 3.32 Qualitative plot of the two probability density functions of the decision signal conditioned on the transmission of a 1 and a 0.

Then the average error probability is equal to

$$P(e) = P_e(1) = P_e(0) = \frac{1}{2}\,\mathrm{erfc}\left[\frac{z_0}{\sqrt{2}[\sigma_n(0) + \sigma_n(1)]}\right] \tag{3.127}$$

The argument of the function erfc in (3.127) can be called the *digital SNR*:

$$\text{digital SNR} \rightarrow Q \overset{\text{def}}{=} \frac{z_0}{[\sigma_n(0) + \sigma_n(1)]} = \frac{1}{\dfrac{1}{\sqrt{\Lambda_0}} + \dfrac{1}{\sqrt{\Lambda_1}}} \tag{3.128}$$

Recalling (3.118), the parameter Q can be interpreted as

$$Q = \frac{\mathrm{E}(v_0|a_0 = 1) - \mathrm{E}(v_0|a_0 = 0)}{\mathrm{RMS}(v_0|a_0 = 1) + \mathrm{RMS}(v_0|a_0 = 0)} \tag{3.129}$$

where RMS denotes the standard deviation (square root of the variance).

According to (3.129), the parameter Q can be evaluated from the measured or simulated eye diagram, by evaluating the mean and the variance of the signal traces in the upper and lower parts of the eye diagram.

Figure 3.33 shows the function

$$f(Q) \overset{\text{def}}{=} \frac{1}{2}\,\mathrm{erfc}\left(\frac{Q}{\sqrt{2}}\right)$$

as a function of Q.

The parameter Q is related to Λ_1 by

$$Q = (1 - \beta_s)\sqrt{\Lambda_1} \tag{3.130}$$

Now we will define the receiver sensitivity with respect to the bit error probability $P(e) = 10^{-9}$. Note that

$$\frac{1}{2}\,\mathrm{erfc}\left(\frac{6}{\sqrt{2}}\right) = 10^{-9}$$

Hence, we need a value of $Q = 6$ or, in dB, $Q_{dB} \simeq 15.6$, for $P(e) = 10^{-9}$.

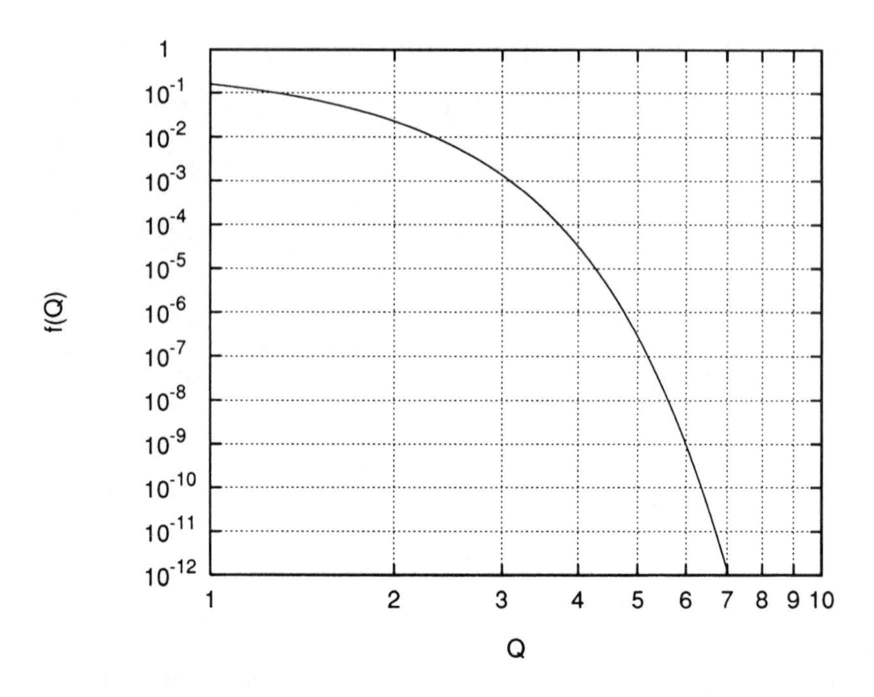

Figure 3.33 Plot of the function $f(Q) = (1/2)\mathrm{erfc}(Q/\sqrt{2})$ versus Q.

However, the condition of symmetry $P_e(0) = P_e(1)$ does not correspond to the minimum error probability. To see this, consider the general expression

$$P(e) = \frac{1}{4}\{\mathrm{erfc}[(1 - \beta)\sqrt{\Lambda_1/2}] + \mathrm{erfc}[\beta\sqrt{\Lambda_0/2}]\} \tag{3.131}$$

Now let us obtain the optimum threshold by minimizing the right side of (3.131) with respect to β. This procedure yields (see Problem 3.28)

$$\beta_{opt} = \frac{1}{\gamma - 1}[(\gamma + (\gamma - 1)\Lambda_1^{-1} \log \gamma)^{\frac{1}{2}} - 1] \tag{3.132}$$

where

$$\gamma \stackrel{\text{def}}{=} \frac{\sigma_n^2(1)}{\sigma_n^2(0)} = \frac{\Lambda_0}{\Lambda_1} \tag{3.133}$$

Figure 3.34 shows the optimal value β_{opt} of the threshold versus the SNR Λ_1 for several values of the ratio γ.

Inspection of Figure 3.34 reveals that $\beta_{opt} \to \beta_s$ for large Λ_1 (which can also be inferred from (3.132)); however, for $\gamma = 1$, $\beta_{opt} = 1/2$. In the latter case, the two SNRs Λ_0 and Λ_1 are equal, meaning that thermal noise is dominant.

In Figure 3.35, the minimum bit error probability is plotted versus the SNR Λ_1. Values of γ greater than 1 mean that the noise at the sampling instant is larger for 1 than for 0, or, equivalently, that the shot noise is significant with respect to the receiver thermal noise.

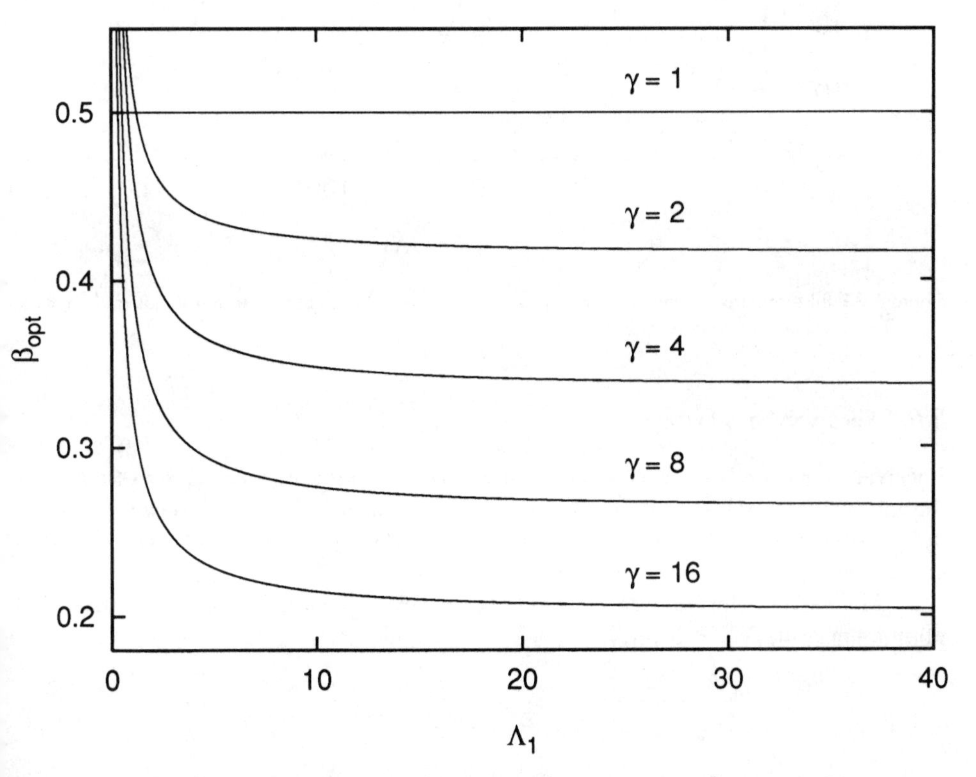

Figure 3.34 Optimum value of the decision threshold as a function of the parameter Λ_1 for an IM-DD transmission system and for three values of the parameter γ.

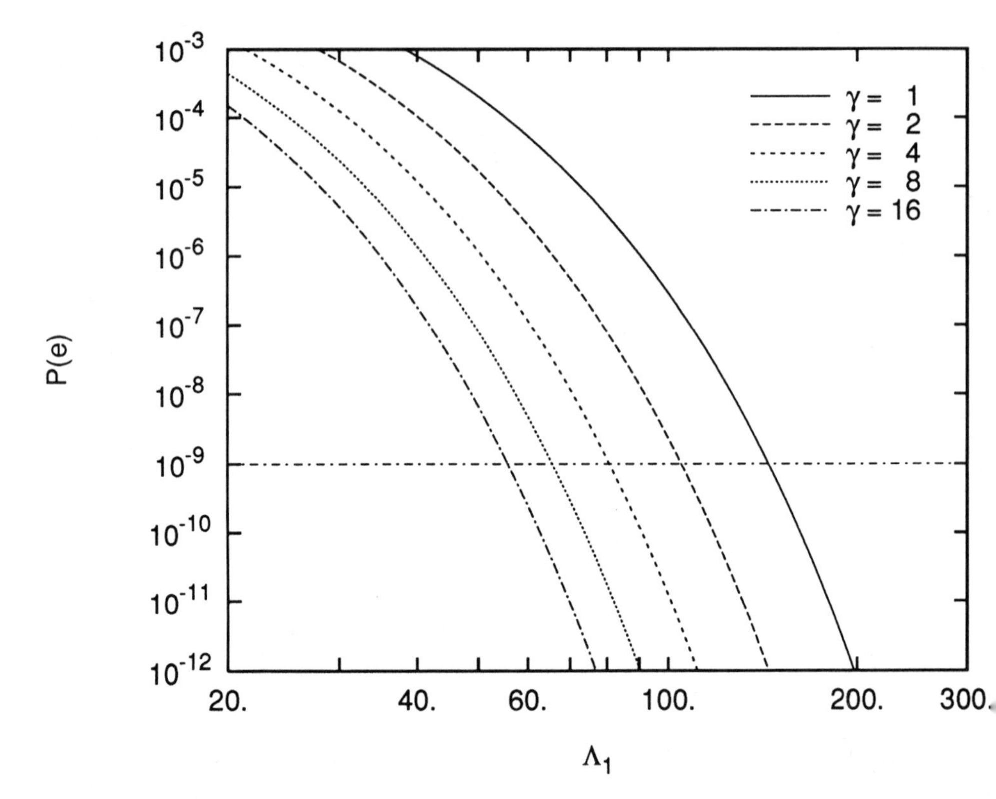

Figure 3.35 Bit error probability as a function of the parameter Λ_1 for optimized thresholds and for three values of the parameter γ.

3.5.3 Receiver Sensitivity

Receivers are usually characterized (and valued) through the receiver *sensitivity*. For digital receivers, the sensitivity is defined as the *minimum average received optical power* (or, equivalently, the *minimum average number of received photons per bit*) that guarantees a bit error probability not greater than 10^{-9}.

The relationship between the average received power, $\overline{P}_R$, and the number of photons per bit, $\overline{N}_R$, is given by

$$\overline{P}_R = \frac{\overline{N}_R h\nu}{T} \tag{3.134}$$

Our goal is to express the receiver sensitivity in terms of $\overline{N}_R$ for a simplified case of no intersymbol interference. To reach that goal, we must derive the relationship between the bit error probability and the average received number of photons per bit.

Example 3.6

Consider a receiver with the threshold V_{th}^* that makes equal the two conditional error probabilities. Then (3.127) holds true. Using a system margin, k_m, to take into account intersymbol interference and other system imperfections, we have

$$P(e) = \frac{1}{2}\,\text{erfc}\left[\frac{1}{k_m\left(\sqrt{\frac{2}{\Lambda_0}} + \sqrt{\frac{2}{\Lambda_1}}\right)}\right] \tag{3.135}$$

Thus, recalling that $\frac{1}{2}$ erfc $(6/\sqrt{2}) = 10^{-9}$ and definition (3.133), we obtain that, for $P(e) = 10^{-9}$,

$$\Lambda_1 = 36k_m^2\left(1 + \frac{2}{\sqrt{\gamma}} + \frac{1}{\gamma}\right) \tag{3.136}$$

For example, for $\gamma = 2$ (which corresponds to the case when the shot and the thermal noise have the same variances) and $k_m = 2$ (a 6-dB margin), we obtain

$$\Lambda_1 \overset{\text{def}}{=} \Lambda_{1min} = 420$$

In Example 3.6, we related the SNR Λ_1 to the bit error probability that defines the sensitivity of the receiver. To find the average received number of photons per bit (and thus the sensitivity), we use (3.105) and (3.106), where Λ_1 and Λ_0 are expressed in terms of $\overline{N}_R$ and the amplifier figure of merit ρ_A.

Thus, to evaluate receiver sensitivity, we can proceed as follows:

1. Evaluate the figure of merit, ρ_A; the quantum efficiency, η; the avalanche average gain, G; the excess avalanche noise factor, G_2; the nonstimulated pairs N_0 due to the dark current; and the parameters A_0 and A_1 as per (3.108) and (3.109) or (3.111) and (3.112). In the definitions (3.105) and (3.106) of Λ_1 and Λ_0, the only unknown parameter is now the received number of photons per bit, $\overline{N}_R$.
2. Substitute the functions $\Lambda_1(\overline{N}_R)$ and $\Lambda_0(\overline{N}_R)$ into (3.132) for the optimum threshold β, to obtain the function $\beta_{opt}(\overline{N}_R)$.
3. Apply the general expression of the error probability (3.131), which depends only on $\overline{N}_R$ through Λ_1, Λ_0, and β, to plot $P(e)$ as a function of $\overline{N}_R$.
4. From the curve of the error probability versus $\overline{N}_R$, read the abscissa

corresponding to $P(e) = 10^{-9}$. It will give the desired sensitivity in terms of $\overline{N}_R$, with the optimum decision threshold.

Previous rigorous procedure requires numerical computations and does not provide an explicit analytical relationship between $\overline{N}_R$ and main receiver parameters. We will develop now an analysis for a simplified situation that will permit a closed-form derivation.

Let us select the threshold β_s yielding symmetry, that is, yielding equal conditional error probabilities: $P_e(1) = P_e(0)$. Then we can write the error probability

$$P_e(1) = \frac{1}{2}\,\mathrm{erfc}\left[(1 - \beta_s)\sqrt{\frac{\Lambda_1}{2k_m^2}}\right] \tag{3.137}$$

where the parameter $k_m^2 \geq 1$ represents the system margin. For $P(e) = 10^{-9}$, we obtain

$$\Lambda_{1min} = \frac{36k_m^2}{(1 - \beta_s)^2} \tag{3.138}$$

From (3.105), we obtain (see Problem 3.29) the sensitivity $\overline{N}_{Rmin}$ as a function of Λ_{1min}:

$$\overline{N}_{Rmin} = \frac{\Lambda_{1min}G^x A_0}{4\eta}\left[1 + \sqrt{1 + \frac{4}{A_0^2\Lambda_{1min}G^x}\left(N_0 A_1 + \frac{\rho_A^2}{G^{2+x}}\right)}\right] \tag{3.139}$$

Now let us examine two important special cases.

3.5.3.1 Integrate-and-Dump Receiver, PIN Photodiode

For a PIN photodiode, $G = G_2 = 1$; for an integrate-and-dump receiver, $A_0 = A_1 = 1$. Hence, (3.139) simplifies to

$$\overline{N}_{Rmin} = \frac{\Lambda_{1min}}{4\eta}\left[1 + \sqrt{1 + \frac{4}{\Lambda_{1min}}(N_0 + \rho_A^2)}\right] \tag{3.140}$$

Neglecting the dark noise contribution and assuming $4\,\rho_A^2/\Lambda_{1min} >> 1$, we obtain the following approximate result:

$$\overline{N}_{Rmin} \simeq \frac{\rho A}{2\eta}\sqrt{\Lambda_{1min}} \tag{3.141}$$

Equation (3.141) shows that $\overline{N}_{Rmin}$ is proportional to ρ_A and $\sqrt{\Lambda_{1min}}$.

Example 3.7

Using the figure of merit $\rho_A = 2{,}497$, as per Example 3.4, the value of $\Lambda_{1min} = 420$, as per Example 3.6, and assuming $\eta = 0.8$ and $N_0 = 0$, we obtain from (3.140) $\overline{N}_{Rmin} \simeq 32{,}115$, which is quite far (35 dB) from the quantum limit.

3.5.3.2 Integrate-and-Dump Receiver, APD

In this case, since $A_0 = A_1 = 1$, (3.139) becomes

$$\overline{N}_{Rmin} = \frac{\Lambda_{1min}G^x}{4\eta}\left[1 + \sqrt{1 + \frac{4}{\Lambda_{1min}G^x}\left(N_0 + \frac{\rho_A^2}{G^{2+x}}\right)}\right] \tag{3.142}$$

Neglecting the dark current contribution and assuming $4\rho_A^2/(\Lambda_{1min}G^{2+2x}) >> 1$, we obtain the following approximate result:

$$\overline{N}_{Rmin} \simeq \frac{\rho A}{2\eta G}\sqrt{\Lambda_{1min}} \tag{3.143}$$

Comparison of this result with (3.141) shows that the APD improves the sensitivity by factor G.

Example 3.8

Using the same numerical values as in Example 3.7 and $G = 50$, $x = 0.3$, (3.142) yields $\overline{N}_{Rmin} \simeq 1{,}192$. This is only 21 dB away from the quantum limit, and 14 dB better than the PIN photodiode case in Example 3.7.

3.5.3.3 Optimizing the Avalanche Gain

The gain, G, of an APD can be modified by adjusting its bias voltage. Thus, it is important to know the optimum value of the gain, G, that yields the highest SNR Λ_1.

Figure 3.36 shows Λ_1 [as per simplified expression (3.116)] versus G for given values of the receiver figure of merit, ρ_A, noise excess factor x, and for several values of $\eta\overline{N}_R$. Initially, Λ_1 increases with G, up to a point where the excess noise becomes dominant. Thus, for each value of $\overline{N}_R$, there is an optimum value of the gain G. That optimum value is given by (see Problem 3.29)

$$G_{opt} = \left[\frac{2\rho_A^2}{x(2\eta\overline{N}_R A_0 + N_0 A_1)}\right]^{\frac{1}{x+2}} \tag{3.144}$$

So far, we have learned how to derive the SNR Λ_{1min} yielding a given bit error probability and how to express the receiver sensitivity (i.e., the minimum received number of photons per bit, $\overline{N}_{Rmin}$) in terms of Λ_{1min}.

To exploit this fundamental relationship, given by (3.139), we need to know the

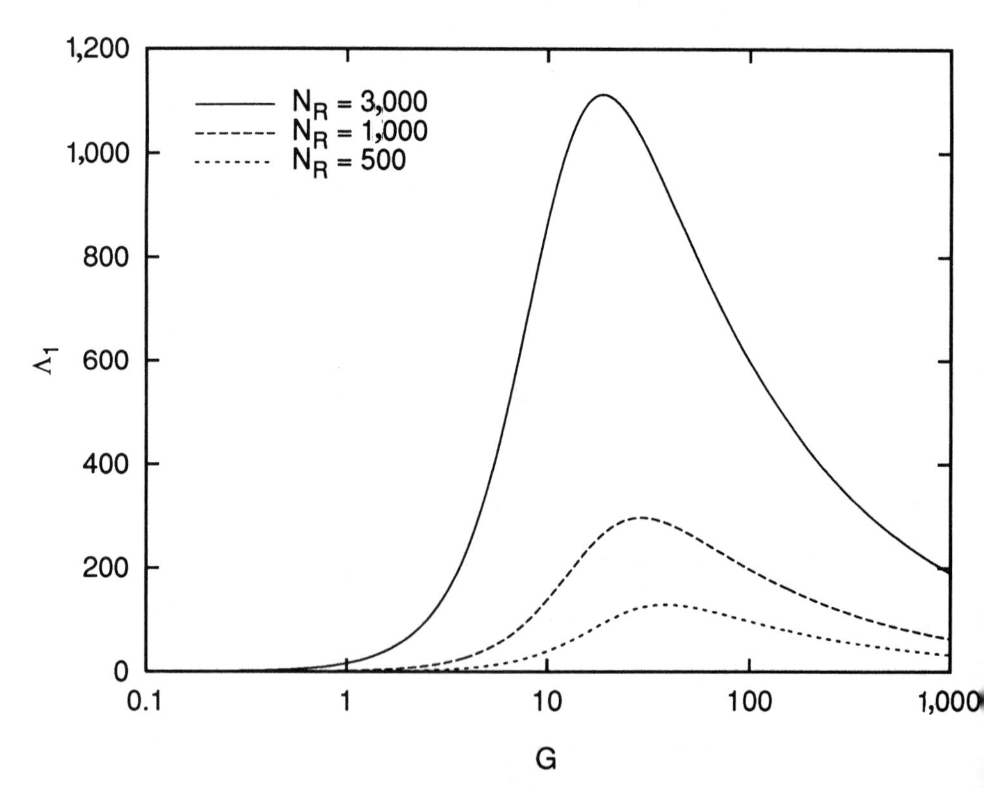

Figure 3.36 Signal-to-noise ratio Λ_1 as a function of the avalanche gain, *G*, of the photodetector for three values of the number of received photons per bit, N_R.

parameters G and G_2 that characterize the APD (when an APD is used), the overall transfer function, $H_L(f) = H_t(f)H_e(f)$, of the linear part of the receiver (amplifiers plus equalizer), and the figure of merit of the preamplifier. In Section 3.6, we will present two important receiver structures, the *high-impedance* receiver and the *transimpedance* receiver, and show how to evaluate the required system parameters.

3.6 HIGH-IMPEDANCE AND TRANSIMPEDANCE RECEIVERS: A COMPARISON

3.6.1 High-Impedance Receiver

A simplified structure of the high-impedance receiver is illustrated in Figure 3.37, and its equivalent circuit is shown in Figure 3.31. To find the receiver figure of merit, ρ_A, we neglect the dark and shot noise currents, i_d and $i_{sh}(t)$. Recalling the definition (3.104) of ρ_A, we need first to evaluate the sampled signal output, r_0.

Consider first the transfer function between $i_s(t)$ and the input voltage, $v_i(t)$, in Figure 3.31. Let us call it $R_i(f)$. Recalling from (3.67) that the shaping waveform of $i_s(t)$ is $p(t)$, we obtain

$$R_i(f) = P(f)R_tH_t(f) = \frac{P(f)R_t}{1 + j\left(\frac{f}{f_t}\right)} \tag{3.145}$$

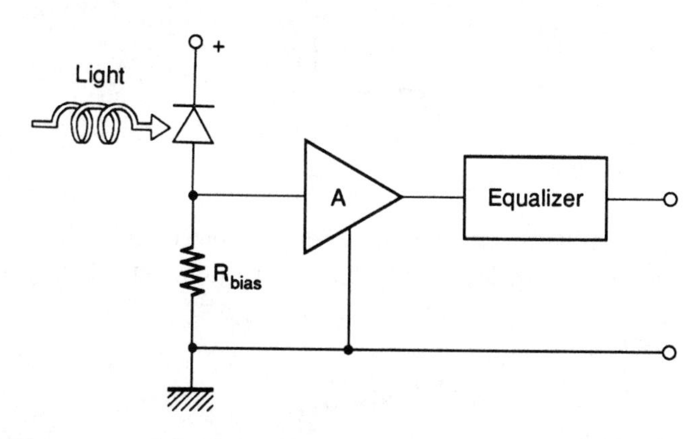

Figure 3.37 The structure of the high-impedance receiver.

where f_t is the cutoff frequency of $H_t(f)$:

$$f_t \overset{\text{def}}{=} \frac{1}{2\pi R_t C_t} \tag{3.146}$$

Now, let us find the thermal noise spectral density. Using (3.84), we obtain

$$G_{r_i}(f) = G_{i_{th}}(f) R_t^2 |H_t(f)|^2 = \frac{2kT_0 R_t F(f)}{1 + \left(\frac{f}{f_t}\right)^2} \tag{3.147}$$

where the noise figure, $F(f)$, is given by (3.85). One can see from (3.145) and (3.147) that the power of $v_i(t)$ is directly proportional to R_t^2, whereas the noise spectral density is proportional to R_t. Thus, to decrease the figure of merit[7], one needs to increase the load resistance, R_t. That is why, in the high-impedance design, R_t is chosen to be large. Consequently, the input time constant, $R_t C_t$, becomes large compared with the bit interval T, so that the cut-off frequency, f_t, is smaller than the bit rate. Because of that fact, the high-impedance receiver is also referred to as an *integrating receiver*.

The insufficient "raw" bandwidth of integrating/high-impedance receivers can lead to strong intersymbol interference. To prevent that from happening, an equalizer in the form of a differentiating network is necessary to extend the receiver bandwidth. The equalizer can be realized as the passive circuit shown in Figure 3.38. Frequency dependence of the transimpedance $Z_t(f) = R_t H_t(f)$ and of the equalizer transfer function are also shown in Figure 3.38. The equalizer transfer function is given by

$$H_e(f) = \alpha_e \frac{1 + j\left(\frac{f}{f_1}\right)}{1 + j\left(\frac{f}{f_2}\right)} \tag{3.148}$$

where α_e is the low-frequency attenuation of the equalizer, and f_1 and f_2 are the equalizer's zero and pole, respectively. The equalizer's parameters are given by

$$\alpha_e = \frac{R_1}{R_1 + R_2} \tag{3.149}$$

7. Recall that a higher figure of merit means a worse receiver.

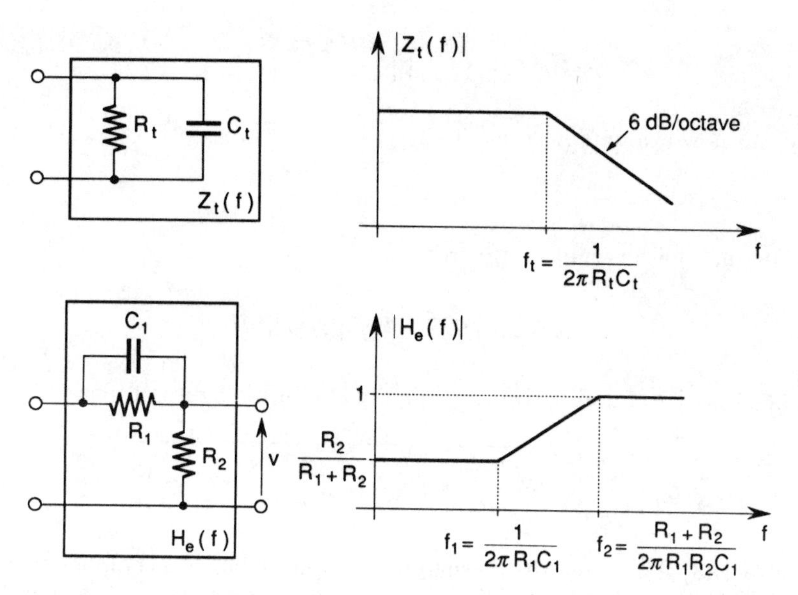

Figure 3.38 Schematic plot of the transfer function of the input impedance, $Z_t(f)$, and of the equalizing circuit, $H_e(f)$, versus the frequency for the high-impedance receiver.

$$f_1 = \frac{1}{2\pi R_1 C_1}$$

$$f_2 = \frac{R_1 + R_2}{2\pi R_1 R_2 C_1}$$

For perfect equalization, the equalizer's zero, f_1, should coincide with the pole frequency, f_t of $H_t(f)$. The bandwidth of the amplifier is then extended to the equalizer pole at f_2.

When an equalizer is used, the receiver gain is reduced by α_e. This can have a negative impact on the receiver dynamic range, since the preamplifier gain must be increased to compensate for the equalizer attenuation.

In the following analysis, we set, for simplicity, $R_1 = R_t$ and $C_1 = C_t$. We also assume that the frequency, f_2, of the equalizer pole is greater than the signal bandwidth, B. At the output of the equalizer, we assume a lowpass filter with a rectangular transfer function having a unit magnitude and bandwidth B identical to the signal bandwidth. Thus, the waveform, $p(t)$, passes through the filter without distortions.

The resulting cascade of $H_t(f)$, $H_e(f)$, and the output lowpass filter has the transfer function given by

$$H_L(f) = \begin{cases} \alpha_e & \text{for } -B \le f \le B \\ 0 & \text{elsewhere} \end{cases} \tag{3.150}$$

We can now evaluate r_0 from (3.73):

$$r_0 = \alpha_e p_0$$

and the thermal noise variance, σ_{th}^2, from (3.94):

$$\sigma_{th}^2 = 2kT_0R_t \int_{-\infty}^{\infty} F(f)|H_L(f)|^2 df \tag{3.152}$$

Using (3.85), (3.104), and (3.150), we finally obtain the figure of merit:

$$\rho_A = \frac{T}{qp_0}\sqrt{2B\left[\frac{2kT_0}{R_t} + G_{i_A} + G_{e_A}\left(\frac{1}{R_t^2} + \frac{4}{3}\pi^2 C_t^2 B^2\right)\right]} \tag{3.153}$$

Equation (3.153) shows that the thermal noise variance has two components related to the input resistance and amplifier current noise generator; both increase with B. The third component, due to the amplifier voltage noise, is proportional to B^3 and is dominant for high bit rates.

Two important conclusions follow from (3.153):

- The figure of merit decreases with the load resistance, R_t, up to the limit

$$\lim_{R_t \to \infty} \rho_A = \frac{T}{qp_0}\sqrt{2B\left(G_{i_A} + \frac{4}{3}\pi^2 G_{e_A} C_t^2 B^2\right)} \tag{3.154}$$

- The performance of high-bit rate systems is limited by the parasitic capacitance C_t:

$$\lim_{R_t, B \to \infty} \rho_A = \sqrt{\frac{8G_{e_A}B}{3}}\frac{\pi C_t}{qp_0} \tag{3.155}$$

where we assumed $B = R_b = 1/T$. In the limit, the figure of merit of the amplifier increases with the square root of the signal bandwidth (and, therefore, of the bit rate).

To use (3.139), we still need to evaluate the coefficients A_0 and A_1 defined by (3.108)–(3.109). For the high-impedance receiver, with the overall transfer function (3.150), we obtain (see Problem 3.30)

$$A_0 \le \frac{2TB}{p_0} \tag{3.156}$$

and

$$A_1 = \frac{2TB}{p_0^2} \tag{3.157}$$

For the equal sign in (3.156) to hold, the ideal filter bandwidth, B, needs to be significantly larger than the signal bandwidth. The exact value of A_0 depends on the received optical power pulse, $p(t)$, and on the receiver transfer function; it can be computed numerically. The largest value of A_0 [corresponding to the equal sign in (3.156)] is a conservative choice for system design, since it yields a lower bound on the SNRs Λ_1 and Λ_0 and thus an upper bound on the error probability. For that reason, we use the largest value of A_0 in the following analysis.

Substituting expressions for ρ_A, A_0, and A_1 into (3.105), we obtain an explicit expression of the SNR Λ_1

$$\Lambda_1 = \frac{(2GR\overline{P}_R p_0)^2}{2B\left[G_2 q(2R\overline{P}_R p_0 + \lambda_0 q) + \dfrac{2kT_0}{R_t} + G_{i_A} + G_{e_A}\left(\dfrac{1}{R_t^2} + \dfrac{4\pi^2 C_t^2 B^2}{3}\right)\right]} \tag{3.158}$$

3.6.2 Transimpedance Receiver

The transimpedance amplifier uses a shunt-feedback amplifier, as illustrated in Figure 3.39. The corresponding equivalent circuit is shown in Figure 3.40. As for the high-impedance receiver, to find the receiver figure of merit, we neglected the dark current and the shot noise, while including the current noise source, $i_F(t)$, due to the

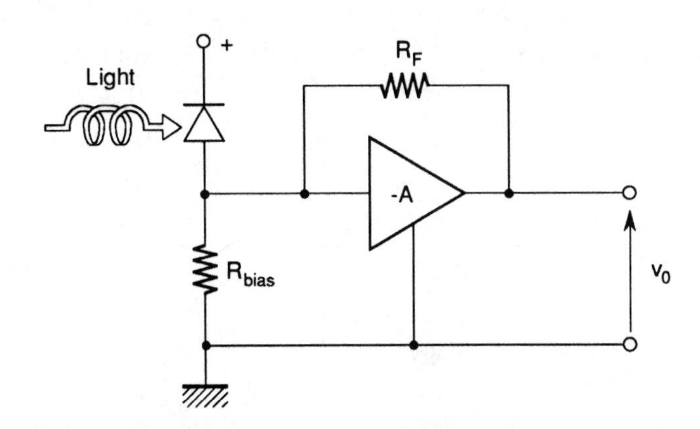

Figure 3.39 The structure of the transimpedance receiver.

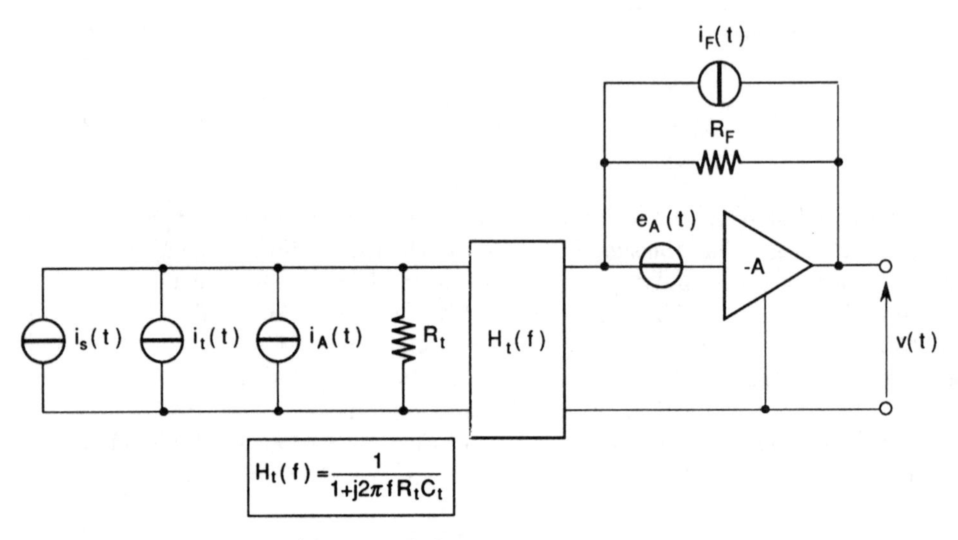

Figure 3.40 Equivalent circuit of the transimpedance receiver.

feedback resistance, R_F. The latter is a white Gaussian random process with spectral density $G_{i_F} = 2kT_0/R_F$. We will evaluate the figure of merit, ρ_A, of this receiver.

Let us consider first the transfer function between the signal current, $i_s(t)$, and the output voltage, $v(t)$.[8] It is given by (see Problem 3.31)

$$Z_1(f) \stackrel{\text{def}}{=} \frac{V(f)}{I_s(f)} = -\frac{A}{A+1} \cdot \frac{R_{eq}}{1 + j\left(\frac{f}{f_{eq}}\right)} \tag{3.159}$$

where A is the amplifier gain, and R_{eq} and f_{eq} are the equivalent input resistance and the pole of the transfer function given by

$$R_{eq} = \frac{R_t R_F}{R_t + \frac{R_F}{A+1}} \tag{3.160}$$

$$f_{eq} = \frac{1+A}{2\pi C_t R_{eq}} \tag{3.161}$$

8. We denote the dimensionless transfer functions by H and the (transimpedance) transfer functions with dimension of Ω by Z.

For large amplifier gains, (3.160) and (3.161) yield

$$R_{eq} \simeq R_F, \qquad f_{eq} \simeq \frac{1 + A}{2\pi C_t R_F} \overset{\text{def}}{=} f_F \tag{3.162}$$

Expression (3.162) shows that the pole of the transfer function has been shifted to higher frequencies by a factor $1 + A$, with a corresponding increase in bandwidth. This phenomenon is known as the Miller effect.

The transimpedance amplifier is normally designed so that its bandwidth is high enough to accommodate the signal bandwidth; thus, no equalization is required. For a given feedback resistance, the bandwidth can be increased by increasing the open-loop gain A. There is a limit to the gain increase, though, mainly due to potential instability of the feedback structure.

The transimpedance amplifier has the advantage of a higher dynamic range than the high-impedance amplifier. The reason is that the absence of the equalizer permits keeping the amplifier gain lower by a factor equal to the equalizer attenuation. The latter is equal to the ratio of the pole frequencies of the input impedance, Z_t, and of the equalizer transfer function, $H_e(f)$. Thus, the transimpedance receiver improves the dynamic range by factor A (amplifier gain) compared to the integrating receiver.

We can now evaluate r_0 as

$$r_0 = \mathcal{F}^{-1}\left[-\frac{A}{A+1} \cdot \frac{P(f)R_{eq}}{1 + j\left(\frac{f}{f_{eq}}\right)} \right]_{|t=t_0} \tag{3.163}$$

To find the thermal noise variance, we first will evaluate the spectral density of the output noise voltage, v_{th}:

$$G_{th}(f) = \left(\frac{2kT_0}{R_t} + G_{i_A}\right) |Z_1(f)|^2 + \frac{2kT_0}{R_F}|Z_2(f)|^2 + G_{e_A}|H_3(f)|^2 \tag{3.164}$$

where $Z_1(f)$ is the transfer function between the the noise currents, $i_t(t)$, $i_A(t)$, and the output voltage; $Z_2(f)$ is the transfer function between the noise current, $i_F(t)$, and the output voltage; and $H_3(f)$ is the transfer function between the noise amplifier voltage, $e_A(t)$, and the output voltage. The variables $Z_2(f)$ and $H_3(f)$ are defined by

$$Z_2(f) \overset{\text{def}}{=} \frac{V(f)}{I_F(f)}$$

$$H_3(f) \stackrel{\text{def}}{=} \frac{V(f)}{E_A(f)}$$

As shown in Problem 3.31, $Z_2(f)$ and $H_3(f)$ are given by

$$Z_2(f) = Z_1(f) \tag{3.165}$$

$$H_3(f) = -\frac{A}{A+1} \cdot \frac{\frac{R_{eq}}{R_F} + \frac{R_{eq}}{R_t} + j2\pi f R_{eq} C_t}{1 + j\left(\frac{f}{f_{eq}}\right)} \tag{3.166}$$

Let us assume that the frequency pole f_{eq} of the transfer functions $Z_1(f)$ and $H_3(f)$ is much greater than the signal bandwidth and that at the output of the amplifier a lowpass filter is used with rectangular transfer function, unit magnitude, and bandwidth B identical to the signal bandwidth. Then the two transfer functions, $Z_1(f)$ and $H_3(f)$, can be approximated as follows:

$$Z_1(f) \simeq \begin{cases} -\frac{A}{A+1} \cdot R_{eq} & \text{for } -B \le f \le B \\ 0 & \text{elsewhere} \end{cases} \tag{3.167}$$

and

$$H_3(f) \simeq \begin{cases} -\frac{A}{A+1} \cdot \left(\frac{R_{eq}}{R_F} + \frac{R_{eq}}{R_t} + j2\pi f R_{eq} C_t\right) & \text{for } -B \le f \le B \\ 0 & \text{elsewhere} \end{cases} \tag{3.168}$$

Substituting (3.167) and (3.168) into (3.164) and integrating the output spectral density from $-B$ to B, we obtain the output thermal noise variance. Using (3.79) and (3.163), we obtain the output voltage, v_{ph}:

$$v_{ph} = -\frac{A}{A+1} \cdot \frac{q p_0 R_{eq}}{T} \tag{3.169}$$

The figure of merit, ρ_A, can now be obtained from (3.104):

$$\rho_A = \frac{T}{q p_0}\sqrt{2B\left\{2kT_0\left(\frac{1}{R_t}+\frac{1}{R_F}\right)+G_{i_A}+G_{e_A}\left[\left(\frac{1}{R_t}+\frac{1}{R_F}\right)^2+\frac{4}{3}\pi^2 C_t^2 B^2\right]\right\}} \quad (3.170)$$

Let us compare the figure of merit in (3.170) of the transimpedance receiver with the figure of merit in (3.153) of the high-impedance receiver. They differ only in the fact that the resistance, R_t, in (3.153) is replaced in (3.170) by two resistances, R_t and R_F, connected in parallel. Yet, for the transimpedance receiver, the equalizer is not required, thanks to the Miller effect.

The parameters A_0 and A_1 needed to use (3.139) are the same as for the high-impedance receiver:

$$A_0 \le \frac{2TB}{p_0} \quad (3.171)$$

and

$$A_1 = \frac{2TB}{p_0^2} \quad (3.172)$$

As explained in Section 3.6.1 for the high-impedance receiver, we will use the equal sign in (3.171).

3.6.3 Summary of the Sensitivity Analysis

In previous subsections, we derived all the ingredients needed to analyze the sensitivity of a given digital IM-DD system. Here, we summarize the main steps of the analysis.

1. The starting point is a high-impedance or transimpedance receiver structure. It is characterized by the values of passive components, like R_t, R_F, C_t, and by the noise characteristics of the active devices. Alternatively, the noise figure, $F(f)$, can be provided. Square roots of the spectral densities, G_{i_A} and G_{e_A}, are usually given in pA/$\sqrt{\text{Hz}}$ and nV/$\sqrt{\text{Hz}}$, respectively. Their values depend on the active device employed, like a field-effect transistor (FET) or a bipolar transistor.
2. Using the data in step 1 and (3.153) or (3.170), we can find the figure of merit of the high-impedance or transimpedance amplifiers and the parameters A_0 and A_1. The photodetector is described by the parameters η, G, x, and i_d. If the gain of the avalanche detector can be controlled, the optimum value can be found using (3.144).

3. Now, we use (3.105) and (3.106) to find the SNRs Λ_1 and Λ_0. Inserting Λ_1 and Λ_0 into (3.132), we find the optimum threshold as a function of $\overline{N}_R$.
4. We then substitute $\Lambda_1(\overline{N}_R)$, $\Lambda_0(\overline{N}_R)$, and $\beta_{opt}(\overline{N}_R)$ into (3.131) for the bit error probability, possibly modified through the insertion of a system margin, k_m^2. Next, $P(e)$ is plotted versus $\overline{N}_R$. The abscissa obtained from the error probability curve at 10^{-9} yields the system sensitivity.

This procedure can be easily implemented in software and allows us to find the sensitivity of optimized systems.

Consider a system with the following characteristics: $r_0 = w_0 = 1$; PIN photodiodes or APDs, with $\eta = 0.8$, $i_d = 3$ nA, $x = 0.77$, and optimized G. The receiver uses a transimpedance amplifier, with $R_t = R_F = 1$ kΩ, $C_t = 2$ pF, $\sqrt{G_{i_A}} = 2$ pA/$\sqrt{\text{Hz}}$, and $\sqrt{G_{e_A}} = 2$ nV/$\sqrt{\text{Hz}}$. The receiver bandwidth is $B = 0.7\,R_b$. The transimpedance amplifier gain is 20 dB. As the bit rate increases, the feedback resistance, R_F is decreased to keep the receiver bandwidth large enough to maintain the relationship $B = 0.7\,R_b$.

Analysis results for the foregoing special case are shown in Figures 3.41, 3.42, and 3.43. In Figure 3.41, the optimized avalanche gain, G, is plotted versus the system bit rate. The optimum gain first decreases with the system bit rate. That is due to the fact that G_{opt} increases with the receiver figure of merit (see (3.144)), which, in turn, decreases with the bit rate for low-medium bit rates (see (3.170)), up to the point at

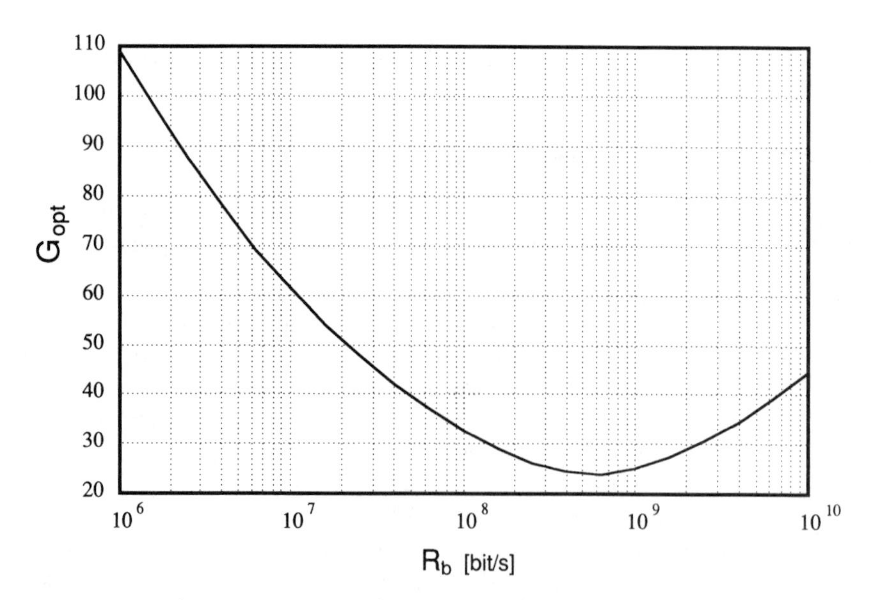

Figure 3.41 Optimized APD gain, G, versus bit rate, R_b.

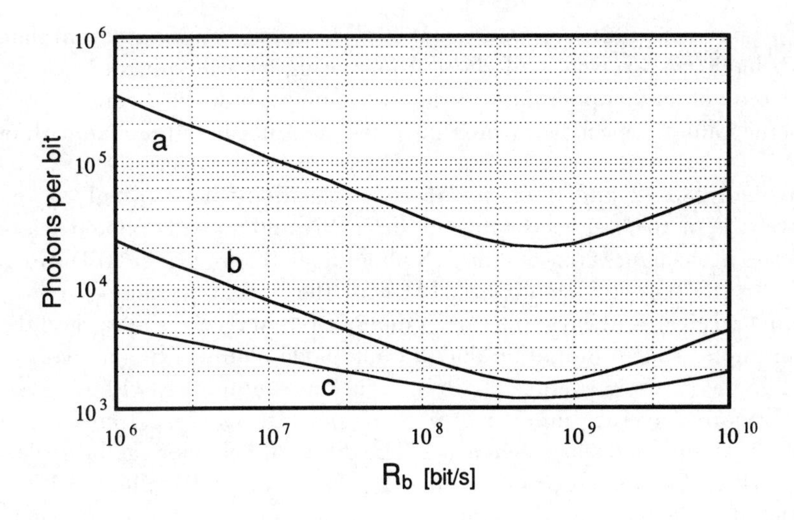

Figure 3.42 System sensitivity expressed in terms of average number of photons per bit, $\overline{N}_R$, versus bit rate, R_b, for a transimpedance receiver employing a PIN (curve *a*), and APD with fixed gain $G = 20$ (curve *b*), and an APD with optimized gain (curve *c*).

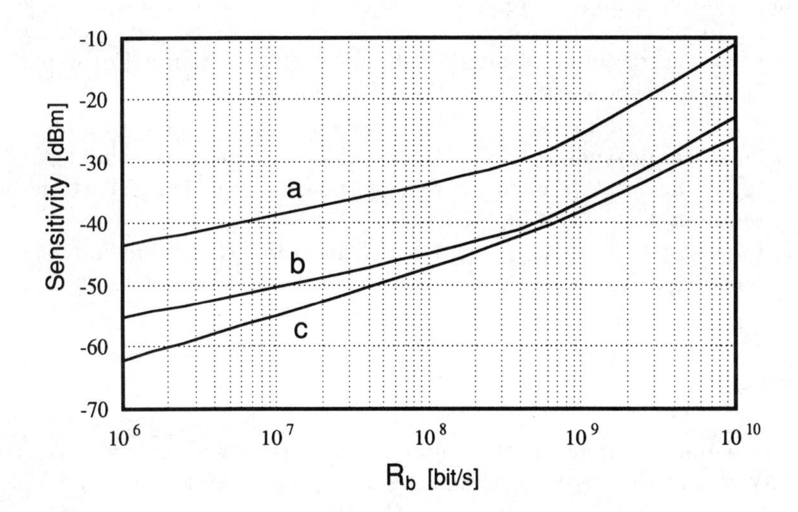

Figure 3.43 System sensitivity expressed in terms of average received power, $\overline{P}_R$, in dBm, versus bit rate, R_b, for a transimpedance receiver employing a PIN (curve *a*), an APD with fixed gain $G = 20$ (curve *b*), and an APD with optimized gain (curve *c*).

which the noise contribution proportional to B^3 becomes dominant. From that point on, the behavior is reversed, and G keeps increasing with the bit rate.

We did not set an upper limit on the APD gain to show the theoretical optimum value; some values, especially at high bit rates, may not be attainable with current APDs.

In Figures 3.42 and 3.43, the receiver sensitivity in terms of $\overline{N}_R$ and $\overline{P}_R$, respectively, is plotted versus the bit rate for a PIN and an APD photodetector. The three curves in the figures refer to the PIN photodiode (curve a), the APD with a fixed gain of $G = 15$ (curve b), and the APD with the optimized gain (curve c). The APD-based receiver is always superior to the PIN-based receiver by several decibels. In this example, system limitation due to photodiode bandwidth is not considered.

While the required average power, $\overline{P}_R$, increases uniformly with the system bit rate, the required average number of photons per bit, $\overline{N}_R$, first decreases with the system bit rate and then starts increasing. The different behavior can be explained as follows. $\overline{N}_R$ is approximately proportional to ρ_A [see (3.141)]. Recalling (3.170) of ρ_A, we see that, for low bit rates, the dominant term is proportional to $1/\sqrt{B}$, whereas for high bit rates the dominant term is proportional to $\sqrt{B}$. At an intermediate bit rate (that depends on the value of C_t), we have a minimum of ρ_A, and, thus, a minimum of $\overline{N}_R$. On the other hand, (3.134), the relationship between $\overline{N}_R$ and $\overline{P}_R$, shows that $\overline{P}_R$ is proportional to the product of $\overline{N}_R$ with the bit rate. Hence, $\overline{P}_R$ increases uniformly with the bit rate.

When a faster analysis is required, we can proceed as follows:

1. Evaluate the figure of merit of the receiver, as explained in step 1 of the analysis at the beginning of the subsection.
2. Fix the system margin k_m^2, the photodiode parameters η, G, x and i_d, and find an approximate threshold value of the symmetric threshold, β_s (3.126) (β_s is close to 0.5 when the thermal noise is prevalent). Then derive the required Λ_{1min} from (3.138).
3. Finally, substitute Λ_{1min}, A_0, A_1, η, G, and x into (3.139) to find the receiver sensitivity, $\overline{N}_R$.

Example 3.9

Consider a 1-Gbps system using a transimpedance receiver with $\sqrt{G_{i_A}} = 2$ pA/$\sqrt{\text{Hz}}$ and $\sqrt{G_{e_A}} = 2$ nV/$\sqrt{\text{Hz}}$. Other receiver parameters are $R_t = R_F = 10$ kΩ and $C_t = 5$ pF. The receiver uses an APD photodetector with $G = 20$ and $G_2 = 200$ (corresponding to $x \simeq 0.77$), a quantum efficiency $\eta = 0.8$, and a dark current $i_d = 3$ nA. These figures are representative of InGaAs PIN diodes and APDs working in the 1,550-nm-wavelength window.

Let us follow the foregoing simplified sensitivity analysis.

1. First, find the amplifier figure of merit by substituting the receiver parameters into (3.170) and assuming a receiver bandwidth $B = (0.7)/T = 0.7$ GHz. For $p_0 = 1$, we obtain:

$$\rho_A = \frac{T}{q p_0}\sqrt{2B}\cdot\left\{2kT_0\left(\frac{1}{R_t}+\frac{1}{R_F}\right)+G_{i_A}+\left[G_{e_A}\left(\frac{1}{R_t}+\frac{1}{R_F}\right)^2+G_{e_A}\frac{4}{3}\pi^2 C_t^2 B^2\right]\right\}^{\frac{1}{2}}$$

$$\rho_A = 6.24\times 10^9\sqrt{1.4\times 10^9}\cdot[1.656\times 10^{-24}+4\times 10^{-24}+(1.6\times 10^{-25}+6.44\times 10^{-22})]^{1/2}$$

We can recognize, under the square root, from left to right, the noise contributions of the input and feedback resistances and of the amplifier current and voltage noises. At high bit rates, the most important noise contribution comes from the amplifier noise voltage; it is proportional to B^3. Completing the calculations yields

$$\rho_A \simeq 5{,}952$$

Then we evaluate the parameters A_1 and A_0 using (3.171) and (3.172). For $p_0 = 1$, we obtain

$$A_0 = A_1 = 1.4$$

2. We now fix a system margin $k_m^2 = 2$, corresponding to 3 dB; a threshold $\beta_s = 0.4$; and, from (3.138), derive the minimum SNR, Λ_{1min}

$$\Lambda_{1min} = \frac{36\cdot 2}{0.36} = 200$$

3. Finally, from (3.139), we obtain the receiver sensitivity:

$$\overline{N}_{Rmin} \simeq 3{,}647 \text{ photons per bit}$$

At the wavelength of 1,550 nm, the corresponding minimum power is

$$P_{Rmin} \simeq -33.3 \text{ dBm}$$

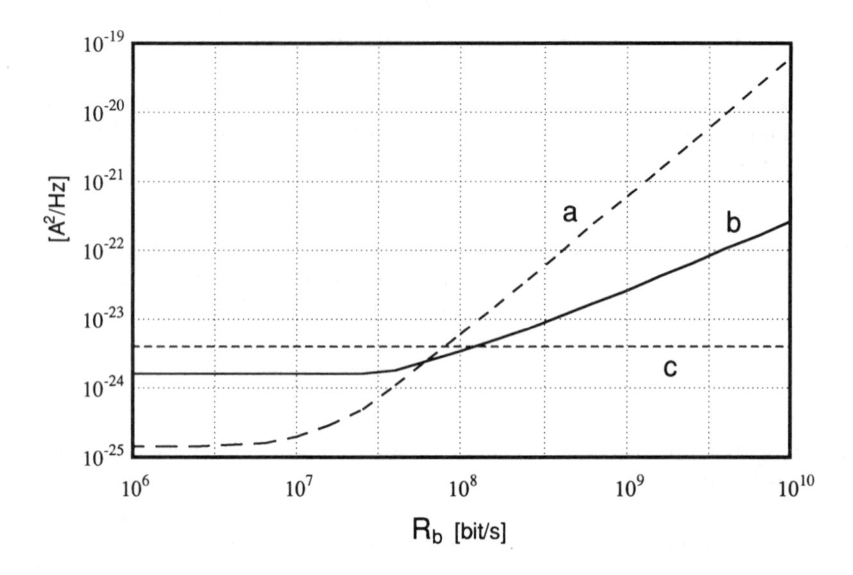

Figure 3.44 Noise spectral densities of receiver resistances (curve *b*), amplifier current generator (curve *c*), and amplifier voltage generator (curve *a*) versus bit rate for system described in Example 3.9.

Now, let us see how various noise contributions and the figure of merit of the amplifier vary with the bit rate. In Figure 3.44, the three noise spectral densities (due to the load-input resistance, amplifier current noise, and amplifier voltage noise) are plotted versus the bit rate. The contribution of the amplifier current noise (curve c) is constant, whereas the other two contributions increase with the bit rate. The contribution of the amplifier voltage noise (curve a) is proportional to B^2 and to the capacitance, C_t. The amplifier gain, A, is fixed at 20 dB. As a result, when the bit rate increases, we need to decrease the feedback resistance, R_F, to keep the amplifier bandwidth large enough. That, in turn, increases the noise contribution due to the feedback resistance, which is proportional to $1/R_F$. As a consequence, the noise contribution due to R_F (curve b) versus bit rate starts increasing from that bit rate at which the feedback resistance must be increased.

In Figure 3.45 we plot the figure of merit of the amplifier versus the bit rate. The curve shows first an improvement and then a progressive deterioration. This behavior can be explained by (3.170), which shows that for low bit rates the dominant term is $1/\sqrt{B}$, whereas for high bit rates the dominant term is proportional to $\sqrt{B}$. At an intermediate bit rate (the value depends on the value of C_t), we have a minimum of ρ_A.

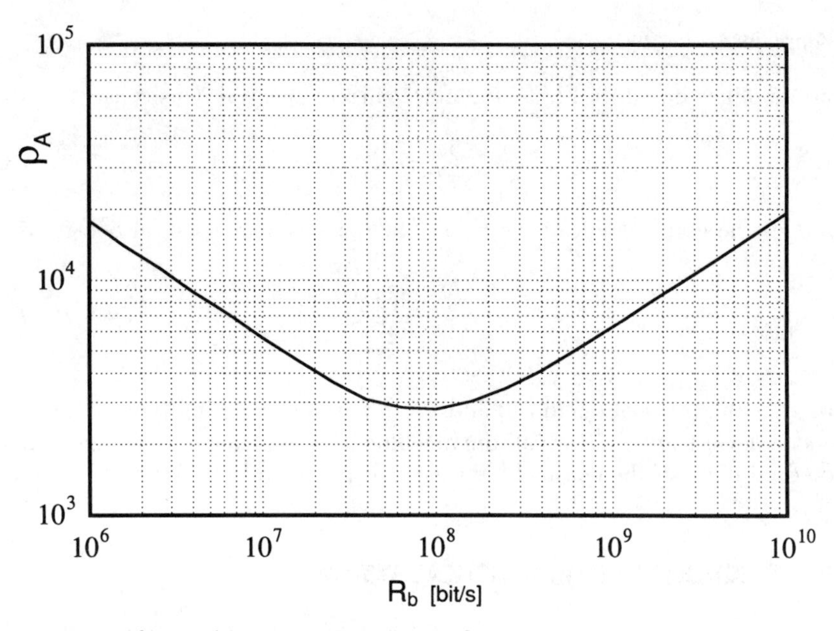

Figure 3.45 Figure of merit of the receiver versus bit rate for system described in Example 3.9.

3.6.3.1 Noise Characteristics of Preamplifiers

A thorough discussion of noise characteristics of low-noise preamplifiers employing FETs and bipolar transistors can be found in [1,Chap. 3;5,Chap. 11]. Here we present a brief simplified characterization of preamplifiers.

Bipolar Transistor Amplifiers

For an amplifier based on bipolar transistors, the noise current spectral density is

$$G_{i_A} = qI_b \tag{3.173}$$

where I_b is the transistor bias base current. The noise voltage spectral density is

$$G_{e_A} = \frac{qI_c}{g_m^2} \tag{3.174}$$

where I_c is the collector bias current, and g_m is the transconductance of the transistor.

FET Amplifiers

For an amplifier based on FETs, the noise current spectral density is

$$G_{i_A} = qI_g \tag{3.175}$$

where I_g is the gate current of the FET. The noise voltage spectral density is

$$G_{e_A} = \frac{2kT_0\Gamma}{g_m} \tag{3.176}$$

where g_m is the transconductance, and Γ is a parameter taking typical values of 0.7 for silicon devices and 1.1 for gallium arsenide devices. The noise voltage stems from the thermal noise of the channel between the source and the drain.

3.7 IMPERFECTIONS IN DIGITAL OPTICAL SYSTEMS

We have seen so far how fiber loss, fiber dispersion, and receiver noise affect IM-DD digital optical communication systems. In particular, receiver analysis permits us to evaluate receiver sensitivity, that is, the minimum average received power necessary to achieve a given bit error probability. Thus, knowing the fiber loss, we can predict the maximum link length for loss-limited systems. In Section 3.3.3 we discussed the SNR penalty associated with fiber dispersion.

Practical systems, however, can suffer from additional impairments that, for the sake of simplicity, we have not considered. Some of the additional impairments are discussed next.

3.7.1 Extinction Ratio

The energy carried by the binary 0 can be larger than zero. For semiconductor lasers, the power associated with the binary 0 depends on the bias current, the threshold current, and the drive current.

The *extinction ratio*, r_e, is defined as the ratio between the power, P_0, associated with the binary 0 and that associated with the binary 1:

$$r_e \stackrel{\text{def}}{=} \frac{P_0}{P_1}, \qquad 0 \le r_e < 1 \tag{3.177}$$

Using (3.177), the average power falling on the photodetector is no longer $P_1/2$. Instead, it is equal to

$$P_{av} = \frac{P_1 + P_0}{2} = \frac{P_1}{2}(1 + r_e) \tag{3.178}$$

The resulting penalty can be evaluated in terms of the parameter Q defined in (3.129). Assuming that the sampled pulse $r_0 = 1$, we obtain

$$Q = \frac{2RP_{av}}{\sigma_n(1) + \sigma_n(0)} \cdot \frac{1 - r_e}{1 + r_e} \tag{3.179}$$

In (3.179), the variances depend on r_e, too, because of the shot noise dependence on the received power. Assuming for simplicity that the thermal noise is dominant, the penalty due to the extinction ratio, Δ_e, is given by:

$$\Delta_e \stackrel{\text{def}}{=} \frac{P_{av}(r_e)}{P_{av}(r_e = 0)} = \frac{1 + r_e}{1 - r_e} \tag{3.180}$$

The penalty Δ_e in decibels is plotted in Figure 3.46 as a function of r_e.

When the laser is biased below the threshold, r_e is very small, and resulting

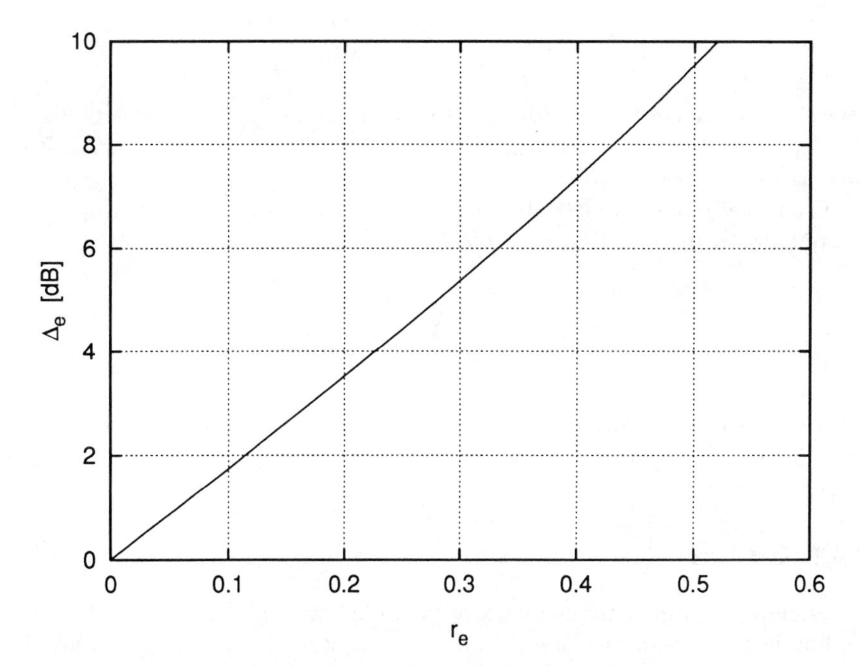

Figure 3.46 Power penalty due to extinction ratio versus extinction ratio, r_e.

penalty is negligible. On the other hand, if the laser is biased above the threshold (this is necessary to extend the laser modulation bandwidth), the penalty can become significant. The analysis was extended to include the dependence of σ_0 and σ_1 from r_e [5,Chap. 12].

3.7.2 Relative Intensity Noise

The impairment due to RIN (*relative intensity noise*) (see Chap. 1) is due to the fact that the transmitted light power (and, consequently, the received light power) fluctuates in time. RIN adds an extra noise to the receiver. As a result, the total noise variance in the receiver becomes

$$\sigma_n^2 = \sigma_{sh}^2 + \sigma_{th}^2 + \sigma_i^2 \tag{3.181}$$

where σ_i reflects the RIN. The value of σ_i^2 depends on the received optical power as follows:

$$\sigma_i^2 = R\overline{P}_R r_i \tag{3.182}$$

where r_i, in turn, can be computed from the RIN of the transmitter:

$$r_i^2 \simeq 10^{RIN/10} \times B \tag{3.183}$$

where B is the system bandwidth, and RIN is the power spectral density of the RIN expressed in decibels per hertz, introduced in Chapter 1. For good lasers, RIN can be as low as -165 dB/Hz.

From the definition of Q, neglecting the extinction ratio and the dark current, we easily obtain the resulting penalty:

$$\Delta_i \overset{\text{def}}{=} \frac{P_{av}(r_i)}{P_{av}(r_i = 0)} = \frac{1}{1 - r_i^2 Q^2} \tag{3.184}$$

Expression (3.184) is plotted in Figure 3.47 for $Q = 6$, necessary to achieve an error probability of 10^{-9}. For RIN below -110 dB, corresponding to $r_i < 0.01$, the penalty is negligible.

3.7.3 Timing Jitter

To eliminate the intersymbol interference from the sampled signal, the optimum sampling instant has to be found. In practice, however, the timing synchronization circuit, shown in Figure 3.24, is affected by noise. As a result, the sampling instants

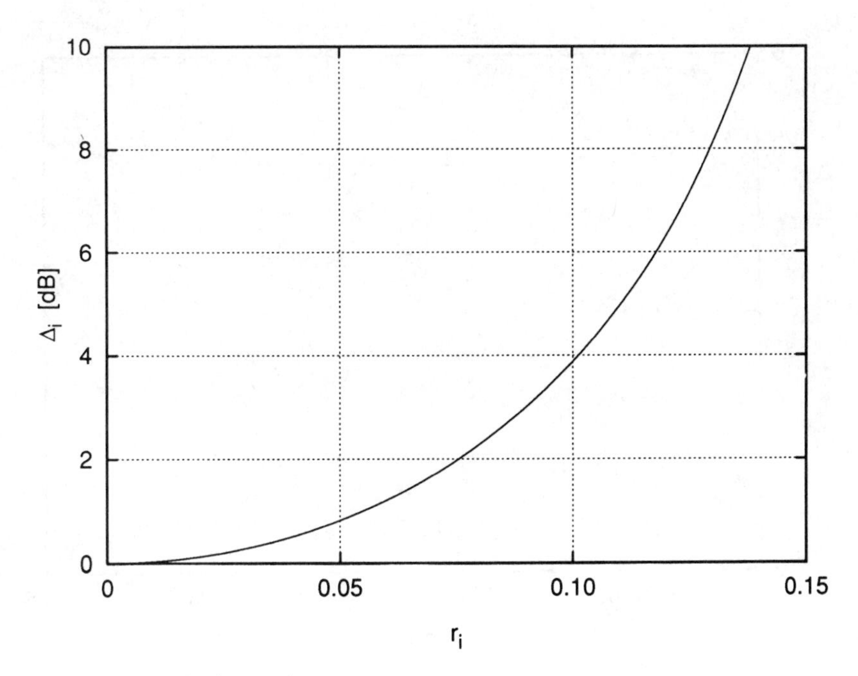

Figure 3.47 Power penalty due to relative intensity noise versus parameter r_i.

fluctuate randomly with time. This phenomenon is known as *timing jitter*. Intersymbol interference can be enhanced by timing jitter.

Let us analyze the impact of timing jitter for a simplified situation. The electrical pulses are assumed to be RZ, and the receiver filter is assumed to be the ideal integrate-and-dump. In this case, the main effect produced by the timing jitter is a sensitivity penalty due to the reduced signal amplitude at the output of the integrator. Let us define the normalized timing jitter as

$$\tau_j \stackrel{\text{def}}{=} \frac{t_0 - \hat{t}_0}{T} \tag{3.185}$$

where $\hat{t}_0$ is the actual sampling instant, and t_0 is the optimum sampling instant. Assuming that τ_j is a zero-mean Gaussian RV[9] with variance σ_j^2, we obtain the timing jitter penalty, Δ_j, by computing first the bit error probability conditioned on a given value of the timing jitter, τ_j, and then averaging with respect to its probability density

9. This is a reasonable assumption for a timing synchronization based on a loop for high SNRs [14,Chap. 6].

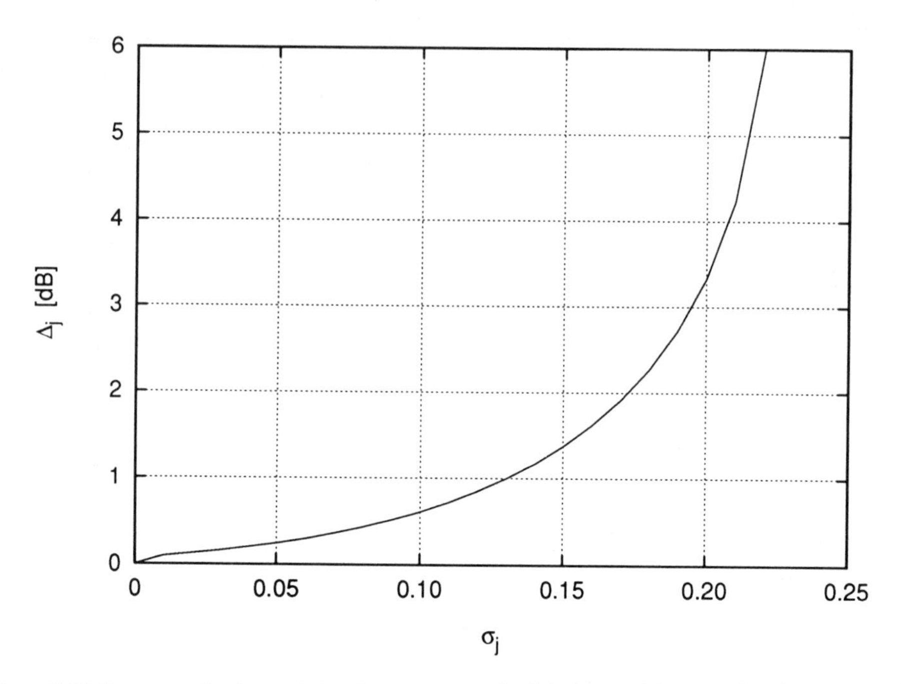

Figure 3.48 Power penalty due to timing jitter versus standard deviation of the normalized timing jitter, τ_j.

function. Then, from the average bit error probability, we derive the penalty at $P(e) = 10^{-9}$.

In Figure 3.48, the penalty, Δ_j, is plotted versus the standard deviation, σ_j. We see that a standard deviation of the normalized timing jitter below 0.1 leads to penalties below 1 dB; the penalty increases sharply above $\sigma_j = 0.15$. Thus, the clock recovery must be designed to keep the standard deviation of the timing error below that value.

3.7.4 Modal Noise

Modal noise is caused by the combination of mode-dependent losses and changes in phase between modes or fluctuations in the distribution of energy among modes. It takes place when the laser light propagates through a multimode fiber link with mode-selective losses. Laser light excites a number of propagation modes, which, at the end of the fiber, produce a *speckle* pattern resulting from constructive and destructive interferences among propagating modes.

Laser chirp, temperature fluctuations, fiber vibrations, and other effects cause the speckle pattern to change with time. When the time-varying speckle pattern travels through system components whose loss is speckle-pattern dependent (like splices, optical connectors, and fiber core discontinuities), the output power varies in time. To minimize modal noise, we can use either multimode fibers with LED transmitters or single-mode fibers with laser transmitters. More recently, multimode fibers with VCSEL transmitters also have been shown to resist the modal noise.

3.7.5 Mode Partition Noise

Mode partition noise stems from the combination of parasitic effects in multilongitudinal-mode semiconductor lasers with a dispersive single-mode fiber.

The optical power of different longitudinal laser modes exhibits large fluctuations even though the total power remains constant. The fiber dispersion desynchronizes the different modes because the group velocity of each mode is different. The net result is additional fluctuations of the received power.

The quantitative analysis of sensitivity penalty due to mode partition noise can be found in [5, Chap. 14].

Mode partition noise can be reduced by the use of single-mode lasers (like DFB lasers).

3.8 GENERAL DESIGN CONSIDERATIONS

As a result of the analysis in Sections 3.1–3.7, we now have equations and relationships that permit us to analyze and design an IM-DD digital optical transmission system. In this section, we add a few heuristic considerations that may help explain basic design choices.

3.8.1 Multimode Versus Single-Mode Fibers

We have seen that modal dispersion limits with multimode fibers are severe. Moreover, multimode fibers give rise to modal noise in conjunction with lasers. As a consequence, the choice of multimode fibers, to be used with LED or VCSEL sources and low-cost receivers, is limited to low bit rate and short-distance systems.

Single-mode fibers with laser transmitters is the universal choice for high-speed communication links. The choice between the 1,300-nm and 1,550-nm range is influenced by EDFAs. They operate in the 1,550-nm window, which is also the most favorable in terms of fiber loss. However, conventional single-mode fibers have a fairly high dispersion at 1,550 nm (a typical value is 16 ps/nm·km). Thus, 1,550-nm systems that employ conventional single-mode fibers are essentially dispersion limited, and dispersion-shifted fibers and/or dispersion-compensation techniques have to be used for long-haul, high-bit-rate links.

3.8.2 PIN Versus APD Photodetectors

The performance of a digital receiver without optical preamplification is bounded by two limits. In one limit, we have shot noise–limited systems, characterized by high received optical powers. The shot noise is then prevalent with respect to the receiver thermal noise. For such systems, a PIN photodetector is a reasonable choice, because it avoids the excess avalanche noise and eliminates the high voltages needed to bias APDs.

In the other limit, we have thermal noise–limited systems. The performance of such systems is improved using an APD photodetector, because its avalanche gain increases the electrical signal power, thus improving the thermal SNR. This advantage must be traded off against the increased shot noise due to the excess avalanche noise, so that there is, in general, an optimum avalanche gain.

A completely different situation occurs when an optical preamplifier is used (see Chap. 5). In that case, the optical power is so high that the thermal noise can be neglected, and the system performance is dominated by the (ASE) noise. For such systems, a better performance is obtained with PIN photodetectors. Similar conclusions can be drawn when the system employs coherent reception (see Chap. 4). Finally, system bandwidth considerations also influence the choice, since PIN photodetectors permit higher bit rates.

3.8.3 High-Impedance Versus Transimpedance Receivers

From the sensitivity point of view, the differences between an equalized high-impedance and transimpedance receivers (when both are optimized with respect to system parameters) are insignificant, with a slight preference toward high-impedance receivers. There are, however, other considerations that may influence the choice. One issue is the receiver dynamic range.

The dynamic range increases when the load (or feedback) resistance decreases. On the other hand, the thermal noise power increases at the same time. Thus, there is a trade-off between sensitivity and dynamic range, and that trade-off is more effective for transimpedance amplifiers. Typical dynamic ranges are 15–20 dB for high-impedance receivers and 35–40 dB for transimpedance receivers. Automatic gain control techniques have been successfully employed to increase those figures up to 60 dB.

Problems

Problem 3.1

Using the basic propagation equations introduced in Section 1.2.4 of Chapter 1, derive the form (3.16) of the received pulse assuming a transmitted Gaussian-shaped pulse as in (3.6).

Problem 3.2

Derive (3.40) and (3.41) of Y_1 and Y_0 defining the worst-case eye opening due to fiber dispersion. Using previous expressions, obtain the SNR penalty (3.42).

Problem 3.3

For an IM-DD system transmitting a power $\overline{P}_T = 0$ dBm at 1,550 nm on a fiber with attenuation of 0.2 dB/km and employing a receiver with sensitivity $\overline{N}_R = 1{,}000$ photons/bit, derive the curves in Figure 3.20 that take into account both attenuation and dispersion. (*Hint:* Use the worst-case bound for dispersion effects, through the minimum eye opening defined by (3.40) and (3.41), and assume that the receiver sensitivity is determined by thermal noise only.)

Problem 3.4

For a single-hop IM-DD system employing optical amplifiers at transmitter and receiving sides, with a transmitted power $\overline{P}_T = 10$ dBm, assuming that ASE noise of the receiver optical amplifier with spectral density $G_n/2$ determines the performance, derive (3.48) of the mean and the variance of the decision variable d_0. Then, under worst-case hypotheses, as in Problem 3.3, and assuming a Gaussian distribution for d_0, derive bound (3.52) to the bit error probability.

Problem 3.5

Repeat computations of Problem 3.4 using the actual χ^2 distribution of the decision variable d_0 and write a computer program to obtain the curves in Figure 3.22 and Figure 3.23 in this case.

Problem 3.6

The distance between the transmitter and the receiver in an optical fiber link is 35 km. The transmission path is made by splicing similar fibers, each of which has a maximum length of 3 km. The attenuation of the fiber is 0.6 dB/km. The losses due to splices average 0.1 dB per splice. The coupling losses are 2 dB at each end, and the design margin is 3 dB. The receiver sensitivity is −30 dBm. What is the needed transmitter power? If the system operates at $\lambda = 1{,}300$ nm, what is the required number of photons per bit?

Problem 3.7

A binary transmission system employs an LED transmitter that launches an average of 300 μW of optical power at a wavelength of 800 nm into the optical fiber. The fiber has an overall attenuation of 4 dB/km. The APD receiver requires 1,200 incident photons per bit to work at $P(e) = 10^{-10}$. Assuming that the effect of dispersion is negligible,

1. Determine the maximum transmission distance provided by the system when the transmission rate is 1 Mbps and 1 Gbps.

2. Plot a graph showing the attenuation limit on transmission distance versus bit rate for this particular system.

Problem 3.8

A binary digital transmission system has the following characteristics:

- The single-mode fiber has a dispersion of 15 ps/(nm·km) and an attenuation of 0.2 dB/km.
- The transmitting GaAs laser operates at $\lambda = 1{,}550$ nm, emits an average optical power of 5 mW, and has a spectral width of 2 nm.
- The APD receiver requires an average of 1,000 photons/bit to operate properly.
- The coupling losses at the transmitter and the receiver amount to 3 dB.

1. Find the maximum transmission distance due to the attenuation limit for data rates of 10 and 100 Mbps.
2. Find the maximum transmission distance due to dispersion limit for data rates of 10 and 100 Mbps.
3. Sketch a graph showing the maximum transmission distance versus data rate for this particular system, including both attenuation and dispersion limits.

Problem 3.9

For the Poisson distribution

$$P(n) = \frac{N^n e^{-N}}{n!}$$

where N is the mean and the variance of the RV n:

1. For what values of N would $P(n)$ decrease monotonically with n?
2. Find the value(s) of n for which $P(n)$ is maximum for $N > 1$.
3. Let n_1 and n_2 be Poisson RVs with means N_1 and N_2, respectively. Derive the distribution of $n_1 + n_2$.

Problem 3.10

Generalize (3.74) and (3.75) for the useful signal and shot noise variance assuming as received optical power the general form (3.55), which includes intersymbol interference at the optical level.

Problem 3.11

On the basis of (3.74) and (3.75) for the useful signal and shot noise variance, for the two cases of

- $p(t) = \delta(t)$
- $P(f) = 1, \ -1/(2T) \leq f \leq 1/(2T)$

propose a solution for the choice of $H_L(f)$ that eliminates intersymbol interference from the signal and variance on samples taken at $t = t_0 + kT$, t_0 being the sampling instant to be chosen appropriately.

Problem 3.12

Evaluate the expression of the impulse response of a raised-cosine filter with transfer function (3.81) and plot it for two values of the roll-off factor $\alpha = 0.5$ and $\alpha = 1$. Compute its noise equivalent bandwidth and plot it as a function of α.

Problem 3.13

On the basis of (3.74) and (3.75) for the useful signal and shot noise variance, in the case of $p(t) = \delta(t)$ and subject to the constraint

$$\int_{-\infty}^{\infty} H_L(f) e^{-j2\pi f t_0} df = 1$$

find the transfer function $H_L(f)$ that maximizes the shot noise SNR defined as $v_s^2(t_0)/\sigma_{sh}^2(t_0)$.

Problem 3.14

A GaAs PIN photodiode generates on average one electron-hole pair per three incident photons. Assuming that all electrons are collected,

1. Calculate the quantum efficiency of the device and its responsivity at $\lambda = 800$ nm.
2. Calculate the mean output photocurrent when the received optical power is 10^{-7} W at $\lambda = 800$ nm.

Problem 3.15

Consider an ideal optical system that transmits a pulse of power P to represent a binary 1 and nothing to represent a 0. The system is operating at $\lambda = 1{,}550$ nm and with a data rate of 1 Mbps.

1. Using Poisson distribution, plot the bit error probability as a function of the average received power assuming that the receiver works at the quantum limit of detection.
2. Repeat the computation and graph in (a) using a Gaussian approximation. What is the inaccuracy due to the Gaussian approximation at $P(e) = 10^{-7}$, 10^{-8}, 10^{-9}?

Problem 3.16

In a quantum-limited digital transmission system, the average received photons per bit is 40 for a binary 1 and 0.15 for a binary 0. Find the optimum decision threshold and the average probability of error.

Problem 3.17

An LED delivers a continuous wave power of 50 μW at $\lambda = 0.85$ μm, into an optical fiber. The total attenuation between transmitter and receiver is 40 dB. Assume that the dispersion is negligible and the photodetector's quantum efficiency is equal to 1. What is the probability that fewer than three electrons will be detected in a time interval of 1 ns?

Problem 3.18

A continuous-wave LED operates at $\lambda = 940$ nm. It couples −20 dBm of power into a fiber, which is 130 km long and has an attenuation of 0.8 dB/km. The fiber output is captured by a detector with quantum efficiency equal to 0.7. There is a 2-dB coupling loss between the fiber and the detector. The detection process is dominated by quantum noise, with a Poisson distribution; the threshold is set between $n = 0$ and $n = 1$.

1. If electrons are counted during 0.5-ns intervals, what is the most probable count?
2. What is the probability of counting 25 electrons in a 5-ns interval?

Problem 3.19

A shot noise–limited PIN photodetector has a responsivity of 0.7 A/W. It receives light of 200 μW average power sinusoidally modulated at 0.5 MHz with modulation index 0.6. The noise bandwidth is 7 MHz.

1. What is the current noise spectral density at the output?
2. What is the SNR at the output?
3. Repeat the computations in (a) and (b) for the case of an APD with $G = 100$ and $x = 0.3$.

Problem 3.20

A quantum-limited 500-Mbps IM-DD system operates with an error probability lower than 10^{-9}. The wavelength is 1,300 nm and the photodetector's quantum efficiency is 0.9. What is the receiver sensitivity in terms of power and received photons per bit?

Problem 3.21

With reference to the equivalent representation of a noisy amplifier in Figure 3.30, derive (3.85) for the noise figure of the amplifier.

Problem 3.22

The lightwave incident on a PIN photodiode is intensity modulated with a sinusoidal signal as

$$P_i = P_0[1 + 0,7 \sin(2\pi f_m t)], \; P_0 = 1 \text{ mW}$$

The modulating frequency is $f_m = 1$ MHz, and the noise-equivalent bandwidth is 1.7 MHz. The responsivity of the photodiode is $R = 0.8$ A/W. The dark current, i_d, is negligible.

1. What is the noise spectral density, in A^2/Hz, at the output of the photodiode?
2. What is the SNR?
3. If the photodiode is replaced with an APD having $G = 25$ and an excess noise factor $x = 0.5$, with the other parameters remaining the same, what is the SNR at the output of the APD?

Problem 3.23

For a value of 10 of the parameter γ defined in (3.133), compute the minimum value of the SNR Λ_1 yielding an error probability equal to 10^{-9}.

Problem 3.24

Devise an experimental setup for the measurement of the amplifier figure of merit, ρ_A, defined in (3.104).

Problem 3.25

Based on (3.105) to (3.110), find the expression of the SNR Λ_1 in the case of $p(t) = \delta(t)$ and $h_L(t)$ of the raised-cosine type.

Problem 3.26

Following the steps for computing the most general expression of the conditional bit error probabilities (3.120) and (3.121), described in Section 3.5.1, evaluate the average bit error probability with the system impulse responses of Problem 3.25, neglecting the thermal noise variance and for $G = G_2 = 1$, $i_d = 0$, $\overline{P}_R = 10^{-6}$ W, $\lambda = 1{,}550$ nm.

Problem 3.27

From (3.124) and (3.125) for the conditional error probabilities, derive the value (3.126) of the threshold β_s that makes equal the two probabilities.

Problem 3.28

Minimize the value of the average bit error probability (3.131) with respect to the parameter β, to obtain (3.132).

Problem 3.29

Derive (3.139) and (3.144) for the sensitivity and optimum avalanche gain.

Problem 3.30

From the definitions in (3.111) and (3.112), derive the results (3.156) and (3.157) for the high-impedance receiver.

Problem 3.31

Derive (3.159) and (3.166) for the transfer functions $Z_1(f)$, $Z_2(f)$, and $H_3(f)$ for the transimpedance amplifier.

Problem 3.32

Consider an optical transmission system limited only by quantum noise, with negligible dark current. The electron number per bit, n, is a Poisson RV with mean N.

1. If $N = 30$, and the threshold is set between $n = 1$ and $n = 2$, what is the bit error probability?
2. With the threshold still set as in (a), what should N be to have $P(e) = 10^{-9}$?

Problem 3.33

Consider a transimpedance receiver with the following parameters: $G_{e_A} = 4 \cdot 10^{-18}$ V^2/Hz, $G_{i_A} = 3 \cdot 10^{-24}$ A^2/Hz, $G = G_2 = 1$, $R_t = R_F = 1$ kΩ, $C_t = 2$ pF, $B = 100$ MHz, $I_{ph} \stackrel{\text{def}}{=} RP_R p_0 = 1$ μA, $i_d = 3$ nA.

1. What is the dominant source of noise?
2. Compute the figure of merit, ρ_A, of the amplifier.
3. Calculate the SNR Λ_1.
4. Compute the average bit error probability for the optimum value of the threshold.
5. By what factor is the SNR in (c) worse than that under the ideal quantum noise-limited condition?
6. How far is the assumed photodetected current from the system sensitivity?

Problem 3.34

An APD photodetector has $R = 0.55$ A/W, $G = 80$, and $G^x = 10$. The received optical signal has a power of 50 nW and is intensity modulated with a sinusoidal waveform $f(t) = 2 \cos 2\pi f_m t$ and modulation index $m = 0.3$. The photodetector load impedance consists of a resistance $R_t = 6$ kΩ and a capacitance $C_t = 2$ pF in parallel. The preamplifier is a transimpedance amplifier with $R_F = 1.2$ kΩ, $B = 300$ MHz (with $f_m << B$), $G_{i_A} = 0$, and $G_{e_A} = 2.2 \cdot 10^{-18}$ V^2/Hz. The temperature of the resistors is 300K.

1. What is the receiver figure of merit, ρ_A?
2. What is the SNR at the output of the preamplifier?

Problem 3.35

A thermal noise–limited transimpedance receiver is used for an optical communication system with a PIN photodiode. The transimpedance amplifier is assumed to have very high input and feedback resistances and noise parameters $G_{e_A} = 4$ $(nV)^2/Hz$ and $G_{i_A} = 4$ $(pA)^2/Hz$.

1. Derive the expression of the photocurrent $I_{ph} \stackrel{\text{def}}{=} RP_R p_0$ required to ensure a specified SNR Λ_1.
2. For a SNR $\Lambda_1 = 21.6$ dB, plot a graph showing the photocurrent versus the bandwidth for $C_t = 2$ and 5 pF.
3. From the graph in (b), in which I_{ph} varies approximately as B^y, find the values of y for different values of B. Identify the break point.

Problem 3.36

A 2-Gbps optical fiber communication system working at a wavelength of 1,300 nm uses a PIN photodetector. The input stage is a transimpedance amplifier with $R_F = 1$ kΩ. Its noise performance can be represented by an equivalent input noise voltage source with $\sqrt{G_{e_1}} = 2$ nV/(Hz)$^{1/2}$ and an equivalent input noise current source with $\sqrt{G_{i_1}} = 0.1$ pA/(Hz)$^{1/2}$. The total input capacitance is $C_t = 2$ pF.

1. Compute the amplifier gain required to ensure that the amplifier stage is equalized over the frequency range 0 to 1 GHz.
2. Evaluate the amplifier noise figure, $F(f)$, and its figure of merit, ρ_A.
3. With a quantum efficiency $\eta = 0.8$, compute the receiver sensitivity assuming a system margin, k_m, of 4 dB.

Problem 3.37

A receiver for an IM-DD optical communication system uses an integrating bipolar front-end amplifier, whose noise sources can be represented as

$$G_{i_A} = qI_b$$

$$G_{e_A} = \frac{(kT)^2}{qI_b\beta}$$

where I_b is the transistor base bias current, and β is the transistor current gain.

1. Find the optimum value of I_b, that is, the value that minimizes the figure of merit, ρ_A, of the amplifier.
2. If the transistor to be used in this receiver has a current gain $\beta = 120$, and $G = G_2 = 1$, $C_t = 5$ pF, calculate the receiver sensitivity in photons per bit and in dBm (for a bit rate of 50 Mbps and an operating wavelength of 1,300 nm), assuming a system margin, k_m, of 6 dB.

Problem 3.38

Consider an IM-DD fiber optic system that employs an equalized high-impedance receiver, characterized by the following parameters:

- Amplifier noises $G_{e_A} = 2.25 \cdot 10^{-18}$ V^2/Hz, $G_{i_A} = 4.0 \cdot 10^{-24}$ A^2/Hz;
- Bit rate $R_b = 2$ Gbps, equalized bandwidth $B = 2$ GHz;
- Resistance temperature $T_0 = 300$K, $R_t = 10$ kΩ, $C_t = 2$ pF;
- PIN photodetector with dark current $i_d = 0$ and $\eta = 0.8$.

Calculate the receiver sensitivity in average number of photons per bit.

Problem 3.39

You are assigned to design (what a burden!) a high-bit-rate underwater fiber optic telecommunications link connecting Europe and the United States. Which of the following devices and techniques would you recommend?

1. Multimode fiber or single-mode fiber?
2. A 1,300-nm or 1,550-nm wavelength of operation?
3. LEDs or semiconductor lasers?
4. APDs or PIN photodiodes?
5. High-impedance or transimpedance receivers?
6. Optical amplifiers (see Chap. 5) or regenerative repeaters?
7. Coherent (see Chap. 4) or DD receivers?
8. Binary or multilevel modulation?

Clearly state all your assumptions for the system and your reasons for choosing one over the other. Use no more than three lines per item in your discussion.

Problem 3.40

Assuming the shot noise variance in Example 3.3, compute the variances of shot and thermal noises as functions of the received optical power for a receiver using an amplifier with noise figure (constant with frequency) of 5 dB. The value of R_t is 1 kΩ, the temperature is 300K, and the receiver uses a PIN photodiode with quantum efficiency 0.7. For what value of the received power are the two variances equal?

References

[1] Personick, S. D., *Optical Fiber Transmission Systems*, Plenum Press, 1981.
[2] Cariolaro, G. F., *Transmissione Numerica (Digital Transmission)* (in Italian), CLEUP, 1987.
[3] Jones, W. B., *Introduction to Optical Fiber Communication Systems*, New York: Oxford University Press, Inc., 1988.
[4] Miller, S. E., and I. P. Kaminow, *Optical Fiber Telecommunications II*, Academic Press, 1988.
[5] Basch, E. E., et al., *Optical-Fiber Transmission*, Howard W. Sams, 1987.
[6] Technical Staff CSELT, *Fiber Optic Communications Handbook*, TAB Books, 1990.
[7] Van Etten, W., and J. Van der Plaats. *Fundamentals of Optical Fiber Communications*, Englewood Cliffs, NJ: Prentice Hall, 1991.
[8] Gowar, J., *Optical Communication Systems*, Prentice-Hall, 2d ed., Englewood Cliffs, NJ: 1991.
[9] Keiser, G., *Optical Fiber Communications*, 2d ed., New York: McGraw-Hill, 1991.
[10] Palais, J. C., *Fiber Optic Communications*, Englewood Cliffs, NJ: Prentice-Hall, 1992.
[11] Senior, J., *Optical Fiber Communications: Principles and Practice*. Englewood Cliffs, NJ: Prentice-Hall, 1992.
[12] Agrawal, G. P., *Fiber-Optic Communications Systems*, New York: John Wiley & Sons, 1992.

[13] Personick, S. D., "Baseband Linearity and Equalization in Fiber Optic Digital Communication Systems," *Bell System Technical Journal,* 52(7):1175–1191, 1973.
[14] Benedetto, S., E. Biglieri, and V. Castellani, *Digital Transmission Theory,* Englewood Cliffs, N.J.: Prentice-Hall, 1987.
[15] Agrawal, G. P., and M. J. Potasek, "Effect of Frequency Chirping on the Performance of Optical Communication Systems," *Optics Letters,* 11(5):318–320, 1986.
[16] Agrawal, G. P., *Nonlinear Fiber Optics,* Academic Press, 1995.
[17] Humblet, P. A., and M. Azizoglu, "On the Bit Error Rate of Lightwave Systems With Optical Amplifiers," *Journal of Lightwave Technology,* 9(11):1576–1582, 1991.
[18] Yamamoto, S., "Undersea Lightwave Systems," in *Proceedings of OFC '96, Tutorial Part,"* 1996.
[19] Tkach, R., "Dispersion Compensation in Nonlinear Systems," in *Proceedings of OFC '96, Tutorial Part,"* 1996.
[20] Cariolaro, G. F., "Error Probability in Digital Fiber Optic Communication Systems," *IEEE Transactions on Information Theory,* 24(2):213–221, 1978.
[21] Dogliotti, R., et al., "Error Probability in Optical-Fiber Transmission Systems," *IEEE Transactions on Information Theory,* 25(2):170–178, 1979.
[22] Helstrom, C. W., "Computation of Photoelectron Counting Distributions by Numerical Contour Integration," *Journal of Optical Society of America,* 2(5):674–681, 1985.
[23] Annovazzi Lodi, V., S. Donati, and S. Merlo, "Squeezed States in Direct and Coherent Detection," *Optical and Quantum Electronics,* 24:285–301, 1992.
[24] Castellani, V., *Rumore (Noise)* (in Italian), Boringhieri, 1982.

Chapter 4

Coherent Systems

4.1 MOTIVATIONS AND BASICS

Despite the rapid progress made in lightwave technology over the past two decades, the basic operations of a digital optical fiber communication link have remained essentially the same. They consist, as seen in Chapter 3, in the direct IM (intensity modulation) of the light source (on-off modulation) and the DD (direct detection) at the receiver site using a PIN or APD (avalanche photodiode). This situation recalls the early stages of electrical modulation and seems to ignore the impetuous development of the modulation schemes and detection processing that have dominated the scenario of electrical digital transmission in the last 30 years.

There are clear motivations for this different attitude in the two fields of electrical and optical applications. The astonishing progress in electrical communication has been motivated by the shortage of power and, mainly, bandwidth caused by the enormous increase of communications demand around the globe. We have witnessed the coming into being of the ultimate promise of Shannon theory, in which system performance is traded for implementation complexity, and that, in turn, has become an increasingly low-cost resource owing to the very large scale integration of digital circuitry.

On the other hand, the huge bandwidth of optical fibers, together with the great improvement of their loss characteristics in the 1970s, led to the success of the simple and reliable IM-DD systems, which, despite their primitiveness and scarce appeal to the sophisticated information theorist, proved to be superior to cable and radio relay systems so as to become the winners for point-to-point high-capacity links. Typically, IM-DD systems may have signal spectra spreading up to 1 THz for information rates a thousand times lower; a waste of the frequency resource, which becomes particularly evident when one tries to send several separate channels on the same fiber using *wavelength division multiplexing* (WDM).[1] Moreover, as we saw in Chapter 3, IM-DD

1. WDM refers to the multiplexing over the same fiber of more than a single channel using conventional DD techniques, which are quite inefficient in channel packing. With the success of the first experiments of coherent transmission, it became clear that the possibility of a much denser channel packing, with

systems ought to use an APD to supply the necessary light amplification, which causes a performance loss of tens of decibels with respect to the theoretical quantum noise limit.

Driven both by the technological improvements in optoelectronic components and techniques (like, for example, single-mode fibers with very low loss and dispersion, semiconductor lasers with spectral linewidths on the order of megahertz or, even, hundreds of kilohertz, and various kinds of high-speed modulation techniques and devices) and by the promise of significantly increasing the repeater spacing through the gains in system sensitivity, the research and development of coherent optical communication experienced a rapid expansion in the 1980s.

Optical coherent communication systems present several advantages over IM-DD. First, they achieve a higher *sensitivity*. The main reason is that, when the power of the local oscillator is sufficiently high, coherent systems are limited only by the quantum noise of the received signal light. This is significantly different from the fact that APD multiplication noise and load-resistance thermal noise dominate the receiver sensitivity in conventional IM-DD systems. We saw in Chapter 3 that several hundreds (and even thousands) of photons per bit are necessary to achieve an error probability of 10^{-9} in IM-DD systems, well above the quantum noise limit of 10 photons/bit. Moreover, use of coherent modulation-demodulation schemes brings about an improvement in sensitivity that is well known in digital communications. As a result, a heterodyne coherent system employing binary PSK modulation requires only 18 photons/bit to achieve an error probability of 10^{-9}. This increase in sensitivity is important in real systems because it permits, for example, an increase in repeater spacing in long-haul point-to-point communication or a way to compensate for bridging losses in passive distribution networks.

This first advantage has been recently shadowed by the great success obtained in the development of erbium-doped fiber amplifiers (see Chap. 5), which have allowed significant increase in the sensitivity of IM-DD systems, up to a point where the application of coherent systems for point-to-point communication is presently a matter for discussions.

The second main advantage of coherent communication lies in the increased *selectivity* of the receivers. In fact, when working in an FDM environment, they perform channel selection in the electrical domain, using sharp microwave filters after heterodyning. DD systems, instead, when used in connection with WDM, must employ optical filters for channel separation, whose selectivity is on the order of a nanometer (hundreds of gigahertz).[2]

As a consequence, coherent communication permits a much denser packing of

channel separations less than 10 times the information rate; for those systems, the same term applied to electrical multiplexing is used, namely, *frequency division multiplexing* (FDM).

2. Multiple-stage Fabry-Perot filters attaining a bandwidth of 0.01 nm (6 GHz) have been reported. However, they require an extremely critical temperature control to prevent detuning and computer-controlled electronics to drive all the stages properly so that the desired channel is selected.

the information channels in the frequency domain, a key feature for the exploitation of the huge bandwidth of optical fibers, particularly in the context of communication networks, where the frequency selectivity of conventional radio systems combined with the immense bandwidth of optical fibers opens up a wide range of new applications for telecommunications. Moreover, coherent technology has the potential of reducing by a factor of up to 4–5 the bandwidth through the use of multilevel transmission.

Further benefits have to do with the possibility of using constant envelope modulation schemes, like PSK and FSK. The former requires an external modulator but yields a reduced impact of stimulated Brillouin scattering (from 3 dBm to 30 dBm), whereas the latter can be obtained by direct modulation of the laser source, like for direct detection ASK modulation. With respect to direct detection, however, it has the advantage of significantly reduced chirp effects.

Optical coherent communication uses the optical field as a very high frequency carrier whose amplitude, phase, frequency, or polarization may be modulated by the information-bearing signal. Although this is very much the same as is commonly done for electromagnetic fields at lower frequencies, the big difference between the carrier frequency and the information signal bandwidth poses in the optical case some peculiar technological problems. Even the term *coherent* has here a different meaning from the standard radio environment. In fact, it is customary in the optical communication community to associate the adjective *coherent* to those systems in which a local oscillator lightwave is added to the incoming signal, even if subsequent processing and demodulation completely ignore the phase and frequency, as is the case of envelope detectors. This contrasts with the meaning of *coherent systems* in the classical communication literature, which require the recovery and use of the phase and frequency of the carrier to perform the demodulation and detection.

The basic configuration of a coherent communication system is shown in Figure 4.1. A laser-emitted light possessing a sufficiently stable frequency (quasi-monochromatic signal) is used as the carrier wave and modulated (in amplitude, frequency, phase, or polarization) by the information signal. At the receiver site, the

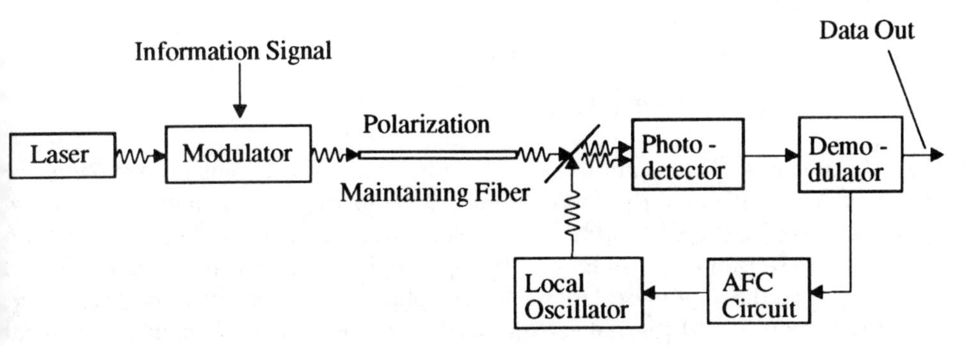

Figure 4.1 Block diagram of a coherent optical communication system.

lightwave of a local oscillator is added to the received signal, to generate at the outpu of the photodetector a suitable IF (intermediate frequency) signal, in the case o heterodyne detection, or directly the BB (baseband) signal, in the case of homodyne detection.

Because coherent detection schemes are sensitive to the polarization states o both the received signal and the local oscillator waves, either a polarization-main taining fiber (as in Fig. 4.1) or a conventional single-mode fiber in conjunction with a polarization controller is needed.

The automatic frequency control circuit is needed in both heterodyne and homodyne receivers to keep the local oscillator frequency relatively stable.

We will first derive the sensitivity of homodyne and heterodyne receivers under the following assumptions:

- The light carrier is monochromatic (no phase noise).
- The modulation is binary, that is, every T seconds we transmit a bit of information, associated with the transmitted waveform.
- The received and local oscillator fields are parallel and with the same polarization.
- The only noise present in the receiver is the shot noise.
- The receiver is able to perform the ideal sum of the electric field of the received and local oscillator lightwaves.

These ideal assumptions will be gradually removed throughout the chapter.

We will consider first the demodulation schemes that exploit the frequency and phase of the carrier to perform the detection; we call these schemes *synchronous*. Successively, we will examine *asynchronous* demodulation, which uses only the envelope of the carrier to detect the transmitted data, and finally we will examine schemes that make use of the phase of the received carrier without requiring, however, its recovery in the receiver. The detection process in the last case is called *differential detection.*

4.1.1 A FIRST QUANTITATIVE INSIGHT

To help understanding quantitatively the advantage of coherent transmission, we will first derive the *signal-to-noise ratio* (SNR) in the case of homodyne binary *phase-shift keying* (PSK). Consider the block diagram in Figure 4.2, which depicts an idealized receiver for coherent binary PSK. In the figure, the received signal is added to the signal, at the same optical frequency and phase, generated by the local oscillator, and the composite light signal is converted to the electronic domain by an ideal photodetector.

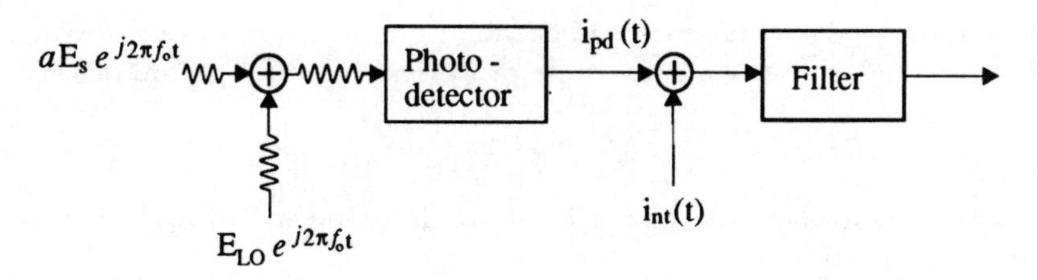

Figure 4.2 Ideal receiver of a coherent binary PSK system.

In analytic signal notation (see Section 2.7), the received signal and the local oscillator in scalar form can be written as

$$\begin{aligned} e_S(t) &= aE_S e^{j2\pi f_0 t} \\ e_{LO}(t) &= E_{LO} e^{j2\pi f_0 t} \end{aligned} \tag{4.1}$$

where a is an RV (random variable) representing the information and assuming the values $+1$ and -1 with equal probabilities.

The power of the sum signal P_t is

$$P_t = |\, aE_S + E_{LO}\,|^2 = E_S^2 + E_{LO}^2 + 2aE_S E_{LO} \stackrel{\text{def}}{=} P_S + P_{LO} + 2a\sqrt{P_S P_{LO}} \tag{4.2}$$

Thus, the output current of the photodetector assumes the form

$$i_{pd}(t) = I_S(t) + i_{ns}(t) \tag{4.3}$$

where $I_S(t)$ is the average current bearing the signal information, with the expression (derived in Section 2.5)

$$I_S(t) = RP_t = R(P_S + P_{LO} + 2a\sqrt{P_S P_{LO}}) \tag{4.4}$$

where R is the responsivity of the photodetector. The current $i_{ns}(t)$ is the shot noise contribution, a white random process with *power spectral density* (PSD) equal to

$$G_{ns}(f) = (RP_t + I_d)q \tag{4.5}$$

where I_d is the dark current. Assuming $P_S << P_{LO}$ and $RP_{LO} >> I_d$, we can approximate G_{ns} as

$$G_{ns}(f) = RP_{LO}q \tag{4.6}$$

The electronic section of the receiver contributes to the noise with the thermal noise $i_{nt}(t)$, considered a white Gaussian noise process with power spectral density

$$G_{ns}(f) = 2kT/R_L \tag{4.7}$$

where k is the Boltzmann constant, T is the absolute temperature of the receiver, and R_L is the load resistance.

The signal $i(t) = I_S(t) + i_{ns}(t) + i_{nt}(t)$ is low-pass filtered to limit the noise power entering the decision part of the receiver. If B is the noise equivalent bandwidth of the filter, the SNR (signal-to-noise ratio) at the output of the filter is easily found to be

$$SNR = \frac{4R^2 P_S P_{LO}}{(2qRP_{LO} + 4kT/R_L)B} \tag{4.8}$$

From (4.8) we can understand the main advantage of coherent detection. Controlling the power of the local oscillator, we can make the first term in the denominator dominant, so as to achieve the shot noise limit for the SNR:

$$SNR = \frac{2RP_S}{qB} \tag{4.9}$$

Thus, use of coherent detection allows us to achieve the shot noise limit even for receivers whose performance is generally dominated by thermal noise, owing to the boosting action of the local oscillator power. Here, we will assume that the receiver is dominated by the shot noise. That will permit us to obtain the performance of coherent systems in terms of the number of photons, N_R, received within a single bit. At the bit rate $1/T$, the average received signal power is $P_S = N_R h\nu/T$. With a filter bandwidth $B = 1/T$ and the relationship $R = \eta q/h\nu$, η being the quantum efficiency of the photodetector, SNR becomes

$$SNR = 2\eta N_R \tag{4.10}$$

We will incorporate the factor η within N_R so that N_R will assume the meaning of *useful*, that is, converted to electrons in the photodetection process, photons received within one bit interval, and $SNR = 2N_R$.

4.2 FUNDAMENTAL RECEIVER SENSITIVITY: HOMODYNE SYSTEMS

4.2.1 PSK Modulation

The received signal and the local oscillator in analytic form are

$$e_S(t) = aE_S\, e^{j\,2\pi f_0 t}$$

$$e_{LO}(t) = E_{LO}e^{j2\pi f_0 t} \tag{4.11}$$

The receiver front-end adds the received and local oscillator lightwaves, so that the resulting signal has a total power P_t;

$$P_t = | aE_S + E_{LO} |^2 = E_S^2 + E_{LO}^2 + 2aE_S E_{LO} = P_S + P_{LO} + 2a\sqrt{P_S P_{LO}} \tag{4.12}$$

According to the power of the local oscillator, P_{LO}, we must distinguish two cases.

4.2.1.1 Super Quantum Limit

Suppose first that $E_{LO} = E_S$. In that case, the optical power P_t equals 0 when $a = -1$ and $4P_S$ when $a = 1$. An ideal photon counter placed after the summation of the two lightwaves would then decide according to an ML (maximum likelihood) criterion by detecting $a = -1$ if it does not count any photons in T seconds, and $a = 1$ if at least one photon shows up in T. This ideal receiver is shown in Figure 4.3. It is then evident that to make an error when transmitting 1 requires that no photons arrive in T, or in formulae:

$$P_e = \frac{1}{2} P[0 \text{ photons} \mid a = 1] \tag{4.13}$$

Let $p_T(1)$ be the RV representing the number of photons counted in an interval of T seconds when transmitting $a = 1$. As seen in Section 2.5, it is a Poisson RV with mean $N_T(1) = 4P_S T/h\nu = 4N_R$, N_R being the average number of photons in T for the received signal. Similarly, $p_T(-1)$ is identically zero. Paralleling the computations made in Chapter 3 to obtain the quantum limit, we obtain

$$P_e = \frac{1}{2} e^{-4N_R} \tag{4.14}$$

which is the so-called *super quantum limit*. From it, we obtain that the sensitivity of PSK modulation with homodyne demodulation and $E_{LO} = E_S$ is equal to 5 photons/bit. It yields an improvement of 3 dB over the quantum limit obtained in Chapter 3.

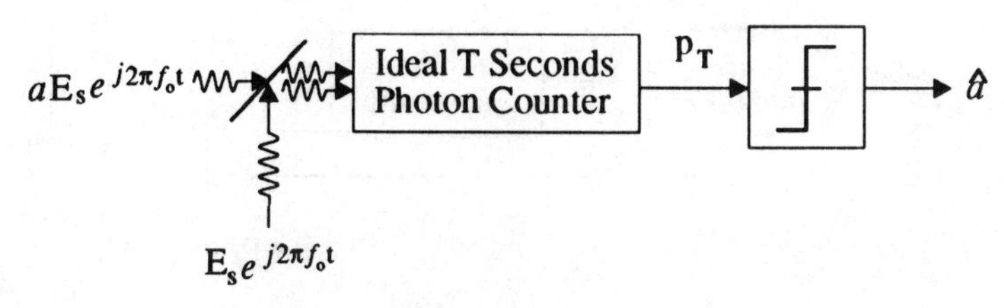

Figure 4.3 Ideal homodyne receiver of a coherent binary PSK system.

4.2.1.2 High-Power Local Oscillator

We now will remove the hypothesis contained in the super quantum limit computation and suppose instead that the power of the local oscillator is much higher than that of the received optical power, that is, $P_{LO} >> P_S$. In this case, we will evaluate the performance of the receiver shown in Figure 4.4, where the photodetector converts the total lightwave into the current $i_S(t) = I_S + i_n(t)$, whose characteristics were derived in Chapter 3. Its average (DC) value is equal to

$$I_S = RP_t = R(P_S + P_{LO} + 2a\sqrt{P_S P_{LO}}) \tag{4.15}$$

where R is the responsivity of the photodiode. The additive shot noise current contribution, $i_n(t)$, has a spectral density equal to

$$G_i = RP_t q \approx RP_{LO} q \tag{4.16}$$

where for the approximation we have used the hypothesis that $P_{LO} >> P_S$. As for the statistics of shot noise, we can recall the conditions described in Section 2.5 needed to invoke the central limit theorem. Essentially, the main requirement is that the rate of photon arrivals be large, which is easily obtained from (4.12) by sufficiently increasing the power of the local oscillator.

We are finally left with a classical problem of detection of signals in white noise, for which the ML receiver (see Section 2.8) assumes the form of Figure 4.4, where after integration over the interval T and ideal high-pass filtering to eliminate the DC term $R(P_S + P_{LO})T$ (operation indicated as a subtraction in Fig. 4.4), we get the RV

$$i_T = 2RTa\sqrt{P_S P_{LO}} + i_{nT} \tag{4.17}$$

where i_{nT} is Gaussian, with zero mean and variance $RP_{LO}qT$.

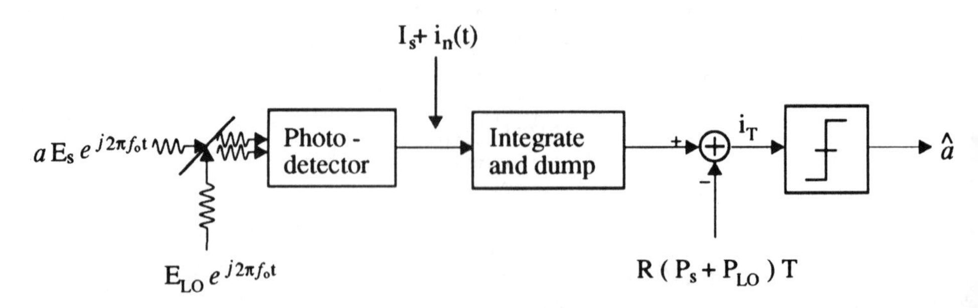

Figure 4.4 Optimum homodyne receiver of a coherent binary PSK system.

The RV i_T, conditioned on the two values of a, has antipodal means, and the ML decision consists in deciding on its sign, so that the resulting bit error probability becomes (see Section 2.8)

$$P_e = \frac{1}{2}\,\text{erfc}\left(\sqrt{\frac{2RP_ST}{q}}\right) = \frac{1}{2}\,\text{erfc}\left(\sqrt{\frac{2\eta P_ST}{h\nu}}\right) = \frac{1}{2}\,\text{erfc}(\sqrt{2N_R}) \qquad (4.18)$$

where N_R represents the average number of "useful" photons (taking into account the quantum efficiency η) in the received lightwave. Asymptotically, using the approximation

$$\text{erfc}(x) \sim 2\text{e}^{-x^2}$$

we obtain

$$P_e \sim \text{e}^{-2N_R}$$

which equals the quantum limit.

Using the exact expression of P_e, the obtained sensitivity is equal to 9 photons/bit.

4.2.2 Synchronous ASK Modulation

The case of amplitude-shift-keying (ASK) modulation differs from PSK only for the values assumed by the RV a, which are now 1 and 0.[3] This obviously prevents the possibility of using a demodulation procedure similar to the one that yielded the super quantum limit in the PSK case. When the local oscillator power is much larger than the received optical power, we can parallel the previous computations (see Problem 4.1) to obtain the following result:

$$P_e = \frac{1}{2}\,\text{erfc}\left(\sqrt{\frac{\eta P_ST}{h\nu}}\right) = \frac{1}{2}\,\text{erfc}(\sqrt{N_R}) \sim \text{e}^{-N_R} \qquad (4.19)$$

where N_R is the number of useful received photons in T averaged with respect to the transmission of a 1 and a 0. Thus, with respect to the average power, ASK modulation

3. In the expressions of the error probability for ASK, we will indicate with P_S the average received signal power. To express everything in terms of the peak power, substitute $P_S = ½\, P_{peak}$ in the expressions of the error probability.

exhibits a 3-dB loss with respect to PSK. If we compare the peak powers, the loss increases to 6 dB.

4.2.3 Frequency-Shift-Keying (FSK) Modulation

In the case of FSK modulation, the transmitter sends a light pulse with frequency $f_0 + af_d$, $a = \pm 1$, where f_d is the frequency deviation. At the receiver side, we need for homodyne detection two local oscillators, one for each possible frequency. Moreover, as the received lightwave must be separated into the two receiver branches, we also have a 3-dB penalty. The optimum receiver is shown in Figure 4.5.

We derive the system performance for orthogonal signaling, which is obtained for $f_d = k/4T$, $k = 1, 2, \ldots$, or, more practically, for very wide frequency deviations. Under the hypothesis of a large local oscillator signal, the optimum receiver in Figure 4.5 yields the same performance of binary orthogonal FSK (see Problem 4.2) over an additive white Gaussian noise channel, that is,

$$P_e = \frac{1}{2}\,\mathrm{erfc}\left(\sqrt{\frac{\eta P_S T}{2h\nu}}\right) = \frac{1}{2}\,\mathrm{erfc}\left(\sqrt{\frac{N_R}{2}}\right) \sim e^{-N_R/2} \tag{4.20}$$

In this case, we loose 6 dB over the homodyne PSK modulation, 3 of which are due to the splitting of the incoming light.

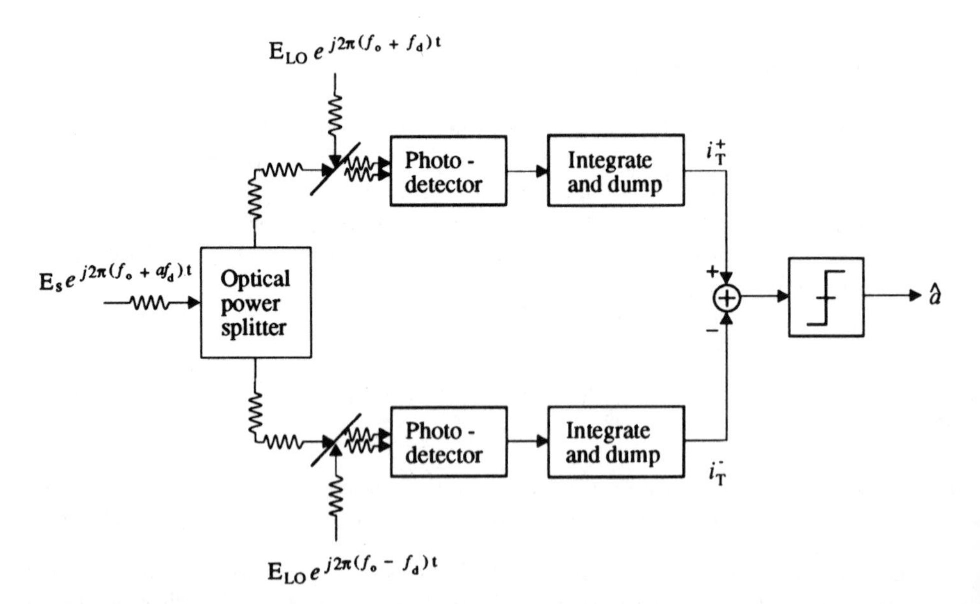

Figure 4.5 Optimum homodyne receiver of a coherent binary FSK system.

4.3 HETERODYNE SYSTEMS: SYNCHRONOUS DETECTION

In heterodyne receivers, the frequency of the local oscillator is different from that of the received carrier, so that the electric signal generated by the photodetector is centered around the IF, which is typically three to six times the bit rate. Thus, to perform the ML detection, we need first to translate the IF signal to BB and then apply to it the ML procedure according to the modulation format.

4.3.1 PSK Modulation

The received signal and the local oscillator in analytic form are

$$e_S(t) = aE_S e^{j2\pi f_0 t}$$
$$e_{LO}(t) = E_{LO} e^{j2\pi f_{LO} t} \tag{4.21}$$

The signal resulting from the sum of the two has power

$$P_t = | aE_S + E_{LO} e^{j2\pi f_{IF} t} |^2 = P_S + P_{LO} + 2a\sqrt{P_S P_{LO}} \cos 2\pi f_{IF} t \tag{4.22}$$

where $f_{IF} = f_0 - f_{LO}$ is the IF.

The ML detection rule requires in this case (Fig. 4.6) the multiplication by $\cos 2\pi f_{IF} t$ followed by integration and dumping. The photodiode current has a time-varying average value equal to

$$I_S = RP_t = R(P_S + P_{LO} + 2a\sqrt{P_S P_{LO}} \cos 2\pi f_{IF} t) \tag{4.23}$$

and an additive white Gaussian shot noise contribution $i_n(t)$ with power spectral density equal to

$$G_i = RP_t q \approx RP_{LO} q \tag{4.24}$$

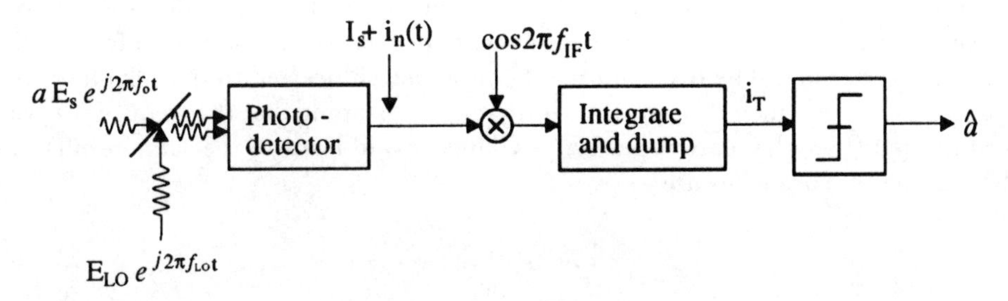

Figure 4.6 Optimum heterodyne receiver of a coherent binary PSK system.

since $P_S << P_{LO}$. After multiplication by $\cos 2\pi f_{IF} t$ and integration, we obtain the RV

$$i_T = Ra\sqrt{P_S P_{LO}}\, T + i_{nT} \tag{4.25}$$

where i_{nT} is Gaussian, with zero mean and variance $RP_{LO}qT/2$. Comparing this with the omologous RV of the homodyne case in (4.17), we see that the only difference consists in a reduction by a factor of 2 the signal component in i_T and of the noise variance. This translates into the following expression for the bit error probability:

$$P_e = \frac{1}{2}\,\mathrm{erfc}\left(\sqrt{\frac{\eta P_S T}{h\nu}}\right) = \frac{1}{2}\,\mathrm{erfc}(\sqrt{N_R}) \sim \mathrm{e}^{-N_R} \tag{4.26}$$

Thus, there is a 3-dB penalty with respect to the homodyne case.

4.3.2 Synchronous ASK Modulation

As in the homodyne case, the only difference between PSK and ASK modulations is a 3-dB penalty in average power (and 6 dB in peak power) due to the unipolar values assumed by the RV a. We have thus the following expression for P_e:

$$P_e = \frac{1}{2}\,\mathrm{erfc}\left(\sqrt{\frac{\eta P_S T}{2h\nu}}\right) = \frac{1}{2}\,\mathrm{erfc}(\sqrt{N_R/2}) \sim \mathrm{e}^{-N_R/2} \tag{4.27}$$

where N_R has the same meaning as for the homodyne case.

4.3.3 Synchronous FSK Modulation

The transmitted signal consists as before of two possible frequencies: $f_0 \pm f_d$, where f_d is the frequency deviation. We still suppose that the frequencies are chosen so as to ensure the orthogonality of the two waveforms. At the receiver side, we have the structure of Figure 4.7, where, after the translation to the IF bandwidth, we need two separate branches in which the correlation with the two possible signals at frequencies $f_{IF} \pm f_d$ is performed. The two resulting RVs are then subtracted from each other, and the decision is taken on the sign of the difference. Assuming that the frequency f_d was transmitted (the other case is perfectly symmetric and leads to the same result), the RVs in the two branches are

$$\begin{aligned} i_T^+ &= R(\sqrt{P_S P_{LO}})T + i_{nT}^+ \\ i_T^- &= i_{nT}^- \end{aligned} \tag{4.28}$$

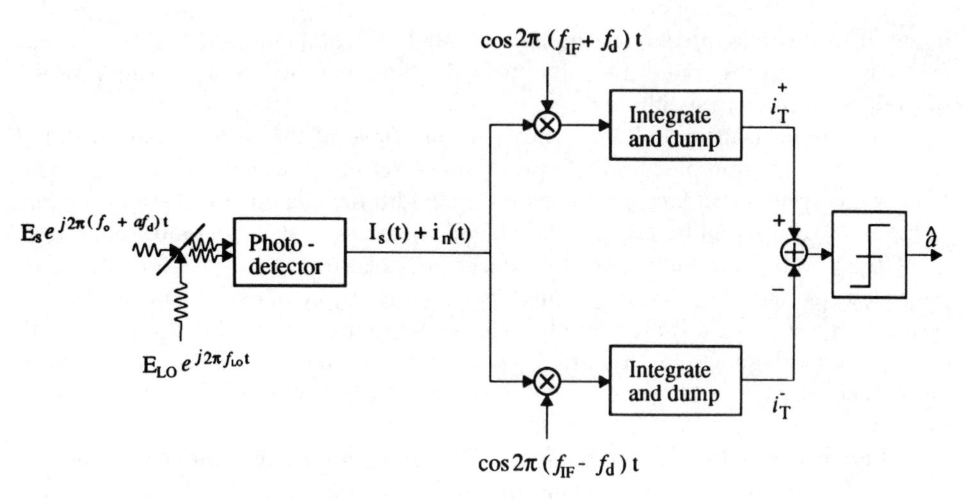

Figure 4.7 Optimum heterodyne receiver of a coherent binary FSK system.

where i_{nT}^+ and i_{nT}^- are independent Gaussian RVs, with zero mean and variance $RP_{LO}qT/2$. The decision variable $i_T^+ - i_T^-$ is thus a Gaussian RV, with mean $R(\sqrt{P_S P_{LO}})T$ and variance $RP_{LO}T$. We have the same situation as for heterodyne PSK, apart from the doubling of the noise variance. The resulting P_e is thus

$$P_e = \frac{1}{2}\,\text{erfc}\left(\sqrt{\frac{\eta P_S T}{2h\nu}}\right) = \frac{1}{2}\,\text{erfc}\left(\sqrt{\frac{N_R}{2}}\right) \sim \text{e}^{-N_R/2} \tag{4.29}$$

which is the same as for heterodyne ASK, with a loss of 3 dB with respect to heterodyne PSK. In this case, however, the average power and the peak power are the same.

4.3.4 Homodyne or Heterodyne?

After evaluating the bit error probability of ideal receivers for synchronous homodyne and heterodyne coherent modulation schemes and before introducing asynchronous detection and some practical problems that arise in receivers, a brief discussion of the relative merits of homodyne and heterodyne detection seems appropriate.

We have seen that heterodyne detection loses 3 dB in sensitivity with respect to homodyne detection. This is different from the case of communications at microwave frequency, where the heterodyning process has no cost in performance. The reason for that is due to the noise-generating mechanism, which is different in the two environments. For microwave communications, noise is additive on the received signal, and thus a bandpass filter at the receiver input can limit the noise spectrum to the signal bandwidth before mixing. This avoids the doubling of the noise spectrum, which is inherent in every frequency translation operated on white noise. In optical communication, on the other hand, the white noise is generated during the mixing process, so prefiltering is not possible.

Another advantage in homodyne detection concerns the required bandwidth of the photodetector and postdetection processing electronics, which is much higher for heterodyne systems. In fact, the detector bandwidth for homodyne receivers is equal to the baseband signal bandwidth, B, whereas heterodyne detectors must operate at $(f_{IF} + B)$, which is normally three to six times B. Seen from the other side, the same photodetector can permit a much higher data rate in homodyne than heterodyne systems. Note, though, that heterodyne systems require a faster bandpass detector, whereas homodyne systems require a slower lowpass detector. Which one of these detectors is easier to implement depends on various factors, which will not be discussed here.

Finally, in favor of heterodyne systems we mention that homodyne detection requires frequency and phase locking (instead of only frequency locking for heterodyne) and is much more sensitive to the optical phase and frequency stability. Moreover, the optical phase-locked loop required for homodyne systems must have a very large bandwidth, and its implementation is not at all trivial (see Section 4.9).

4.4 HETERODYNE SYSTEMS: ASYNCHRONOUS DETECTION

The synchronous receivers previously analyzed are optimum in the ML sense. There are a few good reasons, however, for introducing and analyzing different receivers, characterized by the fact that phase and frequency information at IF are totally disregarded in the detection process. The main reason depends on the fact that the light emitted by the semiconductor lasers is affected by *phase noise*, which produces a broadening of the carrier linewidth. As we will see in Section 4.9, this has a strong impact on the performance of synchronous systems. Thus, the performance of the ideal synchronous receivers cannot be achieved in practice, and detection schemes more robust to phase noise are needed.

Moreover, the receiver section that performs the detection in an asynchronous receiver is simpler to implement and yet only slightly worse in terms of performance.

We will analyze practical asynchronous alternatives to synchronous detection for ASK and FSK modulation schemes. To compare these systems with the synchronous ones, we will still make the hypothesis that the carrier be strictly monochromatic (no phase noise).

4.4.1 Asynchronous ASK

The ML synchronous detector for both PSK (previously shown in Fig. 4.6) and ASK modulations performs a correlation between the IF signal and $\cos 2\pi f_{IF} t$. This correlation is equivalent to a matched-filtering operation followed by a sampler, as shown in Figure 4.8, where the impulse response of the bandpass matched filter is

$$h(t) = \cos[2\pi f_{IF}(T - t)], \qquad 0 \leq t \leq T \tag{4.30}$$

Disregarding the phase of the IF signal in the detection process is equivalent to considering a received carrier affected by a random phase ψ uniformly distributed between 0 and 2π. As a consequence, in the matched-filter receiver, the signal component sampled at $t = T$ would be proportional to a $\cos \psi$ and thus not useful for the decision. On the contrary, the envelope of the signal component at the output of the matched filter is still proportional to the transmitted data, a, and can thus be used for decision. In formulae, with the notation of Figure 4.9, we have

$$i_S(t) = 2a\, R\sqrt{P_S P_{LO}}\, \cos\,(2\pi f_{IF}\, t + \psi) + i_n(t) \tag{4.31}$$

where a is the information-bearing RV assuming the values 0 and 1, ψ is an RV uniformly distributed between 0 and 2π, and $i_n(t)$ is the shot noise contribution with zero mean and spectral density qRP_{LO}.

After matched filtering, the signal becomes

$$i_F(t) = aA\, \cos(2\pi f_{IF}\, t + \psi) + i_{nF}(t) \tag{4.32}$$

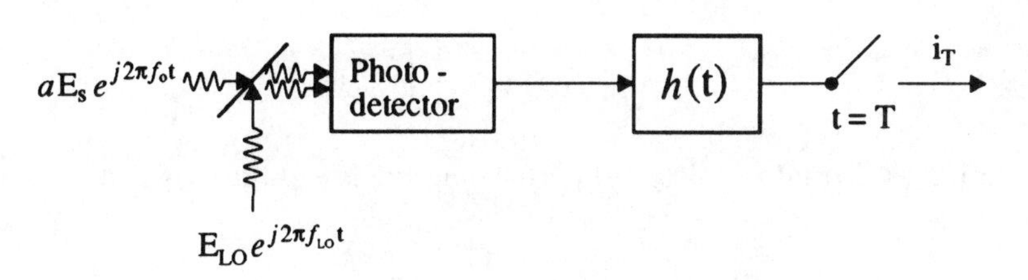

Figure 4.8 Matched filter heterodyne receiver of a coherent binary ASK system.

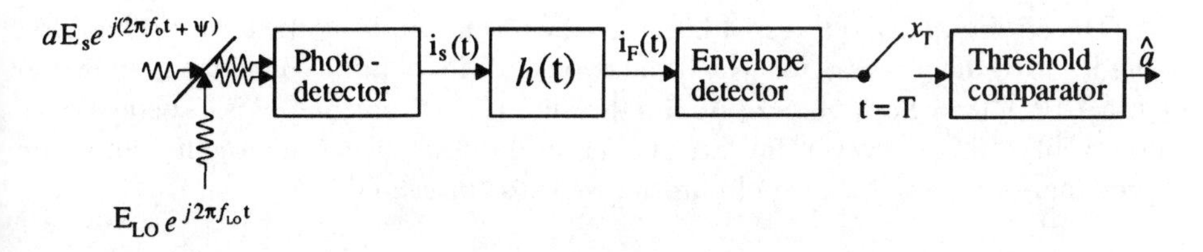

Figure 4.9 Asynchronous heterodyne receiver of a coherent binary ASK system.

where $A = RT\sqrt{P_S P_{LO}}$, and $i_{nF}(t)$ is a bandpass Gaussian noise process. We also included the delay introduced by the filter in the phase term ψ.

The process $i_{nF}(t)$ can be written as

$$i_{nF}(t) = i_{nc}(t)\cos 2\pi f_{IF}t - i_{ns}(t)\sin 2\pi f_{IF}t \tag{4.33}$$

i_{nc} and i_{ns} being two independent Gaussian processes with zero mean and equal variance $\frac{1}{2}qRP_{LO}T$. The envelope of $i_F(t)$ sampled at $t = T$ becomes, with obvious notations,

$$x_T = |\, i_F(T)\,| = \sqrt{(aA\cos\psi + i_{ncT})^2 + (a\,A\sin\psi + i_{nsT})^2} \tag{4.34}$$

To compute the bit error probability, we need the two pdf's (probability density functions) of the RV x_T conditioned on $a = 0$ and $a = 1$. They are (see Section 2.3) the Rayleigh and the Rice pdfs, respectively, with expressions

$$f_0(x) = \frac{x}{\sigma^2}\,e^{-(x^2/2\sigma^2)}$$

$$f_1(x) = \frac{x}{\sigma^2}\,e^{-[(x^2+A^2)/2\sigma^2]}\,I_0\left(\frac{Ax}{\sigma^2}\right) \tag{4.35}$$

where $\sigma^2 = \frac{1}{2}qRP_{LO}T$ is the variance of the RVs i_{ncT} and i_{nsT}, and I_0 is the zeroth-order modified Bessel function of the first kind.

To make a decision about a, we must compare the RV x_T with a threshold, x_{th}. Thus, for equiprobable values of a, the resulting bit error probability becomes

$$P_e = \frac{1}{2}\int_0^{x_{th}} f_1(x)dx + \frac{1}{2}\int_{x_{th}}^{\infty} f_0(x)dx \tag{4.36}$$

Substituting (4.35) into (4.36) and performing the integrations, we obtain

$$P_e = \frac{1}{2}e^{-(x_{th}^2/2\sigma^2)} + \frac{1}{2}\left[1 - Q\left(\frac{A}{\sigma}, \frac{x_{th}}{\sigma}\right)\right] \tag{4.37}$$

where $Q(a, b)$ is the Marcum Q-function (see Section 2.3).

The optimum (ML) threshold, that is, the one minimizing the error probability, coincides with the abscissa for which the two conditional pdf's coincide, as shown in Figure 4.10. A drawback of the optimum threshold (see Problem 4.4) is its dependence on the SNR A^2/σ^2. However, for large values of the SNR, using the rough asymptotic approximation $I_0(x) \sim e^x$, we obtain the constant threshold

$$x_{th,opt} = \frac{A}{2} \tag{4.38}$$

PDASK

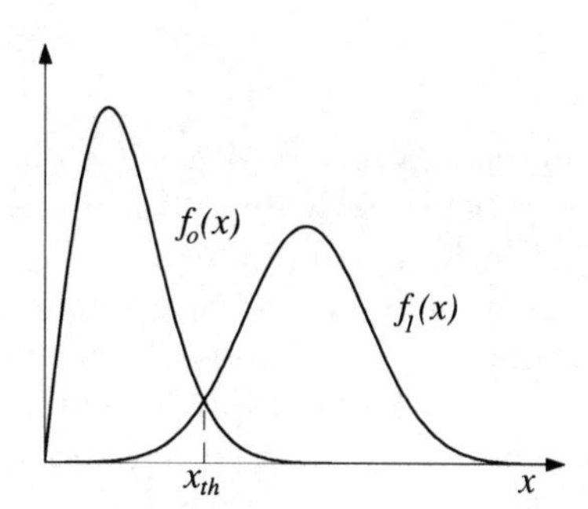

Figure 4.10 Optimum decision threshold for asynchronous binary ASK.

Moreover, making the asymptotic approximations

$$Q(a,b) \approx \frac{1}{2}\,\mathrm{erfc}\left(\frac{b-a}{\sqrt{2}}\right) \tag{4.39}$$

$$\mathrm{erfc}(x) \approx \frac{e^{-x^2}}{\sqrt{\pi}x}$$

we get

$$P_e \approx \frac{1}{2}\left(1+\sqrt{\frac{2\sigma^2}{\pi A^2}}\right)e^{-(A^2/8\sigma^2)} \tag{4.40}$$

which in turn, for large SNRs, can be approximated as

$$P_e \approx \frac{1}{2}e^{-(A^2/8\,\sigma^2)} \tag{4.41}$$

Substituting the values for A and σ, we finally obtain

$$P_e \approx \frac{1}{2}e^{-(RTP_S/4q)} = \frac{1}{2}e^{-(N_R/2)} \tag{4.42}$$

having introduced N_R, the average number of received photons in T.

With respect to the result obtained for synchronous heterodyne ASK, the only difference is the factor 1/2 in the expression of the bit error probability. Its sensitivity turns out to be 40 photons/bit (80 for peak power), compared with the 36 photons/bit required for synchronous heterodyne ASK. We have a 0.46-dB loss, which proves that asynchronous detection is only slightly suboptimal.

4.4.2 Asynchronous FSK

Digital frequency modulation can be detected using asynchronous and differential detection receivers. The choice between the two possibilities depends on the value of the frequency deviation, f_d, which, in turn, influences the signal bandwidth and the required IF bandwidth. We treat here the case of *wide-deviation* FSK, which can be demodulated using asynchronous receivers, and leave for the next section the case of small-deviation FSK leading to a delay-multiply demodulator.

The transmitted signal consists as before of two possible frequencies: $f_+ = f_0 + f_d$ and $f_- = f_0 - f_d$. We suppose that the frequency deviation, f_d, is large enough to permit a complete spectral separation of the two tones centered around f_+ and f_-.

Example 4.1

Using the computational method explained in Section 2.6, the continuous part of the power spectrum of the complex envelope of the binary FSK signal can be found to be

$$G(f) = K\,\{|g(f+f_d)|^2 + |g(f-f_d)|^2 - 2\Re[g(f+f_d)g^*(f-f_d)]\} \quad (4.43)$$

where

$$g(f) = \frac{\sin \pi f T}{\pi f T}\, e^{-j(2\pi f T/2)} \quad (4.44)$$

In Figure 4.11, the spectra of the FSK signal is plotted for various values of the normalized frequency deviation, f_dT, versus the normalized frequency, fT. From it, we learn that a complete separation of the two main lobes requires a normalized deviation, f_dT, larger than 1.

Two asynchronous receivers for wide-deviation binary FSK have been proposed and experimented, that is, the *double-* and *single-filter* receivers. We will analyze first the double-filter receiver shown in Figure 4.12.

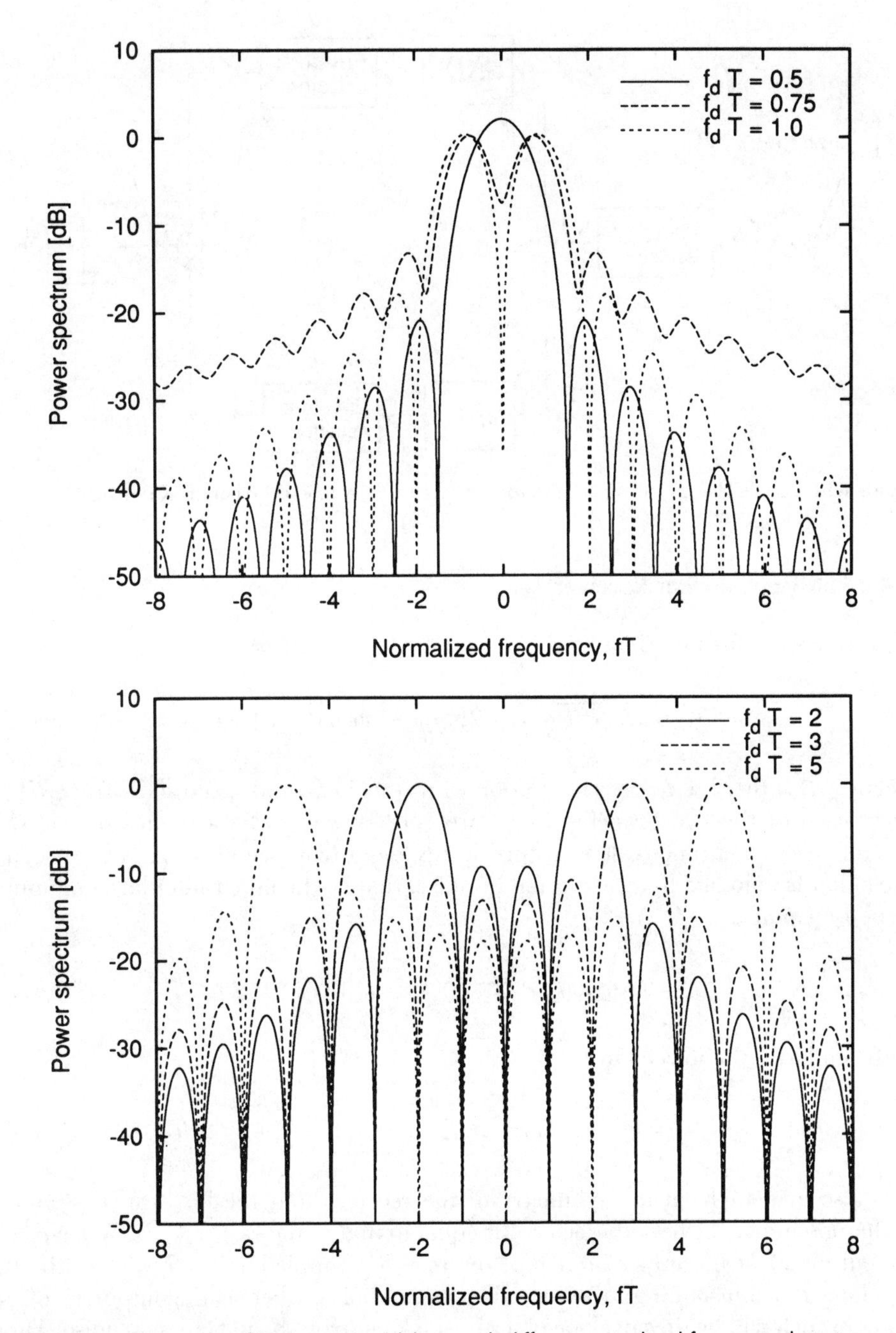

Figure 4.11 Power spectra of binary FSK modulaton with different normalized frequency deviation.

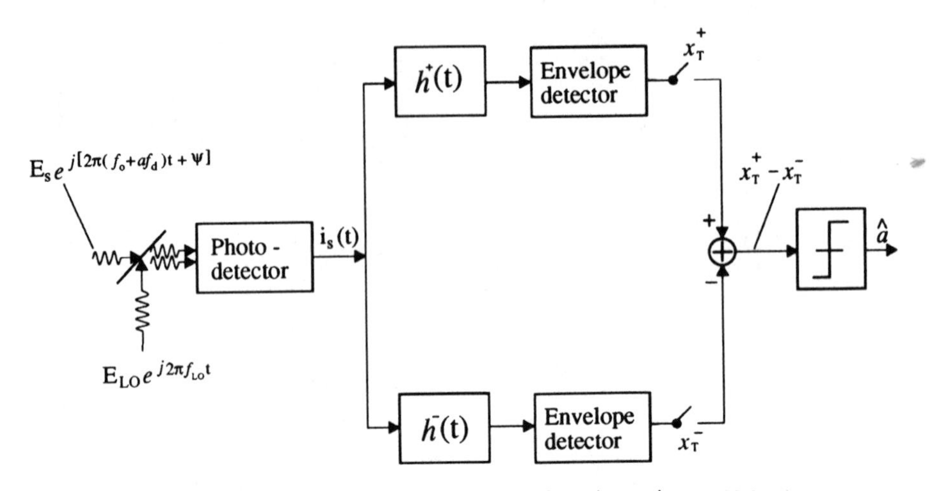

Figure 4.12 Double filter asynchronous heterodyne receiver of a coherent binary FSK system.

4.4.3 FSK Double-Filter Receiver

With reference for notations to the figure, the IF current is

$$i_S(t) = 2R\sqrt{P_S P_{LO}}\cos[(2\pi f_{IF} + 2\pi a f_d)t + \psi] + i_n(t) \tag{4.45}$$

where $i_n(t)$ is the shot noise contribution with zero mean and spectral density qRP_{LO}. Now, each of the two branches in Figure 4.12 is seen to be a replica of the ASK heterodyne asynchronous detector previously described (see also Fig. 4.9), because each filter is matched to the expected IF pulse, that is, the upper filter has an impulse response equal to

$$h^+(t) = \cos[(2\pi f_{IF} + 2\pi f_d)(T - t)], \qquad 0 \le t \le T \tag{4.46}$$

and, similarly, the lower filter

$$h^-(t) = \cos[(2\pi f_{IF} - 2\pi f_d)(T - t)], \qquad 0 \le t \le T \tag{4.47}$$

It is also evident from the symmetry of the receiver that the bit error probability evaluation under the hypothesis $a = 1$ is equal to that in the case $a = -1$. When $a = 1$, the output x_T^+ of the upper branch of the receiver sampled at $t = T$ will still be the envelope of a sinusoid in narrow-band Gaussian noise, whereas the output x_T^- of the lower branch will be the envelope of a zero-mean narrow-band Gaussian noise. Thus, as before, their pdf f^+ and f^- will be

$$f^{+}(x) = \frac{x}{\sigma^2} e^{-[(x^2 + A^2)/2\sigma^2]} I_0\left(\frac{Ax}{\sigma^2}\right)$$

$$f^{-}(x) = \frac{x}{\sigma^2} e^{-(x^2/2\sigma^2)} \tag{4.48}$$

where $A = RT\sqrt{P_S P_{LO}}$ and $\sigma^2 = (1/2)qRP_{LO}T$ is the variance of the two independent lowpass noise components.

To perform an ML decision on the transmitted a, we need to look at the sign of the RV $x_T^+ - x_T^-$, so that the bit error probability when $a = 1$ will be given by

$$P_e = P[x_T^+ - x_T^- < 0] \tag{4.49}$$

From the pdf's of x_T^+ and x_T^-, exploiting their statistical independence, which stems directly from the orthogonality of the two transmitted waveforms (see Problem 4.5), we obtain

$$P_e = \frac{1}{2} e^{-(A^2/4\sigma^2)} = \frac{1}{2} e^{-(N_R/2)} \tag{4.50}$$

which is the same sensitivity (on average power) of 40 photons/bit as the asymptotic approximation of asynchronous ASK, or 3 dB better in terms of peak power.

4.4.4 FSK Single-Filter Receiver

To reduce the complexity of the double-filter FSK receiver, it is common practice to use only one branch (say the upper one, as in Fig. 4.13). In that case, we are throwing away half the signal power, filtered out by the upper matched filter, and the decision is taken in the same way as for asynchronous ASK. All considerations and computations made for ASK about the optimum and asymptotic threshold and error probability evaluation are valid here and yield asymptotically

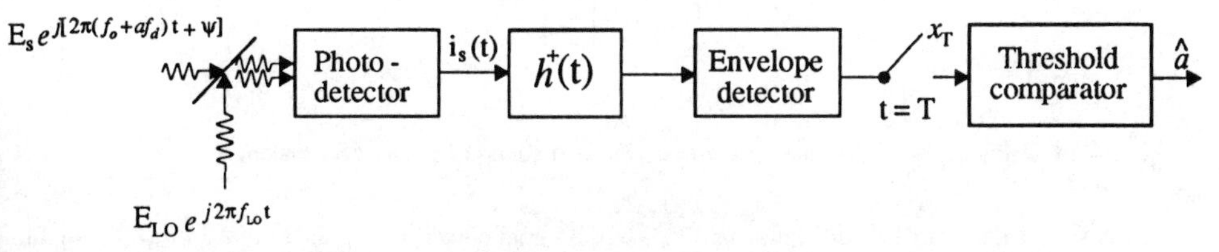

Figure 4.13 Single-filter asynchronous heterodyne receiver of a coherent binary FSK system.

$$P_e \approx \frac{1}{2} e^{-(A^2/8\sigma^2)} = \frac{1}{2} e^{-(N_R/4)} \tag{4.51}$$

which shows an asymptotic 3-dB degradation with respect to the double-filter receiver.

4.5 HETERODYNE SYSTEMS: DIFFERENTIAL DETECTION

In this section, we will analyze two receivers for angle modulation that do not require the recovery of the carrier phase but instead make use of the received signal at different time instants to perform the demodulation. More precisely, the demodulation is performed by multiplying the signal at a given time instant by a delayed version of it and then, after lowpass filtering, deciding on the transmitted bit through comparison of the product signal with a zero threshold. Since the demodulation process is based on the angular difference of two signals received in consecutive intervals, the detection process is called differential detection.

4.5.1 Low-Deviation FSK

To reduce the IF bandwidth required by asynchronous detection of wide-deviation FSK, we can use a smaller frequency deviation (typically $f_d T < 1$) in connection with the delay-and-multiply receiver shown in Figure 4.14. In it, the IF signal is first bandpass filtered by the filter with impulse response $h(t)$, which we assume to let the signal pass undistorted through it[4] while reducing the shot noise by means of an equivalent noise bandwidth, B, and then multiplied by its τ-delayed version. The product signal is lowpass filtered to get rid of the double IF frequency term, sampled, and compared to a zero threshold for the decision.

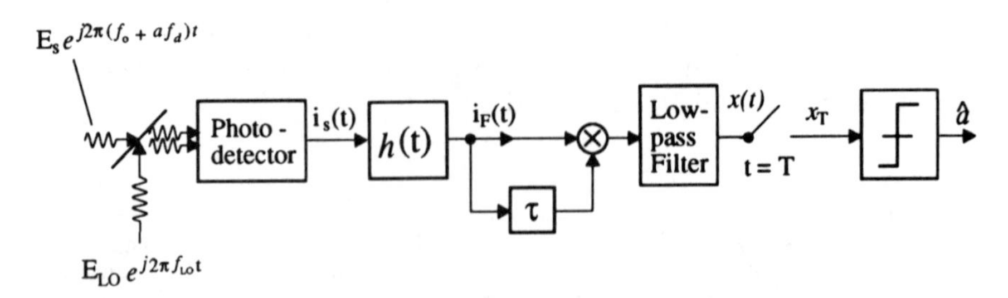

Figure 4.14 Delay-and-multiply heterodyne receiver of a coherent binary FSK system.

4. A widely used approximation states that 95% of the signal power is contained in a bandwidth equal to $f_d + 1/(2T)$.

The received and local oscillator signals in analytic form are

$$e_S(t) = E_S\, e^{j2\pi(f_0 + af_d)t} \tag{4.52}$$

$$e_{LO}(t) = E_{LO} e^{j2\pi f_{LO}t} \tag{4.53}$$

where $a = \pm 1$ is the information-bearing RV. After photodetection and bandpass filtering, the IF photodetected current is

$$i_F(t) = A\cos[2\pi(f_{IF} + af_d)t] + i_{nF}(t) \tag{4.54}$$

where $A = R\sqrt{P_S P_{LO}}$ is the amplitude, and $i_{nF}(t)$ is the shot noise process, a zero mean Gaussian random process with variance $\sigma^2 = qRP_{LO}B$.

After multiplication by its delayed version and lowpass filtering to remove the $2f_{IF}$ term, the signal $x(t)$ (see Fig. 4.14) can be written as

$$x(t) = \frac{A^2}{2}\cos[2\pi(f_{IF} + af_d)\tau + \phi_n(t)] \tag{4.55}$$

where the phase $\phi_n(t)$ is the phase noise due to the contribution of the shot noise.[5]

To show how the demodulator works, let us first assume that $\phi_n(t) = 0$. In that case, it is easily seen (see Problem 4.6) that the two relationships,

$$f_d\tau = \frac{1}{4}$$

$$f_{IF}\tau = \frac{1}{2}\left(k + \frac{1}{2}\right), \qquad k \text{ integer} \tag{4.56}$$

ensure that the amplitude of x behaves as shown in Figure 4.15, that is, that it assumes the maximum and minimum values $\pm(A^2/2)$ corresponding to the values ± 1 of the RV a.

Let us now define the parameter β as

$$\beta = 4f_d\tau = 2m\frac{\tau}{T} \tag{4.57}$$

5. Performing the product using the signal (4.54) yields a signal term, plus noise · noise and signal · noise terms. Equivalently, we can include all noise terms into an equivalent additive phase noise to exploit the knowledge of the statistics of the phase noise random process [2].

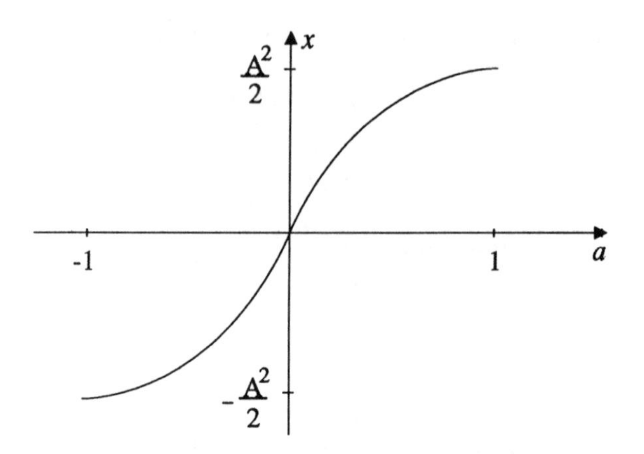

Figure 4.15 Ideal detected signal for the delay-and-multiply heterodyne receiver of a coherent binary FSK system.

where we have used the modulation index $m = 2f_dT$. Equation (4.57) shows that, for a given data rate $1/T$ and for $\beta = 1$ satisfying (4.56), the value of τ can be reduced by increasing the modulation index, m. That will prove to be beneficial when dealing with the system in the presence of laser phase noise.

Data decision is made according to the sign of samples x_T of $x(t)$ taken every T seconds. Due to the complete symmetry of the system, we can assume that $a = 1$ has been transmitted, so that an error is made whenever $x_T < 0$ and the error probability can be written in terms of the sampled noise-induced phase ϕ_{nT} as

$$P_e = P(-2\pi f_d\tau < \phi_{nT} \leq \pi - 2\pi f_d\tau) \tag{4.58}$$

The pdf of the RV ϕ_{nT} has been evaluated in the context of *differential phase shift keying* (DPSK) receivers with delay errors [2], so that integrating it as required in (4.58) gives

$$P_e = \frac{1}{2} - \frac{1}{2}\rho e^{-\rho} \sum_{n=0}^{\infty} \frac{1}{2n+1} [I_n(\rho/2) + I_{n+1}(\rho/2)]^2 \sin\left[(2n+1)\frac{\pi}{2}\beta\right] \tag{4.59}$$

where ρ is the IF SNR $A^2/2\sigma^2$, which is also equal to the received number of photons per bit N_R, and $I_n(\cdot)$ is the modified Bessel function of the first kind of order n.

For $\beta = 1$, (4.59) (see Problem 4.7) yields the same result of DPSK, which will be derived in a different way in Section 4.5.2, that is,

$$P_e = \frac{1}{2} e^{-\rho} \tag{4.60}$$

For $0 < \beta < 1$, the following approximation [2] can also be used:

$$P_e \sim \sqrt{\frac{1 + \cos\frac{\pi}{2}\beta}{1 - \cos\frac{\pi}{2}\beta}} \cdot \frac{e^{-\rho(1-\cos(\pi/2)\beta)}}{\sqrt{8\pi\rho\cos\frac{\pi}{2}\beta}} \tag{4.61}$$

In Figure 4.16 the error probability is plotted for the case of *minimum shift keying* (MSK), corresponding to the particular choice of the modulation index $m = 1/2$ versus N_R for several values of the parameter β, equal, in this case, to τ/T. The curves show that, in the absence of phase noise, the best value of the parameter is $\beta = 1$. The power penalty versus β for an error probability of 10^{-9} are reported in Figure 4.17.

With respect to asynchronous dual-filter FSK, CPFSK achieves a better sensitivity (3 dB for MSK), at the expense of a greater degradation in the presence of

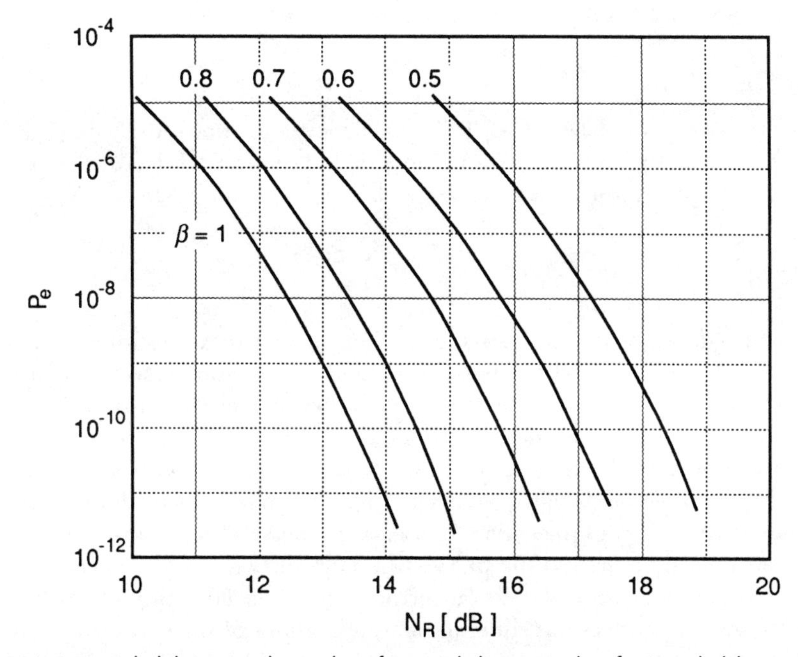

Figure 4.16 Error probability versus the number of received photons per bit of MSK with delay-and-multiply receiver different for values of the parameter β.

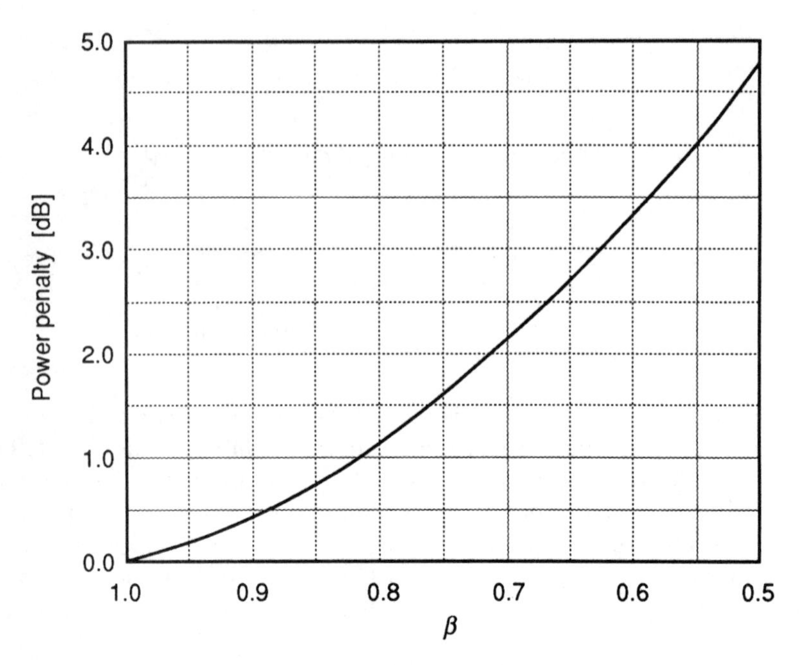

Figure 4.17 Power penalty versus the parameter β for MSK with delay-and-multiply receiver.

phase noise. Compared to DPSK, CPFSK can achieve the same sensitivity (still MSK), and its flexibility in parameters like the modulation index and delay, τ, can be used to decrease the phase noise degradation with respect to *differential* PSK (DPSK).

4.5.2 DPSK

Asynchronous detection described in the previous section allows a simplification of the receiver with a small sensitivity penalty compared with the same modulation schemes when synchronously detected. The best modulation scheme in terms of sensitivity—PSK—however, does not admit asynchronous detection, because it is precisely the phase of the received light that carries information; thus, envelope-based detectors are inappropriate. It is possible, however, to conceive a detection scheme for PSK that does not require the recovery of the carrier phase. It is called DPSK because it recovers the transmitted information from the phase difference of two signals spaced T seconds apart. To avoid propagation of detection errors, it is beneficial in this scheme to encode the data to be transmitted according to phase variations of the carrier instead of the absolute phase.

To send a 0, the transmitter will change the phase of the signal by π with respect to the previous signal; to send a 1, the same phase as in the preceding interval will be transmitted. The transmitted light during the kth signaling interval will be[6]

$$e_S(t) = E_S\, e^{j(2\pi f_0 t + \phi_k)} \tag{4.62}$$

where the phase difference will obey to

$$\phi_k - \phi_{k-1} = 0 \text{ or } \pi \tag{4.63}$$

The DPSK receiver is shown in Figure 4.18. After the IF matched filter, we perform the multiplication of the signal with its replica delayed by T, filter the product by an ideal low-pass filter whose only purpose is to eliminate double-frequency components, and then compare the sampled filter output with a zero threshold, deciding for a transmitted 1 when z_T is positive and for a 0 when it is negative.

With reference to Figure 4.18 notation, we have for the filtered photodetector current, $i_F(t)$:

$$i_F(t) = A \cos\,(2\pi f_{IF} t + \phi_k) + i_{nF}(t) \tag{4.64}$$

where we have defined $A = RT\sqrt{P_S P_{LO}}$ and $i_{nF}(t)$ is a bandpass Gaussian noise process, which can be written as in (4.33). The signal delayed by T is denoted by

$$i_{FD}(t) = A \cos\,(2\pi f_{IF} t + \phi_{k-1}) + i_{nFD}(t) \tag{4.65}$$

where we have supposed, for simplicity of notation, that $f_{IF}T$ is an integer, and where $i_{nFD}(t)$, being originated by a signal in a different interval, is a bandpass Gaussian

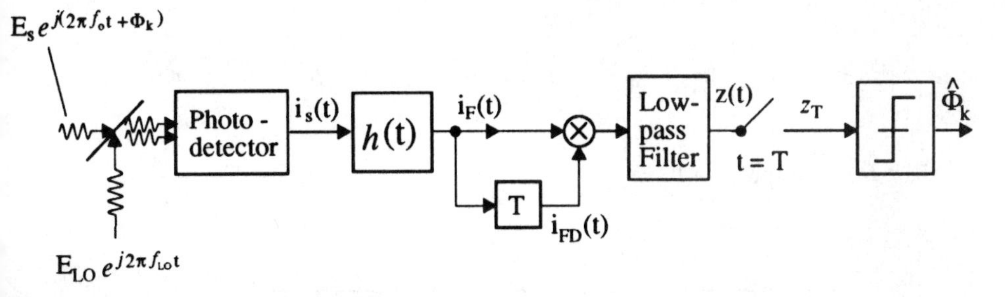

Figure 4.18 Differential detection heterodyne receiver of a coherent binary PSK system.

6. We avoid, for simplicity, the introduction of a random phase, by assuming that it varies slowly with respect to T, so that it can be considered as constant and set to zero.

process independent of $i_{nF}(t)$ and with the same statistical characteristics as $i_{nF}(t)$. Defining

$$x_c(t) = A \cos \phi_k + i_{nc}(t)$$

$$x_s(t) = A \sin \phi_k + i_{ns}(t) \tag{4.66}$$

and similarly, with obvious notations,

$$x_{cD}(t) = A \cos \phi_{k-1} + i_{ncD}(t)$$

$$x_{sD}(t) = A \sin \phi_{k-1} + i_{nsD}(t) \tag{4.67}$$

we obtain, for the low-pass filtered product signal after sampling, the expression

$$z_T = x_c x_{cD} + x_s x_{sD} \tag{4.68}$$

where $x_c = x_c(T)$, and so on. The bit error probability can thus be written as

$$P_e = \frac{1}{2} P(z_T \le 0 \mid \phi_k - \phi_{k-1} = 0) + \frac{1}{2} P(z_T > 0 \mid \phi_k - \phi_{k-1} = \pi) \tag{4.69}$$

To simplify the computations, we make use of the equality

$$z_T = x_c x_{cD} + x_s x_{sD} = \frac{1}{4}[(x_c + x_{cD})^2 + (x_s + x_{sD})^2 - (x_c - x_{cD})^2 - (x_s - x_{sD})^2] \tag{4.70}$$

so that, for example, we have

$$P(z_T < 0 \mid \phi_k - \phi_{k-1} = 0) =$$
$$P\left[\sqrt{(x_c + x_{cD})^2 + (x_s + x_{sD})^2} < \sqrt{(x_c - x_{cD})^2 + (x_s - x_{sD})^2} \mid \phi_k - \phi_{k-1} = 0\right]$$

Under the conditioning value of the phase difference, we have that $x_c + x_{cD}$ and $x_s + x_{sD}$ represent two Gaussian RVs with means $2A \cos \phi_k$ and $2A \sin \phi_k$, respectively, and equal variances $\sigma^2 = qRP_{LO}T$. Thus, as seen in the case of ASK, the RV $\sqrt{(x_c+x_{cD})^2+(x_s+x_{sD})^2}$ is a Rice RV with pdf

$$f(x) = \frac{x}{\sigma^2} e^{-[(x^2+4A^2)/2\sigma^2]} I_0\left(\frac{2Ax}{\sigma^2}\right) \tag{4.71}$$

On the other hand, the RVs $x_c - x_{cD}$ and $x_s - x_{sD}$ are two Gaussian RVs with zero mean and equal variances $\sigma^2 = qRP_{LO}T$, so that the RV $\sqrt{(x_c-x_{cD})^2+(x_s-x_{sD})^2}$ is a Rayleigh RV with pdf

$$f_0(x) = \frac{x}{\sigma^2} e^{-(x^2/2\sigma^2)} \tag{4.72}$$

To compute the error probability conditioned on the transmission of a 1, we must compute the probability that a Rayleigh RV exceeds a Rice RV, which is the same situation already analyzed for double-filter asynchronous FSK detection, with the difference that the signal power is multiplied by 4 and the noise variance by 2, so that the resulting error probability becomes

$$P(z_T < 0 | \phi_k - \phi_{k-1} = 0) = \frac{1}{2} e^{-(4A^2/8\sigma^2)} = \frac{1}{2} e^{-N_R} \tag{4.73}$$

If we consider the transmission of a 0, the role played by the various RVs is reversed, and the final result does not change, so that finally we have

$$P_e = \frac{1}{2} e^{-(4A^2/8\sigma^2)} = \frac{1}{2} e^{-N_R} \tag{4.74}$$

DPSK systems have a sensitivity of 20 photons/bit and asymptotically lose 0.46 dB with respect to coherent heterodyne PSK.

4.6 SUMMARY AND COMPARISON OF FUNDAMENTAL SENSITIVITIES

Table 4.1 lists all the expressions of bit error probability, together with the sensitivity of each system expressed in number of photons per bit necessary to obtain an error probability equal to 10^{-9}. In Figure 4.19 the bit error probabilities of the various modulation schemes are reported versus the number of photons per bit in the received lightwave.

Table 4.1
Bit Error Probabilities and Sensitivity for Different Modulation Schemes

	Bit Error Probability			*Sensitivity, photons/bit*		
	PSK	*ASK*	*FSK*	*PSK*	*ASK*	*FSK*
Homodyne	$\frac{1}{2}\operatorname{erfc}(\sqrt{2N_R})$	$\frac{1}{2}\operatorname{erfc}(\sqrt{N_R})$	$\frac{1}{2}\operatorname{erfc}\left(\sqrt{\frac{N_R}{2}}\right)$	9	Ave. 18 Peak 36	36
Synchronous heterodyne	$\frac{1}{2}\operatorname{erfc}(\sqrt{N_R})$	$\frac{1}{2}\operatorname{erfc}\left(\sqrt{\frac{N_R}{2}}\right)$	$\frac{1}{2}\operatorname{erfc}\left(\sqrt{\frac{N_R}{2}}\right)$	18	Ave. 36 Peak 72	36
Asynchronous heterodyne		$\frac{1}{2}e^{-\frac{N_R}{2}}$	DF: $\frac{1}{2}e^{-\frac{N_R}{2}}$ SF: $\frac{1}{2}e^{-\frac{N_R}{4}}$		Ave. 40 Peak 80	DF 40 SF 80
Differential detection • Heterodyne DPSK	$\frac{1}{2}e^{-N_R}$			20		
• Heterodyne MSK*			$\frac{1}{2}e^{-N_R}$			20

*This line refers to the CPFSK with delay-multiply demodulator for modulation index $m = 0.5$ and delay $\tau = T$, corresponding to MSK.

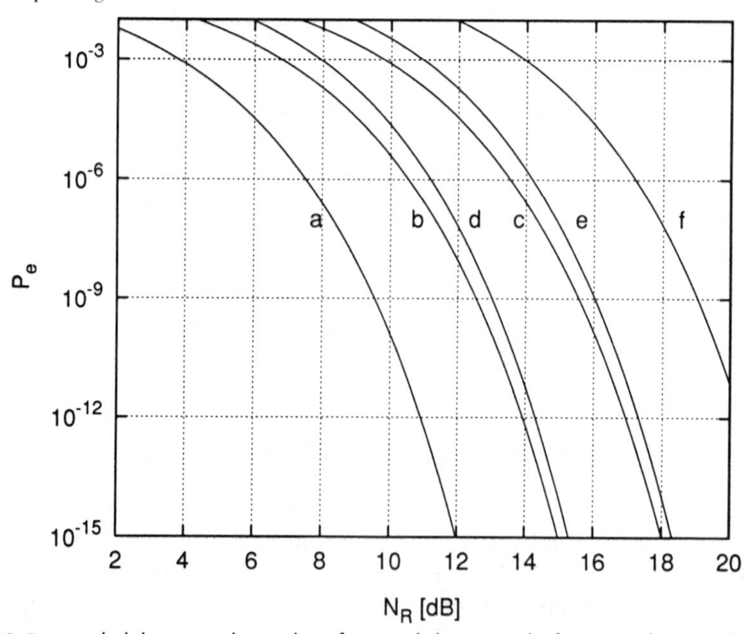

Figure 4.19 Error probability versus the number of received photons per bit for various binary coherent optical communication systems. Curves labeled *a, b, c, d, e,* and *f* refer to homodyne PSK, homodyne ASK, homodyne FSK, heterodyne PSK, double-filter asynchronous heterodyne FSK, and single-filter asynchronous heterodyne FSK, respectively.

Example 4.2

Figure 4.19 uses as abscissa the number of photons per bit in the received lightwave. In shot noise–limited systems, this is the only parameter that influences the performance, and it is independent of the system transmission rate. In the design of an optical transmission system, however, what matters is the required optical power of the transmitting laser, or, given that, the total loss that the system can tolerate from the transmitter to the receiver front end. Both these parameters do depend on the system transmission rate. We derive here a simple relationship among these major system parameters.

We have seen that

$$N_R = \frac{\eta P_S T}{h\nu}$$

so that the received "useful" optical power, including the quantum efficiency, is

$$\eta P_S = \frac{N_R h\nu}{T}$$

With $1/T$ being the system transmission rate, we see that the required received optical power is directly proportional to that rate. In Figure 4.20, the received power necessary

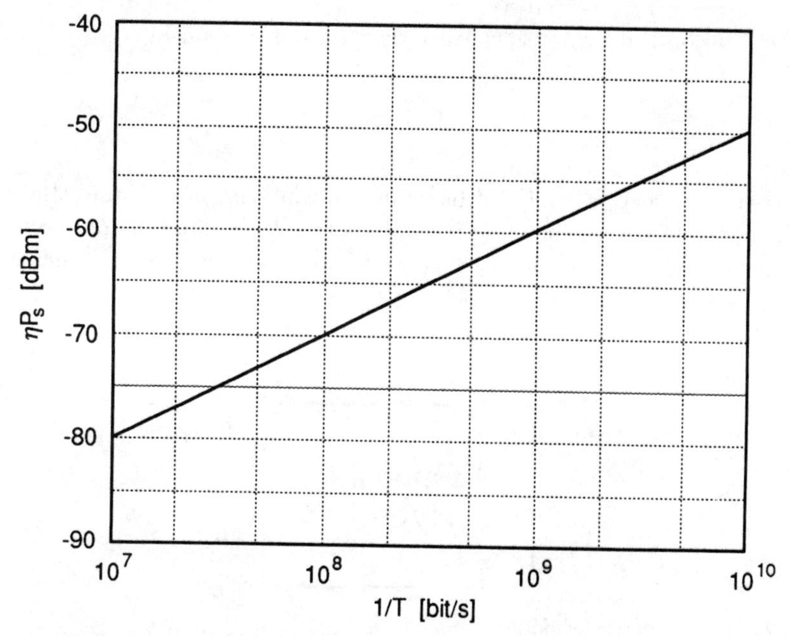

Figure 4.20 Plot of the received power necessary to obtain an error probability of 10^{-9} for binary homodyne PSK versus the transmission rate $1/T$.

to obtain a bit error probability of 10^{-9} for a homodyne PSK system is plotted versus $1/T$ for a system operating in the 1,550-nm range. We read, for example, that a system operating with a transmission rate of 1 Gbps requires about −59 dBm of received power. For a system employing a transmitting laser with 1 dBm power and a fiber with attenuation of 0.25 dB/km, the repeater spacing could be as large as 236 km. Unfortunately, fiber attenuation is not the only detrimental phenomenon limiting the transmitting span, and in this case the fiber dispersion comes into play, as seen in Chapters 1 and 3.

4.7 OPTICAL HYBRIDS

In the previous section, we ideally assumed that the received and local oscillator lightwaves added together before entering the photodetector. That operation, though well defined mathematically, is not physically realizable. We must then resort to directional couplers, which are a particular case of four-port optical devices with two inputs and two outputs, which we call (from microwave applications) *hybrids.* A block diagram of a hybrid is shown in Figure 4.21.

We assume that there are no reflections at the four ports, so that, instead of considering the 4-by-4 scattering matrix to characterize the device, we use a 2-by-2 transfer matrix, which we call **S**.

Thus, an optical hybrid is characterized by the relationship

$$\mathbf{E}_o = \mathbf{S}\mathbf{E}_i \tag{4.75}$$

where (see Fig. 4.21) $\mathbf{E}_i = [E_{1i} E_{2i}]'$ (the symbol $'$ means *transpose*) is the column vector of the fields E_{1i} and E_{2i} at the input ports, $\mathbf{E}_o = [E_{1o}\ E_{2o}]'$ is the column vector of the fields E_{1o} and E_{2o} at the output ports, and **S** is the transfer matrix of the coupler

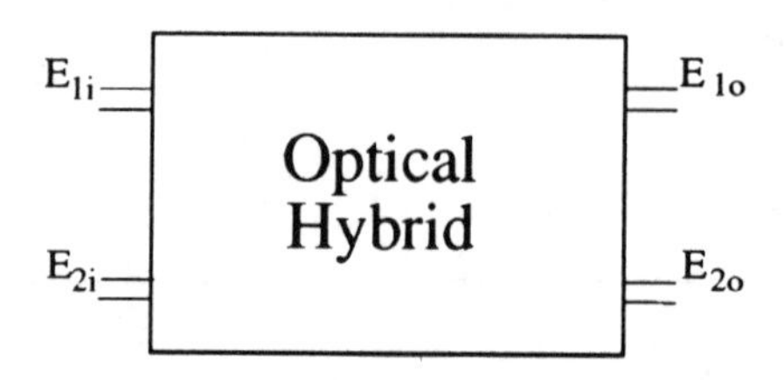

Figure 4.21 Four-port optical hybrid.

$$\mathbf{S} = \begin{bmatrix} s_{11} & s_{12} \\ s_{21} & s_{22} \end{bmatrix} \tag{4.76}$$

We will first derive the conditions that must be fulfilled by the elements of **S** for the device to be lossless. In doing that, we will consider only the input-output relationships and neglect the possibility of using the output ports as inputs.

Without loss of generality, we can assume that three of the four elements of **S** are real positive, namely, s_{11}, s_{12}, and s_{21}. This is obtained simply by adding suitable phase shifts to three of the four ports. As a consequence, we will have $s_{11} = |s_{11}|$, $s_{12} = |s_{12}|$, $s_{21} = |s_{21}|$, and $s_{22} = |s_{22}|e^{j\phi_{22}}$.

4.7.1 Lossless Hybrids

If the device is lossless, the input power must be transferred unaltered to the output. As a consequence, we must have

$$P_{1o} + P_{2o} = P_{1i} + P_{2i} \tag{4.77}$$

or, equivalently,[7]

$$\mathbf{E}_o^{\dagger}\mathbf{E}_o = \mathbf{E}_i^{\dagger}\mathbf{E}_i \tag{4.78}$$

Substituting (4.75) into (4.78) yields

$$\mathbf{E}_o^{\dagger}\mathbf{E}_o = \mathbf{E}_i^{\dagger}\mathbf{S}^{\dagger}\mathbf{S}\mathbf{E}_i \tag{4.79}$$

which, taking into account (4.77), gives the following constraints on the elements of **S** (see Problem 4.8):

$$\begin{aligned} |s_{11}|^2 + |s_{21}|^2 &= 1 \\ |s_{12}|^2 + |s_{22}|^2 &= 1 \\ s_{11}s_{12}^* + s_{21}s_{22}^* &= 0 \end{aligned} \tag{4.80}$$

As a consequence, with the position $s_{11} = \sqrt{1-k}$, from (4.80) we get

7. In the following equations, * means complex conjugate and † means transpose and complex conjugate.

$$s_{21} = \sqrt{k}$$
$$s_{12} = \sqrt{k}$$
$$s_{22} = \sqrt{1-k}e^{j(2m+1)\pi} \quad (4.81)$$

so that the transfer matrix becomes

$$\mathbf{S} = \begin{bmatrix} \sqrt{1-k} & \sqrt{k} \\ \sqrt{k} & -\sqrt{1-k} \end{bmatrix} \quad (4.82)$$

For $k = 1/2$, we have the 3-dB coupler, or π hybrid, whose outputs are proportional to the sum and the difference of its inputs. The value of $k = 1/2$ is the only one allowed by the lossless conditions when we consider reciprocal devices, that is devices for which the input and output functions can be reversed. We will see soon how the 3-dB coupler can be used to build a balanced dual-detector receiver.

When the phase ϕ_{22} is different from π, the hybrid will present some loss. Let us examine the case of $\phi_{22} = -\pi/2$, which is also important for the applications.

4.7.2 The $\pi/2$ Hybrid

A $\pi/2$ hybrid is a four-port device for which the difference between the phase difference at one output port and the phase difference at the other is equal to $\pi/2$. Assuming as before that three of the four s_{ij} are real, the constraint on the output phase difference leads to the following structure for the transfer matrix:

$$\mathbf{S} = \begin{bmatrix} s_{11} & s_{12} \\ s_{21} & s_{22}e^{-j\pi/2} \end{bmatrix} \quad (4.83)$$

We consider two situations with different constraints.

4.7.2.1 Symmetric $\pi/2$ Hybrid

Because the hybrid is symmetric, we assume $|s_{ij}| = 1/\sqrt{L}$, $\forall i,j$. Moreover, let $E_{1i} = |E_{1i}|$ and $E_{2i} = |E_{2i}|e^{j\theta_i}$. Computing the sum of the powers at the output ports, we get

$$\mathbf{E}_o^\dagger \mathbf{E}_o = \frac{2}{L}\left[P_{1i} + P_{2i} + \sqrt{P_{1i}P_{2i}}\left(\cos\theta_i + \sin\theta_i\right)\right] \quad (4.84)$$

Consider now the case of equal input powers, that is, $P_{1i} = P_{2i}$. In this situation, maximization of the output power leads to the condition $\theta_i = -\pi/4$, from which, imposing that the output power does not exceed the input power, leads to the following inequality for the admissible values of L:

$$L \geq 2 + \sqrt{2}$$

In this case, the power from one of the input ports is transferred to the output ports with a minimum achievable loss of $10 \log_{10}(L/2) = 2.32$ dB.

4.7.2.2 Asymmetric π/2 Hybrid

Some system applications, such as the case of homodyne receivers with decision-driven or Costas phase-locked loops (see Section 4.9), require for system optimization a $\pi/2$ hybrid in which the input power is split unevenly between the outputs. Its transfer matrix has the form

$$\mathbf{S} = \frac{1}{\sqrt{L}} \begin{bmatrix} \sqrt{1-k} & \sqrt{1-k} \\ \sqrt{k} & -j\sqrt{k} \end{bmatrix} \tag{4.85}$$

where $0 \leq k \leq 1$ is a parameter that determines the power splitting between the outputs, and L is an excess loss that will prove to be necessary to fulfill the power conservation.

As before, let $E_{1i} = |E_{1i}|$ and $E_{2i} = |E_{2i}|e^{j\theta_i}$. Computing the sum of the powers at the output ports, we get

$$\mathbf{E}_o^{\dagger}\mathbf{E}_o = \frac{1}{L}\{P_{1i} + P_{2i} + 2\sqrt{P_{1i}P_{2i}}\,[(1-k)\cos\theta_i - k\sin\theta_i]\} \tag{4.86}$$

From the conservation of the power, we derive, for the maximum output power (obtained in the case $\theta_i = -\pi/4$), the inequality

$$L \geq 1 + \frac{2\sqrt{P_1 P_2}}{P_1 + P_2} \tag{4.87}$$

We see that in this case a loss is also necessary. The amount of the loss depends on the values of the two input powers, P_1 and P_2. When they are equal, the loss amounts to 3 dB. Fortunately, in many system applications, one of the two inputs has much more power than the other. That is the case when at one input port we have the received lightwave and at the other the local oscillator lightwave. If, for example, the second is 20 dB higher than the first, the loss reduces to 0.78 dB.

In practice, optical hybrids can be fabricated by bringing two optical waveguides, either fibers or integrated optics guides, close together so that their fields overlap. For details on practical implementations, see [3].

4.7.3 Balanced Receivers

If we use the π hybrid to add the fields of the incoming lightwave with that of the local oscillator, and direct to the photodetector the signal of output 1, we lose half the power available at output port 2, making that port useless. With respect to the results obtained with the ideal receivers, the required optical power in the received signal must be increased by 3 dB.

It is possible, however, to take full advantage of the two outputs of the π hybrid, to exploit all the incoming power using the balanced receiver configuration depicted in Figure 4.22. In a balanced receiver, the two outputs are directed to separate photodiodes, which are assumed to be exactly balanced, that is, having the same responsivity and frequency response.

Thus, the currents produced by the two photodiodes are

$$i_1 = R\,|\,E_1\,|^2 = \frac{1}{2}R(P_S + P_{LO} + 2\sqrt{P_S P_{LO}}\cos\theta_S) + i_{n1}(t) \tag{4.88}$$

$$i_2 = R\,|\,E_2\,|^2 = \frac{1}{2}R(P_S + P_{LO} - 2\sqrt{P_S P_{LO}}\cos\theta_S) + i_{n2}(t) \tag{4.89}$$

where θ_S is the phase difference between the received field and the local oscillator field, equal to $f_{IF}t$ for heterodyne and to 0 (we hope) for homodyne receivers. The two shot noise contributions to the current are independent (being produced by separate photodiodes) white Gaussian random processes with spectral densities given by

$$G_{i1} = qR\,|\,E_1\,|^2 \tag{4.90}$$

$$G_{i2} = qR\,|\,E_2\,|^2 \tag{4.91}$$

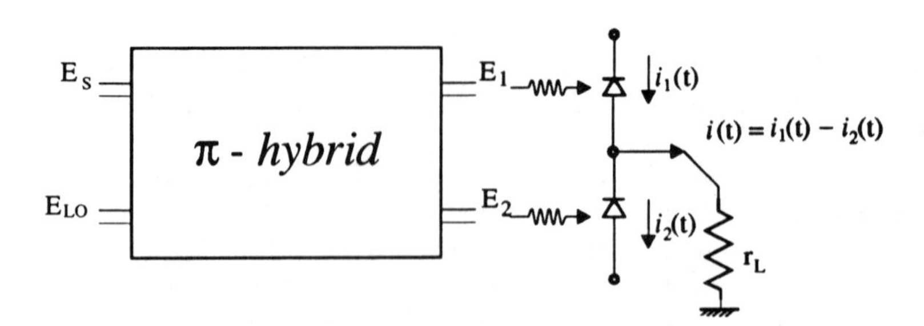

Figure 4.22 Balanced receiver for coherent optical communication.

The load current is thus

$$i(t) = i_1(t) - i_2(t) = 2R\sqrt{P_S P_{LO}} \cos \theta_S + i_n(t) \tag{4.92}$$

where $i_n(t)$ is the overall white Gaussian noise with spectral density

$$G_{in} = G_{i1} + G_{i2} = qR(P_S + P_{LO}) \approx qRP_{LO} \tag{4.93}$$

We have the same signal and noise components as for the homodyne ($\theta = 0$) and the heterodyne ($\theta = 2\pi f_{IF} t$) ideal receivers after DC terms cancellation. As a consequence, without needing a high-pass filter to cancel out the DC currents, the balanced receiver yields the same performance of the ideal receiver.

4.8 PHASE NOISE AND LINEWIDTH

Our analysis of coherent systems presented so far has assumed that the lightwave emitted by the laser in the absence of modulation is an ideal carrier. However, as we have seen in Section 1.2, the signal emitted by a semiconductor laser cannot be modeled as a pure sinewave, because it is affected by phase noise. This is perhaps the main reason why coherent reception techniques, which find so many applications in low-frequency and microwave communications, cannot be easily translated into the optical domain.

The amount of phase noise in an unmodulated lightwave is directly related to its so-called linewidth, that is, the 3-dB bandwidth of its power spectrum. Recent efforts in semiconductor laser technology have achieved the remarkable result of linewidths on the order of hundreds of kilohertz. Relative to the emitted frequency, the purity of a laser source is thus comparable to that of microwave oscillators, whose linewidth may be reduced to a fraction of a hertz.

Unfortunately, what matters in the performance of heterodyne optical receivers is the ratio between the linewidth and the digital information rate, independent of the carrier frequency. This is because the optical signal is down-converted to microwave frequencies, through an operation that leaves the optical linewidth unchanged.

4.8.1 Laser Phase Noise Model

The analytic signal representation of the unmodulated signal emitted by a laser is

$$e_S(t) = E_S \, e^{j[2\pi f_0 t + \phi(t)]} \tag{4.94}$$

where the random process $\phi(t)$ represents the phase noise. The frequency noise, defined as

$$f_n(t) = \frac{1}{2\pi}\dot{\phi}(t) \tag{4.95}$$

has a double-sided spectrum, $G_{fn}(f)$, with the typical form shown in Figure 4.23, where we can distinguish three different regions stemming from different physical phenomena. As a consequence, we can write

$$G_{f_n}(f) = G_{FN}(f) + G_{WN}(f) + G_{RN}(f) \tag{4.96}$$

where the following apply:

- $G_{FN}(f)$ is the spectrum of the *flicker noise*, caused primarily by the laser temperature fluctuations. It is responsible for the increasing values at low frequencies and can be modeled as

$$G_{FN}(f) = k_{FN}/f \text{ [Hz]}$$

  where k_{FN} is a parameter measuring the intensity of the flicker noise.

- G_{WN} is the spectrum of the *white frequency noise*, characterizing the central flat behavior. The white frequency noise represents the most relevant contribution and is modeled as a Gaussian random process with spectrum

$$G_{WN}(f) = \frac{\Delta\nu}{2\pi} \text{ [Hz]}$$

  We shall see that the parameter $\Delta\nu$ is precisely the laser linewidth.

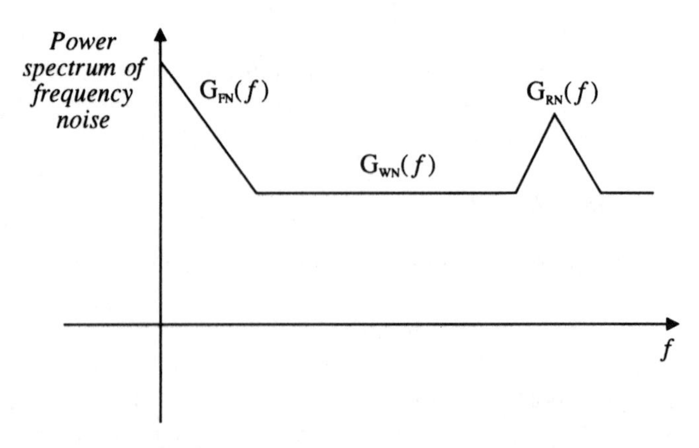

Figure 4.23 Qualitative behavior of the laser frequency noise power spectrum.

- Finally, G_{RN} is the *relaxation frequency noise.* It accounts for the peak appearing at very high frequencies and corresponds to the resonance frequency of the laser cavity.

In a well-designed practical system, a frequency-tracking subsystem will track out the low-frequency components due to the flicker noise. As for the relaxation frequency noise, its spectrum is located at much higher frequencies than those occupied by the system bandwidth. As a consequence, its contribution to the laser phase noise is widely attenuated by the integral relationship between frequency and phase.[8] For that reason, we will consider only the model and the effects of the white frequency noise.

The unmodulated signal emitted by the laser will be modeled as in (4.94), with the phase noise, $\phi(t)$, being a Wiener random process obtained as the integral of the white frequency noise:

$$\phi(t) = \int_0^t \dot{\phi}(\tau)d\tau \tag{4.97}$$

where $\dot{\phi}(t)$ is a zero-mean white Gaussian noise with power spectral density $G_{\dot{\phi}}(f) = 2\pi\Delta\nu$.

We want to compute the power spectral density of $e_S(t)$; to that purpose we first obtain the autocorrelation function

$$R_{e_S}(t_1,t_2) = \mathrm{E}[e_S(t_1)e_S^*(t_2)] = P_S e^{j2\pi f_0(t_1-t_2)}\mathrm{E}[e^{j\Phi(t_1,t_2)}] \tag{4.98}$$

where

$$\Phi(t_1, t_2) = \phi(t_1) - \phi(t_2) = \int_{t_1}^{t_2} \dot{\phi}(z)dz$$

is a zero-mean Gaussian RV with variance

$$\sigma_\Phi^2 = \mathrm{E}\left[\int_{t_1}^{t_2}\int_{t_1}^{t_2} \dot{\phi}(z)\dot{\phi}(v)dzdv\right] = 2\pi\Delta v \int_{t_1}^{t_2}\int_{t_1}^{t_2} \delta(z - v)dzdv = 2\pi\Delta\nu\,|\,t_1 - t_2\,| \tag{4.99}$$

8. Here is a heuristic explanation. Let $A\cos 2\pi f_{RN} t$ represent a single frequency component of the frequency relaxation noise. Its contribution to the phase noise is $\phi(t) = (A/2\pi f_{RN}) \sin 2\pi f_{RN} t$, so that its amplitude is decreased by the factor $2\pi f_{RN}$, being f_{RN} much higher than the highest signal frequencies.

and $P_S = |E_S|^2$ is the signal power. The average in (4.98) is thus the characteristic function of a Gaussian RV with zero-mean and variance $2\pi\Delta\nu|t_1 - t_2|$, which is equal to $e^{-\pi\Delta\nu|t_1 - t_2|}$. Thus, we finally obtain

$$R_{e_S}(t_1, t_2) = P_S e^{j2\pi f_0(t_1 - t_2)} e^{-\pi\Delta\nu|t_1 - t_2|} \tag{4.100}$$

The autocorrelation function depends only on the difference $t_1 - t_2$; the process is then stationary, with PSD equal to the Fourier transform of (4.100)

$$G_{e_S}(f) = \frac{2P_{e_S}}{\pi\Delta\nu} \cdot \frac{1}{1 + \left[\frac{2(f - f_0)}{\Delta\nu}\right]^2} \tag{4.101}$$

The spectral shape (4.101) sketched in Figure 4.24 is referred to as *Lorentzian*. It is easily verified from (4.101) that $\Delta\nu$ represents the linewidth of the lightwave.

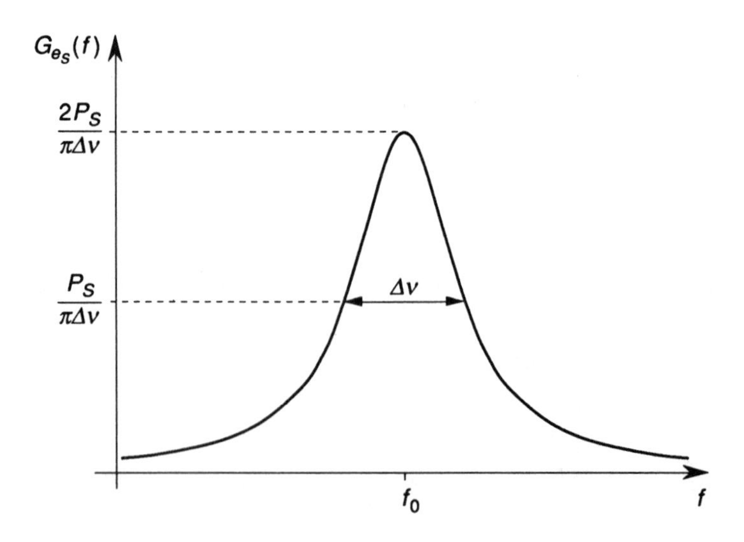

Figure 4.24 Lorentzian shape of the power spectrum of the signal emitted by a laser with phase noise.

4.8.2 System Effects of Laser Phase Noise

The phase noise of the lightwave emitted by a semiconductor laser affects the performance of a coherent digital modulation system in a way that depends on the particular modulation and detection schemes and on the bandwidth (or, equivalently, the information rate) of the system. The latter dependence is reflected in the requirement concerning the parameter *linewidth* × *signaling period* $\Delta\nu T$, and different systems have more or less stringent requirements regarding this parameter.

Synchronous systems make use of the carrier phase in the detection process. It is no surprise then that these systems are highly affected by phase noise, up to the point of exhibiting a *bit error probability floor*, a horizontal asymptote in the error probability versus SNR curves, meaning it is impossible to further decrease the error probability through the use of signal power.[9] As we shall see, this high sensitivity to phase noise requires such low values of the parameter $\Delta\nu T$ that can be satisfied, with the present semiconductor laser technology, only by extremely high signaling rates or by laser with external cavities.

On the contrary, asynchronous modulation schemes are mildly affected by phase noise, in the sense that they suffer from a *power penalty*, that is, they require, to achieve a given error probability, more power than in the absence of phase noise. The amount of power penalty depends on the particular system and on the value of $\Delta\nu T$.

The phenomenon causing this power penalty can be described as follows. The phase noise causes a *broadening* of the signal spectrum. As a consequence, its filtered version is distorted, and that reflects in the signal envelope on which the detection process is based.

Example 4.3

Consider the case of a binary ASK signal modulating a laser with central frequency f_0 and a Lorentzian spectral shape of normalized linewidth $\Delta\nu T$. In Figure 4.25 we plot the continuous part of the power spectral density, $G_x(f - f_0)$ of the modulated signal for several values of the normalized linewidth obtained by simulation. The spectral broadening phenomenon is clearly apparent.

---

9. Actually, this is true for DPSK and CPFSK modulations with delay-multiply receivers. For PSK using synchronous receivers with phase-locked-loops in the optical or electronic domain, the error probability exhibits the error floor as a function of the variance of the residual phase noise after phase tracking. On the other hand, this residual variance can be decreased by increasing the signal power and bandwidth of the phase-locked loop. Also in this case, however, the system becomes completely impractical since its power penalties make it worse than incoherent schemes.

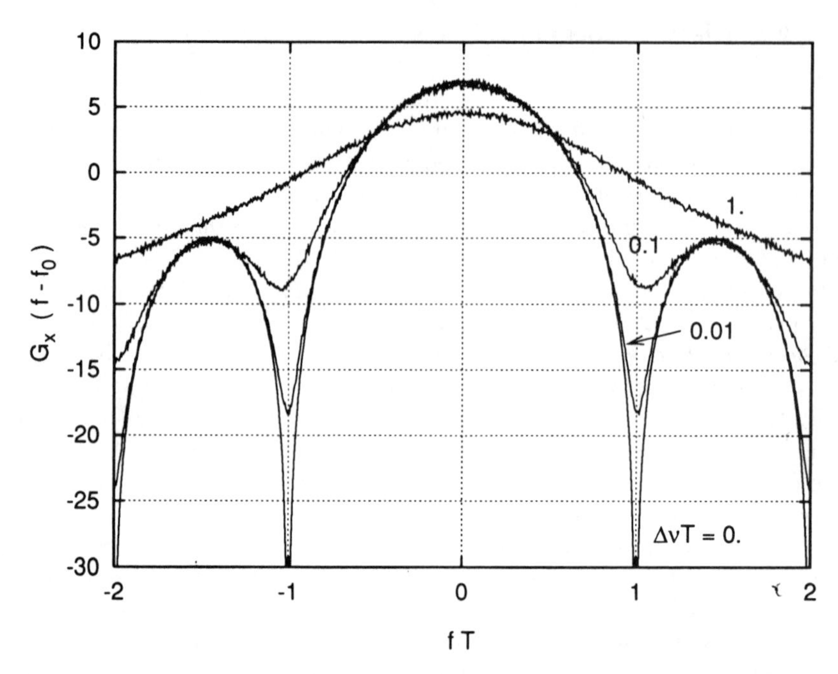

Figure 4.25 Power spectrum of a binary ASK signal modulating a carrier affected by phase noise for different values of the normalized linewidth.

Owing to the spectrum broadening, in the receiver of an asynchronous system an IF matched filter is no longer optimum; a wider bandwidth filter can improve system performance, as we shall see. Enlarging the IF filter bandwidth lets through an increasing amount of noise, which can in turn be limited by a postdetection BB lowpass filter. The optimum filter bandwidth then will be the result of a tradeoff between the signal integrity calling for an IF filter-bandwidth widening and the amount of shot noise entering the IF section of the receiver that increases with the filter bandwidth. Fortunately, thanks to the postdetection filter, the performance degradation due to the excess noise is mild as a function of the bandwidth, so that asynchronous systems can tolerate very high values of $\Delta\nu T$.

Sensitivity between synchronous and asynchronous schemes is provided by differentially detected modulation schemes like DPSK and CPFSK with delay-multiply receivers.

In the following sections, we will revisit the performance of the previously analyzed systems to estimate the effects of phase noise and discuss system optimization. The outcome of the analysis will be a clear assessment of the quantitative dependence of system performance degradation on the parameter $\Delta\nu T$.

4.9 SYNCHRONOUS SYSTEMS

We start our analysis with a synchronous system, in which the detection process requires the acquisition of both the frequency and the phase of the carrier used in the modulation process. This can be done using an *optical phase-locked loop* (OPLL) to track directly the optical carrier (homodyne detection) or an *electrical phase-locked* loop (PLL) in the IF domain (heterodyne detection).

In both cases, system performance is affected by the presence of a phase error in the recovered carrier phase, which is a slow random process with respect to the signaling interval T and can thus be considered as a time-independent RV in each interval T.

The analysis that follows is restricted to the case of binary PSK; the extension to other synchronous modulations previously described is straightforward.

4.9.1 Effects of Phase Error on System Performance

This section investigates the effects of an inaccurate phase of the local oscillator reference on the performance of binary PSK.

If the local carrier reference is affected by a phase error of ϕ_e radians, the output of the integrate-and-dump detector of Figure 4.4 becomes

$$i_T = 2Ra(\sqrt{P_{SP}P_{LO}})T \cos \phi_e + i_{nT}$$

A comparison of this expression with (4.17) shows that the only difference consists in a reduction of the effective signal amplitude by the factor $\cos \phi_e$, where ϕ_e is an RV whose statistics depend on the method used to track the phase of the incoming signal. Thus, when the reference carrier phase is inaccurate by ϕ_e radians, the conditional probability of error $P_e(\phi_e)$ [see (4.18)] is given by

$$P_e(\phi_e) = \frac{1}{2} \operatorname{erfc}(\sqrt{2N_R \cos^2 \phi_e}) \tag{4.102}$$

where we have introduced the average number of received photons per bit in the received lightwave N_R, previously defined in Section 4.2.

To obtain the unconditional value P_e, we must average (4.102) over the statistics of ϕ_e. As we shall see in the following, a good approximation for the statistics of the phase error when the tracking device is a PLL consists in a Gaussian RV,[10] which yields the result

10. This approximation is exact when we use a linearized model for the PLL, as will be done in the following. Without the linear approximation, the true statistics of the phase error is a Tikhonov distribution [4] or an appropriate generalization according to the order of the PLL. Using the true statistics makes the computational effort to get performance results remarkably higher. On the other hand, the obtained results agree quite well with those obtained using the Gaussian approximation for all $\sigma^2_{\phi_e} \leq qR_b/4RP_S$, a condition which is always verified in a well-designed system.

$$P_e = \mathrm{E}_{\phi_e} P_e(\phi_e) = \int_{-\infty}^{\infty} P_e(\phi_e) f(\phi_e) \mathrm{d}\phi_e \tag{4.103}$$
$$= \frac{1}{2\sqrt{2\pi}\sigma_{\phi_e}} \int_{-\infty}^{\infty} \operatorname{erfc}(\sqrt{2N_R \cos^2 \phi_e}) \exp\left(-\frac{\phi_e^2}{2\sigma_{\phi_e}^2}\right) \mathrm{d}\phi_e$$

where σ_{ϕ_e} is the standard deviation of the phase error.

The average in (4.103) can be evaluated numerically using Gauss quadrature rules [5]. The results obtained are plotted in Figure 4.26, as a function of the number of photons per bit, N_R, for various values of the standard deviation of the phase error in degrees.

The already mentioned phenomenon of the floor to the bit error probability is apparent. For any nonzero value of σ_{ϕ_e} and for $N_R \to \infty$, P_e approaches a nonzero constant, the floor. In this situation, an increase in the received signal power becomes less and less effective in terms of decreasing the bit error probability.

Figure 4.26 shows that a penalty of about 0.5 dB corresponds, for an error probability of 10^{-9}, to a standard deviation of the phase error equal to 10 degrees

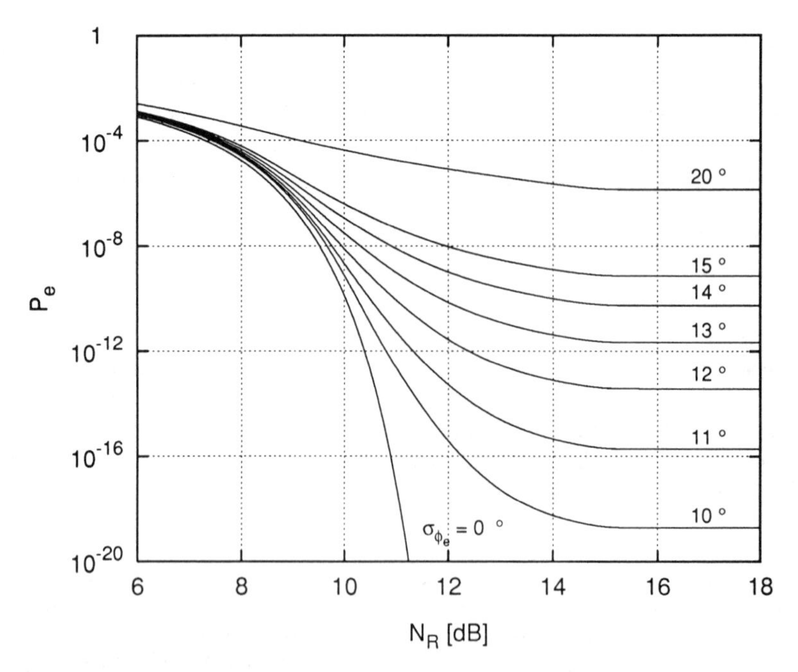

Figure 4.26 Error probability of homodyne binary PSK versus the number of received photons per bit for different values of the phase noise standard deviation.

(0.175 radian). In the following, this value of the standard deviation will be conventionally considered as an upper limit to a well-designed carrier tracking system.

4.9.2 Homodyne Phase Locking

The optical homodyne PSK system, just like its microwave precursor, must incorporate in the receiver a local oscillator phase-locked to the received signal, a major difficulty for the optical system. Phase locking can be achieved either by means of a PLL or through injection locking. Since the latter requires unrealistically high signal power levels, a PLL seems to be a better solution.

In designing a PLL-based PSK homodyne receiver, we encounter the following problems:

- The PSK spectrum is continuous, which contrasts with the PLL requirement of a discrete line to lock to.
- The data signal and the local oscillator signal must be locked *in-phase* for optimum detection; however, a conventional PLL locks the local oscillator *in quadrature* to the reference signal.
- When the data signal and the local oscillator signal are mixed on the photodetector surface, the detector output current contains a DC component in addition to the useful beat product of the two signals. This DC term will interfere with the PLL operation unless it is effectively eliminated.

These three problems can be solved either by a residual carrier (obtained by an incomplete modulation in the transmitter, i.e., using a phase shift of less than $\pi/2$) in connection with a balanced PLL, which tracks the residual carrier, or by a nonlinear loop.

We will analyze in detail the first solution, investigate the behavior of the PLL, which is the heart of both solutions, and, finally, sketch the performance and design guidelines of the second solution.

4.9.2.1 Balanced PLL Receiver

The first solution requires the transmission of a modulated signal that includes a residual carrier in quadrature with respect to the data signal. This can be obtained using a phase shift θ in the modulator less than $\pi/2$, so that the transmitted signal will be given by

$$e_S(t) = E_S \cos[2\pi f_0 t + \phi_{NT}(t) + \theta x(t)] \tag{4.104}$$

where f_0 and $\phi_{NT}(t)$ are the frequency and the phase noise of the transmitting laser, and

$$x(t) = \sum_{k=0}^{\infty} a_k u_T(t - kT) \tag{4.105}$$

In (4.105), $u_T(t)$ is the unit square function of duration T, $a_k = \pm 1$ the information data emitted in the kth interval, and θ is the phase deviation employed by the modulator according to Figure 4.27 ($\theta = \pi/2$ represents the standard binary PSK). Taking into account that $\cos[\theta x(t)] = \cos\theta$ and $\sin[\theta x(t)] = x(t)\sin\theta$, (4.104) can be rewritten as

$$e_S(t) = E_S \cos[2\pi f_0 t + \phi_{NT}(t)]\cos\theta - x(t)E_S \sin[2\pi f_0 t + \phi_{NT}(t)]\sin\theta \tag{4.106}$$

When $\theta = \pi/2$, we have the usual suppressed-carrier PSK modulation; otherwise, an unmodulated carrier in quadrature with the information-bearing signal is present in $e_S(t)$.

So far we have used the actual transmitted signal to clearly separate the two quadrature terms. From now on we will use the complex envelope notation, so that the received optical signal will be written as

$$e_S(t) = E_S\, e^{j\phi_s(t)}$$

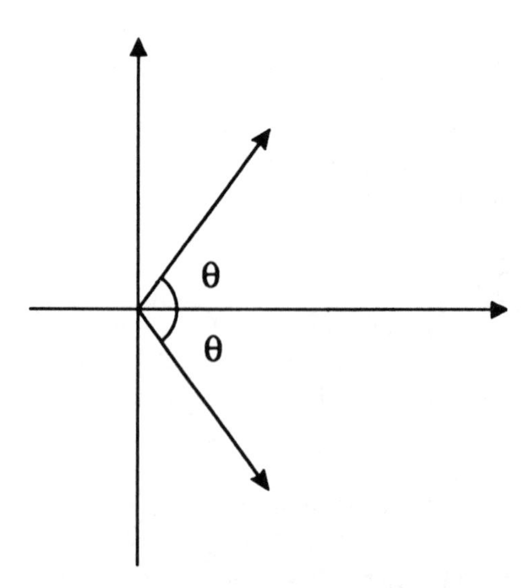

Figure 4.27 Phase deviation θ employed by the modulator for balanced PLL homodyne receiver of binary PSK.

where the phase $\phi_S(t)$ is

$$\phi_S(t) = \theta x(t) + \phi_{NT}(t) + \frac{\pi}{2} \tag{4.107}$$

and the constant $\pi/2$ is introduced only to simplify the analysis.

The block diagram of an optical homodyne receiver based on a balanced PLL is shown in Figure 4.28.

We notice the presence of the signal emitted by the local oscillator, which in complex envelope notation has the form

$$e_{LO}(t) = E_{LO}e^{j\phi_{LO}(t)} \tag{4.108}$$

with $\phi_{LO}(t)$ equal to

$$\phi_{LO}(t) = \phi_C(t) + \phi_{NLO}(t) \tag{4.109}$$

where $\phi_C(t)$ is the controlled phase determined by the input control voltage, $v_C(t)$, to the *voltage-controlled laser oscillator* (VCO) and $\phi_{NLO}(t)$ is the phase noise of the local oscillator.

The received and local oscillator signals are applied to the input ports of a π hybrid (see Section 4.7), whose outputs are linear combinations of inputs according to

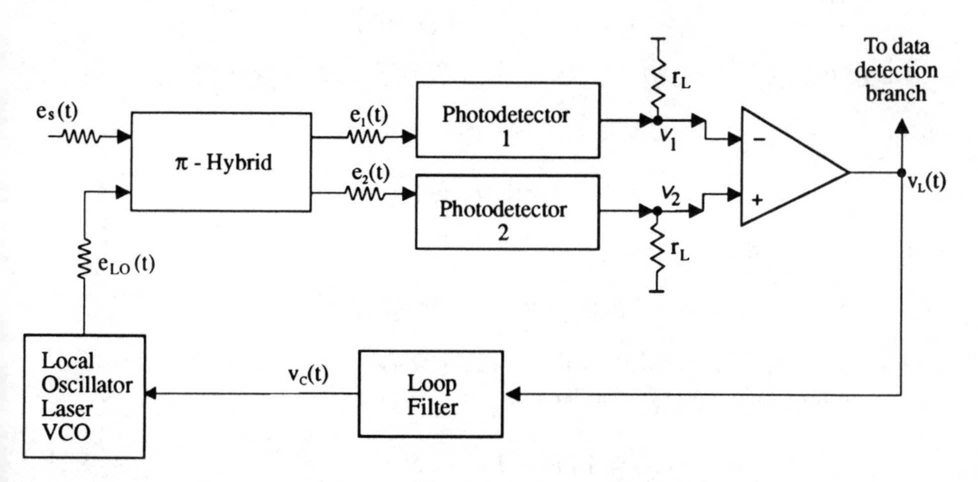

Figure 4.28 Block diagram of the balanced PLL homodyne receiver for binary PSK.

$$e_1(t) = \frac{1}{\sqrt{2}}[e_S(t) + e_{LO}(t)]$$

$$e_2(t) = \frac{1}{\sqrt{2}}[e_S(t) - e_{LO}(t)]$$

The two signals, $e_1(t)$ and $e_2(t)$, are detected by the photodetectors, giving rise to voltages across the load resistances, r_L, equal to

$$v_1(t) = r_L Re_1(t)e_1^*(t) = \frac{1}{2} r_L R[P_S + P_{LO} + 2\sqrt{P_S P_{LO}} \cos(\phi_S - \phi_{LO})] + n_1(t) \quad (4.110)$$

$$v_2(t) = r_L Re_2(t)e_2^*(t) = \frac{1}{2} r_L R[P_S + P_{LO} - 2\sqrt{P_S P_{LO}} \cos(\phi_S - \phi_{LO})] + n_2(t) \quad (4.111)$$

where P_S and P_{LO} are the power of the received and local oscillator signals, R is the responsivity of the two photodetectors, supposed equal for both, and $n_1(t)$ and $n_2(t)$ are the shot noise processes originated in the photodetection process, which, as usual, are assumed to be predominant with respect to the receiver thermal noise.

Taking the difference between the two signal voltages, we obtain the voltage, $v_L(t)$, as

$$v_L(t) = -2r_L R\sqrt{P_S P_{LO}} \cos(\phi_S - \phi_{LO}) + n_T(t) \quad (4.112)$$

where $n_T(t)$ denotes the total shot noise process obtained from the difference $n_2(t) - n_1(t)$. It is a Gaussian random process with spectral density $qr_L^2 RP_{LO}$.

Substitution of expressions (4.107) and (4.109) into (4.112) yields

$$v_L(t) = 2r_L R\sqrt{P_S P_{LO}} \sin\theta\, x(t)\cos[\phi_e(t)] + 2r_L R\sqrt{P_S P_{LO}} \cos\theta \sin[\phi_e(t)] + n_T(t) \quad (4.113)$$

where

$$\phi_e(t) = \phi_N(t) - \phi_C(t) \quad (4.114)$$

being

$$\phi_N(t) = \phi_{NT}(t) - \phi_{NLO}(t)$$

the total transmitter and local oscillator phase noise.

With an easier notation, we can write

$$v_L(t) = A_D(t) + A_{PL} \sin \phi_e(t) + n_T(t) \quad (4.115)$$

where the new symbols mean

$$A_D(t) = 2r_L R\sqrt{P_S P_{LO}} \sin \theta x(t) \cos \phi_e(t) \tag{4.116}$$

$$A_{PL} = 2r_L R\sqrt{P_S P_{LO}} \cos \theta \tag{4.117}$$

In (4.115) we recognize three terms, namely, the information-bearing signal to be further processed by the data detector, the phase error signal to be used by the PLL for locking, and the noise. The effect of a phase deviation θ lower than $\pi/2$ is to reduce the power of the term useful for data detection by a factor $\sin \theta$ and to give rise to a phase error signal that is multiplied by a factor $\cos \theta$.

As a consequence, the power penalty due to residual carrier transmission is

$$\Delta P_{RC} = 10 \log_{10} \frac{1}{\sin^2\theta} \text{ [dB]} \tag{4.118}$$

Since the signal, $v_L(t)$, is employed for both data detection and phase control of the VCO (through the loop filter), we see from these equations that receiver performance is affected by three major sources of impairment: the phase error, $\phi_e(t)$; the noise, $n_T(t)$; and the cross-talk between data and the phase-lock branches of the receiver. The latter interference would disappear if the phase deviation, θ, were equal to $\pi/2$. In that case, however, the phase error signal would equal zero, and no carrier recovery would be possible.

For notational simplicity in the following derivations, we combine the Gaussian noise term and the data noise term into a single additive noise term, $n(t)$:

$$n(t) = A_D(t) + n_T(t) \tag{4.119}$$

The balanced receiver can thus be represented by the synthetic block diagram in Figure 4.29.

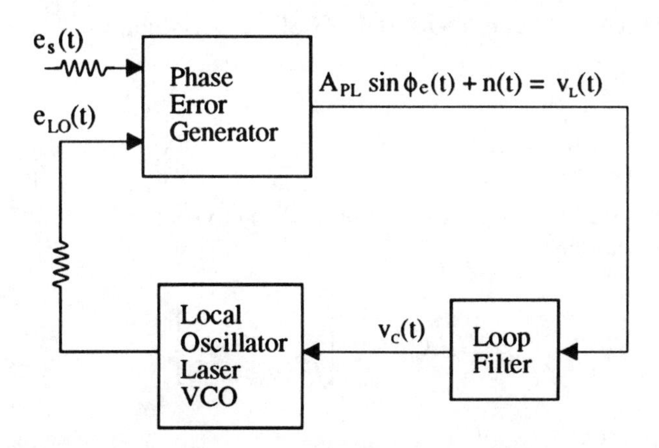

Figure 4.29 Synthetic block diagram of the balanced homodyne receiver for binary PSK.

4.9.2.2 Linear Analysis of the PLL

In the behavior of a PLL, it is customary to distinguish between an acquisition phase and a tracking phase. During the acquisition phase, the phases of the received and local carriers start from a phase difference randomly distributed in $(0, 2\pi)$ to reach a steady-state situation of a very small difference. During that tracking phase, the steady-state situation of small phase error must be kept.

To simplify the analysis, we will assume that the system is in the tracking phase, so that the phase error $\phi_e(t)$ is small, and we can make the following approximation:

$$\sin \phi_e(t) \approx \phi_e(t)$$

so that the phase error signal becomes

$$v_L(t) = A_{PL}\phi_e(t) + n(t) \tag{4.120}$$

4.9.2.3 The Loop Components

Now we will analyze the behavior of the components of the PLL: the VCO and the loop filter.

The VCO

The heart of the PLL is the VCO laser, whose output phase (4.109), apart from the phase noise ϕ_{NLO}, depends linearly on the input voltage, $v_C(t)$, according to

$$\phi_C(t) = G_{VCO} \int_{-\infty}^{t} v_C(\tau) d\tau$$

where G_{VCO} is the VCO gain measured in radians per second volt.

The Loop Filter

The loop filter is a lowpass filter that smooths the phase error signal. Its transfer function, $H_{LF}(f)$, affects the tracking properties of the PLL, as well as its stability. In most applications, the loop filter is a first-order active filter, as shown in Figure 4.30, with transfer function

$$H_{LF}(f) = -\left[\frac{1}{j2\pi f\tau_1} + \frac{\tau_2}{\tau_1}\right] \tag{4.121}$$

where $\tau_1 = r_1 C_1$ and $\tau_2 = r_2 C_1$.

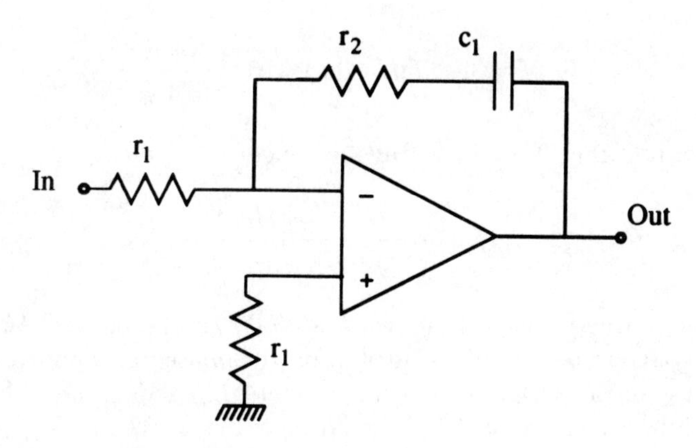

Figure 4.30 First-order active loop filter of a PLL.

4.9.2.4 Linearized PLL Equations

Using the previously obtained linearized equation, together with the one governing the VCO behavior and taking into account (4.114), we obtain the following linear integral equation that relates the input and VCO phases:

$$\phi_C(t) = A_{PL}G_{VCO}\int_{-\infty}^{t} d\tau_1 \int_{-\infty}^{\infty} h_{LF}(\tau_2)\left\{[\phi_N(\tau_1 - \tau_2) - \phi_C(\tau_1 - \tau_2)] + \frac{1}{A_{PL}}n(\tau_1 - \tau_2)\right\}d\tau_2 \tag{4.122}$$

Equation (4.122) is pictorially represented by the block diagram in Figure 4.31, a linear system with feedback in the phase variables.

The linear integral equation (4.122) in the time domain can be turned into a linear algebraic equation in the frequency domain through the Fourier transform, leading to

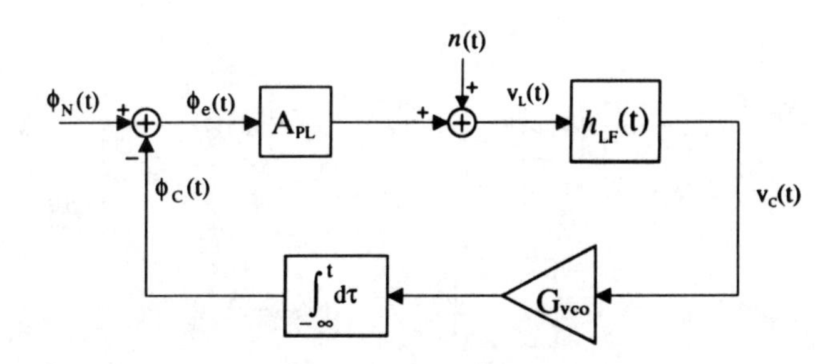

Figure 4.31 Linearized phase model of a PLL.

$$\Phi_C(f) = \Phi_N(f) H_{PLL}(f) + N(f) \frac{H_{PLL}(f)}{A_{PL}} \quad (4.123)$$

where we have introduced the PLL transfer function

$$H_{PLL}(f) = \frac{G_{VCO} A_{PL} H_{LF}(f)}{j2\pi f + G_{VCO} A_{PL} H_{LF}(f)}$$

and the formal Fourier transforms of the phase $\Phi_N(f)$ and shot noise $N(f)$. Equation (4.123) clearly separates the two contributions of phase and additive noise to the controlled phase of the VCO, as from the synthetic block diagram in Figure 4.32.

Taking into account that the phase error, $\phi_e(t)$, is given by $\phi_N(t) - \phi_C(t)$, we obtain the relationship between the phase and additive noises and the phase error:

$$\Phi_e(f) = \Phi_N(f)[1 - H_{PLL}(f)] - N(f) \frac{H_{PLL}(f)}{A_{PL}} \quad (4.124)$$

Recalling (4.119), we can redraw Figure 4.31 as Figure 4.33, where the two contributions of the shot and data noises have been separated. We will now evaluate the effects of the three sources of disturbance on the recovered phase.

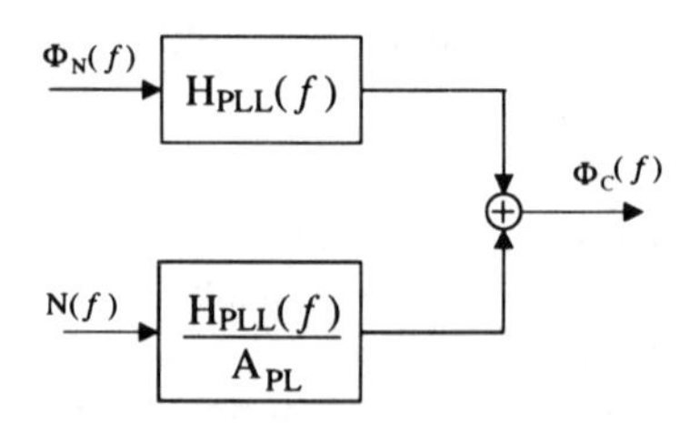

Figure 4.32 Block diagram showing the separate contributions of phase and shot noises for a PLL.

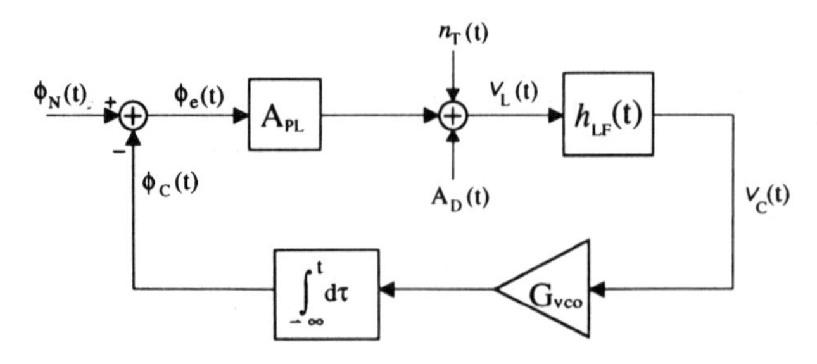

Figure 4.33 Linearized phase model of a PLL with separate contributions of data and shot noises.

4.9.2.5 Phase Error Variance

To estimate quantitatively the effects of noises on the PLL performance, we will refer, as customary, to the variance of the phase error, that is,

$$\sigma_{\phi_e}^2 = E[\phi_e^2(t)]$$

Owing to the linearity of (4.124) and to the independence of the phase and additive noise processes, we can evaluate the two contributions separately. Moreover, recalling that the additive noise is in turn given by the two independent contributions of shot noise and data noise, we can write

$$\sigma_{\phi_e}^2 = \sigma_{PN}^2 + \sigma_{SN}^2 + \sigma_{DN}^2 \tag{4.125}$$

where the three terms on the right side represent the contributions of phase noise, shot noise, and data noise to the variance, respectively.

We will compute the three contributions separately.

Effects of Phase Noise

Using the results of Section 2.6 on linear transformation of random processes, we are able to compute the power spectrum of the phase error due to the phase noise by setting $N(f) = 0$ in (4.124):

$$G_{PN}(f) = G_{\phi_N}(f)\, |1 - H_{PLL}(f)|^2 \tag{4.126}$$

Thus, the phase error variance is given by

$$\sigma_{PN}^2 = \int_{-\infty}^{\infty} G_{\phi_N}(f)\, |\, 1 - H_{PLL}(f)\, |^2 df \tag{4.127}$$

Taking into account that the power spectrum of the total phase noise is

$$G_{\phi_N}(f) = \frac{G_{\dot{\phi}_N}(f)}{(j2\pi f)^2} = \frac{\Delta\nu}{\pi f^2}$$

where we have summed the contributions of the phase noise of the transmitting and local oscillator lasers, assumed to be equal, we obtain

$$\sigma_{PN}^2 = \frac{\Delta\nu}{\pi} \int_{-\infty}^{\infty} \frac{|\, 1 - H_{PLL}(f)\, |^2}{f^2}\, df \tag{4.128}$$

Equation (4.128) cannot be further simplified without assigning a particular form to the closed-loop transfer function, which is equivalent to choosing the loop filter.

Example 4.4

Consider the case of a first-order active loop. For it, the expression $H_{PLL}(f)$ (see Problem 4.15) is the following:

$$H_{PLL}(f) = -\frac{1 + j2\zeta f/f_n}{[1 - (f/f_n)^2] + j2\zeta f/f_n} \tag{4.129}$$

where we have defined the natural frequency, f_n, and the damping coefficient, ζ, as

$$f_n = \frac{\sqrt{A_{PL}G_{VCO}/\tau_1}}{2\pi} \qquad \text{and} \qquad \zeta = \pi f_n \tau_2$$

The noise equivalent bandwidth of $H_{PLL}(f)$ is (see Section 2.6)

$$B_{PLL} = \frac{1}{2}\int_{-\infty}^{\infty} | H_{PLL}(f) |^2 \, df$$

which yields

$$B_{PLL} = \frac{\pi f_n}{4\zeta}(1 + 4\zeta^2) \tag{4.130}$$

Using (4.129) and plugging it into (4.128) we obtain, after easy calculations,

$$\sigma_{PN}^2 = \frac{\Delta\nu}{2\zeta f_n} \tag{4.131}$$

When we recall the relationship among ζ, f_n, and B_{PLL} in (4.130), σ_{PN}^2 can be given the form

$$\sigma_{PN}^2 = \frac{\pi(1 + 4\zeta^2)\Delta\nu}{8\zeta^2 B_{PLL}} \tag{4.132}$$

From (4.132) we see, as anticipated, that the contribution of phase noise to the phase error variance is proportional to the linewidth of the transmitting and local oscillator lasers and inversely proportional to the PLL equivalent bandwidth.

Effects of Shot Noise

As we did for the case of phase noise, we can compute the power spectrum of the phase noise due to shot noise by setting $\Phi_N(f) = 0$ in (4.124) and considering $n(t) = n_T(t)$. It yields

$$G_{SN}(f) = G_{n_T}(f)\,\frac{|H_{PLL}(f)|^2}{A_{PL}^2} \tag{4.133}$$

Thus, the phase error variance is given by

$$\sigma_{SN}^2 = \frac{1}{A_{PL}^2}\int_{-\infty}^{\infty} G_{n_T}(f)\,|H_{PLL}(f)|^2 df \tag{4.134}$$

Assuming the usual expression for the PSD of the shot noise current, valid for a local oscillator power much greater than the received power, the two-sided PSD of the shot noise voltage, $n(t)$, on r_L is

$$G_n(f) = RP_{LO}qr_L^2$$

which leads to

$$\sigma_{SN}^2 = \frac{RP_{LO}\,qr_L^2}{A_{PL}^2}\int_{-\infty}^{\infty} |H_{PLL}(f)|^2\, df = \frac{2RP_{LO}\,qr_L^2}{A_{PL}^2}B_{PLL} \tag{4.135}$$

where we again have used the equivalent loop bandwidth B_{PLL}.

Recalling the expression of the coefficient A_{PL} from (4.117), we get

$$\sigma_{SN}^2 = \frac{qB_{PLL}}{2RP_S\cos^2\theta} \tag{4.136}$$

which, as a function of the number of received photons per bit, N_R, becomes

$$\sigma_{SN}^2 = \frac{TB_{PLL}}{2N_R\cos^2\theta} \tag{4.137}$$

from which we see that the contribution of shot noise to the phase error variance is proportional to the closed-loop bandwidth and inversely proportional to the photodiode responsivity and to the received signal power. As a consequence, the best performance would be obtained with the minimum possible value of B_{PLL}, which is zero. This corresponds to disconnecting the PLL. As we have seen, however, the contributions of phase noise are inversely proportional to the closed-loop bandwidth, so the best choice would be a trade-off between the two effects.

Effects of Data Noise

As we did for the case of shot noise, we can compute the power spectrum of the phase noise due to data noise by setting $\Phi_N(f) = 0$ in (4.124) and considering $n(t) = A_D(t)$. It yields

$$G_{DN}(f) = G_d(f) \frac{|H_{PLL}(f)|^2}{A_{PL}^2} \tag{4.138}$$

where $G_d(f)$ is the power spectrum of the data term $A_D(t)$ in (4.116). Because $A_D(t)$ is a random process representing a binary antipodal digital transmission with symbols a equally likely and independent, its spectrum is given by (see Section 2.6)

$$G_d(f) = 4Tr_L^2 R^2 P_S P_{LO} \sin^2\theta \left(\frac{\sin \pi f T}{\pi f T}\right)^2 \tag{4.139}$$

Thus, the phase error variance is given by

$$\sigma_{DN}^2 = \frac{1}{A_{PL}^2} \int_{-\infty}^{\infty} G_d(f) \, |H_{PLL}(f)|^2 \, df \tag{4.140}$$

Substituting (4.139) and (4.117) into (4.140), we obtain

$$\sigma_{DN}^2 = T \tan^2\theta \int_{-\infty}^{\infty} \left(\frac{\sin \pi f T}{\pi f T}\right)^2 |H_{PLL}(f)|^2 \, df \tag{4.141}$$

Equation (4.141) is exact and valid for an arbitrary loop filter. To obtain a more manageable expression to be used hereafter for design purposes, we suppose now that the closed-loop transfer function, $H_{PLL}(f)$, has a bandwidth much smaller than the signal bandwidth, so that in (4.141) we can make the approximation

$$\left(\frac{\sin \pi f T}{\pi f T}\right) \approx 1$$

This approximation will be justified a posteriori by the values of the PLL bandwidth obtained through the design guidelines. A quantitative measure of the accuracy of the approximation is the objective of Problem 4.18.

Under the previous approximation, we obtain for the phase error variance

$$\sigma_{DN}^2 = 2T \tan^2 \theta \, B_{PLL} \tag{4.142}$$

where B_{PLL} is the noise equivalent bandwidth of $H_{PLL}(f)$.

4.9.2.6 Overall Loop Performance and Design Guidelines

We now have all the ingredients to write the overall variance of the phase error. Using (4.132), (4.137), and (4.142), we obtain from (4.125)

$$\sigma_{\phi_e}^2 = \frac{\pi(1 + 4\zeta^2)\Delta\nu}{8\zeta^2 B_{PLL}} + \frac{T B_{PLL}}{2N_R \cos^2 \theta} + 2T B_{PLL} \tan^2 \theta \tag{4.143}$$

Let us now find the optimum value of the loop noise bandwidth. Taking the first derivative of $\sigma_{\phi_e}^2$ with respect to B_{PLL}, setting it equal to zero, and solving the resulting equation, we obtain

$$B_{PLL,opt} = \sqrt{\frac{[\pi(1 + 4\zeta^2)\Delta\nu]/8\zeta^2}{[T/(2N_R \cos^2 \theta)] + 2T \tan^2 \theta}} \tag{4.144}$$

Substituting (4.144) into (4.143) yields the minimum phase error variance

$$\sigma_{\phi_e,min}^2 = 2\sqrt{\frac{\pi(1 + 4\zeta^2)\Delta\nu T}{8\zeta^2} \cdot \left(\frac{1}{2N_R \cos^2 \theta} + 2 \tan^2 \theta\right)} \tag{4.145}$$

Defining as σ_M^2 the largest value of $\sigma_{\phi_e,min}^2$ that the system can tolerate, we can derive from (4.145) an upper limit to the value of the parameter $\Delta\nu T$:

$$\Delta\nu T \leq \frac{\sigma_M^4}{[\pi(1 + 4\zeta^2)/2\zeta^2] \cdot \{[1/(2 N_R \cos^2 \theta)] + 2 \tan^2 \theta\}} \tag{4.146}$$

The inequality in (4.146) is precisely the result we anticipated at the beginning of this section.

In designing a system, follow these steps:

1. Choose a value for the PLL damping coefficient, ζ.
2. Fix a desired value of the error probability, P_e.
3. Fix a tolerable power penalty, ΔP_S (or equivalently ΔN_R), and split it into the part due to phase error, ΔP_{ϕ_e}, and that due to residual carrier transmission, ΔP_{RC}, so that

$$\Delta P_S = \Delta P_{\phi_e} + \Delta P_{RC}$$

4. From ΔP_{RC} according to (4.118), derive the value of θ.
5. From curves of P_e in Figure 4.26, the value of P_e and the value of ΔP_{ϕ_e}, derive the value of $\sigma_M = \sigma_{\phi_e}$ and $N_R \sin^2\theta$ and, knowing θ, get N_R.
6. Finally, from (4.146), get the value of $\Delta\nu T$.

Example 4.5 illustrates the design procedure.

Example 4.5

We want to limit the power penalty to 1 dB and split it evenly between phase error and residual carrier transmission. Let us perform the steps of the design procedure.

1. We choose, as is customary, for a good compromise between noise bandwidth and acquisition speed, $\zeta = 1/\sqrt{2}$.
2. Our design will be made at a value of $P_e = 10^{-9}$.
3. $\Delta P_{\phi_e} = \Delta P_{RC} = 0.5$ dB.
4. From $\Delta P_{RC} = 0.5$ dB, we obtain $\theta = 1.2347$ rad.
5. From the curves of Figure 4.26 and the values of ΔP_{ϕ_e} we get $\sigma_M = 0.175$ rad and $N_R = 11.22$.
6. Finally, plugging all the numerical values previously obtained into (4.146), we get $\Delta\nu T \le 5.9 \cdot 10^{-6}$.

Previous results were parametric in the data rate $1/T$. For a data rate of 1 Gbps, the laser linewidth must satisfy the inequality $\Delta\nu \le 5.9$ kHz, and the resulting PLL bandwidth is $B_{PLL,opt} = 913.1$ kHz.

As a comment to the numerical results of the example, we are confirmed in our hypothesis that the loop bandwidth is much smaller than the signal bandwidth. Moreover, the obtained value for the laser linewidth is such that binary PSK with homodyne detection is possible, with the present semiconductor laser technology, only with lasers equipped with external cavities.

4.9.2.7 Decision-Driven PLL Receiver

Possible alternatives to the balanced PLL receiver are the Costas loop receiver and the decision-driven loop receiver. Both alternatives employ a fully suppressed carrier transmission, that is, the phase deviation is $\theta = \pi/2$ and the entire transmitted power is used for data transmission. Instead, the received signal power is split at the receiving end of the system. One part of the signal power is sent to the data detector, and the other part is sent to a PLL, so some power penalty is incurred with this approach, too.

Further, the average value of the phase-lock current (produced by the PLL photodetector) will be zero because $\theta = \pi/2$. Thus, the current must be processed nonlinearly before being used for phase locking. Two kinds of processing can be employed:

- The phase-lock current can be multiplied by the signal current produced by the data photodetector. This approach leads to the Costas loop receiver.
- The phase-lock current can be multiplied by the output signal of the data receiver, after decision. This approach leads to the decision-driven loop and provides better performance than the other approach.

In the following text, we will only sketch the analysis of the decision-driven loop, since it follows closely the analysis of the balanced PLL receiver. A block diagram of the receiver is shown in Figure 4.34.

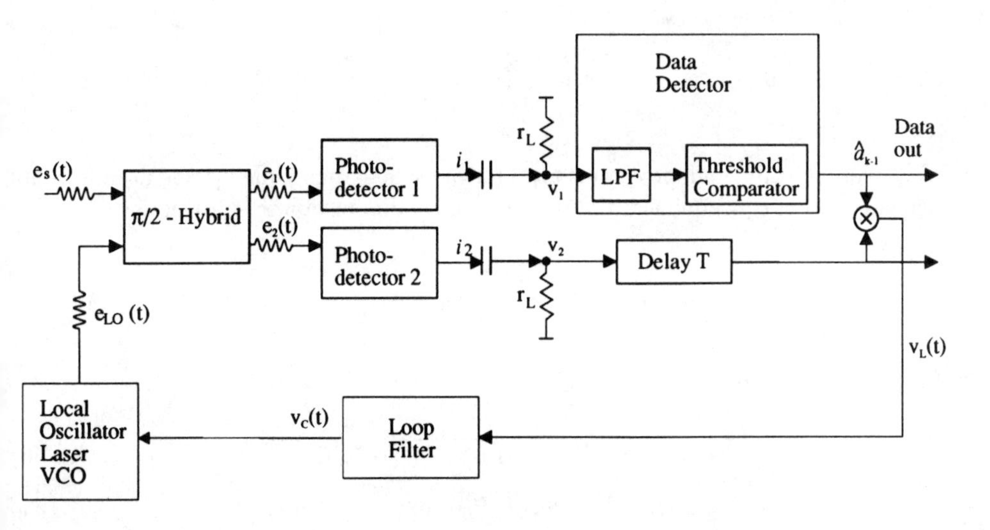

Figure 4.34 Block diagram of the decision-driven PLL homodyne receiver for binary PSK.

The received and local oscillator signals are applied to the input ports of an optical $\pi/2$ hybrid (see Section 4.7), whose outputs are linear combinations of inputs according to

$$e_1(t) = [e_S(t) + e_{LO}(t)]\sqrt{1-k}$$

$$e_2(t) = [e_S(t) + e_{LO}(t)e^{-j(\pi/2)}]\sqrt{k}$$

where k is the power-splitting ratio of the hybrid and determines the percentages of the received power used for data detection and phase locking.

The two signals, $e_1(t)$ and $e_2(t)$, are detected by the photodetectors that produce the currents i_1 and i_2. Their DC components are eliminated by the RC filters, whose output voltages are

$$v_1(t) = 2r_L R(1-k)\sqrt{P_S P_{LO}}\cos(\phi_S - \phi_{LO}) + n_1(t) \tag{4.147}$$

$$v_2(t) = 2r_L Rk\sqrt{P_S P_{LO}}\sin(\phi_S - \phi_{LO}) + n_2(t) \tag{4.148}$$

where P_S and P_{LO} are the power of the received and local oscillator signals, R is the responsivity of the two photodetectors, supposed equal for both, and $n_1(t)$ and $n_2(t)$ are the shot noise processes originated in the photodetection process, which, as usual, are assumed to be predominant with respect to the receiver thermal noise. They have spectral densities

$$G_{n_1}(f) = qr_L^2\, RP_{LO}\,(1-k)$$

$$G_{n_2}(f) = qr_L^2\, RP_{LO}k$$

Note that the voltage $v_2(t)$ is used for phase locking, while $v_1(t)$ is used for data detection. As a consequence, the power penalty due to power splitting is

$$\Delta P_{PS} = 10\log_{10}\frac{1}{(1-k)} \tag{4.149}$$

Proceeding as in the previous case yields

$$v_1(t) = 2ar_L\, R(1-k)\sqrt{P_S P_{LO}}\cos[\phi_e(t)] + n_1(t) \tag{4.150}$$

$$v_2(t) = 2ar_L\, Rk\sqrt{P_S P_{LO}}\sin[\phi_e(t)] + n_2(t) \tag{4.151}$$

where $a = \pm 1$ is the information-bearing RV, and

$$\phi_e(t) = \phi_N(t) - \phi_C(t) \tag{4.152}$$

being

$$\phi_N(t) = \phi_{NT}(t) - \phi_{NLO}(t)$$

the total transmitter and local oscillator phase noise.

Paralleling the previous analysis (see Problem 4.19), we obtain the phase error variance as

$$\sigma_{\phi_e}^2 = \frac{\pi(1 + 4\zeta^2)\Delta\nu}{8\zeta^2 B_{PLL}} + \frac{T\,B_{PLL}}{2k\,N_R} \tag{4.153}$$

which can be rewritten as

$$\sigma_{\phi_e}^2 = \frac{\pi(1 + 4\zeta^2)}{8\zeta^2}\frac{\Delta\nu}{B_{PLL}} + \frac{T\Delta\nu}{2k\,N_R}\frac{B_{PLL}}{\Delta\nu} \tag{4.154}$$

In Figure 4.35 the phase error variance $\sigma_{\phi_e}^2$ in square radians is plotted versus the normalized PLL bandwidth $B_{PLL}/\Delta\nu$ for various values of the parameter *NLR*, representing the ratio between the received number of photons used for phase locking

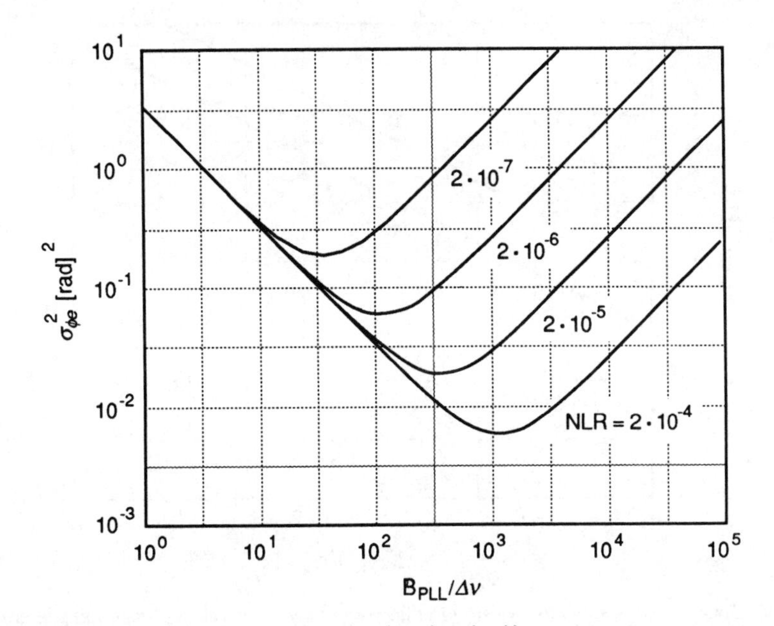

Figure 4.35 Phase error variance versus normalized PLL bandwidth.

and the linewidth $kN_R/\Delta\nu$. The damping coefficient is assumed to be $\zeta = 1/\sqrt{2}$, as usual, and the data rate equal to 1 Gbps.

Inspection of Figure 4.35 reveals that, for small normalized loop bandwidth, $\sigma^2_{\phi_e}$ is large and independent from the received power since the phase error is dominated by the phase noise. For large normalized loop bandwidth, on the contrary, $\sigma^2_{\phi_e}$ is dominated by the shot noise, and its value depends on the received power. The most important conclusion suggested by Figure 4.35 is that a reliable phase locking, in the usual sense of $\sigma_{\phi_e} \leq 0.175$ rad, is possible only if *NLR* is larger than $9 \cdot 10^{-5}$ photons/bit·kHz and if $B_{PLL}/\Delta\nu$ is larger than 90.

In other words, a second-order PLL with active filter and $\zeta = 1/\sqrt{2}$ requires at least $9 \cdot 10^{-5}$ photons/bit/kHz of the laser linewidth. This signal power is used for phase locking and is, therefore, lost from the data detection.

From (4.154) we can also find the optimum value of the loop bandwidth, just as we did for the balanced PLL receiver. With easy calculations (see Problem 4.21), we obtain

$$B_{PLL,opt} = \sqrt{\frac{2\pi\, k\, N_R \Delta\nu(1 + 4\zeta^2)}{8T\zeta^2}} \tag{4.155}$$

In Figure 4.36 the optimum loop bandwidth is plotted as a function of the laser linewidth for various values of the photon number used for phase locking. The

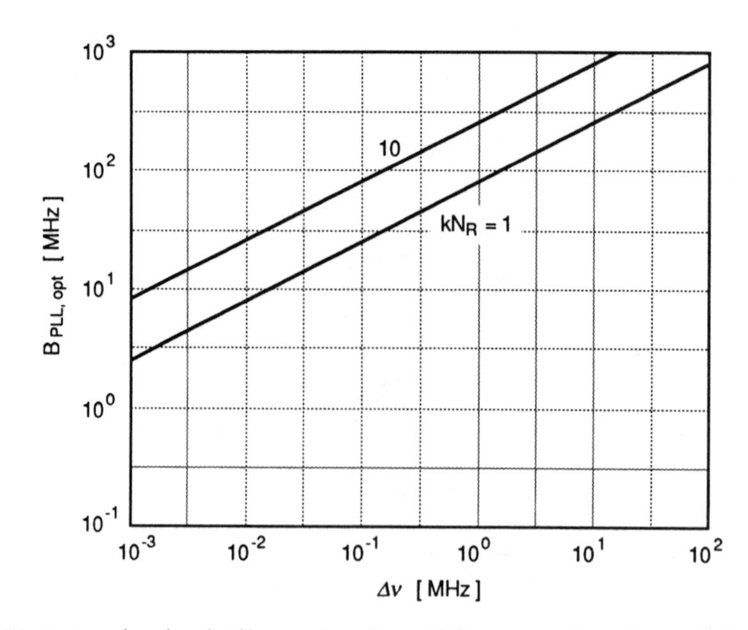

Figure 4.36 Optimum loop bandwidth versus laser linewidth for various values of received photons used for phase locking.

loop filter parameters are as before, and $T = 10^{-9}$, corresponding to a bit rate of 1 Gbps.

Substituting (4.155) into (4.154) we get the minimum phase error variance as

$$\sigma^2_{\phi_e,min} = \sqrt{\frac{\pi T(1 + 4\zeta^2)}{4k\, N_R \zeta^2}\, \Delta\nu} \tag{4.156}$$

Figure 4.37 shows the photon number needed to achieve a standard deviation of 0.175 rad versus $\Delta\nu$, for the same parameter values as before.

Defining as σ^2_M the largest value of $\sigma^2_{\phi_e,min}$ that the system can tolerate, we can derive from (4.156) an upper limit to the value of the parameter $\Delta\nu T$

$$\Delta\nu T \leq \frac{\sigma^4_M}{\pi(1 + 4\zeta^2)/4\zeta^2 k N_R} \tag{4.157}$$

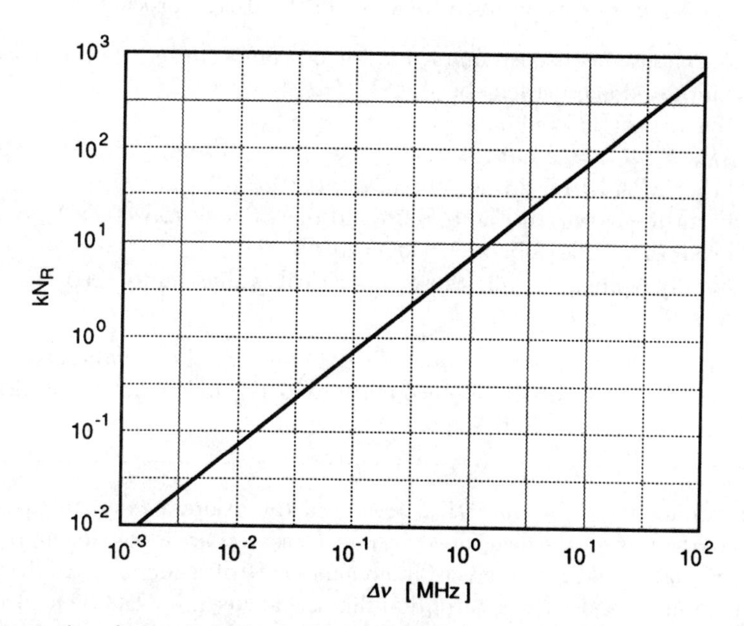

Figure 4.37 Number of received photons used for phase locking needed to achieve a phase noise standard deviation of 0.175 rad versus the laser linewidth.

This relationship can be used in designing a system according to the following procedure:

1. Choose a value for the PLL damping coefficient, ζ.
2. Fix a desired value of the error probability, P_e.
3. Fix a tolerable power penalty, ΔP_S (or, equivalently, ΔN_R), and split it into the part due to phase error, ΔP_{ϕ_e}, and that due to the power splitting for phase locking, ΔP_{PS}, so that

$$\Delta P_S = \Delta P_{\phi_e} + \Delta P_{PS}$$

4. From ΔP_{PS}, according to (4.149), derive the value of k.
5. From curves of P_e of Figure 4.26 and the value of P_e, derive the value of $N_R(1 - k)$ and, knowing k, get N_R.
6. From ΔP_{ϕ_e}, again using the curves in Figure 4.26, derive the value of σ_M.
7. Finally, from (4.157), get the value of $\Delta\nu T$.

Example 4.6 illustrates the design procedure.

Example 4.6

We want to limit the power penalty to 1 dB and split it evenly between phase error and power splitting. Let us perform the steps of the design procedure.

1. We choose, as is customary, for a good compromise between noise bandwidth and acquisition speed, $\zeta = 1/\sqrt{2}$.
2. Our design will be made at a value of $P_e = 10^{-9}$.
3. $\Delta P_{\phi_e} = \Delta P_{PS} = 0.5$ dB.
4. From $\Delta P_{PS} = 0.5$ dB, we obtain $k = 0.1087$.
5. From the curves in Figure 4.26 and the value of k, we get $N_R = 11.21$.
6. From $\Delta P_{\phi_e} = 0.5$ dB, we get $\sigma_M = 0.175$ rad.
7. Finally, plugging all those numerical values into (4.157), we get $\Delta\nu T \leq 1.25 \cdot 10^{-4}$.

Previous results were parametric in the data rate $1/T$. For a data rate of 1 Gbps, the laser linewidth must satisfy the inequality $\Delta\nu \leq 125$ kHz, and the resulting PLL bandwidth is $B_{PLL,opt} = 26.786$ MHz.

As a comment on the numerical results of the example, we can say that the linewidth constraints for the decision-driven PLL receiver are less stringent than those obtained for a balanced PLL receiver (about a factor 10 of reduction). All these results have been obtained under the assumption that the flicker noise can be neglected and that the receiver performance is quantum limited. Problems 4.22 and 4.23 deal with the inclusion of flicker noise in the analysis.

4.9.3 Heterodyne Phase Locking

As an alternative to homodyne phase locking, in which the phase recovery is performed through an electric control signal acting on the local laser oscillator, we can shift the signal at IF and then track its phase completely in the electronic domain. Also in this case, we have the same problems originating from the fact that the received PSK signal has a continuous spectrum, so that the three solutions mentioned for the homodyne case can be applied as well. They consist in the transmission of a residual carrier to lock into, in the decision-driven PLL and in the Costas loop receiver.

We will present here only the final results and design guidelines for the case of the decision-driven receiver shown in Figure 4.38. The other two cases are left as

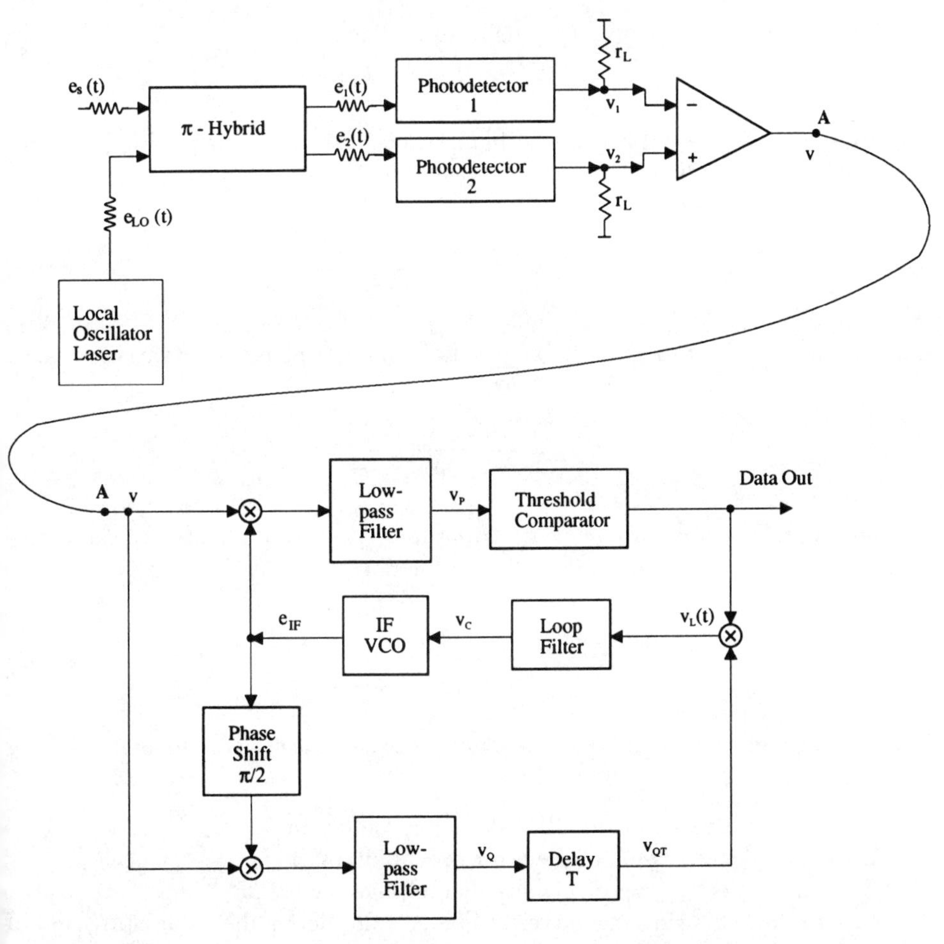

Figure 4.38 Block diagram of the decision-driven PLL heterodyne receiver for binary PSK.

exercises to the reader (see Problem 4.25). Since the derivation follows closely those already explained in detail for the homodyne receivers, we will omit the intermediate steps.

4.9.3.1 Overall Loop Performance and Design Guidelines

The phase error variance is given by

$$\sigma_{\phi_e}^2 = \frac{\pi(1+4\zeta^2)\Delta\nu}{8\zeta^2 B_{PLL}} + \frac{TB_{PLL}}{N_R} \tag{4.158}$$

and can be rewritten as

$$\sigma_{\phi_e}^2 = \frac{\pi(1+4\zeta^2)}{8\zeta^2}\frac{\Delta\nu}{B_{PLL}} + \frac{T\Delta\nu}{N_R}\frac{B_{PLL}}{\Delta\nu} \tag{4.159}$$

From (4.159) we can find the optimum value of the loop bandwidth, just as we did for the balanced PLL receiver. With easy calculations, we obtain

$$B_{PLL,opt} = \sqrt{\frac{\pi\, N_R \Delta\nu(1+4\zeta^2)}{8T\zeta^2}} \tag{4.160}$$

Substituting (4.160) into (4.159), we get the minimum phase error variance as

$$\sigma_{\phi_e,min}^2 = \sqrt{\frac{\pi T(1+4\zeta^2)}{2N_R\zeta^2}\Delta\nu} \tag{4.161}$$

Defining as σ_M^2 the largest value of $\sigma_{\phi_e,min}^2$ that the system can tolerate, we can derive from (4.161) an upper limit to the value of the parameter $\Delta\nu T$:

$$\Delta\nu T \le \frac{\sigma_M^4}{\pi(1+4\zeta^2)/(2\zeta^2 N_R)} \tag{4.162}$$

This relationship can be used in designing a system according to the following procedure:

1. Choose a value for the PLL damping coefficient, ζ.
2. Fix a desired value of the error probability, P_e.
3. Fix a tolerable power penalty due to phase error, ΔP_{ϕ_e}.
4. From ΔP_{ϕ_e} using the curves of Figure 4.26, derive the value of σ_M and of N_P (because the curves of Fig. 4.26 refer to the homodyne PSK, the 3-dB

penalty of the heterodyne receiver, apparent in (4.26), should be taken into account).

5. Finally, from (4.162), get the value of $\Delta \nu T$.

Example 4.7 illustrates the design procedure.

Example 4.7

We want to limit the power penalty to 1 dB. Let us perform the steps of the design procedure.

1. We choose, as is customary, for a good compromise between noise bandwidth and acquisition speed, $\zeta = 1/\sqrt{2}$.
2. Our design will be made at a value of $P_e = 10^{-9}$.
3. $\Delta P_{\phi_e} = 1$ dB.
4. From $\Delta P_{\phi_e} = 1$ dB, we get $\sigma_M = 0.175$ rad and $N_R = 25.2$.
5. Finally, plugging all the numerical values into (4.162), we get $\Delta \nu T \leq 2.5 \cdot 10^{-3}$.

Previous results were parametric in the data rate $1/T$. For a data rate of 1 Gbps, the laser linewidth must satisfy the inequality $\Delta \nu \leq 2.5$ MHz, and the resulting PLL bandwidth is $B_{PLL,opt} = 385$ MHz.

Comparison with results obtained for decision-driven PLL homodyne receiver shows that the heterodyne scheme can tolerate a larger normalized linewidth.

In obtaining these results, we have neglected the one-symbol delay in the decision-driven PLL. Norimatsu and Iwashita [6] have taken into account the delay effects on the system performance.

4.10 ASYNCHRONOUS SYSTEMS

We are already aware that asynchronous detection schemes are less affected by phase noise than synchronous ones, owing to the fact that phase is not directly exploited in the detection process, which is based rather on the envelope of the IF signal.

Nonetheless, phase noise is harmful even in the asynchronous case, although in a less pronounced way, because IF filtering causes a phase-to-amplitude noise conversion, which, in turn, reduces the SNR and results in a power penalty suffered by the system.

In this section, we will derive the power penalty induced by phase noise and show how to design the system in order to minimize it. We will consider in detail the cases of double-filter FSK and asynchronous ASK. Single-filter FSK can be treated in the same way as ASK.

4.10.1 Double-Filter Asynchronous FSK

4.10.1.1 No Postdetection Filtering

The receiver structure for the double-filter asynchronous FSK modulation was shown in Figure 4.12. In the following analysis, we will assume that the IF filters have impulse responses $h^{+}(t) = \cos[2\pi(f_{IF} + f_d)\ (T' - t)],\ 0 \le t \le T'$, and $h^{-}(t) = \cos[2\pi(f_{IF} - f_d)\ (T' - t)],\ 0 \le t \le T'$, that is, they are integrate-and-dump bandpass filters centered around the two tones, $f_{IF} + f_d$ and $f_{IF} - f_d$.

With respect to the ideal case treated in Section 4.4.2, we allow the integration time of the IF filters to be T' seconds, with $0 \le T' \le T$, T being the signaling period. Although the matched filter in the absence of phase noise would require $T' = T$, we will see that a larger IF filter bandwidth may be beneficial.

In complex envelope notation for signals and systems, the receiver structure that follows the photodetector assumes the form of the block diagram shown in Figure 4.39, where the factor $1/T'$ multiplying the integrators is useful for normalization purposes. With reference to the block diagram, the IF signal $i_S(t)$ in a time interval of duration T can be written as

$$i_S(t) = Ae^{j\,[a2\pi f_d t\, +\, \phi(t)]} + i_n(t) \tag{4.163}$$

where $A = 2R\sqrt{P_S P_{LO}}$, $\phi(t)$ represents the phase noise, $a = \pm 1$ is the RV carrying the binary information, and $i_n(t)$ is a white complex Gaussian noise with one-sided

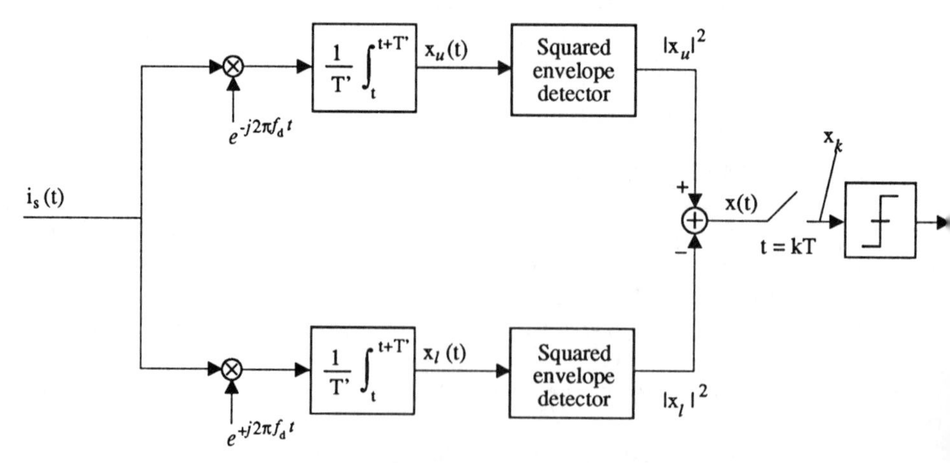

Figure 4.39 Double-filter asynchronous heterodyne receiver of a coherent binary FSK system with phase noise.

spectral density $4N_0$ originated from the receiver thermal noise and the shot noise.[11]

Because of the complete symmetry of the receiver, its performance does not depend on the value of the RV a, so we will assume $a = 1$. In such a case, we have, with reference to Figure 4.39,

$$x_u(t) = \frac{A}{T'} \int_t^{t+T'} e^{j\phi(\tau)} d\tau + n_u(t) \tag{4.164}$$

$$x_l(t) = \frac{A}{T'} \int_t^{t+T'} e^{j[\phi(\tau)+4\pi f_d \tau]} d\tau + n_l(t) \tag{4.165}$$

where $n_u(t)$ and $n_l(t)$ are complex Gaussian random processes with zero mean. The two noise processes are independent only if $1/T'$ is a submultiple of f_d; however, they also can be considered approximately independent when f_d is much greater than $1/T'$. The real and imaginary parts of $n_u(t)$ and $n_l(t)$ are independent with equal variances $\sigma^2 = 2(N_0/T')$. The first term of $x_l(t)$ represents the cross-talk interference of one signal tone on the other. In the absence of phase noise, this term vanishes if $1/T'$ is a submultiple of $2f_d$ and can be neglected when $2f_d$ is much greater than $1/T'$. When phase noise is present, the term is still negligible if $2f_d$ is much greater than $1/T'$ and of the laser linewidth $\Delta\nu$.[12]

We assume that these simplifying hypotheses hold true, so that (4.164) and (4.165) become

$$x_u(t) = \frac{A}{T'} \int_t^{t+T'} e^{j\phi(\tau)} d\tau + n_u(t) \tag{4.166}$$

$$x_l(t) = n_l(t) \tag{4.167}$$

After the square envelope detectors, the adder and the sampler, we obtain

$$x_k \stackrel{\text{def}}{=} x(kT) = |x_u(kT)|^2 - |x_l(kT)|^2 = \frac{A^2}{T'^2} \left| \int_{kT}^{kT+T'} e^{j\phi(\tau)} d\tau + n_{u_k} \right|^2 - |n_{l_k}|^2 \tag{4.168}$$

where n_{u_k} and n_{l_k} are complex Gaussian RVs whose real and imaginary parts are independent, with zero mean and equal variance σ^2.

Because the phase noise random process $\phi(t)$, assumed to obey to the Lorentzian

11. When the shot noise is dominant, $N_0 = qRP_{LO}$.
12. With $\Delta\nu$ we denote here the overall linewidth, which is the sum of the transmitter and local oscillator lasers linewidths.

model previously described, has stationary increments, the RVs x_k preserve their statistical properties from one bit to another. As a consequence, we can consider $k = 0$ without loss of generality, obtaining

$$x \stackrel{\text{def}}{=} x_0 = \frac{A^2}{T'^2} \left| \int_0^{T'} e^{j\phi(\tau)} d\tau + n_u \right|^2 - |n_l|^2 \tag{4.169}$$

The RV x is the decision variable; when it is positive, we decide for $a = 1$, when negative, for $a = -1$. Thus, under the hypothesis that $a = 1$ was transmitted, the error probability is, with obvious notations,

$$P_e = P[x \leq 0] = P[|x_u| \leq |x_l|] \tag{4.170}$$

Defining the RVs

$$z \stackrel{\text{def}}{=} \frac{1}{T'} \int_0^{T'} e^{j\phi(\tau)} d\tau, \qquad Z = |z|^2 \tag{4.171}$$

where Z is equal to 1 in the ideal case of no phase noise, we can compute first the error probability conditioned on $Z = b^2$ and then average with respect to it;

$$P_e = E_Z P[x \leq 0 \mid Z = b^2] \stackrel{\text{def}}{=} E_Z P_e(b) \tag{4.172}$$

The conditional error probability in (4.172) can be computed analytically, as in Section 4.4.3, obtaining

$$P_e(b) = \frac{1}{2} e^{-(A^2 b^2 / 4\sigma^2)} = \frac{1}{2} e^{-(T'/T)(\rho b^2 / 2)} \tag{4.173}$$

having defined the IF SNR[13]:

$$\rho \stackrel{\text{def}}{=} \frac{A^2 T}{4 N_0} \tag{4.174}$$

Equation (4.173) deserves some considerations. We see from it that a reduction of the integration interval, and hence an increase of the IF filter bandwidth, leads to an increased variance of the noise entering the decision process; on the other hand,

13. When the system is shot-noise limited, ρ is equal to the number of photons per bit, N_R, previously used in determining the sensitivities of the various modulation schemes.

decreasing T' reduces the wandering of the phase and makes the values b assumed by the RV Z closer to the ideal value of 1. A trade-off between these contrasting effects leads to an optimum value of T' for a given laser linewidth and SNR.

In Appendix B we show how to obtain the statistical characterization of the RV Z, and in Appendix C we explain how to use the statistics of Z to compute averages like the one in (4.172) using Gauss quadrature rules in the form

$$\mathrm{E}_Z P_e(b) = \int_0^1 P_e(b) f_Z(b) db \approx \sum_{i=1}^{M} w_i P_e(b_i) \tag{4.175}$$

where the integral is performed in the range $(0,1)$ of the RV Z, $f_z(\cdot)$ is the pdf of the RV Z, and $\{w_i\}_{i=1}^{M}$, $\{b_i\}_{i=1}^{M}$ are the weights and abscissa of the quadrature rule. These values can be obtained from the knowledge of the first $2M$ moments of the RV Z through a suitable numerical algorithm, which is explained in Appendix B.

In the results that follow, the number of quadrature points M has been chosen to be large enough to guarantee stable results.

Figure 4.40 plots the error probability, P_e, as a function of the IF SNR, ρ, in

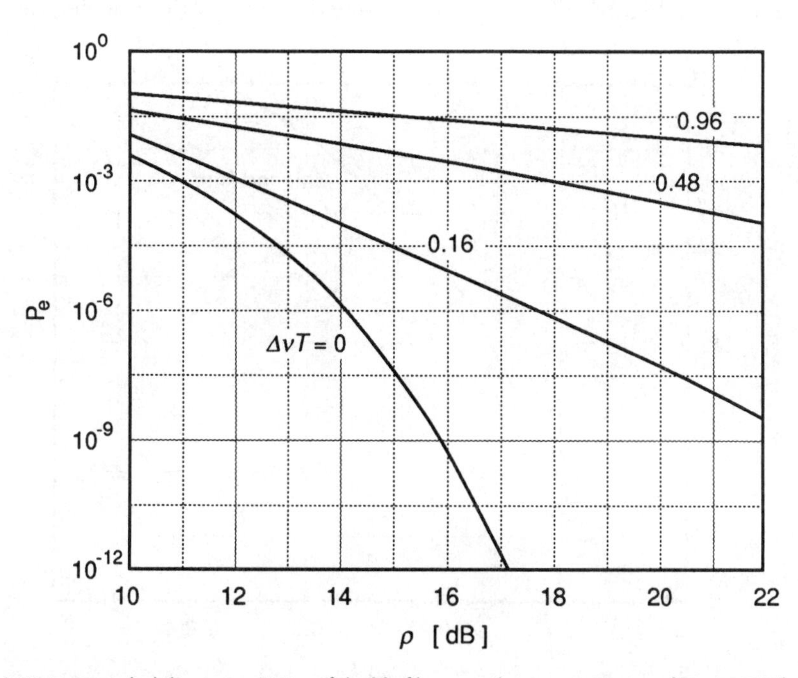

Figure 4.40 Error probability versus SNR ρ of double-filter asynchronous reception of binary FSK for various values of the normalized laser linewidth. The receiver uses a matched IF filter and no postdetection filter.

decibels for some values of the normalized linewidth, $\Delta\nu T$, when the integration time is equal to T, that is, for a matched IF filter. In Figure 4.41, the SNR penalty in decibels for an error probability of 10^{-9} is plotted as a function of the normalized linewidth for the matched filter receiver. Limiting the power penalty to 1 dB requires a normalized linewidth not larger than 0.06.

Let us now vary the parameter T' below the matched filter value, assuming for the sake of simplicity that the ratio $T/T' = N$ is an integer. In particular, for each value of the normalized linewidth and of the SNR, we choose the value of N, which minimizes the error probability.

Figure 4.42 plots the power penalty (with respect to the case of no phase noise) for an error probability of 10^{-9} in this optimized situation. Comparing it with the previous curve, also reported, shows significant improvements, particularly in the region of large normalized bandwidths.

For a given linewidth and SNR, increasing N beyond its optimum value leads to a situation where the power penalty is almost completely due to the increased quantity of noise entering the IF filter. This noise is nonlinearly processed and enhanced in the square envelope detector and enters the decision process directly. That clearly suggests the opportunity of placing a lowpass postdetection filter after the summing device in Figure 4.39, aiming at passing the signal while reducing the noise.

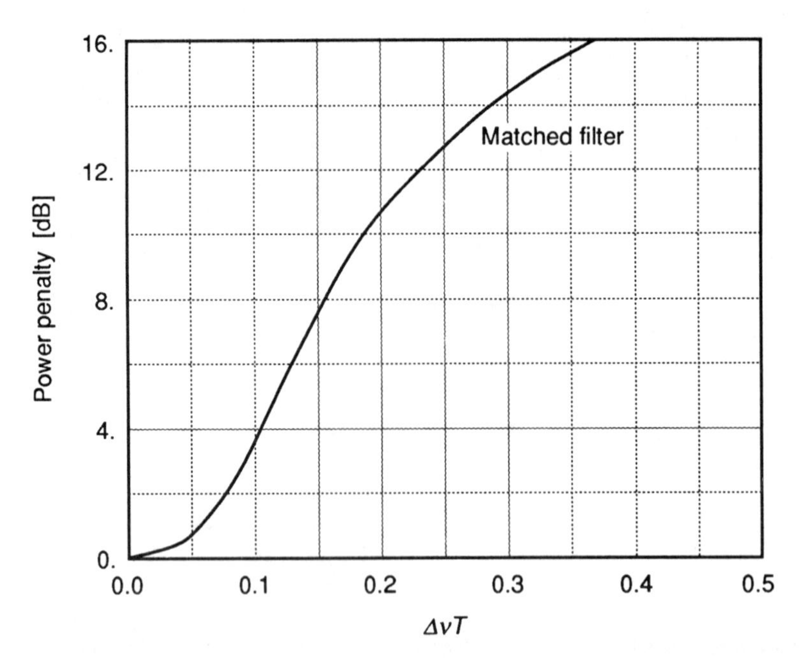

Figure 4.41 Power penalty versus the normalized laser linewidth for double-filter asynchronous reception of binary FSK. The receiver uses a matched IF filter and no postdetection filter.

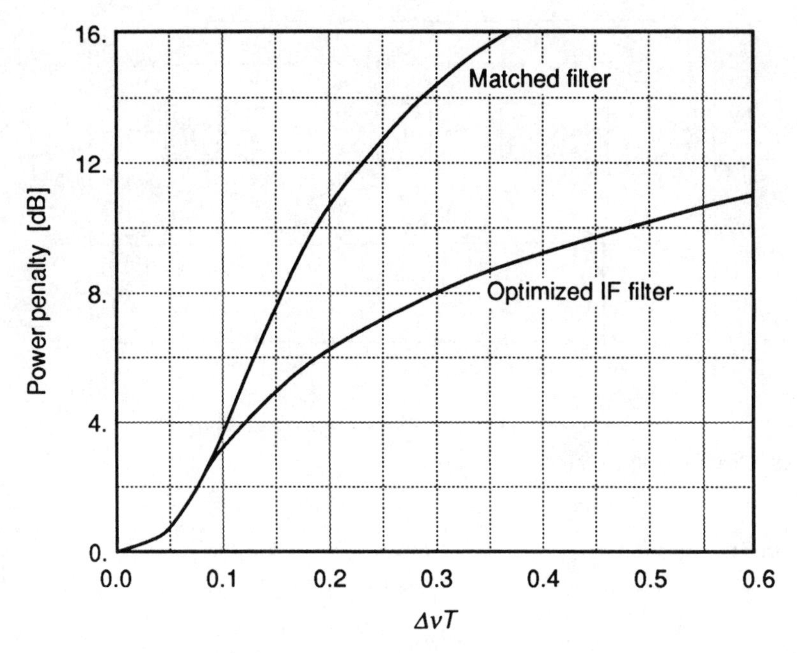

Figure 4.42 Power penalty versus the normalized laser linewidth for double-filter asynchronous reception of binary FSK. The receiver uses an optimized IF filter and no postdetection filter. Also, the curve for matched IF filter is reported.

4.10.1.2 Postdetection filtering

To make the case tractable, we will assume that the lowpass postdetection filter is in fact a numerical filter placed after the sampler, which is now working at a sampling frequency $1/T'$. The output, y, of the filter is simply the sum of the most recent N samples, x_i. Sampler and filter can be implemented by the structure in Figure 4.43. In formulae, we have

$$
\begin{aligned}
y &= \frac{1}{N}\sum_{i=1}^{N} x_i \\
&= \frac{1}{N}\sum_{i=1}^{N}\left(\left|\frac{A}{T'}\int_{(i-1)T'}^{iT'} e^{j\phi(\tau)}d\tau + n_{ui}\right|^2 - |n_{li}|^2\right) \\
&\overset{\text{def}}{=} \frac{1}{N}\sum_{i=1}^{N}\left(|Az_i + n_{ui}|^2 - |n_{li}|^2\right)
\end{aligned}
\tag{4.176}
$$

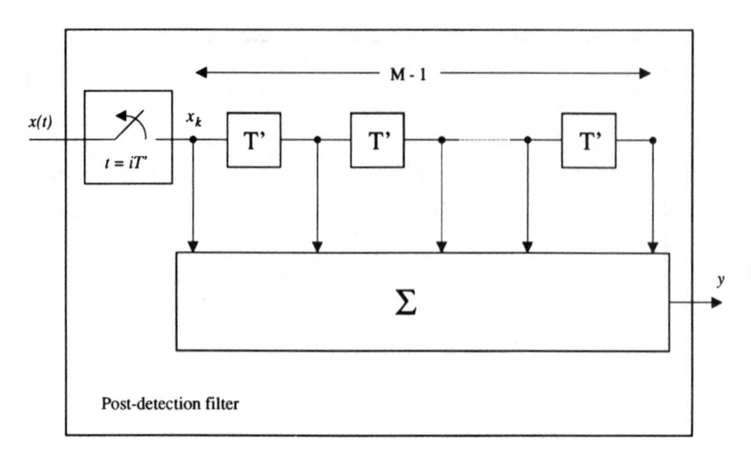

Figure 4.43 Digital postdetection filter.

Defining the RVs

$$u_i \overset{\text{def}}{=} |Az_i + n_{u_i}|^2, \qquad v_i \overset{\text{def}}{=} |n_{l_i}|^2 \tag{4.177}$$

and

$$u \overset{\text{def}}{=} \frac{1}{N}\sum_{i=1}^{N} u_i, \qquad v \overset{\text{def}}{=} \frac{1}{N}\sum_{i=1}^{N} v_i \tag{4.178}$$

we can write the decision variable as

$$y = u - v \tag{4.179}$$

and the error probability, conditioned on the transmission of a 1, becomes

$$P_e = P[y \le 0] = P[u \le v] \tag{4.180}$$

We will proceed now to the characterization of the RVs u and v. First, we observe that v_i is the square of a Rayleigh RV, so that its pdf is exponential, with expression

$$f_{v_i}(b) = \frac{1}{2\sigma^2}\,\mathrm{e}^{-(b/2\sigma^2)}, \qquad b \ge 0 \tag{4.181}$$

on the other hand, u_i, conditioned on $|z_i| = c$, is the square of a Rice RV, with pdf

$$f_{u_i}(b \mid c) = \frac{1}{2\sigma^2} e^{-[(b+c^2A^2)/2\sigma^2]} I_0\left(\frac{cA\sqrt{b}}{\sigma^2}\right), \qquad b \ge 0 \qquad (4.182)$$

where $I_n(\cdot)$ denotes the nth-order modified Bessel function.

The RVs u_i and v_i appearing in the summations in (4.177) are independent (see Problem 4.26), so we can easily derive the pdf of u and of v. In fact, apart from the factor $1/N$, the RV v is the sum of N independent exponential random variables, that is, an Erlang RV of order N. Including the factor $1/N$, it is

$$f_v(b) = \left(\frac{N}{2\sigma^2}\right)^N \frac{b^{N-1}}{(N-1)!} e^{-(Nb/2\sigma^2)}, \qquad b \ge 0 \qquad (4.183)$$

As for u, when conditioned on $W \stackrel{\text{def}}{=} \sum_{i=1}^{N} |z_i|^2 = C$ and apart from the factor $1/N$, it is the sum of the square of N independent Rice RVs, so that its pdf, including the factor $1/N$, is

$$f_u(b \mid C) = \frac{N}{2\sigma^2}\left(\frac{bN}{A^2C}\right)^{(N-1)/2} e^{-[(Nb+A^2C)/2\sigma^2]} I_{N-1}\left(\frac{A\sqrt{bCN}}{\sigma^2}\right), \qquad b \ge 0 \; (4.184)$$

The parameter C is the sum of the squares of the conditional values of the phase noise variables.

To compute the error probability in (4.180), we first condition on the value $u = b$ and then on $W = C$, so as to obtain

$$\begin{aligned} P_e &= \mathrm{E}_u P[v \ge u \mid u = b] \\ &= \mathrm{E}_W \mathrm{E}_u P[v \ge u \mid u = b, W = C] \qquad (4.185) \\ &= \int_0^N f_W(C)dC \int_0^\infty P[v \ge b] f_u(b \mid C)db \end{aligned}$$

A simple but lengthy computation (see Problem 4.27) first gives the expression of the inner probability in the integral in (4.185):

$$P[v \ge b] = e^{-(Nb/2\sigma^2)} \sum_{l=0}^{N-1} \frac{1}{l!}\left(\frac{Nb}{2\sigma^2}\right)^l \qquad (4.186)$$

Plugging (4.186) into (4.185), we obtain

$$\int_0^\infty P[v \ge b] f_u(b \mid C)db$$

$$
\begin{aligned}
&= \int_0^\infty e^{-(Nb/2\sigma^2)} \sum_{l=0}^{N-1} \frac{1}{l!} \left(\frac{Nb}{2\sigma^2}\right)^l f_u(b\,|C)db \\
&= \sum_{l=0}^{N-1} \frac{1}{l!} \left(\frac{N}{2\sigma^2}\right)^l \int_0^\infty e^{-(Nb/2\sigma^2)} b^l f_u(b\,|C)db \\
&= \sum_{l=0}^{N-1} \frac{(-1)^l}{l!} \left(\frac{N}{2\sigma^2}\right)^l \left[\frac{d^l H_u(p\,|C)}{dp^l}\right]_{p=N/2\sigma^2}
\end{aligned} \tag{4.187}
$$

where $H_u(p|C)$ is the moment-generating function associated to the pdf $f_u(b|C)$, that is, its Laplace transform:

$$
H_u(p \mid C) = \int_0^\infty e^{-pb} f_u(b\,|C)db \tag{4.188}
$$

which can be evaluated through (4.184) on the basis of a tabulated transform pair [5] with the following result:

$$
H_u(p \mid C) = \frac{1}{(1 + 2\sigma^2 p/N)^N} e^{-[(2CA^2 p/N)/(1+2\sigma^2 p/N)]} \tag{4.189}
$$

Combining (4.187) and (4.189) (see Problem 4.28), we obtain

$$
\begin{aligned}
P_e(C) &\stackrel{\text{def}}{=} \int_0^\infty P[v \geq b] f_u(b \mid C)db \\
&= \frac{1}{2^N} e^{-(C\rho/2N)} \left[1 + \sum_{j=1}^{N-1} \left(\sum_{l=j}^{N-1} \frac{1}{2^l} \binom{l-1}{j-1}\right) L_j^{(N-1)}\left(-\frac{C\rho}{2N}\right)\right]
\end{aligned} \tag{4.190}
$$

where $\rho = A^2T/4N_0$ is the SNR at the input of the IF stage, and $L_n^{(m)}(\cdot)$ are generalized Laguerre polynomials [5]. For $N = 1$ and $C = b^2$, we find again the result (4.173).

Inserting the closed-form expression (4.190) into (4.185), we are left with an average with respect to the RV W, which can be evaluated using Gauss quadrature rules, leading to

$$
\mathrm{E}_W P_e(C) = \int_0^N P_e(C) f_W(C)dC \approx \sum_{i=1}^M w_i P_e(c_i) \tag{4.191}
$$

where $\{w_i\}_{i=1}^M$, $\{c_i\}_{i=1}^M$ are the weights and abscissa of the quadrature rule, which can be computed, as before, from the knowledge of the first $2M$ moments of the RV W.

In Appendix B we explain how to characterize the RV W derived from the phase noise, and then, in Appendix C, how to use its moments in the computation of averages through Gauss quadrature rules. Here, we will present the results, which show the effectiveness of the postdetection filter.

Figure 4.44 plots the error probability, P_e, as a function of the IF SNR, ρ, in decibels for different values of the normalized linewidth, $\Delta\nu T$. For each value of SNR and $\Delta\nu T$, the value of N is optimized.

Figure 4.45 plots the SNR penalty in decibels for an error probability of 10^{-9} as a function of the normalized linewidth with an optimized postdetection filter. Comparing it with the two curves, also reported, showing the penalty for a matched filter and for an optimized IF filter without postdetection filter, we notice the dramatic improvement induced by the presence of the postdetection filter. A normalized laser linewidth of 1 gives a power penalty slightly greater than 2 dB. In Figure 4.46 we report

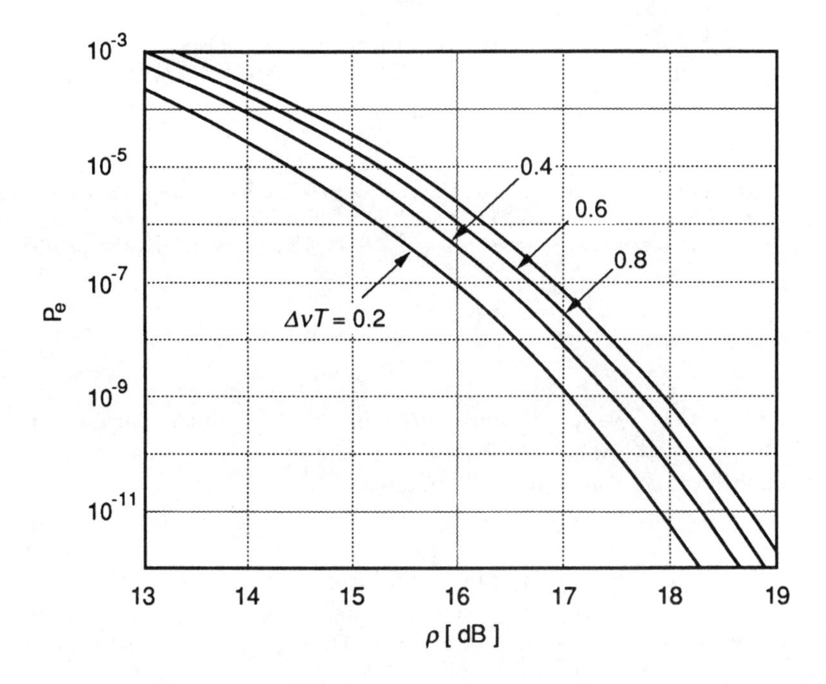

Figure 4.44 Error probability versus SNR ρ of double-filter asynchronous reception of binary FSK for various values of the normalized laser linewidth. The receiver uses an optimized IF filter and the digital postdetection filter.

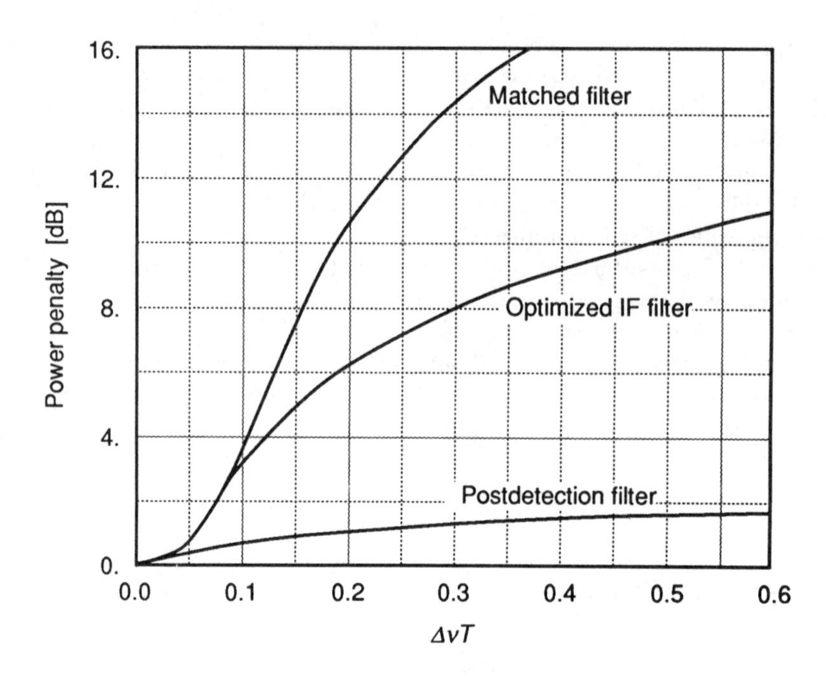

Figure 4.45 Power penalty versus the normalized laser linewidth for double-filter asynchronous reception of binary FSK. The three curves for matched IF filter with no postdetection filter, optimized IF filter with no postdetection filter, and optimized IF filter with postdetection filter are plotted.

the optimum value of N versus the normalized linewidth for an error probability equal to 10^{-9}. The curve shows that for $\Delta\nu T = 1$, the optimum IF filter must have a bandwidth seven times larger than the matched filter.

To show the sensitivity of the system with respect to the optimum postdetection filter, we plot in Figure 4.47 the power penalty versus N for an error probability of 10^{-9}. We notice a mild sensitivity to values of N greater than the optimum one, and an abrupt increase of sensitivity for values of N lower than the optimum. That means that the postdetection filter can cut down very well the "surplus" noise due to the wider IF filter.

As a general conclusion, we can say that asynchronous FSK can tolerate high normalized linewidths, which, for data rates on the order of hundreds of megabits per second, are largely obtainable by current semiconductor laser technology.

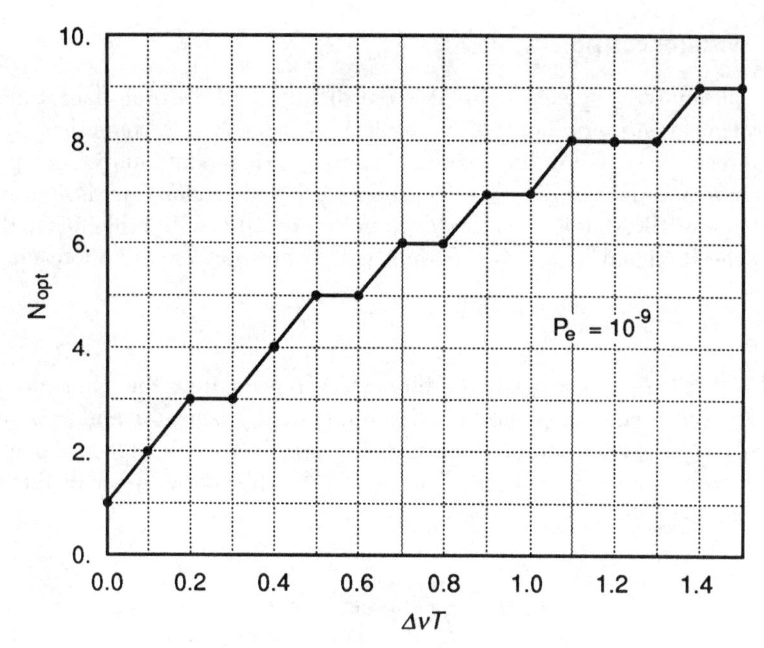

Figure 4.46 Optimum value of the parameter N characterizing the IF filter versus the normalized laser linewidth for double-filter asynchronous reception of binary FSK.

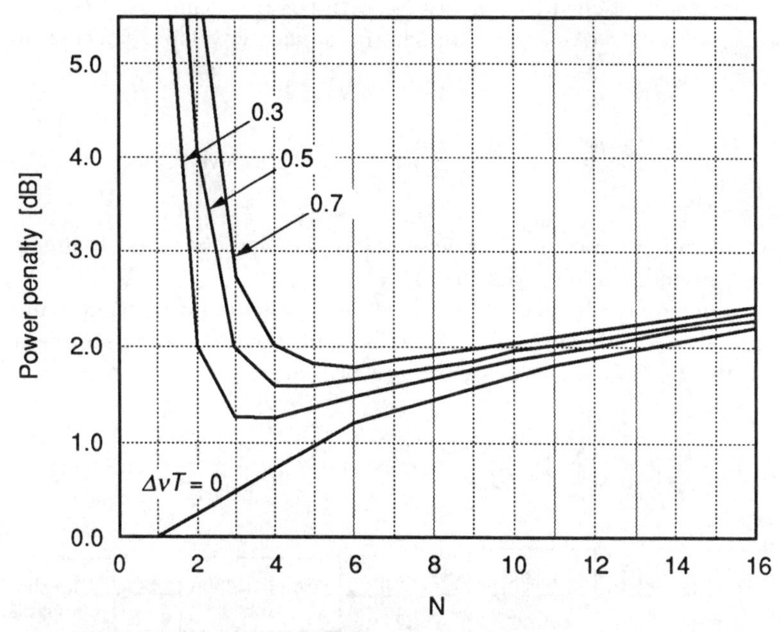

Figure 4.47 Power penalty versus the value of the parameter N characterizing the IF filter for double-filter asynchronous reception of binary FSK for various values of the normalized laser bandwidth.

4.10.2 Asynchronous ASK

Now we will analyze asynchronous ASK paralleling the FSK treatment. Since much of the previous considerations can be duplicated, we will skip some details.

The receiver structure for the asynchronous ASK modulation is shown in complex envelope notations in Figure 4.48, where the postdetection filter is assumed to be the same as for FSK, that is, the one shown in Figure 4.43. With reference to the block diagram, the IF signal $i_S(t)$ in a time interval of duration T can be written as

$$i_S(t) = aAe^{j\phi(t)} + i_n(t) \tag{4.192}$$

where $A = 2R\sqrt{P_S P_{LO}}$, $a = 0{,}1$ is the binary RV representing the transmitted information, $\phi(t)$ represents the phase noise including the transmitting and local oscillator lasers, and $i_n(t)$ is a white complex Gaussian noise with one-sided spectral density $4\,N_0$ originated from the receiver thermal noise and the shot noise. After IF filtering, the signal becomes

$$x_{IF}(t) = \frac{aA}{T'}\int_t^{t+T'} e^{j\phi(\tau)}d\tau + n(t) \tag{4.193}$$

where $n(t)$ is a complex Gaussian random process with zero mean. Its real and imaginary parts are independent processes, with equal variances $\sigma^2 = 2\,(N_0/T')$.

The signal $x_{IF}(t)$ is passed through the square envelope detector and then sampled, yielding

$$x_i \overset{\text{def}}{=} x(iT') = \frac{a^2A^2}{T'^2}\,|\int_{iT'}^{(i+1)T'} e^{j\phi(\tau)}d\tau + n_i\,|^2 \tag{4.194}$$

where n_i is a complex Gaussian RV whose real and imaginary parts are independent, with zero mean and common variance σ^2.

We will not analyze in detail the case of the absence of postdetection filter, since all the considerations made for FSK are still valid. Instead, we pass directly to the analysis of the RV y obtained by summing N samples x_i:

$$y = \frac{1}{N}\sum_{i}^{N} x_i = \frac{1}{N}\sum_{i=1}^{N}\left|\frac{aA}{T'}\int_{iT'}^{(i+1)T'} e^{j\phi(\tau)}d\tau + n_i\right|^2 \overset{\text{def}}{=} \frac{1}{N}\sum_{i=1}^{N}|\,aAz_i + n_i\,|^2 \tag{4.195}$$

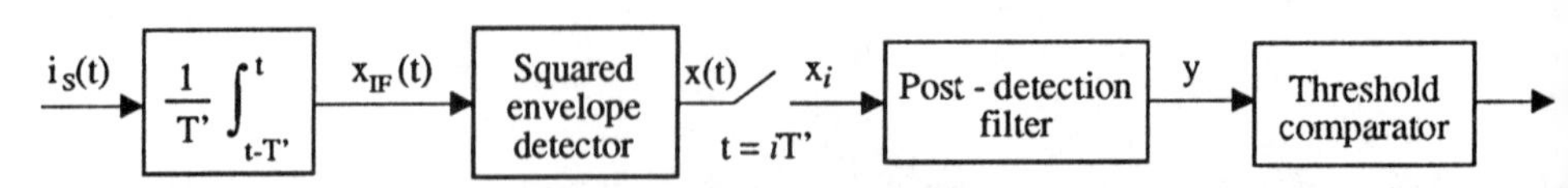

Figure 4.48 Asynchronous heterodyne receiver of a coherent binary ASK system with phase noise.

To make a decision on the transmitted symbol, we compare the RV y with a threshold, y_{th} (to be optimized), so that the error probability can be written as

$$P_e = \frac{1}{2}(P[y \le y_{th} \mid a = 1] + P[y > y_{th} \mid a = 0]) \quad (4.196)$$

We will proceed now to the characterization of the RV y. Consider first the RVs x_i in (4.195); call x_{i0} and x_{i1} the two RVs obtained by x_i under the conditions $a = 0$ and $a = 1$. Proceeding as before in the case of FSK, we easily obtain their pdf's:

$$f_{x_{i0}}(b) = \frac{1}{2\sigma^2}\, e^{-(b/2\sigma^2)}, \qquad b \ge 0 \quad (4.197)$$

and

$$f_{x_{i1}}(b \mid c) = \frac{1}{2\sigma^2}\, e^{-[(b+c^2A^2)/2\sigma^2]}\, I_0\left(\frac{cA\sqrt{b}}{\sigma^2}\right), \qquad b \ge 0 \quad (4.198)$$

where we have conditioned x_{i1} to $|z_i| = c$.

The RVs x_i appearing in the summation of (4.195) are independent (see Problem 4.26), so we can easily derive the pdf of y_0 and y_1, the two RVs obtained from y under conditioning on $a = 0$ and $a = 1$. Proceeding as before, we obtain

$$f_{y0}(b) = \left(\frac{N}{2\sigma^2}\right)^N \frac{b^{N-1}}{(N-1)!}\, e^{-(Nb/2\sigma^2)}, \qquad b \ge 0 \quad (4.199)$$

and

$$f_{y1}(b \mid C) = \frac{N}{2\sigma^2}\left(\frac{bN}{A^2C}\right)^{(N-1)/2} e^{-[(Nb+A^2C)/2\sigma^2]}\, I_{N-1}\left(\frac{A\sqrt{bCN}}{\sigma^2}\right), \qquad b \ge 0 \quad (4.200)$$

where we have conditioned y_1 to $W \overset{\text{def}}{=} \sum_{i=1}^{N} |z_i|^2 = C$.

Using the two pdf's, a simple computation (see Problem 4.29) leads to the following expressions for the two conditional error probabilities appearing on the right side of (4.196). The first one is

$$P[y_0 > y_{th}] = \int_{y_{th}}^{\infty} f_{y0}(b)db = e^{-\rho Y_{th}} \sum_{l=0}^{N-1} \frac{1}{l!}(\rho Y_{th})^l \quad (4.201)$$

where $\rho = A^2T/4N_0$ is the IF SNR, and $Y_{th} = y_{th}/A^2$ is the normalized threshold.

As to the second conditional error probability, we first condition it to $W = C$ and then average it out. We obtain

$$P[y_1 \le y_{th} \mid W = C] = \int_0^{y_{th}} f_{y_1}(b \mid C)db = 1 - Q_N\left(\sqrt{\frac{2\rho C}{N}}, \sqrt{2\rho Y_{th}}\right) \quad (4.202)$$

where $Q_N(\cdot,\cdot)$ is the Nth-order generalized Marcum function.

Averaging with respect to the RV W finally yields

$$P[y_1 \le y_{th}] = \mathrm{E}_W P[y_1 \le y_{th} \mid W = C]$$
$$= \int_0^N P[y_1 \le y_{th} \mid W = C] f_W(C) dC \simeq \sum_{i=1}^{M} w_i P[y_1 \le y_{th} \mid W = C_i] \quad (4.203)$$

where $\{w_i\}_{i=1}^M$, $\{C_i\}_{i=1}^M$ are the weights and abscissas of the quadrature rule, which can be computed as usual from the moments of the RV W.

To obtain error probability results, we need to optimize the value of the normalized threshold Y_{th} for every computation point, that is, for every SNR ρ and for every value of $\Delta\nu T$.

Figure 4.49 plots the error probability, P_e, as a function of the IF

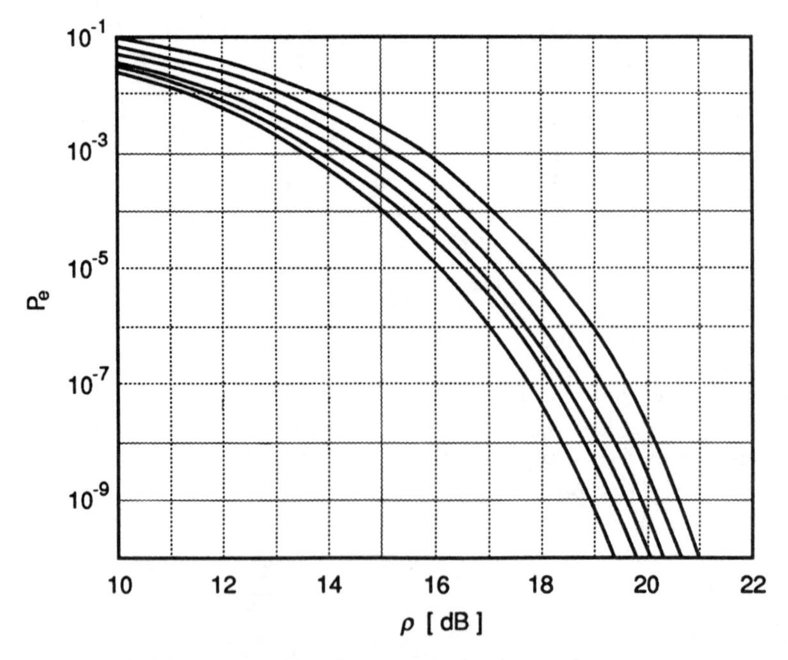

Figure 4.49 Error probability versus SNR ρ of asynchronous reception of binary ASK for various values of the normalized laser linewidth $\Delta\nu T$ (the values, from left to right, are 0, 0.05, 0.1, 0.2, 0.4, and 0.8). The receiver uses an optimized IF filter and the digital postdetection filter.

$\mathrm{SNR} = 10 \log_{10} \rho$ for different values of the normalized linewidth, $\Delta\nu T$. For each value of SNR and $\Delta\nu T$, the value of N and the threshold Y_{th} have been optimized. Figure 4.50 plots the optimum value of N versus the normalized linewidth for an error probability equal to 10^{-9}. In this case, the value $N_{opt} = 7$ is reached for $\Delta\nu T = 0.7$. Figure 4.51 plots the SNR penalty in dB for an error probability of 10^{-9} as a function of the normalized linewidth with optimized postdetection filter and threshold, together with the companion curves obtained in the cases of matched filter and optimized IF filter without postdetection filter. Also in this case we notice the dramatic improvement induced by the presence of the postdetection filter.

Finally, to show the sensitivity of the system with respect to the optimum value of the threshold, we plot in Figure 4.52 the error probability versus Y_{th} for an SNR of 20 dB and a normalized linewidth of 0.2. The curve shows a high sensitivity of the system on both sides of the optimum threshold.

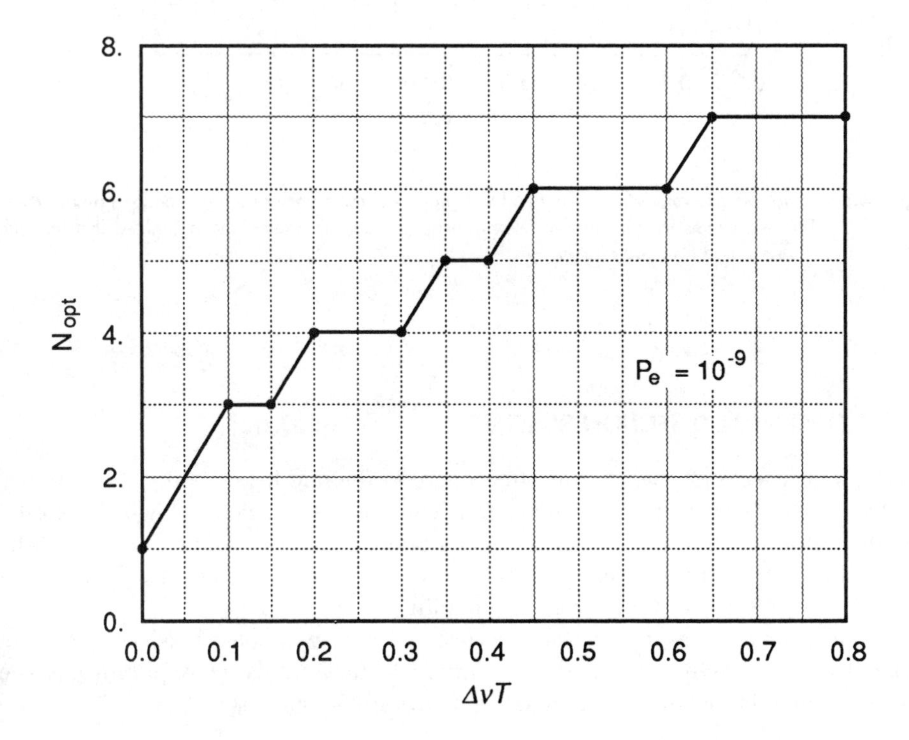

Figure 4.50 Optimum value of the parameter N characterizing the IF filter versus the normalized laser linewidth for asynchronous reception of binary ASK.

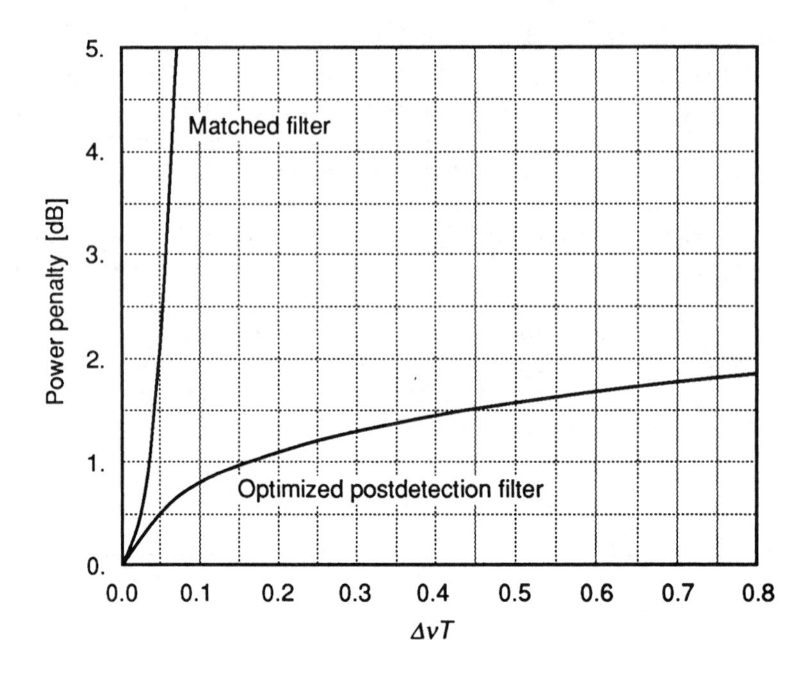

Figure 4.51 Power penalty versus the normalized laser linewidth for asynchronous reception of binary ASK. The two curves for matched IF filter with no postdetection filter and optimized IF filter with postdetection filter are plotted.

4.11 DIFFERENTIAL DETECTION SCHEMES

Differential detection schemes stay somewhere in the middle, in terms of sensitivity to phase noise, between synchronous and asynchronous receivers. This is due to the fact that, although not requiring phase locking, their demodulation strategies use the phase information of the received signal. Unlike asynchronous schemes, differential detected schemes present a bit error probability floor.

The exact analysis of such schemes in the presence of phase noise is even more difficult than for asynchronous demodulation. In fact, the latter required only the statistical characterization of the magnitude of an RV, such as

$$Z = \frac{1}{T}\int_0^T e^{j\phi(t)}dt \tag{4.204}$$

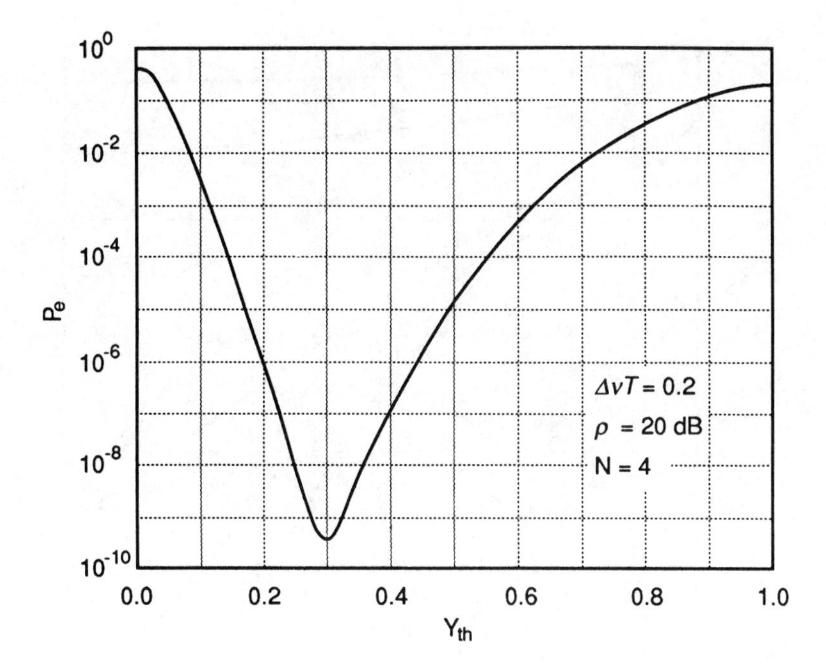

Figure 4.52 Error probability versus the decision threshold value for asynchronous reception of binary ASK with optimized IF filter and postdetection filter.

because the performance degradation due to phase noise depends only on the phase-to-amplitude conversion. A differential detection scheme, however, requires the joint pdf of the magnitude and the phase (or real and imaginary parts) of Z or, equivalently, a certain number of the joint moments of those RVs.

We will present only the most significant results.

4.11.1 DPSK

Kaiser, Shafi, and Smith present an accurate analysis of the DPSK receiver in [7]. Results have been obtained using two methods, that is, the *perturbation* method and the *moments* method, validated through extensive simulations with the importance sampling technique (see Appendix B).

Results of error probability versus the IF SNR, ρ, are shown in Figure 4.53, for several values of the normalized linewidth, $\Delta\nu T$.

The receiver used is the same shown in Figure 4.18, where the IF filter is a matched filter, that is, an integrate (over T)-and-dump filter, and the lowpass filter

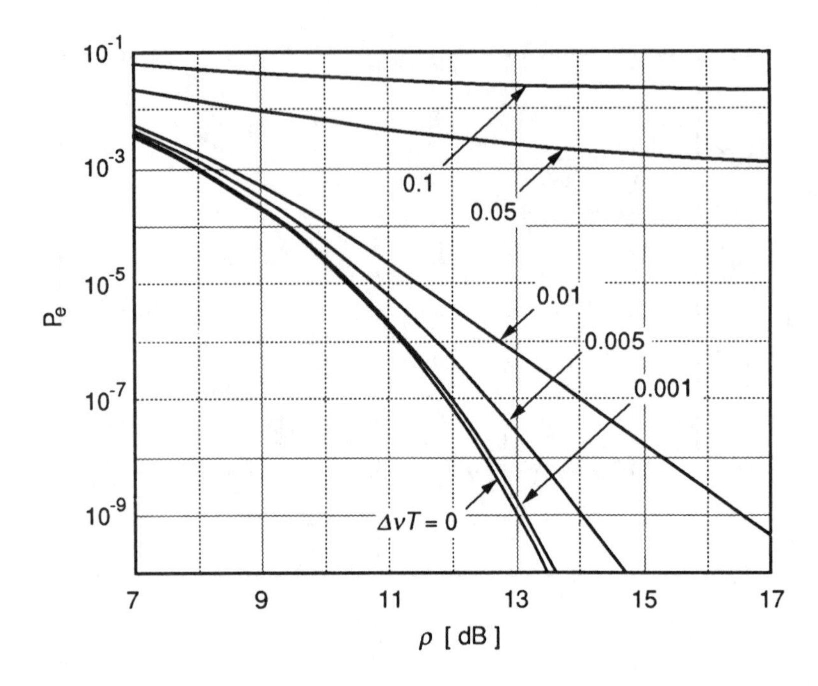

Figure 4.53 Error probability versus the SNR ρ for differential detection of binary PSK with phase noise for various values of the normalized laser linewidth.

simply removes the double IF frequency terms. In fact, it has been proved [8] that, independently of the linewidth, the best choice is an IF matched filter, quite an opposite conclusion with respect to the asynchronous case. The reason is that widening the IF filter has two contrasting effects: It reduces the phase-to-amplitude conversion, thus improving the signal strength, but, on the other hand, it enhances the variance of the phase drift between the two arms of the delay demodulator (and also the shot noise variance). The second effect is more important, so the optimum filter turns out to be the conventional matched filter.

The power penalty at $P_e = 10^{-9}$ with respect to the case of no phase noise versus the normalized linewidth is shown in Figure 4.54. For values of the normalized linewidth slightly larger than 0.01, the error probability cannot be made lower than 10^{-9} (error floor). To limit the power penalty below 1 dB, $\Delta\nu T$ must be less than 0.0034. Finally, Figure 4.55 presents the error floor versus the normalized linewidth.

Kaiser, Smith, and Shafi [9] describe an interesting way of improving the error floor that consists in modifying the receiver.

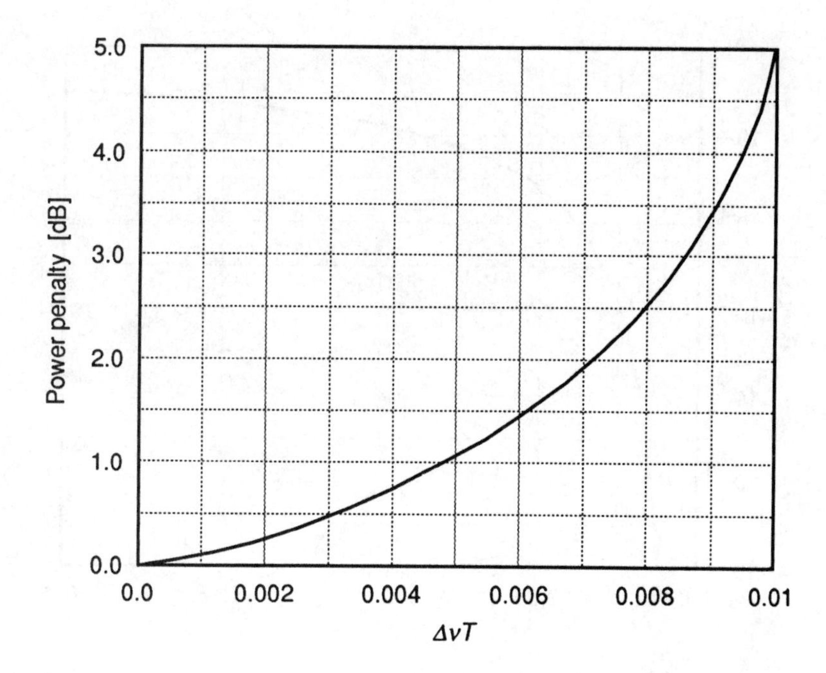

Figure 4.54 Power penalty versus the normalized laser linewidth for differential detection reception of binary PSK with phase noise.

4.11.2 CPFSK

No exact analysis is available for the case of narrow deviation CPFSK with the delay-multiply receiver shown in Figure 4.14. It has been shown [8] that in this case the best IF filter is the one that is optimum in the absence of phase noise. Moreover, since the variance of the phase drift between the two arms of the delay demodulator is proportional to the delay, τ, care should be taken to optimize its value by varying the modulation parameters. As an example, according to (4.57), we can decrease the value of τ, while still keeping β equal to its optimum value of 1, by increasing the modulation index, m.

Iwashita and Matsumoto [10] and Garrett and Jacobsen [11] have evaluated the error probability in the presence of phase noise for CPFSK systems using approximate models for the phase noise.

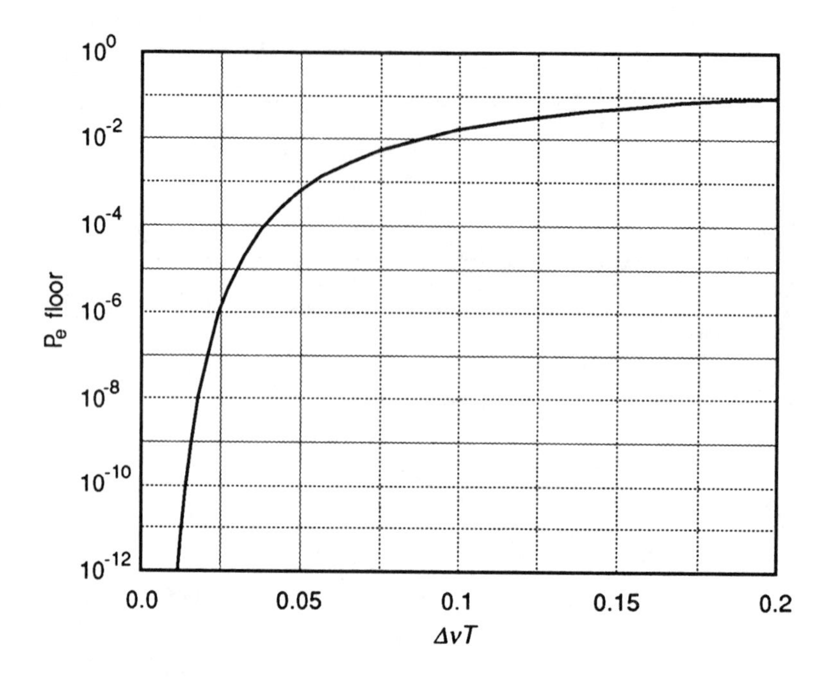

Figure 4.55 Error floor versus the normalized laser bandwidth for differential detection reception of binary PSK with phase noise.

A numerically fitted relationship between the required linewidth and the modulation and receiver parameters that limit the power penalty to 1 dB, useful for design purposes, is as follows [10]:

$$\Delta\nu\tau < 10^{1.53\beta-4} \tag{4.205}$$

As an example, for $\beta = 1$ and, consequently from (4.57), $\tau = T/2m$, the required linewidth must be less than $6.8 \cdot 10^{-3}m$, which, in the case of $m = 0.5$ (MSK) yields a required linewidth less than 0.34% of the bit rate, the same as for DPSK. From (4.205) we see that the sensitivity to laser linewidth can be decreased by simply decreasing the delay, τ. That gives CPFSK an increased degree of freedom with respect to DPSK.

4.11.3 Phase Diversity Receivers

We have seen that asynchronous heterodyne receivers, when optimized using a wider bandwidth of the IF filter and a postdetection filter, can tolerate large amounts of phase noise, up to values of $\Delta\nu T$ of the order of the unity. On the other hand, high-bit-rate heterodyne receivers pose some challenging problems to the designer.

First, extremely large bandwidth optical detectors are required, since the IF frequency is typically equal to three to five times the bit rate, $1/T$. Second, semiconductor lasers frequently exhibit a peak in both amplitude and phase noise spectra located at a frequency of several gigahertz. If the IF spectrum happens to overlap that noise peak, the system performance can deteriorate. Third, in the presence of substantial phase noise, we have seen that some modulation schemes like ASK and FSK require IF filter bandwidth several times larger than the bit rate, so that very wide microwave filters also can be required.

Homodyne receivers can alleviate these problems, because they require only BB filtering and processing. Unfortunately, a conventional synchronous homodyne receiver requires phase locking between the transmitter and local oscillator lasers, which is difficult to achieve and leads to extremely stringent requirements on the laser linewidth (we saw in Section 4.9 that $\Delta\nu T$ must be the order of 10^{-6} for balanced loops and 10^{-4} for decision-driven loops).

Thus, an asynchronous homodyne receiver, that is, a homodyne receiver without phase locking, appears to be a desirable solution. The difference between a synchronous and an asynchronous homodyne receiver can be explained as follows. The signal current produced by a photodetector of a homodyne receiver is equal to $aAcos\phi$, where ϕ is the random phase difference between the transmitted and local oscillator carriers, A is the amplitude of the demodulated signal, and a carries the information. A synchronous receiver tries to keep ϕ as close as possible to zero and, when successful, achieves the highest possible sensitivity, at the expense of extremely stringent constraints on the laser linewidth. An asynchronous receiver makes no attempt to keep ϕ very small. Instead, it uses several detectors with a fixed phase shift between them and processes the obtained signals so as to obtain a decision variable independent of ϕ.

As an example, a two-port receiver would provide two signals equal to $aA \cos \phi$ and $aA \sin \phi$, so that a BB processing squaring and adding them would yield a decision variable $a^2 A^2$, irrespective of the value of ϕ, which is suitable for data detection in the case of ASK (not for PSK, though, where $a = \pm 1$ and $a^2 = 1$).

Since they are not using phase information, asynchronous homodyne multiport receivers are extremely tolerant to phase noise. In whole, they seem to retain the best of the two homodyne and asynchronous heterodyne worlds, because they use only the BB part of the spectrum and yet do not require phase locking. These advantages (nothing comes for free in nature, particularly in engineering) are achieved at the expense of receiver sensitivity (a 3-dB loss with respect to homodyne ASK) and an increased complexity of the optical receiver front end, compensated to some extent by the fact that no IF processing is required.

4.11.3.1 Multiport Receiver Structures

The general structure of a multiport homodyne receiver is shown in Figure 4.56. The received and local oscillator signals are input to an N-port optical hybrid, whose function is to provide N output signals with fixed phase differences $k(2\pi/N)$, $k = 0, 1, \ldots, N-1$ between them. The case $N = 2$ must be treated separately, since it is easily seen that a phase shift of 180 degrees would not eliminate phase noise and have very poor performance. A two-port receiver can be used if the phase shift between the two output ports is designed to be 90 degrees, which is possible using a 90 degrees hybrid (see Section 4.7).

The N signals are then lowpass filtered by filters LPF1 (supposed to be identical in all branches), demodulated, added, and then lowpass filtered again, if required, by lowpass filter LPF2. The structure of the block "DEMODULATOR" depends on the modulation scheme. For ASK signals, it simply consists of a squarer,[14] whereas, for DPSK and CPFSK, it assumes the form of a delay line multiplier; the three cases are shown in Figure 4.57.

In practical applications, for obvious reasons of simplicity, the number of ports, N, is kept quite small, like 2, 3, or 4. Implementations of two-port 90-degree hybrids were discussed in Section 4.7. Several experiments have dealt with three-port receiver; the interested reader can find suitable references in [3].

We analyze the receiver performance in the case of ASK modulation. The N signals at the hybrid outputs (in analytic form) can be written as

$$e_k(t) = \frac{1}{\sqrt{N}}\left[e_S(t)\exp\left(jk\frac{2\pi}{N}\right) + e_{LO}(t)\right] \qquad k = 1, 2, \ldots, N, \qquad N \geq 3 \qquad (4.206)$$

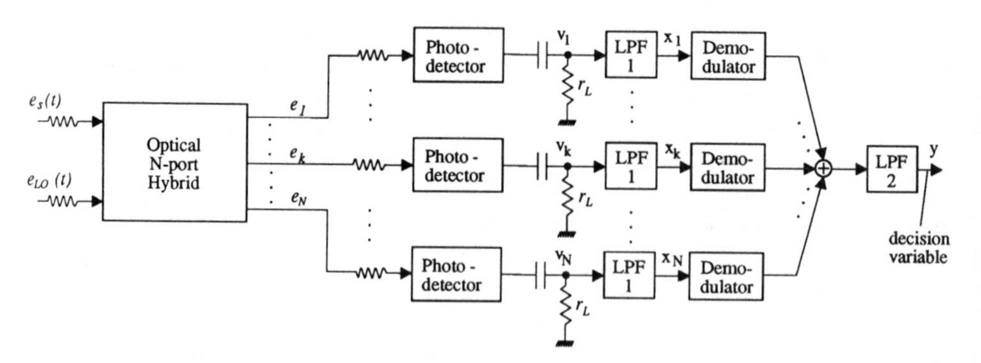

Figure 4.56 Block diagram of a multiport homodyne receiver.

14. It could also be a linear envelope detector. However, performance would be significantly worse, so we will treat here only the case of squarers.

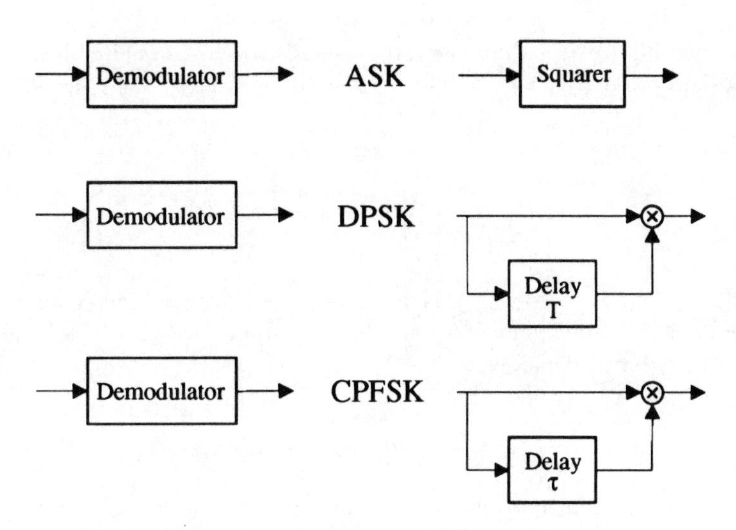

Figure 4.57 Particular demodulators for asynchronous and differential detection receivers.

where e_S and e_{LO} are the received and local oscillator signals:

$$e_S(t) = aE_S e^{j[2\pi f_0 t + \phi_S(t)]}$$

$$e^{LO}(t) = E_{LO} e^{j[2\pi f_0 t + \phi_{LO}(t)]} \qquad (4.207)$$

The frequency, f_0, is the common transmitter and local oscillator frequency, and ϕ_S and ϕ_{LO} are the transmitter and local oscillator laser phase noise, respectively. As for all coherent receivers, this one also requires automatic frequency control of the local oscillator frequency.

The N signals (4.206) at the outputs of the multiport hybrid are detected by N matched photodetectors, which produce N photocurrents like

$$i_k(t) = R|e_k(t)|^2 + i_{nk}(t), \qquad t \in [iT, (i+1)T] \qquad (4.208)$$

where R is the photodetector responsivity, and i_{nk} the current noise process of the kth photodetector. Combining previous equations, we get

$$i_k(t) = \frac{R}{N}\left\{P_{LO} + aP_S + 2a\sqrt{P_S P_{LO}}\cos\left[\phi(t) + k\frac{2\pi}{N}\right]\right\} + i_{nk}(t) \qquad (4.209)$$

where the symbols have the usual meaning, and $\phi(t) = \phi_S(t) - \phi_{LO}(t)$ is the total phase noise. The first term in (4.209) is a DC current and can be rejected by coupling capacitors, as in Figure 4.56. Moreover, the second term has a negligible power under the usual assumption $P_{LO} >> P_S$. Hence, the voltages $v_k(t)$ of Figure 4.56 can be written as

$$v_k(t) = aA_S \cos\left[\phi(t) + k\frac{2\pi}{N}\right] + v_{nk}(t) \quad (4.210)$$

where $A_S = 2Rr_L\sqrt{P_S P_{LO}}/N$ and $v_{nk}(t)$ is the voltage noise process. The N noise processes are independent and, under shot noise–limited conditions, are white Gaussian random processes with spectral densities:

$$G_v = qRr_L^2\, P_{LO}/N \quad (4.211)$$

To proceed with the analysis, we must now distinguish two different situations. The first one refers to the case of small values of the normalized linewidth, $\Delta\nu T$. In that case, the optimal lowpass filters LPF1 are matched filters, that is, integrate (over T)-and-dump filters, and the postdetection lowpass filter LPF2 is not necessary.

The second, and more involved, situation refers to the case in which the normalized linewidth cannot be neglected. In that case, what happens is similar to the situation already analyzed for heterodyne asynchronous ASK reception with phase noise. To reduce the phase-to-amplitude conversion, we may need lowpass filters LPF1 with bandwidth much larger than $1/T$. Thus, a matched postdetection filter LPF2 is necessary because it greatly limits the incurred power penalty by filtering out the extra noise let in by previous filters.

We will analyze only the first situation and offer a few final comments on the second one.

Multiport Receivers With Negligible Phase Noise

Setting $\phi(t) \equiv \phi_0$, ϕ_0 being a constant random phase, in (4.210) and passing the N signals through the integrate-and-dump filters yield the following RVs:

$$x_k = aA_k + nk \quad (4.212)$$

where we have defined

$$A_k \overset{\text{def}}{=} A_S T \cos\left(\phi_0 + k\frac{2\pi}{N}\right), \qquad k = 1, \ldots, N, \qquad N > 3 \quad (4.213)$$

and where n_k are independent Gaussian RVs with zero mean and equal variance

$\sigma^2 = qTRr_L^2 P_{LO}/N$. Recall that the case $N = 2$ requires a different expression of A_k, which is

$$A_k \stackrel{\text{def}}{=} A_S T \cos\left(\phi_0 - k\frac{\pi}{2}\right), \qquad k = 0, 1 \tag{4.214}$$

After squaring and adding the RVs x_k, the receiver of Figure 4.56, without the filter LPF2, takes its decisions by comparing the decision variable

$$y = \sum_{k=1}^{N} x_k^2 \tag{4.215}$$

with a suitable threshold, to be optimized y_{th}.

Defining as y_0 and y_1 the two RVs obtained by y under conditioning on the transmission of $a = 0$ and $a = 1$, respectively, and assuming equally likely transmitted symbols, we can write the error probability as

$$P_e = \frac{1}{2}\{P[y_0 > y_{th}] + P[y_1 \le y_{th}]\} \tag{4.216}$$

The RVs y_0 and y_1 are the sum of N squared Gaussian RVs, with zero mean (for y_0), and means A_k (for y_1) and equal variance σ^2. As a consequence, they are χ-square RVs, with N degrees of freedom, of central type in the case of y_0 and noncentral type in the case of y_1. Their pdf's, derived in Section 2.3, are

$$f_{y_0}(y) = \frac{1}{\sigma^N 2^{N/2}\Gamma(N/2)} y^{(N/2-1)} e^{-(y/2\sigma^2)}, \qquad y \ge 0 \tag{4.217}$$

$$f_{y_1}(y) = \frac{1}{2\sigma^2}\left(\frac{y}{b^2}\right)^{\frac{N-2}{4}} \exp\left(-\frac{b^2 + y}{2\sigma^2}\right) I_{N/2-1}\left(\sqrt{y}\frac{b}{\sigma^2}\right) \tag{4.218}$$

where $\Gamma(\cdot)$ is the gamma function, $I_\alpha(\cdot)$ is the αth-order modified Bessel function of the first kind, and the parameter b is equal to $b = \sum_{k=1}^{N} A_k^2$.

Computing the error probabilities from the pdf's involves only integration over a suitable range. A closed-form expression for the error probability can be given only for $N = 2M$, even. In that case (see Problem 4.30), we obtain

$$P_e = \frac{1}{2}\left[e^{-(y_{th}/2\sigma^2)} \sum_{k=0}^{M} \frac{1}{k!}\left(\frac{y}{2\sigma^2}\right) + 1 - Q_M\left(\frac{b}{\sigma}, \frac{\sqrt{y_{th}}}{\sigma}\right)\right] \tag{4.219}$$

where $Q_M(\cdot,\cdot)$ is the generalized Q function of order M.

It is easily seen that for $N = 2$, that is, $M = 1$, (4.219) yields the same result obtained in Section 4.4.1 for the case of heterodyne asynchronous receiver. As a consequence, use of a two-port receiver does not entail any performance degradations, apart from those related to the use of a 90-degree hybrid and discussed in Section 4.7. As for the case of $N > 2$, we present in Figure 4.58 the results, in terms of error probability versus SNR $A_S^2/4\sigma^2$ (equal to the peak number of photons per bit for shot noise–limited receivers), obtained for $N = 3$, 4 having optimized in each case the threshold y_{th}.

It can be easily verified that the power penalty with respect to the ideal asynchronous heterodyne receiver increases slightly with increasing values of N, and, for $P_e = 10^{-9}$, it is equal to 0.15 and 0.31 for the three- and four-port receivers. These results, which do not seem to have appeared in this exact form previously in the literature, agree with those of Siuzdak and Van Etten [12], based on asymptotic expansion of the pdf f_{y_1}. Kazovsky, Meissner, and Patzak [13] used a Gaussian approximation of the decision RV. Because the SNR for the multiport receiver does not depend on N (what changes is the statistics of the decision RV), the authors were led to the slightly incorrect, although conscious, conclusion that the power penalty is zero, irrespective of N.

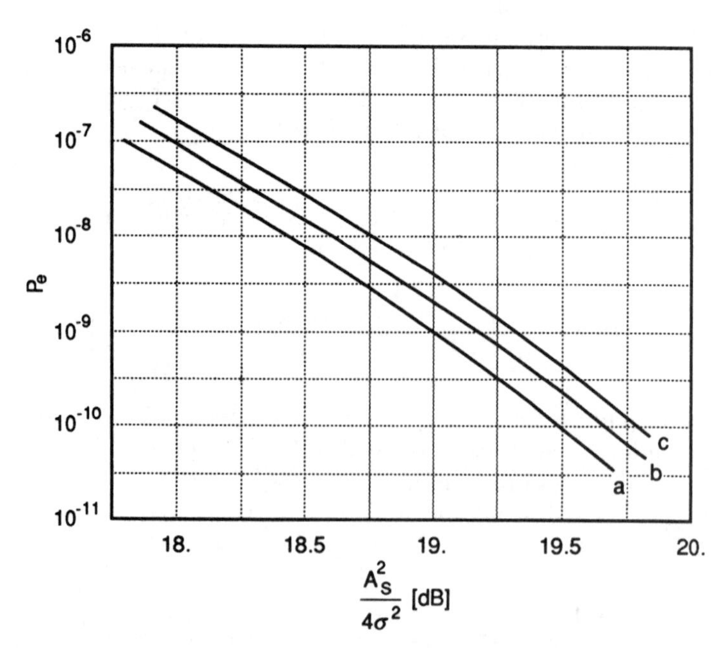

Figure 4.58 Error probability for multiport homodyne receivers and binary ASK modulation versus SNR. The three curves refer to an ideal asynchronous receiver (curve *a*) and three- and four-port receivers (curves *b* and *c*).

A similar analysis can be performed for DPSK and CPFSK multiport receivers, leading to the conclusion that for DPSK no penalty is incurred for a two-port receiver, whereas the penalty amounts to 0.25 dB [14] and 0.45 dB for the three- and four-port receivers, respectively. Almost identical results have been obtained for MSK. For CPFSK, in general, the penalties depend on the modulation index [12].

Multiport Receivers With Phase Noise

When the normalized linewidth of the lasers is not negligible, which happens typically for values of $\Delta\nu$ exceeding a few percent of the bit rate $1/T$, the receiver must be modified as described before and shown in Figure 4.56.

The required analysis follows pretty much the same path as for the case of a heterodyne asynchronous receiver explained in detail in Section 4.10, and the conclusions are qualitatively the same, namely, that increasing the bandwidth of the filter LPF1 with increasing values of $\Delta\nu T$ and inserting a postdetection filter LPF2 leads to a great tolerance to phase noise. Indeed, even a value of $\Delta\nu T = 0.5$ leads to a power penalty as low as 2 dB [13]. These figures are similar to those obtained for the optimized receiver structure of asynchronous heterodyne ASK.

Also, the required widening of the first lowpass filters follows closely what was previously the widening of the IF filters. In fact, for $\Delta\nu T = 0.5$, the optimized filter bandwidth should be six times larger than $1/T$, as it was for the heterodyne receiver.

Kazovsky, Meissner, and Patzak [13] have presented detailed results for the case $N = 3$. As a rule of thumb, we can say that the optimization parameters (and the consequent power penalties) obtained in Section 4.10 can be used as a first design approximation in the case of multiport receivers.

The case of DPSK and CPFSK shows sensitivities to phase noise almost equal to those of the correspondent heterodyne receivers [12].

4.12 HOW TO DEAL WITH PHASE NOISE: A SUMMARY

We have learned how to compute the degradations due to phase noise for the most important modulation schemes and to choose the modulation and receiver parameters so as to limit them to the desired value.

In this section, we will summarize the main results and comment on the possible countermeasures. To make the statements closer to practical needs, we will give all values of the required laser linewidths (transmitter plus local oscillator) with reference to a data rate of 1 Gbps, and such that the incurred power penalty is 1 dB.

4.12.1 Synchronous Schemes

Synchronous schemes show a high sensitivity to phase noise and exhibit an error floor, so that when the linewidth is larger than a certain value, the error probability cannot be decreased no matter what the signal power is.

The best sensitivity is obtained with homodyne PSK. We have analyzed two possible receiver configurations for it, namely, the balanced receiver and the decision directed receiver. The required linewidth is on the order of a few kilohertz for the first receiver and a few hundred kilohertz for the second.

The first figure is out of reach of current semiconductor laser technology, whereas the second is near the border of the best reported achievements using multiquantum-well design for the active region of a single-section DFB laser.

Most viable solutions are external cavity lasers, He-Ne gas lasers or solid-state miniature Nd:YAG lasers. All of them have been experimented with in the laboratory (see [15] and the references therein).

Almost one order of magnitude of insensitivity can be gained using a heterodyne synchronous receiver, where phase locking is performed in the electronic domain. In that case, linewidths on the order of one megahertz are sufficient.

4.12.2 Differential Detection Schemes

Instead of tracking the phase of the received signal, DPSK and CPFSK with delay-line multiplier receivers use as a phase reference the signal received in a previous interval. They are less sensible to phase noise than synchronous schemes. However, they present the phenomenon of the error floor, too. The required linewidth, for CPFSK modulation, depends on the value of the modulation index and of the delay, τ, of the reference signal. In the case of MSK, its sensitivity is the same as for DPSK. The required linewidth for both is around 4 MHz.

4.12.3 Asynchronous Schemes

Asynchronous schemes do not exhibit error floors, only power penalties when affected by phase noise. Also, the power penalty is quite mild when the receiver is optimized with respect to the IF filter bandwidth and postdetection filter. The required linewidth for both ASK and FSK is close to 190 MHz; if a penalty of 2 dB is tolerated, it increases to about 800 MHz. As a consequence, both asynchronous ASK and FSK can be used in connection with commercially available semiconductor lasers for date rate above a few tens of megabits per second.

4.12.4 Phase Diversity Receivers

Phase diversity receivers avoid the burden of IF processing and work well for ASK, DPSK, and CPFSK demodulation. Their sensitivity to phase noise is roughly the same as for the corresponding heterodyne receivers.

Curves showing the required number of photons per bit to obtain an error probability of 10^{-9} versus normalized linewidths are presented in Figure 4.59 for the various modulation schemes. The results for ASK, FSK, and DPSK have been derived from the curves already given in Sections 4.10 and 4.11, whereas for PSK they reflect

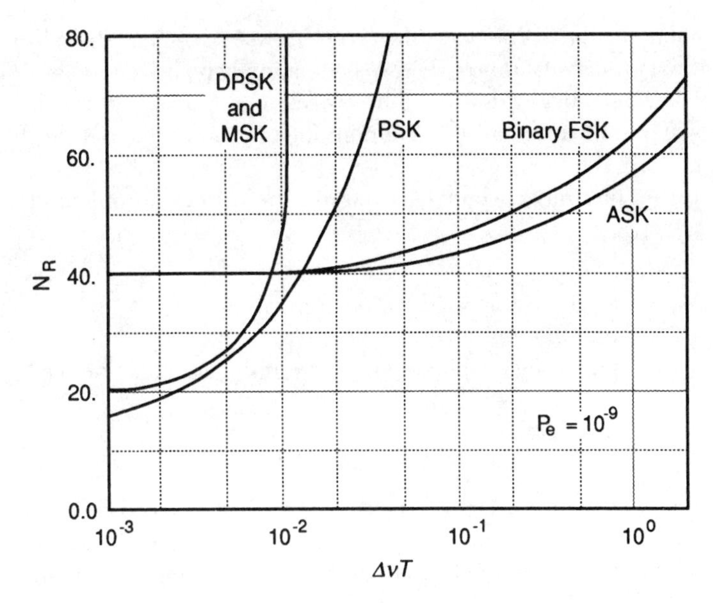

Figure 4.59 Power penalty expressed as number of received photons per bit versus the normalized laser linewidth for various coherent systems with phase noise.

the analysis of homodyne PSK with decision-driven receiver. CPFSK has the same sensitivity as DPSK for a modulation index equal to 0.5; however, its sensitivity can be reduced roughly by a factor of 2 when the delay, τ, of the delay-multiply receiver is reduced and, at the same time, the modulation index is increased.

These curves visualize previous considerations in a synthetic and pictorial way. They also show that the advantages of synchronous and differential detection schemes for zero linewidth are lost very soon as the phase noise starts to become significant.

4.13 POLARIZATION FLUCTUATIONS

Coherent receivers require matching of the SOP (state of polarization) of the local oscillator to that of the received lightwave signal. In all previous performance evaluations, we assumed a perfect matching.

In a practical situation, only the SOP of the local oscillator can be controlled. The SOP of the received signal is different from the one launched into the fiber at the transmitter side and varies with time according to various environmental changes, like temperature and stress. The time constants of these variations can range from seconds to hours, according to the extent and rate of change of perturbation in the ambient conditions.

This phenomenon is known as the *birefringence* of optical fibers. It is caused by different propagation constants of the two orthogonally polarized modes supported by a single-mode fiber and leads to a power exchange between the two polarization components. That, in turn, causes a random fluctuation of the SOP of the lightwave propagating along the fiber.

Four techniques have been conceived to cope with the problem of polarization fluctuations:

- Polarization tracking (or polarization control);
- Polarization-maintaining fibers;
- Polarization spreading (or polarization scrambling or data-induced polarization switching);
- Polarization diversity.

The first two techniques try to keep the SOP of the received signal under control, either by using a fiber fabricated in such a way as to maintain unaltered the SOP launched into it or by modifying the SOP of the received signal so as to match the SOP of the local oscillator.

The third and fourth techniques, on the contrary, aim at modifying the transmitting and/or receiving strategies so the system performance is unaffected by changes in the received SOP.

In this section, we will first see how the differences between the SOPs of the received and local oscillator signals affect the system performance and then describe the four countermeasures.

4.13.1 Effects of the Polarization Mismatch

The fully polarized electromagnetic field launched into the fiber can be written as

$$\vec{E}(t) \equiv \begin{bmatrix} e_1(t) \\ e_2(t) \end{bmatrix} e^{j2\pi f_0 t} \tag{4.220}$$

where $e_1(t)$ and $e_2(t)$ are the complex components of the transversal electric field with respect to two generic orthogonal unit vectors specifying two reference orthogonal SOPs.

Assuming as negligible both polarization dependent losses and depolarization effects, the transit along the fiber can be modeled through a Jones matrix of birefringence, that is, a complex unitary matrix of unit determinant [16]. Dropping the $e^{j2\pi f_0 t}$ and propagation loss terms, the expression of the received field can then be written as

$$\begin{bmatrix} e_1'(t) \\ e_2'(t) \end{bmatrix} = \mathbf{Q}' \begin{bmatrix} e_1(t) \\ e_2(t) \end{bmatrix} \quad (4.221)$$

where $\mathbf{Q}'$ is the Jones matrix of birefringence.

Standard coherent receivers are capable of heterodyning only the received field component with respect to an SOP aligned with the SOP of the local oscillator. In general, that SOP will be a linear combination of the ones used to represent the field in (4.221). Therefore, we introduce another transformation,

$$\begin{bmatrix} e_1''(t) \\ e_2''(t) \end{bmatrix} = \mathbf{Q}'' \begin{bmatrix} e_1'(t) \\ e_2'(t) \end{bmatrix} \quad (4.222)$$

and assume that $e_1''()$ is the heterodyned component, that is, the SOP of the local oscillator, while no heterodyning occurs on $e_2''(t)$. The Jones matrix $\mathbf{Q}''$ is a reference SOP pair transformation matrix.

Thus, we can finally write

$$\begin{bmatrix} e_1''(t) \\ e_2''(t) \end{bmatrix} = \mathbf{Q}_e \begin{bmatrix} e_1(t) \\ e_2(t) \end{bmatrix} \quad (4.223)$$

with $\mathbf{Q}_e = \mathbf{Q}''\mathbf{Q}'$. Being the product of two Jones matrices, $\mathbf{Q}_e$ is itself a Jones matrix.

In evaluating the effect of a polarization mismatch on the system performance, we will make use of the Stokes parameters representation of SOPs [17]. This representation maps every possible SOP into a point with coordinates S_1, S_2, S_3 of the three-dimensional Stokes space identified by the unit vectors $\hat{s}_1$, $\hat{s}_2$, $\hat{s}_3$.

A feature of the Stokes parameters representation is that orthogonal SOPs map onto points that are antipodal on the Poincaré sphere. Therefore, if the SOP of the local oscillator is $\mathbf{S}_{LO} = (S_{1_{LO}}, S_{2_{LO}}, S_{3_{LO}})$ and given $\mathbf{S}_R$, representing the SOP of the received field, we have full heterodyning if $\mathbf{S}_R = \mathbf{S}_{LO}$ and no heterodyning if $\mathbf{S}_R = -\mathbf{S}_{LO}$. We want to know what happens, in terms of power of the heterodyned signal, for all the other possible received SOPs.

The action of a generic Jones matrix of birefringence corresponds to rotating the point representing the launched SOP on the surface of the Poincaré sphere.

Now, every Jones birefringence matrix can be written as:

$$\mathbf{Q}(\alpha,\beta,\gamma) = \mathbf{Q}_1(\alpha)\,\mathbf{Q}_2(\beta)\,\mathbf{Q}_3(\gamma) \quad (4.224)$$

with

$$\mathbf{Q}_1(\alpha) = \begin{bmatrix} e^{j(\alpha/2)} & 0 \\ 0 & e^{-j(\alpha/2)} \end{bmatrix} \quad \mathbf{Q}_2(\beta) = \begin{bmatrix} \cos\frac{\beta}{2} & j\sin\frac{\beta}{2} \\ j\sin\frac{\beta}{2} & \cos\frac{\beta}{2} \end{bmatrix}$$

$$\mathbf{Q}_3(\gamma) = \begin{bmatrix} \cos\frac{\gamma}{2} & \sin\frac{\gamma}{2} \\ -\sin\frac{\gamma}{2} & \cos\frac{\gamma}{2} \end{bmatrix}$$

These matrices, separately, induce an SOP transformation that corresponds to a rotation of the reference axes in the Stokes space around axes $\hat{s}_1$, $\hat{s}_2$, and $\hat{s}_3$, respectively, of angles α, β, and γ, respectively [18].

Consider now a fixed reference vector in the Stokes space, $\vec{S}_{LO}$, representing the SOP $\mathbf{S}_{LO}$ of the local oscillator aligned with $\hat{s}_1$. We can represent the relationship between the received and the local oscillator SOPs in terms of the spherical coordinates (θ, ϕ) with respect to a polar axis coincident with $\mathbf{S}_{LO}$, thus obtaining

$$\begin{bmatrix} e_1'' \\ e_2'' \end{bmatrix} = \mathbf{Q}_1(\phi)\, \mathbf{Q}_2(\theta) \begin{bmatrix} e \\ 0 \end{bmatrix} \tag{4.225}$$

where we have dropped the dependence on t for conciseness.

According to the previous assumptions, the first component on the left side of (4.225) is the heterodyned component. The ratio between the power of the heterodyned component and the total power is then

$$p_f(\theta, \phi) \stackrel{\Delta}{=} \frac{P_{het}}{P_{tot}} = \frac{|e_1''|^2}{|e_1''|^2 + |e_2''|^2} = \frac{|e_1''|^2}{|e|^2} = \left| e^{j(\phi/2)} \cos\frac{\theta}{2} \right|^2 = \frac{1}{2}(1 + \cos\theta) \tag{4.226}$$

where θ is the angle between $\vec{S}_{LO}$ and $\vec{S}_R$. The relationship in (4.226) permits us to evaluate the effect of a polarization mismatch on the heterodyned power.

To explain it more pictorially, if we consider as poles of the Poincaré sphere the points $\pm S_{LO}$, SOPs that lie on parallels of the sphere produce equal received power. The quantity p_f depends only on the "latitude" of the SOP or, in other words, on the θ coordinate of a spherical reference system whose polar axis is $\vec{S}_{LO}$. In general, we do not know the value of θ, which may vary randomly with time.

4.13.2 Polarization Control Systems

The methods proposed and experimentally demonstrated to control the SOP of the received signal so as to match that of the local oscillator make use of a polarization transformer located in the receiver and driven by a feedback loop, as shown in Figure 4.60. Commonly, a controller dithers the driving signals of the device, by detecting the changes in the IF or BB signal power of the receiver to gather information about the extent of SOP matching. The polarization transformer can be located either in the local oscillator path or, in reversed order, in the signal path. It is generally preferable to place it in the local oscillator path because the involved attenuation of the local oscillator power does not decrease the receiver sensitivity as much as attenuation of the received signal.

The differences between the various proposed schemes lie essentially in the choice of the endless *polarization transformer*, that is, the device that permits a continuous SOP matching, irrespective of variations of the received signal SOP. The endless polarization transformer consists of one or more birefringent devices, which introduce a phase shift (or a retard) between the two orthogonal polarizations traveling into the fiber. We will explain here only the common principle to all proposed experimental schemes. For more detailed information, refer to Walker and Walker's excellent tutorial [19].

Polarization transformers exploit the property that if a pressure is applied to a conventional fiber, the induced change in refractive index in the direction of the stress will, in turn, induce a phase change between the components of the propagating electromagnetic wave parallel and orthogonal to the direction of the stress. The phase shift or retardation angle can be varied by changing the squeezing force.

For instance, if light were propagating through the fiber with a linear polarization forming a $\pi/4$ angle with the vertical, and an increasing stress were applied to the fiber in the vertical direction, then an increasing phase difference would be introduced between the horizontal and vertical components of the electric vector of the lightwave, and the SOP would evolve from linear, through elliptical, to circular polarization, then through elliptical to linear polarization oriented at $-\pi/4$ to the vertical, and so on.

If we remember that the equator of the Poincaré sphere carries all linear SOPs, like horizontal H and vertical V, or P and Q, which are inclined by $\pm\pi/4$ to the

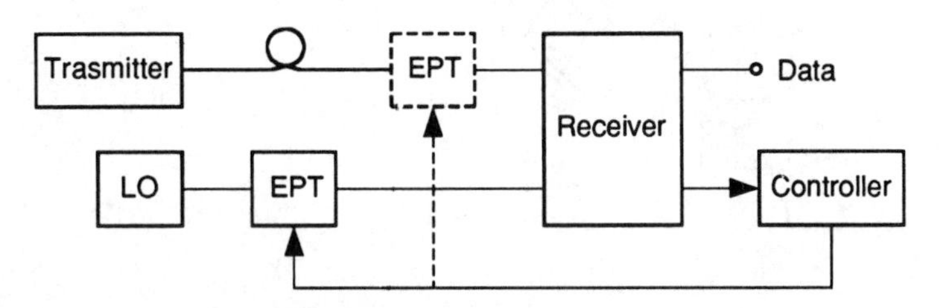

Figure 4.60 Block diagram of a coherent receiver with polarization control system.

horizontal, whereas the poles refer to the right, R, and left, L, circular polarization (Fig. 4.61), the point representing the SOP would describe a circle on the Poincaré sphere around the axis $V - H$, starting from the point P, through the pole L, to the point Q, then via the circular state R, back to P, and so on.

Exploiting that principle, an input SOP $\mathbf{S}_1$ can then be transformed into an output SOP $\mathbf{S}_2$ lying anywhere on the Poincaré sphere by a series of transducers that apply stress alternately at $\pi/4$. This produces individual transformations on the Poincaré sphere along arcs of circles about axes lying at right angles to each other in the equatorial plane.

A configuration with four fiber squeezers is shown in Figure 4.62. We can identify

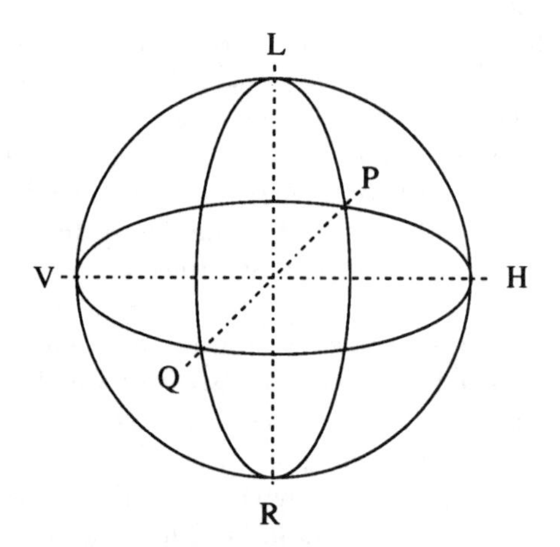

Figure 4.61 Main points on the Poincaré sphere.

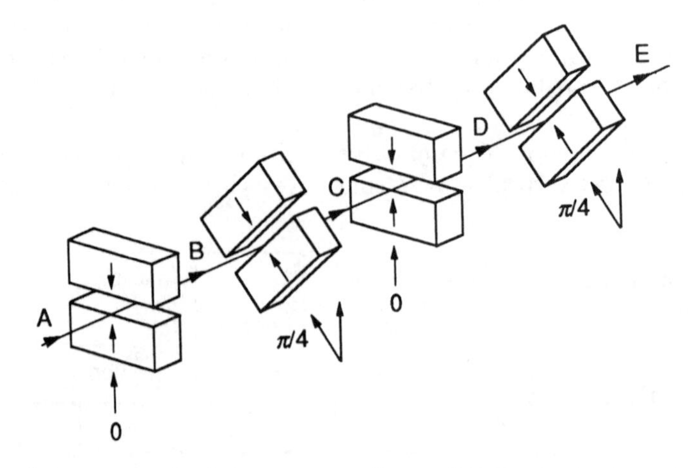

Figure 4.62 Polarization control with four squeezers. (*Source:* [19]. Reprinted with permission.)

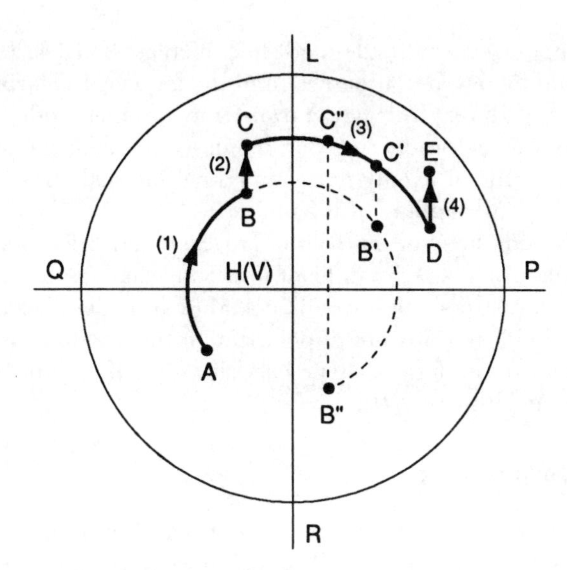

Figure 4.63 How to move on the Poincaré sphere through the four-squeezers configuration.

the transformation operated by the first (and third) squeezer, applied vertically, as a rotation along the axis $H - V$ of the Poincaré sphere, whereas the transformation of the second (and fourth) squeezers, applied at an angle of $\pm\pi/4$ with respect to the vertical, is a rotation along the axis $P - Q$.

By changing the squeezing forces on the squeezers, it is possible to reach all points of the Poincaré sphere, irrespective of the starting point representing the initial SOP. An example is shown in Figure 4.63, where all movements have been projected onto the plane $P - Q$, $L - R$ of the Poincaré sphere.

In principle, only two squeezers are required to achieve SOP matching. In practice, however, since its range is limited, a squeezer must be reset when it reaches a range limit. The reset operation can cause momentary signal loss in a polarization control using only two squeezers. For that reason, practical schemes employ three or four squeezers.

4.13.3 Polarization-Maintaining Fibers

Polarization-maintaining fibers are the most straightforward and simplest solution to the problem of matching the SOPs of the received and local oscillator lightwaves. They are single-mode fibers, in which a large amount of birefringence is intentionally introduced through design modifications so that small random birefringence fluctuations do not affect the light polarization significantly.

In fact, when the propagation constant difference of the two orthogonally polarized modes supported by a single-mode fiber increases, so does the birefringence; the two polarization modes are decoupled, and the launched polarization mode will travel along the fiber with very little power transfer to the other mode. The propagation constant difference can be produced either by an axially nonsymmetrical refractive index distribution, or by an axially nonsymmetrical internal stress, as induced, for example, by an elliptical cladding of the fiber.

In spite of the simplicity of this solution to the polarization matching problem, polarization-maintaining fibers are difficult to produce and expensive, and their loss is higher than for standard single-mode fibers. Moreover, the fiber already installed (tens of million of kilometers) is not of polarization-maintaining type.

Interested readers can find a detailed classification of all kinds of polarization-maintaining fibers in [20, Chap. 9].

4.13.4 Polarization Spreading

A completely different solution to the problem of polarization matching is given by *polarization spreading*, or *scrambling* (PS). Rather than trying to vary the SOP of the local oscillator to match the SOP of the incoming lightwave, PS spreads the signal power over different SOPs during the transmission of every bit, so that heterodyning of at least half the optical incident bit energy is always ensured. An inherent 3-dB penalty is incurred, but in principle this technique permits use of very simple heterodyne receivers, where no care is taken of the incoming SOP.

Hodgkinson et al. [21] first proposed and experimentally demonstrated PS by switching the SOP of the launched light between orthogonal polarizations several times within each bit interval. The high polarization switching rate causes in this case a large broadening of the IF signal spectrum.

Marone, Poggiolini, and Vezzoni [22], further developed the concept by proposing SOP switching at only twice the symbol rate, together with a technique for varying smoothly the SOP within the symbol interval to limit the spectral broadening. Caponio et al. presented experimental evidence [23].

In this subsection, we deal with the principle of PS in a rigorous way, following closely the treatment of Poggiolini and Benedetto [16,24], to which the reader is referred for some of the analytical details and proofs here avoided for brevity.

There is also a recent discovery that makes PS an important technique beyond its applications to coherent schemes. It is the so-called phenomenon of *polarization hole burning* in EDFAs (erbium-doped fiber amplifiers). This phenomenon results from an anisotropic saturation created when a polarized saturating signal is launched into the erbium-doped fiber and has been described for the first time [25] as an excess of noise accumulation in a chain of saturated EDFAs. In a long chain of saturated EDFAs, like in a transoceanic optical link, polarization hole burning can cause amplified spontaneous emission noise to accumulate in the polarization orthogonal to the signal faster than along the parallel axis, thus producing losses up to several decibels.

The deleterious effects of polarization hole burning can be avoided by PS of the signal at a rate faster than the EDFA can respond. Thus, PS also becomes an important issue in connection with long-haul IM-DD systems.

4.13.4.1 The Energy Constraint on PS Systems

We saw in Section 4.13.1 that the normalized fraction, p_f, of the total incident power that is heterodyned converted is given by

$$p_f = \frac{1}{2}(1 + \cos\theta) \qquad p_f \in [0, 1], \theta \in [0, \pi] \tag{4.227}$$

where θ is the angle between the two vectors, $\vec{S}_{LO}$ and $\vec{S}_R$, representing the SOPs of the received and local oscillator signals in the Stokes space, respectively.

When the SOP of the received signal is constant during a bit interval, p_f also represents the normalized fraction of the signal energy that is heterodyned converted in a bit interval.

Now we want to vary the SOP of the received signal, and thus, rather than the time-dependent instantaneous value of the received power fraction, $p_f(t)$, the quantity of interest is the energy of the heterodyned fraction of the optical signal, divided by the total energy of the incident signal, in a given time span, which we call the energy fraction:

$$\varepsilon_f(t, \tau) \stackrel{\text{def}}{=} \frac{\int_{t-\tau}^{t} p_f(t')dt'}{\tau} = \frac{\varepsilon_{het}(t, \tau)}{\varepsilon_{tot}(t, \tau)} \tag{4.228}$$

The principle on which PS operates is that of spreading the signal over different SOPs, so that heterodyning of at least part of the optical incident energy is always ensured, in a fixed time span.

A desirable condition is that of having a constant energy detected in each symbol time T and that it is the highest possible, compatibly with the hypothesis of a complete ignorance of $\mathbf{S}_R$. We therefore impose

$$\varepsilon_{f_T} \stackrel{\Delta}{=} \varepsilon_f(kT, T) = \frac{\int_{(k-1)T}^{kT} p_f(t')dt'}{T} = \frac{1}{2} \qquad \forall k \tag{4.229}$$

In fact, it is easily understood that ε_{f_T} exceeding 1/2 implies at least some knowledge about the received SOP S_R.

Therefore, the problem is that of varying the angle θ between $\vec{S}_R$ and $\vec{S}_{LO}$ in such a way to satisfy (4.229). PS can be performed at either the transmitter or the receiver sites. In the latter case, it can be applied on either the received signal or the local oscillator signal. Because of its greater practical interest, our development lies on the fact that PS is performed during transmission, but all the results hold unaltered if PS is applied at reception.

We assume that all the points of the Poincaré sphere spanned by S_{TR}, the SOP of the transmitted field, belong to a maximum circle of the sphere. In other words, we assume that

$$\vec{S}_{TR}(t) = S_0\{\hat{s}_a \sin[\phi(t)] + \hat{s}_b \cos[\phi(t)]\} \tag{4.230}$$

where S_0 is the radius of the Poincaré sphere, and $\hat{s}_a$, $\hat{s}_b$ are orthogonal unit vectors that lie on the plane of a maximum circle. We call $\phi(t)$ the spreading waveform. Due to the intrinsic features of the transformation (4.223), the received field SOP vector $\vec{S}_R$ also moves on a maximum circle of the Poincaré sphere:

$$\vec{S}_R(t) = S_0\{\hat{s}''_a \sin[\phi(t)] + \hat{s}''_b \cos[\phi(t)]\} \tag{4.231}$$

where $\hat{s}''_a$, $\hat{s}''_b$ are orthogonal unit vectors, in general different from $\hat{s}_a$, $\hat{s}_b$ (Fig. 4.64).

It can be proved (see Problem 4.32) that a sufficient condition for the fulfillment of the energy constraint (4.229) is

$$\int_{(k-1)T}^{kT} \cos[\phi(t) + \phi_0]\, dt = 0 \qquad \forall k \tag{4.232}$$

where ϕ_0 is an unknown random parameter that depends on the polarization transformation (4.223).

There exists an infinite number of functions, $\phi(t)$, that satisfy (4.232), independent of the value of ϕ_0. We will propose some of those solutions.

4.13.4.2 Classes of Solutions of the Energy Constraint

The most interesting spreading waveforms that satisfy the energy constraint (4.232) can be grouped into three classes:

- Staircase PS: $\phi(t)$ is a discontinuous staircase function.
- Linear PS: $\phi(t)$ is a linear function of time.
- Sinusoidal PS: $\phi(t)$ is a sinusoidal function of time.

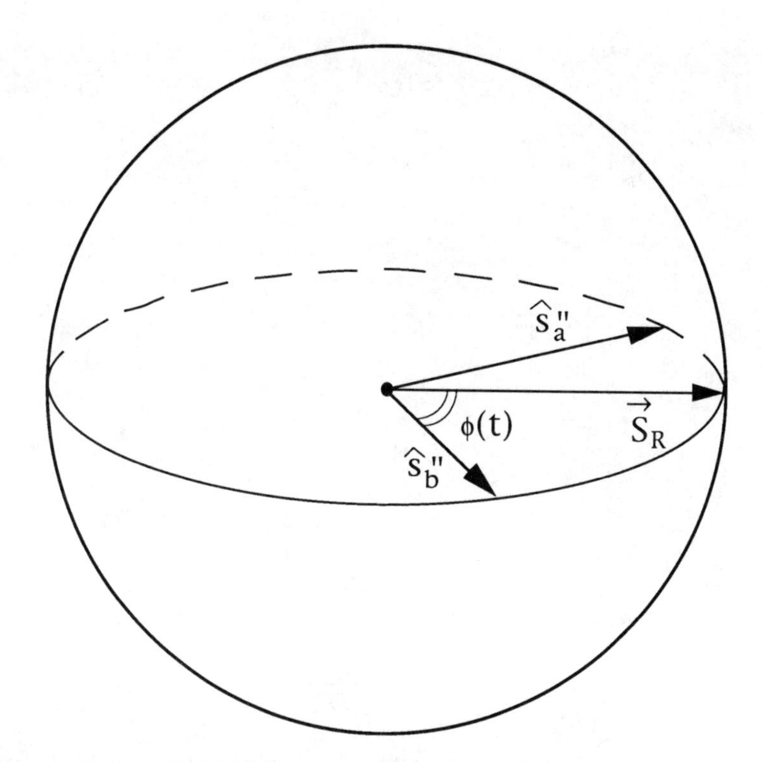

Figure 4.64 State of polarization spanning a maximum circle on the Poincaré sphere.

Staircase PS

We examine the two spreading laws of different specimens belonging to this class and shown in Figure 4.65. The first one consists in transmitting the first half of the bit on one SOP and the second half on the orthogonal SOP. All first halves are sent on the same SOP, so that the period of the spreading waveform is identical to the bit rate. We call this waveform *staircase full* (STF) rate, because its period is equal to the bit duration.

A variation of the same principle consists in still sending the first and the second half of the bit on orthogonal polarizations. This time, though, we do not switch back the polarization at the completion of the bit interval, so the second half of a bit is sent on the same polarization as the first half of the following bit. We call this waveform *staircase half* (STH) rate, because its period is equal to twice the bit duration.

Linear SOP Spreading

Linear spreading waveforms move the SOP vector along a maximum circle, in the Stokes space, at a constant angular speed. In other words, $\phi(t)$, which is the angle

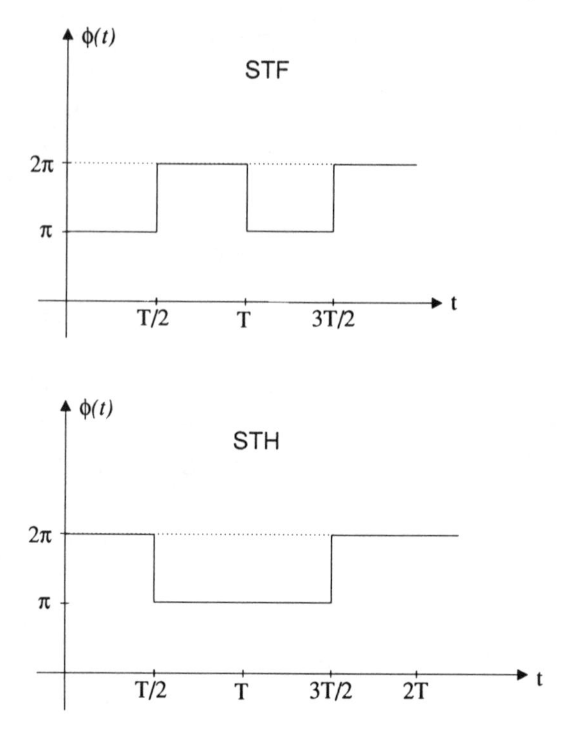

Figure 4.65 Two staircase spreading functions (STF, STH) satisfying the energy constraint. (*Source:* [16]. Reprinted with permission.)

subtended by this movement, grows or decreases linearly with time. We consider two of them, shown in Figure 4.66. The first is called *linear half* (LH) rate because its period is equal to twice the bit duration. A completely equivalent function is obtained by letting $\phi(t)$ grow indefinitely in time, without ever resetting it:

$$\phi(t) = 2\pi \frac{t}{T}$$

To do so, however, a truly endless polarization modulator is needed.

The linear spreading waveform LH has the drawback that it entails either endless polarization modulation or very deep modulation followed by an abrupt return to zero. Both these conditions are not easy to obtain from practical modulators. A different solution is that of driving $\phi(t)$ up to 2π during a bit interval and then linearly back

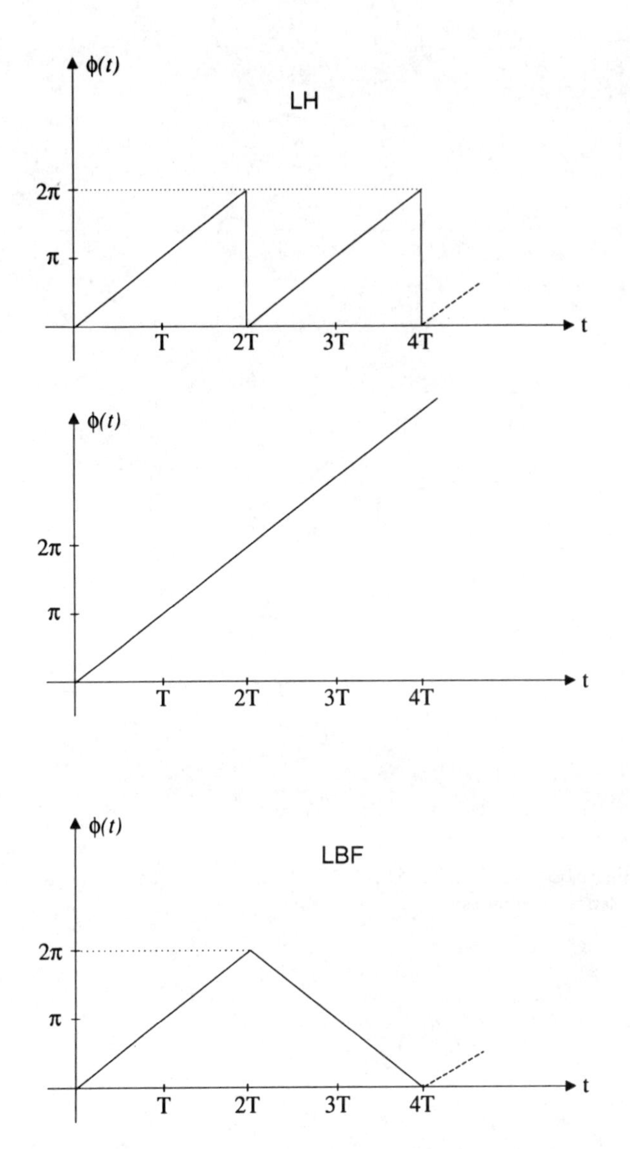

Figure 4.66 Two linear spreading functions (LH, LBF) satisfying the energy constraint. (*Source:* [16]. Reprinted with permission.)

from 2π to zero during the following bit interval. We call this waveform *linear back and forth* (LBF).

Sinusoidal SOP Spreading

We will consider the two best performing members of this family, shown in Figure 4.67. In the first instance, $\phi(t)$ spans a complete period of a sinusoidal function:

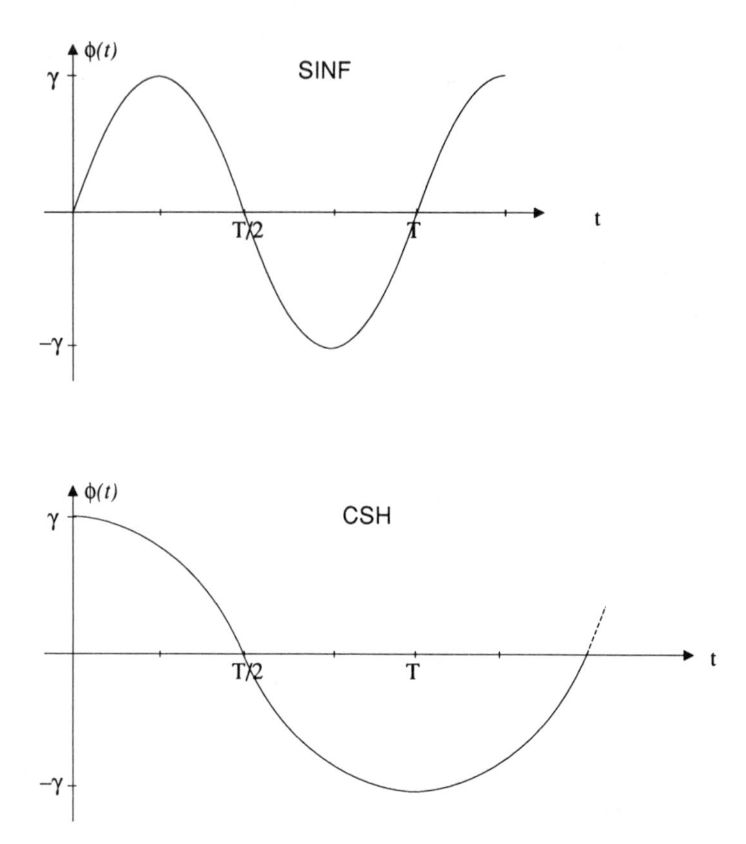

Figure 4.67 Two sinusoidal spreading functions (SINF, CSH) satisfying the energy constraint. (*Source:* [16]. Reprinted with permission.)

$$\phi(t) = \gamma \sin\left(2\pi \frac{t}{T} + \beta\right) \qquad \beta \in [0, 2\pi] \tag{4.233}$$

We call this waveform *sinusoidal at full* (SINF) rate.

Recalling that γ must be one of the zeroes of the Bessel function, J_0, we obtain the best result (in terms of spectral occupancy) by choosing the first zero $\gamma = 2.405$.

The value of β in (4.233) is not constrained. Any initial phase satisfies the energy constraint. In particular, β can be 0, so that ϕ would follow a cosine law, or $\pi/2$, resulting in a sine function of time.

SINF is a compromise between the excellent performance of the linear waveforms and ease of implementation. An interesting point to be made is that with linear laws the Stokes vector has to swing a full 360 degrees (2π) to meet the energy constraint. Here, the full swing amounts to 235 degrees (i.e., 2γ). Since in any modulator structure the applied peak-to-peak voltage is proportional to the peak-to-peak swing of $\phi(t)$, these waveforms require less voltage than the linear ones. In addition, they have the very nice property that the applied voltage strictly follows a sinusoidal waveform, so that a very narrow bandwidth is needed to apply this type of spreading and slow modulators can be used. No precise synchronization with the bit stream is required, since β can be any value. It could be even allowed to drift, provided the drift is slow compared to T.

The second alternative slows down the movement of the Stokes vector so that only half a period of the sinusoidal function is spanned, in every bit interval. This solution still meets the energy constraint, provided that the initial phase β is 0:

$$\phi(t) = \gamma \cos\left(\pi \frac{t}{T}\right) \tag{4.234}$$

We call this waveform CSH, because it must be a cosine and the rate is halved.

From the viewpoint of practical implementation of this spreading technique, the advantages pointed out for the sinusoidal law SINF still hold, except for synchronization with the bit stream, which is now required since β must be kept zero.

4.13.4.3 Spectral Analysis of PS Modulation

We will soon see that the performances of the optimum receiver do not depend on the choice of the spreading waveform, provided it conforms to the energy constraint, whereas, for the suboptimum receiver, the bandwidth of the spread signal can indirectly influence the performance through the additive noise let through by the IF filter. This bandwidth is strictly related to the spreading waveform, and its quantitative evaluation represents a crucial step guiding the choice of $\phi(t)$ in the applications.

We will now analyze the spectral behavior of spread signals and try to establish a performance hierarchy among spreading waveforms according to the spectral broadening that they bring about.

We focus our attention on the IF spectrum in the optical receiver. As a reference modulation format we will use ASK.

The received IF signal depends on the polarization fluctuations on the fiber through the elements q_{ij} of the Jones matrix, to be defined in (4.238). For ASK modulation, it has the following structure

$$v(t) = \sum_{n=0}^{\infty} a_n u_T(t - nT)\{[q_{11}e^{j\phi(t)} + q_{12}e^{-j\phi(t)}]e^{j\,2\pi f_{IF}\,t}\} \tag{4.235}$$

We have to compute the power spectrum of $v(t)$:

$$\Xi\,(f;\,q_{11},q_{12}) \stackrel{\text{def}}{=} [\text{power spectrum of } v(t)] \tag{4.236}$$

which is a function of the random polarization transformations due to the fiber, through q_{11} and q_{12}.

Benedetto and Poggiolini [24] have presented the full mathematical expressions of the power spectra for ASK, FSK, and DPSK as a function of a generic Jones matrix for the various spreading laws. Here, we will base our considerations on two more compact (and useful in practical applications) spectral quantities.

The first quantity, $\Xi(f)$, consists in averaging the power spectrum (4.236) over all the possible polarization transformations that may occur in the channel. The main features of $\Xi(f)$ are as follows:

- Wherever a notch (a zero) is present in Ξ, that notch occurs in all possible realizations of the actual power spectrum.
- Wherever Ξ is not zero, the actual value of the power spectrum can vary between zero and twice (+3 dB) the value of Ξ.

Even though some information is lost, Ξ carries what is needed to design an actual receiver, that is, it shows where signal power can be present and where it never is.

Most of the power spectra that we show include spectral lines. They are displayed as well, but only the relative level between different spectral lines is preserved.

Power spectra are computed from the BB complex envelope of the IF signal. Therefore, they are centered around $f = 0$. This conventional zero frequency corresponds to the IF frequency of the actual receiver. Power spectra are normalized so that the peak transmitted power is equal to 1.

The $\Xi\,(f)$s corresponding to different spreading laws permit us to get a general understanding of the overall spectral behavior of the spread signal, but they are not enough to clearly assess a hierarchy regarding the actual widening of the signal spectra due to spreading. To extract this additional information, we will use mainly a second spectral quantity, the *power confinement function* (PCF), $\Psi(B)$, which is defined as follows:

$$\Psi(B) \stackrel{\text{def}}{=} \frac{\int_{-B/2}^{+B/2} \Xi\,(f)\,df}{\textit{total signal power}} \tag{4.237}$$

and gives the fraction of the total signal power in a bandwidth B between $[-B/2, B/2]$.

Plotting together the Ψs originating from different spreading laws, a power confinement hierarchy among spreading laws is obtained.

The PCF correctly integrates the power carried by spectral lines, resulting in discontinuities in its plot. From the extent of those discontinuities, it is possible to find the power carried by each spectral line.

Let us now compare the previously introduced spreading laws with respect to their spectral occupancy.

Staircase Spreading Laws

The power spectrum of the STF spreading waveform is shown in Figure 4.68, together with unspread ASK. The PCF is shown in Figure 4.69.

The spread signal shows a main lobe, which is twice as wide as the unspread one. So is for the side lobes. The doubling of the lobe size can be heuristically ascribed to the fact that discontinuities in the IF signal occur every T seconds, due to ASK modulation, for the unspread signal, while they occur every $T/2$ seconds for the spread one.

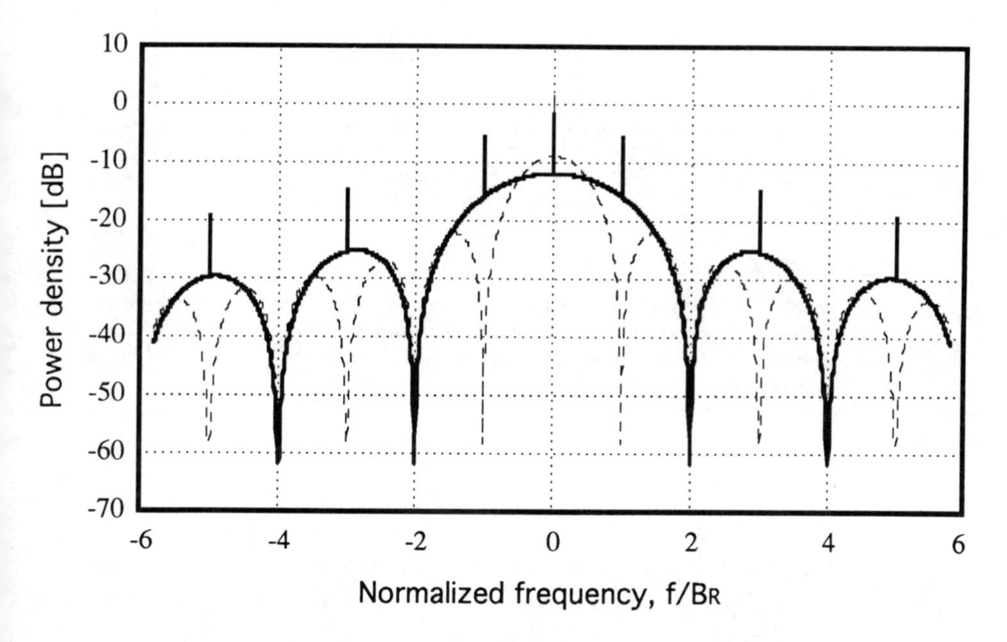

Figure 4.68 Plot of the average power spectrum of polarization spreading ASK with spreading waveform STF (solid line) and of unspread ASK (dashed line) versus normalized frequency. (*Source:* [24]. Reprinted with permission.)

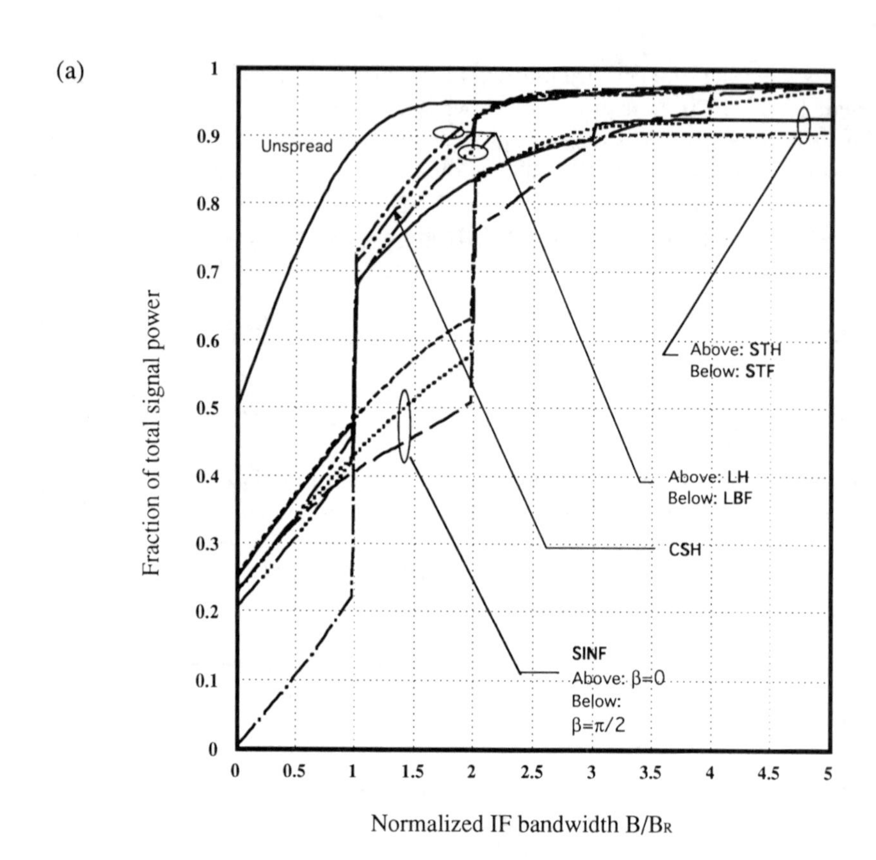

Figure 4.69 (a) Plot of the power confinement function representing the fraction of the total ASK polarization spreading signal power contained within a bandwidth, B, around IF, for various spreading waveforms. The abscissa is the normalized bandwidth, BT. Vertical jumps in the curves correspond to spectral lines in the power spectrum. (b) The zoomed upper part of (a). (*Source:* [24]. Reprinted with permission.)

This spreading law is one of the simplest conceivable, but it is also the poorest in terms of spectral performance. From Figure 4.69, we see that all other spreading laws considered here perform better than this one.

The PCF of the STH law is also shown in Figure 4.69. A certain faster convergence of its PCF with respect to STF can be seen in the figure, but this scheme still performs rather poorly.

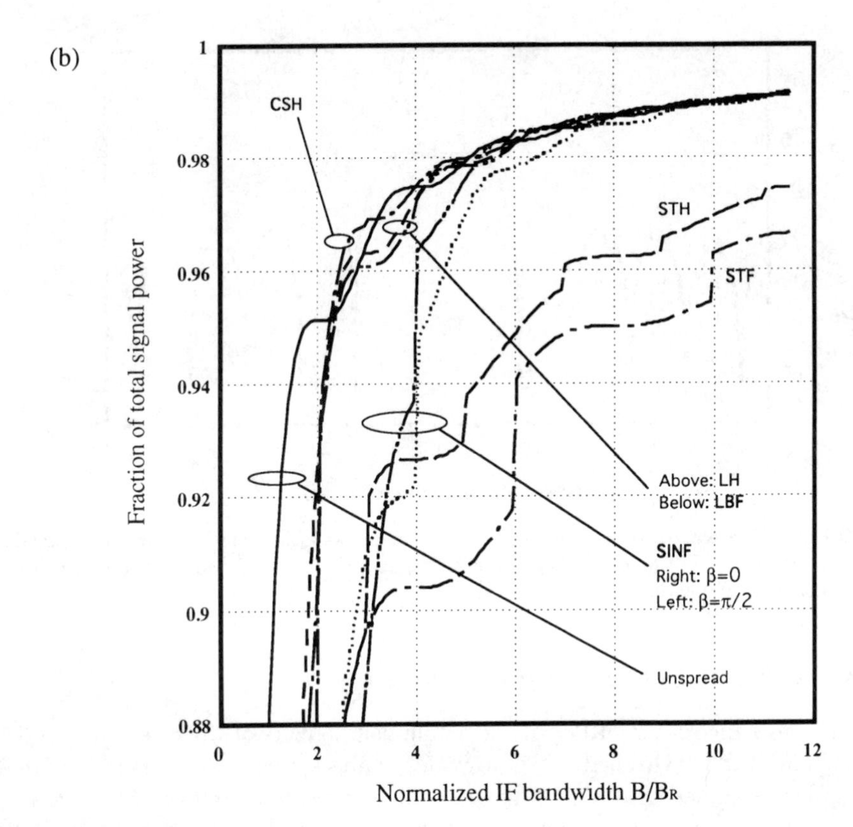

Figure 4.69 (continued)

Linear Spreading Laws

The power spectrum of the LH waveform is reproduced in Figure 4.70. It is much narrower than that of the staircase waveforms previously shown. The main lobe has a full width equal to $3/T$, while the side lobes have the customary $1/T$ width of the unspread ASK signal. In fact, this spreading waveform is one of the best performing, as it can be verified through the PCF plot in Figure 4.69. Its PCF catches up with the PCF of unspread ASK at $2.5/T$, and from that point on the two curves proceed with similar values.

The power spectrum of the LBF law differs from that of LH only in the distribution of spectral lines. So does its PCF in Figure 4.69.

Sinusoidal Spreading Waveforms

The PCFs of the two spreading laws SINF and CSH are plotted in Figure 4.69. In particular, it can be verified that the resulting power containment performance of CSH is very good, and in some spectral regions it even slightly outperforms linear laws.

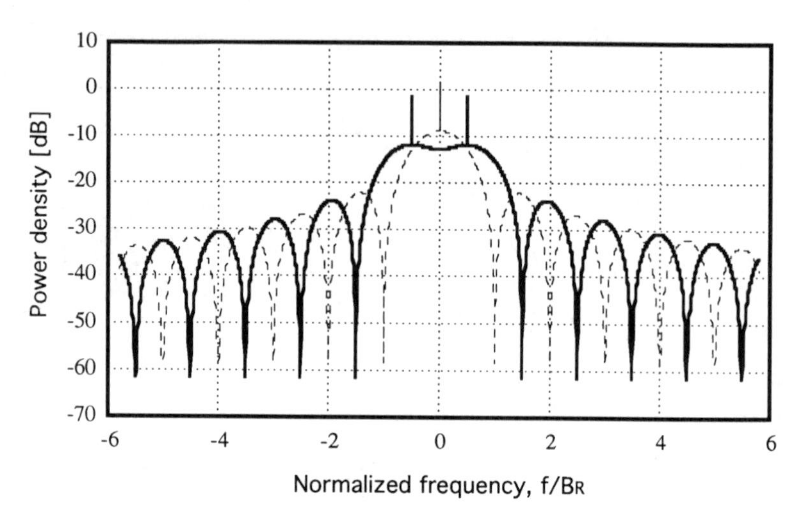

Figure 4.70 Plot of the average power spectrum of polarization spreading ASK with spreading waveform LH (solid line) and of unspread ASK (dashed line) versus normalized frequency. (*Source:* [24]. Reprinted with permission.)

4.13.4.4 Field Encoding of Spreading Waveforms

We examine now the effects of PS on the actual components of the electromagnetic field defined in (4.220). On the basis of previous analysis, the maximum circle of the Poincaré sphere spanned by the spread SOP, defined in (4.230), can be whatever. We need, therefore, to write a Jones matrix that describes a generic rotation of an SOP vector over a generic maximum circle. One possible solution is the following

$$\begin{bmatrix} e_1'' \\ e_2'' \end{bmatrix} = \mathbf{Q}_e \begin{bmatrix} e^{j\phi(t)/2} & 0 \\ 0 & e^{-j\,\phi(t)/2} \end{bmatrix} \begin{bmatrix} 1 \\ 1 \end{bmatrix} \tag{4.238}$$

where the rightmost Jones vector represents the launched optical signal (with transmitted peak power equal to 1) whose SOP is linear, and aligned with the $\hat{s}_2$ axis in the space of the Stokes parameters. The square matrix rotates this SOP around axis $\hat{s}_1$ by an angle $\phi(t)$. The matrix $\mathbf{Q}_e$

$$\mathbf{Q}_e \stackrel{\Delta}{=} \begin{bmatrix} q_{11} & q_{12} \\ q_{21} & q_{22} \end{bmatrix}$$

represents the polarization transformation matrix due to the optical channel. By assumption, the heterodyned signal component is e_1.

The operation described by (4.238) can be performed in practice by a polarization modulator, that is, a device capable of moving the SOP point in the Stokes space when driven by an appropriate voltage waveform. As an example, it can be realized using $LiNbO_3$ devices, as described in [26].

4.13.4.5 Performance of PS Modulation Schemes

Poggiolini and Benedetto [16] have derived the optimum receiver structures and the performance of asynchronous PS modulation schemes, such as ASK, FSK (single and dual filter), and DPSK. The optimum receivers yield performances that are invariant with respect to both the polarization transformations induced by the fiber and the spreading waveform employed.

PS ASK Modulation

The optimum receiver structure is shown in Figure 4.71. It consists of two IF branches, filtering the IF signal with lowpass equivalent impulse responses $\tilde{h}_a(t)$ and $\tilde{h}_b(t)$, which depend on the spreading waveform $\phi(t)$ through

$$\tilde{h}_a(t) = \tilde{a}^*(T - t), \qquad \tilde{h}_b(t) = \tilde{b}^*(T - t) \tag{4.239}$$

with

$$\tilde{a}(t) = \frac{1}{\sqrt{T}} \exp\left[j\frac{\phi(t)}{2}\right], \qquad \tilde{b}(t) = \frac{1}{\sqrt{T}} \exp\left[-j\frac{\phi(t)}{2}\right] \tag{4.240}$$

We will express the performance as a function of the parameter $A = RT\sqrt{P_S P_{LO}}$, which is the square root of the IF signal energy, and the noise variance σ^2, which, for the shot noise–limited case, is equal to $\frac{1}{2}qRP_{LO}T$. This choice implies that the intrinsic 3-dB penalty due to the energy constraint will not appear in the following performance calculations. This penalty in fact occurs during the heterodyning process, and therefore has nothing to do with the subsequent demodulation accomplished by the receiver.

After some algebra performed on the IF noisy signal, it can be shown (see Problem 4.34) that the BB decision variable d, obtained by sampling the signal $d(t)$ in Figure 4.71 at the optimum sampling instant, is

$$d = (aA + n_1)^2 + n_2^2 + n_3^2 + n_4^2 \tag{4.241}$$

where the n_is are Gaussian, independent zero-mean RVs of variance σ^2, and a is the information-bearing RV taking the value 0 or 1.

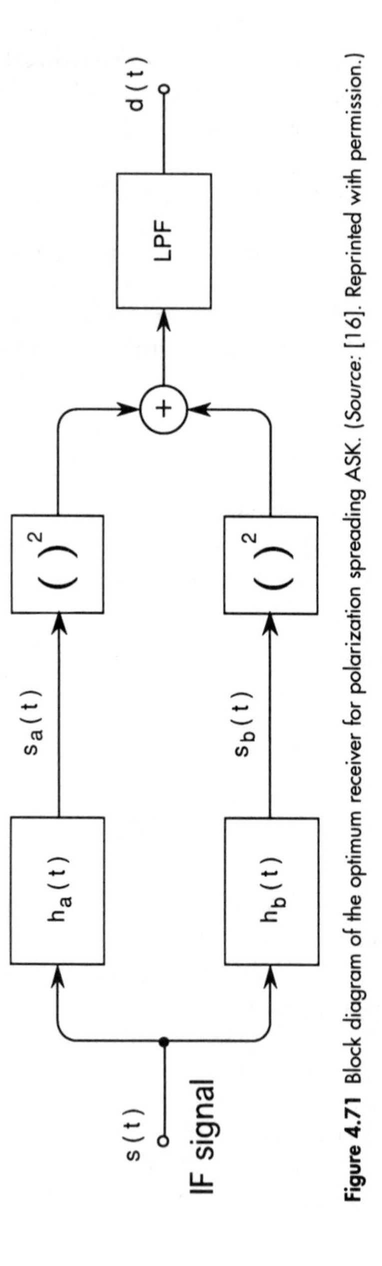

Figure 4.71 Block diagram of the optimum receiver for polarization spreading ASK. (*Source:* [16]. Reprinted with permission.)

Thus, the error probability with respect to a decision threshold x_{th}^2 turns out to be (see Problem 4.35)

$$P_e = \frac{1}{2}\left[e^{-(x_{th}^2/2\sigma^2)}\left(1 + \frac{x_{th}}{\sigma^2}\right) + 1 - Q_2\left(\frac{A}{\sigma}, \frac{x_{th}}{\sigma}\right)\right] \quad (4.242)$$

where Q_2 is a generalized Marcum Q function of order 2. This expression must be compared with the error probability without PS, already obtained in Section 4.4.1, repeated here for completeness:

$$P_e = \frac{1}{2}\left[e^{-(x_{th}^2/2\sigma^2)} + 1 - Q\left(\frac{A}{\sigma}, \frac{x_{th}}{\sigma}\right)\right] \quad (4.243)$$

In both cases, the threshold, x_{th}, must be adjusted according to the ratio $A \neq \sigma$ to get the best result. For $P_e = 10^{-9}$, the penalty between ASK-PS and ASK, each one with its optimized threshold, is approximately 0.31 dB. A plot of P_e for both the PS and the non-PS case is shown in Figure 4.72.

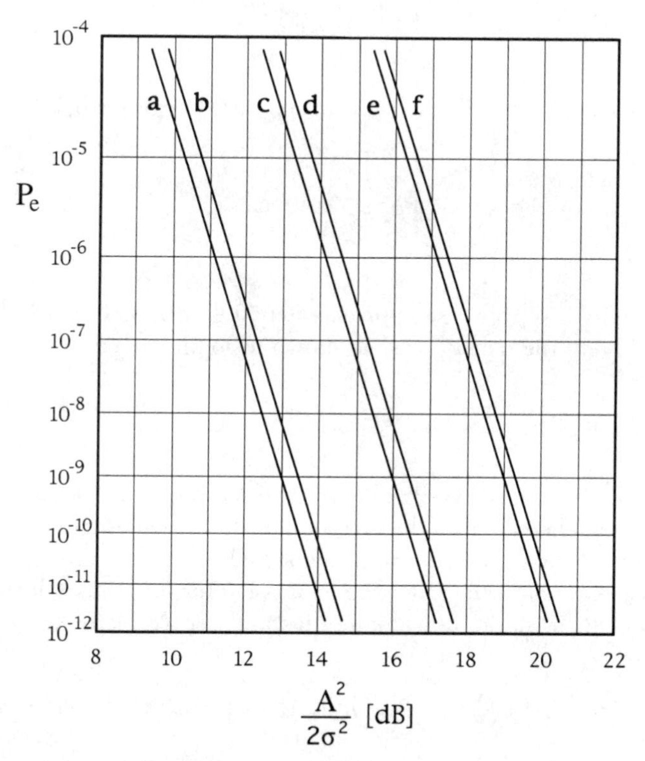

Figure 4.72 Error probability versus SNR for various synchronous modulations with and without PS. Curves *a* and *b* refer to DPSK, curves *c* and *d* refer to FSK and *e* and *f* refer to ASK. (*Source:* [16]. Reprinted with permission.)

PS FSK Modulation

FSK-PS single-filter gives exactly the same results of ASK-PS as FSK single-filter does with respect to ASK. The only necessary assumption is that the interference of the rejected tone be negligible. FSK-PS dual-filter detection can be implemented using the optimum receiver structure in Figure 4.73, where the different pairs of matched filters are centered around the frequencies of the two tones. It can be shown (see Problem 4.36) that the BB decision variable takes on the form

$$d = (aA + n_1)^2 + n_2^2 + n_3^2 + n_4^2 - [(1 - a)A + n_5]^2 - n_6^2 - n_7^2 - n_8^2 \quad (4.244)$$

where all the symbols have the same meaning as before.

The resulting error probability expression turns out to be (see Problem 4.37)

$$P_e = \left(\frac{1}{2} + \frac{1}{16}\frac{A^2}{2\sigma^2}\right) e^{-(A^2/4\sigma^2)} \quad (4.245)$$

This should be compared with the already computed error probability without PS

$$P_e = \frac{1}{2} e^{-(A^2/4\sigma^2)} \quad (4.246)$$

The excess penalty is in this case approximately 0.45 dB for $P_e = 10^{-9}$. A plot of P_e for both the PS and the non-PS case is shown in Figure. 4.72.

PS DPSK Modulation

The optimum receiver for DPSK-PS requires two delay lines with multipliers, one after each of the two matched filters of the optimum IF stage, as shown in Figure 4.74. Then an adder performs the sum of the resulting signals.

In this case, we have the following expression for the demodulated BB decision signal, sampled at the optimum sampling instant (see Problem 4.38):

$$d = (2a - 1)[(\sqrt{2}A + n_1)^2 + n_2^2 + n_3^2 + n_4^2 - n_5^2 - n_6^2 - n_7^2 - n_8^2] \quad (4.247)$$

where all the symbols have the same meaning as before.

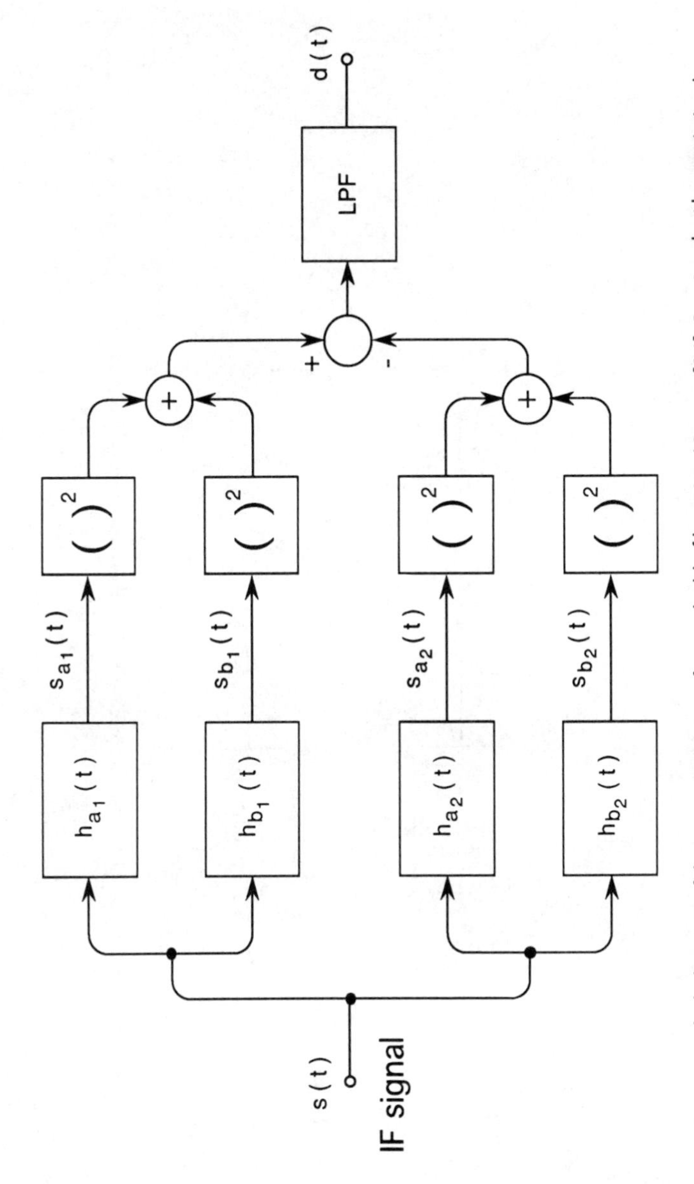

Figure 4.73 Block diagram of the optimum receiver for PS double-filter FSK. (*Source:* [16]. Reprinted with permission.)

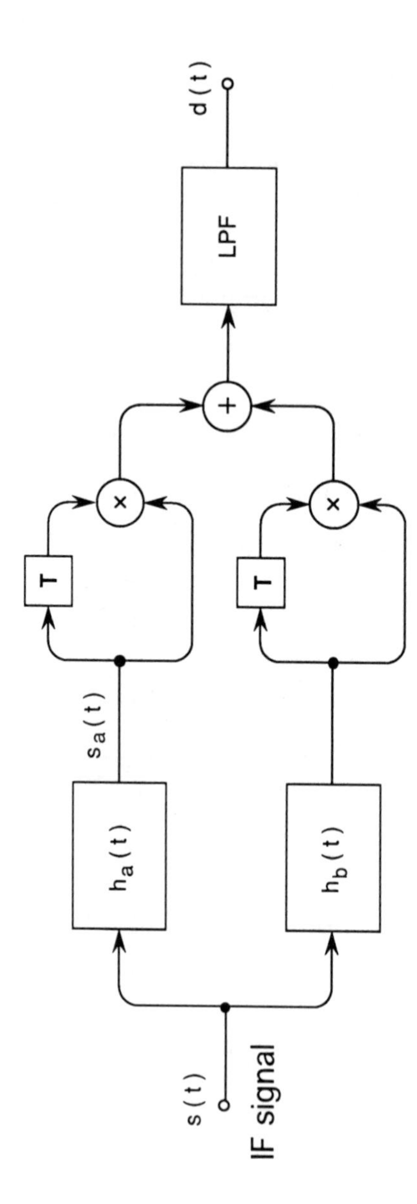

Figure 4.74 Block diagram of the optimum receiver for PS differential detection PSK. (*Source:* [16]. Reprinted with permission.)

The resulting P_e expression is

$$P_e = \left(\frac{1}{2} + \frac{1}{16}\frac{A^2}{\sigma^2}\right) e^{-(A^2/2\sigma^2)} \tag{4.248}$$

This should be compared with the already obtained error probability of the unspread system:

$$P_e = \frac{1}{2} e^{-(A^2/2\sigma^2)} \tag{4.249}$$

The excess penalty is approximately 0.45 dB for $P_e = 10^{-9}$. A plot of P_e for both the PS and the non-PS case is shown in Figure 4.72.

4.13.4.6 Simplified Receiver Structures

The comparison between PS and non-PS schemes with optimum receivers has shown that the penalties, apart from the inherent 3 dB, are very low. These results are important because they establish a firm performance reference.

On the other hand, optimum PS receivers appear unrealistic in the field of optical communications. This is due not only to implementation problems; in fact, even if a matched IF filter is the optimal choice with Gaussian additive noise, we have seen that it may be highly suboptimum in the presence of phase noise, especially for the uncoherent modulation schemes analyzed here.

Fortunately enough, it can be proved [24] that independence of the statistics of the decision variable from both polarization fluctuations and chosen spreading waveform can also be obtained by ASK, FSK, and DPSK spread modulation schemes using a simplified receiver structure like the one shown in Figure 4.75 for ASK. In it, the IF filter labeled "WIDE BPF" is simply a noise-limiting filter, letting the spread IF signal pass unaltered through it. To do so, its transfer function must be substantially flat over the IF signal bandwidth. The postdetection filter, assumed as an integrate (over T) and dump lowpass filter in one analysis [24], removes the effects of PS from the baseband signal.

In this case, however, if we have two spreading waveforms, say, $\phi_1(t)$ and $\phi_2(t)$, of which the former causes a less pronounced spectral broadening, we will be able to

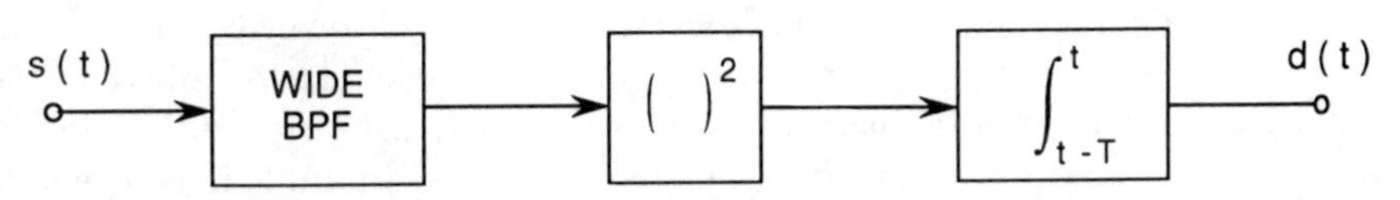

Figure 4.75 Block diagram of the simplified suboptimum receiver for polarization spreading ASK. (*Source:* [24]. Reprinted with permission.)

use a narrower IF filter with ϕ_1 than with ϕ_2. Hence, even though the decision variable has the same statistics with both spreading waveforms, the noise variance can be made lower with ϕ_1, because it uniquely depends on the IF filter noise equivalent bandwidth. This aspect is peculiar to the simplified receiver, the use of which establishes a performance hierarchy among spreading waveforms not only in required IF bandwidth but also, indirectly, on potential sensitivity of the receiver.

Another important remark is that, under the above hypothesis of an IF filter that is "wide enough," the statistics of the decision variable turn out to be identical to those of a heterodyne receiver, demodulating non-PS signals with an ideal polarization control and using the same (in type and bandwidth) IF and postdetection filters. In that case, the penalty between the PS system and the polarization-control system would be exactly the inherent 3 dB.

However, since for the unspread signal we can narrow the IF filter considerably with respect to that which must be used for the spread signal, it turns out that an additional penalty between PS and non-PS does exist, deriving from the larger noise power entering the necessarily wider IF filter.

To sum up, the simplified receiver behaves essentially as a heterodyne receiver with a wide IF filter and a narrow postdetection one.

We analyzed such receivers in Section 4.10 and saw that their performance penalty is quite mild with increasing IF bandwidth. As an example, with an IF filter as wide as 20 times the symbol rate, overall penalties with respect to matched IF filter receivers can be kept within 1.5–2 dB, provided that a tight postdetection filter is used. Therefore, we can expect the penalty between a PS system using the simplified receiver and a polarization-control system (which can use a narrower IF filter) to be approximately bounded within the same figures. The same result helps estimate the penalty deriving from the use of a simplified PS receiver instead of an optimum PS one, which should not exceed the same figures.

The important indication stemming from these results is that the required IF bandwidth is the key parameter to assess how good a PS scheme is, since very wide IF bandwidths are difficult to obtain in practical receivers and they cause more noise to enter the system.

This conjecture has been checked in [22], where a single-tone FSK-PS with simplified receiver was extensively analyzed through computer simulations and compared to single-tone FSK with ideal polarization recovery and postdetection filtering. The spreading waveforms used were staircase and sinusoidal. Using sinusoidal spreading waveforms, the PS-scheme was found to operate with penalties very close to the inherent 3 dB, even if the postdetection filter was of Bessel type instead of the optimum integrate-and-dump. These results have been recently confirmed by an experiment [23] in which the minimum penalty was around 4 dB (including the inherent 3 dB).

To take into account the laser phase noise, we only need to add the constraint of choosing an IF filter wide enough to suppress phase-to-amplitude noise conversion so harmful in ASK and FSK modulation schemes, in much the same way as it was shown for non-PS ASK and FSK.

In conclusion, we can say that a simplified receiver for ASK, FSK, and DPSK with PS, consisting essentially of a conventional heterodyne receiver, with a large enough IF filter and an integrate-and-dump BB one, can closely approach the performance of the optimum receiver.

As for the IF stage, the "large enough" IF filter theoretically should be designed so that the IF signal may go through totally undistorted. This is obviously impossible, but the PCF plots in Figure 4.69 give a precise idea of how large this filter should be, assuming as a target that a certain amount of the signal power falls within its bandwidth.

For unspread ASK, as a rule of thumb established through common experience and extensive simulations, it has been found that an IF bandwidth (at −3 dB) of 2 to $2.5/T$ ensures a relatively undistorted transit of the IF signal. This corresponds to 95% of the signal power. We will take this power figure as our reference target in evaluating spreading law performance.

The PCF's of the spreading laws LH, LBF, and CSH reach 95% immediately before $2.5/T$, where they actually merge with the PCF of unspread ASK, as shown in Figure 4.69. As a result, with this kind of receiver configuration, the necessary IF filter broadening with respect to unspread ASK, due to the use of any of these three spreading laws, turns out to be very small, on the order of 25–30%.

A totally different picture emerges when the other spreading laws are used. With the full-rate sinusoidal SINF, 95% of the power is reached beyond $4/T$, suggesting that an IF filter of bandwidth greater than that amount is needed. With staircase spreading STF, the situation is even worse, indicating that an IF bandwidth of at least $7/T$ is needed not to distort the signal substantially.

If IF filters of lower bandwidth are used, the signal is distorted and some of the power is cut off. At the same time, however, less noise enters the system, so that the optimum bandwidth from the viewpoint of system performance is, for SINF and STF, less than the figures given here. For the specific case of SINF and STF, an extensive simulation study was conducted, and some of the main results were reported in [22]. The bandwidths for best sensitivity were approximately $4/T$ and $5/T$ for SINF and STF, respectively. With these bandwidths, excess penalties are minimized to 0.5 and 1.5 dB, respectively.

In conclusion, we have seen that PS in conjunction with ASK, DPSK, and wide-deviation FSK, solves the problem of polarization fluctuations for coherent receivers, at the expense of a power penalty slightly over 3 dB, excess bandwidths that can be reduced to less than 30%, and a manageable increase of complexity in the transmitter. These results suggest that PS can represent an interesting alternative to polarization control methods, in particular for coherent frequency-division multiplexed point-to-multipoint distribution networks. In that environment, in fact, it would be advantageous to remove complexity from the many users' terminals and concentrate the increased complexity at the few head ends.

4.13.5 Polarization Diversity

Insensitivity with respect to polarization fluctuations can be achieved in coherent optical communication if the receiver derives two demodulated signals stemming from two orthogonal polarizations of the received signal. The two signals are then combined (generally at BB) so as to obtain a decision signal virtually independent of the SOP of the received lightwave.

The structure of the *polarization diversity* (PD) receiver is shown in Figure 4.76 for an unbalanced receiver.[15] The polarization beam splitter is used to obtain the orthogonally polarized components, which are then processed by separate branches of the receiver and added. The way the signals are processed in the block labeled "DEMODULATOR" depends on the modulation format. As an example, for ASK modulation, each of the two signals is bandpass filtered and envelope detected before being added to the other. The resulting decision signal is polarization independent.

PD is one of the best candidates for obtaining polarization fluctuation insensitivity in coherent receivers. As for the performance analysis, it turns out (see Problem 4.40) that the statistics of the decision variable, as well as the noise variance, for the ASK, FSK, and DPSK modulation schemes are identical to those previously evaluated for polarization spreading signals.

An intuitive justification of this result is that both PS and PD schemes work by demodulating signals traveling on two orthogonal channels. In PD, these channels are simply two orthogonal polarizations on the fiber, and separation is performed in the optical front end through a polarization beam splitter. In PS receivers, the two channels are orthogonal in the time domain, and they are separated by means of the IF matched filters through scalar products with the time-orthogonal functions $\tilde{a}(t)$, $\tilde{b}(t)$, defined in (4.239). In fact, after the IF filters, the layout of PS and PD receivers is identical.

As a consequence, the performance of a PD receiver is the same as that of an optimum PS receiver, apart from the inherent 3 dB lost by PS. The curves shown in

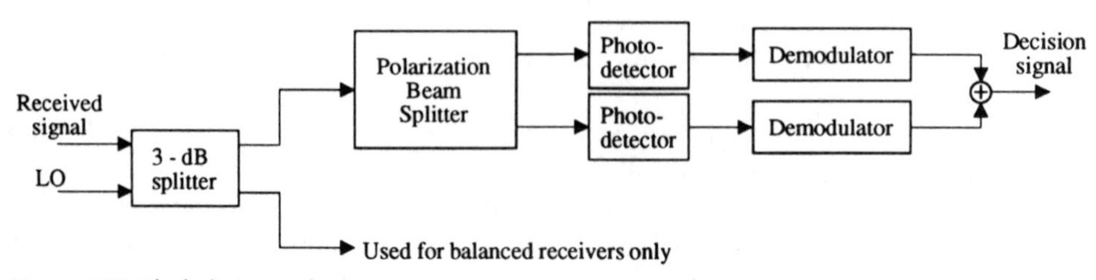

Figure 4.76 Block diagram of a PD receiver.

15. A similar configuration can be easily derived in the case of a balanced receiver, at the expense of doubling the structure complexity.

Figure 4.72 give then the exact error probability performance of PD receivers and show that the incurred penalties are very low, below 1 dB, for all three modulation schemes.

This is confirmed by the results obtained analytically by Okoshi [27] for a PD DPSK and numerically by Enning et al. [28] for ASK. No comparison is possible for FSK since, to our knowledge, no exact calculation of the performance of PD applied to those schemes has been published so far.

As is clearly apparent from Figure 4.76, the polarization diversity receiver presents a complexity almost doubled with respect to a standard receiver. However, the structure is suitable for integration in an *optoelectronic integrated circuit* (OEIC), as shown in Figure 4.77, where the structure of an OEIC based on InP, described in Heidrich [29], is shown. The OEIC receiver includes a DFB-type local oscillator laser, an optical hybrid for PD, four photodiodes for a balanced receiver, and, finally, a preamplifier stage.

Several experiments using PD and various modulation schemes have been reported, whose sensitivities confirm the theoretical results here described. Some of them, together with a tutorial treatment of the subject, are referenced and described in [3].

Combining the concepts of phase diversity asynchronous homodyne receivers, described in Section 4.11.3, and of PD, receivers insensitive to polarization fluctuations and with the advantages of homodyne receivers can be obtained. The basic scheme shown in Figure 4.78 derives from the PD receiver in Figure 4.76 in which each of the two demodulators was replaced by a phase diversity receiver shown in Figure 4.56.

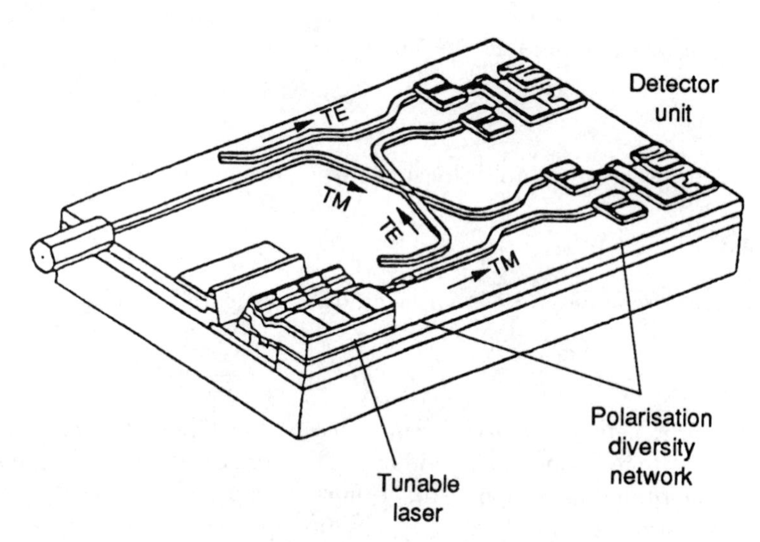

Figure 4.77 Optoelectronic integrated circuit implementing a PD receiver.

The overall complexity depends on the particular structure chosen for the phase diversity receiver but is, in any case, rather high. A few experiments have been performed using this form of combined receiver [3,30]. As for the performance of such schemes, a theoretical analysis using accurate asymptotic approximations [14] has shown that power penalties for combined receivers that employ two- and three-port phase-diversity receivers are 0.25 and 0.45 dB for ASK modulation and 0.45 and 0.75 for DPSK, respectively.

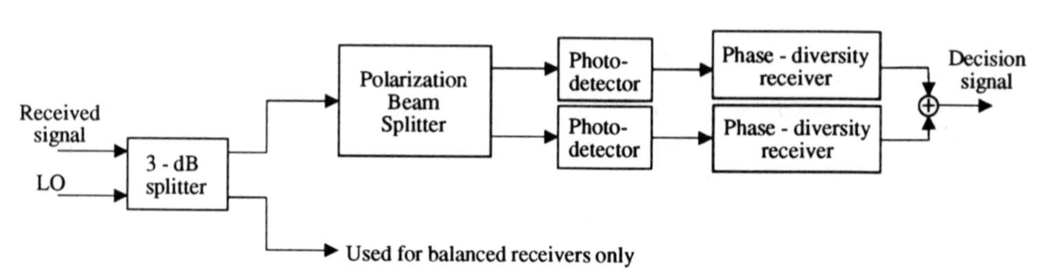

Figure 4.78 Block diagram of a multiport receiver incorporating both phase and polarization diversity.

Problems

Problem 4.1

Derive the error probability (4.19) for homodyne ASK.

Problem 4.2

Derive the error probability (4.20) for homodyne orthogonal FSK.

Problem 4.3

Derive the error probability (4.37) for heterodyne asynchronous ASK.

Problem 4.4

Consider the ASK heterodyne asynchronous receiver in Figure 4.9. Find the exact value of its sensitivity from (4.37) without any approximations and optimize the threshold. (Algorithms to compute the Q-function can be found in the paper by Brennan, *IEEE Trans. on Inf. Theory*, April 1965, pp. 312–313.) Compare with the approximate value $N_R = 80$ and compute the inaccuracy in decibels.

Problem 4.5

Derive the error probability (4.50) for heterodyne asynchronous double-filter FSK.

Problem 4.6

Explain the constraints (4.56) that relate the frequency deviation and intermediate frequency to the delay in the demodulator for CPFSK receivers with reference to Figure 4.15.

Problem 4.7

Show that the error probability (4.59) of CPFSK with delay-multiply receiver reduces to (4.60) for the value of the parameter β equal to 1.

Problem 4.8

Derive the constraints (4.80) and compare them with those obtained by representing the hybrid with the 4-by-4 scattering matrix (the difference consists in considering the device reciprocal, i.e., usable also from the output ports as inputs).

Problem 4.9

An optical hybrid has a transfer matrix with

$$s_{11} = \frac{1}{\sqrt{L}} e^{j\pi/4}$$

$$s_{12} = s_{11}$$

$$s_{21} = \frac{1}{\sqrt{L}} e^{-j\pi/4}$$

$$s_{22} = s_{11}$$

Show that this device is a $\pi/2$ hybrid. Denoting by θ the phase difference at the input ports and under the hypothesis of equal power at the input ports, find the value of θ that maximizes the overall output power and the corresponding power loss, in decibels, at the output.

Problem 4.10

Consider an ideal 3-by-3 coupler. Assume that the input power is equally split among the three output ports and that the loss is negligible.

1. Applying the signals $E_{1i}e^{j\theta_1}$ and $E_{2i}e^{j\theta_2}$ at input ports 1 and 2 (and no signal at input port 3), show that

$$\phi_1 - \phi_2 = \pm 120°$$

$$\phi_3 - \phi_1 = \pm 120°$$

where ϕ_1 is the phase difference at output port 1 and ϕ_2 at output port 2.

2. Applying the signals at other input-port pairs, derive the transfer matrix, S, of an ideal coupler.

Problem 4.11

For the balanced receiver in Figure 4.22, R_1 and R_2 are the responsivities of the two photodetectors. Let SNR be the quantum-limited homodyne SNR and let $P_{LO} >> P_S$.

1. Find SNR in terms of R_1 and R_2.
2. Calculate the degradation, in decibels, of SNR when the second photodiode is disconnected, that is, $R_2 = 0$, with respect to the case $R_1 = R_2 = R$.
3. Suppose $R_1 = R_2 - \Delta R$; sketch the SNR as a function of ΔR.

Problem 4.12

Consider the signal $x(t) = A\cos[2\pi f_0 t + \phi(t)]$, where $\phi(t)$ is the phase noise due to white frequency noise. The signal $x(t)$ is passed through the system of Figure 4.79, where the low-pass filter completely removes the second harmonic of f_0.

1. Derive the autocorrelation function and the power spectral density of $y(t)$ and sketch them.
2. Find the ratio of power between $x(t)$ and $y(t)$.

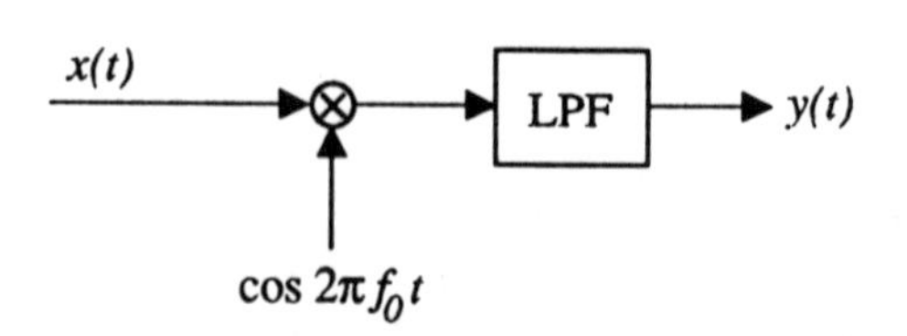

Figure 4.79

Problem 4.13

For a certain quantum-limited homodyne PSK receiver, the phase error ϕ_e between the received signal and the local oscillator is a zero-mean Gaussian random process slowly varying with time (i.e., it remains constant during a 1-bit interval).

1. Compute and plot the error probability, P_e versus $10 \log_{10}(N_R)$ for $\sigma_{\phi_e} = 0$ degrees, 5 degrees, 10 degrees, 12 degrees, 15 degrees and N_R in the interval (5,20).
2. Find the maximum permitted value of σ_{ϕ_e} for 0.5- and 1-dB sensitivity penalty.
3. Find and plot the error probability curve versus σ_{ϕ_e} in the case of absence of quantum noise.

Problem 4.14

Compute the transfer function of the active first-order loop filter in Figure 4.30.

Problem 4.15

Derive the form (4.129) of the closed-loop transfer function of a second-order PLL with active filter.

Problem 4.16

Rederive the transfer function of the closed-loop and its noise equivalent bandwidth when the loop filter is the passive RC filter shown in Figure 4.80.

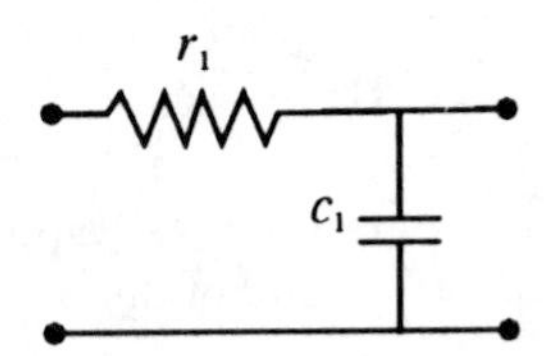

Figure 4.80

Problem 4.17

Consider a PSK optical homodyne communication system with the following characteristics:

- Transmitter and local oscillator lasers:
  1. Wavelength = 1.55 μm;
  2. Power output equal to 0 dBm for both local oscillator and transmitter lasers;
  3. Laser linewidth equal to 10 KHz for both lasers.
- The phase modulator runs at 4 Gbps and produces a phase modulation of $\theta = \pm 80$ degrees with a 6-dB insertion loss.
- The balanced receiver uses photodiodes with quantum efficiency equal to 0.8; the noise current of the preamplifier is 10 pA/$\sqrt{\text{Hz}}$. It uses a PLL with $\zeta = 0.707$; and the PLL operates with $\sigma_{\phi e} \leq 10$ degrees. Our goal is a sensitivity of 90 photons/bit.

Assuming that the fiber dispersion is negligible and that its attenuation is 0.2 dB/km, evaluate the largest achievable transmission distance.

Problem 4.18

In deriving the effects of data noise on the phase error variance for balanced PLL receivers, we have made the approximation

$$\left(\frac{\sin \pi f T}{\pi f T}\right) \simeq 1$$

Evaluate numerically the exact contribution of data noise to the phase error variance and plot the relative error as a function of the normalized parameter $f_n T$.

Problem 4.19

For the decision-driven PLL system shown in Figure 4.34, derive the phase error variance of (4.153).

Problem 4.20

For the decision-driven PLL, compute the noise equivalent bandwidth B_{PLL}, as a function of the loop natural frequency f_n and of the normalized parameter $\alpha = 2\pi f_n T$, taking into account the delay T.

Defining as B_{PLL0} the noise equivalent bandwidth in the absence of the delay T, plot the relative error, $\delta = |B_{PLL} - B_{PLLO}|/B_{PLL}$, as a function of the parameter α and find its value for the practical case of $\alpha = 0.03$.

Problem 4.21

Derive the optimum PLL bandwidth and the minimum phase error variance [(4.155) and (4.156)] for the decision-driven PLL receiver.

Problem 4.22

Including in the analysis the flicker frequency noise, modeled as a random process with spectral density $G_{FN}(f) = k_{FN}/f$ [Hz], derive the overall phase error variance $\sigma_{\phi_e}^2$ for the balanced PLL receiver and sketch a design procedure as it was done in the absence of flicker noise. Find conditions on the parameter k_{FN} allowing neglect of the effect of flicker noise.

Problem 4.23

Including the flicker frequency noise in the analysis, modeled as a random process with spectral density, $G_{FN}(f) = k_{FN}/f$ [Hz], derive the overall phase error variance $\sigma_{\phi_e}^2$ for the decision-driven PLL receiver and sketch a design procedure as it was done in the absence of flicker noise.

Problem 4.24

Evaluate the two contributions of shot noise and phase noise to the noise error variance (4.158) for the decision-directed heterodyne PLL receiver.

Problem 4.25

Derive the expression of the overall phase error variance $\sigma_{\phi_e}^2$ in the case of the balanced PLL heterodyne receiver shown in Figure 4.81, including the effects of shot noise, phase noise, and data noise.

Problem 4.26

Prove that the RVs u_i defined in (4.177) are independent for different values of i.

Problem 4.27

Using (4.183) of the pdf $f_v(b)$ of the RV v, compute the probability $P[v \geq b]$ appearing in the integral on the right side of (4.185).

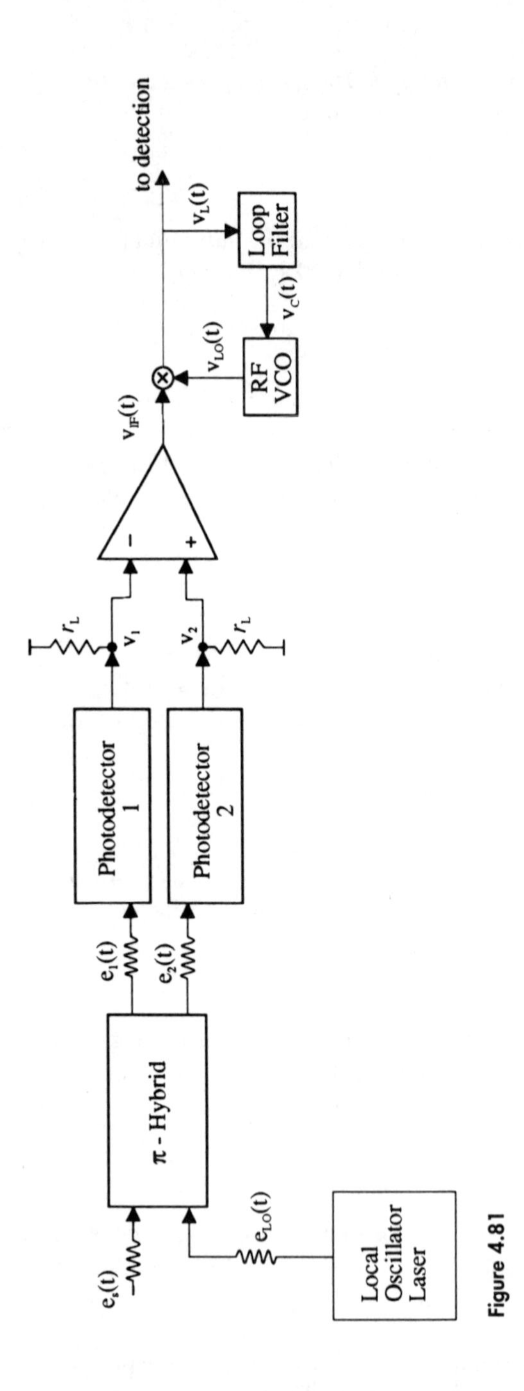

e_s(t)
π - Hybrid
e_LO(t)
Local
Oscillator
Laser
e_1(t)
e_2(t)
Photodetector
1
Photodetector
2
r_L
v_1
v_2
r_L
−
+
v_IF(t)
v_LO(t)
RF
VCO
v_C(t)
Loop
Filter
v_L(t)
to detection

Figure 4.81

Problem 4.28

Derive the result in (4.190).

Problem 4.29

Derive the two conditional error probabilities (4.201) and (4.202).

Problem 4.30

Using the pdf's (4.217) and (4.218), compute the error probability for the N-port phase diversity receiver for $N = 2M$.

Problem 4.31

Derive the expression of the error probability for a DPSK multiport receiver for the cases $N = 2,3,4$ and plot the curves of the error probability versus the SNR $A_{\hat{S}}^2/4\sigma^2$. Evaluate the power penalty with respect to the ideal heterodyne DPSK receiver.

Problem 4.32

Prove that (4.232) is a sufficient condition for a spreading waveform to meet the energy constraint (4.229).

Problem 4.33

Prove that the three classes of spreading waveforms (staircase, linear, and sinusoidal) described in Section 4.13.4 meet the energy constraint (4.229).

Problem 4.34

Using the expression of the polarization spread ASK IF signal (4.235) and the structure of the matched receiver shown in Figure 4.71, prove that the decision variable is independent of the Jones matrix of birefringence and of the spreading waveform and derive its expression (4.241).

Problem 4.35

Using the expression (4.241) of the decision variable, derive the error probability (4.242) for the PS ASK optimum receiver.

Problem 4.36

Suitably adapting the expression of the PS ASK IF signal (4.235) to the FSK dual-filter case and the structure of the matched receiver shown in Figure 4.73, prove that the decision variable is independent of the Jones matrix of birefringence and of the spreading waveform and derive its expression (4.244).

Problem 4.37

Using (4.244) of the decision variable, derive the error probability (4.245) for the PS FSK dual-filter optimum receiver.

Problem 4.38

Suitably adapting the expression of the PS ASK IF signal (4.235) to the DPSK case and the structure of the matched receiver shown in Figure 4.74, prove that the decision variable is independent of the Jones matrix of birefringence and of the spreading waveform and derive its expression (4.247).

Problem 4.39

Using (4.244) of the decision variable, derive the error probability (4.245) for the PS FSK dual-filter optimum receiver.

Problem 4.40

Prove that the decision variables (as well as the error probabilities) for ASK, dual-filter FSK, and DPSK employing PD receivers according to the structure of Figure 4.76 are identical to those of the correspondent schemes using PS.

References

[1] Benedetto, S., E. Biglieri, and V. Castellani, *Digital Transmission Theory*, Englewood Cliffs, NJ: Prentice-Hall, 1987.

[2] Blackman, N., "The Effect of Phase Error on DPSK Error Probability," *IEEE Transactions on Communications*, 29(3):364–365, 1981.

[3] Kazovsky, L. G., "Phase and Polarization-Diversity Coherent Optical Techniques," *Journal of Lightwave Technology*, 7(2):279–292, 1989.

[4] Stiffler, J. J., *Theory of Synchronous Communications*, Englewood Cliffs, N.J.: Prentice-Hall, 1971.

[5] Abramowitz, M., and I. A. Stegun, *Handbook of Mathematical Functions*, Dover Publications, 1972.

[6] Norimatsu, S., and K. Iwashita, "Linewidth Requirements for Optical Synchronous Detection Systems With Nonnegligible Loop Delay Time," *Journal of Lightwave Technology*, 10(3):341–349, 1992.

[7] Kaiser, C. P., M. Shafi, and P. J. Smith, "Analysis Method for Optical Heterodyne DPSK Receivers Corrupted by Laser Phase Noise," *Journal of Lightwave Technology*, to be published.

[8] Jacobsen, G., et al., "Bit Error Rate Floors in Coherent Optical Systems With Delay Demodulation," *Electronics Letters*, 25(21):1425–1427, 1989.
[9] Kaiser, C. P., P. J. Smith, and M. Shafi, "An Improved Optical Heterodyne DPSK Receiver To Combat Laser Phase Noise," *Journal of Lightwave Technology*, 13(3):525–533, 1995.
[10] Iwashita, K., and T. Matsumoto, "Modulation and Detection Characteristics of Optical Continuous Phase FSK Transmission System," *Journal of Lightwave Technology*, 5(4):452–460, 1987.
[11] Garrett, I., and G. Jacobsen, "Theory for Optical Heterodyne Narrow-Deviation FSK Receivers With Delay Demodulation," *Journal of Lightwave Technology*, 6(9):1415–1422, 1988.
[12] Siuzdak, J., and W. Van Etten, "BER Performance Evaluation for CPFSK Phase and Polarization Diversity Coherent Optical Receivers," *Journal of Lightwave Technology*, 9(11):1583–1592, 1991.
[13] Kazovsky, L. G., P. Meissner, and E. Patzak, "ASK Multiport Homodyne Receivers," *Journal of Lightwave Technology*, 5(6):770–791, 1987.
[14] Siuzdak, J., and W. Van Etten, "BER Evaluation for Phase and Polarization Diversity Optical Homodyne Receivers Using Noncoherent ASK and DPSK Demodulation," *Journal of Lightwave Technology*, 7(4):584–599, 1989.
[15] Kazovsky, L. G., "A 1320-nm Experimental Optical Phase-Locked Loop: Performance Investigation and PSK Homodyne Experiments at 140 Mc/s and 2 Gb/s," *Journal of Lightwave Technology*, 8(9):1414–1422, 1990.
[16] Poggiolini, P., and S. Benedetto, "Theory of Polarization Spreading Techniques—Part I," IEEE Transactions on Communications, 42(5):2105–2118, 1994.
[17] Born, M., and E. Wolf, *Principles of Optics*, Pergamon Press, 1964.
[18] Takenaka, H., "A Unified Formalism for Polarization Optics by Using Group Theory," *Nouvelle Revue d'Optique*, 4:37–42, 1973.
[19] Walker, N. G., and G. R. Walker, "Polarization Control for Coherent Communications," *Journal of Lightwave Technology*, 8(3):438–458, 1990.
[20] Okoshi, T., *Coherent Optical Fiber Communications.*, Boston: Kluwer Academic Publisher, 1988.
[21] Hodgkinson, T. G., et al., "Polarisation Insensitive Heterodyne Detection Using Polarisation Scrambling," *Electronics Letters*, 23(10):513–514, 1987.
[22] Marone, G., P. Poggiolini, and E. Vezzoni, "Polarization Independent Detection by Synchronous Intra-Bit Polarization Switching," in *Proceedings of ICC '90*, 1990.
[23] Caponio, N., et al., "Polarisation Insensitive Coherent Transmission by Synchronous Intra-Bit Polarisation Spreading," *Electronics Letters*, 27(4), 1991.
[24] Benedetto, S., and P. Poggiolini, "Theory of Polarization Spreading Techniques—Part II," *IEEE Transactions on Communications*, 42(6):2291–2304, 1994.
[25] Taylor, M. G., "Observation of New Polarization Dependence Effect in Long Haul Optically Amplified System," in *Proceedings of OFC '93*, 1993.
[26] Benedetto, S., et al., "A $LiNbO_3$ Modulator for Binary and Multilevel Polarization Modulation," In *Proceedings of OFC '94*, 1994.
[27] Okoshi, T., "Simple Formula for Bit-Error Rate in Optical Heterodyne DPSK Systems Employing Polarisation Diversity," *Electronics Letters*, 24(2):120–122, 1988.
[28] Enning, B., et al., "Signal Processing in an Optical Polarization Diversity Receiver for 560-Mbit/s ASK Heterodyne Detection," *Journal of Lightwave Technology*, 7(3):459–464, 1989.
[29] Hamacher, M., et al. "Coherent Receiver Front-End Module Including a Polarization Diversity Waveguide OIC and a High-Speed In GaAs Twin Dual p-i-n Photodiode OEIC Both Based on In P," *IEEE Photonics Technology Letters*, 4(11):1234–1237, 1992
[30] Okoshi, T., and Y. H. Cheng, "Four-port Homodyne Receiver for Optical Fiber Communications Comprising Phase and Polarisation Diversity," *Electronics Letters*, 23:377–378, 1987.

Chapter 5

Optical Amplifiers

5.1 INTRODUCTION

Optical amplifiers have recently stolen center stage in the optical telecommunications world due to rapid device progress and revolutionary systems results [1]. These devices enable new and exciting optical systems to be conceived and demonstrated. In fact, many of the most relevant recent advances in optical communications (i.e., long-distance NRZ [nonreturn to zero] and soliton systems and wide-area and broadcast multichannel systems) can be traced to the incorporation of optical amplifiers.

As a simple introduction, optical amplifiers can be thought of as a laser (gain medium) with a low feedback mechanism and whose excited carriers amplify an incident signal but do not generate their own coherent signal [2]. Similar to electronic amplifiers, optical amplifiers can be used to compensate for signal attenuation resulting from distribution, transmission, or component-insertion losses [3,4]. As shown in Figure 5.1, both types of amplifiers provide signal gain, G, but also introduce additive noise (variance = σ^2) into the system. Each amplifier requires some form of external power to provide the energy for amplification. A voltage source is required for the electrical amplifier, and a current or optical source is required for the optical amplifier. The current or optical source for the optical amplifier is used to pump carriers into a higher energy level, which can then decay and emit a photon at the input signal wavelength. One figure of merit for both amplifiers is a low noise figure, NF, to be defined later. Additionally, the amplifier's design, input parameters, and position along a channel must all be optimized for a given system. In this chapter, we will discuss fundamental characteristics, systems issues, and potential applications associated with implementing optical amplifiers in an optical communication system.

5.1.1 Benefits of Direct Amplification

Optical amplifiers were relatively unknown in the systems community before 1980. The original motivation for the recent widespread research was to replace the regenerators in long-haul transoceanic fiber-optic systems, which were placed every

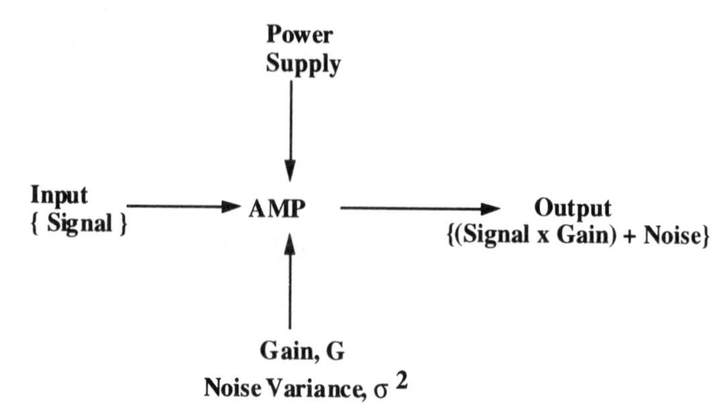

Figure 5.1 Basic amplifier characteristics.

~ 50 km along the entire span. Such regenerators corrected for fiber attenuation and chromatic dispersion by detecting an optical signal and then retransmitting it as a new signal using its own internal laser, as illustrated in Figure 5.2. Regenerators (hybrids of optics and electronics) are expensive and bit-rate and modulation-format specific and waste power and time converting from photons to electrons and back again to photons. In contrast, the optical amplifier is ideally a transparent box that provides

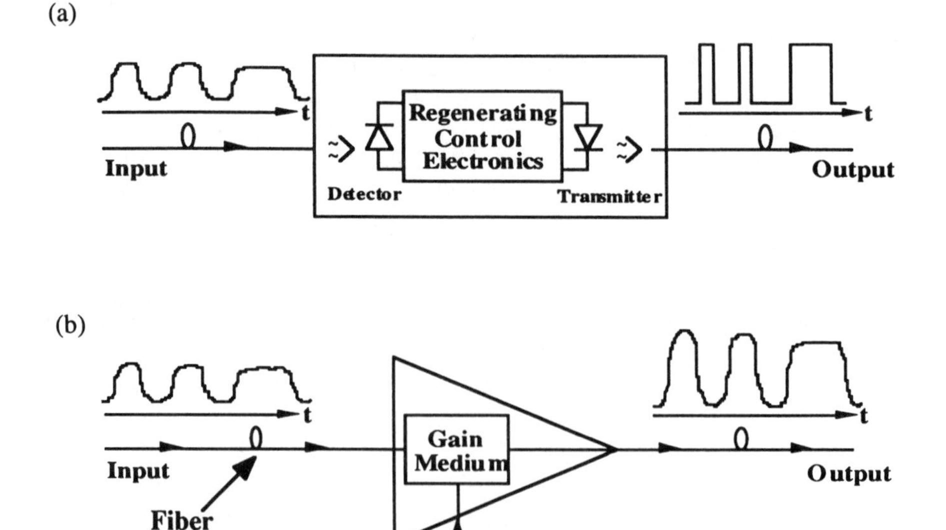

Figure 5.2 Schematic of (a) optoelectronic regenerator and (b) optical amplifier.

gain and is also insensitive to the bit rate, modulation format, power, and wavelengths of the signal(s) passing through it. The signals remain in optical form during amplification, and optical amplifiers are potentially cheaper and more reliable than regenerators. However, the optical amplifier is not an ideal device for the following reasons:

- It can provide only a limited amount of output power before the gain diminishes due to a finite number of excited carriers available to amplify an intense input signal.
- The gain spectrum is not necessarily flat over the entire region in which signals may be transmitted.
- The additive noise causes a degradation in receiver sensitivity.
- Fiber dispersion and nonlinear effects are allowed to accumulate unimpeded.

We note here that for long-haul systems (1) the signal wavelength must be near 1.55 μm for lowest attenuation, and (2) the fiber must be dispersion-shifted, with its dispersion parameter having a value close to zero at the signal wavelength since the optical amplifier does not correct for chromatic dispersion along the fiber span.

5.1.2 Basic Amplifier Configurations

The three basic system configurations envisioned for the incorporation of optical amplifiers are shown in Figure 5.3 [5]. The first configuration is to place the amplifier immediately following the laser transmitter to act as a power or postamplifier. This

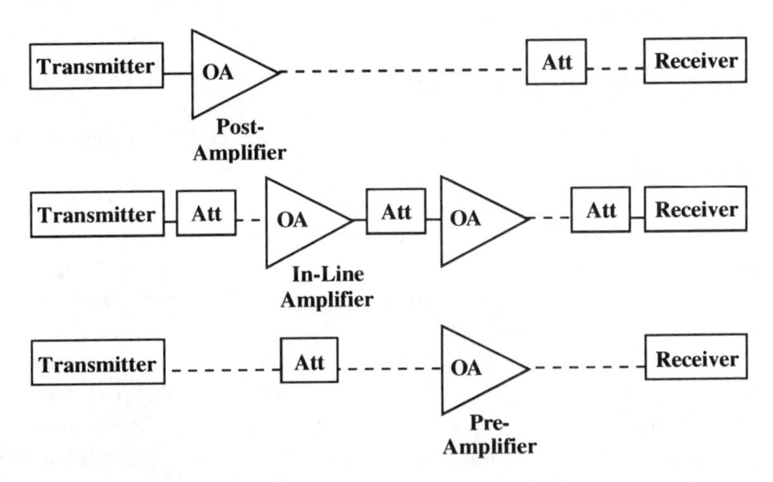

Figure 5.3 Three generic configurations for incorporating optical amplifiers into transmission or distribution systems (AH = attenuator, OA = optical amplifier).

boosts the signal power so the signal is still above the thermal noise level of the receiver even after attenuation. Furthermore, any noise introduced by the power amplifier will be similarly attenuated together with the signal as they are transmitted through the lossy system. Since the signal power input to the power amplifier is typically large (0.1–1.0 mW), the key parameter for the power amplifier will be to maximize the saturation output power and not necessarily the absolute gain. If the amplifier can supply no more power than the original laser transmitter, this configuration produces little advantage.

The second configuration is to place the amplifier in-line and perhaps incorporated at one or more places along the transmission path. The in-line amplifier corrects for periodic signal attenuation due either to fiber absorption or to network distribution splitting losses [6]. The in-line amplifier may exist in a cascade form, with a given amplifier's output signal and noise feeding into a subsequent amplifier with some attenuation between the amplifiers. Issues such as optical filtering and isolation must be considered.

The third possibility is to place the amplifier directly before the receiver, so it functions as a preamplifier. In that case, the signal has already been significantly attenuated along the transmission path. The main figures of merit are high gain and low amplifier noise, because the entire amplifier output is immediately detected. As we shall see later in this chapter, the receiver will be limited by the amplifier noise and not by the receiver thermal noise.

5.1.3 Types of Optical Amplifiers

Before discussing details of specific amplifier systems, we will introduce briefly the main classes of optical amplifiers.

5.1.3.1 Semiconductor and Fiber Amplifiers

Semiconductor optical amplifiers (SOAs) [7–11] and EDFAs (erbium-doped fiber optic amplifiers) [12–15] consist of an active medium that has its carriers inverted into an excited energy level, thus enabling an externally input optical field to initiate stimulated emission and achieve coherent gain. The population inversion is achieved by the absorption of energy from a pump source, and an external signal must be efficiently coupled into and out of the amplifier. Figure 5.4 depicts the basic amplifier building blocks for an SOA and an EDFA.

An SOA is nothing more than a semiconductor laser, with or without facet reflections (the antireflection coating reduces the reflections). An electrical current inverts the medium, by transferring electrons from the valence to the conduction band, thereby producing spontaneous emission (fluorescence) and the potential for stimulated emission if an external optical field is present. The stimulated emission yields the signal gain. On the other hand, the spontaneous emission is itself amplified and is

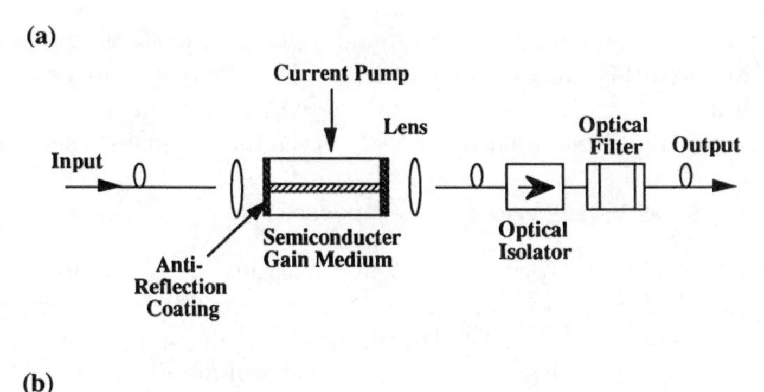

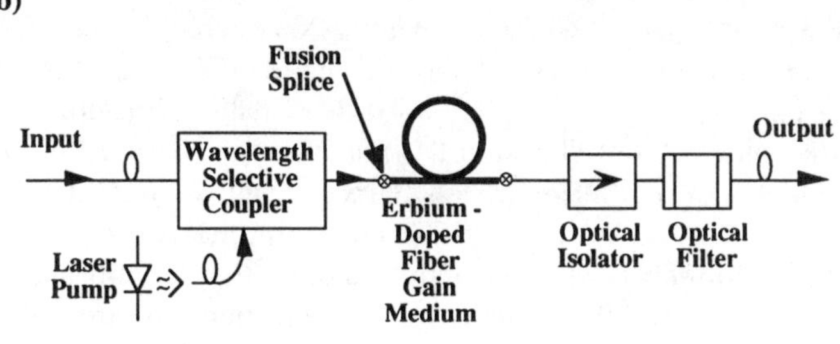

Figure 5.4 Block diagram of (a) a semiconductor amplifier and (b) an erbium-doped fiber amplifier. The optical isolator and optical filter are included, although they may not be required under all circumstances.

considered the randomly fluctuating amplifier noise, called the ASE (amplified spontaneous emission). If we are dealing with a circular-waveguide fiber-based communication system [16], an external signal must be coupled into and out of the amplifier's rectangular active region, producing a mode field mismatch and, consequently, insertion losses.

The fiber amplifier is a length of glass fiber that has been doped with the ions of a rare-earth metal, such as erbium. The ions act as an active medium with the potential to experience inversion of carriers and emit spontaneous and stimulated emission light near a desirable signal wavelength. The pump typically is another light source whose wavelength is preferentially absorbed by the ions. The pump and the signal must be combined, typically by a wavelength-selective coupler, and may co- or counterpropagate with respect to each other inside the doped length of fiber. Therefore, light absorbed by the doped fiber at the pump wavelength will produce gain for a signal at a different wavelength. Because the transmission and the active medium are both fiber based, the insertion losses are minimal.

Both types of amplifiers are susceptible to external reflections that adversely affect the stimulated and spontaneous emission rates and the frequency selectivity of the cavity. As a result, an optical isolator, which permits light to pass in one direction only and prevents reflections back into the amplifier, typically is required for both amplifiers, although they may not be needed under all circumstances.

We will first address the fundamentals of SOAs. Subsequently, we will discuss the basics of fiber amplifiers and differences between the two main types of amplifiers.

5.1.3.2 Narrowband Versus Wideband Amplifiers

Before discussing the principles of amplification, it is important to further classify amplifiers into narrowband and wideband categories, depending on the wavelength range for which gain is provided uniformly to an input signal. Narrowband amplifiers have a bandwidth < 0.1 nm, and wideband amplifiers can perform adequately over as much as tens of nanometers. Narrowband amplifiers contain an active medium that has frequency-selective feedback in a cavity. Such feedback can be of the *Fabry-Perot* (FP) etalon [17] or DFB (distributed feedback) grating variety [2]. Another narrowband fiber amplifier involves *stimulated Brillouin scattering* (SBS) [18], which is a nonlinear process derived from the nonlinearity of the fiber medium. SBS produces narrowband gain from a pump wave to a signal.

Wideband amplifiers contain an active medium that has a broad gain spectrum and that has nearly all the frequency-selective reflections suppressed. Such an amplifier is commonly known as a *traveling-wave* (TW) amplifier, in which a wave simply experiences amplification as it traverses the gain medium from the input to the output [9]. In addition to SOAs and EDFAs, the wideband amplifier category also includes fiber amplifiers based on another nonlinear amplification phenomenon due to *stimulated Raman scattering* (SRS) [17]. As with SBS, SRS transfers energy from a pump wave to a signal wave as they exist side by side in the fiber.

This chapter will deal mainly with TW amplifiers, with some early discussion of FP amplifiers. We will address SBS and SRS amplifiers near the end of the chapter, but only briefly, because their usefulness in optical communication systems is limited given the attractive qualities of fiber and semiconductor TW amplifiers.

5.2 SEMICONDUCTOR AMPLIFIERS

We will derive many basic amplifier characteristics within the context of semiconductor amplifiers. Where there is a discrepancy between SOAs and EDFAs, the analysis will be repeated later for the fiber amplifiers.

5.2.1 Rate Equations and Gain

The two most important quantities of optical amplifiers are the gain and the noise, since these directly affect the SNR (signal-to-noise ratio) of the receiver. We will describe the gain here and address the noise in a later section.

The semiconductor active medium is a rectangular waveguide that provides gain to an optical signal propagating through the amplifier [19]. Signal gain occurs when carriers are excited from the valence to the conduction band in this quasi-two-level energy system and when an external signal initiates stimulated emission of these conduction band carriers back down to the valence band in this two-level system. We can approximate the unsaturated (i.e., enough excited carriers exist to provide gain to a small signal) spectral gain coefficient, $g(\omega)$, of the active medium to be homogeneously broadened and to have a Lorentzian lineshape (Fig. 5.5), as given by

$$g(\omega) = \frac{g_o}{1 + (\omega - \omega_o)^2 \tau_R^2} \tag{5.1}$$

where g_0 is the maximum gain coefficient, τ_R is the dipole relaxation time of the carriers, and ω_0 is the optical frequency at which the gain is a maximum. (Note that in this chapter we will interchangeably use *frequency* and *wavelength,* since wavelength equals the speed of the light in a medium divided by the optical frequency.) This center wavelength is typically designed to be near either 1.3 or 1.55 μm, the two fiber-loss minima most useful for optical systems. The gain coefficient represents the likelihood of the occurrence of a spontaneous emission event. Such a spontaneous event means that a carrier makes a transition from the conduction band to the valence band, thus causing a photon to be emitted over a broad range of wavelengths. (It should be emphasized that the gain is really not Lorentzian-shaped, because the semiconductor is not, strictly speaking, a two-level system. In fact, the conduction and valence bands both have a distribution of carriers over a wide energy range.) For comparison, see the real measured gain spectrum in Figure 5.4. This real spectrum is asymmetric and difficult to describe analytically [20]. There are empirical formulae that

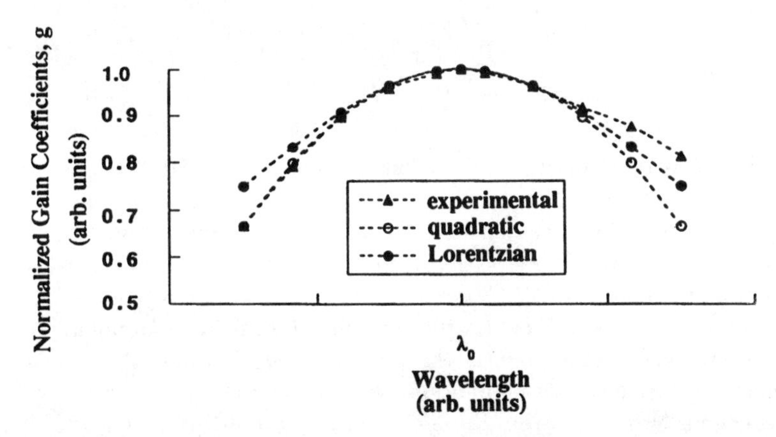

Figure 5.5 Lorentzian-shaped and inverse-quadratic-shaped gain coefficient, $g(\omega)$, in an SOA. Also shown is a measured gain coefficient curve.

describe the central portion of the curve well, such as the following quadratic approximation [21,22]:

$$g = a(n - n_0) - \gamma(\lambda - \lambda_0)^2 \tag{5.2}$$

in which n_0 is the carrier density required to achieve transparency, and a and γ are fitting constants. λ_0 is the gain peak wavelength and is dependent on carrier density according to the relation such that the peak wavelength shifts by an amount ($\Delta\lambda_0 = C_1\Delta n$) corresponding to a change in carrier density. A full comparison, including the possibility of a cubic formula, is left as a problem at the end of the chapter.

In general, the same gain coefficient in (5.1) is a function of the carrier population inversion from the valence to the conduction band, as described by [22]

$$g = \gamma(J - J_0) = \delta(n_e - n_0) \tag{5.3}$$

where γ and δ are the differential-gain constants $\Delta g/\Delta J$ and $\Delta g/\Delta n$, respectively, J is the current density in the active area, n_e is the carrier population in the excited state, and J_0 and n_0 represent quantities to achieve transparency (i.e., gain equals medium losses). The rate equation (5.4) and the spatial equation (5.5) that completely describe this quasi-two-level system are

$$\frac{dn_e}{dt} = \frac{J}{q} - \frac{CT}{(c/n)}(n_e - n_o)\frac{P}{h\nu} - \frac{n_e}{\tau_s} \tag{5.4}$$

and

$$\frac{dP}{dz} = \frac{C\Gamma}{(c/n)}(n_e - n_o)P \tag{5.5}$$

where C is a constant of the semiconductor (typically $\sim 10^{-6}$ cm^3/s for 1.3- or 1.55-μm SOAs), (c/n) is the speed of light in a medium of refractive index n, P is the optical signal power, $h\nu$ is the energy of a photon at frequency ν, τ_s is the carrier (spontaneous emission) lifetime, q is the charge of an electron, and Γ is the overlap fill factor (i.e., the confinement factor) between the active area (containing the electrical carriers) and the optical field, which accounts for the fact that not all the photons interact with the active gain material. Essentially, (5.4) describes the time-dependent change in the electrical carrier population from current injection, stimulated emission, and spontaneous emission. From those two rate equations, we arrive at a descriptive equation that contains the maximum amount of optical power or saturated optical intensity, P_{sat}, that can be output from the active medium:

$$\frac{P_{sat}}{h\nu}\ln(G_o) = \Gamma L\left(\frac{J}{q} - \frac{n_o}{\tau_s}\right) = \frac{P_{sat}}{h\nu}\ln(G) + \frac{P_{in}}{h\nu}(G-1) \qquad (5.6)$$

where

$$P_{sat} = \frac{h\nu(c/n)}{\tau_s C\Gamma} \qquad [W/m^2] \qquad (5.7)$$

and τ_s is the spontaneous emission lifetime, L is the length of the active medium, P_{in} is the input optical intensity (magnitude of the square of the field), as an external optical field traverses the amplifier from left to right, and $G = (P_{out}/P_{in})$ represents the total amplifier gain. Note that $G = G_0$ when no input signal is present.

Many communication functions are directly affected by the value of P_{sat}, and so we generally desire a high value. Note that there are two types of gain in (5.6), the total gain, G, and the unsaturated small-signal gain, G_0. The unsaturated small-signal gain represents the gain that an amplifier can theoretically provide to a small signal and that does not consume all the excited carriers available in the amplifier. However, the actual gain provided by the SOA depends on the input-signal intensity. This gain can be much lower if the inverted gain-producing carriers become too depleted to amplify the entire input signal, resulting in amplifier saturation. Rearranging (5.6), we obtain the general expression for the unsaturated single-pass gain, G_0, through the amplifier medium:

$$G_o = \exp\left[\frac{L\Gamma C\tau_s}{v}\left(\frac{J}{q} - \frac{n_o}{\tau_s}\right)\right] = \exp\left[\frac{Lh\nu}{P_{sat}}\left(\frac{J}{q} - \frac{n_0}{\tau_s}\right)\right] \qquad (5.8)$$

These equations provide the basic gain expression but ignore the fact that there may be reflections at the right and left boundaries of the amplifier due to either an index-of-refraction mismatch or a reflective grating. These reflections have a profound impact on the gain achievable from the amplifier. We can determine the effect reflections have on the amplifier gain by considering the wave equations of the optical fields inside and outside the amplifier. As shown in Figure 5.6, the externally input signal field, E_{in}, enters the amplifier, propagates from left to right, and exits the amplifier as the output signal field, E_{out}.

The right, $A(z)$, and left, $B(z)$, propagating waves are represented by [23]

$$A(z) = A_o \exp\left(-j\beta z + \frac{gz}{2}\right) \qquad (5.9)$$

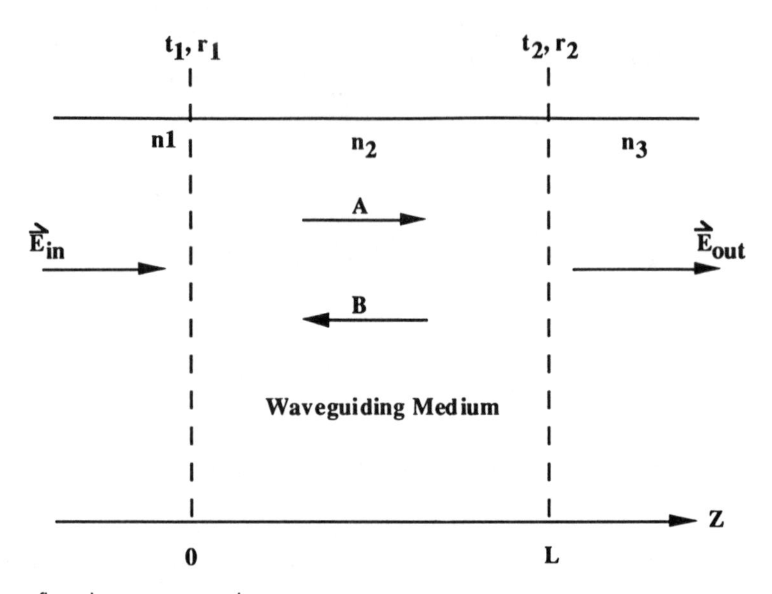

Figure 5.6 Reflected waves in a medium.

$$B(z) = B_o \exp\left(+j\beta z + \frac{gz}{2}\right) \tag{5.10}$$

where β is the propagation constant of the wave. The boundary conditions for continuity of the fields at an interface, which must be satisfied, are

$$t_1 E_{in} + r_1 B(z = 0) = A(z = 0) \tag{5.11}$$

$$B(z = L) = r_2\, A(z = L) \tag{5.12}$$

where r_1 and r_2 are the field reflection coefficients at the input ($n_1 \| n_2$) and output ($n_2 \| n_3$) boundaries, respectively, and t_1 and t_2 are the field transmission coefficients at the input and output boundaries, respectively. These boundary conditions represent the accounting for transmission and reflection by equating fields A and B at a given boundary and requiring continuity. Solving for the constants of A_0 and B_0 yields a relationship between E_{in} and E_{out} in terms of the boundary reflections at each semiconductor facet:

$$E_{out} = t_2 A(L) = \frac{E_{in} t_1 t_2 \exp\left(-j\beta L + \frac{gL}{2}\right)}{1 - r_1 r_2 \exp(-j2\beta L + gL)} \tag{5.13}$$

After some algebraic rearranging and assuming that the reflection coefficients are real, the power gain, G_r, resulting from traversing a gain medium containing facet reflectivities is then [7]

$$G_r = \frac{(1 - R_1)(1 - R_2)G_o}{1 + R_1 R_2 G_o^2 - 2G_o \sqrt{R_1 R_2} \cos(2\beta L)} \tag{5.14}$$

where

$$R_1 = |r_1|^2 = 1 - |t_1|^2 = \left| \frac{n_1 - n_2}{n_1 + n_2} \right|^2 \tag{5.15}$$

$$R_2 = |r_2|^2 = 1 - |t_2|^2 = \left| \frac{n_2 - n_3}{n_2 + n_3} \right|^2 \tag{5.16}$$

are the intensity reflectivity coefficients. G_0, the maximum unsaturated spectral gain, can be found by rearranging (5.8) or in terms of the gain coefficient of (5.1):

$$G_0 = \exp(g_0 L) \tag{5.17}$$

Returning to (5.14), we can rearrange the gain expression into a more useful form by noting that a simple wave relation exists:

$$\beta L = \frac{(\omega - \omega_o)L}{(c/n)} \tag{5.18}$$

A more manageable frequency-dependent gain expression than (5.14) can then be derived:

$$G_r = \frac{(1 - R_1)(1 - R_2)G_o}{(1 - G_o \sqrt{R_1 R_2})^2 + 4G_o \sqrt{R_1 R_2} \sin^2 \left[\frac{(\omega - \omega_o)L}{(c/n)} \right]} \tag{5.19}$$

The denominator in (5.19) is periodic, producing periodic FP resonances in the gain spectrum. This equation clearly illustrates the difference between two reflection-dependent types of amplifiers. If the reflections are suppressed and $R_1 = R_2 = 0$, then this is a wideband TW amplifier, also known as a one-pass amplifier. If there are reflections and $R_1 = R_2 > 0$, then this is a narrowband FP amplifier with narrow FP ripples in the wide-gain spectrum. The signal travels multiple times between the two facets in the FP amplifier. The gain is plotted in Figure 5.7 for $R_1 = R_2 > 0$ and $R_1 = R_2 \sim 0$. The maximum TW gain is still G_0, and the maximum FP gain, G_r^{max}, is

$$G_r^{max} = \frac{(1 - R_1)(1 - R_2)G_o}{(1 - G_o\sqrt{R_1R_2})^2} \tag{5.20}$$

(Note that G_0 would be replaced by the actual gain, G, in (5.14) to (5.20) if the amplifier is saturated, as will be discussed soon.) The maximum FP gain is clearly larger than the maximum TW single-pass gain as long as the signal wavelength is at one of the FP resonances of the cavity. The crossover point between considering an amplifier an FP or a TW amplifier is the reflectivity that will produce a 3-dB sinusoidal ripple in the gain. Section 5.2.2 will discuss methods to reduce the reflections to achieve TW characteristics.

Because FP amplifiers are narrowband, they are not considered practical as TW amplifiers for implementation into optical systems. In fact, TW amplifiers can provide high gain for signals covering many possible wavelengths covering tens of nm. As such, the discussion in this chapter will now deal exclusively with the characteristics of TW amplifiers.

We discussed the unsaturated gain in (5.8) without defining specifically what is meant by saturation. Essentially, an incoming small optical signal can undergo a

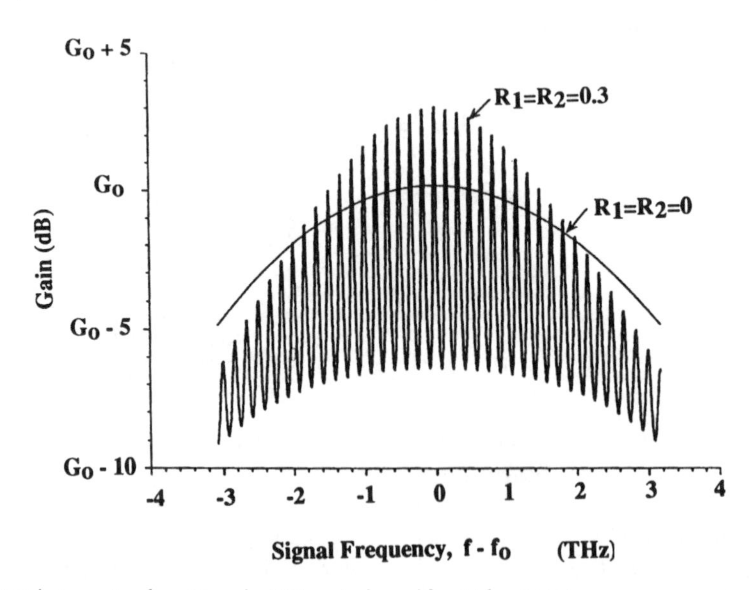

Figure 5.7 Relative gain of an FP and a TW optical amplifier (*After:* [24].)

certain amount of gain by initiating stimulated emission of the inverted carriers. However, if an incoming signal is so large that there simply are an insufficient number of carriers in the inverted state to allow stimulated emission to occur for all the incoming photons, then the total gain for this intense signal will be less than in the small-signal case. For intense signals, the amplifier gain is diminished and the amplifier itself is considered saturated. Another way of thinking about saturation is that the pump can only provide a maximum number of excited carriers; therefore, an incoming large signal cannot be amplified to the point where more power is output from the amplifier than was initially provided by the pump source. Furthermore, the saturation input (or output) power is usually considered that input (or output) power that will produce a reduction in the small-signal gain by 3 dB. This is an important amplifier figure of merit, since we desire (1) the amplifier to operate uniformly over a wide dynamic range and (2) to provide as large a "boost" to the signal if the configuration is as a power amplifier. The true gain of a TW amplifier under saturated and unsaturated conditions can be expressed as a transcendental equation in several forms. By modifying (5.1) to include gain saturation, the expression for signal power along the length of the amplifier is [4]

$$\left.\frac{dP}{dz}\right|_{\omega=\omega_o} = gP = \frac{g_o P}{(1 + P/P_{sat})} \tag{5.21}$$

and the resulting equivalent transcendental equations are

$$G = G_o \exp\left(-\frac{(G-1)}{G}\frac{P_{out}}{P_{sat}}\right) = \exp(gz) \tag{5.22}$$

$$\frac{P_{in}}{P_{sat}} = \frac{\ln(G_o/G)}{G-1} \tag{5.23}$$

where g is the gain coefficient, including saturation. Figure 5.8 shows how the gain is reduced for an increase in the input signal power for a given initial gain and P_{sat}.

Example 5.1

An optical amplifier has the following properties: g = 250 dB/cm, L = 400 μm, and P_{sat} = +1 dBm. Find G_0. If an input signal results in a saturated gain of 10 dB, what is the input signal power? What is the total small-signal gain if the loss in the amplifier is 50 dB/cm? If the right and left reflectivities are −20 dB, what is G^{max} and G^{min} near the center of the gain peak?

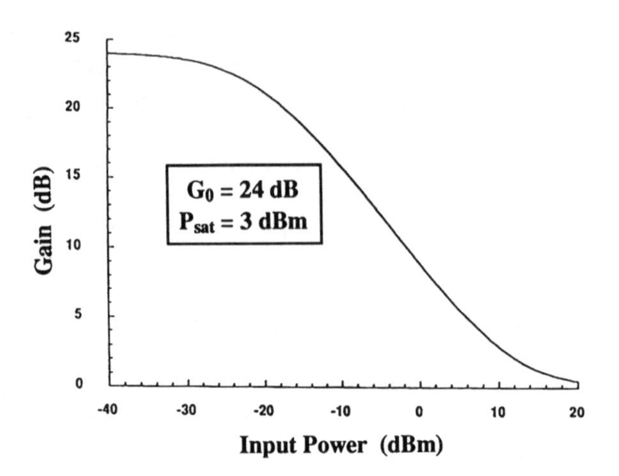

Figure 5.8 TW-amplifier gain versus input signal power demonstrating the effects of gain saturation.

Solution: The total gain is $G_0 = \exp(g_0 L) = 22026.466 = 43.43$ dB. The input power that compresses the gain to 33.43 dB is derived from the following equation:

$$\frac{P_{in}}{P_{sat}} = \frac{\ln(G_o/G)}{G - 1}$$

Rearranging this equation results in

$$P_{in} = 1.077 \text{ mW} = 0.32 \text{ dBm}$$

If the loss = 50 dB/cm, then

$$G_0 = \exp[(g_0 - \alpha)L] = 2980.96 = 34.74 \text{ dB}$$

To find G^{max} and G^{min}, we use (5.19). By inserting $\sin^2$ (*expression*) = 0 and $\sin^2$ (*expression*) = 1 for *max* and *min*, respectively, we have:

$$G_r^{max} = \frac{(1 - R_1)(1 - R_2)G_o}{(1 - G_o\sqrt{R_1R_2})^2} = 12.1 = 10.8 \text{ dB}$$

$$G_r^{min} = \frac{(1 - R_1)(1 - R_2)G_o}{(1 - G_o\sqrt{R_1R_2})^2 + 4G_o\sqrt{R_1R_2}} = 8.1 = 9.08 \text{ dB}$$

5.2.2 Correcting for Facet Reflections

We have established that it is advantageous for optical communication systems to employ TW and not FP amplifiers. As such, we must reduce or eliminate the reflections at the amplifier boundaries. In a semiconductor, these reflections occur at the facets. One method to reduce the reflections is simply to cover the facets with *antireflection* (AR) coatings. Such coatings can be understood by considering the waves at the input (or output) facet (Fig. 5.9). The various transmitted and reflected waves can be equated at the boundaries to ensure continuity [23]:

$$E_{refl} = r_1 E_{in} + t_1 B(z = 0) = 0 \tag{5.24}$$

The following two conditions must be met for the reflected wave to be negligible and for the two counterpropagating waves within the coating layer to cancel each other by destructive interference (180 degrees out-of-phase):

$$n_2^2 = n_1 n_3 \tag{5.25}$$

and

$$\frac{L_{AR}}{(\lambda/n_2)} = \frac{1}{4} + \frac{k}{2} \qquad k = 1, 2, \ldots \tag{5.26}$$

where k is an integer, and L_{AR} is the thickness of the AR coating. The second condition is commonly referred to as a quarter-wave layer. This AR coating is bidirectional and thus identical for both facets.

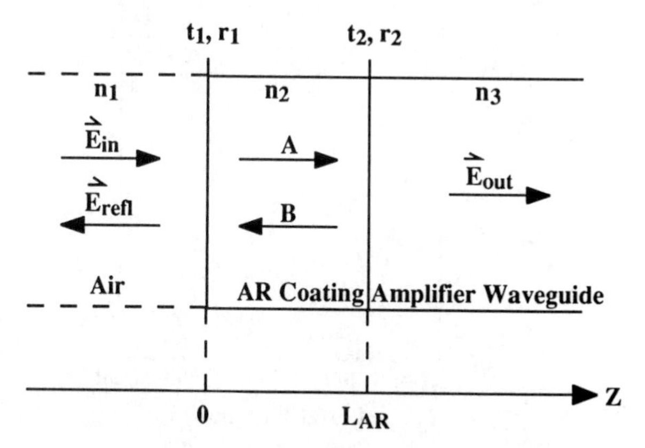

Figure 5.9 Transmitted and reflected waves propagating through an antireflection coating.

Using an AR coating can substantially reduce the reflections at the boundaries back into the FP cavity to a value $\sim 10^{-4}$. However, two polarizations propagate inside the semiconductor amplifier: TE and TM waves. The two polarizations experience different reflection coefficients at the AR-coated facets even at normal incidence. Therefore, the AR coatings reduce one polarization reflection more than the other, which is one reason for the polarization-dependent gain characteristics of an SOA; this dependence can be significantly reduced by a modified double-layer AR coating [24]. (Note that the gain coefficient inside the active medium is also polarization dependent; we will address this issue later.) Figure 5.10 is a typical reflectivity curve showing the wavelength and polarization dependence of a single-layer AR coating [25].

Because efficient AR coatings are difficult to fabricate and never reach a value of $R = 0$, two other methods commonly employed to reduce reflections (usually in conjunction with the AR coating) are depicted in Figure 5.11. The first method is simply to angle the guiding active region in relation to the facet normal [26]. Under that scenario, the optical reflection occurring as the wave hits the semiconductor-air interface on exiting the cavity will reflect back into the waveguide at the designed angle. If that angle is large enough, that is, greater than the critical angle necessary for achieving total internal reflection on reflection back into the guiding cavity, then the reflection will not be coupled into the amplifier waveguide and thus not affect the gain characteristics. The second method incorporates a buried passive nonguiding window region between the end of the central active layer and each of the facets [27]. As the wave exits the guiding active region, it reaches a nonguiding semiconductor region of the same index of refraction. Therefore, no reflection occurs at the interface,

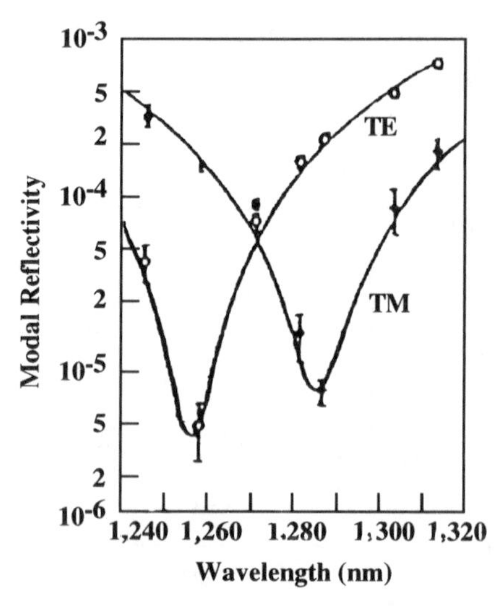

Figure 5.10 Reflectivity of an AR coating, showing dependence on wavelength and polarization (*After:* [24].)

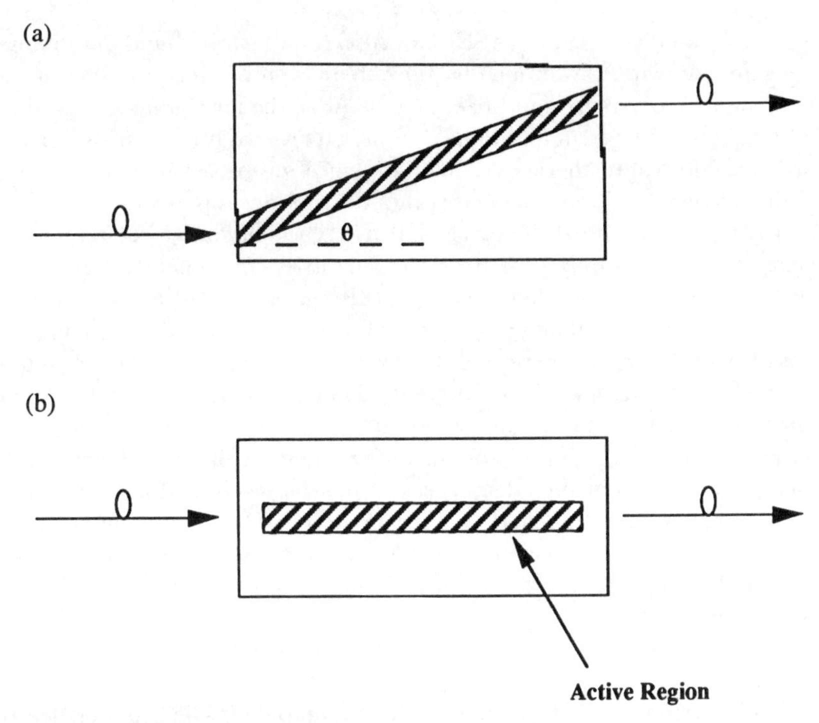

Figure 5.11 Two methods for reducing reflections in an optical amplifier cavity: (a) Angled waveguide and (b) buried passive nonguiding window region.

and the beam diverges. The divergence continues until the beam reaches a facet, at which point a reflection occurs. Since there is still no guiding, the reflected wave continues to diverge, eventually coupling very little light back into the guiding active region.

5.2.3 Amplifier Noise and SNR

In addition to gain, understanding the effect noise has on the implementation of optical amplifiers in optical communication systems is of paramount importance. We will now discuss the generation of the amplifier noise and its effect on receiver sensitivity.

The noise in an amplifier is inherently due to the random incoherent spontaneous emission events of excited carriers. (Recall that the stimulated emission that is coherent with an input field was responsible for the signal gain.) Each spontaneously decaying carrier can radiate in any solid angle. The fraction of the spontaneous emission that is emitted within the critical angle, allowing it to be coupled into the optically guiding region, will itself cause further stimulated emission. Therefore, the spontaneously emitted light that gets coupled into the beam propagation path is subsequently amplified and in total is called the ASE. This ASE is quite broadband, occurring over the entire gain bandwidth. Additionally, since there is only a finite number of excited carriers, the more carriers being utilized for the ASE, the fewer that are available to provide signal gain. This is an additional reason why we want to suppress forward or backward reflections into the active medium, since any reflected nondesired wave depletes the available gain and increases the ASE noise component.

A fairly typical amplified channel is shown in block diagram form in Figure 5.12. The signal, P_{sig}, may initially pass through some lossy components, with a lumped insertion loss of L_{in}. The amplifier then provides gain, G, and adds noise, having variance σ^2. Since the modulated laser-generated signal is typically single mode (and narrowband) and the ASE spectrum of the TW amplifier is quite broad (tens of nm), there will probably be an optical filter present, of bandwidth B_0, to pass the signal and block much of the noise. We can also lump any insertion losses at the output of the amplifier in one term, L_{out}. (We will not include L_{in} and L_{out} in the following analysis; the reader should note that the effect as seen from the receiver of L_{in} is to attenuate the optical signal power, P_{sig}, and L_{out} attenuates both P_{sig} and the spontaneous emission power, P_{sp}.) The optical detector has a characteristic resistance, Ω_r, and the entire receiver has an electrical bandwidth, B_e, to pass the modulated BB (baseband) signal and again block higher-frequency noise terms. Although we will discuss this point later, note that the signal is in one polarization but the ASE occurs in both polarizations.

We begin by finding the ASE generated by the amplifier. The equation describing the light power, P, in a solid angle that is guided in the active medium is [2]

$$\frac{dP}{dz} = gP + \left(\frac{N_2}{N_2 - N_1}\right) g\, h\nu\, (\Delta\nu) \tag{5.27}$$

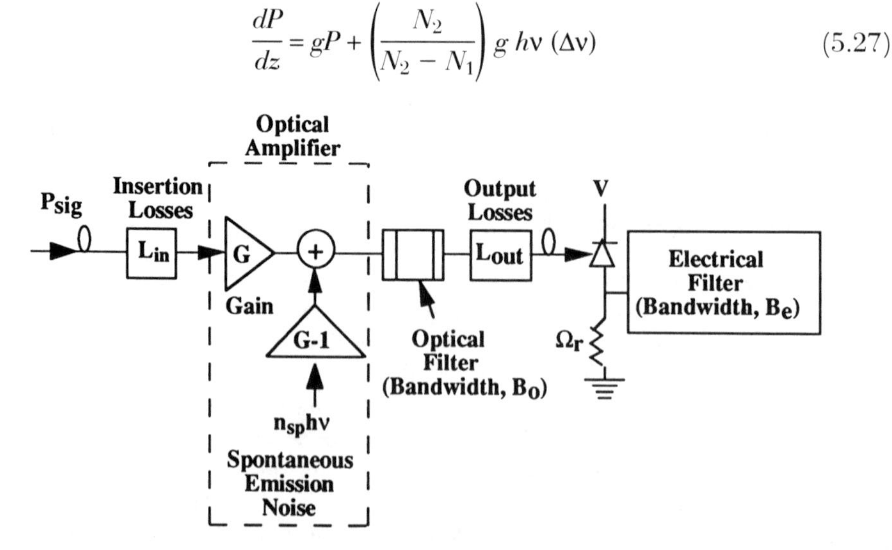

Figure 5.12 Block diagram of a signal passing through a typical optical amplifier and then being detected.

where the first term describes the signal, the second term describes the noise, z is the length along the amplifier, N_2 is the population in the carrier conduction band (higher energy level), N_1 is the carrier population in the valence band (lower energy state), $h\nu$ is the photon energy, and $(\Delta\nu)$ is the optical bandwidth of the active medium. Solving (5.27) for P given $P(0) = P_{sig}$ (that is, input signal power) reduces to

$$P(z) = P_{sig}\, e^{gz} + n_{sp} h\nu\, (\Delta\nu)\, (e^{gz} - 1) \tag{5.28}$$

for which we have defined the spontaneous emission factor, n_{sp}, as a measure of the carrier population inversion in terms of [28,29]

$$n_{sp} = \frac{N_2}{N_2 - N_1} \tag{5.29}$$

in which the carrier population in the lower energy level is N_1 and the population in the excited energy level is N_2. The higher the n_{sp}, the more ASE noise is generated in the amplifier. Based on the definition, the quantum-limited minimum for n_{sp} in any amplifier is equal to 1. Note that the first term in (5.28) describes the total signal and the second term describes the total noise (in one polarization) exiting an amplifier of length z. This second term is more clearly understood when defining the total ASE power, P_{sp}, over the gain bandwidth $(\Delta\nu)$ in one polarization as being

$$P_{sp} = n_{sp}\, (G - 1)\, h\nu\, (\Delta\nu) \tag{5.30}$$

If we were to consider both polarizations, a simple factor of 2 is added before n_{sp} in (5.30). We approximate the spectral density of the ASE power to be nearly uniform over the entire bandwidth and equal to $n_{sp}\ (G - 1)h\nu$. Thus, the optical SNR emanating from an individual amplifier is

$$SNR = \frac{P_{sig} G}{n_{sp}\, h\nu(\Delta\nu)(G - 1)} \tag{5.31}$$

Note that we have included only the ASE noise from one polarization, because, as we will see later, only the noise that is in the same polarization as the signal will affect the ultimate system sensitivity.

5.2.4 Electrically Equivalent SNR

We have thus far derived the spontaneously generated optical noise power in the amplifier. However, we are more interested in the electrical noise that is ultimately generated in the optical detector and that governs the overall sensitivity of the system [30]. Two noise terms common to all detectors are the shot noise and the thermal noise. The shot noise, having a variance σ_{sh}^2, excluding the dark current, is [31]

$$\sigma_{sh}^2 = 2\eta q\left(\frac{q}{h\nu}\right)(GP_{sig} + P_{sp})B_e \tag{5.32}$$

where q is the electric charge constant, and η is the detector responsivity or quantum efficiency (typically close to a value of 1 and which will be dropped henceforth from the analysis). The term $(q/h\nu)$ (A/W) converts optical power into electrical photocurrent. The shot noise is essentially a quantum-mechanical phenomenon due to the generating of electrons at random times and is thus proportional to the incoming optical power. The electrical thermal noise, having variance σ_{th}^2, is [31]

$$\sigma_{th}^2 = \frac{4k\,T\,B_e}{\Omega_r} \tag{5.33}$$

where k is Boltzmann's constant, Ω_r is the resistance of the detector, and T is the receiver temperature. The thermal noise is due to the thermal generation of carriers in the semiconductor detector or in any resistive element in the receiver circuitry, even in the absence of incoming light. In unamplified systems, it is typically the thermal noise that dominates the electrical SNR.

The incoherent ASE generated in the amplifier is very broadband and can be considered as white noise. The noise components due to the ASE impinging on the detector is very dependent on the optical and electrical bandwidths of the corresponding filters. The spectral shape of these filters can take many forms. The most common optical filter shape is a Lorentzian-shaped FP filter [32,33]. The most common electrical filter shapes are single-pole or Nyquist functions [32]. However, for this analysis, we can approximate the filter shapes to be uniform with a certain bandwidth, resembling a rectangular form, which allows easy overlapping calculations with the approximately rectangular shape of the ASE spectrum.

We can think of the broadband ASE as being composed of an infinite number of incoherent thin spectral slices, each containing some ASE power. This helps us understand the ASE-generated noise. Because the detector is inherently a square-law device that responds to the square of the incoming optical field, a "beat" term is produced if two different signals are incident. We represent these two arbitrary field waves as $A(\omega_1)$ and $B(\omega_2)$, each at a specific frequency. Specifically, the field would be represented by

$$A(\omega_1) = A(t)\cos(\omega_1 t + \phi) \tag{5.34}$$

where A is the amplitude and can even be a time-varying data function, ω_1 is the optical frequency, and ϕ is the phase, which we will drop for simplicity. When $A(\omega_1)$ and $B(\omega_2)$ are summed and squared by the detector, the resulting photocurrent is

$$[A(\omega_1) + B(\omega_2)]^2 = A^2(\omega_1) + B^2(\omega_2) + 2A(\omega_1)B(\omega_2)$$

$$= A^2(\omega_1) + B^2(\omega_2) + A(t)\,B(t)\,\{\cos[(\omega_1 + \omega_2)t] + \cos[(\omega_1 - \omega_2)t]\} \quad (5.35)$$

The $2AB$ term is the beat term between A and B. Because the frequencies in (5.34) and (5.35) are ultra-high optical frequencies (THz) undetectable by the photodetector, the sum frequency term will not be detected and only the difference frequency will appear at the electrical output.

In a simple sense, our amplified signal impinging on a detector has two waves as well, namely, the signal and the ASE noise. Therefore, based on (5.35), A^2 represents the signal power, B^2 represents the ASE noise power, and $2AB$ represents the signal-spontaneous electrical beat noise. One would expect signal-spontaneous beat noise to produce a beat term at a given frequency. However, the situation is more complicated because the ASE does not exist at a specific wavelength but is broadband consisting of an infinite number of waves, each at a different frequency within the gain spectrum. Therefore, we must integrate over the entire ASE noise passing through the optical and electrical filters and beat (i.e., multiply) each thin slice of ASE with the approximately single-frequency signal term. The resulting signal-spontaneous beat noise, $\sigma^2_{sig\text{-}sp}$, that falls within the optical filter and electrical detector bandwidths is then [28]

$$\sigma^2_{sig\text{-}sp} = 4q\left(\frac{q}{h\nu}\right)P_{sig}\,G(G-1)\,n_{sp}\,B_e \quad (5.36)$$

Note that only one polarization of the ASE is considered here, since beating only occurs in the same polarization as the signal. The B^2, or ASE noise power, term also must be evaluated since it is not at a single frequency and also produces beat terms between one part of the ASE spectrum and another. After integration and convolution and considering only the noise passing through the various filters, the spontaneous-spontaneous electrical beat noise, $\sigma^2_{sp\text{-}sp}$, is [28]

$$\sigma^2_{sp\text{-}sp} = 4q^2\,(G-1)^2\,n^2_{sp}\,B_e\,B_o \quad (5.37)$$

Here each separate polarization of the ASE noise beats with itself, producing twice as much noise as a single polarization. Both the ASE beating-noise terms depend critically on filtering. However, since the optical and electrical filters must be at least as wide as the modulated signal bandwidth for undistorted transmission of the signal, the ASE that produces the noise must also pass through the filters at the same spectral position as the signal itself. In other words, $\sigma^2_{sig\text{-}sp}$ and $\sigma^2_{sp\text{-}sp}$ can be reduced but not eliminated, since some ASE must pass through the filter within the same bandwidth as the signal. Additionally, the majority of the signal-spontaneous beat noise is generated from the small frequency portion (within a few gigahertz) of the ASE immediately surrounding the signal. This happens because the ASE farther away in wavelength will inevitably produce electrical beat terms with such high frequencies (for example, when $\omega_1 - \omega_2$ is large) so as to be outside the electrical bandwidth of the receiver. These noise terms are the only ones present in an IM-DD system; a coherent system, which we do not concentrate on here but which is discussed at the end of this section, would additionally contain some local oscillator-generated beat noise terms and an enhanced shot noise component.

We are now ready to define the main figure of merit for a typical IM-DD system, that being the detected electrical SNR. Recall from Chapter 1 that the SNR is typically thought of as the signal mean squared divided by the sum of the noise terms squared (i.e., noise variances). Therefore,

$$SNR = \frac{(GP_{sig})^2}{\sigma^2_{sh} + \sigma^2_{th} + \sigma^2_{sig\text{-}sp} + \sigma^2_{sp\text{-}sp}} \tag{5.38}$$

We can simplify this expression by making three generally valid assumptions: (1) The shot noise is small in comparison to all other terms; (2) the receiver noise will be dominated by the ASE noise and not by the thermal noise (typically true for systems employing a pre-amplifier); and (3) $G >> 1$. After making the appropriate substitutions, the SNR can be approximated as

$$SNR = \frac{P^2_{sig}}{[2P_{sig}\, n_{sp}\, h\nu + 2n^2_{sp}\, (h\nu)^2\, B_o]\; 2\, B_e} \tag{5.39}$$

Given all our assumptions, the SNR does not change much with an increase in gain! Our approximation that the receiver noise is dominated by the ASE noise and relatively unaffected by thermal noise may not be true for systems employing only power amplifiers. In such a scenario, the ASE noise is significantly attenuated before it reaches the detector and may be smaller than the receiver thermal noise. Determining the SNR gives us the ability to evaluate the system performance in terms of the BER (bit error rate) and system power penalties.

Another important parameter is the NF (*noise figure*) of an optical amplifier, which is defined by

$$NF = \frac{SNR_{in}}{SNR_{out}} \tag{5.40}$$

where SNR_{in} is the electrically equivalent SNR of the optical wave going into the amplifier if it were to be detected, and SNR_{out} is the electrically equivalent SNR of the optical wave coming out of the amplifier (see Figure 5.13 for the detection circuit). These quantum-limited quantities (the absolute maximum possible SNRs given a large P_{sig} and $G >> 1$) are

$$SNR_{in} = \frac{P_{sig}}{2q\,B_e} \tag{5.41}$$

which is dominated by the shot noise, and

$$SNR_{out} = \frac{P_{sig}}{4n_{sp}\,q\,B_e} \tag{5.42}$$

which is dominated by the signal–spontaneous beat noise. These formulae give rise to an expression for NF, assuming a large enough signal power and gain, of

$$NF \cong 2n_{sp} \tag{5.43}$$

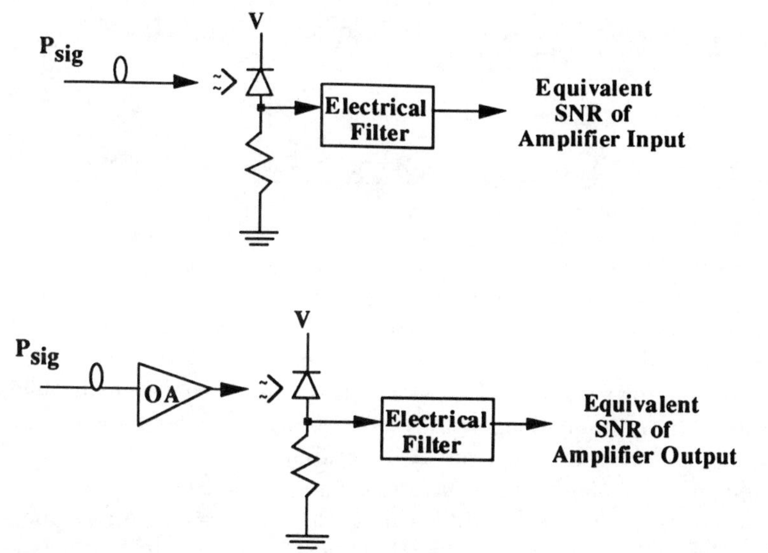

Figure 5.13 Schematic of detection circuits for deriving the input and output equivalent electrical SNRs of an optical wave into and out of an optical amplifier.

Because the minimum n_{sp} is 1 for complete inversion, the quantum-limited NF for an amplifier given the approximations above is 2 or 3 dB. Essentially, the noise must be increased by at least 3 dB, but the system performance is enhanced significantly because the signal has undergone enormous gain to overcome the losses. The typical NF for a semiconductor amplifier is 6–8 dB [4,34].

Now that we have the SNR, we can easily extend our analysis to include the calculation of the probability of error, or BER, for a random digital bit stream [35,36]. We define the Q parameter as [28]

$$Q = \frac{\sqrt{S(1)} - \sqrt{S(0)}}{\sqrt{N(1)} + \sqrt{N(0)}} \tag{5.44}$$

in which $S(1)$ and $S(0)$ are the signal powers, and $N(1)$ and $N(0)$ are the noise powers for the 1 and 0 bits, respectively. For the case of optical amplification in which $G >> 1$ and the ASE-generated noise dominates the thermal noise, Q can be further defined as [37]

$$Q = \frac{2\sqrt{\dfrac{B_o}{B_e}}}{\sqrt{4\left(\dfrac{P_{ASE}}{P_S}\right) + \left(\dfrac{P_{ASE}}{P_S}\right)^2 + \left(\dfrac{P_{ASE}}{P_S}\right)}} \tag{5.45}$$

As with any direct detection system, the BER can be found directly [28]:

$$BER = \frac{1}{\sqrt{2\pi}} \frac{\exp\left(\dfrac{-Q^2}{2}\right)}{Q} \tag{5.46}$$

Based on this analysis, we emphasize again that an amplified system (excluding a postamplifier configuration) should be operated so that the receiver SNR is spontaneous noise limited and not thermal-noise limited. We want to increase the signal gain to the point where an increase in the signal gain will also increase proportionally the signal-spontaneous beat noise, which then provides us with the highest SNR possible for our system. As long as the system is thermal-noise limited (i.e., constant noise term), we haven't reached our SNR maximum.

Another interesting point is the importance of the coupling losses at the semiconductor facets. If there is a 3-dB loss at each facet, system performance is highly dependent on whether the loss occurs at the input or output facet. A loss at the input facet degrades the signal, whereas loss at the output facet degrades both the signal and the spontaneous emission noise. The relative differences will depend on the dominating noise factor in the receiver and is left as a problem at the end of the chapter.

Based on the equations, it is also obvious that the optical and electrical bandwidths have a significant effect on the system SNR. For instance, if the signal-spontaneous beat noise is dominant, then limiting the optical bandwidth will not help the SNR much, since the signal-spontaneous noise is generated within a very close bandwidth to the signal itself. It is left as a problem at the end of the chapter to determine the relative dependencies.

For the sake of completeness, we can easily and briefly describe how the preceding analysis would change for a coherent detection system. In a coherent heterodyne optical detector, a local oscillator is used to beat with the incoming signal and bring the beat term down to low frequencies, which can be filtered and detected (see Chapter 2 for more details on coherent systems). This local oscillator is typically very intense, and the signal–shot noise will dominate in unamplified systems. The additional noise terms when amplifiers are used are also present, with the addition of a local oscillator-ASE beat noise term, $\sigma^2_{lo\text{-}sp}$ [38]:

$$\sigma^2_{lo\text{-}sp} = 2\left(\frac{q}{h\nu}\right)^2 P_{sp}\, P_{LO}\, B_e \tag{5.47}$$

where P_{LO} is the power in the local oscillator. The resulting SNR, given a very intense local oscillator, will have the numerator (signal) equal to the input signal multiplied by both the gain and the P_{LO}, and the denominator (noise) dominated by $\sigma^2_{lo\text{-}sp}$

This analysis of noise is mostly valid for fiber-based amplifiers. Any slight discrepancies will be visited later. We will discuss later how the ASE noise and NF affect system performance given a cascade of in-line fiber amplifiers, a subject more appropriately connected to fiber amplifiers as opposed to semiconductor amplifiers.

Example 5.2

An optical amplifier is used as a preamplifier for a detector. The signal power = −38 dBm, the signal wavelength = 1.55 μm, n_{sp} = 1.4, the gain = 30 dB, the full optical bandwidth of the amplifier = 30 nm, and the detector electrical bandwidth = 2 GHz. Find P_{sp}, $\sigma^2_{s\text{-}sp}$, $\sigma^2_{sp\text{-}sp}$, and σ^2_{th} if an optical filter is placed between the amplifier and the detector: filter bandwidth = 0.3 nm, 2 nm, and ∞.

Solution: The signal power, in watts, is P_s = −38 dBm = 0.158 μW, and the gain in linear units is G = 30 dB = 1,000. Therefore, the total generated ASE power is

$$P_{sp} = n_{sp}\,(G-1)\,h\nu(\Delta\nu)$$

$$= 0.672 \text{ mW} \qquad (\Delta\nu = \infty)$$

$$= 0.0448 \text{ mW} \qquad (\Delta\nu = 2 \text{ nm})$$

$$= 6.72\ \mu\text{W} \qquad (\Delta\nu = 0.3 \text{ nm})$$

Assuming room temperature and a 50-Ω resistance, the thermal noise is the same for all three cases since it is a function of the detector only:

$$\sigma_{th}^2 = \frac{4\,k\,t\,B_e}{\Omega_r} = 6.624 \cdot 10^{-13}\,A^2$$

The signal-spontaneous beat noise is also the same for all cases, because the electronic bandwidth is much smaller than the electrical bandwidth:

$$\sigma_{sig-sp}^2 = 4q\left(\frac{q}{h\nu}\right) P_{sig}\,G\,(G-1)\,n_{sp}\,B_e = 3.54 \cdot 10^{-10}\,A^2$$

The spontaneous-spontaneous beat noise changes for the three cases are

$$\sigma_{sp-sp}^2 = 4q^2(G-1)^2\,n_{sp}^2 B_e B_o$$

$$= 1.5 \times 10^{-9}\ A^2 \quad (\Delta\nu = \infty)$$

$$= 1.0 \times 10^{-10}\ \mathrm{A}^2 \quad (\Delta\nu = 2\ \mathrm{nm})$$

$$= 1.5 \times 10^{-11}\ \mathrm{A}^2 \quad (\Delta\nu = 0.3\ \mathrm{nm})$$

For our cases, we will assume that the shot noise is negligible. The SNR at the detector, then, is

$$SNR = \frac{(GP_{sig})^2}{\sigma_{sh}^2 + \sigma_{th}^2 + \sigma_{sig-sp}^2 + \sigma_{sp-sp}^2}$$

$$= 8.10 \qquad (\Delta\nu = \infty)$$

$$= 54.8 \qquad (\Delta\nu = 2\ \mathrm{nm})$$

$$= 67.5 \qquad (\Delta\nu = 0.3\ \mathrm{nm})$$

The SNR is dominated by the spontaneous-spontaneous beat noise when no filter is used, and the signal-spontaneous beat noise dominates when a narrowband optical filter is used. The thermal noise is negligible.

5.2.5 Methods for Correcting the Effects of the Polarization-Dependent Semiconductor Gain Medium

Unlike fiber amplifiers, the semiconductor gain medium, being rectangular (not square) and having different crystal planes, has a polarization-dependent gain inherent in the medium. Furthermore, the confinement fill factor (Γ) of the two polarizations is different, further producing a difference in gain. This differential gain between the TE and TM waves can be as large as 8 dB. Operating an SOA with significant polarization dependence in the gain may require using unwanted polarization controllers to control the input signal polarization from the input fiber. There is recent work that attempts to minimize this differential by creating the cross-sectional active area to be nearly square. Thus, the TE and TM fill factors would be similar. The square shape could be obtained by either enlarging the active area thickness or reducing the active area width. Additionally, the AR coatings on the facets must be designed to have a similar response for both polarizations. A difference of <1 dB has been produced by combining these methods plus *multiple-quantum-well* (MQW) and strained-layer material [39].

Another type of systems-based solution is to pass the two polarizations through two different amplifiers. We will examine only one possible method, as depicted in Figure 5.14. When two identical amplifiers are placed in series and the second amplifier is rotated axially 90° in relation to the first, any wave passing through both amplifiers will experience the same combined gain for each polarization, assuming the second amplifier has not been saturated by the first and residual reflections do not couple the two cavities [40]. Other solutions exist but with different advantages and disadvantages.

Using semiconductor amplifiers in optical fiber systems requires attention to their polarization dependence, since the polarization of any wave along the fiber typically is unknown and can vary. However, if the semiconductor amplifier is integrated on the same chip following a fixed linear-polarization laser transmitter, then only a single polarization passes through the amplifier, and the polarization

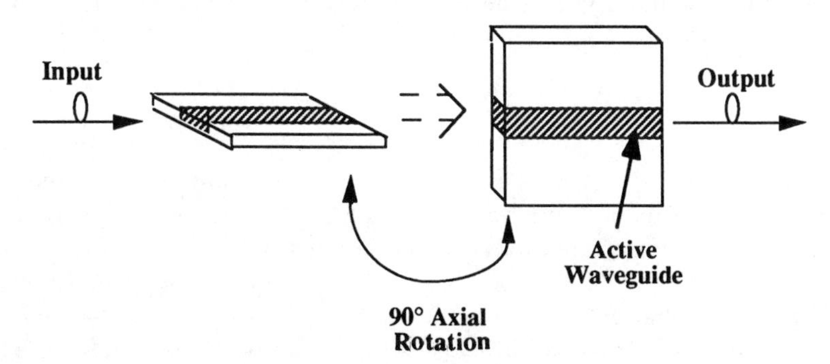

Figure 5.14 Two semiconductor amplifiers connected in series, with the second one axially rotated by 90 degrees in relation to the first.

dependence is not a problem. Note that, since the fiber-based amplifiers are circula and the wave in a fiber typically is circularly polarized, the gain medium and the signa gain are negligibly dependent on the polarization of the incoming signal.

5.2.6 Multichannel Intermodulation and Saturation-Induced Cross-talk

In multichannel systems, the ideal amplifier will provide gain equally to all channel over a broad wavelength range without any other effects. However, a few factors mak *wavelength-division-multiplexed* (WDM) multichannel systems more difficult to im plement than single-channel systems. These factors include nonuniform spectral gain intermodulation distortion, and saturation-induced cross-talk. We will not be con cerned with nonuniform gain in semiconductor amplifiers, since the broad gain profil can provide approximately uniform gain to many channels located near the center o the gain spectrum. On the other hand, fiber amplifiers are more prone to nonuniforn gain problems, which will discussed in Chapter 7. Intermodulation distortion an cross-talk both are nonlinear effects and will be the main focus of this section.

When two channels are incident in a closed amplifier system and their combine powers are near the amplifier saturation power, nonlinear effects occur that generat beat frequencies at the cross-product of the two optical carrier waves [see (5.35)]. Th carrier density (i.e., gain) will be modulated by the interference between any tw optical signals, and this modulation occurs at the various beat (i.e., sum and differ ence) frequencies generated by the possible combinations of input channels. Thi carrier density modulation at the beat frequencies produces additional signals that ca interfere with the original desired signals. Figure 5.15 illustrates this scenario for tw input signals that produce four output waves. Therefore, this nonlinear effect is calle four-wave mixing or, alternatively, *intermodulation distortion* (IMD) [41,42,43]. Th modulated gain will modulate both the amplitude and the phase of all signals and ca be described by the third-order optical susceptibility of the semiconductor crystal.

We can determine the population fluctuations and generated signals if we con sider the population rate equations as describing a time-varying carrier number. I a simple sense, the amplitude of these fluctuations in a two-channel system is pro portional to the angular beat frequency, Ω [41]:

$$\Omega = (\omega_1 - \omega_2)\tau_s \quad (5.48$$

where ω_1 and ω_2 are the optical frequencies of the two channels. In semiconducto amplifiers, τ_s is in the nanosecond range. In general, when τ_s and $(\omega_1 - \omega_2)$ are small significant power is transferred to the signal products. Conversely, if $(\omega_1 - \omega_2)$ an τ_s are large (i.e., $\Omega >> 1$), then the beat frequency modulation for widely space channels is so great that the carrier population will be too slow to respond and transfe power to the products; this can be understood as a lowpass filter effect. Thus, fo semiconductor amplifiers, a spacing greater than several gigahertz typically i

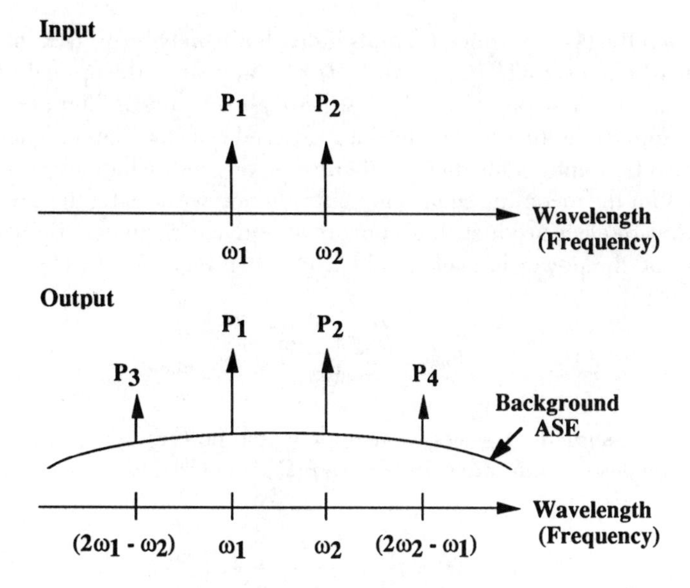

Figure 5.15 Example of four-wave mixing with two original signals and two newly produced signals.

sufficient to quench any IMD production. As we will see, an EDFA has such a long lifetime such that IMD is of no real consequence.

Intermodulation distortion includes two-wave products (as in Fig. 5.15) and three-wave products (Fig. 5.16). The two-wave products appear at frequencies of $(2\omega_i - \omega_j)$ and $(2\omega_j - \omega_i)$, and the three-wave products appear at $(\omega_i + \omega_j - \omega_k)$, $(\omega_i - \omega_j + \omega_k)$, and $(-\omega_i + \omega_j + \omega_k)$. The number of two-wave and three-wave products is proportional to the square and the cube, respectively, of the number of total signals.

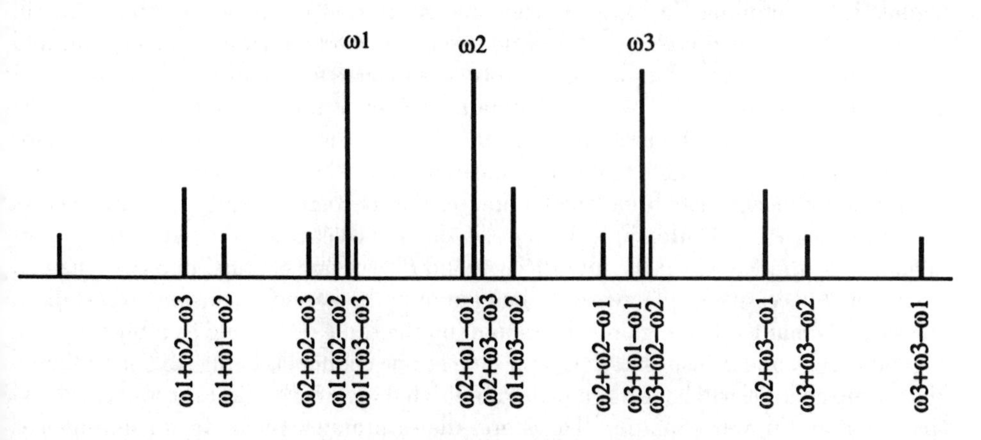

Figure 5.16 Example of three-wave products generated due to intermodulation distortion.

Although the IMD-produced signals individually may be quite weak when compared to the original signals, many products may appear in the optical bandwidth of an original signal. These products will pass through any optical filter at a receiver and give rise to amplitude fluctuations in the recovered selected signal. Therefore, these products must be summed for their total effect on any multichannel system. Since we can assume that the incoming signals are not coherent with each other, we can simply add the power of each product that appears at a given frequency. By using a small-signal analysis, the power in each product, P_{imd} can be derived [41]:

$$P_{imd} = \frac{P_{out}}{4}\left(\frac{P_{out}}{P_{sat}}\right)^2 f(\Omega) \tag{5.49}$$

where we have assumed a lossless waveguide, equal powers of the input signals, and a high-gain single-pass amplifier. In (5.49), $f(\Omega)$ is defined for two-wave products as

$$f(\Omega) = \left| \frac{1}{1 + i(\omega_i - \omega_j)\,\tau_s} \right|^2 \tag{5.50}$$

and for three-wave products as

$$f(\Omega) = \left| \frac{1}{1 + i(\omega_j - \omega_k)\,\tau_s} + \frac{1}{1 + i(\omega_i - \omega_k)\,\tau_s} \right|^2 \tag{5.51}$$

These IMD products must be summed to establish if any nonlinear distortion will take place for a given wavelength channel.

Cross-talk can also occur in a gain-saturated amplifier. As the intensity of the signals in an amplifier increases beyond the saturation input power, the gain will decrease. When the input-signal intensity drops, the gain increases to its unsaturated value. Thus, the gain and input signal power are inverse functions of each other when the amplifier is saturated. This gain fluctuation occurs as rapidly as determined by the carrier lifetime, τ_s, of the amplifier, again being $\sim$ 1 ns in a typical semiconductor amplifier and is comparable to the bit-time in a gigabit-per-second data stream. If we assume a homogeneously broadened amplifier that becomes equally saturated across the entire gain bandwidth independent of the wavelength of the saturating input signal, then an increase in the intensity beyond P^{in}_{sat} of one channel in a two-channel system will necessitate a decrease in the gain of both channels, causing cross-talk in the second channel. If the gain can respond on the same time scale as a bit time in a gigabit per-second transmission system, then as one channel is being ASK modulated, the second channel will have its gain also modulated within a bit time, producing signal distortion and power penalties. Therefore, the saturated semiconductor amplifier is bit-rate dependent. This scenario is depicted in Figure 5.17.

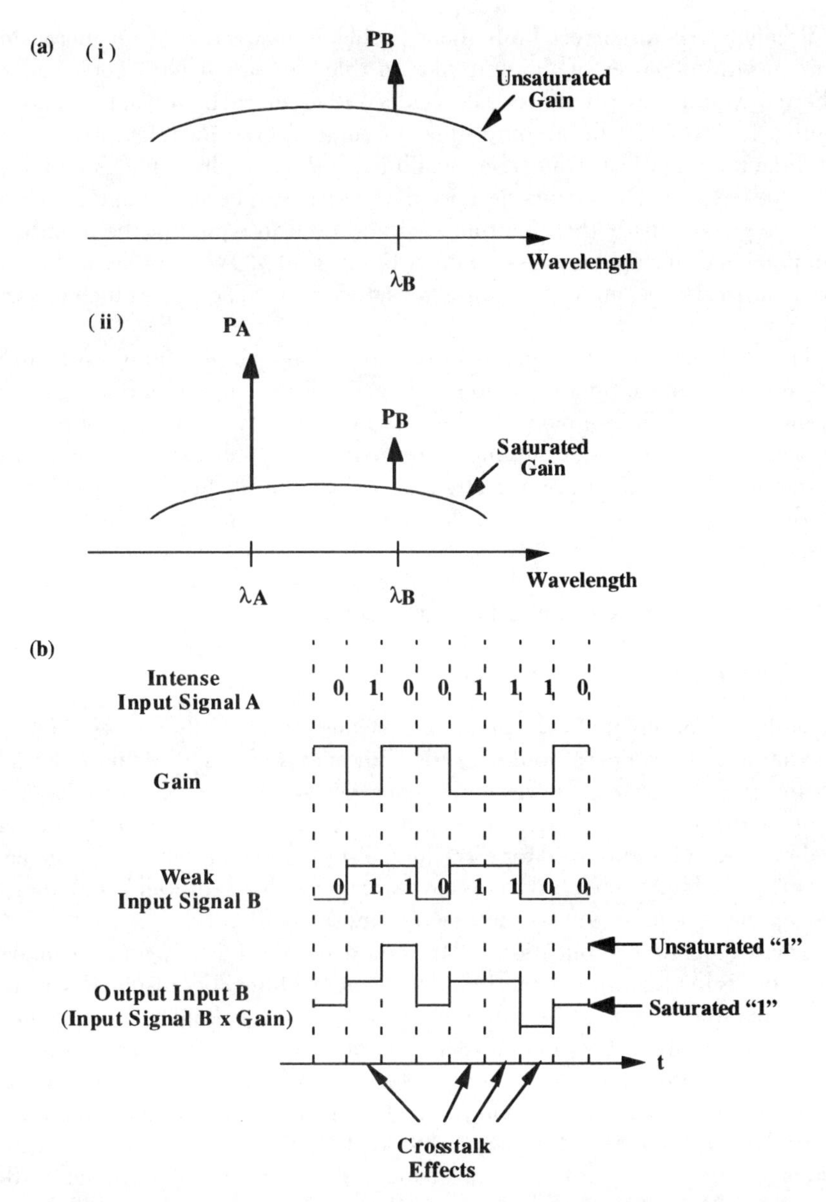

Figure 5.17 (a) Signal and gain spectra given (i) a weak signal B "ON" or (ii) an intense signal A and a weak signal B "ON" simultaneously, producing gain saturation. (b) Bit stream sequences for two signals propagating through a semiconductor amplifier. All pulse transitions are sharp, because we assume the response time of the amplifier gain to be much greater than the bit rate. If they are comparable, then pulse-rounding effects will occur.

Recent work to correct both these problems has centered on increasing the saturation output power of the amplifier so that saturation does not occur easily. Furthermore, four-wave mixing can be significantly reduced by reducing τ_s. Both these corrective features can be accomplished to some degree by fabricating an MQW semiconductor amplifier, which has exhibited a P_{sat}^{out} (3 dB) of ~ 40 mW and a τ_s ~ 0.2 ns [44,45]. This occurs because the carriers can be stored in the barrier and cladding regions outside the quantum well and used to replenish the quantum-well population via diffusion. One disadvantage of using an MQW amplifier is that its gain is very polarization dependent, because the active region's height is much less than its width.

This section has addressed nonlinearities that exhibit significantly different effects for the semiconductor as opposed to the fiber amplifiers due to the vastly different carrier lifetimes involved. The two nonlinear effects are negligible for fiber amplifiers, since their carrier lifetime is approximately 10 ms, far too long to produce any intermodulation distortion for any reasonably spaced channels and far too long to produce saturation-induced gain fluctuations on the time scale of an individual high-speed bit.

5.2.7 Applications of Semiconductor Amplifiers

5.2.7.1 Long-Distance Communications

Semiconductor amplifiers have been demonstrated successfully as power amplifiers, in-line amplifiers, and preamplifiers [46]. However, because of their superior characteristics (i.e., higher P_{sat}, lower NF, higher gain, and fiber compatibility), EDFAs are more desirable for all three of these applications at 1.55 μm. Certainly, dispersion-shifted fiber must be employed for ultra-long distance systems, although conventional fiber can be used for 1.55-μm systems covering up to a few hundred kilometers. These fiber-amplifier qualities will be discussed in more detail later.

The real possible application of semiconductor amplifiers in long-distance communications is for 1.3-μm systems [47]. Since much of the fiber installed worldwide is conventional fiber with the dispersion zero at 1.3 μm, it is still quite possible that some systems will operate at 1.3 μm and require periodic amplification, postamplification, and preamplification. Moreover, there is currently no practical fiber-based amplifier alternative in the 1.3-μm range. Alternatively, it is worthwhile to mention that recent research has allowed long-distance (>300 km) transmission at 1.55 μm with conventional fiber by using chromatic inversion and erbium-doped amplifiers [48]. It is still unclear which method is preferred for long-distance conventional fiber systems.

5.2.7.2 Optoelectronic Integrated Circuits (OEICs)

The main advantages of using a semiconductor amplifier as opposed to fiber amplifiers include its small size, potential low cost, and integratability on a chip containing many

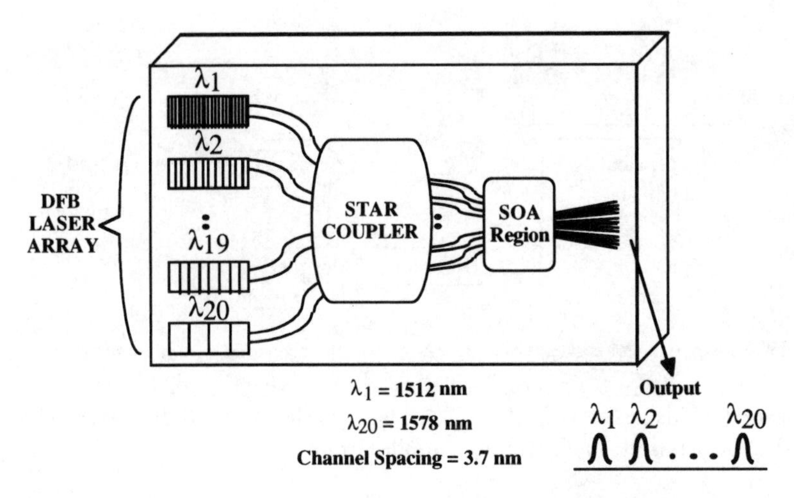

Figure 5.18 Twenty-wavelength laser array combined in a single waveguide and then amplified.

other optoelectronic components (i.e., lasers and detectors). For instance, a semiconductor amplifier can be integrated in a photonic integrated circuit (PIC) [49]. If an array of N lasers is combined in a single monomode output waveguide, then $1/N$ optical splitting losses occur. If a power amplifier is placed following this lossy combining, then the polarization is still well defined and linear. Thus, the semiconductor amplifier can adequately compensate for the splitting loss without polarization dependence. Note that some residual bit-rate dependence may occur if the amplifier is saturated, as it almost always is in the postamplifier mode. Such a PIC is shown in Figure 5.18 [50].

5.2.7.3 Photonic Switching Gates and Modulators

Beyond providing simple gain, a semiconductor amplifier can be used as a high-speed switching element in a photonic system, because the semiconductor will amplify if pumped and absorb if unpumped. The operation is simply to provide a current pump when an optical data packet is to be passed and discontinue the pump when a data packet is to be blocked. Thus, the passed packet is amplified by the population inversion, and the blocked packet is absorbed by the semiconductor because it now lacks a population inversion. This can be in an integrated or bulk-component configuration. A demonstration of a data packet being switched between two different possible outputs with two amplifiers in a 1-by-2 photonic switch is described in [51] and is shown in Figure 5.19. Note that high-speed switching can be performed only in a semiconductor amplifier, not in a fiber-based amplifier; the lifetimes are too long in fiber amplifiers.

Additionally, there is much work on using *quantum-well* (QW) semiconductor amplifiers as modulators, whereby an incident wave is passed or blocked by

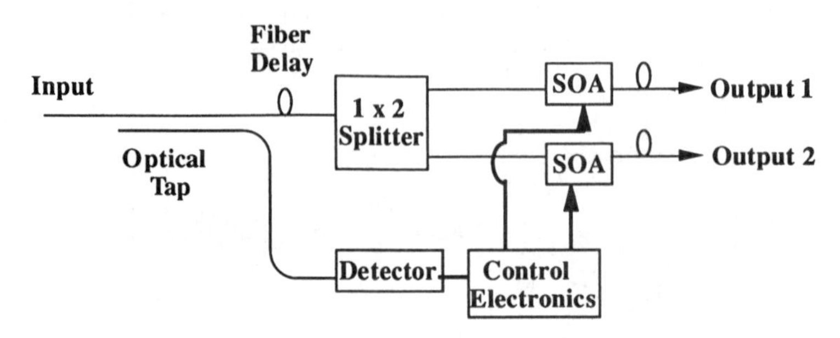

Figure 5.19 An experimental example of a semiconductor amplifier used as a photonic switching gate.

modulating the bias on the amplifier [52]. These modulators would compete with more mature $LiNbO_3$-based Mach-Zehnder modulators [53].

5.2.7.4 Broadband All-Optical Wavelength Shifter

In WDM switching network routing nodes, the ability to shift the wavelength (i.e., convert the frequency) of an incoming packet from one wavelength to another may be of critical importance for network efficiency. A brute force method is to detect a signal at one wavelength and then retransmit it on another wavelength using another laser transmitter. That technique would eventually suffer from an electronic speed bottleneck. Recent research has demonstrated the ability to shift the wavelength of a bit stream by several nanometer and as quickly as several nanoseconds in an all-optical manner [54]. One method is shown in Figure 5.20 in which the experimental setup includes two beams copropagating through the amplifier: one intense ASK-modulated signal at wavelength 1 (λ_1) and one weak continuous wave (CW) probe beam at λ_2 [55–58].

The operation of a wavelength shifter is based on the ability of an intense signal to reduce the gain of an optical amplifier. If the intense signal at wavelength 1, having power P_1 is amplitude modulated and the gain can respond on fast time scales, then the gain of the amplifier will also be modulated. Since $P_1(t)$ is a function of time, then the gain is now a function of time as well. Specifically, as the input λ_1 signal increases, the gain decreases. Therefore, the gain modulation is an inverse function of the original modulated signal at λ_1. If an unmodulated (CW) signal located at λ_2 is then input to the amplifier, the output of the amplifier on this λ_2 is simply the input signal at λ_2 multiplied by the gain. Since the gain has the inverse modulation from $P_1(t)$, then the output $P_2(t)$ on λ_2 also now has the inverse modulation that had formerly appeared on λ_1, but with a different amplitude.

5.2.7.5 Filters

Subsequent to discussing the effect of facet reflections on amplifier gain, we have been exclusively discussing TW amplifiers. However, the FP and DFB amplifiers containing

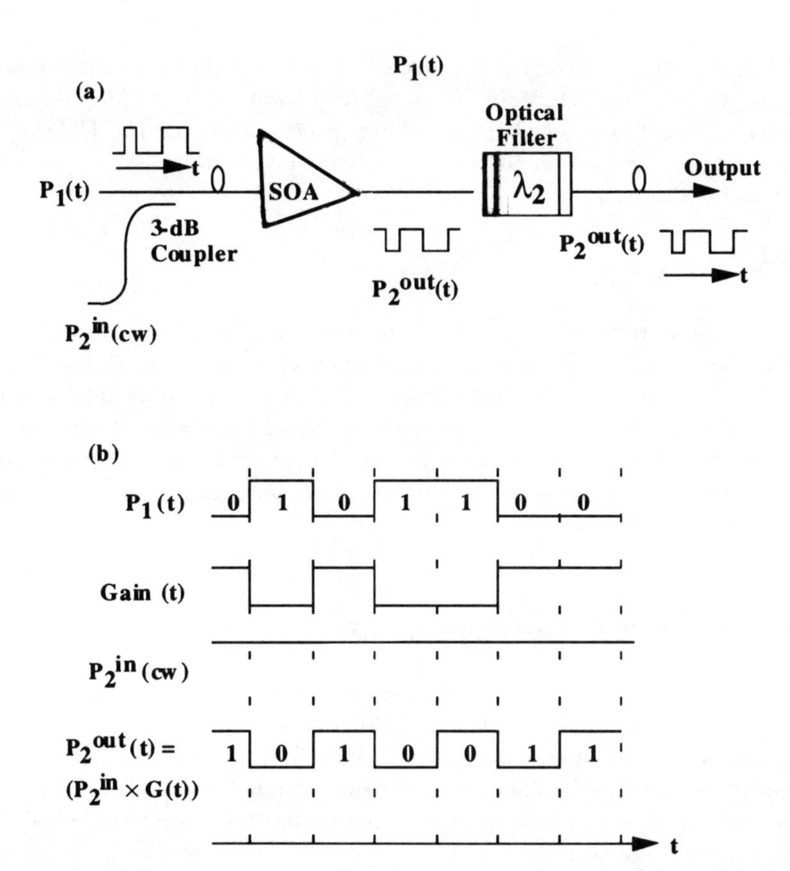

Figure 5.20 (a) Experimental block diagram of a semiconductor amplifier acting as a broadband wavelength shifter and (b) various illustrative power levels, assuming an infinitely fast SOA gain.

feedback have some interesting properties that may be useful for WDM multichannel systems. Because the FP or DFB cavity is by nature wavelength selective, it can act as an optical filter to amplify one channel at a given wavelength and block all others. The FP cavity periodic frequency passbands are spaced apart by the free spectral range, $\Delta\nu$, by

$$\Delta\nu = \frac{c}{2nL} \tag{5.52}$$

and is limited by the length of the semiconductor. The passband of an individual resonance is the free spectral range divided by the cavity finesse, itself a quantity determined by the reflectivities of the facets. This filter is wavelength tunable over a narrow range, because a change in the bias will also change the carrier population and the index of refraction of the cavity. [59–64] The advantage of this filter is that it can be wavelength tuned at a very high switching speed (nanoseconds). The major disadvantage of this filter is that the passband finesse and transmissivity changes for varying bias conditions.

5.2.7.6 Optical Tapping

A semiconductor amplifier can also be used to tap optical energy from a signal passing through the amplifier [65]. This is accomplished in an unpumped semiconductor amplifier because the material will absorb part or all of the incoming light, which then generates electron-hole pairs and a measurable current. The partial detection is useful to tap off some data that will be used for control information concerning the data packet. Full detection can be accomplished by reverse biasing the semiconductor amplifier.

5.3 ERBIUM-DOPED FIBER AMPLIFIERS (EDFAs)

Recently, the EDFA has generated excitement in the telecommunications world [1,14,66]. The main reasons include the fact that erbium ions (Er^{3+}) emit light in the 1.55-μm loss-minimum band of optical fiber and that a circular fiber-based amplifier is inherently compatible with a fiber optic system. As stated in Section 5.1, the original motivation for work on optical amplifiers was to insert a transparent optical pipe in long-haul transoceanic systems, replacing expensive regenerators with relatively inexpensive optical amplifiers. We have discussed the main disadvantages of semiconductor amplifiers for long-haul optical communication systems. The EDFA has relatively few disadvantages, making it an almost ideal critically important component for long-haul communications.

The basic operation of the fiber amplifier (see Fig. 5.2) is similar to that of the semiconductor amplifier. The fiber amplifier contains a gain medium that must be inverted by a pump source [67]. A signal initiates stimulated emission resulting in gain, and spontaneous emission occurs naturally, which results in noise. There are fundamental differences, however, between the two amplifiers. These differences can be traced to four attributes.

- The semiconductor amplifier is, essentially, a two-energy-level system, whereas the EDFA is a three-energy-level system [68]. In the EDFA, carriers are excited from the ground state (population N_0) into an excited state (N_2), which quickly decays to the metastable level (N_1), and from which both stimulated and spontaneous emission occurs. Additionally, an obvious difference is that the population inversion in the semiconductor amplifier is achieved by a current source, whereas the fiber amplifier is pumped by an optical source.

- The length of the fiber amplifier is meters whereas the length of the semiconductor amplifier is ~ 1 mm. This dramatic difference in length makes the assumption of uniform inversion along the length of the amplifier valid only for the semiconductor case and not for the fiber amplifier. Furthermore, since the doping level is much lower in the fiber than in the semiconductor, the ion cross-section that defines the area of interaction between a pump photon and an ion is of critical importance to fiber amplifier operation.
- The fiber amplifier is circular, not rectangular, thus eliminating significant attenuation when it is coupled to a standard optical fiber as well as removing any polarization dependence in the gain.
- The carrier lifetime of erbium ions is milliseconds, whereas that of semiconductor carriers is nanoseconds. That difference reduces significantly the two nonlinear problems in multichannel systems of intermodulation distortion (four-wave mixing) and bit-rate-dependent cross-talk due to gain saturation. (We will assume that the EDFA gain changes only on millisecond time scales, although we will note later in this chapter that even microsecond time changes can occur in cascaded amplifier systems [69].)

Unless otherwise stated, the gain and noise analyses for the semiconductor amplifier will be true for the EDFA. We will concentrate on the unique attributes of the EDFA in this section.

5.3.1 Basic Spectral Characteristics

To produce the amplifier gain medium, the silica fiber core of a standard single-mode fiber is doped with erbium ions. Because of the many different energy levels in erbium, several wavelengths will be absorbed by the ions. Figure 5.21 shows the energy levels and some of the key wavelengths that can be absorbed [70]. In general, absorption corresponds to a photon being absorbed and causing a carrier (ion) to jump to a higher energy level of energy difference, $\Delta E = h\nu$, that roughly matches the energy of the photon. Different wavelengths can cause either *ground-state absorption* (GSA) or *excited-state absorption* (ESA) [3]. GSA corresponds to a photon exciting a carrier from the ground state to a higher state, and ESA corresponds to a photon exciting a carrier from one of the non-ground-state energy levels to an even higher level. Since the population is greatest in the ground state, the probability of GSA occurring is much greater than ESA. Once a photon is absorbed and a carrier is excited to a higher energy level, the carrier decays rapidly to the first excited level. Once the carrier is in the first excited state, it has a very long lifetime of ~ 10 ms [12], thereby enabling us to consider

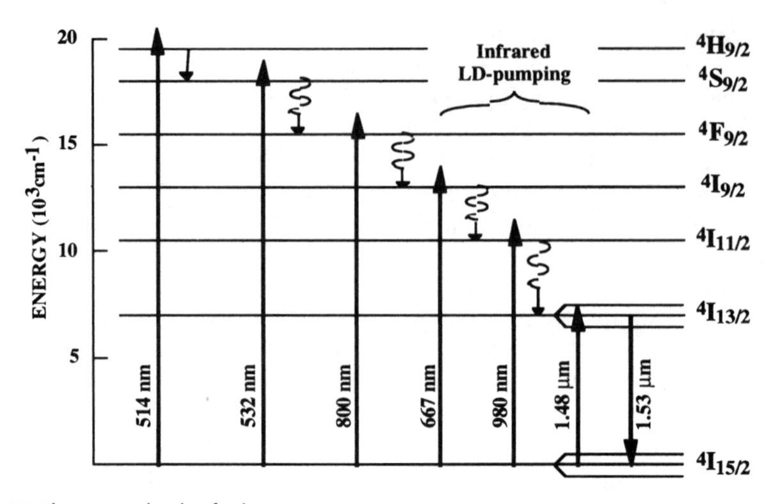

Figure 5.21 The energy levels of erbium.

the first excited level to be metastable. Depending on the external optical excitation signal, this carrier will decay in a stimulated or spontaneous fashion to the ground state and emit a photon. The emission in erbium is fortuitously near the 1.55-μm loss-minimum of standard silica optical fiber.

The absorption is not as strong for all the possible wavelengths, governed critically by the tendency of a pump photon to be absorbed as determined by the cross-section of the erbium ion with that photon. The wavelength having the strongest absorption coefficient is 0.98 μm, the second strongest is at 1.48 μm, and weaker absorptions occur at 0.53 and 0.8 μm. Fortunately, fabricating high-power multimode laser diodes for the 0.98 and 1.48-μm wavelengths can be done by using strained-layer QW material, with output powers >100 mW achievable and commercially available [71]. Laser diode pumps are attractive sources because they are compact, reliable, and potentially inexpensive.

Both the absorption and the emission spectra have an associated bandwidth. These bandwidths depend on the spread in wavelengths that can be absorbed or emitted from a given energy level, allowing multimode multiwavelength diode laser light to be absorbed. Such a spread in wavelengths is caused by Stark-splitting of the energy levels, allowing a deviation from an exact wavelength [72]. This is highly desirable because (1) the exact wavelength of the pump laser may not be controllable and is impossible for a multimode laser, and (2) the signal may be at one of several wavelengths, especially in a WDM system; that makes implementation of the amplifier very flexible. Figure 5.22 shows the bandwidth in the 1.48-μm absorption and 1.55-μm fluorescence spectrum of a typical erbium-doped fiber [73].

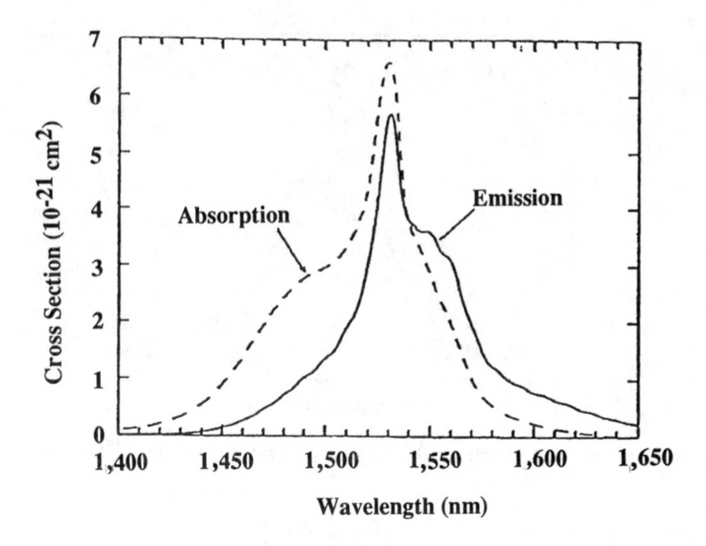

Figure 5.22 The absorption and fluorescence spectra for erbium near 1.5 μm. (*After:* [71]. © 1991 IEEE.)

The fluorescence spectrum in Figure 5.22 was taken from a fiber that not only contained erbium but was co-doped with another material [73]. Co-doping the erbium fiber with another material is important for two reasons. First, erbium ions are quite large in comparison to silica atoms and are not very soluble in silica, thereby making it difficult to highly dope the fiber so as to achieve high gain in a short length. Therefore, another material that is more soluble in silica but similar in size to erbium is required to help incorporate more erbium in the fiber core region. A concentration of 1,000 parts per million can be attained by using co-dopants.

The second reason is related to systems performance. Because we want the amplifier to be as versatile as possible in many different applications, the gain bandwidth should be as broad and as uniform as possible. Semiconductor amplifiers are quite adequate in this regard since the bandwidth is enormous, covering ~ 200 nm, and have a well-behaved and slowly varying Lorentzian profile. Unfortunately, erbium is not as well behaved. The bandwidth is only on the order of 10 nm and exhibits a definite nonuniform structure. Incorporating a co-dopant enhances the Stark-splitting of the various levels, thereby broadening and smoothing the gain profile. Figure 5.23 shows the fluorescence spectra of an EDFA with three different co-dopants. It has been found that aluminum achieves the best performance as a co-dopant, although a peak at 1.53 μm and a bandwidth of only ~ 30 nm remain [70].

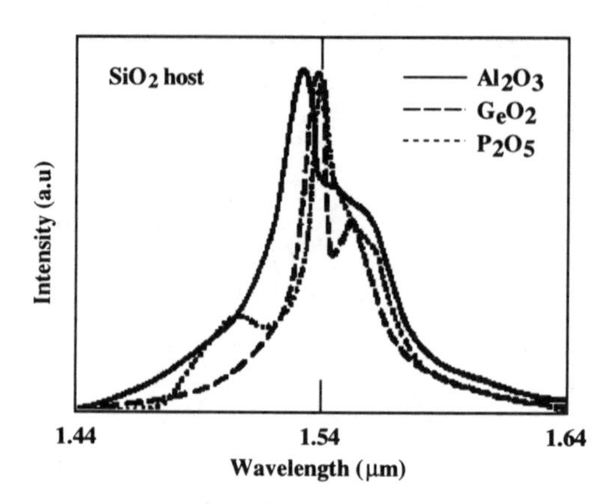

Figure 5.23 Effect of co-doping an erbium-doped fiber with other materials. (*After:* [68]. © 1991 OSA.)

5.3.2 Required Components in a Functional EDFA

Section 5.3.1 described the fundamental mechanisms of the erbium gain medium. Placing this active element in an amplifier configuration is relatively straightforward. Figure 5.4(b) showed the main EDFA components.

The erbium-doped fiber core diameter may be either quite small (~ 4 μm) or a standard 8-μm single-mode size. If the core size is small, then there is a mismatch between conventional fiber and the erbium-doped fiber, causing a loss in both the signal and the pump. Care must be taken to carefully splice together the two fibers. One method to minimize the loss (<0.5 dB) has been to place the two cores flush against each other, heating them up to expand the cores, and then fusion-splicing them together.

Because the erbium-doped gain medium must be pumped optically, the 0.98- or 1.48-μm pump light (usually in the form of a diode laser) and the 1.55-μm signal must be combined in the doped fiber to achieve medium pumping and signal gain. Pigtailed grating-based three-port WDMs can perform this coupling function with <0.5 dB of insertion loss and >40 dB of return loss, even when combining wavelengths as close as 1.48 and 1.53 μm. Naturally, the direction of signal propagation is fixed. However, the pump light, which is absorbed by the gain medium as it propagates along the fiber, can either copropagate or counterpropagate, for which the WDM would be required at the input or output of the erbium-doped fiber, respectively. (Co- and counterpropagating the pump with the signal have slightly different gain and NF properties, as will be discussed in Section 5.3.3.)

Two other components that are not, strictly speaking, essential to an EDFA but that may be required to prevent system degradation are an optical filter and an optical isolator. The broadband ASE emanating from the amplifier will produce spontaneous-spontaneous beat noise. This must be limited in some way to achieve a reasonable SNR at the receiver. A bandpass filter of ~ 1–2 nm or narrower may be placed at the output of any EDFA. Alternatively for WDM systems that incorporate a cascade of amplifiers and many channels located across the gain bandwidth, the wavelength filter can be placed only at the receiver, which will minimize noise as well as demultiplex the many channels. Additionally, we want to prevent reflections back into the amplifier, which may cause the NF of the EDFA to increase and may even cause the EDFA to lase if the gain is high enough (i.e., gain greater than loss in a cavity with reflections). Note that the noise figure may increase in the presence of reflections due to the inverted carriers amplifying an unwanted reflected field and *not* the desired signal. An isolator would prevent those deleterious properties. Furthermore, an isolator may be necessary on the input side to prevent backward-traveling ASE from propagating back into reflection-sensitive components, such as a laser transmitter. (Isolators and bandpass filters may be required in semiconductor amplifiers for essentially the same reasons as for EDFA's.)

5.3.3 Gain and Noise Dependencies

As stated previously, the three major differences between the semiconductor amplifier and the EDFA with regard to an analysis of gain are (1) the semiconductor is essentially a two-level system and the EDFA is a three-level system; (2) the semiconductor carrier density can be considered uniform along its short length, whereas the carrier density in the meters-long EDFA will vary along its length; and (3) the gain of the EDFA is highly nonuniform, whereas the gain of the SOA is fairly well behaved with respect to wavelength. In general, the gain in the EDFA will depend on the amplifier doping, length, pump power, pump wavelength, and the emission and absorption cross-sections for the signal and the pump. In a simple sense, the higher the doping, the more gain per unit length. The gain's wavelength dependence corresponds closely to the ASE spectrum of the erbium-doped amplifier, as was shown in Section 5.3.2. We will now concentrate on the wavelength-dependent gain and noise, as well as the evolution of the signal, ASE, and pump within the fiber. For this analysis, we will assume that the pump and the signal are copropagating (unless otherwise stated). Issues not addressed here should be considered similar to the analysis for the semiconductor gain.

In general, the wavelength-dependent gain, g, and loss, α, coefficients of the EDFA can be described as follows [70]:

$$g(\lambda) = \sigma_a(\lambda)\Gamma(\lambda)N_t \tag{5.53}$$

and

$$\alpha(\lambda) = \sigma_e(\lambda)\Gamma(\lambda)N_t \tag{5.54}$$

where σ_e and σ_a are the wavelength-dependent emission and absorption cross-sections of the erbium ions, Γ is the wavelength-dependent overlap integral between the optical mode and the erbium ions, and N_t is the density of erbium ions in the fiber. The total wavelength-dependent gain is the integral of the gain and loss over the entire length of the fiber:

$$G_k = \exp\left[\int_0^L \left(g_k \frac{N_1(z)}{N_t} - \alpha_k \frac{N_0(z)}{N_t}\right) dz\right] \tag{5.55}$$

where the subscript k denotes that we consider k possible wavelengths; L is the erbium-doped fiber length; and N_1 and N_0 are the carrier densities in the first metastable and ground states, respectively. Two equations that fully describe the gain and the signal power along the erbium-doped fiber in a spectrally resolved, spatially integrated model developed by Giles and Desurvire are [72]:

$$\frac{\overline{N}_1}{\overline{N}_t} = \frac{\sum_k \left(\frac{P_k(z)\, \alpha_k\, \tau_{10}}{\pi h \nu_k\, b_{eff}^2 \overline{N}_t}\right)}{1 + \sum_k \left(\frac{P_k(z)(\alpha_k + g_k)\, \tau_{10}}{\pi h \nu_k\, b_{eff}^2 \overline{N}_t}\right)} \tag{5.56}$$

and

$$\frac{dP_k}{dz} = \left[u_k(\alpha_k + g_k)\frac{\overline{N}_1}{\overline{N}_t} P_k(z)\right] + \left[u_k g_k \left(\frac{\overline{N}_1}{\overline{N}_t}\right) m h \nu_k\, (\Delta\nu_k)\right] - \left[u_k(\alpha_k + l_k) P_k(z)\right] \tag{5.57}$$

where τ_{10} is the metastable carrier lifetime; $(\Delta\nu)$ is the amplifier homogeneous optical bandwidth); $P_k(z)$ is the optical power at each wavelength slice along the length of the fiber; l_k is the additional loss due to fiber absorption; m^{-1} is the linear density of ions; u equals either 1 or -1, depending on whether the optical beam is traveling forward or backward, respectively; $\overline{N}_1$ and $\overline{N}_t$ are the average respective densities; and b_{eff} represents the equivalent radius of the doped region and is given by

$$b_{eff} = \left[\frac{1}{\pi}\int_0^{2\pi}\int_0^{\infty} \frac{N_t(r)}{N_t(0)}\, r\, dr\, d\varphi\right]^{\frac{1}{2}} \tag{5.58}$$

in which r and ϕ are simply radial coordinates. Note that the factors of

$$(\pi b_{eff} \overline{N}_t) = \frac{A}{\Gamma_k} \frac{(\alpha_k + g_k)}{(\sigma_{ak} + \sigma_{ek})} \tag{5.59}$$

are related to experimental measurements of the cross-sections, emission and fluorescence, and saturation powers. A is the mode area in the fiber. These equations uniquely determine the operation of the EDFA.

The normalized carrier densities as they depend on the pump power can be expressed as [66]

$$N_0(z) = \frac{Pp^{th}}{Pp^{th} + Pp(z)} \tag{5.60}$$

and

$$N_1(z) = \frac{Pp(z)}{Pp^{th} + Pp(z)} \tag{5.61}$$

where $P_p(z)$ is the pump power at any point along the length of the fiber, and $P_p{}^{th}$ is the pump power necessary for the gain to compensate for the losses, that is, for the fiber to be transparent.

There are many interesting phenomena in amplifiers that depend on the time dynamics of the carrier populations. The rate equation describing the carrier dynamics between the various levels is [66]

$$\frac{dN_0}{dt} = -\frac{dN_1}{dt} = -\left(\frac{\sigma_{02}P_p}{h\nu_p} + \frac{\sigma_{01}P_{sig}}{h\nu_s}\right)N_0 + \left(\frac{\sigma_{10}P_{sig}}{h\nu_s} + \frac{1}{\tau_s}\right)N_1 \tag{5.62}$$

where σ_{02}, σ_{01}, and σ_{10} represent the cross-sections for transitions from the first numbered level to the second-numbered level, and ν_p and ν_s are, respectively, the pump and signal optical frequencies. The four terms on the right side of (5.62) represent, starting from the leftmost term, the pumping, the stimulated absorption, the stimulated emission, and the spontaneous emission. These transitions are shown in Figure 5.24. Note that the carriers pumped to a higher excited state (i.e., N_2) decay rapidly to the metastable state, because σ_{21} is much larger than σ_{10}. Also note that (5.62) is not that dissimilar from (5.4), which described the carrier dynamics in a semiconductor amplifier.

Three equations that define, in space and time, what the evolution of the pump, the signal, and the ASE along the fiber are [37]

$$dP_p^{\pm}(z,t) = \mp P_p^{\pm}\Gamma_p(\sigma_{pa}N_0 - \sigma_{pe}N_1 - \sigma_{pe}N_2) \mp \alpha_p P_p^{\pm} \tag{5.63}$$

$$dP_s(z,t) = P_s\Gamma_s(\sigma_{se}N_1 - \sigma_{sa}N_0) - \alpha_p P_s \tag{5.64}$$

$$\frac{dP_{ASE}^{\pm}(z,t)}{dz} = \pm P_{ASE}^{\pm}\,\Gamma_s(\sigma_{se}\,N_1 - \sigma_{sa}\,N_0) \pm 2\sigma_{se}\,N_1\,\Gamma_s\,h\,\nu_s\,(\Delta\nu) \mp \alpha_s\,P_{ASE}^{\pm} \tag{5.65}$$

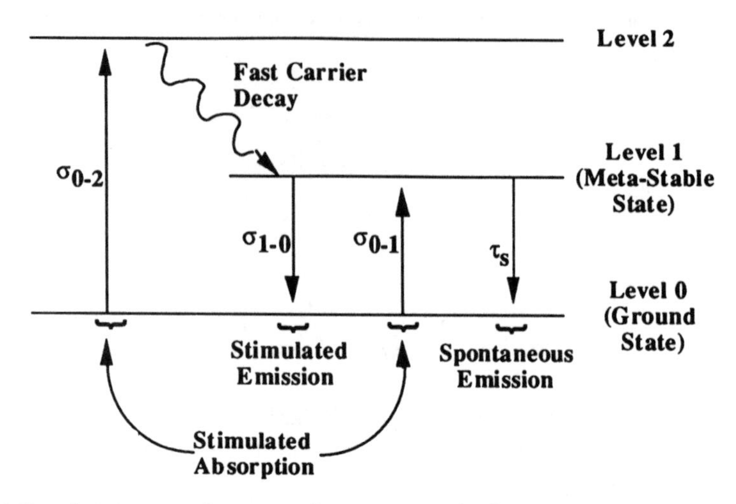

Figure 5.24 The relative rates and transitions between energy levels in an EDFA.

where + and − represent propagation in the forward or backward direction, respectively, and the subscripts p and s represent pump and signal, respectively, for the various power and absorption/emission cross-section terms. Note that the signal propagates only in the forward direction, the ASE propagates in both directions, and the pump can co- or counterpropagate. Equations (5.63), (5.64), and (5.65) all account for the relevant absorption and emission of the wave in question, as well as the accompanying gain and intrinsic erbium-doped fiber losses. By conservation of carriers, the following equality exists:

$$N_1 = N_2 + N_1 + N_0 \tag{5.66}$$

One can even find the total output flux power of an amplifier by examining the input power fluxes, Φ, at each wavelength value [70]:

$$\Phi^{out} = \sum_k \Phi_k^{in} \exp\left\{\frac{(\alpha_k + g_k)\tau_s}{\pi b_{eff}^2 \overline{N}_t}(\Phi^{in} - \Phi^{out}) - \alpha_k L\right\} \tag{5.67}$$

There exists an optimum length of fiber to produce maximum gain in a system. Figure 5.25 depicts the relative amplitudes for the pump and signal powers and for the carrier populations. The pump is absorbed as it travels along the EDFA and thereby inverts the carriers, increasing the N_1 carrier population and reducing the N_0 carrier population. After some length, however, there is insufficient pump remaining to adequately invert the erbium to provide gain. Therefore, as the signal propagates through the amplifier, the signal will undergo gain as long as the pump has inverted

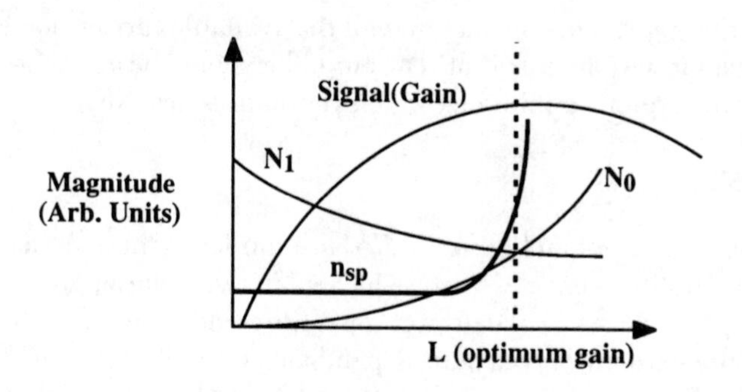

Figure 5.25 Relative amplitudes for the pump and signal powers and for the carrier populations as they all traverse a length of erbium-doped fiber.

the medium. At the point along the fiber where the pump has not adequately inverted the medium, the signal will no longer experience gain but will be attenuated by absorption. The point at which the signal gain ceases would be considered close to the optimum length, although the noise properties may be better served if the length was even shorter to prevent n_{sp} from rising (recall the definition for n_{sp}, which we will revisit in the next section).

A typical experimental gain dependence on pump power is shown in Figure 5.26 [70]. A minimum pump power is required for the gain to overcome the losses, that is, achieve transparency. The gain then increases rapidly, followed by a

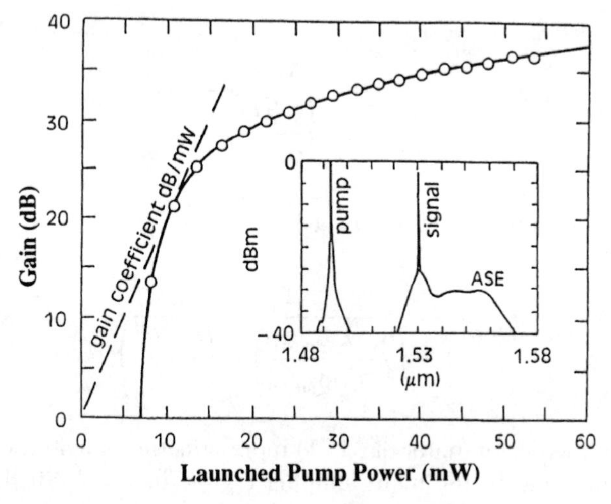

Figure 5.26 EDFA gain as function of 1.48-μm pump power. (*After:* [68].)

plateau region that represents the fact that all the available carriers have already been inverted throughout the gain medium. The amplifier should be operated near this fully inverted (plateau) region to minimize n_{sp} and obtain a low NF.

5.3.4 EDFA Noise

The noise considerations of an EDFA [74,75] are similar to those for a semiconductor amplifier. Both amplifiers have ASE noise located in two polarizations, and that noise generates beat-noise terms in a detector. Two differences between the two types of amplifiers are the extreme wavelength dependence on the NF of the EDFA and the relatively subtle effect of co- versus counterpropagating the pump light with the signal in an EDFA.

If we assume Poisson statistics for the ASE, we can derive the NF and spontaneous emission factor for the erbium-doped amplifier [70,76–78]. The NF is, by definition and by approximation [76],

$$NF(\lambda) = \frac{SNR_{in}}{SNR_{out}} \approx \frac{2P_{sp}(\lambda) + 1}{G(\lambda)} = \frac{2n_{sp}(\lambda)[G(\lambda) - 1] + 1}{G(\lambda)} \tag{5.68}$$

where

$$n_{sp} = \frac{P_{sp}}{h\nu(\Delta\nu)(G - 1)} \tag{5.69}$$

Just as in the SOA, the noise figure of the EDFA is approximately equal to $2n_{sp}$ if $G >> 1$. When factoring in the specific properties of the EDFA, then n_{sp} can be defined as

$$n_{sp} = \frac{N_1}{N_1 - \left(\frac{\sigma_{sa}(\lambda)}{\sigma_{se}(\lambda)}\right)N_0} \tag{5.70}$$

and becomes, when proper substitutions are made [76,79],

$$n_{sp} = \frac{1}{1 - \left(\frac{\sigma_{sa}(\lambda)\sigma_{pe}(\lambda)}{\sigma_{se}(\lambda)\sigma_{pa}(\lambda)}\right) - \left(\frac{\sigma_{sa}}{\sigma_{se}}\frac{P_p^{th}}{P_p}\right)} \tag{5.71}$$

The third term in the denominator of (5.71) represents the contribution to noise due to inadequate pumping. In the high-pumping regime in which all the carriers are inverted along the length of the amplifier, this third term can be neglected. The contribution to noise of this third term is that n_{sp} increases either when pumping is too low or when the amplifier is in deep saturation.

The individual spectra for the absorption and emission (fluorescence) of light near 1.55 μm in an EDFA was shown to be quite nonuniform but also to overlap with each other. That means that if a signal is incident on the amplifier at a wavelength containing significant absorption and emission cross-sections, then the signal will experience both gain and absorption, thereby contributing to some additional ASE. This overlap occurs on the short wavelength end of the gain spectra, and so n_{sp} and the NF will be higher at this end of the spectrum than in the case where the wavelength is higher and the overlap is much less. Figure 5.27 shows the NF as a function of wavelength and how the NF is clearly lower at longer wavelengths [80]. Thus, it is advantageous to operate the signal wavelength not at the gain peak of the amplifier but at a higher wavelength with a lower NF. Note that the NF is very close to 3 dB, which is the quantum-limited value, as determined earlier in the chapter.

The noise issue of whether to co- or counterpropagate the signal with the pump is more subtle. Certainly, if the two beams are counterpropagating, then the pump does not appear at the detector but must be blocked in the reverse direction by an isolator so as not to propagate back into the signal source. However, the pump should still be blocked from the detector even in the copropagating case by an in-line optical filter. The subtle noise issue revolves around the accumulation of spontaneous emission noise. The two pump-propagation cases differ when the medium is not fully inverted. This occurs when the pump does not contain sufficient power to fully invert the entire length of the amplifier. An incomplete inversion will then occur at the fiber end in the copropagating case and at the fiber beginning in the counterpropagating case. The emission dynamics suggest that the better scenario is for copropagation, since in any amplifier system the NF is determined by the first amplifier and not by the last one. However, practical counterpropagation systems will not be affected too adversely,

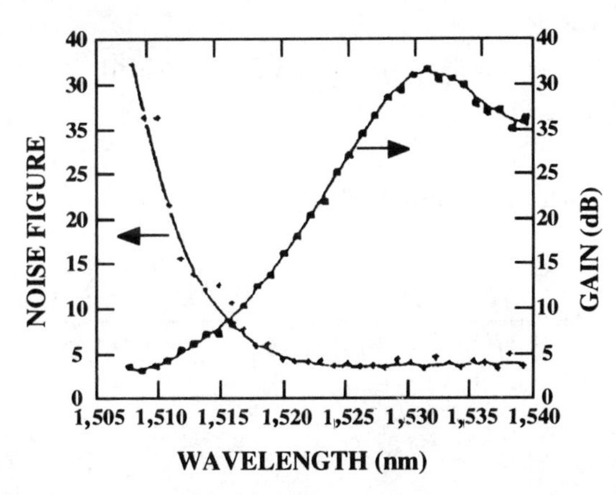

Figure 5.27 Noise figure as a function of wavelength in an EDFA pumped with 1.48-μm light. (*After:* [78]. © 1989 IEEE.)

since most of these systems will have a high medium inversion along the entire amplifier due to the availability of high-power pump diodes. The slight difference in NF probably will not be significant. Moreover, some systems will employ a combination of both types of pumping schemes to provide high pump powers for high gain and low noise figure.

Example 5.3

An EDFA has an input 1.48-μm pump power of 25 mW and a pump threshold of 8 mW. The gain at 1.53 and 1.55 μm is, respectively, 29 and 20 dB. The *normalized* energy transition cross-sections are as follows:

Normalized Cross-Sections	1.48 μm	1.53 μm	1.55 μm
Absorption	2.5	6.5	2.7
Emission	0.7	5.8	4.0

Find n_{sp} and the NF for signals at 1.53 and 1.55 μm.

Solution: We can use the following equation to find the spontaneous emission factor:

$$n_{sp} = \frac{1}{1 - \left(\frac{\sigma_{sa}(\lambda)\sigma_{pe}(\lambda)}{\sigma_{se}(\lambda)\sigma_{pa}(\lambda)}\right) - \left(\frac{\sigma_{sa}}{\sigma_{se}}\frac{P_p^{th}}{P_p}\right)}$$

By substitution using the numbers from the table provided:

n_{sp} (1.53 μm) = 3.052

n_{sp} (1.55 μm) = 1.680

The noise figure is derived from the following equation:

$$NF(\lambda) = \frac{SNR_{in}}{SNR_{out}} \approx \frac{2P_{sp}(\lambda) + 1}{G(\lambda)} = \frac{2n_{sp}(\lambda)[G(\lambda) - 1] + 1}{G(\lambda)}$$

NF(1.53 mm) = 6.09
NF(1.55 mm) = 3.34

5.3.5 Saturation Characteristics and Cross-talk

As with semiconductor amplifiers, the EDFA gain can be compressed to a small value [81] if the amplifier is saturated by an intense input signal [82]. An amplifier figure of merit is to have a high output saturation power, which is especially desirable for power amplifier applications in which we want to boost the output of a laser diode to a value higher than what a semiconductor laser or amplifier can provide. The saturation output power is a function of the erbium concentration, fiber length, and pump power, because more carriers inverted in the medium will result in a higher output power. Of course, the limit is reached when all the erbium ions in the fiber length have been inverted. For the semiconductor amplifier, we plotted signal gain as a function of input signal power. For the case of the EDFA, we alternatively plot in Figure 5.28 the signal gain as a function of signal output power, showing the saturation output power of the amplifier [83]. This example is for a given length of fiber and pump power. Saturation output powers as high as ~ 20 dBm have been measured with extremely high pumping and an extremely large number of erbium ions.

We have previously discussed cross-talk issues in a multichannel system with regard to semiconductor amplifiers. Recall that two problems existed, intermodulation distortion caused by four-wave mixing and gain saturation–induced crosstalk. The relative effects of each of these can be traced to the carrier lifetime (i.e., gain response time) of an optical amplifier. Semiconductor amplifiers have lifetimes on the order of nanosecond, whereas EDFA lifetimes are much longer, specifically in the millisecond range. Both of these two deleterious nonlinear effects are considered negligible in the EDFA due to its extremely long gain response time. For instance, four-wave mixing, which is proportional to the product of the lifetime and the frequency difference

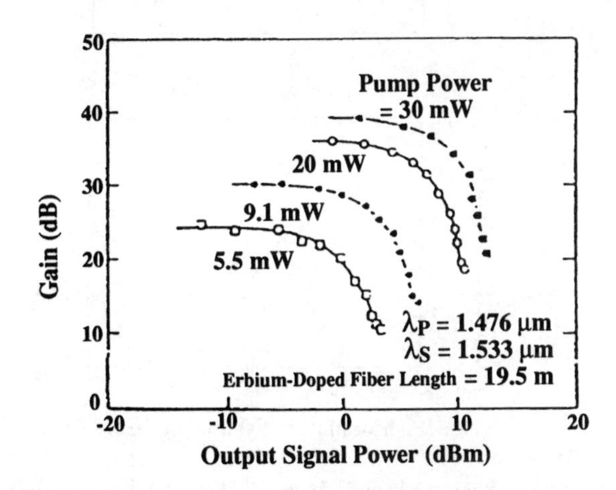

Figure 5.28 EDFA gain versus output signal power. (*After:* [81]. © 1991 OSA.)

between any two channels, would be significant only if the two channels would b extremely close to each other, an unlikely prospect given high-bandwidth channels an the need for channel spacings to be large enough to negate the adverse effects o nonideal optical filtering at a direct-detection receiver.

Gain saturation-induced cross-talk is also considered negligible. If the power o signal A fluctuates, then the amplifier gain will fluctuate as well, impressing a undesirable modulation on signal B. This scenario occurs only if the gain can respon on the time scale of a data bit. Since the EDFA gain can respond only on milliseco time scales, then a modulation from signal A will not affect signal B. The gain transien response and the effect of cross-talk are shown in Figure 5.29, the major result bein that any signal faster than kilohertz speeds will not induce cross-talk effects in anothe signal when both are passing through an EDFA [84]. (We again emphasize tha cross-talk effects on the order of megahertz can occur in EDFA cascades [69]. See not at end of Section 5.3.5.)

We can determine analytically what effects a strong input signal into the EDF would have on the gain and carrier populations, as was shown in Figure 5.29(b). Th time-dependent population inversion and the saturation and recovery of the gain ca be determined by the following two equations [85]:

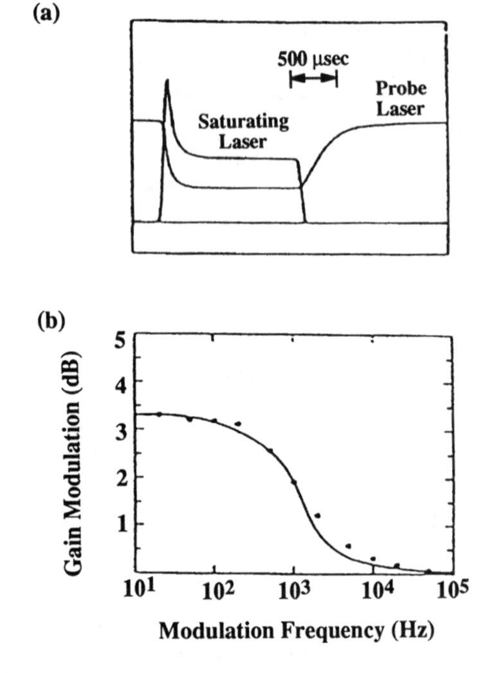

Figure 5.29 (a) Signal and gain transients in an EDFA and (b) cross-talk between two channels as a functio of the rate of amplitude modulation on one of the channels. (*After:* [82]. © 1989 OSA.)

$$\Delta N_{01}(0,t) = \left(\frac{\alpha - 1}{1 + \alpha + 2\beta}\right)\left\{1 + \frac{2\beta}{1+\alpha}\exp\left[-\left(\frac{1+\alpha-2\beta}{\tau_{10}}\right)t\right]\right\}, \qquad 0 < t < \Delta \quad (5.72)$$

and

$$\Delta N_{01}(0,t) = \left(\frac{\alpha - 1}{1 + \alpha}\right)\left[1 - \frac{2\beta}{1+\alpha+2\beta}\exp\left(-\frac{1+\alpha}{\tau_{10}}(t - \Delta T)\right)\right], \qquad t > \Delta T \quad (5.73)$$

where α and β describe the pump absorption and signal emission, respectively, and are defined by:

$$\alpha = \frac{P_p(0)}{P_p^{th}} = \frac{P_p(0)\sigma_{pa}\tau_{10}A_p}{h\nu\Gamma_p} \quad (5.74)$$

and

$$\beta = \frac{P_s(0)}{P_{sat}} = \frac{P_s(0)\sigma_{se}\tau_{10}A_s}{h\nu\Gamma_s} \quad (5.75)$$

and where ΔN_{01} is the carrier inversion from the ground state to the metastable state and is directly related to the gain, and ΔT is the time for which the intense input signal is incident. We have assumed that at time $t = 0$ the gain is at its small-signal value and that we are investigating the scenario at the input to the amplifier ($z = 0$). Equation (5.72) describes the saturation of the amplifier (i.e., carrier transitions from the metastable state to the ground state) and (5.73) describes the recovery of the gain back to its small-signal value (i.e., carrier transitions from the ground state back to the metastable state). The higher excited state, N_2, does not appear, since we have assumed that the lifetime is so short as to be inconsequential.

Important Note: A late-breaking result by researchers at Bell Labs has shown that the gain in a cascade of EDFAs can fluctuate much faster than millisecond due to saturated conditions along a cascaded-amplifier link. In fact, the gain can change as fast as even a few microseconds [69]! Microsecond gain transients will affect system performance in which the total signal input power dynamically changes. However, an unsaturated single EDFA can still be considered to have gain that fluctuates on the order of millisecond.

5.3.6 Noise in Lumped and Distributed Amplifiers

Amplifiers can be cascaded in series to provide periodic gain to compensate for fiber attenuation losses of a signal traveling a long distance. An important noise consideration involves the trade-off between the gain per amplifier and the number of in-line amplifiers, assuming that a given total gain must be achieved to propagate a certain distance. In other words, the question from a system SNR perspective is the following: Is it better to have one very high gain amplifier or many low-gain amplifiers distributed periodically along the propagation path (Fig. 5.30) [86]? The argument for having many low-gain amplifiers proceeds as follows. Based on (5.30), an amplifier produces noise in proportion to the gain, and the gain is chosen to compensate for the link losses. For a large lumped-gain amplifier, the amplification typically is placed near the receiver (1) to provide enough gain to the signal to overcome the receiver thermal noise and (2) to provide the highest gain without being saturated by an intense signal. The high ASE noise generated will be detected in totality. However, if this large gain is divided by 2 and two amplifiers are used, one placed midway along the transmission span and one at the receiver, then the ASE from the first amplifier will experience attenuation in the length of fiber as it traverses the distance between the two amplifiers. Concurrently, the signal still experiences the same total gain. In the extreme case, if there are an infinite number of very low gain amplifiers placed all along the length of

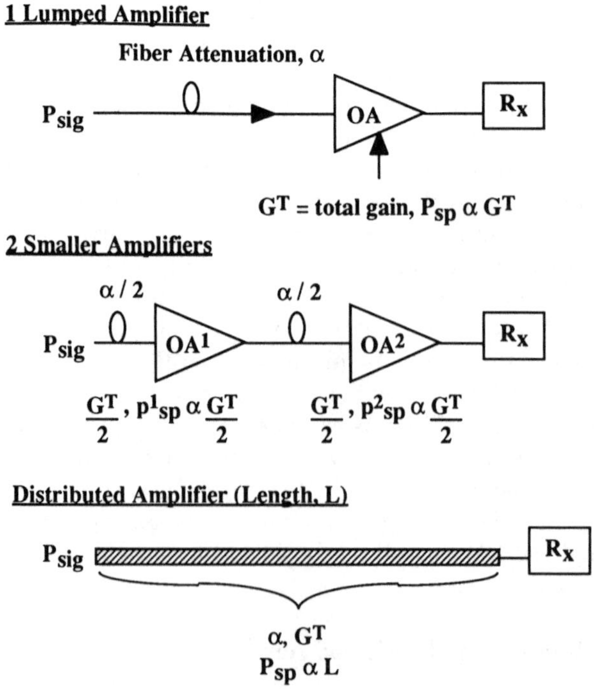

Figure 5.30 ASE accumulation in high-gain, low-gain, and distributed EDFAs. G^T is the total gain required to compensate for losses (α) along the entire fiber span.

the transmission fiber, then the ASE from the first amplifier will be severely attenuated by the time it reaches the detector, the second amplifier's ASE will be attenuated but less so, and so on. The ASE will accumulate much more slowly in the case of many low-gain amplifiers than in the case of one high-gain amplifier. Of course, it is more expensive to have many small low-gain amplifiers than few high-gain ones, and component insertion losses will affect what the optimum gain should be.

The noise argument concerning ASE accumulation can be extended to include a distributed fiber amplifier [84,87,88], which is nothing more than a low-doped EDFA (see Fig. 5.30). The gain of such an amplifier would almost exactly compensate for the loss along a transmission fiber path, with the distributed EDFA forming the transmission path itself. The ASE would grow quite slowly. The expression for P_{sp} in a distributed EDFA, in which the loss equals the gain, is proportional to the fiber length, L, and can be expressed as [88]

$$P_{sp} = n_{sp} h\nu(\Delta\nu)L \tag{5.76}$$

This is in contrast to a standard EDFA, whose ASE grows exponentially with fiber length [see (5.28)]. The exponential gain of an amplifier is directly compensated by the exponential loss in a fiber. Furthermore, a 1.48-μm laser would be necessary for pumping, since the pump must now travel without normal fiber attenuation down a long length of distributed EDFA, one that is perhaps as long as 50 km. Such fibers may be very expensive to make and difficult to implement, without much performance benefit (see Section 5.3.7 for the results involving lumped amplification). Therefore, whereas low-gain (<20 dB) EDFAs may be employed in long-distance amplifier cascades, providing quite good performance, it seems unlikely that distributed EDFAs will be used in the near future.

5.3.7 EDFA Applications

EDFAs have highly desirable system qualities that make it virtually certain that they will be implemented in actual systems in the very near future. This section outlines the basic applications of EDFAs within optical communication systems. We specifically leave the subject of amplification of WDM and soliton systems for later chapters.

5.3.7.1 Amplifier Cascades and Long-Distance Communications

Probably the one implementation of critical importance is for long-distance terrestrial and transoceanic communications. The EDFA is an adequate substitute for expensive regenerators, even though EDFAs do not compensate for intramodal chromatic dispersion. Figure 5.31 shows a typical EDFA cascade with emphasis on the periodic introduction of gain, noise, and fiber attenuation losses. In this example, an optical isolator is placed after each EDFA to prevent any backward propagating ASE from

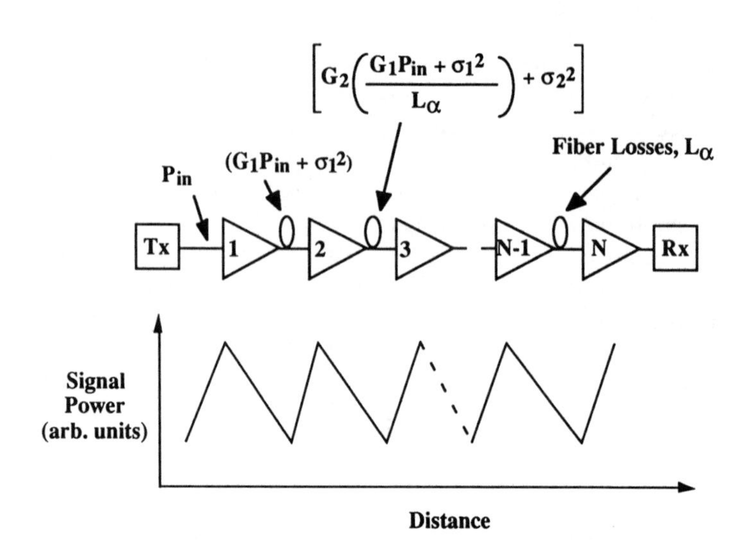

Figure 5.31 Cascade of EDFAs with periodic addition of gain, noise, and fiber losses. In this example, an isolator is placed after each EDFA.

accumulating and also to limit nonlinear scattering effects. Furthermore, a bandpass filter of bandwidth B_o is used to limit the ASE propagating from one stage to the next.

The operation of a cascade can be understood by considering the propagation of the signal and ASE as they traverse the system. An extremely simplified model can be employed to understand the gross effects in a cascade. The input of the first amplifier is simply the signal, whereas the input to all subsequent amplifiers is the output of the previous amplifier, which includes both signal and ASE power. The output signal power, $P_{s,i}$ output ASE power, P_{ASE}; and the total output power, $P_{out,i}$, of the ith amplifier in a cascade is [37]:

$$P_{s,i} = LG_iP_{s,i-1} \tag{5.77}$$

$$P_{ASE,i} = LG_iP_{ASE,i-1} + \left(\frac{B_o}{\Delta\nu}\right)P_{ASE,i} \tag{5.78}$$

$$P_{out,i} = LG_iP_{out,i-1} + 2n_{sp}(G_i - 1)h\nu B_o \tag{5.79}$$

where L is the interamplifier fiber attenuation and insertion losses, and the gain at each amplifier is not necessarily the small signal gain (G_o) but is described by the following transcendental equation:

$$G_i = G_o \left[(1 - G_i) \frac{P_{in}}{P_{sat}} \right] \tag{5.80}$$

The spontaneous emission factor is assumed to be constant from one amplifier to the next, and only the forward propagating ASE is considered, since optical isolators are placed along the cascade.

Figure 5.32 shows the signal and ASE noise progression when a cascade of many EDFAs is traversed [37]. In this example, the system is allowed to operate without constraints and naturally equilibrate and self-regulate. As the signal propagates along the cascade, three phenomena occur: (1) the signal slowly decays, (2) the ASE noise slowly accumulates, and (3) the SNR decreases slowly. A fundamental principle of cascaded amplifiers is that the signal and the ASE are both amplified from one amplifier to the next. In fact, since both the signal and ASE from one amplifier are input to the following amplifier, amplifier saturation will quickly result, causing the total output power from each EDFA to equilibrate to the same value.

An interesting effect of a freely operating cascade is the self-regulating effect of the signal power. If the interamplifier losses are such that the amplifiers are operated in a slightly saturated mode, then even if the signal experiences a boost or attenuation, the chain of amplifiers will cause the signal power to return close to its original value. This power-regulating effect is illustrated in Figure 5.33, for which the gain is compressed by ~ 3 dB at each amplifier. If nothing unexpected happens, the signal power will generally be constant from one EDFA to the next. If a signal undergoes an unexpected power boost, this will cause more gain compression in the following EDFAs, and so the signal power will return to its original lower level. On the other hand, if the signal experiences an unexpected power loss, the following amplifiers will not be as saturated, their gains will be higher, and the signal will be amplified until it again reaches its original power level.

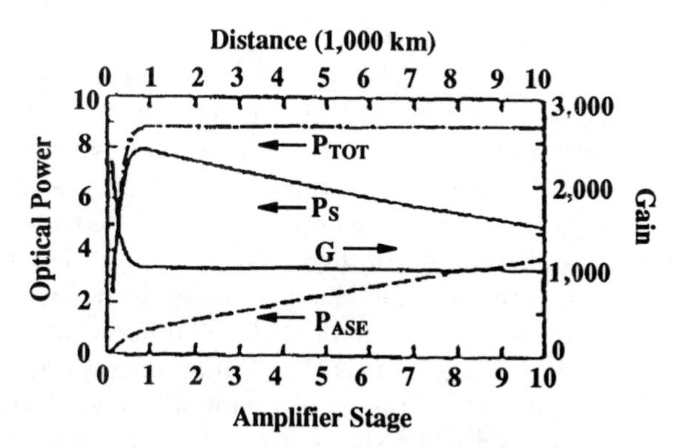

igure 5.32 Signal and ASE noise powers along an EDFA amplifier cascade. (*After:* [36]. © 1991 IEEE.)

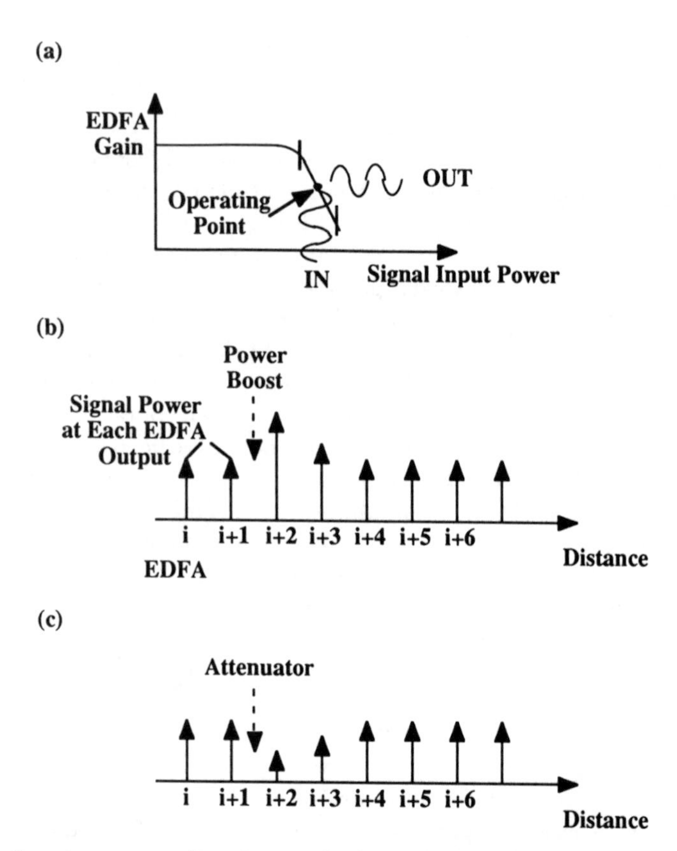

Figure 5.33 Self-regulating power effect of a cascade of EDFAs in which the gain is slightly saturated at each amplifier: (a) saturated amplification, (b) power boost, and (c) power attenuator.

Recent results for 5-Gbps and 10-Gbps NRZ transmission with 274 cascaded EDFAs have shown that nearly error-free transmission can be achieved for distances of 9,000 km, effectively conquering the barrier of repeaterless transmission for any possible distance around the globe (e.g., trans-Pacific distances are ~ 9,000 km) [89,90]. Figure 5.34 shows the BER curve for the results of this experiment [89]. Due to the relative positions and strengths of the absorption and emission cross-sections, the spectrum at the end of the cascade changes significantly from a typical single amplifier; the 1.531-μm peak has been attenuated, and the gain and ASE have shifted to longer wavelengths near 1.560 μm. As is clear from these results, EDFAs will have a remarkable impact on future long-distance communications. (Even more impressive results have been achieved using soliton transmission; see Chapter 6 for more information).

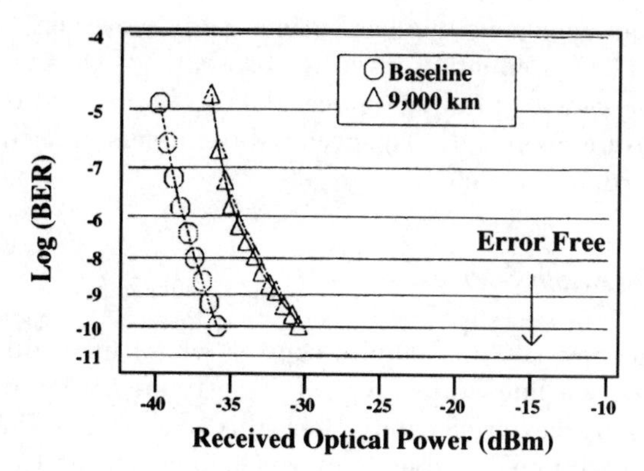

Figure 5.34 BER curve of the 9,000-km 274-EDFA cascaded system (*After:* [87]).

It should be mentioned that dispersion-shifted fiber must be used in these ultra-long-distance systems such that the transmitted channel experiences a dispersion factor very close to 0. However, EDFAs can be of considerable use by extending the unrepeatered distances even using conventional single-mode fiber with a dispersion minimum near 1.3 μm; results have shown 318-km spans using EDFAs and conventional fiber [91]. Also note that many extremely important nonlinear effects occur in ultra-long-distance links when several channels are transmitted on different wavelengths [92,93]. These effects will be discussed in Chapter 7.

5.3.7.2 Preamplifiers and Power Amplifiers

EDFAs have been successfully used as preamplifiers in receivers [94,95]. The results are quite impressive, with a sensitivity of −46 dB at 2 Gbps for a single-stage set-up and −40 dB at 10 Gbps for a double-stage EDFA set-up to achieve 10^{-9} [80,96]. It now seems clear that direct detection with EDFAs can rival the high sensitivity of coherent detection with a high-power local oscillator.

Additionally, EDFAs have shown their ability for impressive power boosting at the output of a laser transmitter. Results showing +27-dBm output power are unheard of with semiconductor amplifiers [97].

EDFAs are commercially available, so implementing them anywhere in an optical system has become advantageous and quite easy to achieve.

5.3.7.3 Recirculating Loops

One method for storing optical data is to construct an optical memory or delay line. It is simply a loop of fiber that holds optical data as long as the data propagate around the loop, perhaps an enormous number of times [98]. However, each circular prop agation results in fiber attenuation losses. Incorporating EDFAs in the recirculatin loop can overcome losses and allow optical data to be delayed or stored for man revolutions around the loop [99]. The eventual usefulness of such loops is not clea and is awaiting further research.

5.3.7.4 Narrow-Linewidth Sources

Certain communication systems require sources whose linewidth is quite narrow perhaps ~ 100 kHz. The linewidth of typical single-frequency lasers is >1 MHz, unles they are connected to an external cavity. Since this arrangement is complex, a simple solution may be an erbium-doped fiber laser, which lases near 1.55 μm. As with an laser, the basic requirements include (1) a gain medium, (2) frequency-selectiv feedback, and (3) gain greater than loss inside the cavity. As shown in Figure 5.35 a circular piece of erbium-doped fiber with a frequency-selective filter incorporate into the loop acts as a laser whose cavity is several meters long [100]. Since the cavit is circular, the feedback is provided not by mirrors but by circular return. An isolato

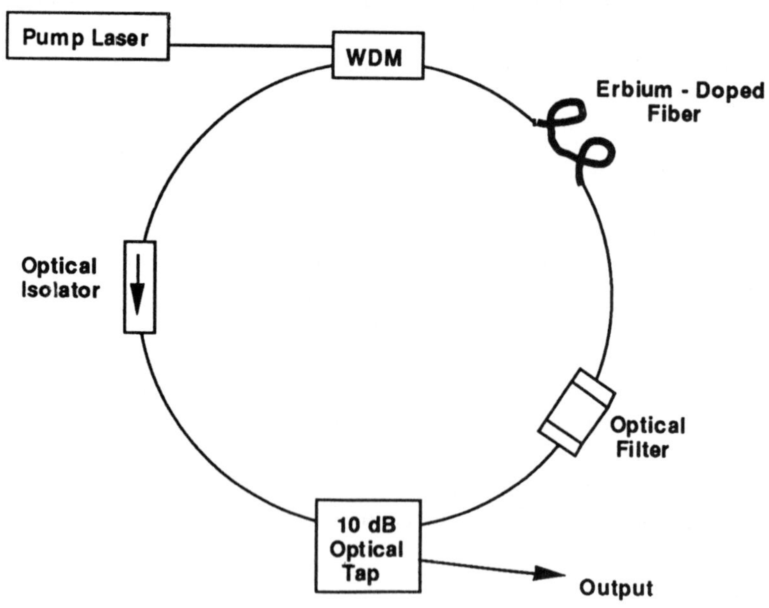

Figure 5.35 Erbium-doped fiber laser.

is required to prevent backward reflections into the optical path, and light is coupled out of the loop by a passive optical tap. Linewidths of 100 kHz have been achieved. This scheme is quite straightforward to implement, is fiber compatible, and can be tuned over the entire gain bandwidth of the EDFA. In addition, extremely short pulses can be generated in a fiber ring laser [101]. This may be a viable signal source in the future.

5.3.8 Erbium-Doped Integrated Waveguide Amplifiers

Given that erbium has so many desirable qualities and that an integrated erbium-doped amplifier would be occasionally preferred for low cost, compact design, and easy OEIC integration, research has progressed on doping integrated silica waveguides with erbium. Gain as high as 15 dB has been achieved, and the doping densities are much higher than for a fiber [102–104]. More results are expected in the near future, but the fiber amplifier will not be replaced by an integrated waveguide amplifier, only augmented.

5.4 COMPARISON OF MAJOR SOA AND EDFA CHARACTERISTICS

For easy reference, Table 5.1 compares many of the fundamental amplifier characteristics of the SOA and the EDFA.

Table 5.1
Major SOA and EDFA Characteristics

Type of Amplifier	*Energy Level*	*Pumping Source*	*Lifetime*	*Inserstion Losses (per facet)*	*Center Wavelength (Gain Bandwidth)*	*Length*	*Fiber-to-Fiber Gain*	*Typical Noise Figure*
SOA	~2 levels	Electrical (~100 mA)	~1 ns	~3 dB	1.3 or 1.55 μm (~50–75 nm)	~500 μm	10–15 dB	8–12 dB
EDFA	3 levels	Optical (~20–50 mW)	~1 ms	~0.2 dB	1.55 μm (~20–30 nm)	~10m	10–40 dB	4–5 dB

5.5 OTHER TYPES OF FIBER AMPLIFIERS

The field of optical amplifiers is overwhelmingly dominated by the mechanisms discussed in previous sections. However, three other types of fiber amplifiers deserve mention and a brief discussion.

5.5.1 Praseodymium Amplifiers

Because most of the installed optical fiber is conventional with a dispersion zero at 1.3 μm, it is highly desirable to have a fiber-based optical amplifier at 1.3 μm, which has similar characteristics to the 1.55-μm EDFA. Doping a fiber with praseodymium will enable a fiber to experience fluorescence and thus amplify in the 1.3-μm range. However, because praseodymium is a four-level energy system and not a three-level system (as in erbium), the pump efficiency to achieve high gain is extremely low. In fact, hundreds of mW of pump power are required to achieve 20 dB gain, whereas <20 mW is required for EDFAs [105]. The future of praseodymium-doped fiber amplifiers (PDFAs) hinges on the future of extremely high-power and inexpensive pump sources. Moreover, as mentioned with semiconductor amplifiers, using chromatic inversion and EDFAs will allow for long-distance (>300 km) 1.55-μm transmission even with conventional fiber, making PDFAs even less attractive for system implementation.

5.5.2 Brillouin and Raman Amplifiers

It is important to briefly mention two other types of fiber amplifiers because of their historic significance in preceding erbium by many years. These two amplifiers are Brillouin and Raman amplifiers. The fundamental mechanisms involve molecular interactions of light with the fiber, causing the molecules to vibrate. Energy can be transferred from a propagating pump beam to a propagating signal beam. Relative to the amplifiers examined in this chapter, these amplifiers are much less efficient, requiring high pump power and extremely long lengths of fiber (kilometers).

The fundamental principles are as follows [18,82]. In the Brillouin amplifier, an incident photon interacts with the material forming an acoustic wave (phonon) and a reradiated photon at a lower frequency (Stokes frequency). This photon is shifted ~ 10 GHz down in frequency, or about 1 angstrom, since the acoustic wave is uniform in a fiber. The odd part of the Brillouin amplifier is that the pump and the signal must be counterpropagating, with no copropagating interaction possible.

In the Raman amplifier, the pump light photon is absorbed and sets the fiber molecules into mechanical vibrations. A photon is again radiated at the Stokes frequency, but since mechanical vibrations are not uniform in a fiber, the Stokes frequency is not a set number. Furthermore, the pump and the signal may co- or counterpropagate in the fiber. The Raman amplifier is much less efficient than the Brillouin amplifier.

It seems improbable that future optical communication systems will incorporate Brillouin or Raman amplifiers. Furthermore, the fundamental gain mechanism for these two amplifiers is fundamentally different than the semiconductor and erbium amplifiers. The reader is encouraged to seek further information pertaining to these amplifiers in texts dealing with quantum electronics. Brillouin and Raman scattering effects will be discussed in Chapter 7.

5.6 SUMMARY

This chapter covered a wide range of topics involving optical amplifiers. These topics include the fundamental gain and noise mechanisms, important device parameters, systems performance, and applications. Furthermore, basic issues were compared between semiconductor amplifiers and EDFAs.

It should be evident to the reader that enormous research progress has taken place in just the last few years, with mind-boggling transmission distances effectively conquered. This explosion was fueled by the almost ideal characteristics of EDFAs for incorporation into optical systems. EDFAs are commercially available, with the future cost expected to drop well below the present bench mark of $20,000. They are also embarrassingly straightforward to implement. We will revisit optical amplifiers in Chapters 6 and 7 because of their significant impact on multichannel and soliton communication systems. There is little doubt that optical amplifiers will enable global and local optical systems within the near future.

Problems

Problem 5.1

We described a method by which the polarization-dependent gain of a semiconductor amplifier can be compensated by concatenating two amplifiers, with each one rotated axially by 90 degrees. Describe a scenario whereby the same kind of gain polarization dependence can be overcome by a *parallel* amplifier approach.

Problem 5.2

a. If we want the facet reflectivities in the semiconductor amplifier to be 10^{-4} without using AR coatings, how far should the waveguiding region be from the facet if we use a buried-window technique? *Hint:* Assume Gaussian beam divergence. (The rectangular waveguide has dimensions 10 μm wide by 2 μm high, and the material is GaAs.)
b. What would this distance be if we also had an AR coating applied that provided 10^{-3} facet reflectivity?

Problem 5.3

In a typical GaAs-based FP semiconductor amplifier (length = 400 μm, gain peak = 1.55 μm, gain bandwidth = 40 nm), plot the output gain as a function of signal wavelength when one facet has an AR coating.

Problem 5.4

We discussed that the gain coefficient in a semiconductor amplifier is only *approximated* by a Lorentzian lineshape. An empirical quadratic formula was also given that replicated the gain only in the center of its gain bandwidth, since the gain is asymmetric at the wings. Arrive at a generic cubic equation that would approximate the empirical shape of the gain over most of the gain spectrum. *Hint:* The constant in front of the quadratic term accounts for the gain bandwidth, and the constant in front of the cubic term accounts for asymmetry in the gain spectrum.

Problem 5.5

Plot how the gain coefficient would change in an SOA as (a) the bias current and (b) the carrier concentration is changed. Assume typical values (i.e., carrier lifetime ~0.5 ns, P_{sat} ~+1 dB).

Problem 5.6

Optical filtering plays a critical role in limiting the ASE-generated noise. Plot the relevant noise powers (i.e., N_{th}, N_{s-sp}, and N_{sp-sp}) as a function of optical filter bandwidth if the EDFA is used as a preamplifier. Assume signal power = .1, 1, 10, and 100 μW, n_{sp} = 1.4, λ = 1.55 μm, electrical bandwidth = 2 GHz, and the thermal noise = 10^{-22} A/$\sqrt{\text{Hz}}$. Next, plot the SNR versus the optical filter bandwidth and comment on the results.

Problem 5.7

We have assumed that the ASE spectrum can be considered uniform with frequency over the optical filter bandwidth. However, real filters have a Lorentzian-shaped passband. If we keep the total ASE power that passes through the filter as a constant and just change the filter shape from flat-top to Lorentzian, how will this affect the ASE-generated noise terms and the ASE? Derive the approximate new formulae for N_{s-sp} and N_{sp-sp}. Assume optical filter bandwidth = 1 nm, electrical bandwidth = 2 GHz, n_{sp} = 1.4, λ = 1.55 μm, P_{sig}^{in} = 0.1 μW.

Problem 5.8

Find the SNR in a distributed erbium-doped amplifier and for a lumped-EDFA cascade (as a function of the number of amplifiers) in which the amplifiers are placed every $1/\alpha$ meters, where α is the loss coefficient per meter.

Problem 5.9

Derive an analytical solution for the effect that an optical isolator has on output SNR when the isolator is placed somewhere along the span of the erbium-doped fiber. Repeat this analysis for the case of placing a Lorentzian-shaped optical transmission filter (of 3-dB passband f_{BW}) along the amplifier span. *Hint:* Evaluate the ASE accumulation in the forward and backward directions.

Problem 5.10

Numerically compare the system performance (SNR) of an SOA and an EDFA. Assume $\lambda = 1.55$ μm, the SOA has coupling losses of 3-dB per facet, n_{sp} for the SOA and EDFA is 3.5 and 1.4, respectively, electrical bandwidth is 2 GHz, signal power is 1 μW, and all other variables have typical values.

Problem 5.11

a. What is the noise level on the 1 and 0 bits given EDFA gain = 30 dB, $\lambda = 1.55$ μm, optical filter bandwidth = 1 nm, electrical bandwidth = 2 GHz, $n_{sp} = 1.4$, and signal power = .1 μW. Give all noise terms separately.
b. A typical receiver places its decision level when recovering bits at the half amplitude between the 1 and 0 levels. Given the answer derived in part (a), find the optimum decision amplitude level in this system.

Problem 5.12

The EDFA gain and noise figure is highly nonuniform with regard to wavelength. Given high amplifier pumping conditions, compare the detected SNR for signals at 1.531 and 1.560 μm when considering that the gain is 40 and 30 dB and the n_{sp} is 2.5 and 1.2, respectively. Signal power = 0.1 μW, electrical bandwidth = 2 GHz, and the optical filter bandwidth = 1 nm.

Problem 5.13

Optical amplification solves the issue of fiber attenuation. However, dispersion remains unchecked. In an ultra-long-distance system of 2,000 km and using dispersion-shifted fiber (DSF), some dispersion still accumulates. It was mentioned in this chapter that we can use the positive dispersion in a length of conventional fiber to compensate for the negative dispersion in the DSF. How long a piece of conventional fiber is necessary to compensate for the dispersion from the DSF? Dispersion at 1.55 μm for DSF and conventional fiber is -1 and $+17$ ps/nm·km, respectively. If we have a 2.5-Gbps signal, after how many kilometers of propagation in DSF alone does dispersion cause a 20% spread in a transmitted bit?

Problem 5.14

What is the total gain needed for a 10,000-km fiber link (fiber attenuation = 0.2 dB/km)? Since the gain of the EDFAs will be saturated due to ASE and signal accumulation, the small-signal gain will be much higher than the actual gain provided to the signal. How many 20-dB gain amplifiers will be needed if their saturated gain is only 10.5 dB? What would be the effect if we now placed an optical filter of 1-nm bandwidth after each amplifier in the cascade (assume the EDFA ASE bandwidth to be ~30 nm)? Show the result quantitatively using typical EDFA values.

References

[1] Li, T., "The Impact of Optical Amplifiers on Long-Distance Lightwave Telecommunications," *IEEE Proc.* vol. 81, pp. 1568–1579, 1993.

[2] Yariv, A., *Optical Electronics*, 4th Ed., App. D, Philadelphia: Holt, Rinehart & Winston, 1991.

[3] Green, P. E., Jr., *Fiber Optic Networks*, chap. 6, Englewood Cliffs, NJ: Prentice Hall, 1993.

[4] Agrawal, G. P., *Fiber-Optic Communication Systems*, Chapter 8, New York: Wiley, 1992.

[5] Giles, C. R., "System Applications of Optical Amplifiers," Conference on Optical Fiber Communications '92, Tutorial Sessions, TuF, Washington, D.C.: Optical Society of America, 1992.

[6] Hill, A. M., et al., "Million-Way WDM Broadcast Network Employing Two Stages of Erbium-Doped Fibre Amplifiers," *Electronics Letters*, Vol. 26, 1990, pp. 1882–1884.

[7] O'Mahony, M. J., "Semiconductor Laser Optical Amplifiers for Use in Future Fiber Systems," *J. of Lightwave Technology*, Vol. 6, 1988, pp. 531–544.

[8] Saitoh, T., et al., "Recent Progress in Semiconductor Laser Amplifiers," *J. of Lightwave Technology*, Vol. 6, 1986, pp. 1656–1664.

[9] Saitoh, T., and T. Mukai, "1.5 μm GaInAsP Traveling-Wave Semiconductor Laser Amplifier," *IEEE J. of Quantum Electronics*, Vol. 23, 1987, p. 1010.

[10] Simon, J. C., "GaInAsP Semiconductor Laser Amplifiers for Single Mode Fiber Communication," *J. of Lightwave Technology*, Vol. 5, 1987, pp. 1286–1295.

[11] Fye, D. M., "Practical Limitations on Optical Amplifier Performance," *J. of Lightwave Technology*, Vol. 2, 1984, pp. 403–406.

[12] Mears, R. J., L. Reekie, I. M. Jauncy, and D. N, Payne, "Low Noise Erbium-doped Fiber Amplifier Operating at 1.54 µm," *Electronics Letter*, Vol. 23, 1987, p. 1026.
[13] Desurvire, E., J. R. Simpson, and P. C. Becker, "High-Gain Erbium-Doped Traveling-Wave Fiber Amplifier," *Optical Letter*, Vol. 12, 1987, pp. 888–890.
[14] Desurvire, E., *Erbium-Doped Fiber Amplifiers: Principles and Applications*, New York: John Wiley & Sons, 1994.
[15] Bjarklev, A., *Optical Fiber Amplifiers*, Norwood, MA: Artech House, 1993, p. 215.
[16] Stone, J., and C. A. Burrus, "Neodymium Doped Silica Lasers in End-Pumped Fiber Geometry," *Applied Physics Letter*, Vol. 23, 1973, p. 388.
[17] Yamamoto, Y., "Characteristics of AlGaAs Fabry-Perot Cavity Type Laser Amplifiers," *J. Quantum Electronics*, Vol. QE-16, 1980, pp. 1047–1052.
[18] Yariv, A., *Quantum Electronics*, 2nd Ed., New York: John Wiley & Sons 1975, chap. 18.
[19] Siegman, A. E., *Lasers*, Mill Valley, CA: University Science Books, 1986, chap. 30.
[20] Eisenstein, G., and R. M. Jopson, "Measurements of the Gain Spectrum of Near Traveling-Wave and Fabry-Perot Semiconductor Optical Amplifiers at 1.5 µm," *Int. J. Electronics*, Vol. 60, 1986, pp. 113–121.
[21] Henning, I. D., M. J. Adams, and J. V. Collins, "Performance Prediction From a New Optical Amplifier Model," *J. Quantum Electronics*, Vol. 2, 1985, pp. 609–613.
[22] Agrawal, G. P., and N. K. Dutta, *Long-Wavelength Semiconductor Lasers*, New York: Van Nostrand Reinhold, 1986, chap. 2.
[23] Born, M., and E. Wolf, *Principles of Optics*, 6th Ed., New York: Pergamon Press, 1980, chap 1.
[24] Vassallo, C., "Gain Ripple Minimization and Higher-Order Modes in Semiconductor Optical Amplifiers," *Electronics Letter*, Vol. 25, 1989, pp. 789–790.
[25] Simon, J. C., et al, Proc. European Conference on Optical Communications, 1986.
[26] Zah, C. E., et al., *Electronics Letter*, Vol. 23, 1987, p. 990.
[27] Olsson, N. A., et al., "Polarization Independent Optical Amplifier With Buried Facets," *Electronics Letter*, Vol. 25, 1989, p. 1048.
[28] Olsson, N. A., "Lightwave Systems With Optical Amplifiers," *J. Lightwave Technology*, Vol. 7, 1989, pp. 1071–1082.
[29] Henry, C. H., "Theory of Spontaneous Emission Noise in Open Resonators and Its Application to Lasers and Amplifiers," *J. Lightwave Technology*, Vol. 4, 1986, pp. 288–297.
[30] Yamamoto, Y.,"Noise and Error Rate Performance of Semiconductor Laser Amplifiers in PCM-IM Optical Transmission Systems," *IEEE J. of Quantum Electronics*, Vol. QE-16, 1980, pp. 1073–1081.
[31] Personick, S. D., "Applications for Quantum Amplifiers in Simple Digital Optical Communications Systems," *Bell Syst. Tech. J.*, Vol. 52, 1973, pp. 117–133.
[32] Willner, A. E., "SNR Analysis of Crosstalk and Filtering Effects in an Amplified Multi-Channel Direct-Detection Dense-WDM System," *IEEE Photonics Technology Letter*, Vol. 4, 1992, pp. 186–189.
[33] Steele, R. C., G. R. Walker, and N. G. Walker, "Sensitivity of Optically Preamplified Receivers With Optical Filtering," *IEEE Photonics Technology Letter*, Vol. 3, 1991, pp. 545–547.
[34] Suzuki, Y., et al., "High-Gain, High-Power 1.3 µm Compressive Strained MQW Optical Amplifier," *IEEE Photonics Technology Letter*, Vol. 4, 1993, pp. 404–406.
[35] Marcuse, D., "Derivation of Analytical Expressions for the Bit-Error Probability in Lightwave Systems with Optical Amplifiers," *J. Lightwave Technology*, Vol. 8, 1990, pp. 1816–1823.
[36] Humblet, P. A., and M. Azizoglu, "On the Bit Error Rate of Lightwave Systems With Optical Amplifiers," *J. Lightwave Technology*, Vol. 9, 1991, pp. 1576–1582.
[37] Giles, C. R., and E. Desurvire, "Propagation of Signal and Noise in Concatenated Erbium-Doped Fiber Optical Amplifiers," *J. Lightwave Technology*, Vol. 9, 1991, pp. 147–154.
[38] Tonguz, O. K., and R. E. Wagner, "Equivalence Between Preamplified Direct Detection and Heterodyne Receivers," *IEEE Photonics Technology Letter*, Vol. 3, 1991, pp. 835–837.

[39] Newkirk, M. A., et al., "1.5 μm Multiquantum-Well Semiconductor Optical Amplifier With Tensile and Compressively Strained Wells for Polarization-Independent Gain," *IEEE Photonics Technology Letter*, Vol. 4, 1993, pp. 406–408.

[40] Grosskopf, G., R. Ludwig, and H. G. Weber, "Crosstalk in Optical Amplifiers for Two-Channel Transmission," *Electronics Letter*, Vol. 22, 1986, pp. 900–902.

[41] Jopson, R. M., and T. E. Darcie, "Calculation of Multicarrier Intermodulation Distortion in Semiconductor Optical Amplifiers," *Electronics Letter*, Vol. 24, 1988, pp. 1372–1374.

[42] Jopson, R. M., et al, "Measurement of Carrier Density Mediated Intermodulation Distortion in an Optical Amplifier," *Electronics Letter*, Vol. 23, 1987, pp. 1394–1395.

[43] Agrawal, G. P., "Four-Wave Mixing and Phase Conjunction in Semiconductor Laser Media," *Optical Letter*, Vol. 12, No. 4, 1987, pp. 260–262.

[44] Wiesenfeld, J. M., et al., "Gain Spectra and Gain Compression of Strained-Layer Multiple Quantum Well Optical Amplifiers," *Applied Physics Letter*, Vol. 58, 1991, pp. 219–221.

[45] Wiesenfeld, J. M., A. H. Gnauck, G. Raybon, and U. Koren, "High-Speed Multiple-Quantum-Well Optical Power Amplifier," *IEEE Photonics Technology Letter*, Vol. 4, 1992, pp. 708–711.

[46] Jopson, R. M., et al., "8 Gbit/s, 1.3 μm Receiver Using Optical Preamplifier," *Electronics Letter*, Vol. 25, 1988, pp. 233–235.

[47] Ryu, S., et al., *Electronics Letter*, Vol. 25, 1989, p. 1682.

[48] Gnauk, A. H., R. M. Jopson, and R. M. Derosier, "10-Gb/s 360km Transmission Over Dispersive Fiber using Midsystem Spectral Inversion," *IEEE Photonics Technology Letter*, Vol. 5, 1993, pp. 663–666.

[49] Koch, T. L., and U. Koren, "Semiconductor Photonic Integrated Circuits," *IEEE J. Quantum Electronics*, Vol. 27, 1991, pp. 641–653.

[50] Zah, C. E., et al., "Monolithic Integration of Multiwavelength Compressive-Strained Multiquantum-Well Distributed-Feedback Laser Array with Star Coupler and Optical Amplifiers," *Electronics Letter*, Vol. 28, 1992, pp. 2361–2362.

[51] Fortenberry, R., A. J. Lowery, W. L. Ha, and R. S. Tucker, "Photonic Packet Switch Using Semiconductor Optical Amplifier Gates," *Electronics Letter*, Vol. 27, 1991, pp. 1305–1307.

[52] Zucker, J. E., "Multiple-Quantum-Well Modulators and Switches," Optical Society of America Annual Meeting '93, *Technical Digest*, Washington, DC: Optical Society of America, 1993, tutorial MM.

[53] Alferness, R. C., in *Guided-Wave Optoelectronics*, T. Tamir, ed., New York: Springer Verlag, 1989.

[54] Barnsley, P. E., and P. J. Fiddyment, "Wavelength Conversion From 1.3 to 1.55 μm Using Split Contact Optical Amplifiers," *IEEE Photonics Technology Letter*, Vol. 3, 1991, pp. 256–258.

[55] Glance, B., et al., "High Performance Optical Wavelength Shifter," *Electronics Letter*, Vol. 28, 1992, pp. 1714–1715.

[56] Wiesenfeld, J. M., and B. Glance, "Cascadability and Fanout of Semiconductor Optical Amplifier Wavelength Shifter," *IEEE Photonics Technology Letter*, Vol. 4, 1992, pp. 1168–1171.

[57] Durhuus, T., B. Mikkelsen, and K.E. Stubkjaer, "Detailed Dynamic Model for Semiconductor Optical Amplifiers and Their Crosstalk and Intermodulation Distortion," *J. Lightwave Technology*, Vol. 10, 1992, pp. 1056–1065.

[58] Joergensen, C., et al., "4 Gb/s Optical Wavelength Conversion Using Semiconductor Optical Amplifiers," *IEEE Photonics Technology Letter*, Vol. 5, 1993, pp. 657–660.

[59] Choa, F. S., and T. L. Koch, "Static and Dynamical Characteristics of Narrow-Band Tunable Resonant Amplifiers as Active Filters and Receivers," *J. Lightwave Technology*, Vol. 9, 1991, pp. 73–83.

[60] Koch, T. L., F. S. Choa, F. Heismann, and U. Koren, "Tunable Multiple-Quantum-Well Distributed-Bragg-Reflector Lasers as Tunable Narrowband Receivers," *Electronics Letter*, Vol. 25, 1989, pp. 890–892.

[61] Chawki, M. J., et al., "Two-Electrode DFB Laser Filter Used as a Wide Tunable Narrow-Band FM Receiver: Tuning Analysis, Characteristics and Experimental FSK-WDM System," *J. Lightwave Technology*, Vol. 10, 1992, pp. 1388–1396.
[62] Magari, K., H. Kawaguchi, K. Oe, and M. Fukuda, "Optical Narrow-Band Filters Using Optical Amplification With Distributed Feedback," *IEEE J. Quantum Electronics*, Vol. 24, 1988, pp. 2178–2190.
[63] Kazovsky, L. G., M. Stern, S. G. Menocal, Jr., and C. E. Zah, "DBR Active Optical Filters: Transfer Function and Noise Characteristics," *J. Lightwave Technology*, Vol. 8, 1990, pp. 1441–1451.
[64] Goldstein, E. L., and M. C. Teich, "Noise Measurements on Distributed-Feedback Optical Amplifiers Used as Tunable Active Filters," *IEEE Photonics Technology Letter*, Vol. 3, 1991, pp. 45–46.
[65] Koai, K. T., and R. Olshansky, "Simultaneous Optical Amplification, Detection, and Transmission Using In-Line Semiconductor Laser Amplifiers," *IEEE Photonics Technology Letter*, Vol. 4, 1992, pp. 441–443.
[66] Sunak, H. R. D., "Fundamentals of Erbium-Doped Fiber Amplifiers," Society of Photo-Instrumentation Engineers OE/Fibers '92 Conference, Short Course, SC9, Bellingham, WA: SPIE, 1992.
[67] Koester, C. J., and E. Snitzer, "Amplification in a Fiber Laser," *Applied Opt.*, Vol. 3, 1964, p. 1182.
[68] Armitage, J. R., "Three-Level Fiber Laser Amplifier: A Theoretical Model," *Applied Optical*, Vol. 27, 1988, pp. 4831–4836.
[69] Zyskind, J. L., et al., "Fast Power Transients in Optically Amplified Multiwavelength Optical Networks," Conference on Optical Fiber Communications '96, San Jose, CA, *Technical Digest*, Washington, DC: Optical Society of America, 1996, Paper PD-31.
[70] Desurvire, E., "Erbium-Doped Fiber Amplifiers," Conference on Optical Fiber Communications '91, Tutorial Sessions, Washington, DC: Optical Society of America, 1991.
[71] OKI Corp., Product brochure, 1992.
[72] Giles, C. R., and E. Desurvire, "Modeling Erbium-Doped Fiber Amplifiers," *J. Lightwave Technology*, Vol. 9, 1991, pp. 271–283.
[73] Miniscalco, W. J., "Erbium-Doped Glasses for Fiber Amplifiers at 1500 nm," *J. Lightwave Technology*, Vol. 9, 1991, pp. 234–250.
[74] Li, T., and M. C. Teich, "Bit-Error Rate for a Lightwave Communication System Incorporating an Erbium-Doped Fibre Amplifier," *Electronics Letter*, Vol. 27, 1991, pp. 598–599.
[75] Saleh, A. A. M., R. M. Jopson, J. D. Evankow, and J. Aspell, "Modeling of Gain in Erbium-Doped Fiber Amplifiers," *IEEE Photonics Technology Letter*, Vol. 2, 1990, pp. 714–717.
[76] Desurvire, E., "Spectral Noise Figure of Er^{3+}-Doped Fiber Amplifiers," *IEEE Photonics Technology Letter*, Vol. 2, 1990, pp. 208–210.
[77] Olshansky, R., "Noise Figure for Erbium-Doped Optical Fibre Amplifiers," *Electronics Letter*, Vol. 24, 1988, pp. 1363–1365.
[78] Walker, G. R., D. M. Spirit, D. L. Williams, and S. T. Davey, "Noise Performance of Distributed Fibre Amplifiers," *Electronics Letter*, Vol. 27, 1991, pp. 1390–1391.
[79] Giles, C. R., and D. Di Giovanni, "Spectral Dependence of Gain and Noise in Erbium-Doped Fiber Amplifiers," *IEEE Photonics Technology Letter*, Vol. 2, 1990, pp. 797–800.
[80] Giles, C. R., E. Desurvire, J. L. Zyskind, and J. R. Simpson, "Noise Performance of Erbium-Doped Fiber Amplifier Pumped at 1.49 μm, and Application to Signal Preamplification at 1.8 Gbit/s," *IEEE Photonics Technology Letter*, Vol. 1, 1989, pp. 367–369.
[81] Way, W. I., et al., "Noise Figure of a Gain-Saturated Erbium-Doped Fiber Amplifier Pumped at 980 nm," Topical Meeting on Optical Amplifiers and Their Applications '91, *Technical Digest*, Washington, DC: Optical Society of America, 1991, Paper TuB3.
[82] Willner, A. E., and E. Desurvire, "Effect of Gain Saturation on Receiver Sensitivity in 1 Gb/s Multichannel FSK Direct-Detection Systems Using Erbium-Doped Fiber Preamplifiers," *IEEE Photonics Technology Letter*, Vol. 3, 1991, pp. 259–261.

[83] Zyskind, J. L., et al., "High-Performance Erbium-Doped Fiber Amplifier Pumped at 1.48 μm and 0.97 μm," Topical Meeting on Optical Amplifiers and Their Applications '91, Monterey, CA, *Technical Digest*, Washington, DC: Optical Society of America, 1991, Paper PDP6.

[84] Giles, C. R., E. Desurvire, and J. R. Simpson, "Transient Gain and Cross Talk in Erbium-Doped Fiber Amplifiers," *Optics Letter*, Vol. 14, 1989, pp. 880–882.

[85] Desurvire, E., "Analysis of Transient Gain Saturation and Recovery in Erbium-Doped Fiber Amplifiers," *IEEE Photonics Technology Letter*, Vol. 1, 1989, pp. 196–199.

[86] Yariv, A., "Signal-to-Noise Considerations in Fiber Links With Periodic or Distributed Optical Amplification," *Optics Letter*, Vol. 15, 1990, pp. 1064–1066.

[87] Simpson, J. R., et al., "Performance of a Distributed Erbium-Doped Dispersion-Shifted Fiber Amplifier," *J. Lightwave Technology*, Vol. 9, 1991, pp. 228–233.

[88] Goldstein, E. L., "Noise Performance of Bus-Configured Optical Networks with Distributed Fiber Amplification," Topical Meeting on Optical Amplifiers and Their Applications '91, *Technical Digest*, Washington, DC: Optical Society of America, 1991, Paper FB5.

[89] Bergano, N. S., et al., "9000 km, 5 Gb/s NRZ Transmission Experiment Using 274 Erbium-Doped Fiber Amplifiers," Topical Meeting on Optical Amplifiers and Their Applications '92, Santa Fe, *Technical Digest*, Washington, DC: Optical Society of America, 1992, Paper PD-11.

[90] Taga, H., et al., "10 Gb/s, 9000 km IM-DD Transmission Experiment Using 274 Erbium-Doped Fiber Amplifier Repeaters," Conference on Optical Fiber Communications '93, Washington, DC: Optical Society of America, 1993, Paper PD-1.

[91] Park, P. K., et al., "318-km Repeaterless Transmission Using Erbium-Doped Fiber Amplifiers in a 2.5-Gb/s IM/DD System," Conference on Optical Fiber Communications '92, *Technical Digest*, Washington, DC: Optical Society of America, 1992, Paper ThK1.

[92] Chraplyvy, A. R., "Limitations on Lightwave Communications Imposed by Optical-Fiber Nonlinearities," *J. Lightwave Technology*, Vol. 8, 1990, pp. 1548–1557.

[93] Tkach, R. W., "System Implications of Optical Fiber Nonlinearities," Optical Society of America Annual Meeting '92, *Technical Digest*, Washington, DC: Optical Society of America, 1992, Tutorial.

[94] Gabla, P. M., E. Leclerc, and C. Coeurjolly, "Practical Implementation of a Highly Sensitive Receiver Using an Erbium-Doped Fiber Preamplifier," *IEEE Photonics Technology Letter*, Vol. 3, 1991, pp. 727–729.

[95] Jacobs, I., "Effect of Optical Amplifier Bandwidth on Receiver Sensitivity," *IEEE Trans. on Communications*, Vol. 38, 1990, pp. 1863–1864.

[96] Laming, R. I., et al., "High Sensitivity Optical Pre-Amplifier at 10 Gbit/s Employing a Low Noise Composite EDFA With 46 dB Gain," Topical Meeting on Optical Amplifiers and Their Applications '92, *Technical Digest*, Washington, DC: Optical Society of America, 1992, Paper PD-13.

[97] Massicott, J. F., R. Wyatt, B. J. Ainslie, and S. P. Craig-Ryan, "Efficient, High Power, High Gain Er^{3+} Doped Silica Fibre Amplifier," *Electronics Letter*, Vol. 26, 1990, pp. 1038–1039.

[98] Eiselt, M., et al., "One Million Pulse Circulations in a Fiber Ring Using a SLALOM for Pulse Shaping and Noise Reduction," *IEEE Photonics Technology Letter*, Vol. 4, 1993, pp. 422–424.

[99] Spring, J., and R. S. Tucker, "Photonic 2 by 2 Packet Switch With Input Buffers," *Electronics Letter*, Vol. 29, 1993, pp. 283–284.

[100] Zyskind, J. L. et al., "Diode-Pumped, Electrically Tunable Erbium-Doped Fiber Ring Laser With Fiber Fabry-Perot Etalon," *Electronics Letter*, Vol. 27, 1991, p. 1950.

[101] Nakazawa, M., E. Yoshida, and Y. Kimura, "Generation of 98 fs Optical Pulses Directly from an Erbium-Doped Fibre Ring Laser at 1.57 μm," *Electronics Letter*, Vol. 29, 1993, pp. 63–65.

[102] Becker, P., et al., "Erbium-Doped Integrated Optical Amplifiers and Lasers in Lithium Niobate," Topical Meeting on Optical Amplifiers and Their Applications '92, *Technical Digest*, Washington DC: Optical Society of America, 1992, Paper ThB4.

[103] Miliou, A. N., X. F. Cao, R. Srivastava, and R. V. Ramaswamy, "15-dB Amplification at 1.06 μm

in Ion-Exchanged Silicate Glass Waveguides," *IEEE Photonics Technology Letter,* Vol. 4, 1993, pp. 416–418.

[104] Nykolak, G., "Systems Evaluation of an Er^{3+}-Doped Planar Waveguide Amplifier," *IEEE Photonics Technology Letter,* Vol. 5, 1993, pp. 1185–1187.

[105] Shimizu, M., et al., "28.3 dB Gain 1.3 μm-Band Pr-Doped Fluoride Fiber Amplifier Module Pumped by 1.017 μm InGaAs-LD's," *IEEE Photonics Technology Letter,* Vol. 5, 1993, pp. 654–657.

Chapter 6

Soliton Systems

6.1 INTRODUCTION

Coherent systems, optical amplifiers, and WDM (wavelength division multiplexing) greatly enhance the capacity of optical fiber communication systems. The common theme in all these technologies is their similarity to conventional *radio frequency* (RF) systems. Fundamentally, all three just transfer the familiar RF principles to optical communication systems.

Soliton systems are fundamentally different: They rely on fiber properties and therefore have no analog in RF systems. The two relevant fiber properties are dispersion and nonlinearity. Taken separately, each property can be detrimental to the performance of a communication system. However, under certain conditions, the impact of nonlinearity cancels the impact of dispersion. When that happens, a pulse of light—a so-called soliton—can propagate over extremely long distances (thousands of kilometers) without any distortion. Thus, soliton systems are potentially important for transoceanic and transcontinental applications.

Because solitons rely on the exact compensation of dispersion by nonlinearities, soliton systems must sustain a fixed pulse energy throughout the link. That is possible only if fiber attenuation is compensated for by the use of optical amplifiers. Thus, optical amplifiers are absolutely necessary for soliton systems.

The goal of this chapter is to explain the principles of solitons and to examine their use in communication systems. We focus on the physical insight, applications, and an understanding of limitations, as opposed to rigorous derivations. Thus, mathematics is kept simple and complemented by simulation and experimental results.

We begin with an intuitive explanation of solitons in Section 6.2 and then examine their advantages for long-distance transmission in Section 6.3. A formal derivation of solitons from the Schrodinger equation is contained in Section 6.4. In Section 6.5, the interrelationship among soliton amplitude, duration, energy, and power is discussed, while Section 6.6 is devoted to higher-order solitons. Section 6.7 contains a qualitative physical explanation of solitons in terms of chirp and compression. In Section 6.8, we estimate the peak power needed for soliton transmission and discuss what happens when the power condition is not satisfied. In Section 6.9, the fiber loss and its compensation are discussed. In Section 6.10, we discuss the compensation of fiber loss using lumped amplifiers. Sections 6.11 and 6.12 are devoted to two potentially detrimental phenomena in soliton systems, fiber polarization dispersion and spontaneous emission noise stemming from optical amplifiers. Error rates of soliton systems are discussed in Section 6.13. In Section 6.14, we review soliton experiments. Section 6.15 is devoted to soliton WDM systems. Section 6.16 examines sources of soliton pulses. Finally, Section 6.17 is devoted to time-domain and frequency-domain filtering extending the performance of soliton systems beyond the Gordon-Haus limit.

Most of the material in this chapter (except for Sections 6.15–6.17) can be found in [1] and [2]. A more general (and more complicated) treatment of solitons can be found in [3].

6.2 INTUITIVE EXPLANATION OF SOLITONS

Figure 6.1 shows a simple intuitive explanation of solitons. In the figure, photons are represented by runners. Heavier runners are assumed to run faster than lighter ones—this corresponds to fiber dispersion (photons having different wavelengths and, therefore, different energies travel at different speeds through the fiber). As a result, when a group of runners runs on a hard surface (and a low-energy group of photons travels through fiber), different runners (and different photons) arrive at the destination at different times even if they leave the starting point at the same time. Thus, when a low-energy light pulse is injected into a fiber, it becomes longer as it propagates through the fiber.

However, when a high-energy group of photons is injected into fiber, its energy changes the effective refractive index of the fiber and, therefore, the group's velocity.

Figure 6.1 Intuitive explanation of solitons. Photons of different wavelengths are represented by runners, and fiber is represented by a mattress. (*Source:* [1]. © 1991 Optical Society of America.)

In Figure 6.1, this phenomenon corresponds to running on a mattress. The mattress's soft surface is distorted by the runners, and the resulting dip changes the runners' speed: lighter and normally slower runners begin to run faster (they are "helped" by the dip), while heavier and normally faster runners begin to run slower (they are "slowed down" by the dip). As a result, the group of runners tends to stay together while running on a mattress even though it would disperse on a hard surface. Similarly, a high-energy group of photons tends to stay together in the fiber (even though it would disperse at lower energies), because the light pulse creates a moving valley of higher dielectric constant. In other words, fiber nonlinearity caused by the packet's high energy (refractive index "dip") cancels the impact of dispersion.

A high-energy group of photons that tends to stay together in the fiber is called a *soliton.*

6.3 ADVANTAGES OF SOLITONS FOR LONG-DISTANCE TRANSMISSION

Figure 6.2 shows a classic long-distance transmission system. When the transmitted signal is distorted by fiber attenuation or dispersion, the signal is detected, reshaped, and retransmitted by a repeater. While this technique has served long-distance transmission needs for years, it suffers from the following drawbacks:

- Repeaters are inherently bit-rate limiting. Thus, if a link is upgraded to a higher bit rate, all old repeaters have to be pulled out and replaced by newer ones—an expensive operation, especially for underwater links.
- Repeaters are not truly compatible with WDM. Because each repeater processes only one wavelength, an n-wavelength WDM system has to have a WDM demultiplexer, n repeaters, and a WDM multiplexer for each repeater of a comparable single-wavelength system (Figure 6.3).
- Repeaters are inherently unidirectional. Since long-distance systems normally carry bidirectional traffic, each repeater location has to be served by two repeaters even in a single-wavelength system.
- As a result of these factors, repeater systems tend to be expensive.

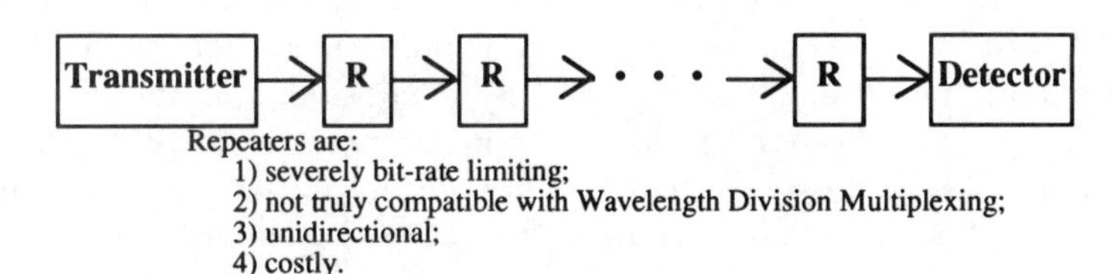

Figure 6.2 Repeated system. (*Source:* [1]. © 1991 Optical Society of America.)

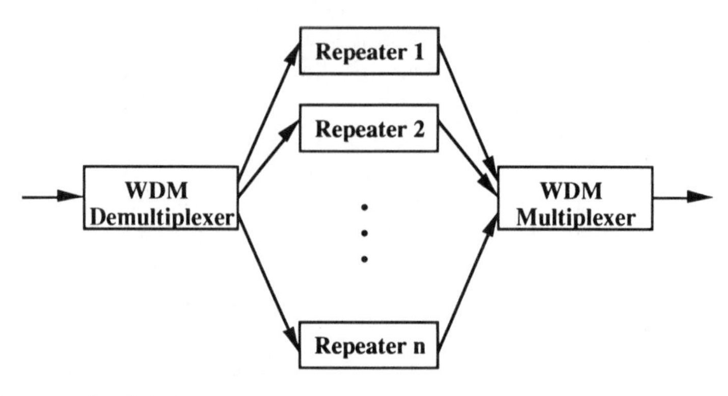

Figure 6.3 An *n*-wavelength WDM system.

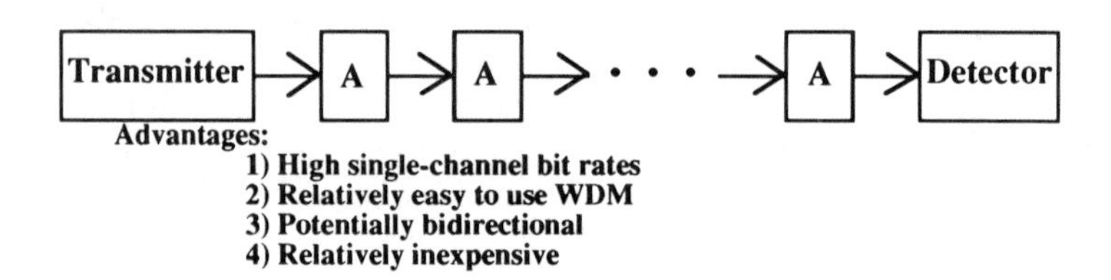

Figure 6.4 Optically amplified system. (*Source:* [1]. © 1991 Optical Society of America.)

Figure 6.4 shows an alternative long-distance system design. Instead of repeaters, this system utilizes optical amplifiers. Compared to the repeated system in Figure 6.2, the amplified system in Figure 6.4 offers the following advantages:

- Since optical amplifiers are bit-rate independent, the system can support very high bit rates. Most important, the amplifiers can remain in place when the system bit rate is increased. This advantage is particularly important for underwater systems, where access to repeaters and amplifiers is extremely difficult and expensive.
- It is relatively easy to use WDM as long as the combined WDM spectrum falls within the optical amplifier bandwidth. On the negative side, the optical amplifier bandwidth is substantially smaller than the fiber bandwidth (see Chap. 4), so that optical amplifiers, once installed, do limit the overall system bandwidth.
- Optical amplifiers are, in principle, bidirectional, so that, in theory, one optical amplifier can serve both traffic directions. In practice, this advantage is not likely to materialize: To ensure system stability and to prevent noise accumulation, optical amplifier modules have to contain optical isolators. The isolators are unidirectional and prevent optical amplifiers from being used as bidirectional devices.

- Probably the most important advantage of optical amplifiers is their potentially low cost. A typical optical amplifier unit contains a WDM coupler, a laser diode pump, a piece of erbium-doped fiber, and one or two isolators. This unit is much simpler (and, therefore, less expensive) than a repeater.

At the moment, the cost of an amplifier unit is dominated by the cost of the laser diode pump.

Once a decision is made to use amplifiers (and not repeaters) in a long-distance system, the following arguments point toward a soliton-based system:

- Alternative modulation formats—PSK (phase-shift keying) and FSK (frequency-shift keying)—succumb to the ASE (amplified spontaneous emission) noise introduced by optical amplifiers long before ASK (amplitude-shift keying). Thus, long-distance system designers prefer ASK over PSK and FSK.
- Once an ASK modulation format is selected, there are two ways to achieve long-distance operation: transmission at the zero-dispersion wavelength, λ_0, and soliton transmission. The first technique—transmission at λ_0—is plagued by the effects of fiber nonlinearity and by practical variations in λ_0 (it is difficult to make laser diodes that exactly match the λ_0 wavelength of the fiber and to maintain the match over the range of operating temperatures and over the very long lifetime of underwater systems).
- Finally, all experimental and theoretical evidence to date shows that solitons are extremely robust against the known defects of practical transmission systems. In other words, once a soliton system is built, it can be expected to provide a stable operation in spite of aging and temperature effects.

6.4 DERIVATION OF SOLITONS

Fibers are slightly nonlinear. The small nonlinearity, in conjunction with some linear dispersion, permits solitons to establish themselves. In a lossless dispersive fiber, solitons would travel arbitrary distances without changing shape. Solitons must carry a given amount of energy that depends on the pulse duration, the fiber nonlinearity, and dispersion. Solitons do not solve the problem of fiber losses, but they do conquer the limits set by fiber dispersion. For solitons to be useful, optical amplifiers are needed to counter fiber losses and thus maintain the critical pulse energy.

We derive the differential equation for solitons in three stages:

- Nonlinear material, no dispersion;
- Dispersive material, no nonlinearity;
- Combination of nonlinear dispersive material.

To simplify the derivation, the fiber mode is approximated by a plane wave. Then the connection with the actual fiber mode is made by introducing an effective area covered by the mode field.

6.4.1 Nonlinear Nondispersive Medium

A [plane] wave propagating in z-direction can be described by the wave equation

$$\frac{\partial^2 E}{\partial z^2} = \frac{1}{c^2} \frac{\partial^2 (n^2 E)}{\partial t^2} \tag{6.1}$$

where c is the speed of light in vacuum. The refractive index, n, depends on the light intensity:

$$n = n_0 + n_2 |E|^2 \tag{6.2}$$

The electric field, E, has a slowly varying envelope, $\phi(z,t)$:

$$E(z,t) = \phi(z,t) \exp[j(\omega_0 t - \beta_0 z)] \tag{6.3}$$

where

$$\beta_0 = n_0 \omega_0 / c \tag{6.4}$$

The units of the envelope $\phi(z,t)$ are volts per meter, the same as those of the electric field.

Substitution of (6.2) and (6.3) into (6.1) yields a fairly complex equation that can be simplified using the following approximations:

- The second derivatives of ϕ can be neglected since they are much smaller than the terms $\omega_0^2\, \phi$ and $\beta_0^2\, \phi$.
- n_2^2 can be neglected since n_2 is very small.
- Only the leading term, $-\omega^2 |\phi|^2 \phi$, of the second time derivative, $\partial^2[|\phi^2|\phi]/\partial t^2$ needs to be kept, since the products of all additional terms with n_2 are negligibly small.

Using the foregoing approximations, the result of the substitution of (6.2) and (6.3) into (6.1) can be reduced to the following form (the detailed derivation is Problem 6.3):

$$\frac{\partial \phi}{\partial z} + \dot{\beta}_0 \frac{\partial \phi}{\partial t} = -j \frac{n_2}{n_0} \beta_0 \,|\, \phi \,|^2\, \phi \tag{6.5}$$

where

$$\beta_0 \equiv \frac{\partial \beta_0}{\partial \omega_0} = n_0/c$$

6.4.2 Linear Dispersive Medium

A plane wave with angular frequency ω can be described as

$$E_p = A \exp[j(\omega t - \beta z)] \tag{6.6}$$

A pulse is a superposition of plane waves:

$$E = \int_{-\infty}^{\infty} A(\omega) \exp[j(\omega t - \beta z)]d\omega \tag{6.7}$$

Because the medium is dispersive, β depends on ω. A good approximation (except for the zero-dispersion point) contains just the first three forms of the power expansion series:

$$\beta = \beta_0 + \dot{\beta}_0(\omega - \omega_0) + \frac{1}{2}\ddot{\beta}_0(\omega - \omega_0)^2 \tag{6.8}$$

where

$$\dot{\beta}_0 \equiv \partial\beta/\partial\omega \Big|_{\omega=\omega_0} \text{ and } \ddot{\beta}_0 \equiv \partial^2\beta/\partial\omega^2 \Big|_{\omega=\omega_0}$$

Substitute (6.8) into (6.7):

$$E = \phi(z,t) \exp[j(\omega_0 t - \beta_0 z)] \tag{6.9}$$

where

$$\phi(z, t) \equiv \int_{-\infty}^{\infty} A(u) \exp\left\{j\left[(t - \dot{\beta}_0 z)u - \frac{1}{2}\ddot{\beta} z u^2\right]\right\}du \tag{6.10}$$

where $u \equiv \omega - \omega_0$. The envelope in (6.10) satisfies the following differential equation (it can be verified by substitution):

$$\frac{\partial\phi}{\partial z} + \dot{\beta}_0 \frac{\partial\phi}{\partial t} = \frac{j}{2}\ddot{\beta}_0 \frac{\partial^2\phi}{\partial t^2} \tag{6.11}$$

Differential equation (6.11) is exact, provided the medium exhibits only the first-order dispersion as per (6.8).

6.4.3 Nonlinear Dispersive Medium

The left side of (6.5) and of (6.11) are identical. Each describes a traveling pulse of an arbitrary shape $\phi(t,z) = f(t - \dot{\beta}_0 z)$. The right side of (6.5) modifies the wave to account for the nonlinearity, while the right side of (6.11) modifies it to account for dispersion. Since both effects are weak in typical optical fibers, the two effects can be assumed to be additive:

$$\frac{\partial \phi}{\partial z} + \dot{\beta}_0 \frac{\partial \phi}{\partial t} = \frac{j}{2} \ddot{\beta}_0 \frac{\partial^2 \phi}{\partial t^2} - j \frac{n_2}{n_0} \beta_0 |\phi^2| \phi \tag{6.12}$$

Equation (6.12) is only approximately valid provided that nonlinearity and dispersion are weak. The equation does, however, yield the soliton solutions. Let us convert (6.12) to a normalized form by introducing dimensionless variables:

$$t' = \frac{1}{\tau}(t - \dot{\beta}_0 z) \quad \text{and} \quad z' = \frac{|\ddot{\beta}_0|}{\tau^2} z \tag{6.13}$$

and a normalized envelope function

$$u(z',t') = \tau \left(\frac{n_2 \beta_0}{n_0 |\ddot{\beta}|} \right)^{1/2} \phi \tag{6.14}$$

The new quantity τ in (6.13) and (6.14) will be shown to be related to the soliton pulse width. With the normalized variables t', z' and u, (6.12) can be rewritten as the nonlinear Schrödinger equation:

$$\frac{\partial u}{\partial z'} = \frac{j}{2} \cdot \frac{\ddot{\beta}_0}{|\ddot{\beta}_0|} \cdot \frac{\partial^2 u}{\partial t'^2} - j|u|^2 u \tag{6.15}$$

The new equation has an infinite number of solutions. The simplest solution is a soliton of first order (see Problem 6.4):

$$\phi(z,t) = \phi_0 \cdot \frac{\exp(jaz)}{\cosh\left(\dfrac{t - \dot{\beta}_0 z}{\tau}\right)} \tag{6.16}$$

6.5 AMPLITUDE, DURATION, ENERGY, AND POWER

In this section, we will derive the relationship among the amplitude, duration, energy, and power of a soliton. We begin with substitution of (6.16) into (6.12). The result shows that (6.16) is indeed a solution of (6.12), provided that

$$a = \frac{\ddot{\beta}_0}{2\tau^2} \tag{6.17}$$

and the amplitude of the soliton is

$$|\phi_0|^2 = -\frac{n_0}{n_2}\frac{\ddot{\beta}_0}{\beta_0\tau^2} \tag{6.18}$$

Equation 6.16 shows that $|\phi|^2$ does not change its shape or width as the soliton travels in the dispersive fiber.

The amplitude equation (6.18) shows the following:

- The amplitudes of the solitons are not arbitrary; they are uniquely specified by the nonlinearity represented by n_2, by the dispersion represented by $\ddot{\beta}_0$, and by the pulse width represented by τ.
- Solitons can exist only if $\ddot{\beta}_0/n_2 < 0$.

Thus, nonlinearity and dispersion must cooperate in the right direction to produce solitons. Since n_2 of silica is positive, the fiber must be operated at wavelengths longer than the zero dispersion wavelength if we want to excite solitons. The full width of the soliton power, $|\phi|^2$, at half maximum is obtained as

$$\Delta t = 1.76\tau \tag{6.19}$$

To find the energy of the soliton W_e, we integrate $|\phi|^2$ over $t \in [-\infty,\infty]$ and multiply the result by A_{eff}, the effective area of the fiber mode, and by the admittance of free space, $\sqrt{\varepsilon_0/\mu_0}$. (The quantity $\sqrt{\mu_0/\varepsilon_0} = 377\Omega$ is known as the characteristic impedance of the free space.) The result is

$$W_e = \sqrt{\frac{\varepsilon_0}{\mu_0}}A_{eff}\frac{n_0^2\,|\ddot{\beta}_0|}{n_2\beta_0\tau} \tag{6.20}$$

The peak power of the soliton can be shown to be

$$P_{max} = \frac{W_e}{2\tau} \sim \frac{1}{\tau^2} \tag{6.21}$$

Equation (6.20) shows that the product $W_e \cdot \tau$ is constant for a given fiber; thus, shorter solitons require more energy. The reason is that for a given dispersion, a short pulse spreads more rapidly than a longer pulse. To counteract the faster spread requires a larger contribution from the nonlinearity, that is, more energy.

Table 6.1 is a summary of soliton units [1]; z_0 characterizes soliton length and corresponds to the soliton phase shift of 45 degrees [see (6.16)]. In other words, $az_0 = \pi/4$.

To get a numerical estimate of soliton parameters, consider the following example. Assume that solitons need to be generated with the duration Δt = 50 ps at λ = 1,550 nm in the fiber with the effective mode area A_{eff} = 78.5 μm^2, $n_2 = 6.1 \cdot 10^{19}$ cm^2/V^2, and D = 1 ps/nm·km. The resulting soliton parameters are time τ = 28.4 ps; length, z_0 = 1,000 km; power, P_{max} = 1.3 mW; and energy, W_e = 73.8 fJ (see Section 6.8 for further discussion).

6.6 HIGHER-ORDER SOLITONS

The soliton equations (6.12) and (6.15) have an infinite number of solutions yielding higher-order solitons. Each has a definite amplitude and changes its shape periodically along the fiber. Thus, higher-order solitons do not spread continuously due to dispersion, but their widths pulsate and their shapes are complicated.

To excite the first-order soliton only we need to inject into the fiber a pulse with the correct power and duration specified in Table 6.1. If the energy injected is too high, higher-order solitons can be excited.

Table 6.1
Summary of Soliton Units

Parameter	*Expression*
Time, τ	$\dfrac{\Delta t}{1.763}$
Length, $\dfrac{2}{\pi} z_0$	$\dfrac{2}{\pi} z_0 = 0.322 \dfrac{2\pi c}{\lambda^2} \dfrac{(\Delta t)^2}{D}$ (For Δt = 50 ps, D = 1 ps/nm · km, and λ = 1,550 nm, z_0 = 1,000 km.)
Power, P_{max}	$\sqrt{\dfrac{\varepsilon_0}{\mu_0}} = \dfrac{A_{eff}}{4n_2} \dfrac{\lambda}{z_0} \sim \dfrac{D}{(\Delta t)^2}$

Source: [1]. © 1991 Optical Society of America.

6.7 QUALITATIVE PHYSICAL EXPLANATION OF SOLITONS

If the pulse is chirped (f_0 changes so that it is lower at one end of the pulse and higher at the other end), the pulse may either spread more rapidly than an unchirped pulse or contract temporarily until a minimum pulse width is reached. The direction depends on the direction of the chirp and on the sign of $\ddot{\beta}$. Compression of chirped pulses has been used to generate extremely short light pulses starting with longer pulses.

In soliton systems, the chirp is generated continuously by the fiber nonlinearity. This phenomenon is known as self-phase modulation. To understand it, consider again the nonlinear equation (6.5):

$$\frac{\partial \phi}{\partial z} + \dot{\beta}_0 \frac{\partial \phi}{\partial t} = -j \frac{n_2}{n_0} \beta_0 \, |\phi|^2 \, \phi \tag{6.22}$$

A solution of (6.22) is

$$\phi(t, 3) = f(t, 3) \cdot \exp\left[-j \cdot \frac{n_2}{n_0} \cdot \beta_0 \, z \cdot f^2(t, 3)\right] \tag{6.23}$$

where $f(t, 3)$ is an arbitrary real function, $f(t, 3) = f(t - \dot{\beta}_0 z)$.

Exercise 6.1

By substitution, show that (6.23) is a solution of (6.22).

Combining (6.3) with (6.23), we obtain

$$E(t, z) = f(t, z)\exp\left[j(\omega_0 t - \beta_0 z) - j\frac{n_2}{n_0} \beta_0 f^2 z\right] \tag{6.24}$$

The instantaneous angular carrier frequency of (6.24) is the time derivative of the phase angle:

$$\omega = \omega_0 - 2\frac{n_2}{n_0} \beta_0 f z \frac{\partial f}{\partial t} \tag{6.25}$$

Thus, the pulse has become chirped by the fiber nonlinearity.

When self-chirping and fiber dispersion act in the proper direction and have the proper magnitudes, they cancel each other. As a result, the pulse maintains its width exactly in first-order solitons.

In higher-order solitons, self-chirping and pulse compression are not perfectly balanced at every point along the fiber (and in time). Instead, the pulse shrinks and expands periodically, depending on whether the mechanisms causing pulse compression or pulse spreading momentarily have the upper hand.

6.8 ESTIMATE OF PEAK PULSE POWER REQUIRED FOR SOLITONS

The power needed for soliton transmission depends on the nonlinear coefficient, n_2, and on fiber dispersion; Figure 6.5 shows typical fiber dispersion curves. Let us consider a numerical example: For fused silica, the nonlinear coefficient is $n_2 = 6.1 \cdot 10^{-19}$ cm^2/V^2. Assume the effective mode area has a radius $r = 5$ µm so that $A_{eff} = \pi r^2 = 7.85 \cdot 10^{-7}$ cm^2. Further, let dispersion be $\ddot{\beta} = -20$ ps^2/km.

With $n_0 = 1.46$ and $\lambda = 1.5$ µm, $\beta = 6.12 \cdot 10^4$ cm^{-1}. If we let the pulse width be $\Delta t = 10$ ps, (6.20) and (6.21) yield a peak pulse power of 375 mW. The required peak power is inversely proportional to the square of the pulse width; thus, a 100-ps pulse requires only 3.75 mW peak power. Also, the required peak power is directly proportional to $|\ddot{\beta}|$. Thus, placing the operating λ closer to λ_0 reduces the required peak power: $\ddot{\beta}$ of -1 ps^2/km yields the required peak pulse power of only 19 mW for $\Delta t = 10$ ps.

If the input pulse has the shape and amplitude described by (6.16) to (6.18), a first-order soliton is excited. If the input pulse energy is too low, no soliton will form and the pulse will spread due to dispersion. If the input pulse energy is too high (or

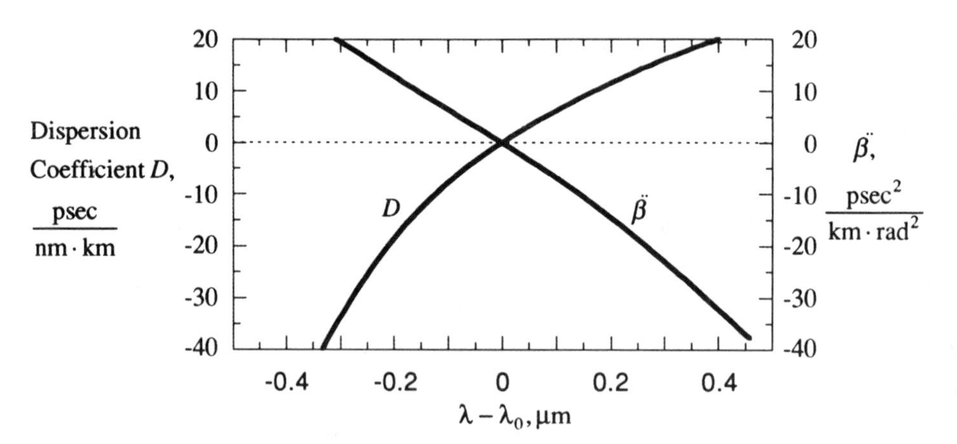

Figure 6.5 Dispersion coefficient, *D*, and the second derivative of the propagation constant, $\ddot{\beta}$ versus light wavelength, λ. Here, λ_0 is the zero-dispersion wavelength. The relationship between $\ddot{\beta}$ and *D* is given by (1.222) in Chapter 1. (*Source:* [4,5]. © 1981 IEEE.)

the pulse is too broad), a soliton (possibly of higher order) will form with the proper energy and pulse shape, but the remaining energy will be spun off as a spurious pulse (its presence leads to an undesirable interference).

The shape of the input pulse does not have to comply exactly with the ideal soliton shape in (6.16). All that is necessary is to have the ideal soliton described by (6.16) to (6.18) contained in the input pulse. The rest of the input pulse (i.e., the difference between the actual input pulse and the ideal soliton pulse) will be spun off in the form of undesirable interference. That interference generally leads to performance degradation (and, therefore, should be minimized); however, its impact can be greatly attenuated (essentially, eliminated) using sliding frequency filters [6,7] described in Section 6.17.

6.9 FIBER LOSS AND ITS COMPENSATION

A soliton is attenuated by the fiber loss mechanisms. Consider the quantity z_0, first introduced in Table 6.1:

$$z_0 \equiv \frac{\pi}{2} \cdot \frac{\tau^2}{|\ddot{\beta}|} \cong 0.5 \frac{(\Delta t)^2}{|\ddot{\beta}|} \quad (6.26)$$

It is closely related to $1/a$ of (6.16) and has the meaning of spatial length of the soliton (check units: Δt has units of seconds and $\ddot{\beta}$ has units of $1/(\text{m}\cdot\text{Hz}^2)$ so that z_0 has units of meters). If the fiber loss is small—$1/\alpha > z_0$—the soliton will adjust its width so the product, $W_e\Delta t$, remains constant per (6.19) and (6.20). If the fiber loss is large—$1/\alpha < z_0$—the soliton becomes an ordinary light pulse subject to dispersion spreading. Thus, soliton transmission over large distances requires repeated amplification.

At one time, Raman amplifiers were considered to compensate for fiber attenuation in soliton systems. The Raman effects involves three processes: signal, pump, and vibrations of the fiber molecules. The difference between the pump frequency, ω_p, and the signal frequency, ω_s, corresponds to a vibrational frequency of the molecules. Phase matching of the two optical waves is not required.

The Raman effect converts a usual communications fiber into a distributed amplifier *if* a CW optical pump is provided with the power 30–100 mW. To amplify a signal at 1.55 μm requires a pump at 1.45 μm.

Table 6.2 lists several proposed soliton systems.

Computer simulations [8] show that a resonance instability develops if $z_0 = L/8$. Then the soliton energy and width fluctuate widely. This condition is avoided in all the designs in Table 6.2.

With the advent of EDFAs (erbium-doped fiber amplifiers), the idea of using Raman amplifiers has been abandoned in favor of EDFAs. However, many important

Table 6.2
Design Examples of Soliton-Based, Single-Channel, High-Bit-Rate Systems

Design Number	*L** *(km)*	z_0 *(km)*	Δt *(ps)*	P_{max} *(mW)*	*R* *(GHz)*	*Z* *(km)*
1	30	6	5.5	50	18	1,600
2	40	11	7.5	27	13	2,200
3	50	17	9.4	18	10.6	2,700
4	(~50)	30	12.3	10	8.1	3,600
5	(~50)	50	16	6.1	6.2	4,700
6	(~50)	100	22.6	3	4.4	6,600

Source: [2], © 1988 Academic Press.
Note: $A_{eff} = 25\ \mu m^2$; $\beta = -2.4\ ps^2/km$.
*Spacing between injection points of the bidirectional Raman pump waves.

design features hold true for both EDFAs and Raman amplifiers. For example, Table 6.2 suggests that a reasonable value of L is between 30 and 50 km; that conclusion is valid for EDFA systems scheduled to be installed soon.

6.10 LUMPED AMPLIFIERS IN SOLITON SYSTEMS

Because EDFAs are lumped rather than distributed amplifiers, the energy of a soliton is not constant over the length of an amplified system. Instead, a soliton's energy fluctuates: It is maximum at amplifiers' outputs, decays as the soliton propagates along the fiber, and reaches its minimum at amplifiers' inputs. Therefore, the simplified theory presented in Sections 6.4–6.8, strictly speaking, does not apply.

Intuitively, we can expect the soliton to propagate if the amplification period, L, is sufficiently small (L is the distance between adjacent amplifiers). Indeed, simulations and experiments support that intuition: They show that if $L << z_0$, where z_0 is defined by (6.26) and has the meaning of the spatial length of the soliton, nothing happens to the pulse shape and width over one period L. The nonlinear effects in the fiber are determined by the corresponding path-averaged power (and energy) of the soliton, and, if the path-averaged power (over each period of L meters) is equal to the soliton power required by (6.21), well-behaved solitons propagate along the fiber.

An additional, closely related, and very similar practical problem arises when a long-distance amplified system is constructed. The problem is that fiber dispersion varies along the link length because the link consists of many fiber segments possibly drawn from different preforms or even manufactured by different factories. The answer to that problem is similar to the answer to the lumped-amplifiers problem: As long as the path-averaged dispersion coefficient, D, remains constant from one period to the next, solitons can tolerate considerable variation in D over short (comparatively to z_0) distances. In the rest of this section, we will review simulation results that support those conclusions [1].

Figure 6.6 illustrates the behavior of dispersion and pulse power in a system that employs lumped amplifiers and dispersion-shifted fiber. In that hypothetical system, the amplifiers are placed every 100 km (L = 100 km), which is several times greater than would be chosen in a practical system (in a practical system, L would be 30 km or so to minimize the amount of accumulated ASE). Thus, the test is especially rigorous, because a practical system would behave better. The dispersion variation selected is also rigorous: The dispersion is assumed to vary periodically, with a period of 100 km.

Figure 6.7 shows pulse shapes and spectra in the system in Figure 6.6. Solid lines show what happens after propagation through 9,000 km of fiber, and dotted lines show, for comparison, the initially launched pulses. Inspection of Figure 6.7 shows that the pulses at 9,000 km are virtually indistinguishable from those at the input. This conclusion is valid not only for τ = 25 ps (z_0 = 250 km) but for τ = 35 ps (z_0 = 480 km) and for τ = 50 ps (z_0 = 980 km) as well.

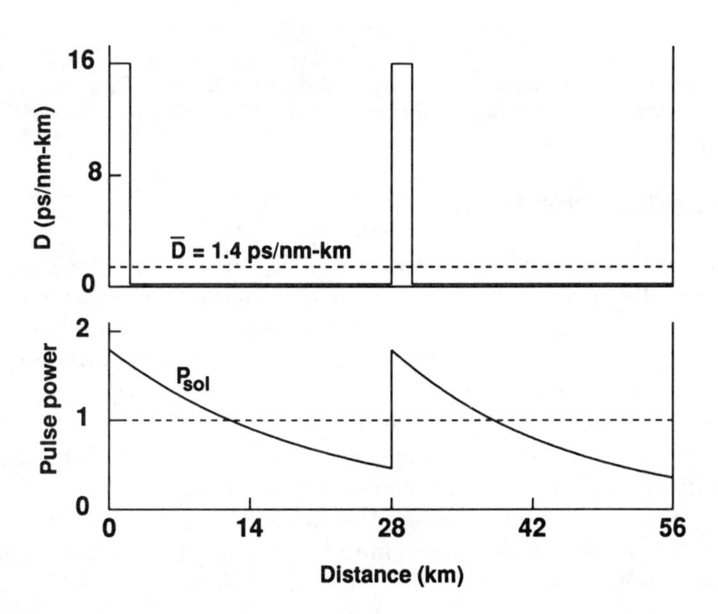

Figure 6.6 Segment of model system for test of soliton propagation through chain of lumped amplifiers and dispersion-shifted fiber. (*Source:* [1]. © 1991 Optical Society of America.)

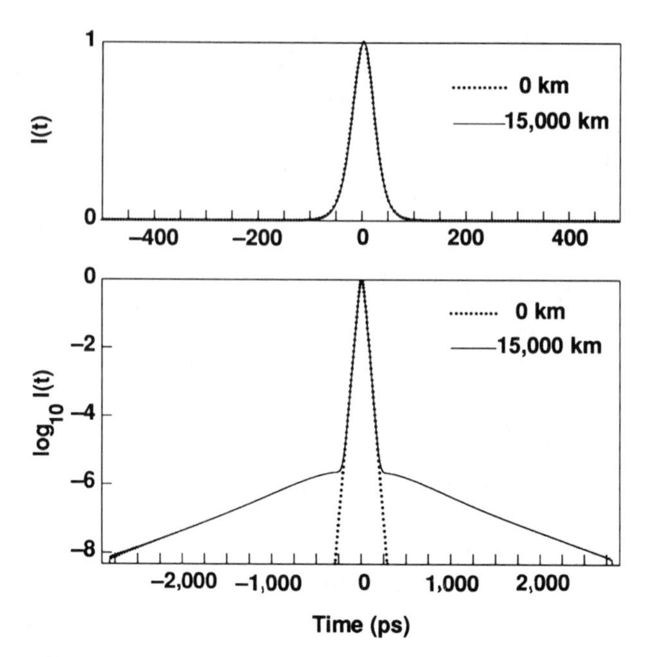

Figure 6.7 Numerically simulated soliton transmission over 9,000 km through a chain of lumped amplifiers and dispersion-shifted fiber at 1,550 nm, $t = 25$ ps and $D = 1$ ps/nm · km [5]; $z_0 = 250$ km; $L = 100$ km; D steps back and forth every 50 km between 0.5 and 1.5 ps/nm·km. Dotted curve: initial pulses; solid curve: after 9,000 km. (*Source:* [1]. © 1991 Optical Society of America.)

6.11 POLARIZATION DISPERSION

A fiber with a perfectly circular core and a perfectly circular cladding should exhibit no polarization dispersion if there is no stress or variation in angular material composition. In practice, a cabled fiber does exhibit some polarization dispersion due to manufacturing imperfections and stress. The amount of polarization dispersion is characterized by $\Delta\beta/\sqrt{h}$, where $\Delta\beta$ is the difference between the propagation constants of the two eigenmodes, and h is the fiber length.

The existence of polarization dispersion again poses a practical imperfection problem: Will solitons propagate through an imperfect fiber with a realistic amount of polarization dispersion? That problem has been investigated via computer simulations, with the following result [1]: Solitons resist polarization dispersion and propagate virtually undistorted as long as the polarization dispersion is sufficiently small:

$$\frac{\Delta\beta}{\sqrt{h}} \leq \sqrt{\frac{D}{10}} \tag{6.27}$$

where $\Delta\beta/\sqrt{h}$ is in ps/$\sqrt{\text{km}}$, and D is in ps/(nm · km). Dispersion-shifted fiber with $\Delta\beta/\sqrt{h}$ of less than 0.2 ps/$\sqrt{\text{km}}$ is readily available. Then, for $D \geq 0.5$ ps/(nm · km) (which is used for solitons' transmission), (6.27) is well satisfied. Experiments to date are consistent with this prediction.

6.12 ASE NOISE IN SOLITON SYSTEMS

In an amplified system like the one shown in Figure 6.8, soliton power at the receiver is nearly identical to that at the transmitter because fiber loss is compensated for by optical amplifiers. Because received signal power is so high, detector noise tends to be insignificant compared to ASE noise.

We can easily show that the ASE spectral density (power per unit bandwidth), $P(\nu)$, at the output of the last amplifier in the chain, is

$$P(\nu) = \alpha Z h\nu n_{sp} \frac{(G-1)}{\ln G} \tag{6.28}$$

where α is the fiber loss coefficient, Z the system length, $h\nu$ the photon energy, and n_{sp} the amplifier excess spontaneous emission factor.

For solitons where the path-average signal power is fixed, however, the SNR (signal-to-noise-ratio) is most conveniently calculated by comparing with the path-averaged $P(\nu)$.

Because the ratio of path-average to peak power in each fiber span is $(G-1)^2/(G \ln G)$, the path averaged $P(\nu)$, is

$$\overline{P}(\nu) = \alpha Z h\nu n_{sp} F(G) \tag{6.29}$$

where the function $F(G) = (G-1)^2/G(\ln G)^2$ represents an important noise penalty. Figure 6.9 shows $F(G)$ versus amplifier gain. Because amplifier spacing is uniquely determined by the amplifier gain for a given fiber loss, the top axis in Figure 6.9 shows the corresponding amplifier spacing.

The ASE noise adds to the signal and increases the system BER (bit error rate) through two different mechanisms: (1) the conventional additive noise effect and

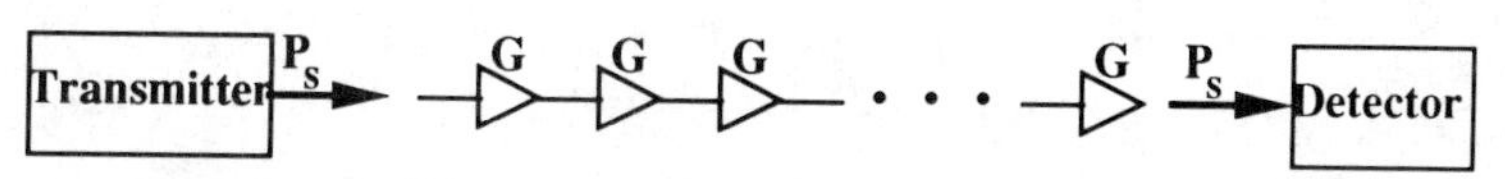

Figure 6.8 All-optical system with N amplifiers of gain G each preceded by $1/G$ fiber loss factor. *Source:* [1]. ©1991 Optical Society of America.

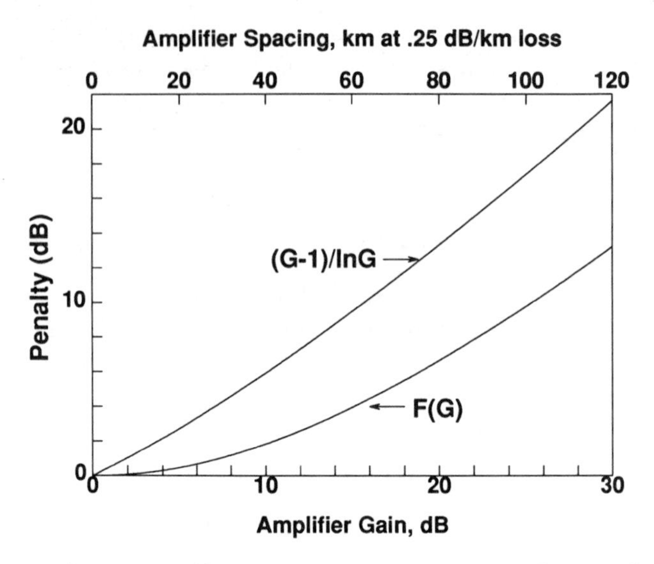

Figure 6.9 Noise penalty versus amplifier gain. *Source:* [1]. © 1991 Optical Society of America.

(2) perturbation of the soliton velocities. The second effect leads to a jitter in pulse arrival times and is known as the Gordon-Haus effect.

Theory of the Gordon-Haus effect shows that the ASE perturbs the soliton velocities such that at the end of a system of length Z, there is a Gaussian distribution in arrival times with the following variance:

$$\sigma^2 = 4138 n_{sp} F(G) \frac{\alpha_{loss}}{A_{eff}} \frac{D}{\tau} Z^3 \tag{6.30}$$

where α_{loss} is in km^{-1}, D is in ps/nm·km, Z is in thousands of km, τ is in ps, A_{eff} in square microns, G and n_{sp} are the amplifier gain and excess spontaneous emission factor, respectively, and $F(G)$ is the parameter used in (6.29):

$$F(G) = \frac{(G-1)^2}{G(\ln G)^2} \tag{6.31}$$

To estimate the numerical value of σ, consider the following example. Let $\tau = 50$ ps, $D = 1$ ps/nm·km, $A_{eff} = 35\ \mu\text{m}^2$, $\alpha_{loss} = 0.0576$ 1/km (0.25 dB/km), $F = 1.24$ (span between amplifiers = 28 km), and $n_{sp} \sim 1.5$. Then, for 9,000 km, we obtain

$$\sigma = 13.6 \text{ ps} \tag{6.32}$$

In Section 6.13, we will see what impact the Gordon-Haus effect has on the system BER.

6.13 ERROR RATES IN SOLITON SYSTEMS

In a conventional lightwave system, the SNR improves monotonically as the received signal power increases. Simultaneously, the BER improves, too, provided the decision threshold is adjusted to be optimum, as illustrated in Figure 6.10. This mechanism stems from very simple physics: When the signal becomes stronger and stronger, the noise becomes less and less important and leads to fewer and fewer errors.

In soliton systems, this effect is present as well. However, it is complicated by the Gordon-Haus effect. To understand the resulting BER, consider the plot of BER versus τ/D shown in Figure 6.11. Because soliton power is inversely proportional to τ^2 [see (6.21)], and the soliton energy is inversely proportional to τ [see (6.20)], both signal energy and signal power vary in Figure 6.11: As we move from left to right, both energy and power decrease, not increase as in conventional BER plots.

As we move from left to right in Figure 6.11, at first the BER falls due to the decrease of the Gordon-Haus effect [(6.30) shows that the pulse arrival time variance due to Gordon-Haus effect is inversely proportional to τ]. The resulting BER increases

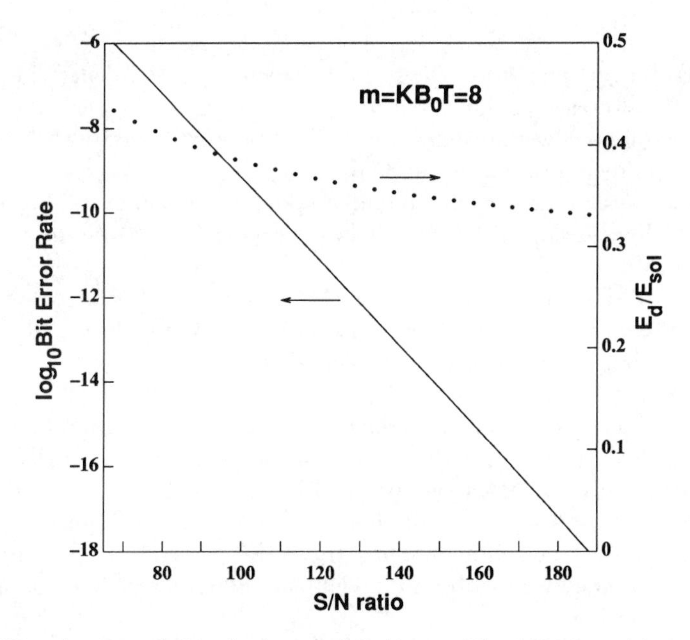

Figure 6.10 BER and optimum decision level versus SNR. (*Source:* [1]. ©1991 Optical Society of America.)

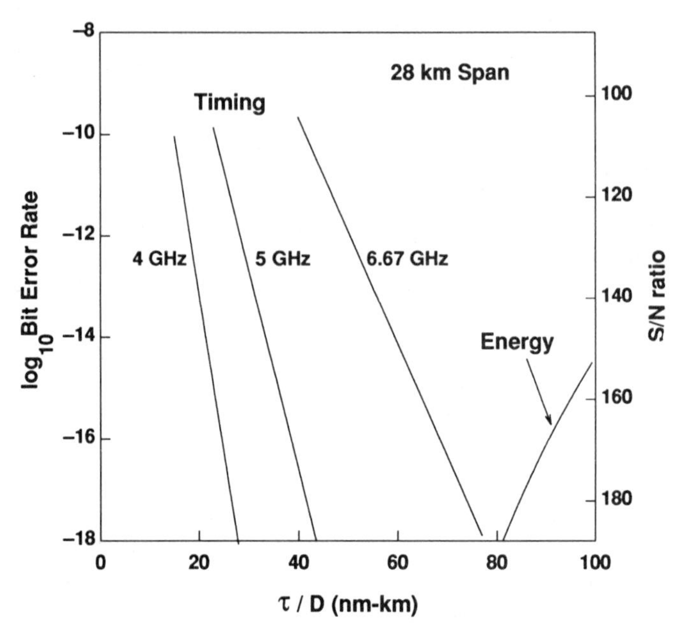

Figure 6.11 Error rates at 9,000 km assuming 0.21 dB/km loss, $\tau_s \sim 1550$ nm, and $n_{sp} = 1.1$ (erbium amplifiers). (*Source:* [1]. ©1991 Optical Society of America.)

as the system bit rate increases. The second branch of the BER curve in Figure 6.11 stems from the decrease of the signal power and energy at higher values of τ: As τ increases, the soliton power and energy decrease [see (6.20) and (6.21)], and the BER increases due to deterioration of the SNR.

Optimum system design corresponds to the intersection of the two branches in Figure 6.11. For example, for 6.67-Gbps systems, the optimum value of τ/D is 80 nm-km.

Figure 6.12 also shows the BER versus τ/D but for different parameter values: larger fiber loss and larger amplifier noise. Two cases are illustrated: the span of 80 km (broken lines) and the span of 28 km (solid line). The case of $L = 28$ km leads to excellent results: The two branches do not cross (down to BER = 10^{-18}) at 2.5 Gbps, and even at 4 Gbps the BER can be driven to below 10^{-18} with optimum system design (optimum τ). However, with $L = 80$ km, the two branches cross at BER = 10^{-8}. That means a system with $L = 80$ km will have a BER floor of 10^{-8}, even under optimum conditions (this BER is unacceptable for most applications). Figure 6.12 explains, therefore, why ultra-long soliton systems are likely to operate with L around 30 km (and not more) and a bit rate around 2.5 Gbps (and not more than 4 Gbps) and use τ/D of 20–50 nm-km.

Recently, it has been shown that soliton systems can operate well beyond the

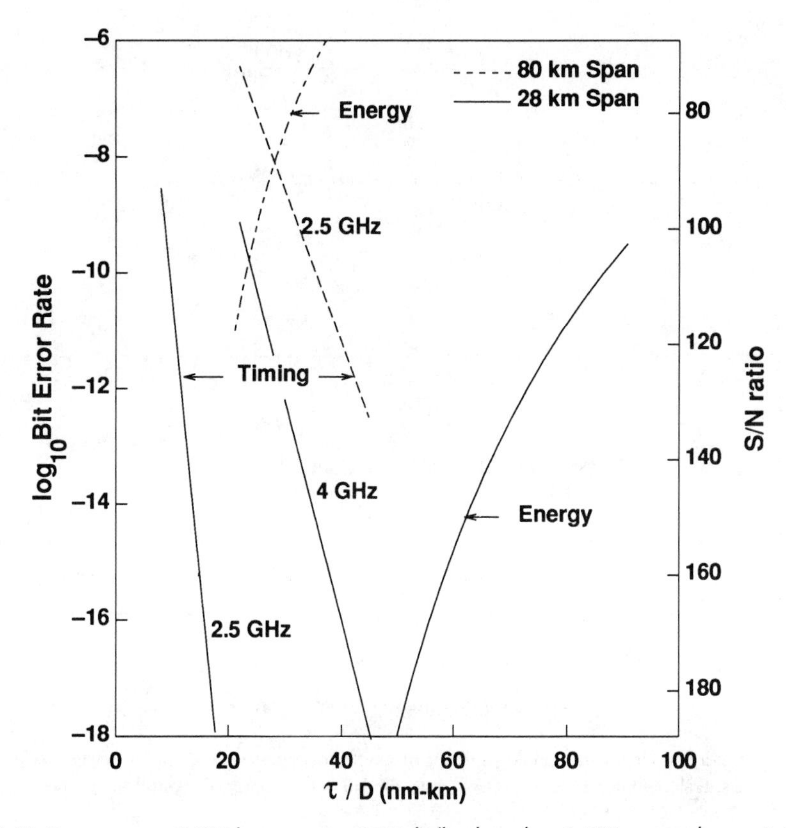

Figure 6.12 Error rates at 9,000 km assuming 0.25-dB/km loss, $l_s \sim$ 1,532 nm, and n_{sp} = 1.5 (erbium amplifiers). (*Source:* [1]. ©1991 Optical Society of America.)

Gordon-Haus limit using frequency-domain or time-domain filtering, as will be discussed in Section 6.17.

6.14 SOLITON EXPERIMENTS USING RECIRCULATING LOOPS

Early research toward transoceanic soliton systems faced an interesting problem: Convincing long-distance (many thousands of kilometers) soliton experiments had to be constructed before large-scale investments in soliton system development would be made. But how do you construct a convincing long-distance experiment without a large-scale investment in thousands of kilometers of fiber?

Figure 6.13 shows an ingenious solution—a recirculating fiber loop. The loop is only 75 km long; yet it allows propagation experiments over some 10,000 km. This

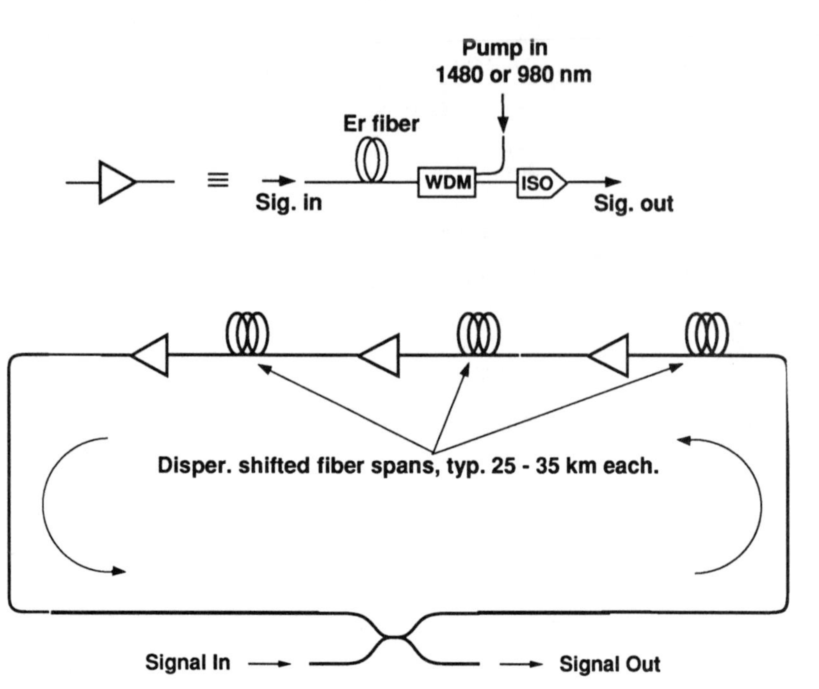

Figure 6.13 Experiment involving running solitons around a recirculating loop. In this particular experiment, dispersion-shifted fiber with $D = 1.4$ ps/nm·km at the signal wavelength was used. (*Source:* [1]. © 1991 Optical Society of America.)

Table 6.3
Experimental Parameters for the Soliton Experiment Using Recirculating Loop (Figure 6.13)

Soliton Duration, τ	*Fiber Dispersion, D*	*Path-Averaged Power, P_{sol}*	*Soliton Peak Power, Pk*	*Average Power Into One Bit Period at 2.5 Gbps*	*Measured ASE Noise Power in 2.5 GHz*	*SNR*
50 ps	1.38 ps/nm·km	660 μW	1.8 mW	250 μW	0.8 μW	250

Source: [1]. © 1991 Optical Society of America.
Note: At amplifier output.

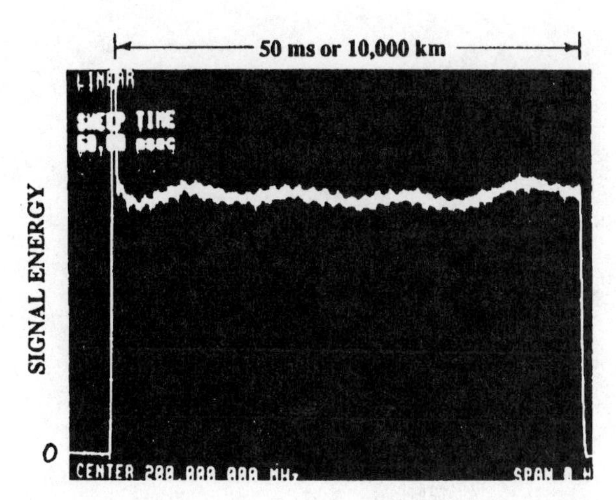

Figure 6.14 Measured signal energy versus distance. (*Source:* [1]. © 1991 Optical Society of America.)

goal is achieved by running light pulses around the loop many times, much in the same way runners run long distances by running many loops in a stadium.

The particular experiment of Figure 6.13 was conducted at AT&T Bell Labs (a number of similar experiments have been reported). Table 6.3 gives key experimental parameters.

Light pulses in Figure 6.13 are injected into the loop using a 20% coupler and travel counterclockwise. To minimize BER degradation due to ASE, the signal is amplified every 25 km (see Section 6.13) by EDFAs pumped both codirectionally and counterdirectionally with the signal. By monitoring signal pulse evolution as a function of the number of runs around the loop, we can predict the signal evolution as a function of distance traveled in an actual long-distance link.

Figure 6.14 shows the measured signal energy versus distance traveled for the experiment in Figure 6.13. The signal energy is seen to remain fairly constant over 10,000 km, illustrating the tight control over gain and signal energy possible with erbium amplifiers.

Figure 6.14 indicates that the energy—not the shape—of the pulses remains the same. To evaluate signal shape, consider signal spectra observed after 6,000 km, 8,000 km, and 10,200 km (Fig. 6.15). Inspection of Figure 6.15 shows that the signal spectrum (and, therefore, signal shape) does change as a function of distance.

The pulse spectrum after propagation through 9,000 km is shown in Figure 6.16. The dashed line corresponds to an unjittered pulse, while the solid curve is obtained by multiplying the dashed curve by the Gaussian weight function

$$e^{-1/2(2\pi f \sigma)^2} \tag{6.33}$$

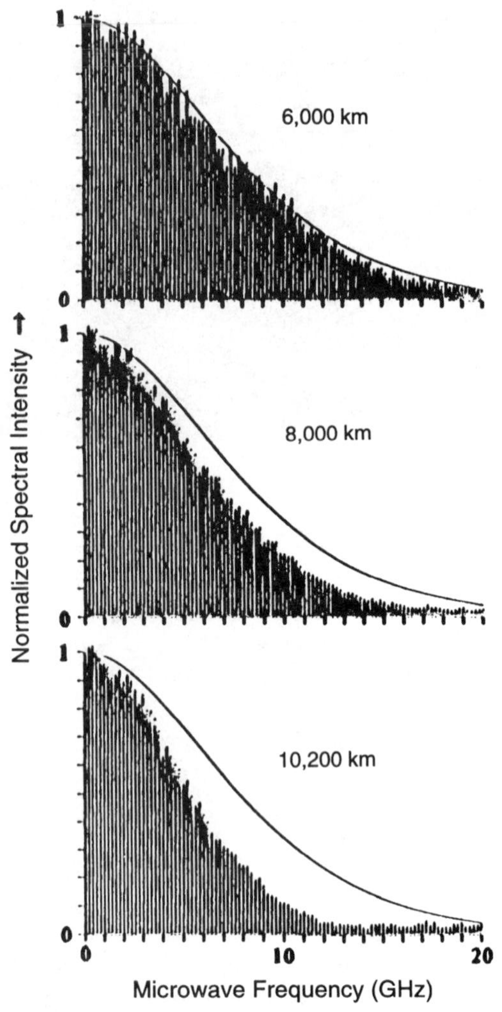

Figure 6.15 Microwave spectra of soliton pulse trains. (*Source:* [1]. © 1991 Optical Society of America.)

where σ is chosen for best fit to the experimental data. Inspection of Figure 6.16 shows that the jitter distortion is nicely approximated by the Gaussian weight function.

The variance σ in (6.33) is plotted versus path length in Figure 6.17. The squares and dots correspond to experimental data, while the dashed lines correspond to the Gordon-Haus expression (6.30). Comparison of the two sets shows that (6.30) provides a good prediction of experimental results; however, the accuracy of the prediction seems to deteriorate at large transmission distances (~10,000 km).

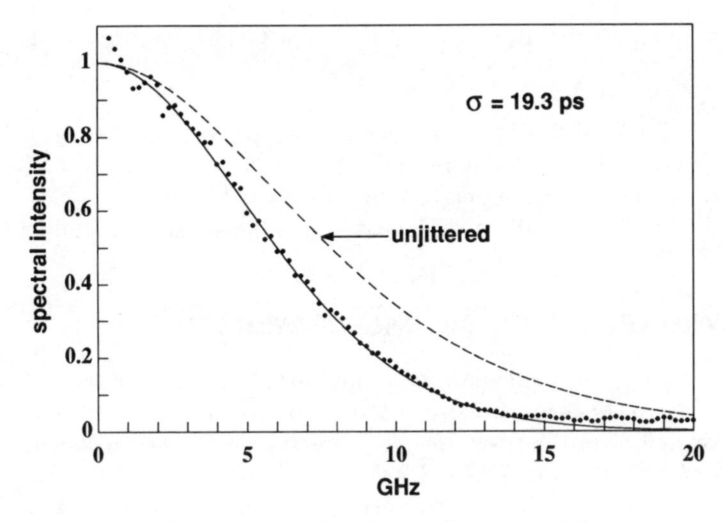

Figure 6.16 Soliton spectrum after propagation through 9,000 km. (*Source:* [1]. © 1991 Optical Society of America.)

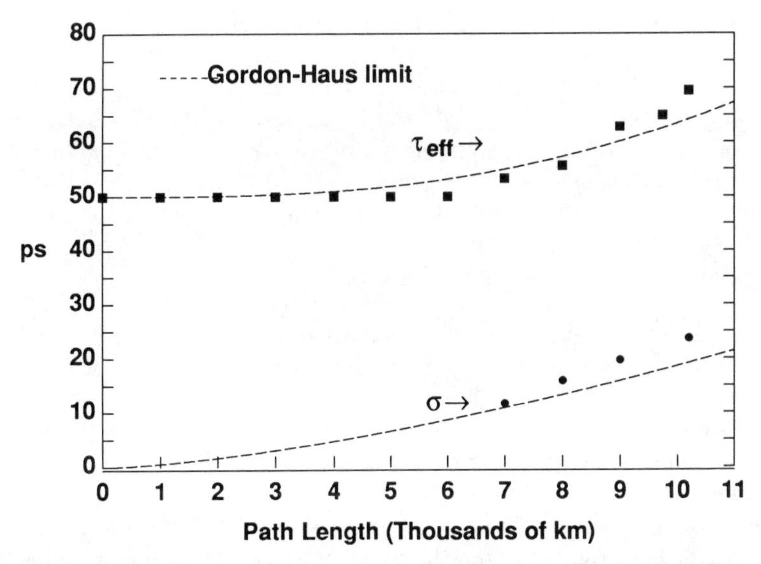

Figure 6.17 Results of the soliton transmission experiment in Figure 6.13: effective pulse width, τ, and jitter, σ, versus distance. The theoretical curves are for the conditions of experiment, namely, $b - 1.5$, 0.25-dB/km loss, 25-km amplifier spacing, and so on. (*Source:* [1]. © 1991 Optical Society of America.)

To summarize, the 10,000-km experiment reviewed in this section shows the following:

- Measured timing jitter agrees well with that predicted theoretically by (6.30).
- Measured SNR ~250 is more than adequate for digital transmission.
- There is no measurable polarization dispersion.
- There is no measurable interaction between pulses separated by ≥5 τ.

6.15 WAVELENGTH DIVISION MULTIPLEXING WITH SOLITONS

Can WDM be used in conjunction with solitons to increase the overall system capacity? That is not a trivial question, because if and when two solitons of different wavelengths overlap ("collide") in the fiber, their total power no longer satisfies the power requirement in (6.20) and (6.21); this leads to a perturbation of the propagation conditions and, therefore, to a system performance deterioration. And solitons having different wavelengths *will* overlap every so often, because fiber dispersion causes them to travel at different speeds.

The net result is that WDM soliton systems behave quite differently from conventional WDM systems. For example, the channel spacing, $\Delta\lambda$, in conventional WDM systems must be larger than a certain critical value for satisfactory performance. But soliton systems impose an additional requirement: $\Delta\lambda$ must be smaller than another critical value. In this section, we will examine the reasons for this behavior and the physics involved in WDM soliton systems. We begin by examining the collision of WDM solitons. Then we consider the acceleration and the velocity shift of colliding solitons, and finally, we deal with soliton WDM design rules and channel counts.

6.15.1 Collision of Solitons in WDM Systems

When two light pulses (solitons) have different wavelengths, they travel at different speeds because of fiber dispersion. Soliton systems must operate at wavelengths that are longer than the zero dispersion wavelength (see Section 6.5) so that $\ddot{\beta}_0 < 0$. Therefore, solitons of a higher-frequency channel travel faster than those of a lower-frequency channel. They will overtake lower-frequency solitons and periodically pass through them, as shown in Figure 6.18.

How long, in time and space, does the collision last? That is governed by the collision length, L_{coll}. By definition, the collision length, L_{coll}, begins and ends where solitons overlap at their half-power points. The collision length can be found as follows [1]:

$$L_{coll} = 0.6298 \frac{z_0}{\tau \Delta f} = \frac{2\tau}{D\Delta\lambda} \tag{6.34}$$

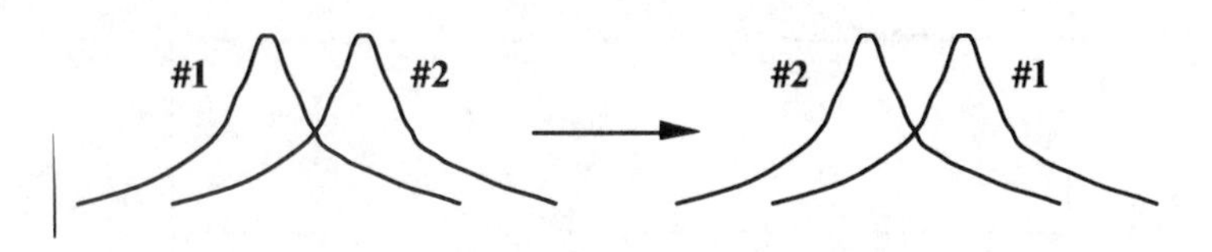

Figure 6.18 Soliton-soliton collisions in WDM. (*Source:* [1]. © 1991 Optical Society of America.)

For example, for $\tau = 50$ ps, $D = 1$ ps/nm·km, and $\Delta f = 0.125$ THz ($\Delta\lambda = 1$ nm at 1,550 nm), $L_{coll} = 100$ km. The corresponding quantity in the time domain is called the collision time, t_{coll}.

6.15.2 Acceleration and Velocity Shift of Colliding Solitons

When two (or more) solitons collide, their combined power exceeds the nominal soliton power, given by (6.20) and (6.21). As a result, their velocity is perturbed: Both solitons travel somewhat faster at the collision segment. The change of velocity can be characterized by acceleration (that is nominally equal to zero in an unperturbed soliton system). Colliding solitons will, therefore, travel somewhat faster than noncolliding solitons. The net result is an additional dispersion in the soliton arrival time (somewhat similar to the Gordon-Haus effect) and additional system performance deterioration.

Due to the nonlinear nature of the physical effects involved, theoretical analysis is difficult. We will review simulation results [9] that shed light on the quantitative limitations arising from soliton collisions.

Figure 6.19 shows the acceleration and velocity shift during the collision of solitons. For simplicity, the fiber is assumed to be lossless and otherwise unperturbed. On the horizontal axis, we have time (normalized to the collision time, t_{coll}) or distance (normalized to the collision length, L_{coll}). The velocity change is maximum when the two solitons exactly overlap. As a result of collision, solitons experience time shift, which for $\tau\Delta f > 1$ (where Δf is channel spacing) is given by

$$\Delta t = 0.1786 \frac{1}{\tau(\Delta f)^2} \tag{6.35}$$

For $\tau = 50$ ps and $\Delta f = 0.03$ THz ($\Delta\lambda \cong 0.25$ nm at 1,550 nm), $\Delta t = 4$ ps.

In practical soliton systems, amplifiers are used to boost soliton power periodically (see Section 6.10). Thus, the soliton power will vary along the fiber; it will be

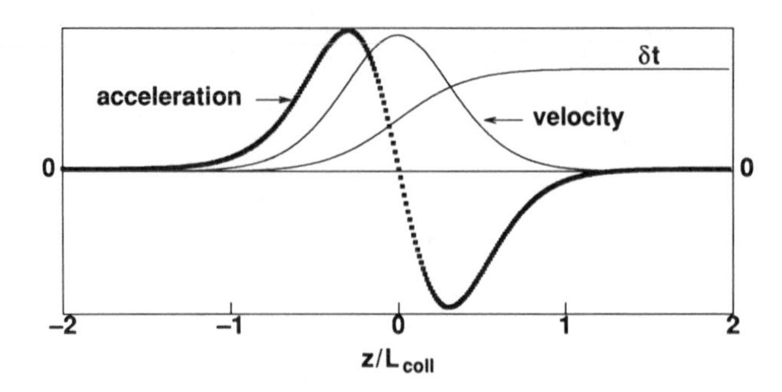

Figure 6.19 Acceleration and velocity shift during collision of solitons in lossless and otherwise unperturbed fiber. (*Source:* [1]. © 1991 Optical Society of America.)

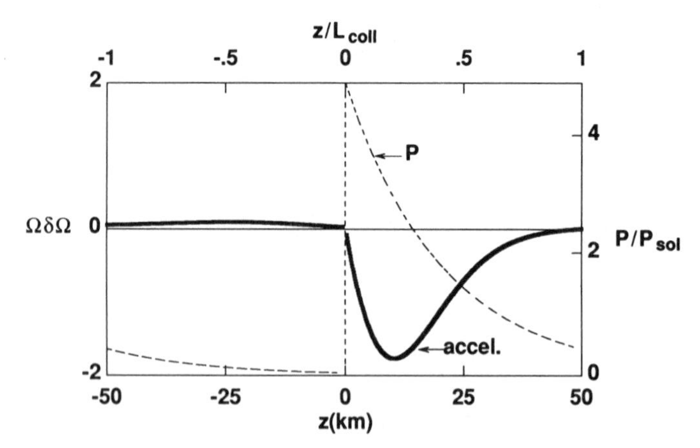

Figure 6.20 Acceleration of colliding solitons with collision centered at amplifier (1 amp/100 km). (*Source:* [1]. © 1991 Optical Society of America.)

maximum at amplifier output, decrease as solitons propagate to the next amplifier, and reach its minimum at amplifier input. Because the perturbation caused by solitons' collision increases with soliton power, the impact of a collision depends on where, along the link, the collision took place. The worst case is when the collision is centered around an amplifier output, where solitons' power is maximum; the best case (least perturbation) is when solitons collide just before an amplifier's input, where their power is minimum. However, the exact location of the collision point is not critical, since the collision length is typically several tens of kilometers.

Figures 6.20 and 6.21 show the worst-case acceleration and the worst-case velocity, respectively, that take place when two solitons collide at an amplifier's output. The distance between adjacent amplifiers is assumed to be 100 km, yielding a

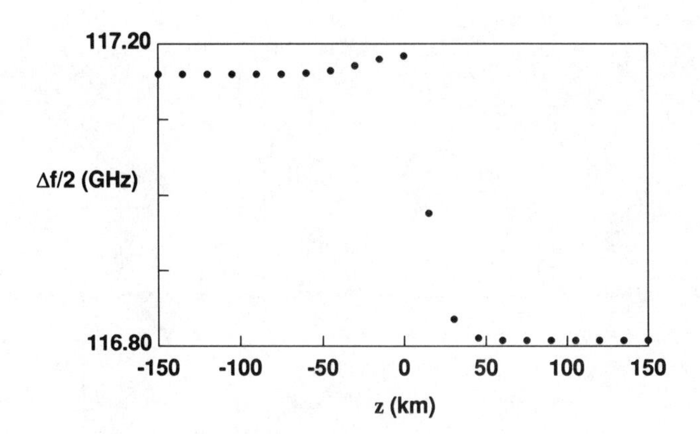

Figure 6.21 Velocity of colliding solitons with collision centered at amplifier, 1 amp/100 km. (*Source:* [1]. © 1991 Optical Society of America.)

P_{max}/P_{min} ratio of 4.7. Inspection of Figure 6.21 shows that the resulting peak-to-peak relative velocity change is 0.3%.

The change of 0.3% may seem insignificant, but this change is per collision. Solitons traveling through many thousands of kilometers of fiber may undergo many collisions and may accumulate a substantial arrival time deviation.

6.15.3 WDM Design Rules and Channel Count

Figure 6.22 is similar to Figures 6.20 and 6.21 in that it shows the acceleration and velocity shift in a system with lumped amplifiers. However, Figure 6.22 corresponds to an amplifier spacing of 20 km versus 100 km for Figures 6.20 and 6.21. In Figures 6.20 and 6.21, $L_{coll} = L_{pert}$, whereas in Figure 6.22 $L_{coll} >> L_{pert}$, where L_{pert} is the amplifier spacing. Comparison of Figure 6.22 with Figures 6.20 and 6.21 shows that the impact of collisions is much weaker when $L_{coll} >> L_{pert}$. A more accurate analysis [9] reveals that the condition preventing serious system degradation can be written as

$$L_{coll} \geq 2L_{pert} \tag{6.36}$$

In turn, the requirement (6.36) determines the maximum allowable channel spacing:

$$\Delta f_{max} = 0.31 \frac{z_0}{\tau L_{pert}} \quad \text{or} \quad \Delta\lambda_{max} = \frac{\tau}{DL_{pert}} \tag{6.37}$$

Note carefully that (6.37) specifies the maximum, not the minimum, channel spacing. This is quite different from conventional WDM systems, where only the minimum channel spacing is set by basic physics. However, the minimum channel

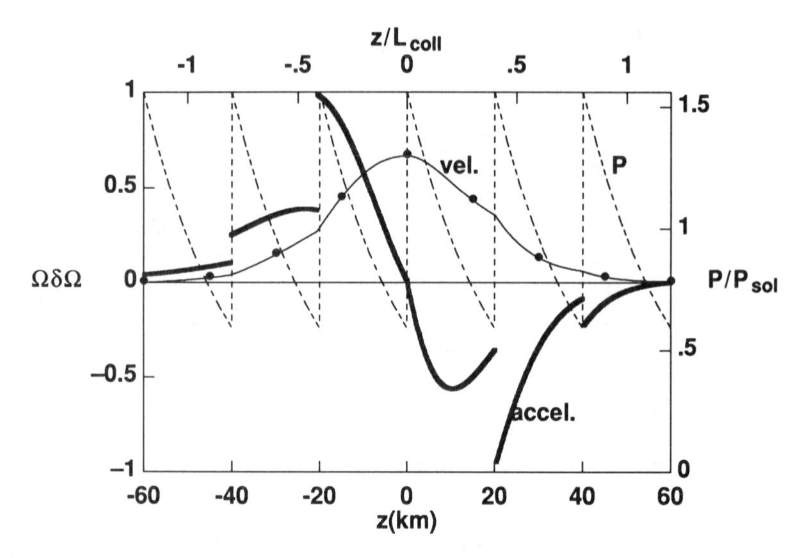

Figure 6.22 Acceleration and velocity shift of solitons colliding in system with lumped amps every 20 km. (*Source:* [1]. © 1991 Optical Society of America.)

spacing is also fixed in soliton systems. There are two basic mechanisms. The first one stems from the basic Fourier transform inequality:

$$\tau \Delta f_{spec} > 1 \tag{6.38}$$

where τ is the pulse duration, and Δf_{spec} is the resulting spectrum width. Because the channel spacing, Δf, must be larger than the spectrum width, Δf_{spec}, (6.38) leads to minimum channel spacing for both soliton and conventional systems.

There is an additional minimum channel-spacing limitation specific to soliton systems. That limitation stems from the fact that pulses of the *i*th channel tend to suffer a range of collisions with the *i*th channel, from none to a maximum given by $N_{ij} = Z\tau/(L_{coll}^{ij} T)$, where Z is the total system length, and T is the bit period. As a result, there will be a spread of arrival times about the mean. The spread can be found by multiplying the time shift per collision by $N_{ij}/2$ and summing over all channels $i \neq j$:

$$\Delta t_i = \pm 0.1418 \frac{Z}{z_0} \frac{\tau}{T} \sum_{j \neq i} \frac{1}{(\Delta f)_{ij}} \tag{6.39}$$

Thus, the maximum allowable Δt sets a limit on the minimum allowable Δf. Of course the minimum Δf may instead be determined by the requirement $\tau \Delta f_{spec} > 1$ as per (6.38).

These factors considerably limit the freedom of a WDM system designer. They

Table 6.4
WDM System Design

Fiber Dispersion, $\overline{D}$ *(ps/nm·km)*	*Minimum Channel Spacing,* $\Delta\lambda_{min}$ *(nm)*	*Arrival Time Spread,* Δt *(ps)*	*Maximum Channel Spacing,* $\Delta\lambda_{max}$ *(nm)*	*Number of Channels*	$\overline{P}_{sig}$*/channel at amplifier outputs (μW)*	*Total Bidirectional Capacity (Gbps)*
1.0	0.27	±19.0	1.08	5	125	40 (20 in each direction)

Source: [1].
Note: Soliton duration = 50 psec, link length = 9,000 km, and bit rate = 4 Gbps/channel.

seem to point to the overall speed of 20 Gbps in each direction for the whole WDM system, as illustrated by Table 6.4. However, the capacity limits in Table 6.4 can be exceeded by use of frequency-domain and time-domain filtering, as described in Section 6.17.

6.16 SOURCES OF SOLITON PULSES

Ideally, soliton pulses should have well-defined shape, width, power, and energy (see (6.18) to (6.21) and the discussion in Section 6.8). Several solutions to this problem have been proposed.

6.16.1 Fiber Ring Soliton Lasers

Figure 6.23 shows one elegant solution successfully demonstrated in a laboratory. It is based on the observation that a soliton ring transmission loop (such as the one shown in Fig. 6.13) will become unstable if the gain around the loop exactly compensates the loss. With a few additional elements to adjust the loop length and to control the soliton repetition rate, the loop will become a good source of soliton pulses.

The loop solitons' source shown in Figure 6.23 has the following attractive features:

- It is self-mode-locking (no modulator required).
- Repetition rate is set by etalon free spectral range (FSR) (T = etalon round-trip time).
- It is dispersion tuned.
- It has high output power (≥10 mW time averaged).
- Its pulse width scales inversely with $\overline{P}(\overline{P} \sim E_{sol} \sim D/\tau)$.
- It yields transform-limited, sech^2-shaped pulses.

The last feature is illustrated in Figure 6.24, where the experimentally obtained

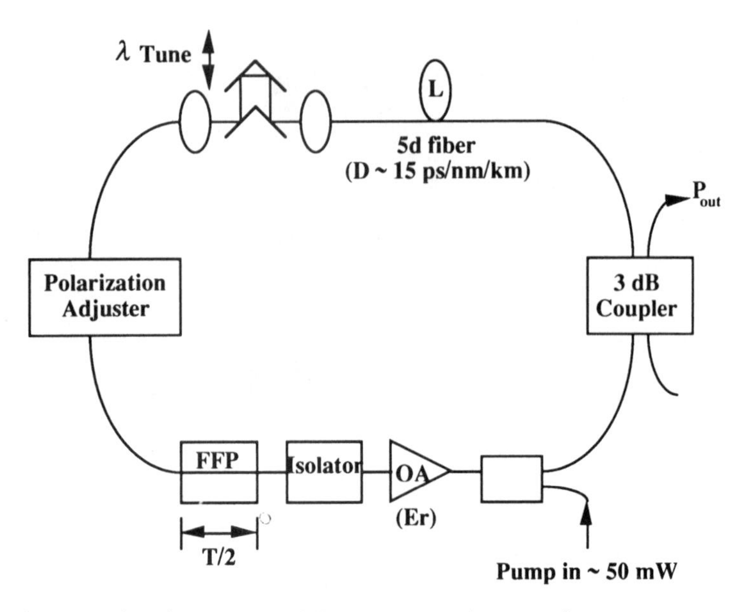

Figure 6.23 Fiber ring soliton laser. (*Source:* [1] © 1991 Optical Society of America.)

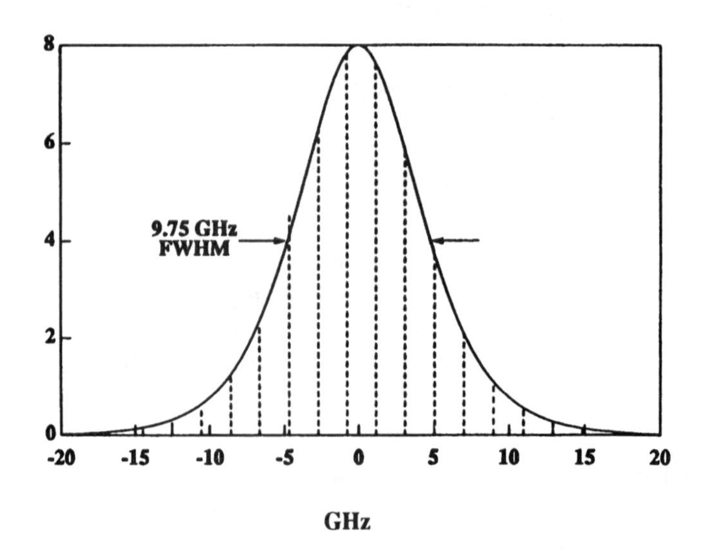

Figure 6.24 Frequency spectrum of soliton laser (dots) compared with sech^2 function L = 10 km. τ (from autocorrelation) = 32.5 ps. $\tau\Delta f$ = 0.317. (*Source:* [1]. © 1991 Optical Society of America.)

frequency spectrum of the soliton laser is compared with the ideal $sech^2$ function. The comparison confirms that the soliton ring laser indeed generates nearly ideal $sech^2$ pulses.

The advantages of the soliton ring laser have contributed to its popularity with researchers of soliton systems and have made this laser an excellent experimental tool. However, for practical field installations, a more rugged and compact source of soliton pulses may be needed, and the search for a pragmatic solution is an important research topic at the time of this writing.

Several alternatives are discussed next.

6.16.2 Fabry-Perot Fiber Soliton Lasers

Figure 6.25 shows a block diagram of a passively mode-locked Fabry-Perot fiber soliton laser reported in [10]. The fiber used in the experiment had numerical aperture, $NA = 0.12$, dispersion, $D = 17$ ps/nm·km, effective mode area = 124 μm^2, and beat length of more than 10m. Pumping was provided by a Ti:sapphire laser operating at 980 nm.

The experimental laser has a CW threshold of 25 mW launched pump power. When the pump power is increased to above the mode-locking threshold of 450 mW, the laser enters a mode-locking regime and generates narrow mode-locked pulses. Once initiated, mode-locked operation is sustained even if the pump power is lowered to within a few milliwatts of the 25-mW CW threshold. The mode-locking threshold of 450 mW is fairly high; it is much lower (some 20–30 mW) in other configurations, such as figure-eight and ring lasers.

Depending on the pump power and on the setting of the polarization controller, two mode-locking regimes were observed. In the "square-pulse" regime, long (7,500 ps) square pulses at the cavity round-trip frequency (2 MHz) were observed. In the soliton regime and at high pump powers (around the 450-mW mode-locking threshold), the laser generates tightly packed bunches of solitons exhibiting chaotic

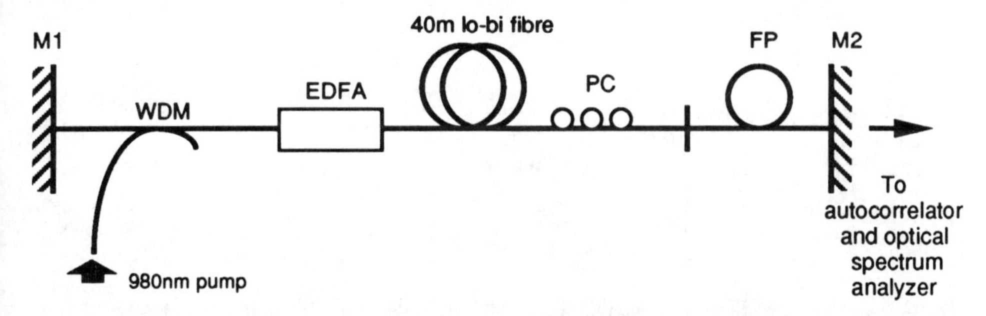

Figure 6.25 A block diagram of a passively mode-locked Fabry-Perot fiber soliton laser. Notation: WDM = WDM coupler; PC = polarization controller; FP = fiber polarizer. (*Source:* [10]. © 1993 IEEE.)

behavior. As the pump power is gradually decreased, the bunches break up to produce stable bunches of randomly spaced fundamental solitons; the bunches are repeating with the cavity round-trip period.

Figure 6.26 shows the autocorrelation trace of the pulses obtained in the soliton regime. The trace has the width of 2.5 ps, corresponding to a pulse duration of 1.6 ps. While the pulse width of this device is exceptionally short, it is unstable and the pulse frequency is difficult to control. Thus, it seems to be ill-suited for commercial communication applications.

6.16.3 DFB Laser/External Modulator Soliton Source

The simplest and the most pragmatic (at the moment) source of soliton pulses is shown in Figure 6.27.

Essentially, the external modulator (a $LiNbO_3$ device in [11]) carves out pulses of desired length (30 ps) and desired period (100 ps). The modulator also provides pulses with a controlled and desired chirp. This chirp is used to compress the pulse

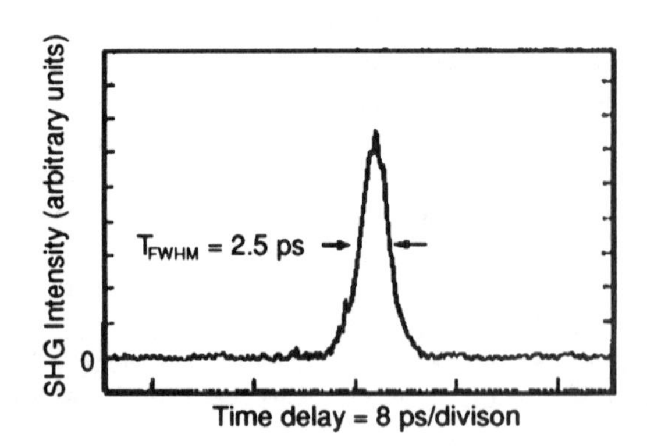

Figure 6.26 Autocorrelation trace of pulses generated by the laser in Figure 6.27 in the soliton regime. (*Source:* [10]. © 1993 IEEE.)

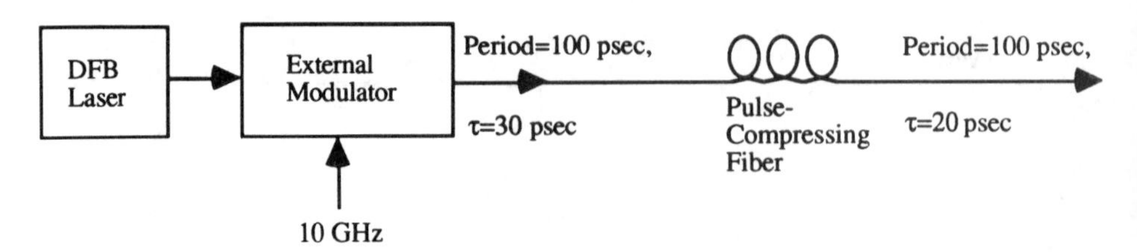

Figure 6.27 A simple source of soliton pulses for 10-GHz systems consisting of a CW-driven DFB laser, 10-GHz-driven external modulator, and a pulse-compressing fiber. (*Source:* [11]. Reprinted with author's permission.)

length from the original 30 ps to 20 ps, which is better suited for soliton transmission. The compression and unchirping are performed by a 4-km-long conventional single-mode fiber with the dispersion coefficient $D \sim 17$ ps/nm·km at 1,557 nm.

6.16.4 DFB Laser/Integrated Modulator Soliton Source

Ideally, the source of soliton pulses should be integrated into a single chip. One report [12] says that 7-ps transform-limited pulses at the rate of 20 GHz have indeed been obtained using a single semiconductor device—an integrated MQW (multiple-quantum-well) DFB laser/modulator.

Figure 6.28 shows a schematic of the laser used in [12]. The modulator section consists of a PIN structure with an i-region of eight periods of an 89-Å InGaAsP well and a 50-Å InGaAsP barrier; p-doped and n-doped InP cladding layers; and a P-InGaAsP capping layer. The laser section consists of two MQW guide layers: a laser active layer and a modulator core layer. The upper active layer consists of four periods of a 67-Å InGaAsP well and a 151-Å InGaAsP barrier. The facet of the modulator is coated with a 0.1% antireflection coating.

The CW threshold of the device is 20 mA; it oscillates in a single longitudinal mode with a side-mode suppression of more than 40 dB. The output power is 5 mW at an injection current of 100 mA when the modulator is biased at 0V; the output power decreases to 0.5 mW as the reverse bias increases.

To generate short pulses, the laser was driven at a DC injection current of 60 mA, with an output power of some 0 dBm. The modulator was driven by a 20-GHz RF signal, 3.2V peak-to-peak. The output light was coupled into a single-mode fiber amplifier. The resulting pulses are shown in Figure 6.29; they have the width of 14 ps and the duty ratio of 7.

This device is compact, stable, and simple, and the repetition rate is easily controlled by adjusting the RF frequency. It is, therefore, a highly promising source for future systems.

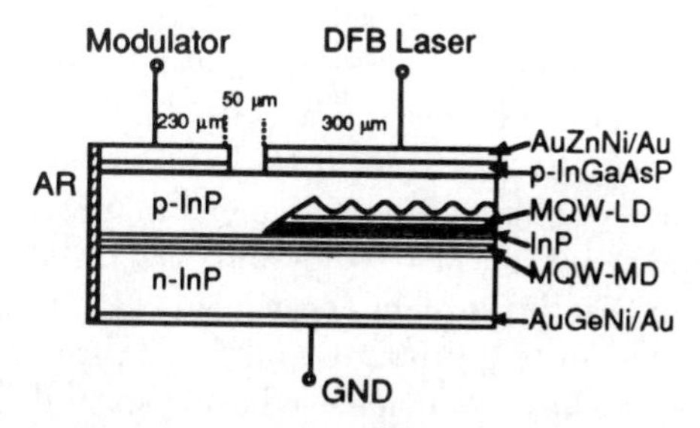

Figure 6.28 Schematic of a monolithically integrated light source consisting of a strained InGaAsP MQW DFB laser and a strained InGaAsP MQW electroabsorption modulator. (*Source:* [12]. © 1993 IEEE.)

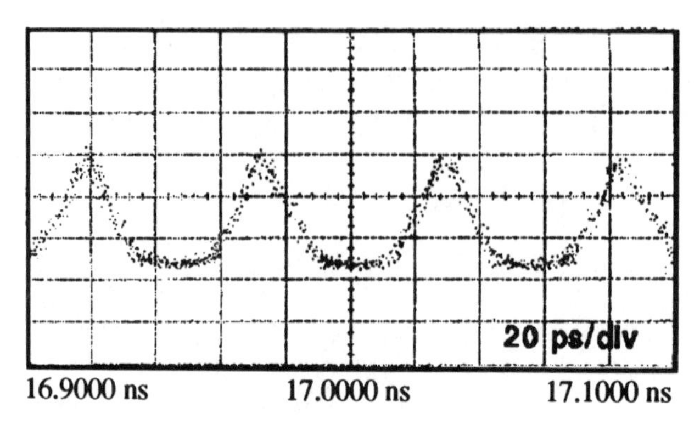

Figure 6.29 Optical pulses generated by the device in Figure 6.28. The repetition frequency is 20 GHz, and the pulse width is 14 ps. (*Source:* [12]. © 1993 IEEE.)

6.17 BEYOND THE GORDON-HAUS LIMIT

As discussed in Sections 6.12 and 6.13, the Gordon-Haus effect limits the performance of soliton systems. Recently, two techniques have been developed [6,7,13–18] to overcome the Gordon-Haus limit: time-domain and frequency-domain filtering. Both techniques improve system performance well beyond the Gordon-Haus limit and make the transmission distance virtually unlimited.

6.17.1 Frequency-Domain Filtering

Arrival-time jitter (stemming from the ASE) can be reduced by using a conventional guiding frequency filter [13]. However, a filter causes an additional loss. To compensate for that loss, more amplification is needed, which in turn leads to more ASE.

A major advance overcame this limit by using sliding frequency filters, whose frequency is slightly shifted after each amplification step [6,7,19,20]. As a result, the transmission line becomes transparent to solitons (since they adjust to the frequency shift) but opaque to ASE. Sliding-frequency filters also greatly attenuate (essentially, eliminate) the potentially harmful interference caused by the difference between the ideal soliton pulse [described by (6.16) to (6.18)] and the actual input pulse. Using that approach, Mollenauer and his colleagues demonstrated error-free propagation of solitons over 20,000 km at a bit rate of 10 Gbps and over 13,000 km at 20 Gbps using two 10-Gbps WDM channels. Subsequent experiments [11] have reached even more impressive results: 60 Gbps over 11,400 km and 70 Gbps over 9,400 km error free (BER $< 10^{-9}$). These results have been achieved using six and seven WDM channels, respectively, each running at 10 Gbps. These records are being improved on every year, and 100-Gbps systems operating over trans-Pacific distances are already on the horizon. Even faster transmission rates will undoubtedly be demonstrated in the future.

The achievement of multi-10-Gbps transmission bit rates over transoceanic distances required, in addition to solitons and sliding-frequency guiding filters, an additional innovation: dispersion-tapered fiber spans between amplifiers [11,21,22]. Without dispersion tapering, pseudo–phase matching allows four-wave mixing products from soliton-soliton collisions to grow uncontrollably and to cause severe timing and amplitude jitter. Dispersion tapering has an additional advantage of lifting the restriction on the maximum allowable channel spacing, discussed in Section 6.15.

Frequency-domain filtering is implemented using simple Fabry-Perot filters. Because their transfer function is periodic, many WDM channels can be transmitted simultaneously without any additional equipment (except for terminal equipment, of course), a powerful advantage. Opponents of this approach argue that the many slightly different Fabry-Perot filters needed for transoceanic applications will be difficult to manufacture, install correctly, and maintain. Proponents of frequency-domain filtering respond that (1) solitons adjust themselves to filters even if the filters' frequencies are imprecise or drifting; (2) highly stable Fabry-Perot filters can be manufactured using gas-filled sealed cavities, and (3) the necessary variety of filters is easy to obtain using the natural spread of filters' resonant frequencies due to manufacturing inaccuracies, by measuring and "binning out" Fabry-Perot filters at the output of a production line.

The performance of soliton WDM systems can be further improved by moving the acceptance window of a time-division demultiplexer in approximate synchronism with the correlated time shifts due to acoustic effects, soliton-soliton collisions, and collision-induced polarization scattering [23] (the latter effect is described in [24,25]). This technique is a close relative of time-domain filtering.

6.17.2 Time-Domain Filtering

Time-domain filtering [16–18] is the optical analog of conventional repeaters. Essentially, at each repeater/amplifier, the gain is synchronously modulated to retime the position of the soliton and to remove the ASE-induced jitter. It has been proved theoretically that soliton data transmission is possible over unlimited distances with this technique, because periodic synchronous modulation reduces the ASE to a very low level [14].

This technique is also being vigorously pursued experimentally. Efficient timing clock extraction circuits have been developed, and combined time-domain/frequency-domain techniques led to repeater spacings as long as 50 km with 24- to 30-ps solitons [14]. Transmission distances as long as 1,000,000 km at 10 Gbps have been demonstrated with no appreciable degradation [14].

Proponents of this technique anticipate reaching transmission rates exceeding 100 Gbps over distances longer than 10,000 km. Its critics, however, point to what appears to be a fundamental disadvantage: This technique is fairly complex and operates on one channel only. If several—say, N—WDM channels are transmitted, N repeaters are needed at each repeater node, along with WDM demultiplexers and multiplexers (see Figure 6.3).

Whichever technique—time-domain or frequency-domain filtering—emerges as the winner in this competition, intensive research in the field of 10- to 100-Gbps transmission over transoceanic distances will continue and is very likely to be successful.

Problems

Problem 6.1

Fiber loss can lead to soliton broadening. The impact of fiber loss can be analyzed by adding a loss term to the nonlinear Schrödinger equation:

$$\frac{\partial u}{\partial z'} + \frac{j}{2}\frac{\partial^2 u}{\partial t'^2} + j|u|^2 u = -2\Gamma u \tag{6.40}$$

where

$$\Gamma = \frac{\alpha}{2}\frac{\tau^2}{|\ddot{\beta}_0|}$$

Here, α is the attenuation constant of the fiber, and z' and t' are defined in Section 6.4.3.

a. Show that, for $e^{2\Gamma z'} >> t'$,

$$u(z',t') = \frac{\exp\left[-\frac{j}{8\Gamma}(1 - e^{-4\Gamma z'})\right]}{\cosh(t' e^{-2\Gamma z'})} e^{-2\Gamma z'} \tag{6.41}$$

is a solution to the nonlinear Schrödinger equation (6.40).

b. Show that the time width of the soliton increases exponentially with z and find the distance after which the soliton time width will double.

c. Show that the amplitude of the soliton decays exponentially with z.

d. If a soliton with energy E_{in} is launched into the fiber, what is its energy after 1 km of transmission?

Problem 6.2

The length of a soliton pulse is 1.5 cm assuming an effective mode radius of 6 μm,

$$n_2 = 6.1 \cdot 10^{19}\ \mathrm{cm}^2/V^2,\ \frac{\partial^2\beta}{\partial\omega^2} = -2\ \mathrm{ps}^2/\mathrm{km},$$

and a free space wavelength of 1.55 μm. Find the energy in the given soliton. Assume that the group velocity is approximately equal to the phase velocity. See Figure 6.30 for the definition of the length of the soliton.

Problem 6.3

Derive the following equation:

$$\frac{\partial\phi}{\partial z} + \dot{\beta}_0 \frac{\partial\phi}{\partial t} = -j\frac{n_2}{n_0}\beta_0\,|\phi|^2\,\phi$$

Problem 6.4

a. Show (by substitution) that

$$\phi(z,t) = \phi_0 \cdot \frac{\exp(jaz)}{\cosh\left(\dfrac{t-\dot{\beta}_0 z}{\tau}\right)}$$

where $a = \ddot{\beta}_0/2\tau^2$ and $|\phi_0|^2 = (-n_0/n_2)\cdot(\ddot{\beta}_0/\beta_0\tau^2)$ is a solution of the nonlinear Schrödinger equation.

b. Plot $|\phi(z,t)|^2$ versus t for $z = 0$, $\ddot{\beta}_0 = -20\mathrm{ps}^2/\mathrm{km}$, $n_0 = 1.54$, $n_2 = 6 \cdot 10^{-19}\ \mathrm{cm}^2/V^2$, $\lambda = 1.5\ \mu\mathrm{m}$.

c. Plot three cases: (i) $\tau = 5$ ps; (ii) $\tau = 10$ ps; (iii) $\tau = 15$ ps. Make sure the same scale is used for all three cases.

Problem 6.5

A soliton has the wavelength of 1.55 μm and the energy of 50 fJ. Find the length of the soliton pulse in centimeters assuming the mode diameter of 10 μm, $n_2 = 6 \cdot 10^{-19}\ \mathrm{cm}^2/\mathrm{V}^2$, $\ddot{\beta}_0 = -2\ \mathrm{ps}^2/\mathrm{km}$ (dispersion-shifted fiber), and group velocity ≈ phase velocity. See Figure 6.30 for the definition of the length of the soliton.

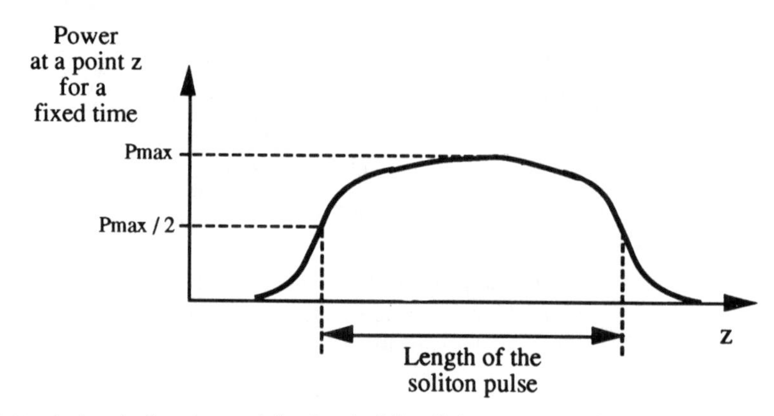

Figure 6.30 The length of a soliton is defined as the full width between the half-maximum power points along the direction of propagation *z* for a fixed time.

References

[1] Mollenauer, L. F., "Solitons in Ultra Long Distance Transmission," *OFC '91*, San Diego, CA, 1991.

[2] Marcuse, D., Miller, S. E. and I. P. Kaminow, Editors, *Optical Fiber Telecommunications II*, London: Academic Press, 1988.

[3] Agrawal, Govind P., *Fiber-Optic Communications Systems*, New York: John Wily & Son, 1992.

[4] Marcuse, D., and C. Lin, "Low Dispersion Single-Mode Fiber Transmission—The Question of Practical Versus Theoretical Maximum Transmission Bandwidth," *IEEE J. Quantum Electronics*, Vol. QE-17, No. 6, 1981, p. 870.

[5] Gemelos, S. M., unpublished.

[6] Mollenauer, L. F., J. P. Gordon, and S. G. Evangelides, "The Sliding-Frequency Guiding Filter. An Improved Form of Soliton Jitter Control," *Optics Letters*, Vol. 17, 1992, pp. 1575–1577.

[7] Mollenauer, L. F., E. Lichtman, M. J. Neubelt, and G. T. Harvey, "Demonstration, Using Sliding-Frequency Guiding Filters, of Error-Free Soliton Transmission Over More Than 20Mm at 10 Gbit/s, Single Channel, and Over More Than 13 Mm at 20 Gbit/s in a Two-Channel WDM," *Electronics Letters*, Vol. 29, 1993, pp. 910–911.

[8] Mollenauer, L. F., J. P. Gordon, and M. N. Islam, "Soliton Propagation in Long Fibers With Periodically Compensated Loss," *IEEE J. Quantum Electronics*, Vol. QE-22, January 1986, pp. 157–173.

[9] Mollenauer, L. F., S. G. Evangelides, and J. P. Gordon, "Wavelength Division Multiplexing With Solitons in Ultra Long Distance Transmission Using Lumped Amplifiers," *IEEE J. Lightwave Technology*, Vol. 9, March 1991, pp. 362–367.

[10] Matsas, V. J., W. H. Loh, and D. J. Tichardson, "Self-Starting, Passively Mode-Locked Fabry-Perot Fiber Soliton Laser Using Nonlinear Polarization Evolution," *IEEE Photonics Technology Letters*, Vol. 5, No. 5, May 1993, pp. 492–494.

[11] Mollenauer, L. F., P. V. Mamyshev, and M. J. Neubelt, "Demonstration of Soliton WDM Transmission at 6 and 7 × 10 Gbit/s Error Free Over Transoceanic Distances," Submitted to *Electronics Letters*.

[12] Wakita, K., K. Sato, I. Kotaka, M. Yamamoto, and M. Asobe, "Transform-Limited 7-ps Optical Pulse Generation Using a Sinusoidally Driven InGaAsP/InGaAsP Strained Multiple-Quantum-Well DFB Laser/Modulator Monolithically Integrated Light Source," *IEEE Photonics Technology Letters*, Vol. 5, No. 8, August 1993, pp. 899–901.

[13] Glass, A. M., "Fiber Optics," *Physics Today*, October 1993, pp. 34–38.
[14] Nakazawa, M., et al, "Nonlinear Optics in Optical Fibers and Future Prospects for Optical Soliton Communication Technologies," NTT R&D 42, (1993), pp. 1317–1326.
[15] Mecozzi, A., J. D. Moores, H. A. Haus, and Y. Lai, "Soliton Transmission Control," *Optics Letters*, Vol. 16, 1991, pp. 1841–1842.
[16] Kodama, Y., and A. Hasegawa, "Generation of Asymptotically Stable Optical Solitons and Suppression of the Gordon-Haus Effect," *Optics Letters*, Vol. 17, 1992, pp. 31–33.
[17] Nakazawa, M., E. Yamada, H. Kubota, and K. Suzuki, "10 Gbit/s Soliton Transmission Over One Million Kilometers," *Electronics Letters*, Vol. 27, 1990, pp. 1270–1272.
[18] Nakazawa, M., H. Kubota, E. Yamoda, and K. Suzuki, "Infinite-Distance Soliton Transmission With Soliton Controls in Time and Frequency Domains," *Electronics Letters*, Vol. 28, 1992, pp. 1099–1100.
[19] Mamyshev, P. V., and L. F. Mollenauer, "Stability of Soliton Propagation With Sliding-Frequency Guilding Filters," *Optics Letters*, Vol. 19, 1994, pp. 2083–2085.
[20] Mollenauer, L. F., P. V. Mamysev, and M. J. Neubelt, "Measurement of Timing Jitter in Filter-Guided Soliton Transmission at 10 Gbits/s and Achievement of 375 Gbits/s-Mm, Error Free, at 12.5 and 15 Gbits/s," *Optics Letters*, Vol. 19, 1994, pp. 704–706.
[21] Mamyshev, P. V., and L. F. Mollenauer, "Pseudo Phase Matched Four-Wave Mixing in Soliton WDM Transmission," *Optics Letters*, to be published.
[22] Mollenauer, L. F., "Soliton Transmission Speeds," *Optics and Photonics News*, April 1994, pp. 15–19.
[23] Mollenauer, L. F., "Method for Nulling Non-Random Timing Jitter in Soliton Transmission," *Optics Letters*, to be published.
[24] Mollenauer, L. F., J. P. Gordon, and F. Heismann, "Polarization Scattering by Soliton-Soliton Collisions," *Optics Letters*, Vol. 20, 1995, pp. 2060–2062.
[25] Mollenauer, L. F., and J. P. Gordon, "Birefringence-Mediated Timing Jitter in Soliton Transmission," *Optics Letters*, Vol. 19, 1994, pp. 375–377.

Selected Bibliography

Andrekson, P. A., et al., "Observation of Multi-Wavelength Soliton Collisions in Optical Systems With Fiber Amplifiers," *Applied Physics Letters*, Vol. 57, 1990, p. 1715.
Andrekson, P. A., et al., "Soliton Collision Interaction Force Dependence on Wavelength Separation in Fiber Amplifier Based Systems," *Electronics Letters*, Vol. 26, 1990, p. 1499.
Andrekson, P. A., et al., "Observation of Collision Induced Temporary Soliton Carrier Frequency Shifts in Ultra Long Fiber Transmission Systems," *IEEE J. Lightwave Technology*, Vol. 9, September 1991, pp. 1132–1135.
Gordon, J. P., and L. F. Mollenauer, "Effects of Fiber Nonlinearities and Amplifier Spacing on Ultra Long Distance Transmission," Joint Issue on Optical Amplifiers of *IEEE J. Quantum Electronics* and *J. Lightwave Technology*, Vol. 9, February 1991, pp. 170–173.
Gordon, J. P., and L. F. Mollenauer, "Phase Noise in Photonic Communications Systems Using Linear Amplifiers," *Optics Letters*, Vol. 15, 1990, p. 1351.
Mollenauer, L. F., and K. Smith, "Demonstration of Soliton Transmission Over More Than 4000 km in Fiber With Loss Periodically Compensated by Raman Gain," *Optics Letters*, Vol. 13, 1988, p. 675.
Mollenauer, L. F., K. Smith, J. P. Gordon, and C. R. Menyuk, "Resistance of Solitons to the Effects of Polarization Dispersion in Optical Fibers," *Optics Letters*, Vol. 14, 1989, p. 1219.
Mollenauer, L. F., S. G. Evangelides, and H. A. Haus, "Long Distance Soliton Propagation Using Lumped Amplifiers and Dispersion Shifted Fiber," Joint Issue on Optical Amplifiers of *IEEE J. Quantum Electronics* and *J. Lightwave Technology*, Vol. 9, February 1991, pp. 170–173.

Mollenauer, L. F., et al., "Experimental Study of Soliton Transmission Over More Than 10,000 km in Dispersion Shifted Fiber," *Optics Letters*, Vol. 15, 1990, p. 1203.
Smith, K., and L. F. Mollenauer, "Experimental Observation of Adiabatic Compression and Expansion of Soliton Pulses Over Long Fiber Paths," *Optics Letters*, Vol. 13, 1989, p. 751.
Smith, K., and L. F. Mollenauer, "Experimental Observation of Soliton Interaction Over Very Long Fiber Paths: Discovery of a Long Range Interaction," *Optics Letters*, Vol. 14, 1989, p. 1284.

Chapter 7

Multichannel Systems

7.1 INTRODUCTION

Optical communications has experienced many revolutionary changes since the days of short-distance multimode transmission at 0.8 μm [1]. We have seen how, with the advent of EDFAs (erbium-doped fiber amplifiers), single-channel repeaterless transmission at 5 Gbps across 9,000 km has been accomplished [2]. Therefore, we can consider single-channel point-to-point links to be state of the art and an accomplished fact, albeit with many improvements possible. (Soliton transmission, which has the potential for much higher speeds and longer distances, was discussed in Chapter 6.) For short distances, transmission of a 100-Gbps single channel has been achieved using some very sophisticated modulation and demodulation techniques [3]. However, although these single-channel results are quite impressive, they do have two disadvantages: (1) They take advantage of only a very small fraction of the enormous bandwidth available in an optical fiber, and (2) they connect two distinct end points, not allowing for a multiuser environment. Because the required rates of data transmission among many users have been increasing at an impressive pace for the past several years, it is a highly desirable goal to eventually connect many users with a high-bandwidth optical communication system. A simple multiuser system may be a point-to-point link with many simultaneous channels. A more complicated system can take the form of a local-, metropolitan-, or wide-area network with either high bidirectional connectivity or simple unidirectional distribution [4]. Electrical networks that exist today will eventually exhibit the severe limitations of reduced speed and increased power consumption for high-speed (~Gbps/user), multiuser (hundreds), or medium-distance-link (exceeding ~1 km) networks. Therefore, it seems quite likely that multiuser optical systems will be needed in the not-too-distant future.

Multichannel optical systems were relatively unknown in 1980, but much technological progress has been achieved since then. The applications can include a multiplexed high-bandwidth library resource system, simultaneous information sharing, supercomputer data and processor interaction, a myriad of multimedia services, video applications, and many undreamed-of services. As demands for more network bandwidth increase, the need will become apparent for multiuser optical networks, with issues such as functionality, compatibility, and cost determining which systems will eventually be implemented. This chapter will deal with the many technical issues, possible solutions, and recent progress in the exciting and relatively young area of multichannel systems.

7.1.1 Fiber Bandwidth

The driving force motivating the use of multichannel optical systems is the enormous bandwidth available in the optical fiber. When the single-mode silica fiber was described in Chapter 2, we concentrated on its low-loss, low-dispersion, low-cost, and high-bandwidth waveguiding properties. The attenuation curve as a function of optical carrier wavelength is shown in Figure 7.1 [5]. There are two loss minima, one near 1.3 μm and an even lower one near 1.55 μm. Let's discuss the one surrounding 1.55 μm, which is approximately 25,000 GHz wide. (Due to the extremely desirable characteristics of the EDFA [see Chapter 5], which amplifies only near 1.55 μm, the conventional wisdom is that future systems will use EDFAs and therefore not use the dispersion-zero 1.3-μm band of the existing embedded conventional fiber base. Of course, dispersion-shifted fiber in which the dispersion-zero point is located near 1.55 μm may be widely installed in the future, thereby totally circumventing the need to consider the 1.3-μm band.) The high-bandwidth characteristic of the optical fiber implies that a single optical carrier at 1.55 μm can be baseband modulated at ~25,000 Gbps, occupying 25,000 GHz surrounding 1.55 μm, before transmission losses of the optical fiber would limit transmission. Obviously, this bit rate is impossible for present-day optical devices to achieve, given that heroic lasers, external modulators, switches, or detectors have bandwidths <100 GHz; note that practical data links today would be significantly slower, perhaps no more than 10 Gbps per channel. As

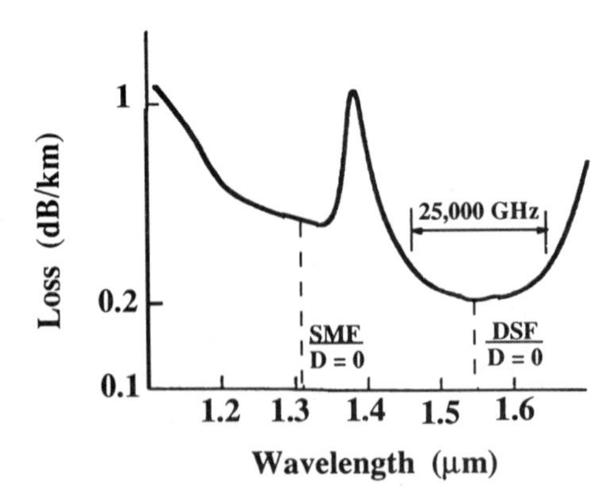

Figure 7.1 Signal attenuation as a function of wavelength in a conventional single-mode silica fiber (*After:* [5]).

such, a single high-speed channel takes advantage of an extremely small portion of the available fiber bandwidth. In this chapter, we will discuss the various multiplexing methods that attempt to utilize as much of the fiber bandwidth as possible.

7.1.2 Popular Multiplexing Methods

This section will briefly introduce the various multichannel systems that will be covered in detail throughout this chapter.

The 10-Gbps channel mentioned in the previous section probably will be a combination of many lower-speed signals, since very few individual applications today utilize this high bandwidth. These lower-speed channels are multiplexed together in time to form a higher-speed channel. This *time-division multiplexing* (TDM) can be accomplished in the electrical or optical domain, with each lower-speed channel transmitting a bit (or a collection of bits known as a packet) in a given time slot and then waiting its turn to transmit another bit (or packet) after all the other channels have had their opportunity to transmit [6]. Figure 7.2 shows the basic TDM concept with interleaving (i.e., multiplexing) bits from several users. TDM is quite popular with today's electrical networks and is fairly straightforward to implement in an optical network at <10-Gbps speeds. This scheme by itself cannot hope to utilize the available bandwidth because it is limited by the speed of the time-multiplexing and -demultiplexing components. Moreover, ultra-high-speed transmission becomes severely limited by fiber dispersion and nonlinearities [7].

To exploit more of the fiber's THz bandwidth, we seek solutions that complement or replace TDM. One obvious choice is WDM (wavelength-division multiplexing), in which several baseband-modulated channels are transmitted along a single fiber but with each channel located at a different wavelength (see Figure 7.3) [8–11]. Each of N different wavelength lasers is operating at the slower Gbps speeds, but the aggregate system is transmitting at N times the individual laser speed, providing a significant capacity enhancement. The WDM channels are separated in wavelength to avoid cross-talk when they are (de)multiplexed by a nonideal optical filter. The wavelengths can be individually routed through a network or individually recovered by wavelength-selective components. WDM allows us to use much of the fiber bandwidth, although various device, system, and network issues will limit the utilization of the full fiber

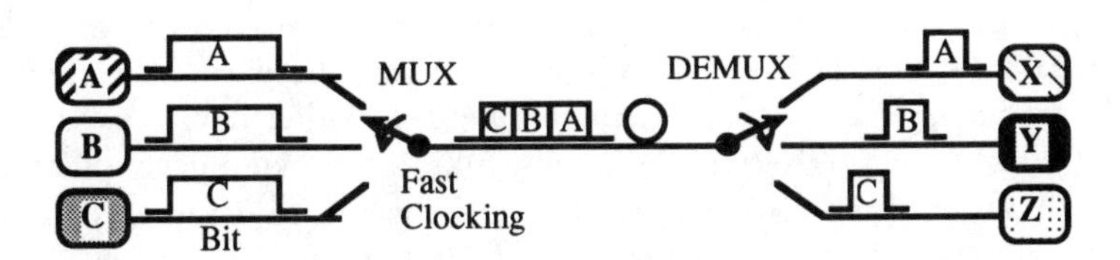

Figure 7.2 Several TDM channels with bit-interleaved multiplexing (mux = multiplexer, demux = demultiplexer).

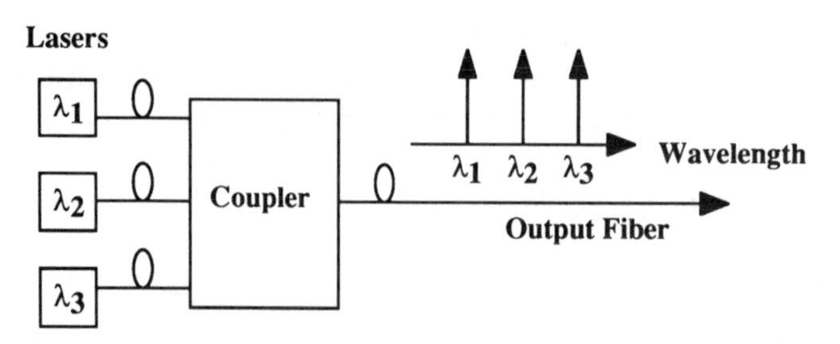

Figure 7.3 Many WDM channels propagating in a single optical fiber.

bandwidth. Note that **each** WDM channel may contain a set of even slower time-multiplexed channels.

Another method conceptually related to WDM is *subcarrier multiplexing* (SCM) [12]. Instead of directly modulating a ~terahertz optical carrier wave with ~100s Mbps baseband data, the baseband data are impressed on a ~gigahertz subcarrier wave that is subsequently impressed on the THz optical carrier. Figure 7.4 illustrates the situation in which each channel is located at a different subcarrier frequency, thereby occupying a different portion of the spectrum surrounding the optical carrier. SCM is similar to commercial radio, in which many stations are placed at different RF (radio frequency) such that a radio receiver can tune its filter to the appropriate subcarrier RF [13]. The multiplexing and demultiplexing of the SCM channels is accomplished electronically, not optically. The obvious advantage for cost-conscious users is that several channels can share the same expensive optical components; electrical components typically are less expensive than optical ones. Just as with TDM, SCM is limited in maximum subcarrier frequencies and data rates by the available bandwidth of the electrical and optical components. Therefore, SCM must be used in conjunction with WDM if we want to utilize any significant fraction of the fiber bandwidth, but it can be used effectively for lower-speed, lower-cost multiuser systems.

We will additionally discuss a method known as *code-division multiplexing* (CDM) [14] (Fig. 7.5). Instead of each channel occupying a given wavelength, frequency, or time slot, each channel transmits its bits as a coded channel-specific sequence of pulses. This coded transmission typically is accomplished by transmitting a unique time-dependent series of short pulses. These short pulses are placed within chip times within the larger bit time. All channels, each with a different code, can be transmitted on the same fiber and asynchronously demultiplexed. One effect of coding is that the frequency bandwidth of each channel is broadened, or "spread." If ultra-short (<100 fs) optical pulses can be successfully generated and modulated, then a significant fraction of the fiber bandwidth can be used. Unfortunately, it is difficult for the entire system to operate at these speeds without incurring enormous cost and complexity.

The final optical multiplexing scheme we will discuss in this chapter is called

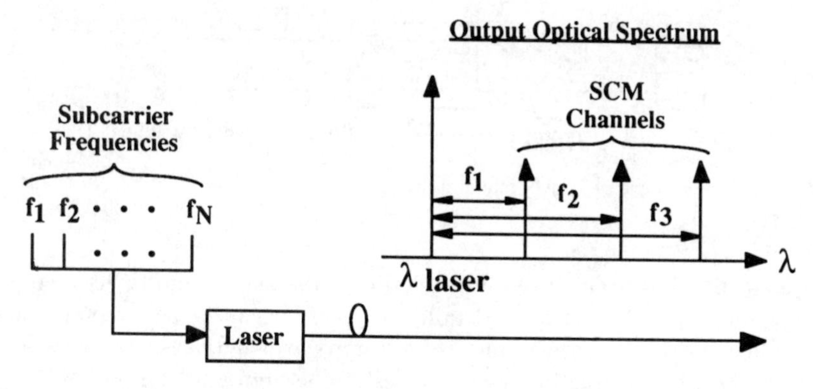

Figure 7.4 Frequency spectrum of several SCM channels transmitted from a single laser.

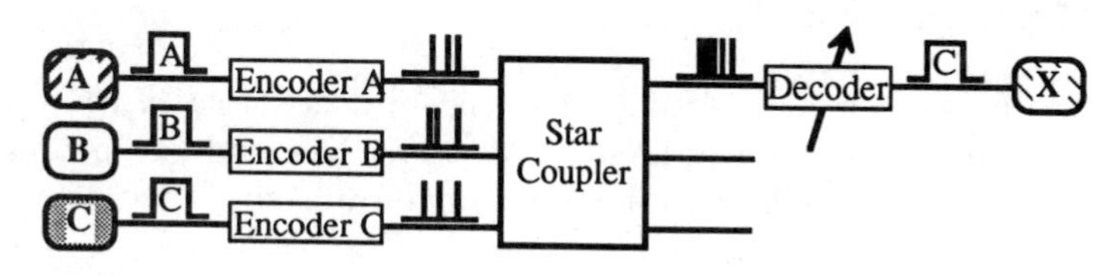

Figure 7.5 The basic concept of a coded pulse sequence for CDM, with each pulse located in a chip time and the entire code occupying a larger bit time slot.

space-division multiplexing (SDM), in which the channel-routing path is determined by a different spatial position (i.e., a different output fiber) [15]. A simple example of this is shown in Figure 7.6, in which the optical output of a fiber is split into N different and parallel optical beam paths. Each of the N output beams is passed through a light-modulating switch and then coupled to a different output fiber. By controlling the transmissivity of each optical modulator, a signal on the input fiber can be routed to any fiber output port. By extending this scenario, N input fiber ports can be fully interconnected with N output fiber ports by an array of N^2 optical switches. The technology for implementing moderate-speed systems is already commercially available. In contrast to all the other methods, however, each channel occupies its own spatial coordinate, and all other channels cannot be transmitted simultaneously on the same fiber. In other words, we are not more fully utilizing the high bandwidth of the fiber, but we are creating a high-bandwidth space-switching matrix, with the result that a high overall switching capacity can be realized.

Irrespective of the specific multiplexing scheme, the signal routing through a network can be performed actively, by reading information encoded on the optical signal, or passively, in which the specific time slot, wavelength, frequency, code, or spatial position of the signal dictates the network path. These routing schemes and multiplexing techniques can be used individually or collectively in a hybrid system.

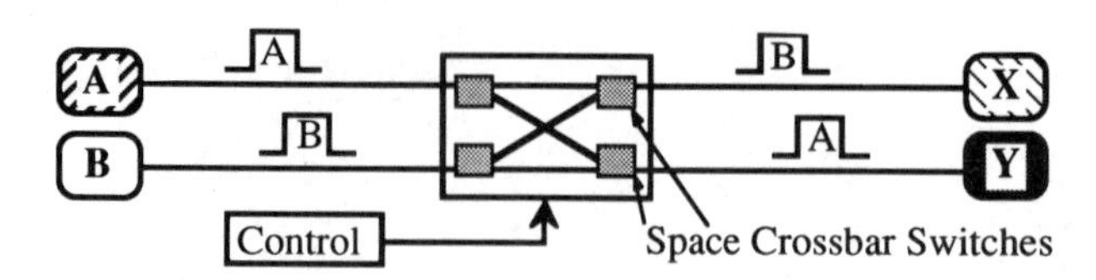

Figure 7.6 A simple example of 1-by-N spatial multiplexing.

This section has been a brief introduction to the various multiplexing methods, each of which we will revisit in detail. However, because the preponderance of multichannel work is currently in the area of multiwavelength systems, we will devote special attention to the enabling technologies and system guidelines of WDM.

7.1.3 Basic Topologies

The multiplexing schemes described in Section 7.1.2 are used to route data and connect N users in a high-bandwidth optical network. One multichannel implementation is to have many channels multiplexed together, transmitted over some distance, and then demultiplexed at a common destination. This represents the simplest type of topology and is known as a point-to-point system. In addition, several common network topologies exist for electrical networks, each of which can be modified to facilitate optical communications networks. These are the ring, bus, star, and tree configurations, which are shown in Figure 7.7 [16]. The names are descriptive of the topology, and each type is conceptually straightforward.

The ring structure has nodes (i.e., users) that are periodically connected in a closed ring formation. The single ring is unidirectional, but a second inner ring can be added that would facilitate bidirectional communication, with each ring representing a different direction of propagation. Furthermore, the inner ring provides an alternative protection path in case of a single-link outage, enabling all nodes to still communicate with each other even if a circuitous route must be taken [17]; such protection path provides for a survivable ring network.

A similar topology to the ring is the bus network, which is simply a ring that has failed to close on itself and represents a situation in which the network nodes are connected to a common "backplane." Shown in Figure 7.7 is a single bus. A dual bus can also be implemented, with one "rail" (i.e., fiber) used for transmitting data to nodes upstream and the other rail used for transmitting data to nodes downstream. The bus is quite easy to build given almost any geographical configuration of the network, but a break in the bus will isolate a node or nodes from communicating with the rest of the network.

Historically, the ring and the bus have been highly favored for electrical TDM networks. One drawback in the implementation of an optical bus or ring is that a passive optical tap typically would be required at each node for signal injection and/or

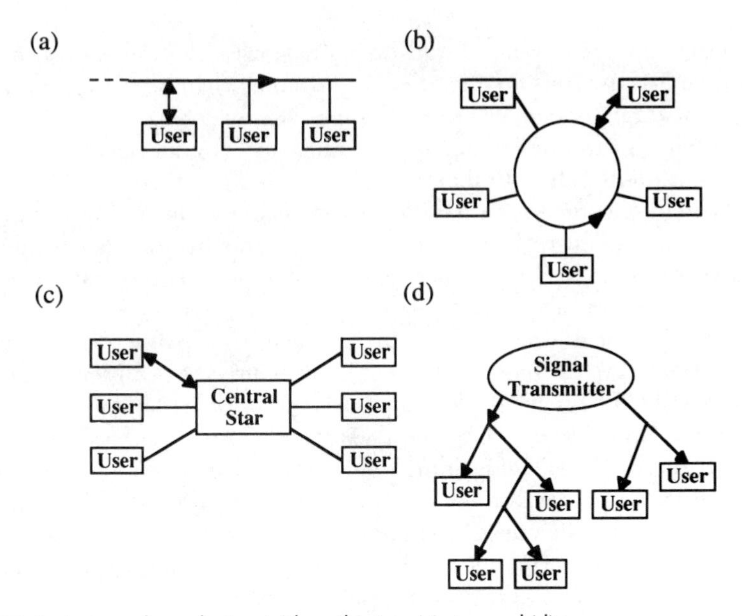

Figure 7.7 Basic network topologies: (a) bus, (b) ring, (c) star, and (d) tree.

recovery; therefore, the optical losses would become intolerable for a large number of users. The optical power available at the Mth user node in a bus, P^M_{bus}, for an M node network is given by [18]

$$P^M_{bus} = P_T C[\beta(1 - C)]^{M-1} \tag{7.1}$$

where C is the ratio of the amount of optical power tapped off at each node, β is the excess loss at each tapping node, and γ is the excess loss in an M-by-M star.

Another network topology is the star, which has some advantages for an optical network. The star is configured such that a central device interconnects many nodes and each node can transmit to every other node through the central star. The star may involve active routing, but generally it is thought of as a passive element, such as a passive star coupler [19]. This passive star coupler has N input fibers (connected to each node's transmitter) and N output fibers (connected to each node's receiver). The passive star splits the input light from any given input port equally among the N output ports. The optical power available at each of M output ports in a star, P^M_{star}, is given by [18]

$$P^M_{star} = \frac{P_T \gamma}{M} \tag{7.2}$$

The inherent optical loss in a star grows more slowly with the number of users, M, than does the loss in a bus or ring, and the difference can be significant in large networks (optical amplifiers can be used to compensate for some of these losses). Any node in a star can transmit to and access any other node, and a break in a link will disrupt communications only with that node, as opposed to a bus, for which a break could disrupt communications for a significant portion of the entire network. Since all transmission must pass through the central star, the star has the disadvantages of having a higher propagation delay, requiring more fiber, and requiring the geography to accommodate a central device.

The final topological example is the tree, which is a favorite of broadcast, or distribution, systems. At the base of the tree is the source transmitter, from which emanates the signal to be broadcast throughout the network. From that base, the tree splits many times into different branches, with each branch either having nodes connected to it or further dividing into subbranches. This continues until all the nodes in the network can access the base transmitter. Whereas the other topologies are intended to support bidirectional communication among the nodes, this topology is useful for distributing information unidirectionally from a central point to a multitude of users. This is a straightforward topology and is in use in many electrical systems, most notably cable television (CATV). We will revisit CATV when we discuss SCM.

The network users being referred to in this chapter can take many forms and require vastly different bit rates. The nodes may be video stations, supercomputers, library mainframes, personal computers, multimedia centers, personal communications devices, or gateways to other networks. Moreover, the selection as to which topology to implement depends on many factors, including the multiplexing scheme, geography, power budget, and cost. In fact, it is quite likely that a hybrid topology will be used in a large network, as is depicted in Figure 7.8. In the example shown in that figure, smaller rings, buses, and stars are connected in larger networks.

In Figure 7.8, in which a larger network is composed of smaller ones, we have introduced the subject of the architecture of the network, which depends on the network's geographical extent. The three main architectural types are the local-, metropolitan-, and wide-area networks (LAN, MAN, and WAN, respectively) [20]. Although no rule exists, the generally accepted understanding is that a LAN interconnects a small number of users covering a few kilometers (i.e., intra- and inter-building), a MAN interconnects users in a city and its outlying regions, and a WAN interconnects significant portions of a country (hundreds of kilometers). Based on Figure 7.8, the smaller networks represent LANs, the larger ones MANs, and the entire figure would represent a WAN. In other words, a WAN is composed of smaller MANs, and a MAN is composed of smaller LANs. Hybrid systems exist, and typically a WAN will consist of smaller LANs, with mixing and matching between the most practical topologies for a given system. For example, stars and rings may be desirable for LANs, whereas buses may be the only practical solution for WANs. It is, at present, unclear

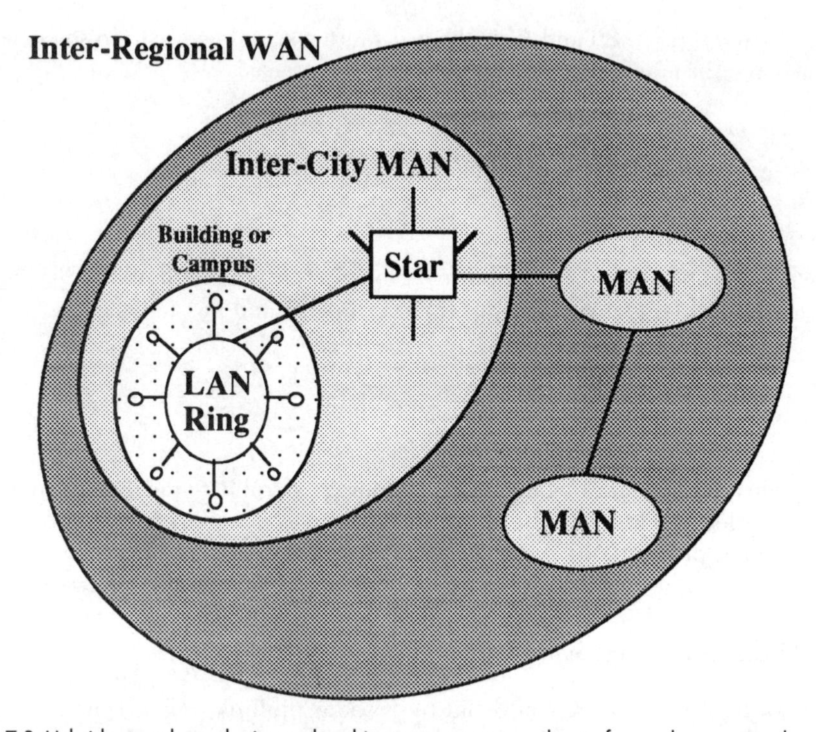

Figure 7.8 Hybrid network topologies and architectures woven together to form a large network.

which network topology and architecture will ultimately and most effectively take advantage of high-capacity optical systems.

Example 7.1

Let's compare the optical losses in a bus and in a star configuration. Assume ther are $M = 50$ users (i.e., nodes). In the bus, the tapping ratio is 10% and the excess loss is 0.5 dB. In the central star, the excess loss is 3 dB. What is the power ratio at a given receiving node when comparing the bus to the star?

Solution

Using (7.1), power P_{Bus} at the $M = 50$ user is

$$\begin{aligned} P_{\text{Bus}}^{M} &= P_T C[\beta(1 - C)]^{M-1} \\ &= 0.0000019 P_T \end{aligned}$$

where $C = 0.01$, $\beta = 0.89$ and P_T is the transmitted signal power from the first node. Similarly for the star topology, power at the M_{th} user is:

$$P^{M}_{\text{Star}} = \frac{P_T \gamma}{M} = 0.01 P_T$$

where $\gamma = 0.5$ and P_T is the total power transmitted from any input node. The ratio between the available powers at the M_{th} user in star and bus topology is

$$\frac{P_{\text{Star}}}{P_{\text{Bus}}} = 5260.00$$

It is obvious that the star topology is much more perferable to the bus (or ring) when considering losses. However, the optical amplifier has, for most purposes, solved this power-related problem.

7.1.4 Circuit and Packet Switching

Before we proceed to discuss in detail the various multiplexing schemes, the reader must be familiar with the distinction between the two main methods of switching transmitted information across a network: circuit switching and packet switching [21]. The distinction is extremely relevant in appreciating the switching requirements on the enabling technologies. Figure 7.9 shows the scenario for both switching schemes in a simple six-node network. To begin with, it is important to realize that circuit switching is widely used and is the method typically employed in the public-switched telephone network. In circuit switching, if user A wants to communicate with user E, then user A sends a request signal to user E to initiate data transmission. If user E is available for interconnection, it will acknowledge user A's original request, and then user A can begin data transmission. This process of request and acknowledgment is known as *handshaking*. Once transmission commences, a fixed path, or circuit, along the network is created between the two users until such time that the communications link is terminated by either user. The fiber medium on which the A-E circuit resides has a certain total bandwidth potential, and only a portion of that bandwidth is dedicated (i.e., reserved) for these two users. Many circuit switched links can coexist on the same transmission medium by using one of the multiplexing schemes, such as TDM. Because the medium bandwidth is reserved for the A-E circuit independent of whether a stream of bits is actually being transmitted (i.e., a pause), bandwidth is being unused, and the links over which the circuit has been created are underutilized.

A circuit-switched network is easy to implement and control. Circuit switching

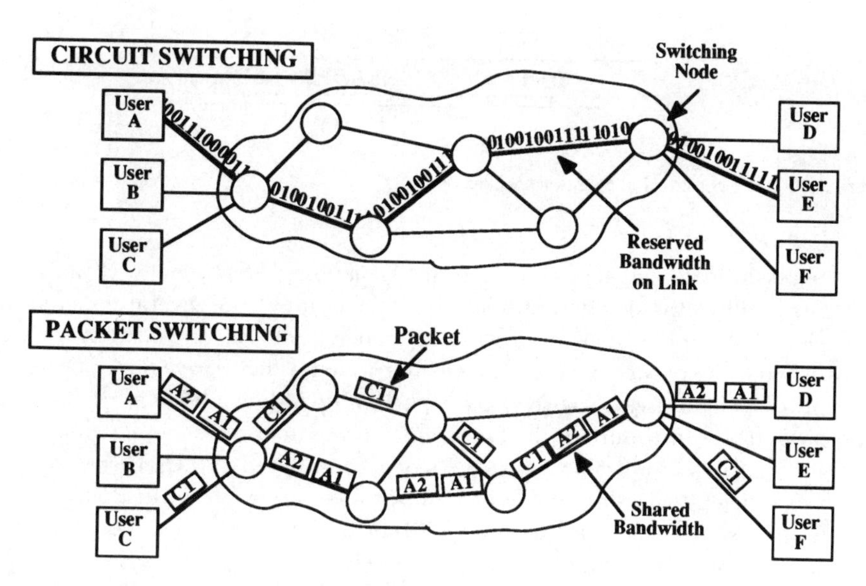

Figure 7.9 (a) Circuit and (b) packet switching systems for a simple six-node network.

emphasizes a transparent data pipe with a long set-up and tear-down time and is more appropriate for low-speed transmission for a long transmission time, such as a telephone call. Therefore, it has low component-switching-speed requirements (~milliseconds), since the communications path is intended for minutes of transmission, as in a telephone call. In that case, the handshaking process between widely separated nodes may take a short time compared to the slow data transfer speed. However, if the handshaking process takes a long time compared to the data transfer rate, the link overhead due to handshaking is undesirably large. (The switching speed discussed is the time it takes for a switch to change the routing of a data stream or packet, i.e., determining which output port of the switch is enabled for data arriving at a given input port.)

However, for high-speed (gigabits per second) optical transmission in which a user may require only a few seconds of transmission time, the more likely scenario to maximize system throughput is to implement a packet-switched network, which does not require a link with dedicated bandwidth. Each user in such a network transmits data in the form of packets, as depicted in Figure 7.9. A data packet in a generic network is composed of a flag (i.e., a unique set of bits), which is typically at the beginning of the packet to synchronize data recovery and alert a receiver that a packet is arriving: header bits, which contain destination and routing information for switching through the network; and the user-generated end-to-end data payload (Fig. 7.10) [22].

A packet may be required to traverse many intermediate switching nodes before

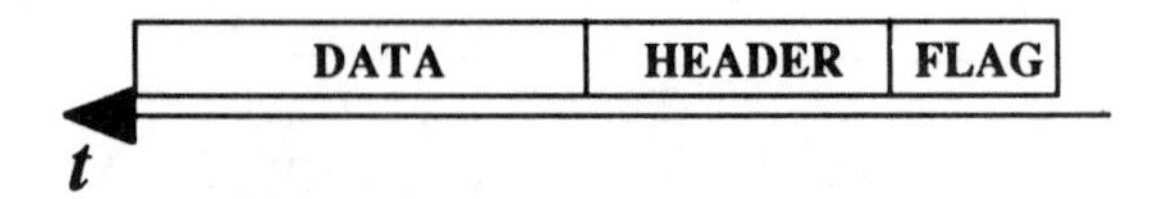

Figure 7.10 Basic construction of a data packet.

it is successfully routed to its destination. Depending on the network and on the available bandwidth on different lines, any packet may take several possible routes to reach a destination. In such a case, a sequence number is included in the packet to ensure that the packets will not be misinterpreted if they arrive out of order. Each transmitted packet uses some fraction of the bandwidth available in the fiber links. If no packets are transmitted, the bandwidth in the fiber links can be allocated to a different transmission of packets, such as between users B and D. Therefore, packet switching efficiently uses the available bandwidth—bandwidth is used only for transmitting high-speed data, and the line will not be forced idle if a given user is not transmitting data. Thus, bandwidth can be freely allocated in this system to where it is needed and is used efficiently, making it ideal for either bursty or uniform high-speed traffic. However, packet switching has much more stringent requirements on the switching technologies since switching must occur on the time scale of an individual packet. The component-switching-speed requirement is severe, with the range being microsecond to nanosecond, depending on the specific system being implemented. We will discuss SONET and ATM later, but a SONET frame is 125 μs in length and a 1-Gbps ATM packet is only ~500 ns in length.

To underscore the importance of minimizing the switching speed in a high-speed network, we can compare the total transmission time required for a circuit-switched and a packet-switched network for cases involving low-speed and high-speed transmission. The total time for a communications link for circuit and packet switching, $T_{circuit}$ and T_{packet}, respectively, is

$$T_{initial} = 2T_{prop} + T_{ack} + T_{switch} + \left(\frac{\#\ of\ transmitted\ bits}{B\,R}\right) \tag{7.3}$$

$$T_{initial} = 2T_{prop} + T_{ack} + T_{switch} + \left(\frac{\#\ of\ transmitted\ bits}{B\,R}\right) \tag{7.4}$$

where T_{ack} is the time for handshaking (i.e., request and acknowledgment) in circuit switching, T_{prop} is the signal propagation delay, BR is the data bit rate, and T_{switch} is the switching time of the system components. The typical value for T_{ack} in circuit-switched systems is several milliseconds, negating the requirement that component switching times be less than milliseconds: As an example, if 100,000 bits are being transferred at 1 Gbps and the set-up time is 1 ms, then the total time is 1.1 ms and is dominated by the long set-up time. On the other hand, high-data-rate packet-switched networks are typically limited by the switching speed of the components, requiring switching on the order of microseconds or less.

Because of the potential for higher network throughput when packet switching is used, the remainder of the chapter will discuss the technological issues in accomplishing such a system. It will be noted when circuit switching is being discussed, as in the case of a tree network in CATV systems.

7.2 TIME-DIVISION MULTIPLEXING

At its most basic, optical networks can imitate electrical networks in which TDM is overwhelmingly used for digital data transmission. A fiber can carry many time-multiplexed channels, in which each channel can transmit its data in an assigned time slot [23,24]. That time slot also identifies either the sender's or the receiver's address, but we will assume it is the sender's address, for simplicity. Furthermore, it should be emphasized that since the network will allow only one user to transmit during a given time slot, no output-port contention problems exist with TDM. (There will be a more complete discussion on contention problems later in this chapter.)

A typical TDM link is shown in Figure 7.11, in which N transmitters are sequentially polled by a fast multiplexer to transmit their data. The time-multiplexed data are transmitted along a fiber link and then sequentially and rapidly demultiplexed at the receiving node. If a time slot represents the sender's address, then the demultiplexing is accomplished so that a given receiver node can access only that specific

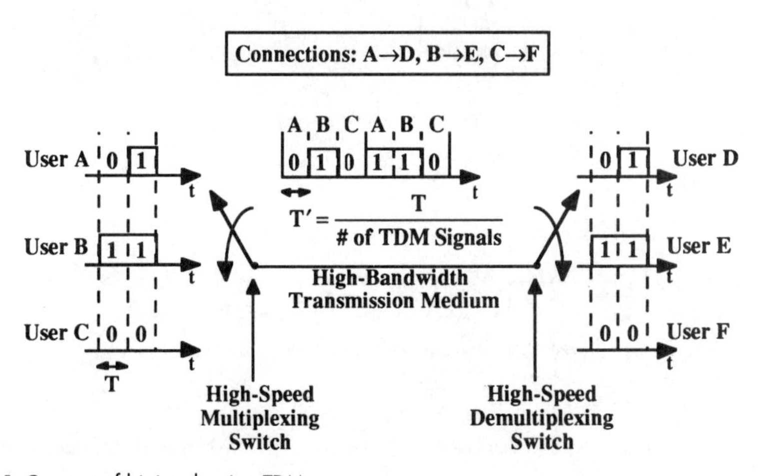

Figure 7.11 Concept of bit-interleaving TDM.

time slot associated with the desired sender. The multiplexing can be of the bit- or packet-interleaved form, although only bit-interleaving is shown in Figure 7.11. In bit interleaving, the multiplexer polls each node for a single bit and then polls the next node. If N users wish to share the same high-bandwidth optical medium, then we must divide each transmitter's bit time, T, into N slots for (de)multiplexing. In packet interleaving, the multiplexer polls a node for an entire packet time and then polls the next node. Although either method is valid, we will limit ourselves to the bit-interleaving case, which has more stringent requirements and which is the more traditional method of TDM.

Depending on the bit rate of each channel, BR_{ch} and the number of users, N, it is possible that the multiplexer and demultiplexer may not be sufficiently fast. Specifically, switching (multiplexing and demultiplexing) must be performed at the time-multiplexed speed. The bit rate of the TDM transmission (BR_{TDM}), and, consequently, the speed requirement of the (de)multiplexer is

$$BR_{TDM} = N \cdot BR_{ch} \tag{7.5}$$

It is obvious that optical components do not exist that accommodate very large N if each channel operates at moderate speeds, with even the fastest and most complicated TDM switches limited to <100 GHz.

The crucial time multiplexing and demultiplexing functions can be performed in the electrical or the optical domain, as illustrated in Figure 7.12. Electrical

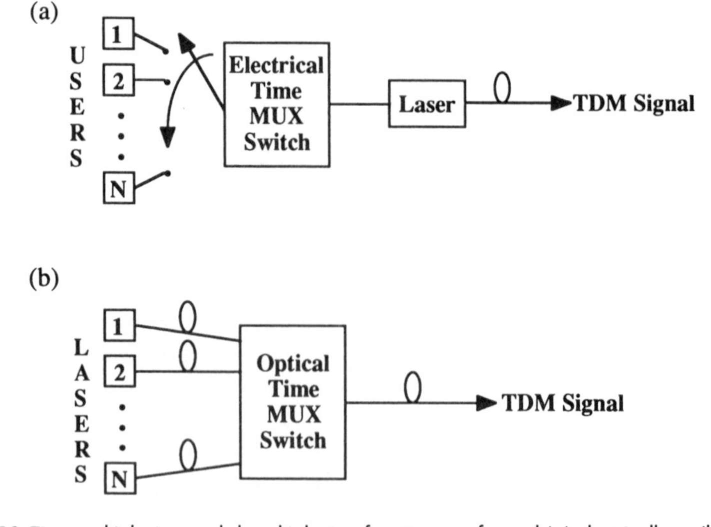

Figure 7.12 Time multiplexing and demultiplexing functions performed (a) electrically or (b) optically.

multiplexing essentially time multiplexes several lower-speed channels into one high-speed channel, which is then used to modulate a single optical signal. Electrical demultiplexing follows the same argument, in which one optical signal is detected and then electrically time demultiplexed. The only optical procedure is the transmission along the fiber.

Electrical TDM is efficient and will almost certainly form the backbone for many high-speed optical signals, since few users will individually generate gigabit-per-second signals. However, the forefront of research in TDM is in performing the (de)multiplexing in the optical domain. Several lower-speed optical signals are time (de)multiplexed by a fast optical switch; this optical switch, in which the data path remains in optical form, can be controlled by an external electrical input signal. The importance of this optical TDM technology is that the many lower-speed channels can originate and terminate at locations far away from the TDM switch. With electrical TDM, the lower-speed channels must be in close proximity to the TDM switch, since even the lower-speed signals still can be quite fast and thus incur high transmission losses across an electrical coaxial transmission line. In networks, optical TDM incorporates the ability to perform routing and switching functions based on the specific time slot of a particular optical bit.

Another important issue in TDM is the critical requirement of synchronizing the incoming bits. Since bits may be arriving at the multiplexer from sources located at some distance away, it is crucial to have the bits synchronized so that the multiplexer is not polling an individual input during a bit transition but only in the middle of a bit. Additionally, the demultiplexer must also know precisely the time slots of the high-speed bits. Therefore, for almost any section of a high-speed optical TDM system, it is critical to recover the transmitted signal's clock within a few clock cycles. We will discuss the issue of network synchronization, clock distribution, and clock recovery in a later section.

There are two major advantages of TDM: (1) There is no output-port contention problem, since each data bit occupies its own time slot and there is only a single high-speed signal present at any given instant, and (2) the implementation for low-speed photonic networks is quite straightforward and similar to electronic networks. One major disadvantage of TDM is that the scheme requires ultra-high-speed switching components if the individual signals are themselves high-speed and if there are many users. For example, if there are 10 users, each transmitting at 2.5 Gbps, the photonic switches must have a 25-GHz bandwidth. It is obvious that this method will experience a capacity limitation since the bandwidth of photonic switches will not exceed tens of gigahertz in the near future. Another difficulty of TDM is that network control, stability, and electronic processing become difficult (and expensive) to perform efficiently at very high speeds. Furthermore, the transmission of short pulses is extremely prone to fiber dispersion and nonlinear effects, unless soliton pulses are being propagated.

Some device technologies that may be critical to performing high-speed TDM include the following.

- *High-speed switches.* These could be high-speed lithium-niobate switches in which many 2-by-2 switches are cascaded to form any N-by-N combination [25].
- *Generation of high-speed pulses.* A mode-locked laser could produce picosecond pulses at gigahertz repetition rates. These pulses could be modulated to produce a high-speed TDM signal, and such a signal would be in RZ format. Additionally, a single high-power short pulse can be passed through several parallel delay lines (Fig. 7.13), which can produce a series of lower-power short pulses, each delayed by one bit time [24].

Much of the recent work concerning optical TDM has centered on reducing the speed bottleneck imposed by the multiplexer and the demultiplexer. By using a high-speed lithium-niobate switch and a mode-locked laser to generate short optical pulses, time multiplexing was performed at 72 GHz, allowing transmission of two 36-Gbps channels [26]. Slower experiments using conventional NRZ-modulated lasers and lithium-niobate switches have showed 16-Gbps system capacity for either four 4-Gbps channels or sixteen 1-Gbps channels [25]. Another interesting area of research is the use of an interference-type *nonlinear optical loop mirror* (NOLM) to enable ultra-high-speed multiplexing and demultiplexing [27]. Results have been demonstrated using the NOLM in which individual bits have been switched at speeds >100 GHz, an astounding achievement. This ultra-high-speed device is still in the research phase and has yet to be embraced as a practical technology.

Although TDM is conceptually straightforward with no output-port contention problems, it is obvious that high-speed implementation is difficult and that by no means will the ~THz fiber bandwidth be adequately utilized. However, future systems

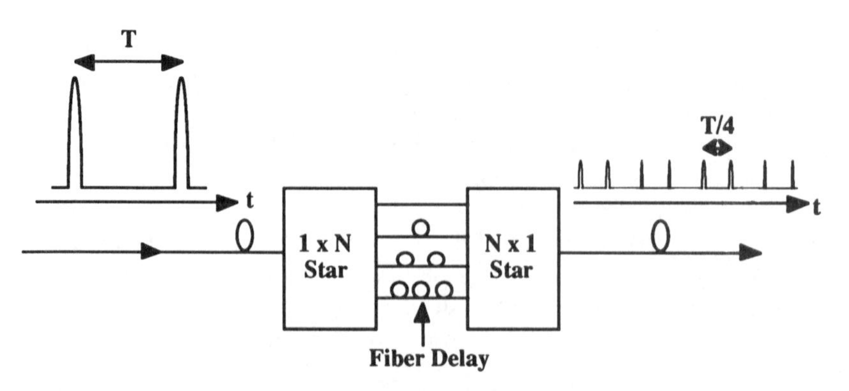

Figure 7.13 Generation of a series of short pulses from a single short pulse.

will most probably use some form of TDM (optical or electrical) in conjunction with another type of multiplexing scheme, since it is unlikely that individual users will require much greater than gigabit-per-second transmission. Therefore, many users would share a single laser or a single optical switch input port.

7.3 WAVELENGTH-DIVISION MULTIPLEXING

Until the late 1980s, optical fiber communications was mainly confined to transmitting a single optical channel. Due to fiber attenuation, this channel required periodic regeneration, which included detection, electronic processing, and optical retransmission. Such regeneration causes a high-speed optoelectronic bottleneck, is bit-rate specific, and can handle only a single wavelength. The need for these single-channel regenerators (i.e., repeaters) was replaced when the EDFA was developed, enabling high-speed repeaterless single-channel transmission. We can think of this single ~Gbps channel as a single high-speed lane in a highway in which the cars are packets of optical data and the highway is the optical fiber. However, the ~25 THz optical fiber can accommodate much more bandwidth than the traffic from a single lane. It seems natural to dramatically increase the system capacity by transmitting several different independent wavelengths simultaneously down a fiber to more fully utilize this enormous fiber bandwidth [28–30]. Therefore, the intent was to develop a multiple-lane highway, with each lane representing data traveling on a different wavelength. The WDM system enables the fiber to carry more throughput. By using wavelength-selective devices for the on and off ramps, independent signal routing also can be accomplished. This highway cartoon scenario is illustrated in Figure 7.14. It should be emphasized that a major enabling technology to the practical vision of this multiwavelength system is the EDFA, which can provide gain to many channels simultaneously over a ~THz wavelength range.

At present, it is expected that WDM will be one of the methods of choice for future ultra-high-bandwidth multichannel systems. Of course, this sentiment could change as technology evolves. Since the vast majority of multichannel optical systems research involves the technology and systems management of WDM, this section will be dramatically larger than the other sections and will discuss many aspects involved in WDM systems.

7.3.1 Basic Operation

We explained in Section 7.1.2 how WDM enables the utilization of a significant portion of the available fiber bandwidth by allowing many independent signals to be transmitted simultaneously on one fiber, with each signal located at a different wavelength. These signals can be routed and detected independently, with the wavelength determining the communication path by acting as the signature address of the origin, destination, or routing. Components are therefore required that are wavelength

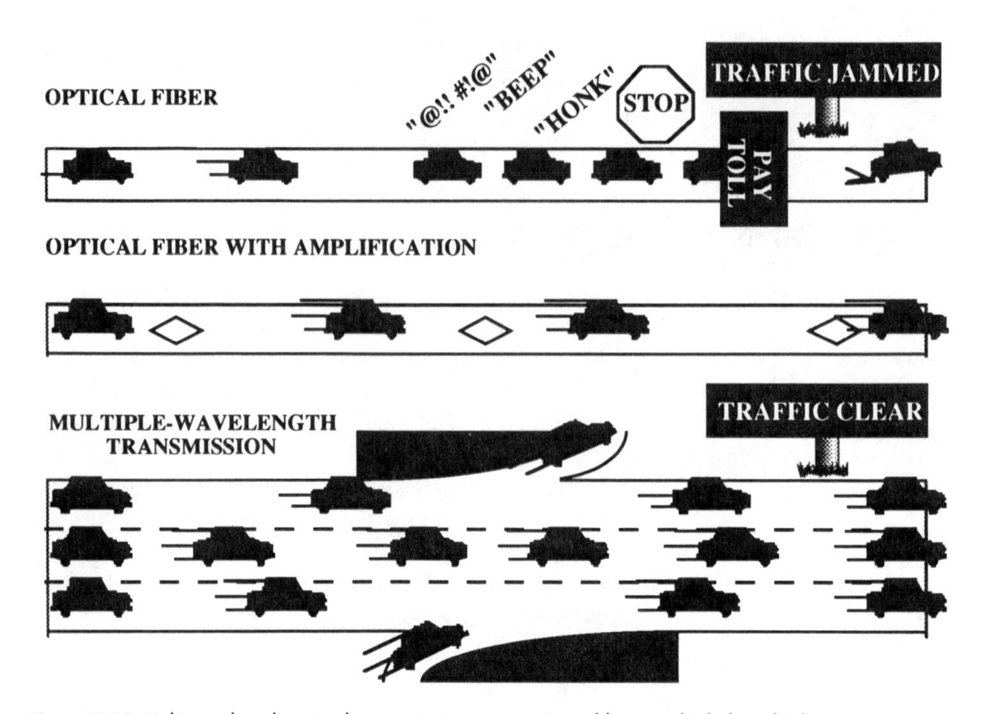

Figure 7.14 Multiwavelength optical transmission as represented by a multiple-lane highway.

selective, allowing for the transmission, recovery, or routing of specific wavelengths. As shown in the simple system in Figure 7.15, each laser transmitter must emit light at a different wavelength, with all the lasers' light multiplexed together onto a single optical fiber. After being transmitted through a high-bandwidth optical fiber, the combined optical signals must be demultiplexed at the receiving end by distributing the total optical power to each output port and then requiring that each receiver selectively recover only one wavelength by using a tunable optical filter. Each laser is modulated at a given speed, and the total aggregate capacity being transmitted along the high-bandwidth fiber (or high-bandwidth optical switching matrix) is the sum total of the bit rates of the individual lasers. One example of the system capacity enhancement is the situation in which ten 2.5-Gbps signals can be transmitted on one fiber, producing a system capacity of 25 Gbps. This wavelength-parallelism circumvents the problem of typical optoelectronic devices, which do not have bandwidths exceeding a few gigahertz unless they are exotic and expensive. The speed requirements for the individual optoelectronic components (i.e., lasers, detectors, modulators) are, therefore, relaxed, even though a significant amount of total fiber bandwidth is still being utilized.

Figure 7.16 illustrates the concept of wavelength demultiplexing using an optical

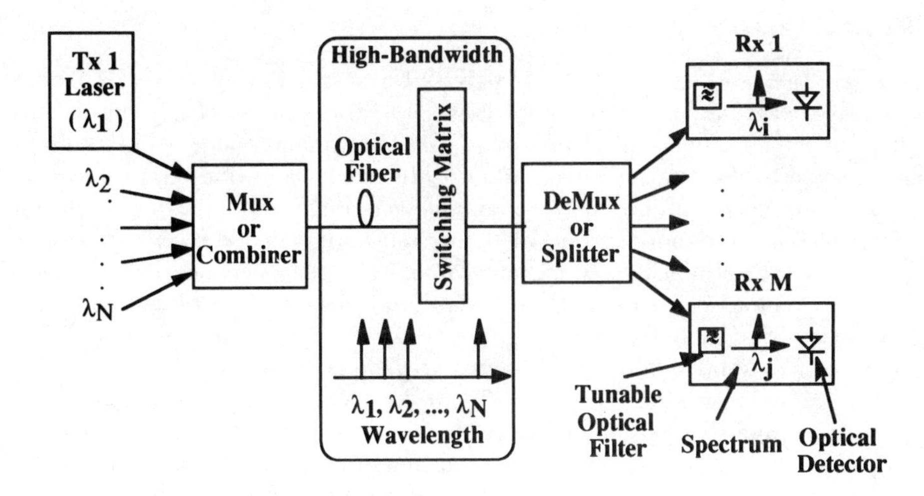

Figure 7.15 Diagram of a simple WDM system.

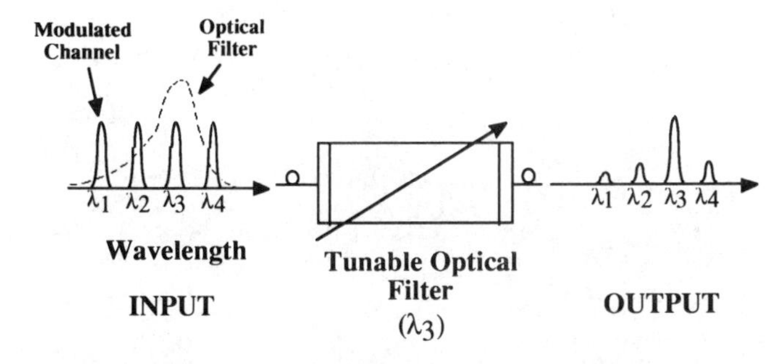

Figure 7.16 Optical WDM channels being demultiplexed by an optical filter.

filter. In the figure, four channels are input to an optical filter that has a nonideal transmission filtering function. The filter transmission peak is centered over the desired channel, in this case, λ_3, thereby transmitting that channel and blocking all other channels. Because of the nonideal filter transmission function, some optical energy of the neighboring channels leaks through the filter, causing interchannel, interwavelength cross-talk. This cross-talk has the effect of reducing the selected signal's contrast ratio and can be minimized by increasing the spectral separation between channels. Although there is no set definition, a nonstandardized convention exists for defining optical WDM, dense WDM, and *frequency-division multiplexing* (FDM) as encompassing a system for which the channel spacing is approximately 10 nm, 1 nm, and 0.1 nm, respectively. However, we will not make any distinction among those system labels in this book.

7.3.2 Topologies and Architectures

Figure 7.17 (a) shows a simple point-to-point WDM system in which several channels are multiplexed at one node, the combined signals are transmitted across some distance of fiber, and the channels are demultiplexed at a destination node. This facilitates high-bandwidth fiber transmission. Additionally, high-bandwidth routing can be facilitated through a multiuser network. As shown in Figure 7.17(b), the wavelength becomes the signature address for either the transmitters or the receivers, and the wavelength will determine the routing path through an optical network. Because nodes will want to communicate with each other, either the transmitters or the receivers must be wavelength tunable to facilitate the proper link set-up; we have arbitrarily chosen the transmitters to be tunable in this network example.

The simple illustration in Figure 7.18 shows two common network topologies that can use WDM, namely, the star and the ring networks [31–33]. Each node in the star has a transmitter and a receiver, with the transmitter connected to one of the central passive star's inputs and the receiver connected to one of the star's outputs. WDM networks can also be of the ring variety, as shown in Figure 7.18(b). Rings are popular because so many electrical networks use this topology and because rings are easy to implement for any network geographical configuration. In this example, each node in the unidirectional ring can transmit on a specific signature wavelength, and each node can recover any other node's wavelength signal by means of a wavelength-tunable

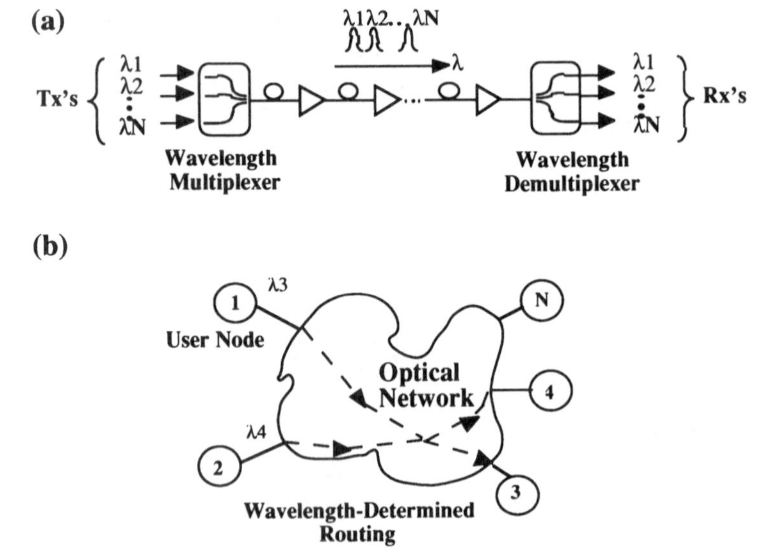

Figure 7.17 (a) A simple point-to-point WDM transmission system. (b) A generic multiuser network in which the communications links and routing paths are determined by the wavelengths used within the optical switching fabric.

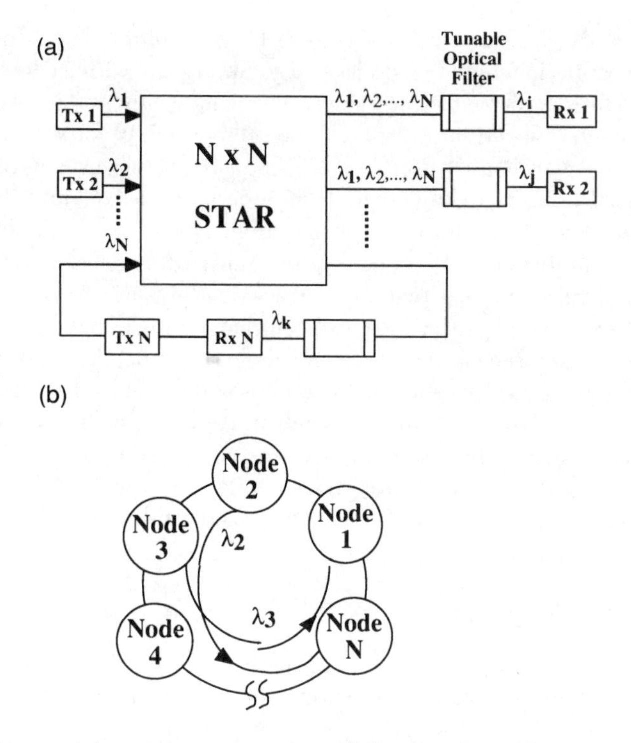

Figure 7.18 (a) Diagram of a simple star network in which WDM is used for routing and multiplexing purposes. (b) An example of a WDM unidirectional-ring network.

receiver. Although not shown in the figure, each node must recover a desired channel, which can be performed in one of two ways: (1) Only a small portion of all the signals is tapped off by a passive optical coupler, thereby allowing a tunable filter to recover a specific channel, or (2) a channel-dropping filter completely removes only the desired signal and allows all other channels to continue propagating around the ring.

In both the star and the ring scenarios, each node has a signature wavelength, and any two nodes can communicate with each other by transmitting on that wavelength. This implies that we require N wavelengths to connect N nodes. The obvious advantage is that data transfer occurs with an uninterrupted optical path between the origin and the destination, known as a single-hop network. The optical data start at the originating node and reach the destination node without stopping at any other intermediate node. A disadvantage of a single-hop WDM network is that the network and all its components must accommodate N wavelengths, which may be difficult (or impossible) to achieve in a large network. Current fabrication technology cannot provide and transmission capability cannot accommodate 1,000 distinct wavelengths for a 1,000-user network.

An alternative to requiring N wavelengths to accommodate N nodes is to have a multihop network, in which two nodes can communicate with each other by sending data through a third node, with many such intermediate hops possible [34]. A dual-bus multihop eight-node WDM network is shown in Figure 7.19 for which each node can transmit on two wavelengths and receive on two other wavelengths. The logical connectivity is also shown. As an example, if node 1 wants to communicate with node 5, it transmits on wavelength λ_1 and only a single hop is required. However, if node 1 wants to communicate with node 2, it first must transmit to node 5, which then transmits to node 2, incurring two hops. Any extra hops are deleterious in that they (1) increase the transmit time between two communicating nodes, since a hop typically requires some form of detection and retransmission, and (2) decrease the throughput, since a relaying node can transmit its own data while it is in the process of relaying another node's data. However, multihop networks do reduce the required number of wavelengths and the wavelength tunability range of the components. (The concept of wavelength reuse throughout a limited-wavelength network will be addressed later in this chapter when integrated frequency routers and passive/active routing are discussed.)

7.3.3 Enabling Technologies

Many interesting challenges face the eventual implementation of WDM systems and networks. Several of these challenges involve the control and management of the data

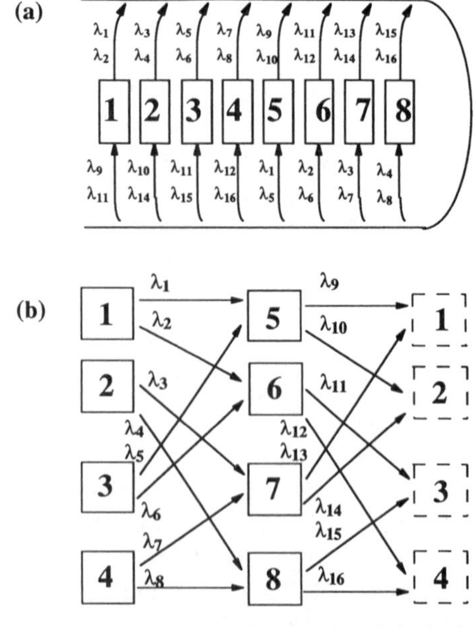

Figure 7.19 (a) A dual-rail WDM bus multihop 8 node network and (b) the logical network connectivity.

through this novel high-speed network. However, as was clear at the outset when research groups were envisioning WDM systems, the required devices simply were not available commercially and were hardly more than prospective research projects. Therefore, the most fundamental problems initially were in the area of providing the unique wavelength-selective devices necessary for good system performance. After several years of research, much progress has been made in the laboratory with several WDM-related components just now becoming commercially available. With all the recent progress, devices still place limits on the design of a WDM network. Moreover, cost will always be a factor in determining which type of network will be implemented, and the cost of these WDM components is still quite high. Future work will center on making the devices more affordable and practical for individual users. This section will deal exclusively with the critical components necessary for a WDM system. For the sake of brevity, we will include only a limited descriptive discussion of each technology. It should be emphasized that much of what distinguishes an optical network from an electrical one is in the physical layer implementation, which mainly includes the device and transmission technologies.

Figure 7.20 shows a small subset of critical component technologies typically required for a WDM system, including multiple-wavelength transmitters, multiport star couplers, passive and active wavelength routers, EDFAs, and tunable optical filters. Before discussing the specifics of the individual components, it is important to familiarize the reader with some generic laudable goals that a WDM-device technologist aims to achieve. These goals include

- Large wavelength tuning range;
- Multiuser capability;
- Wavelength stability and repeatability;
- Low cross-talk;
- High extinction ratio;

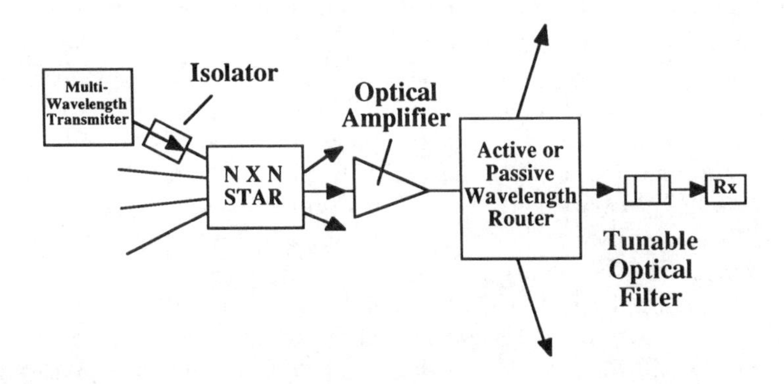

Figure 7.20 Schematic of a small subset of enabling device technologies for a WDM system.

- Minimum excess losses;
- Fast wavelength tunability (especially for packet switching);
- High-speed modulation bandwidth;
- Low residual chirp;
- High finesse;
- Low noise;
- Robustness;
- High yield;
- Potential low cost.

These device goals enable a system that can accommodate a large number of users and achieve high-speed and packet-switching operation. Of course, the list is not exhaustive, and not all goals apply to each specific component.

7.3.3.1 Wavelength-Tunable Lasers

Multi-wavelength transmitters are required if the wavelength of the transmitter must be tuned to different values corresponding to different destinations. It is quite possible for a fixed-transmitter WDM system to use ordinary nontunable single-frequency (distributed feedback (DFB)) [35] lasers that have been hand-selected to be at the desired wavelengths with the appropriate wavelength spacings [36]. However, it is impractical to design a robust system around such discrete-wavelength lasers because of the unpredictability in the exact wavelength of a processed laser, the limited thermal tuning range of DFBs (~0.1 nm/°C), and the inherent long-term laser wavelength drift. We will concentrate in this section on describing wavelength-tunable lasers to be used in a WDM system. In general, the figures of merit for these devices are large wavelength-tuning range, high-speed wavelength tunability, high data modulation speed for ASK or FSK transmission, rigid wavelength stability and repeatability, and low potential cost.

The specific tunable laser to be discussed will be the multiple-section DBR (distributed Bragg reflector) laser [37], with mention made of the two-section DFB laser [38]. It is assumed that the reader is already familiar with the operation of a basic laser diode, as outlined in Chapter 2. To briefly review, carriers that are injected into a frequency-selective cavity can experience stimulated emission and lasing if the following conditions are met: (1) there exists a gain medium, (2) the gain is greater than the losses in the cavity, and (3) the cavity provides frequency-selective feedback by having, for example, two separated parallel mirrors providing periodic wavelength-selective transmission resonances.

The two main physical processes underlying wavelength tuning are (1) Bragg reflection [39] and (2) the carrier-dependent refractive index. Bragg gratings can be formed by a periodic longitudinal variation in the real or imaginary part of the index of refraction of a medium. These gratings are quite efficient and highly wavelength selective in their reflectivity, for which the laser light and the gratings must match the Bragg condition [39]:

$$\Lambda = \frac{m\lambda}{2n(\cos\theta)} \tag{7.6}$$

where Λ is the periodic distance between adjacent grating peaks, λ is the wavelength of the emitted light, n is the refractive index of the medium, m is the order of the grating, and θ is the angle between the incident light and the grating normal (e.g., $\theta = 90$ degrees for light incident parallel to the grating).

We limit our discussion to enabling only the fundamental Bragg condition ($m = 1$), in which incident light is almost totally reflected back toward the direction of the incident wave. This Bragg wavelength condition should also correspond to the wavelength peak of the laser gain medium to simultaneously have efficient gratings and efficient gain for lasing. These Bragg gratings can replace end mirrors to provide feedback in a laser cavity, in which the gain-medium waveguide is followed by a waveguiding region with a periodic index grating. The Bragg cavity is more selective than a Fabry-Perot (FPE) cavity formed by two parallel dielectric mirrors, with the periodic wavelength-selective resonances spaced farther apart in frequency (i.e., high finesse). These widely-spaced resonances ensure that only one reflectance resonance peak will fall within the gain-greater-than-loss bandwidth, thus providing single-mode operation at one cavity resonance.

Tuning of the laser wavelength output of a laser given a Bragg reflective grating for feedback can be accomplished by tuning the parameters that govern the condition for Bragg reflection [see (7.6)]. One method is to change the index of refraction, which is a function of the carrier concentration [40]. If the Bragg region is simply a passive (i.e., undoped and no gain) waveguiding semiconductor region, then we can inject current into that grating region, thereby changing the refractive index, the frequency-selective Bragg reflection function, and ultimately the laser output wavelength. This tuning can be accomplished relatively fast, since it depends on carrier dynamics.

The second tuning mechanism relates to the refractive index inside the cavity in which the light propagates. As stated before, the refractive index increases as the carrier concentration increases. In a simple sense, injecting more current into a semiconductor will increase its refractive index, thereby changing the effective length of and the speed of light within a semiconductor cavity. If the effective cavity length is changed, then the wavelength-selective cavity resonances will shift, and the lasing mode will appear at a different wavelength or even a different cavity mode will lase. Therefore, changing the carrier concentration in a waveguiding phase region will change the wavelength of the emitted light. If the laser light is propagating in a passive section, then the carrier concentration can be controlled by controlling an injected current. However, the situation is more complicated if the laser light is traversing an active gain section of the cavity. In an ideal active section of a laser, the carrier concentration increases until the lasing threshold has been reached (i.e., gain greater than loss for one of the cavity resonances). Once lasing starts, any additional generated carriers should contribute to stimulated emission of photons and not to increasing the carrier concentration. Consequently, it is difficult to change the refractive index and the lasing wavelength by varying the injected current of any active section of a laser, such as in a standard DFB laser.

Figure 7.21 shows a three-section DBR laser, an illustration of the tuning mechanisms, and a tuning curve as a function of Bragg current [41]. The three sections are the active, phase, and Bragg sections, with each section electrically isolated from the others and current-pumped through individual independent electrodes. The optical cavity includes all three sections making an entire unit. Feedback is provided by two "mirrors," one being the cleaved facet providing reflectivity due to a refractive-index discontinuity (i.e., Snell's law) and the other being the Bragg grating section. The active section is highly doped and provides the gain necessary for lasing, whereas the phase and Bragg sections are passive and lightly doped. Two tuning processes exist, both of which rely on changing the refractive index through injecting current into a below-threshold passive semiconductor section.

The phase and Bragg sections, being passive and electrically isolated from the active section, are both below-lasing-threshold semiconductors for which it is

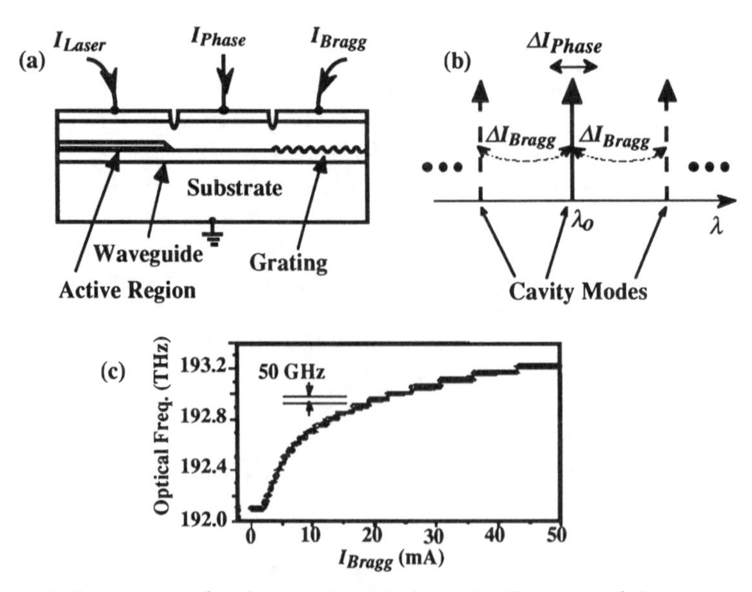

Figure 7.21 (a) The structure of a three-section DBR laser, (b) illustration of the tuning mechanisms, and (c) frequency tuning as a function of the Bragg bias current.

straightforward to change their refractive index with injected current. If the refractive index of the phase section is changed, then (1) the effective length of the phase section is changed, (2) the effective length of the entire cavity is changed, and ultimately (3) the output wavelength is tuned. Injecting current into the passive phase section will fine tune the wavelength of the cavity's existing lasing resonance. Alternatively, by changing the refractive index of the Bragg section, we change the condition for Bragg reflection, which changes the cavity resonances themselves. Coarse tuning can be achieved by causing hopping from one resonance mode to another. This wavelength hopping is shown clearly in the tuning curve of Figure 7.21, in which the mode spacing is 50 GHz (~0.5 nm) and there are ~22 total modes. Such structures have tuning ranges of several (5–10) nanometers. However, because two sections are passive (i.e., low carrier concentration and long carrier lifetime), the wavelength tuning speed is typically limited to ~200 MHz [42]. Because this stucture is complicated, wavelength predictability and repeatability are difficult, and each device must be individually well characterized.

A similar structure to the three-section DBR is the two-section DFB laser [38]. In this structure, both sections are active and are electrically isolated from each other. As is the case for a standard DFB laser, placing the Bragg reflectors within the active gain cavity provides a frequency-selective reflection that is distributed throughout the active region of the laser. In such a structure, the carrier concentration for one section can be maintained above threshold, and the carrier concentration in the second section can be maintained slightly below threshold, allowing us to change the refractive index of the below-threshold passive section. The tuning range is smaller than that of the DBR and is typically ~2 nm, but the optical-modulation and wavelength-tuning speeds are both >1 GHz, owing to the fact that both sections are highly doped and exhibit small carrier lifetimes. As with the tunable DBR laser, predictability and reliability are significant issues that must be addressed.

For the sake of a mature, commercially available source technology, we will also discuss the external-cavity laser, which cannot be modulated or tuned rapidly. This laser is composed of a typical FP semiconductor laser with one facet antireflection coated. The light output from this low-reflectivity facet propagates until it hits an angle-tuned Bragg grating, which forms the second mirror for the frequency-selective cavity. The light output from the uncoated facet is the laser light output. By angle-tuning the Bragg grating, the Bragg condition for the cavity changes, and the output wavelength is changed. This cavity can be quite long, perhaps even several centimeters. Since this cavity is quite long, small changes in angle can produce large changes in wavelength. Tunability on the order of ~50 nm is common. However, since the round-trip propagation time of the cavity is so long, it is not possible to modulate or tune the output wavelength at speeds of more than tens of megahertz. Such a long laser cavity also produces a very small laser linewidth, useful in coherent systems. Therefore, this laser is suitable only as a highly tunable source in a fixed-transmitter WDM system.

Because the modulation speed of the external-cavity laser is extremely small, the output laser light must pass through a high-speed external optical modulator to transmit an optical bit stream. An example of an external modulator is a lithium-niobate voltage-controlled *Mach-Zehnder* (MZ) interferometer, as shown in Figure 7.22 [43]. The operation of the modulator is as follows. After light is coupled into a single input waveguide, the single waveguide is split into two equivalent waveguides, with half the light taking each of the two possible optical paths. These two arms are then recombined into an output waveguide. The two arms of the interferometer produce either constructive or destructive interference at the output waveguide corresponding, respectively, to a 0 or π phase difference between the optical path lengths of the two arms. Modulation is provided by placement of an electrode over one arm of the interferometer. An applied voltage causes an electric-field-induced change in the refractive index and therefore a change in the effective length of that arm, producing as much as a π phase shift. Constructive interference corresponds to light being passed (a 1 or a mark), and destructive interference corresponds to light being blocked (a 0 or a space). These external modulators can be quite fast, with common bandwidths exceeding 10 GHz. The external-cavity laser coupled with the external modulator may be functional, but they are also quite bulky and expensive. We emphasize that an external modulator does not induce any wavelength shift in the light, thus having the desirable effect of practically eliminating any residual laser chirp resulting from direct laser modulation; we will discuss chirp again in a later section.

7.3.3.2 Multiple-Wavelength Laser Arrays

Tunable lasers, in which one laser can tune to many wavelengths, suffer most from their unpredictability in their wavelength-tuning characteristics. Recent advances have been made in developing an array of lasers, with each individual laser emitting light at a different wavelength [44]. We trade one tunable device for an array of fixed devices. All lasers in the array are coupled to a single output fiber, and the fiber output

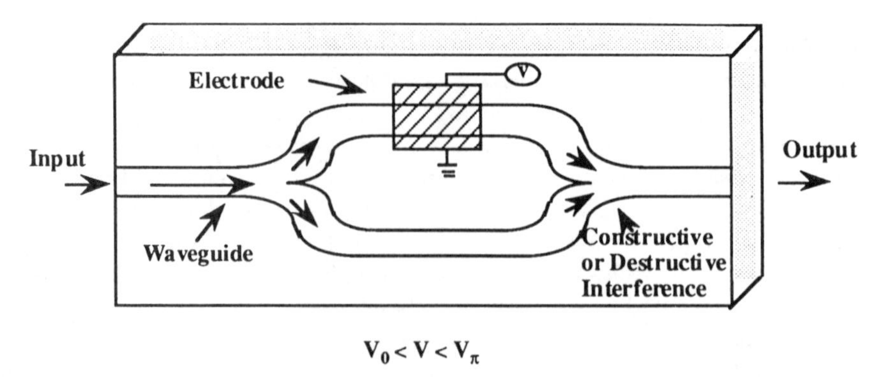

Figure 7.22 Lithium-niobate Mach-Zehnder interferometer representing an external modulator.

wavelength can be changed by simply changing which single laser is biased ON, while leaving all the other lasers unbiased and OFF. (Under certain broadcasting or control circumstances, it may be advantageous to have more than one fiber at the output or to have more than one laser on simultaneously.) The basic fabrication design process of creating a series of different-wavelength lasers is to gradually change from element to element either the optical length of the laser cavities or the frequency selectivity of the cavities (i.e., change the Bragg grating period in a DFB or DBR laser array). The two major categories of multiple-wavelength laser arrays are the edge-emitting one-dimensional array and the surface-emitting vertical-cavity two-dimensional array [45].

Figure 7.23 shows a one-dimensional edge-emitting 20-wavelength DFB laser array. The wavelength is varied from one laser to the next by a slight change in the pitch (i.e., spacing) of the Bragg grating, thereby slightly changing the output wavelength. The wavelength spacing is ~3.7 nm covering a range of 10 nm. In the complete integrated device, the 20-laser waveguides are coupled in an integrated star coupler and amplified in an SOA (semiconductor optical amplifier) region to compensate for the optical splitting losses of the star. A fiber is coupled to one output waveguide. Wavelength uniformity is quite good and wavelength predictability is improving, but the yield is still low and these are very complex devices to fabricate. Other DBR-based arrays have used the same technique of varying the Bragg spacing, but with the added ability to tune the individual lasers' wavelengths [46].

A multiple-wavelength two-dimensional *vertical-cavity surface-emitting laser* (VCSEL) array containing 140 different wavelengths is shown in Figure 7.24 [45].

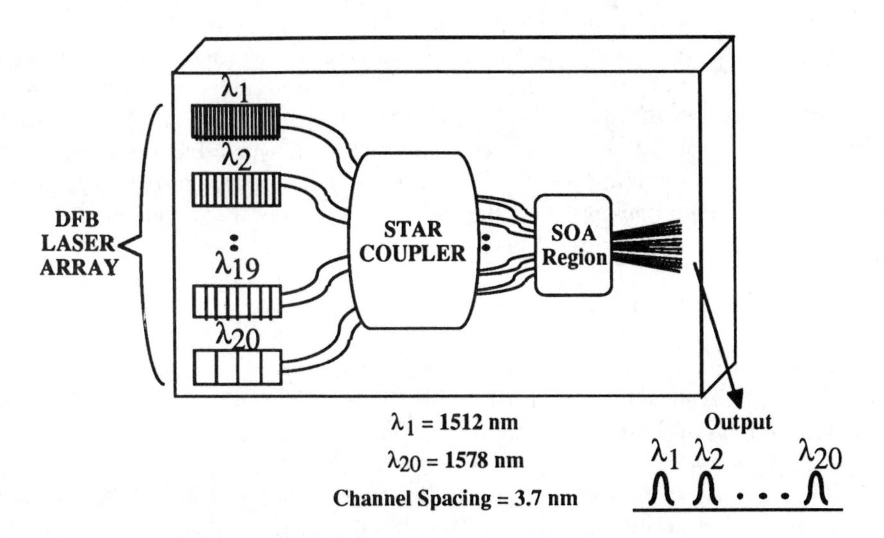

Figure 7.23 An edge-emitting 20-wavelength DFB laser array.

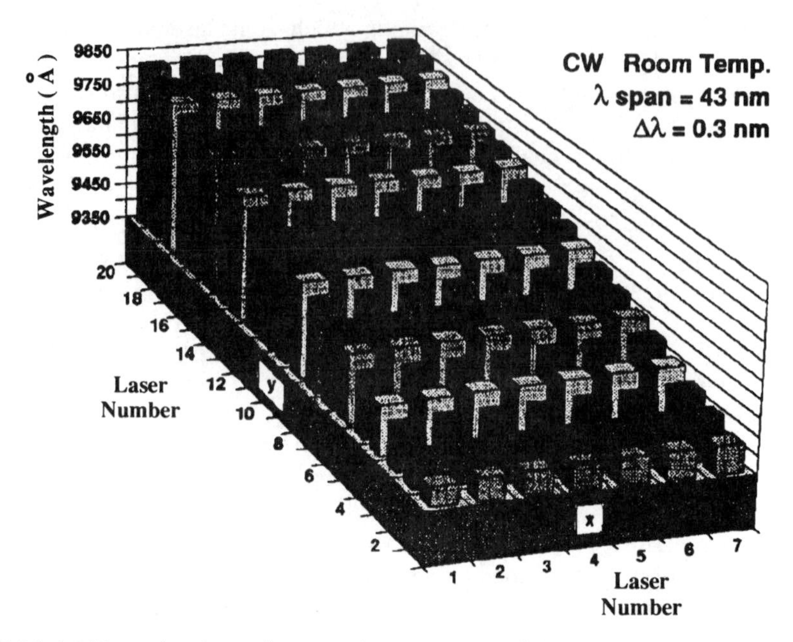

Figure 7.24 A 140-wavelength two-dimensional VCSEL array. (*After:* [95]. © 1992 IEEE.)

Each laser is composed of an MOW (multiple quantum well) active vertical cavity region and upper and lower mirrors made from dielectric stacks. The wavelength of each laser is determined by the vertical spacing between the upper and lower mirrors. For this special array, a spacer layer was added into the vertical laser cavity. This spacer layer can be varied in thickness across the face of the semiconductor wafer during processing. As the spacer layer thickness changes, so does the output wavelength of the individual lasers. This two-dimensional structure has the potential for producing an extremely large number of wavelengths in one integrated package, but there is still significant packaging challenges to solve in coupling an array of vertical-cavity lasers into a single output fiber.

7.3.3.3 Tunable Optical Filters

A key functionality required in most WDM systems is the wavelength selectivity afforded at any receiver by use of a tunable optical filter. Such a filter allows one wavelength to pass and blocks all others, thus enabling unambiguous data recovery. If a system requires wavelength-tunable receivers, then the optical filters must be tunable. Even if the system allows for receivers to be fixed at a specific wavelength, some tunability is desirable to center the passband over the correct spectral location and compensate for any long-term frequency drift. Optical filters are also frequently used to limit the very broadband optical ASE (amplified spontaneous emission) noise generated in an optical amplifier.

(Although we will discuss only optical filters in this section, the reader should bear in mind that wavelength-selective routing and recovery can also be performed by grating-based wavelength (de)multiplexers. These will be discussed in a later section.)

One of the most common optical filters is the FP etalon, which operates in much the same manner as does the FP resonant laser cavity [47]. Figure 7.25 (a) shows three media separated by two partially reflecting mirrors, thereby forming a wavelength-selective FP optical wavelength filter cavity. The filtering is achieved as follows. An incident right-propagating optical wave at angle θ to the normal of the left mirror is partially transmitted at that mirror. The right-propagating transmitted wave will then experience partial transmission, T_1, and reflection at the right mirror. Subsequently, the reflected wave from the right mirror will experience reflection and transmission at

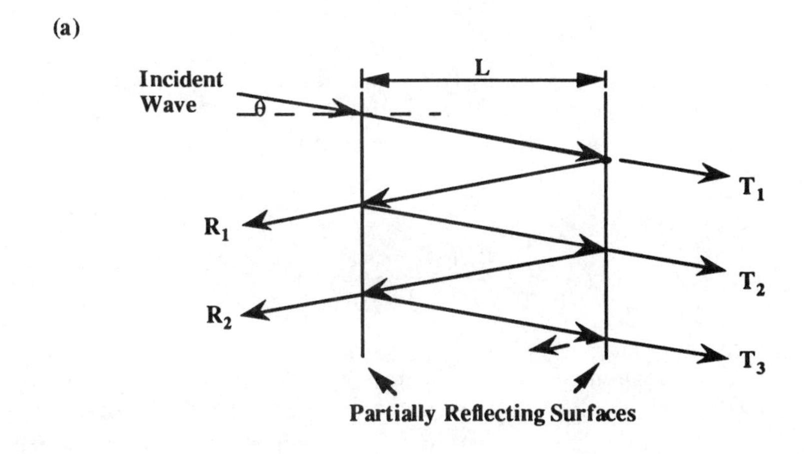

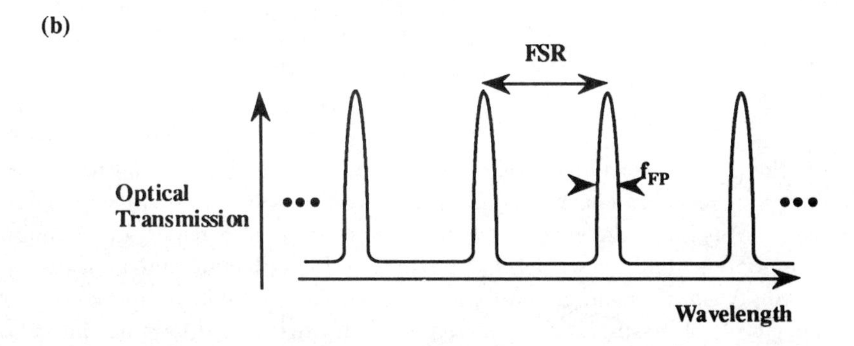

Figure 7.25 (a) Basic operation of an FP filter and (b) wavelength-selective optical transmission of an FP filter.

the left mirror. The new reflected wave from the left mirror will propagate to the right mirror and experience partial transmission, T_2, and reflection. This process continues such that there are an infinite number of transmitted waves at the right mirror. Since the distance between the mirrors is typically small in comparison to the laser light coherence length, all these waves can be considered to originate from the same coherent optical laser source and will therefore interfere with each other as they are transmitted. If these waves are in-phase with each other (i.e, phase difference equals 0, 2π, 4π, . . .), then they will add constructively, and all the waves will be transmitted. If they are even slightly out of phase, then they will add destructively and very little effective optical power will be transmitted through the optical filter. Therefore, all the waves must be in phase, implying that the round-trip wave propagation distance inside the cavity be an integral number of wavelengths [47]

$$L = \frac{m\lambda}{2n\cos\theta}, \qquad m = 1,2,3, \ldots \tag{7.7}$$

An infinite number of integers will satisfy the in-phase requirement; therefore, the transmission passband of the filter is periodic with wavelength (see Fig. 7.25(b)). The optical filter has a transmission resonance bandwidth, f_{FP}; a free spectral range between any two resonances, FSR; and a filter finesse, $\Im$ [48]:

$$\Im = \frac{FSR}{f_{FP}} = \frac{c}{2nLf_{FP}} \tag{7.8}$$

where c is the speed of light in a vacuum. The finesse is a figure of merit for the optical filter with a high finesse corresponding to a more efficient filter. The reflectivity of the feedback mirrors affects the finesse according to the following relationship [48]:

$$\Im = \frac{\pi\sqrt{r}}{1 - r} \tag{7.9}$$

where r is the power reflectivity at each mirror and is considered equal for the right and left mirrors. As is evident from (7.7), wavelength selectivity depends on the angle of incidence of the incoming wave since that defines the round-trip distance required for phase matching of all the wave reflections. Therefore, this filter can be wavelength tuned by simply tilting the angle of the filter in relation to the incoming wave. Angle tuning effectively shifts the periodic resonances in unison to different transmitting wavelengths.

Because the FP filter has a periodic transmission function, two channels can simultaneously pass through the filter if each channel is located at a different transmission resonance. Therefore, it is important to locate all the WDM channels within one FSR of the filter to prevent unambiguous data recovery. Figures of merit for wavelength-tunable optical filters include large finesse to accommodate many channels; narrow bandwidth to allow for closely spaced channels; wide wavelength tunability; fast wavelength switching to allow for more rapid channel switching, which is critically important in packet switching; wavelength stability, repeatability, and accuracy; and potential low cost. Five possible tunable filters, both FP and non-FP, are summarized next.

Semiconductor Filter

A semiconductor waveguide with either an FP cavity or a Bragg grating can act as a wavelength-selective filter. In the case of the intracavity Bragg grating Figure 7.26, changing the current will change the index of refraction and therefore the wavelength, which will satisfy the Bragg condition. Tuning speed is extremely fast (nanoseconds), but the tuning range is limited, and it is difficult to maintain a uniform passband over the entire wavelength range [49].

Fiber Fabry-Perot (FFP)

Highly-reflective mirrors are deposited on the ends of two pieces of fiber. These mirrors define a wavelength-selective FP cavity with the fiber acting as an intracavity waveguide. The wavelength tunability is achieved by mechanically varying the distance between the two mirrors (Fig. 7.27) by either separating two pieces of fiber or by stretching the same piece of fiber [50]. The tuning range is large (>10 nm), but the switching speed is slow (milliseconds). This filter is commercially available.

Liquid-Crystal Fabry-Perot

Liquid crystal is used to fill the FP cavity formed between two mirrors. The orientation of the molecules in the liquid crystal can be realigned based on an externally applied

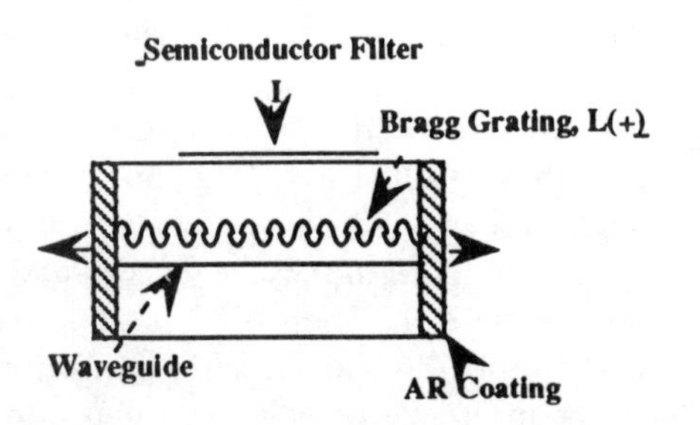

Figure 7.26 The semiconductor filter using an intracavity Bragg grating.

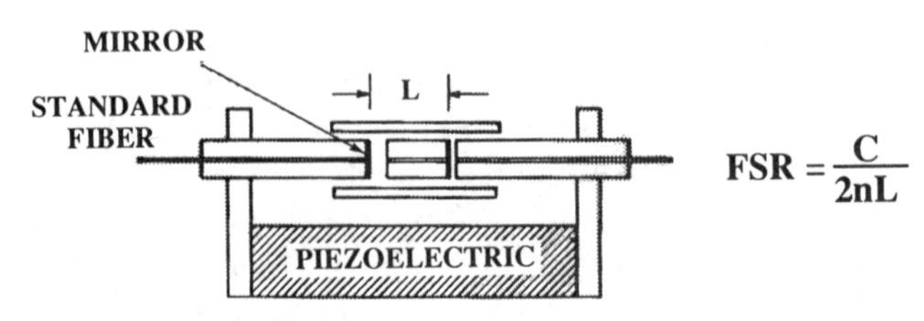

Figure 7.27 An FFP tunable optical filter.

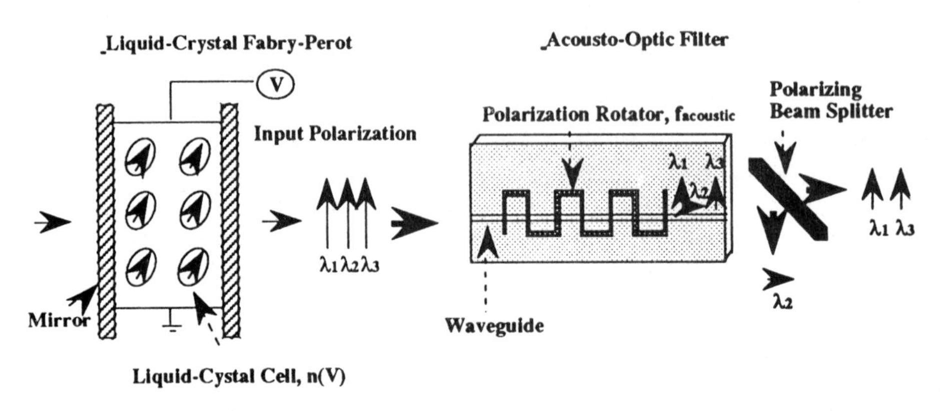

Figure 7.28 Schematic of a liquid-crystal FP tunable optical filter.

voltage. It is that molecular alignment that determines the index of refraction of the crystal and thereby the effective length of the FP cavity (Fig. 7.28) [51]. Tuning range is large (>10 nm) but switching speed is slow (milliseconds).

Cascaded Mach-Zehnder Interferometers

We described in Section 7.3.3.1 the operation of an MZ interferometer for providing external optical modulation. The same principle of constructive and destructive interference can be applied to wavelength-selective filtering. A single MZ interferometer has an input waveguide that is split into two parallel waveguide arms. The light takes these two different optical paths and is then recombined at an output waveguide. If the path-length difference between the two arms is an integral number of wavelengths, the two beams will add constructively at the output waveguide. If the path lengths are an integral number of half-wavelengths, the two beams will add destructively. The MZ is clearly wavelength selective, with the transmission being a raised cosine-squared function of input wavelength and the transmission bandwidth being critically dependent on the path-length difference between the two interferometer arms. If a voltage is applied to an electrode over one arm or if the temperature of the waveguide arms is changed, the effective path lengths are changed and the filter can be tuned. Such a filter is not very wavelength selective (i.e., it has a low finesse) due to its simple raised cosine-squared function. However, if several MZ interferometers of decreasing bandwidths can be cascaded, the overlapping transmission function of several passbands will effectively allow transmission through only the narrowest passband, resulting in a high finesse filter (Fig. 7.29) [52]. Temperature tuning is slow (milliseconds), but this kind of filter has the potential for integration on lithium-niobate and fast electronic tuning using several electrodes.

Acousto-Optic Tunable Filter (AOTF)

Wavelength-selective Bragg gratings can be formed on a lithium-niobate substrate by inducing a radio frequency (RF) acoustic wave in the material. As an input optical wave propagates along a waveguide that has the acoustic wave inside, a shift in the light's polarization is induced only if that laser light wavelength matches the Bragg acoustic grating. An output polarizing beamsplitter can be used to filter out only that wavelength that has had its polarization changed, thus providing wavelength selectivity (Fig. 7.30) [53]. The AOTF can accommodate multiple acoustic waves and, therefore, multiple independent passbands. This filter is very input-polarization sensitive, but two such waveguides can be fabricated on the same substrate, in which one waveguide filters the transverse electric (TE) waves and the other filters the transverse magnetic (TM) waves, all followed by a polarization recombiner. This arrangement will provide polarization diversity and a polarization-independent AOTF. Switching speed is in the microsecond, regime, and the tuning range is large (>10 nm). This filter is commercially available.

The choice of which filter to use depends on the specific requirements of the WDM system being implemented. At present, the FFP and the AOTF are commercially available and can satisfy the needs of many types of systems. However, no filter can achieve nanosecond switching speeds and tens of nanometers tuning range, so the search continues for the ultimate filter for WDM use.

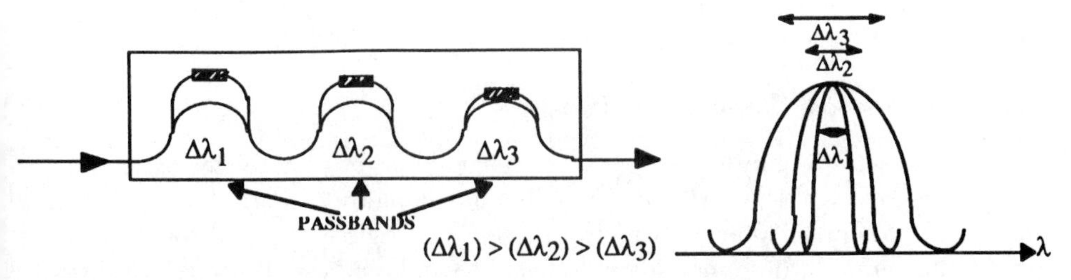

Figure 7.29 Cascaded MZ interferometer filters and their overlapping passbands.

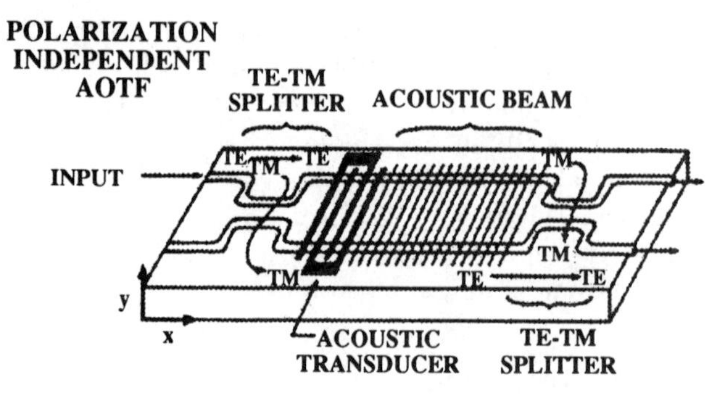

Figure 7.30 Illustration of the operation of a tunable AOTF filter and the accompanying polarization-independent AOTF.

7.3.3.4 Passive Star Couplers

In star network topology, passive stars are necessary to allow all users to communicate with all other users. Each output port has access to all the input ports, with the input power divided among all the outputs. Small N-by-N passive stars typically are formed by cascading 2-by-2 directional couplers. However, it would be impractical to have a 128-by-128 passive star composed of cascaded directional couplers. In fact, the number (Q) of 2-by-2 couplers necessary to connect N users in a star configuration is [54]

$$Q = \frac{N(\log_2 N)}{2} \tag{7.10}$$

One clever planar solution is to have all input waveguides radiate their light into a free-space region, which is then coupled into all the output waveguides (Fig. 7.31) [55]. By tailoring the shape of the free-space region according to antenna theory and Fourier optics, the light from any input waveguide will radiate and couple equally into each of the output waveguides. This is known as the integrated star coupler. Large stars of 128-by-128 input and output ports have been fabricated exhibiting low (<3 dB) excess loss [56]. Furthermore, efficient fiber connectorization of the waveguides has been achieved for a 20-by-20 star [57].

7.3.3.5 Wavelength-Division Multiplexers

Wavelength-selective multispatial routing of signals may be required for signal routing or signal recovery in a WDM network. (This functionality is more advanced than is the case for the optical filters for which only one wavelength is passed through the main output port and all the remaining wavelengths can be accessed through the second remaining output port.) If passive wavelength routing is required in a network, then

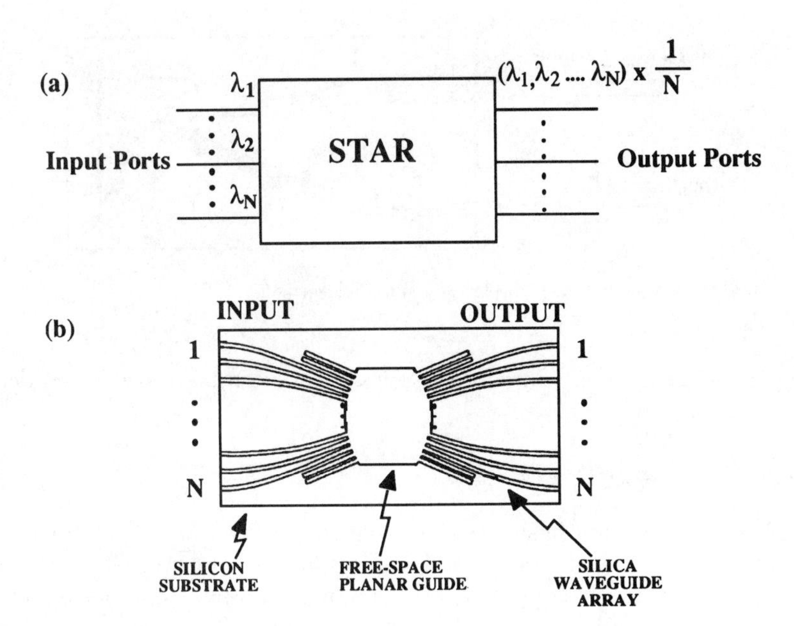

Figure 7.31 Integrated planar N-by-N star coupler: (a) star functionality and (b) integrated device. (*After:* [55]. © 1989 IEEE.)

M-by-N port devices are needed, facilitate the selective routing of a signal from a given input port to a given output port according to the input signal's wavelength. These devices are most commonly fabricated from some type of wavelength-selective grating that spatially deflects an optical beam depending on its wavelength. As examples, we will discuss three wavelength (de)multiplexers.

Spectrometer on a Chip

Figure 7.32 shows an N-by-1 wavelength multiplexer [58], N input fibers are coupled to N input planar waveguides, for which a different wavelength would be input to each different fiber. The light from each waveguide would reflect off an integrated Bragg grating. We have already discussed Bragg reflection, and it was shown how the angular reflection off a grating is dependent on the incident wavelength. The multiplexing can occur because each input waveguide is at a different spatial location, and each wavelength has a different angular deflection. When these two mechanisms are combined, the different waveguides with different wavelengths can be arranged so that all light will be focused to the same focal position. Each input waveguide has a specific wavelength associated with it would satisfy that condition. An output waveguide can be placed at the common focal position and will collect all the light. This multiplexer

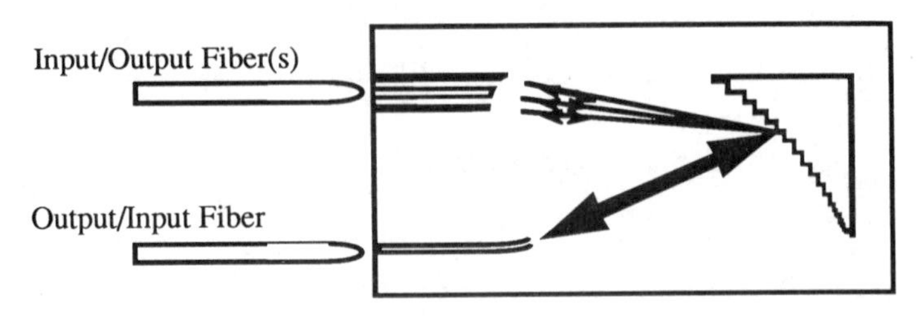

Figure 7.32 Integrated grating-based wavelength (de)multiplexer (*After:* [58]).

can be operated in reverse, with one input fiber containing N wavelengths demultiplexed into N output fibers.

Integrated Channel Add/Drop Filter

Figure 7.33 shows a grating-based structure that acts as a two-input-port/two-output-port device, with one input port and one output port representing the through data path and the other ports corresponding to the input and output from a local node [59]. The gratings act to provide wavelength-selective coupling of the optical waves between two adjacent waveguides, and the device essentially functions as a wavelength-selective directional coupler. The device can be used by a local node to inject a given wavelength onto a fiber (i.e., add new data) and remove that same wavelength from the fiber (i.e., remove old data). This functionality can be very useful in a ring or bus network configuration.

Integrated Frequency Router

The highly acclaimed frequency-selective integrated router has a different functionality than either a simple wavelength multiplexer or a passive star coupler [60]. Figure 7.34 shows both the new functionality as well as the integrated device itself. The device has N input ports and N output ports, and the same set of N different wavelengths can be input into each input port. We will designate the different ports as A, B, . . . , N and the wavelength set as 1, 2, . . . , N. The functionality is as follows. At a given input port i, λ_1 is output to the same numbered output port i, λ_2 is output to the next higher-numbered output port $(i + 1)$, λ_3 is output to the second higher-numbered output port $(i + 2)$, and so on. This happens for each input port and for each wavelength. In a simple 3-by-3 example, different signals on λ_1, λ_2, and λ_3 are input at independent input ports A, B, and C, totaling nine different combinations of signals. The outputs are as follows: λ_1^A, λ_3^B, λ_2^C at output A; λ_2^A, λ_1^B, and λ_3^C at output B; and λ_3^A, λ_2^B, and λ_1^C at output C. There is no possibility for two input ports to transmit to the same output port on the same wavelength! Note that such output-port contention on the same wavelength would occur for a simple star coupler and would

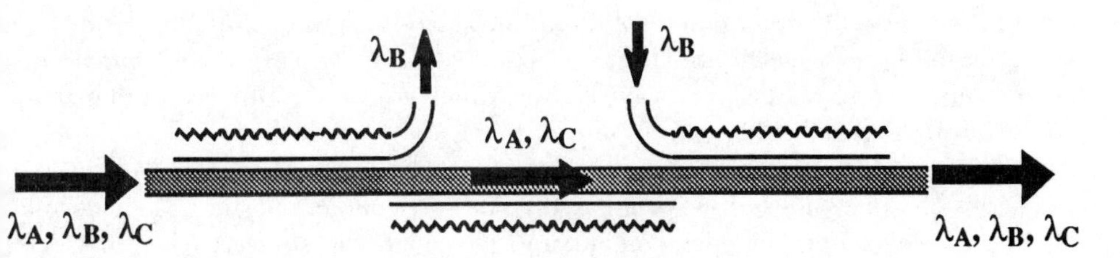

Figure 7.33 An integrated grating-based channel add/drop filter.

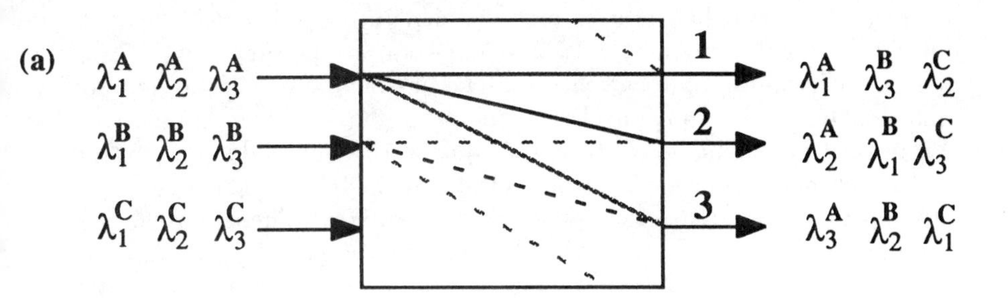

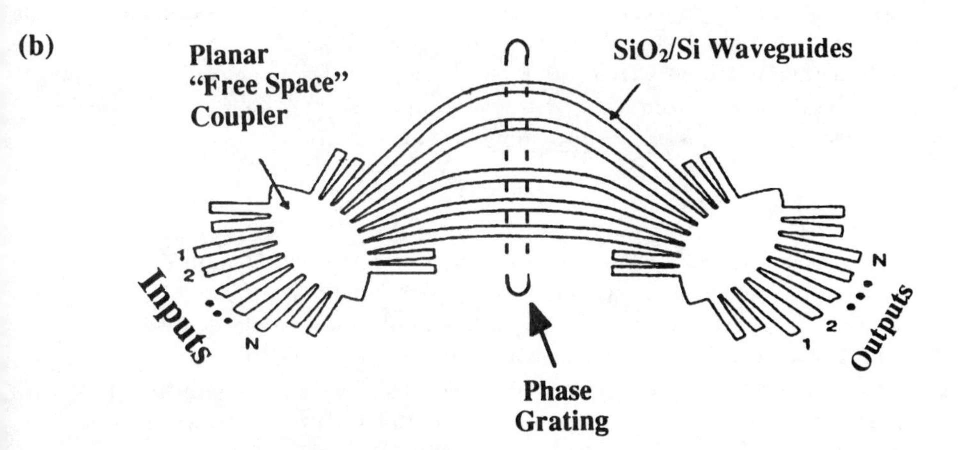

Figure 7.34 The (a) functionality and (b) device structure of an integrated frequency router. (*After:* [60]. © 1992 IEEE.)

require some form of contention resolution, since an optical filter cannot separate two signals on the same wavelength. However, in our scenario, two input ports can access the same output port on different wavelengths, which is a valid routing condition given an optical filter demultiplexer at an output port.

This frequency router is composed of two integrated star couplers interconnected by a series of waveguides of varying length. The star couplers perform their normal function of wavelength-independent mixing. However, the staggered-length waveguides between the two stars form a kind of phase grating, which is the basis of the wavelength selectivity. A simplistic physical mechanism is as follows. A given wavelength input into the first star coupler is output to all the interconnecting waveguides. This wavelength thus propagates through many different-length parallel paths, thereby experiencing a phase grating and a wavelength-dependent deflection. All these parallel beams are then input to and mixed in the output star, producing an excitation only at a single appropriate output port. The passbands can be <0.5 nm and the number of input and output wavelengths/ports can be >20. The routing performed here is passive routing based solely on the input wavelength and input port number.

An extremely important point for WDM systems is that the system capacity is greatly enhanced by an integrated router. In a conventional system in which each user has a signature wavelength on which it either transmits or receives data, N wavelengths produce only N possible communication paths and can accommodate N users. Alternatively, an integrated frequency router can accommodate N^2 possible connection paths, because the same wavelength set can be reused at different input ports:

$$[\#\ of\ connections] = [\#\ of\ wavelengths] \cdot [\#\ of\ input\ ports] \tag{7.11}$$

Therefore, N wavelengths can accommodate N^2 users! This added capacity is at the expense of the ultimately-simple network control and scalability afforded by a passive star coupler and a unique wavelength for each user. The frequency router has a higher capacity because we are in some sense combining the benefits of WDM and space multiplexing in a single device, providing N^2 possibilities.

7.3.3.6 Receiver Arrays

We have already discussed the use of optical filters and wavelength demultiplexers allow a receiver to recover data on a single wavelength from a multitude of input wavelengths. However, a receiver array will not only allow single-wavelength demultiplexing but also the simultaneous recovery of several wavelengths at a given receiver. Each receiver in the array will recover a single transmitted wavelength, with the entire receiver array detecting all the possible wavelengths. Figure 7.35 shows the schematic of an input multiple-wavelength optical beam reflected off a Bragg grating [61]. Each wavelength is reflected at a different wavelength-dependent angle and is then incident on a different receiver in a receiver array. This data reception converts wavelength

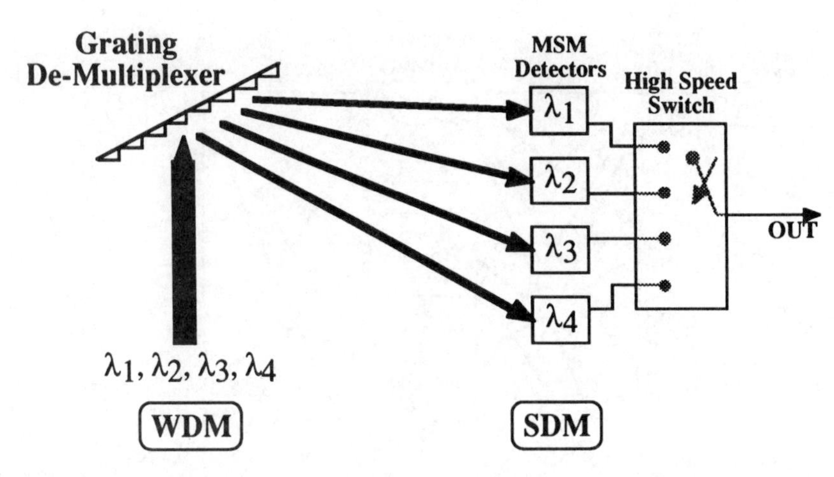

Figure 7.35 A receiver array in which each receiver can recover a different wavelength signal (MSM = metal-semiconductor-metal).

multiplexing on the optical fiber into space multiplexing at the receiver. By biasing one receiver or many receivers simultaneously, one or many different wavelength signals can be recovered.

7.3.3.7 WDM/SDM Cross-Connects

We have been discussing how a grating or wavelength demultiplexer acts to separate wavelengths spatially and can add a new dimension of flexibility by introducing SDM. To take advantage of hybridizing a system using two types of multiplexing, we can combine WDM and SDM in a cross-connect configuration, as shown in Figure 7.36 [62]. In so doing, we can reduce the complexity of each type of multiplexing scheme since, typically, a low-N multiplexing system is much simpler to implement than a high-N system. In this hybrid example, there are four input ports, four output ports, and four possible wavelengths at each input port. Each input port has a wavelength demultiplexer that routes each wavelength to a different crossbar switch. Each space crossbar switch corresponds to a specific input wavelength and has four inputs derived from each of the four major input fibers. Subsequently, each space switch independently switches a single input wavelength from any of the four original input ports to the appropriate destination output port. The switching is achieved by use some local (electronic) control signals that determine the complete optical path. If there is any output-port contention, the electronic control will reroute one of the signals to a different output port, known as deflection routing. This is considered active routing since is some control necessary within each of the space switches. The outputs from the individual space switches are distributed to four output-port wavelength

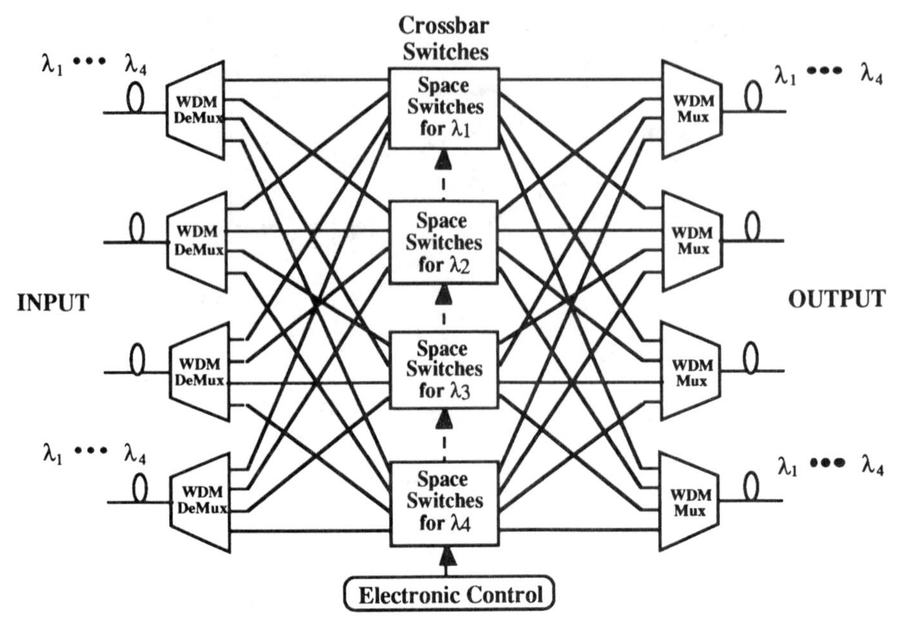

Figure 7.36 A typical WDM/SDM cross-connect.

multiplexers. In this active WDM/SDM routing, N wavelengths can potentially form N^2 connections, but control signals are necessary for high throughput and effective routing.

Subsequent to introducing this WDM/SDM example, it is instructive to briefly discuss implementing an optical space switch. A simple diagram of a space crossbar switch is shown in Figure 7.37 for a 2-input port and 2-output port device:[15] Two input waveguides and two output waveguides are coupled based on a voltage-controlled coupling region. An external control voltage will determine if the light from one spatial input port will appear at spatial output port 1 or spatial output port 2. Many such 2-by-2 crossbar switches can be cascaded in stages to form a crossbar switch of any size. One advantage of this WDM/SDM approach is that small crossbar switches are commercially available and a fairly mature technology and therefore can be implemented in the near future. We will revisit the concept of SDM and its generic advantages and disadvantages in Section 7.6.

We have been extensively discussing many of the enabling technologies for a WDM system. We have discussed only a few examples of each technology, with many other types left untreated. We will now begin to discuss actual WDM system design criteria. It must be emphasized that different WDM systems will require different technologies and operating parameters for implementation.

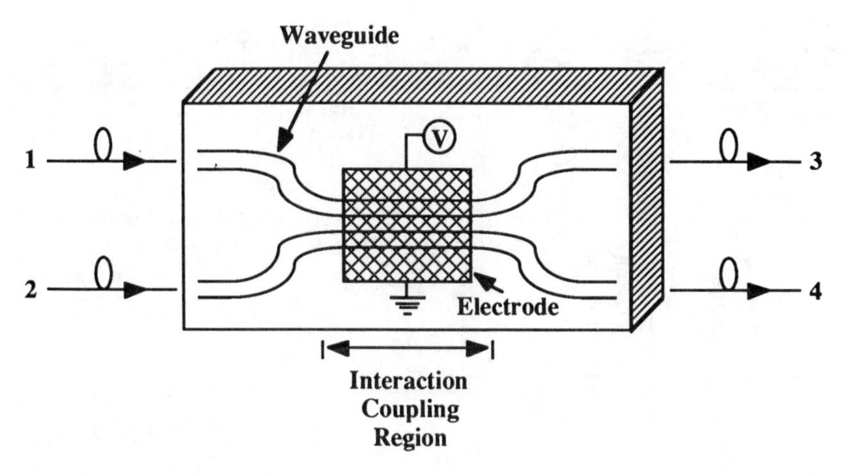

Figure 7.37 Illustration of a simple 2 by 2 optical crossbar space switch.

7.3.4 System Guidelines

7.3.4.1 Demultiplexing, Filtering, and Cross-talk

Several critical issues must be addressed in a WDM system. The first issue involves the most basic function of demultiplexing many WDM channels by use of an optical filter [63]. Figure 7.38 shows the spectrum of many WDM channels that must be demultiplexed by an optical filter so that only one channel is received and the other channels blocked. If we assume that the optical filter is etalon, then the periodic transmission function of the filter has a Lorentzian-shaped passband: [64].

$$T(f) = \frac{1}{1 + \left[\frac{2(f - f_0)}{f_{FP}}\right]^2} \tag{7.12}$$

where f_{FP} is the *full-width half-maximum* (FWHM) of the filter transmission, f_0 is the center frequency of a single resonant passband where the transmission is a maximum, and the optical wavelength can be related to optical frequency by $\lambda = c/f$. Such a slowly diminishing lineshape with long "tails" will allow some interwavelength cross-talk since not all the optical power from the other channels will be blocked. The total cross-talk power, P_{cr}, from $(N - 1)$ zero-bandwidth rejected channels is: [65]

$$P_{cr} = \sum_{i=1}^{N-1}\left[\frac{P}{1 + \left[\frac{2if_d}{f_{FP}}\right]^2}\right] \tag{7.13}$$

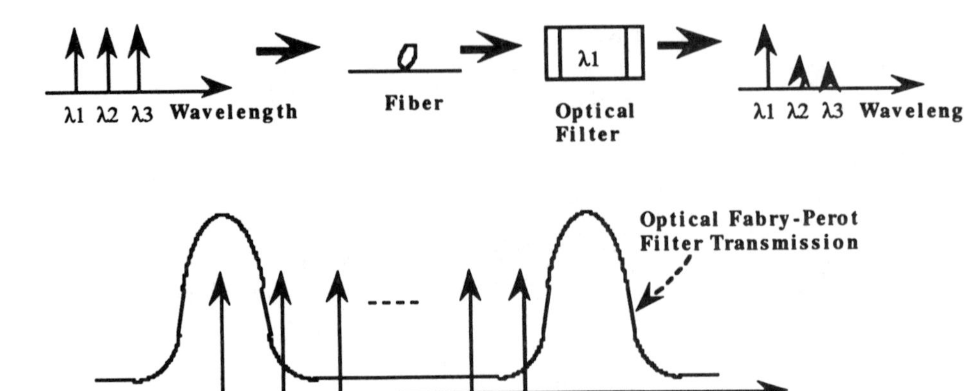

Figure 7.38 Spectrum of many WDM channels being demultiplexed by an FP optical filter.

where P is the prefiltered power in each rejected channel, the channel spacing is fixed at f_d and the selected channel is located at the extreme left or right of the collection of wavelengths. We are always concerned about the SNR (signal-to-noise ratio) of the recovered data in the system, and this interchannel cross-talk will act to reduce the power contrast ratio of the selected recovered signal:

$$S = P_S - P_{cr} \tag{7.14}$$

where S is the effective recovered signal power, and P_S is the power in the selected wavelength that passes through the filter. The cross-talk due to power leakage from the other channels is deterministic and affects the signal mean (i.e., signal power) and not the signal variance (i.e., noise power). Although the exact tolerable amount of cross-talk depends on the WDM system, cross-talk should not exceed a few percent of the selected channel power to maintain good system performance. Note that a narrow filter is desirable, because it will limit the amount of power leaking through the filter from neighboring channels. Furthermore, devices also must have a wide wavelength-tuning range for the channels to be spaced far apart, thereby reducing cross-talk.

Because the filter has a periodic passband, two different wavelength channels could be transmitted through the filter at different passbands, causing ambiguous channel recovery. Therefore, all the channels must be located within one free spectral range of the FP filter. The maximum number of channels that can be accommodated in a simple WDM system is [32]

$$N = \frac{FSR}{\Delta f} \tag{7.15}$$

where N is the number of channels, FSR is the free spectral range, and Δf is the channel spacing that produces negligible power leakage through the filter by the neighboring channels.

The previous discussion assumed that the signals had zero bandwidth and acted as if they were delta functions in comparison to the transmission function of the optical filter. In reality, the individual WDM channels have a broadened bandwidth in wavelength, which will significantly affect the filtering and demultiplexing in a WDM system. A broadened channel will require a wider optical filter to recover the entire signal and a larger interchannel spacing. Therefore, fewer channels would be accommodated if the entire system is of limited overall bandwidth. Three major mechanisms that cause channel broadening are information bandwidth, laser linewidth, and laser chirp.

Information Bandwidth

The channel broadening that is the most fundamental and almost impossible to reduce is the information bandwidth of the transmitted signal. Any modulated signal is spectrally broadened due to the inherent information capacity of that channel [66]. For intensity-modulated channels, the 3-dB FWHM bandwidth is roughly equal to the modulation bit rate, with a 1-Gbps signal having a 1-GHz bandwidth. For very fast modulated signals at 20 Gbps, the information bandwidth is ~1.5 Å and will affect the required width of the optical filter passband that must be used to recover the entire transmitted signal.

Laser Linewidth

Due to phase fluctuations in a laser cavity and the quantum-mechanical energy-level fluctuations of the stimulated emission events, the spectral linewidth of the laser optical output has a nonzero value [67]. The fluctuations produce a broadening of the output spectra even when the light is not being modulated. Typical DFB lasers have output-power-dependent laser linewidths of 10–100 MHz, with most lasers being closer to 10 MHz. Such a value for the laser linewidth would have a severe impact on coherent systems, because the input signal and local oscillator must be mixed together and fluctuations inherently reduce the coherence time and length of an optical signal (see Chap. 4). For direct-detection systems, the laser linewidth broadens the signal spectrum, requiring a wider optical filter bandwidth to pass the selected channel. However, since most direct-detection optical transmission systems operate at speeds greater than hundreds of megabits per second, the laser linewidth is usually considered negligible compared to the signal information bandwidth.

Laser Chirp

One of the most troubling broadening mechanisms is the chirp (i.e., wavelength broadening) of the laser output when the injected current into the laser is directly modulated to produce a modulation in the output light [68]. Unfortunately, directly modulating the laser current is also considered the simplest way to modulate the light. In an ideal semiconductor laser, the number of carriers will increase with increasing current only until the threshold level is reached, at which point lasing occurs when there is more gain than loss. After lasing is initiated, any additional injected carriers would only contribute to additional photon generation, since the total number of carriers in the cavity will be fixed and clamped at the gain-equals-loss value. However, since nonradiative transitions and undesirable current leakage will occur in a nonideal laser, the carrier density will vary with a change in the injected current even after lasing has begun. A change in the current density causes (1) a slight change in the refractive index, (2) a change in the effective length of the cavity, (3) a change in the wavelength that will satisfy the cavity lasing conditions (as determined by the facet mirrors or the distributed Bragg reflections), and, ultimately, (4) a change in the final output lasing wavelength. Additionally, both nonradiative recombinations and a current-density variation will cause a temperature change in the cavity, thereby changing the length of the cavity and the wavelength that will be output. Direct current injection can cause as much as a ~1 Å (~15 GHz) spectral broadening (Fig. 7.39).

One straightforward method to practically eliminate the effects of chirp is to avoid directly modulating the current of the laser. In Section 7.3.3.1 we discussed using an external optical modulator, which does not produce any chirp, to modulate the output of a tunable laser having a slow modulation speed. The example given of an external modulator was an MZ interferometer, in which the phase difference between the two arms can be electrically changed from 0 to π radians, resulting in, respectively, constructive and destructive interference at the modulator output. An external modulator is effective but can be more expensive than the laser itself. There has been some recent work to integrate an external modulator on the same chip as the laser to reduce the cost and complexity of having a separate bulk device. The modulator is external

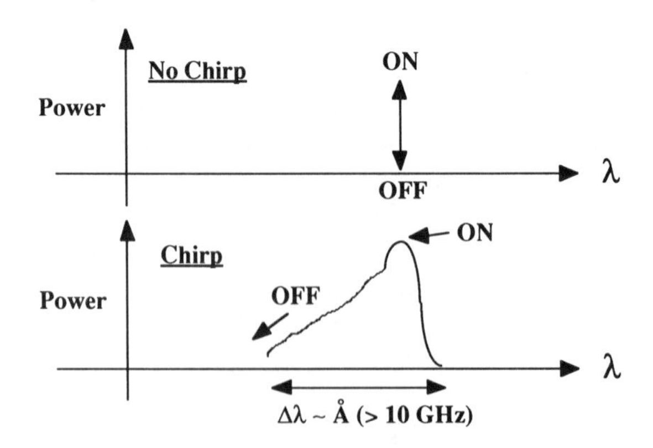

Figure 7.39 Channel spectral broadening due to direct modulation of the laser's injected current.

to the lasing cavity but is still located on the same chip. The example in Figure 7.40 shows a wavelength-tunable DBR laser with a modulating region external to the Bragg reflecting region [69]. The modulating region controls the optical output by changing the transmissivity and electroabsorption within a quantum-well waveguiding region.

In passing, it should be emphasized that the issue of channel spectral broadening not only affects the channel spacing and filter width in a WDM system but also critically affects the capacity in a long-distance transmission system. Different frequencies travel down an optical fiber at different speeds due to a wavelength-dependent refractive index, so an optical pulse that is spectrally broad will suffer severe pulse spreading and dispersion as it propagates (i.e., intramodal material dispersion). Therefore, external modulators are used exclusively for transmitting high-speed optical signals over long-distance optical links.

The analysis of filtering and demultiplexing in a WDM system can be extended to include the issue of channel broadening. For the sake of simplicity in deriving closed-form solutions, we will assume that the channel broadening is well behaved according to a normalized Lorentzian lineshape power spectrum. We will analyze the scenario as portrayed in Figure 7.41, in which there is a broadened selected and rejected channel of spectral width, respectively, of f_S and f_R [70]. The systems goals include maximizing the transmission through the filter of the selected channel, minimizing the transmission through the filter of the rejected channel, and minimizing the channel spacing but still maintaining a low system power penalty. Examination of the eye diagram in Figure 7.42, representing the recovered eye diagram generated from these two demultiplexed signals, shows that (1) transmitting less than the entire selected channel lowers the 1's ("on") rail of the eye and (2) transmitting some interwavelength cross-talk from a rejected channel raises the 0's ("off") rail. Both these effects will reduce the contrast ratio of the recovered signal and act to subtract

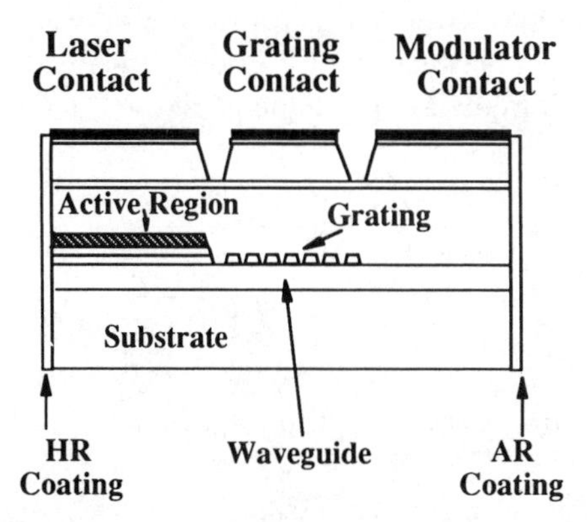

Figure 7.40 A diagram of a wavelength-tunable DBR laser with an external optical modulator integrated directly on the same chip (HR = high reflection, AR = antireflection).

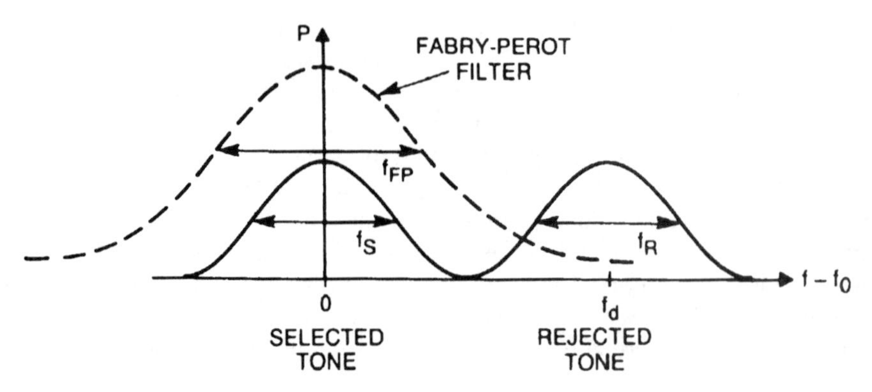

Figure 7.41 Spectrally broadened selected and rejected channels being demultiplexed by an FP optical filter. (*After:* [70]. © 1991 IEEE.)

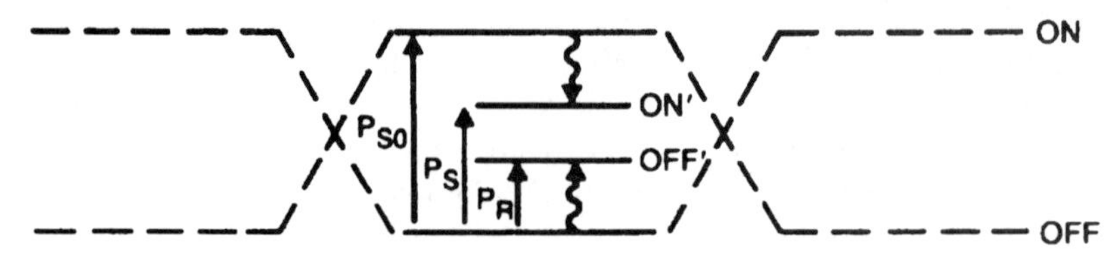

Figure 7.42 An idealized eye diagram showing various powers transmitted through the FP optical filter. (*After:* [71]. © 1991 IEEE.)

power from the numerator in the SNR calculation. The selected channel's power before filtering is P_{So}, the selected channel's power after being transmitted through the filter is P_S, the rejected channel's power before the filter is P_{Ro}, and the rejected channel's power after filtering is P_R.

We can calculate the power that is transmitted through the FP optical filter by performing a convolution between the FP Lorentzian-lineshaped transmission function and the assumed Lorentzian lineshape of the selected or rejected channel [70]:

$$P_{S,R} = \left(\frac{\pi f_{FP} f_{S,R}}{2(f_{FP} + f_{S,R})}\right)\left[\frac{1}{1 + \left(\frac{2f_d}{(f_{FP} + f_{S,R})}\right)^2}\right](P_{So,Ro}) \tag{7.16}$$

The total power penalty due to optical filtering is

$$PP = 10 \log\left[\frac{P_{So}}{P_S - P_R}\right] = 10 \log\left\{\left(1 + \frac{f_{TONE}}{f_{FP}}\right)\left[1 + \left(\frac{f_{FP} + f_{TONE}}{2f_d}\right)^2\right]\right\} \tag{7.17}$$

where the first part of the right side of the equation is due to imperfect transmission of the selected channel and the second part of the right side is due to imperfect blockage of the rejected channel. If there are many rejected channels, a simple summation can be performed, although the majority of the power penalty is clearly due to the most adjacent channel. Figure 7.43a, shows the power penalty due to imperfect transmission of the selected channel, imperfect blockage of the rejected channel, and a combination of the two. System guidelines include that (1) the FP should be at least three times the width of the channel broadening and (2) the channels should be spaced apart at least 1.5 times the FP width. Given a fixed channel spacing, a minimum exists in the curves in Figure 7.43(c), in which a too-narrow filter doesn't transmit enough selected channel and a too-wide filter has too much cross-talk. Although this analysis was performed by simplifying the channel broadening shape, the trends and guidelines are valid and correspond to experimental data.

Example 7.2

A multiple wavelength system has an optical filter at each receiver to demultiplex the incoming channels (i.e., wavelengths). The filter 3-dB transmission passband is 0.5 nm, and the 2.5 Gbps channels are chirp-free except for the inherent information bandwidth. Approximately how far apart must the channels be separated in order to have the detected power through the filter due to the nearest neighbor channel be only 1% of the selected channel? Assume that the optical filter is a Fabry-Perot Lorentzian-lineshaped filter. Furthemore, how many channels will fit in the free-spectral-range given a filter finesse = 40? Given the number of wavelengths, what is the total system penalty due to optical power crosstalk through the filter?

Solution

If we substitute $T(f) = 0.01$ and $f_{FP} = 0.5$ nm in (7.12), we have:

$$T(f) = \frac{1}{1 + \left[\frac{2(f - f_0)}{f_{FP}}\right]^2} = 0.01$$

$$\Rightarrow f - f_0 = \Delta f \cong 2.5\text{nm}$$

The free-spectral-range is determined by (7.8):

$$\begin{aligned} FSR &= \Im f_{FP} \\ &= 20\text{nm} \end{aligned}$$

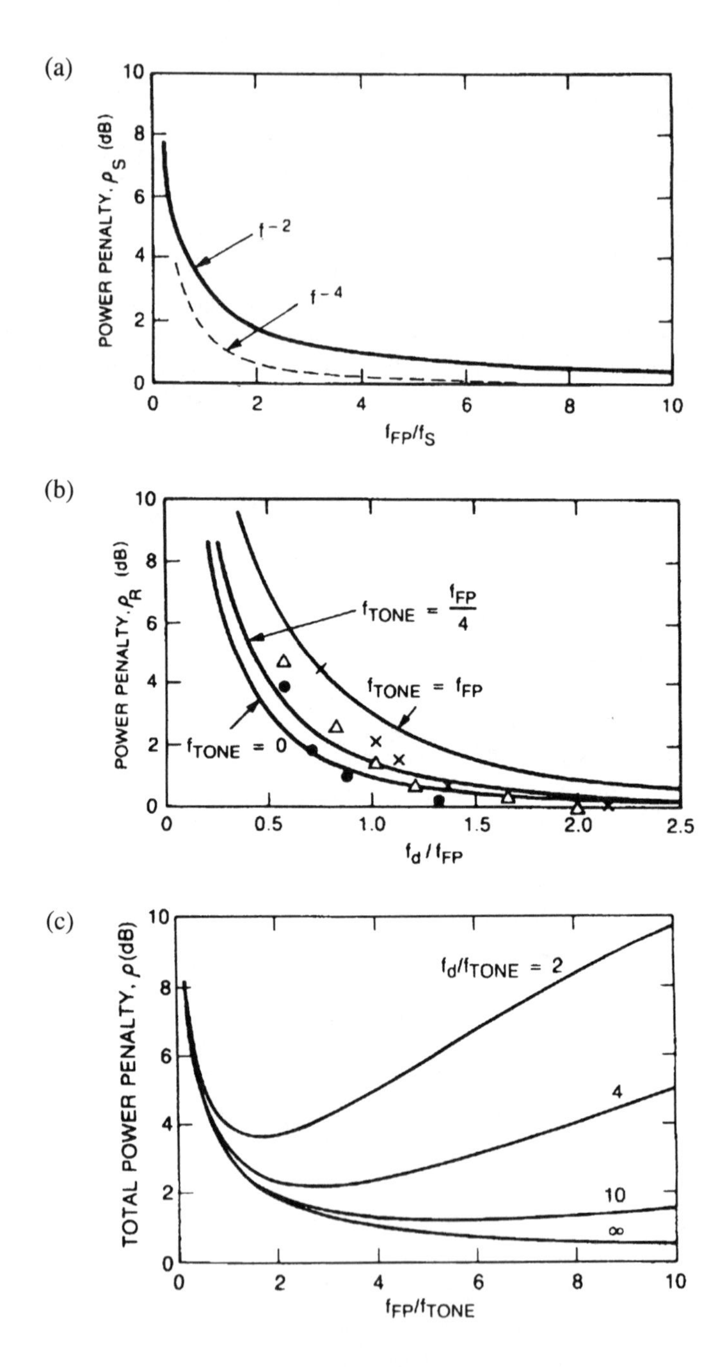

Figure 7.43 Power penalty due to filtering and demultiplexing of a WDM system: (a) penalty from the selected channel versus filter width; (b) penalty from cross-talk versus channel spacing; and (c) total penalty versus filter width. The experimental data points in (b) are taken from [31,65,71]. (*After:* [70, 71]. © 1991 IEEE.)

Therefore, based on (7.15), the total number of channels that can be accommodated in one free-spectral-range is

$$N = \frac{FSR}{\Delta f}$$

$$N + \frac{20}{2.5} = 8$$

When all these eight channels are present, the total crosstalk-induced system penalty P_{er} is given by (7.17):

$$P_{cr} = \sum_{i=1}^{N-1} \left[\frac{P}{1 + \left[\frac{2if_d}{f_{FP}} \right]^2} \right]$$

$$\Rightarrow P_{cr} \cong 0.015 P_s$$

Where P_s is the power in each of the signals before the filter. The system power penalty can be determined using (7.17):

$$PP\ 10 \log \left[\frac{P_{SO}}{P_S - P_R} \right] = 0.066\text{dB}$$

This is a negligable penalty due to the small power leaking through the filter from the neighboring channels.

7.3.4.2 Wavelength Stability

Although experimental demonstrations of WDM systems can be performed over the course of a few minutes, real systems must be stable over the course of years. A critical issue in the long-term physical-layer performance of a WDM system is the stability, repeatability, and accuracy of the wavelength characteristics of the various components. The wavelength characteristics of lasers, filters, and multiplexers will change slightly with time due to many factors, chief of which are temperature changes, device aging, and hysteresis. An uncontrolled drift in the wavelength characteristics of the network would wreak havoc and bring this ultra-high-capacity, multichannel, high-speed system to a grinding halt.

Wavelength stability can be achieved by various locking or control, methods. Four basic representative examples follow:

- *Lock channel to an atomic transition.* Figure 7.44 shows the basic concept of locking a DFB laser to an atomic transition, in this case using argon as the atom [72]. A small gas-filled lamp is used to generate the atomic line, and the laser can be rigidly locked to any desirable wavelength in relation to the atomic line. This method ensures that a signal channel will not drift.
- *Lock filter to a selected channel.* An optical filter will drift in wavelength. However, by using a straightforward feedback scheme, the peak of the optical filter passband can be rigidly locked to the peak of the recovered channel [65]. Figure 7.45 shows this locking technique, in which a small fraction of the filtered optical signal is detected and used in a feedback loop to control the filter passband wavelength. The feedback loop tunes the filter to transmit a maximum of the selected channel. The center of the filter passband is then locked to the same wavelength as the transmitted signal, thereby ensuring that the filter will not drift in relation to the recovered signal. This method requires that the selected signal already be partially inside the passband of the filter for the capturing and locking to occur.
- *Lock each channel to a different resonance of an FP.* Figure 7.46 depicts, without detail, the concept of locking each of several WDM signal lasers to a different periodic transmission resonance of etalon [73]. The FP resonances must be separated by the required WDM channel spacing. Appropriate and individual optical filtering and feedback mechanisms must be used to ensure that each laser is locked to a different resonance of a centralized FP, and some form of

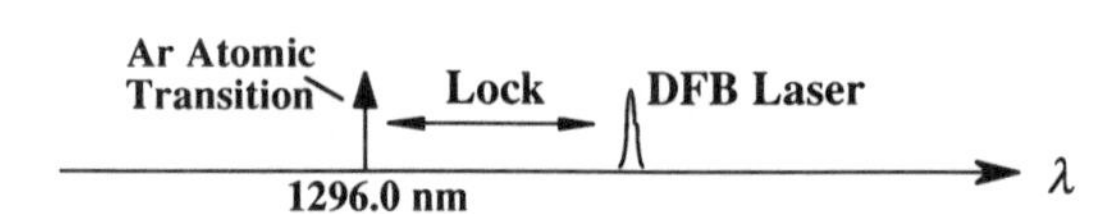

Figure 7.44 Locking a DFB to a wavelength relative to the atomic transition line of argon.

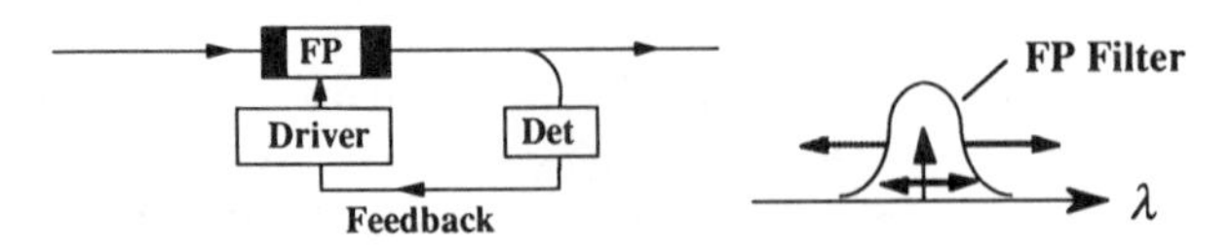

Figure 7.45 Locking the wavelength of the optical filter transmission peak to the peak of a transmitted and selected signal.

Figure 7.46 Several WDM lasers each locked to a different periodic transmission resonance of an FP etalon.

centralized control is required. This method does not guarantee that any component will not drift in wavelength, but it does provide for the entire WDM system drifting in unison, thereby preserving the relative channel spacing.
- *Reference individual wavelength to a central source.* Figure 7.47 shows a scenario in which the signature wavelength at an isolated network node is periodically updated in comparison to a stable reference wavelength located at a central control node. An individual node would transmit its wavelength to the central control node. Subsequently, the central node would transmit back to the isolated node a signal that represents the wavelength difference between the isolated node's wavelength and the stable reference control wavelength [74].

Of course, the employment of any type of external control techniques in a WDM system will add to its complexity and cost. Furthermore, these methods can themselves be unreliable over a long period of time. Therefore, it is perhaps more desirable to design a system that is itself robust to any slight changes in wavelength characteristics. The most straightforward way to have a robust WDM system is to have the wavelengths spaced very far apart, to have optical filters with large transmission bandwidths, and so on. If the optical filter passband is large compared to the channel spacing, the system will be fairly tolerant to any slight wavelength drift. However, there will still be the requirement that the lasers or filters will not drift significantly over the course of years. Some recent studies of free-running standard 1.55-μm DFB lasers has produced hope that the long-term wavelength stability may be good enough for a WDM system. In one study, 13 DFB lasers drifted by less than 300 MHz over the course of 1 year [75]. A second study measured the wavelength of more than 100 DFB lasers over one year, and the wavelength drift for 90% of the lasers was <13 GHz (1 Å). This second study measured the laser under elevated temperatures that replicated a 17-year aging process.[76] It must be emphasized that a real system may have many wavelength-dependent components cascaded over any given communications path, thereby exacerbating any wavelength-drift problems which may exist (i.e., the effective

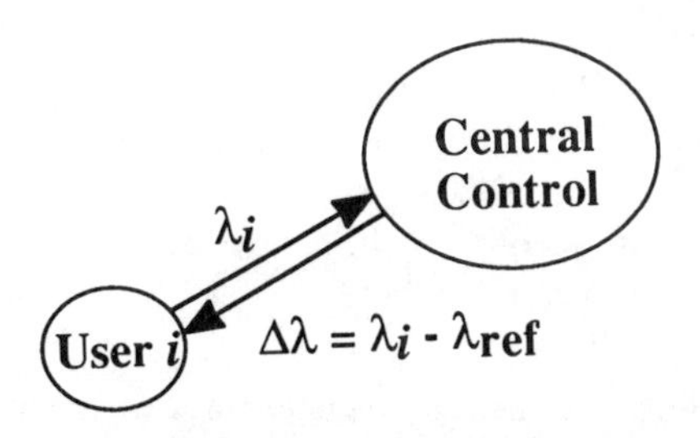

Figure 7.47 Referencing the output wavelength at a network node to the stable wavelength reference source at a central control location.

wavelength passband for the entire link is much narrower for a cascade than for a single element).

Wavelength stability problems, in addition to channel spectral broadening, have generated the conventional wisdom that near-term WDM systems will probably not be operated with ultra-dense channel spacings on the order of <1 Å (i.e., on the order of the information bandwidth), but rather more in the range of ≥ 0.5 nm.

7.3.4.3 Direct-Detection FSK Systems

We have been discussing exclusively amplitude-shift keyed (ASK) systems, in which the intensity of the optical WDM signals is modulated and then directly detected. However, there has also been some work on using frequency-shift keying (FSK) as the modulation scheme for a WDM system [11]. (The basic concepts of FSK modulation were introduced in earlier chapters.)

FSK is a modulation scheme in which "1" and "0" bits in the data stream are defined as two different frequencies, f_1 and f_0. These two frequencies, or tones, can be considered as two separate and complementary ASK signals if the frequency separation is large (>3) compared to the modulation frequency; the frequency separation will necessarily be large since we will be using nonideal optical filtering to demultiplex the two tones when using direct detection. It is necessary to use a wavelength-tunable laser for which two different injected-current levels will generate the two separate FSK tones. A desirable characteristic is that the frequency modulation will be accompanied by a minimal amplitude modulation so as not to disrupt system performance. Additionally, two other figures of merit for the lasers that generate the FSK tones are a wide and uniform modulation bandwidth and a high *frequency-modulation* (FM) response as measured in units of frequency deviation per milliampere of injected current (MHz/mA). These two characteristics typically are inversely related, with an increase in one giving rise to a decrease in the other. Three-electrode DBR lasers have very high FM responses (>10 GHz/mA) but have modulation speeds of only ~200 MHz [77]; two-electrode DFB lasers have moderate FM responses (~1 GHz/mA) but can be modulated at ~1 GHz speeds [78]; and single-electrode DFB lasers have low FM responses (~200 MHz/mA) but can be modulated at several GHz [79]. Single-electrode DFB lasers have phase reversals at kilohertz modulation speeds and may cause significant demodulation problems for bit streams containing long strings of 1s or 0s. Figure 7.48 shows the FM characteristics of a two-electrode DFB laser.

Two potential advantages of using FSK over ASK modulation for a WDM system are the following:

- Smaller modulation currents may be necessary to generate two tones as opposed to turning the laser completely on and off for ASK modulation.
- The bandwidth that an FSK-modulated channel occupies can be significantly

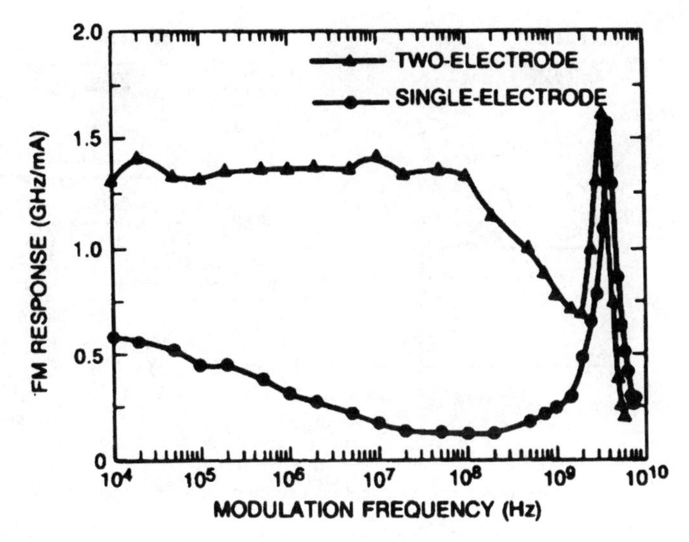

Figure 7.48 The FM response of a two-electrode DFB laser. (*After:* [78]. © 1989 IEEE.)

smaller than the ASK chirp that is generated by a directly-modulated laser. We can, in theory, place more WDM channels within one free spectral range of the optical filter. Of course, external ASK modulation produces the smallest channel bandwidth possible.

The technique for demodulating an FSK channel is shown in Figure 7.49. An electrical bit stream modulates the injected current on the wavelength-tunable laser, which, consequently, produces the two tones. The passband of an optical filter is placed over one tone, allowing that tone to pass and blocking the other tone. Only one tone is output from the optical filter, representing for example the 1 tone. This tone is high for a 1 but low for a 0, since the 0 tone is now blocked. This single-tone output is almost exactly the same modulation scheme as a typical ASK-modulated signal and can therefore be directly detected. The entire WDM system with FSK-modulated channels can use an optical filter to simultaneously demodulate one channel and demultiplex many WDM channels.

The main advantage of using FSK (i.e., its smaller generated channel bandwidth) may unfortunately be offset by the more primary concern about wavelength drift in a WDM system. If large channel spacings are required to ensure system robustness, it may be unnecessary to use FSK transmission. As components become more wavelength stable, FSK transmission may become important for ultra-dense WDM systems. Unless otherwise noted, we will be discussing ASK, and not FSK, systems for the remainder of this chapter.

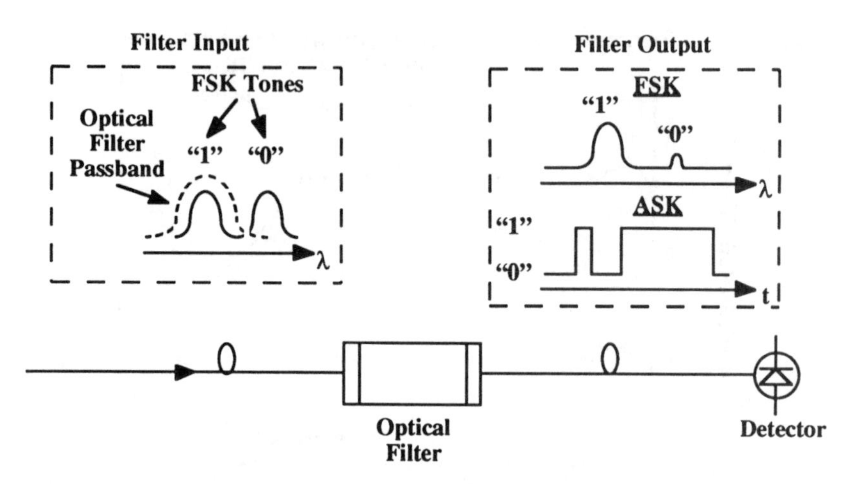

Figure 7.49 FSK-to-ASK conversion using an optical filter to demultiplex the two FSK tones.

7.3.4.4 Coherent Systems

Coherent optical systems were discussed extensively in Chapter 4. We will now describe briefly some concepts is multichannel coherent WDM systems that are appropriate at this point in the book.

One fundamental difference between coherent and direct-detection WDM systems is that coherent detection performs the demodulating and demultiplexing in the electrical domain using a wavelength-tunable local oscillator and electrical filters, whereas direct-detection systems have these functions performed in the optical domain. The optical and electrical spectra for a coherent multichannel system are shown in Figure 7.50 [80]. Two stages of electrical filters are required in the data recovery, one filter stage to demultiplex the many WDM channels and the second stage to demodulate the remaining selected channel if it is FSK or PSK modulated.

Figure 7.51 shows the cross-talk power penalty as a function of the channel spacing normalized to the bit rate [81]. This power penalty is plotted for the cases of ASK, FSK, and PSK modulation. A matched electrical filter is used for demodulation, and the modulation index for FSK is 1. A power penalty less than 1 dB can be achieved for an electrical domain channel spacing that is at least four times the bit rate.

Figure 7.50 also shows one of the best examples of an experimental coherent WDM system, one in which six FSK-modulated channels were used [80]. The sensitivity of this ~200-Mb/s system was <3 dB away from the shot-noise receiver sensitivity limit. The WDM channels were combined in a star configuration, and a wavelength-tunable DBR laser was used as the local oscillator.

Some of the advantages of using coherent detection for multichannel WDM systems include the following.

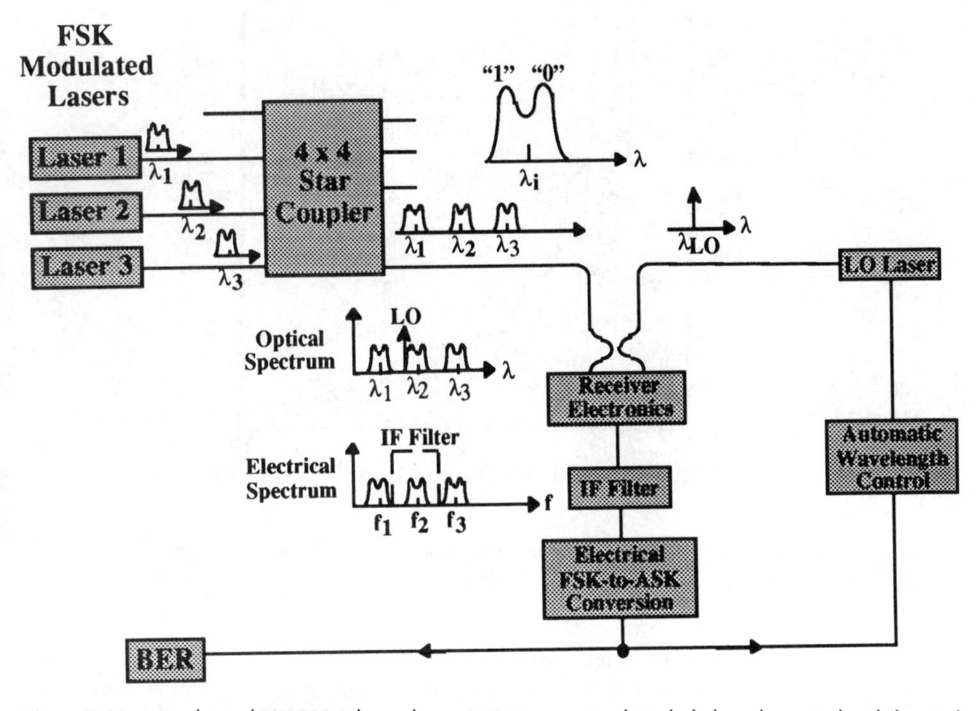

Figure 7.50 A six-channel FSK 200-Mbps coherent WDM star network. Included are the optical and electrical spectra of the WDM system.

- The sensitivity of coherent systems is higher than direct-detection systems that do not use optical amplifiers. Therefore, more channels can be accommodated by a network limited by the optical splitting losses of multiusers. (The sensitivity of coherent detection is equivalent to direct detection using optical amplifiers, and the decision as to which technology to use may very well rest on whether the amplifier or the local oscillator is more convenient to use.)
- Using a local oscillator to down-convert the selected channel into the IF regime allows for electrical filtering. Because electrical filters can be made with more efficient and steeper cut-off frequencies than optical filters, we can place the WDM channels closer to each other and conserve system bandwidth. Thus, coherent detection allows for greater selectivity. Additionally, since smaller system bandwidth is an advantage of using coherent detection, many coherent WDM experiments have used FSK modulation.

However, the disadvantages of using standard coherent detection systems also apply to coherent WDM systems, namely, (1) increased complexity due to using a local oscillator, (2) polarization sensitivity requiring polarization diversity receivers, and

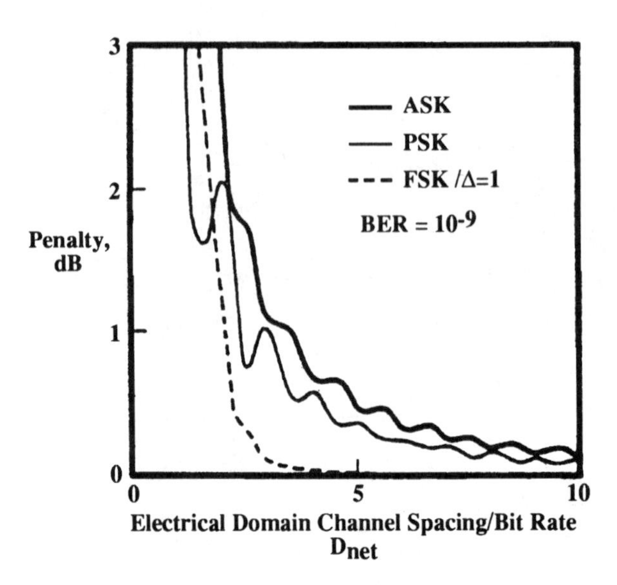

Figure 7.51 Cross-talk power penalty versus normalized electrical domain channel spacing for ASK, FSK, and PSK modulation. (*After:* [7,8]. © 1988 IEEE.)

(3) greater sensitivity to component wavelength drift. The reason behind the greater sensitivity to wavelength drift in a coherent system compared to a direct detection system is because an optical wavelength drift is usually smaller in comparison to an optical filter bandwidth than an electrical frequency drift is in relation to the electrical filter bandwidth of a coherent receiver.

7.3.4.5 Sample Systems

There are several examples of demonstrated experimental WDM systems. Figure 7.52 and Figure 7.53 show two early examples, a 100-channel experiment by NTT and the LambdaNet project at Bellcore. The 100-channel NTT star experiment used standard DFB lasers, which were each FSK-modulated at 622 Mbps with a total throughput of 62.2 Gbps [36]. The optical filter was a temperature-tuned multistage MZ interferometer, which adequately demultiplexed the WDM channels. The LambdaNet star experiment consisted of 16 users, with each user having a transmitter and a receiver [82]. Additionally, the Rainbow project at IBM [83] represents an advance in the state of the art of WDM system technology. We will discuss in Section 7.7.2 some more recent examples of WDM networks.

7.3.5 Optical Amplification

We discussed optical amplifiers in great detail in Chapter 5. In this section, optical amplifiers will be discussed with regard to their implementation in WDM systems. We

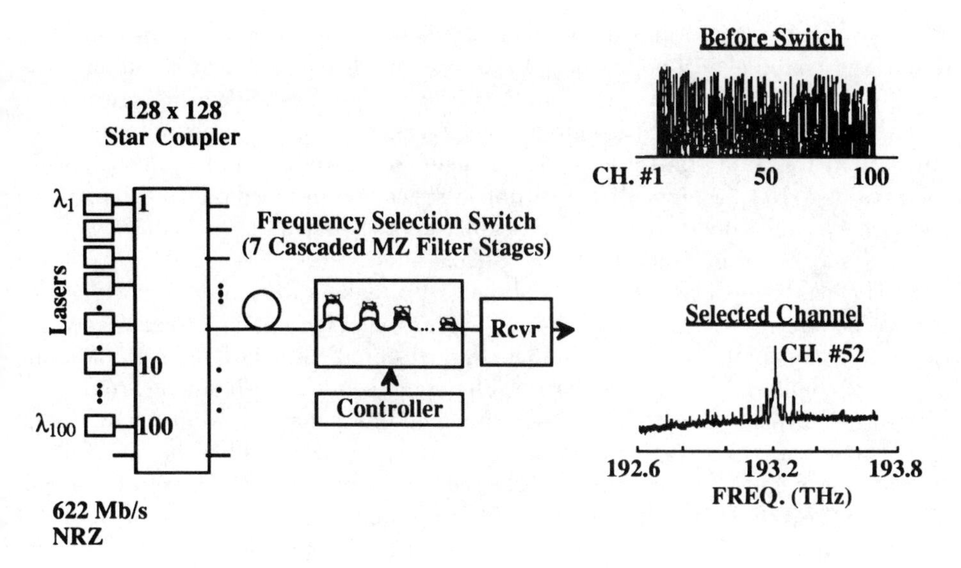

Figure 7.52 The 100-channel WDM NTT experiment. (*After:* [36]. © 1991 IEEE.)

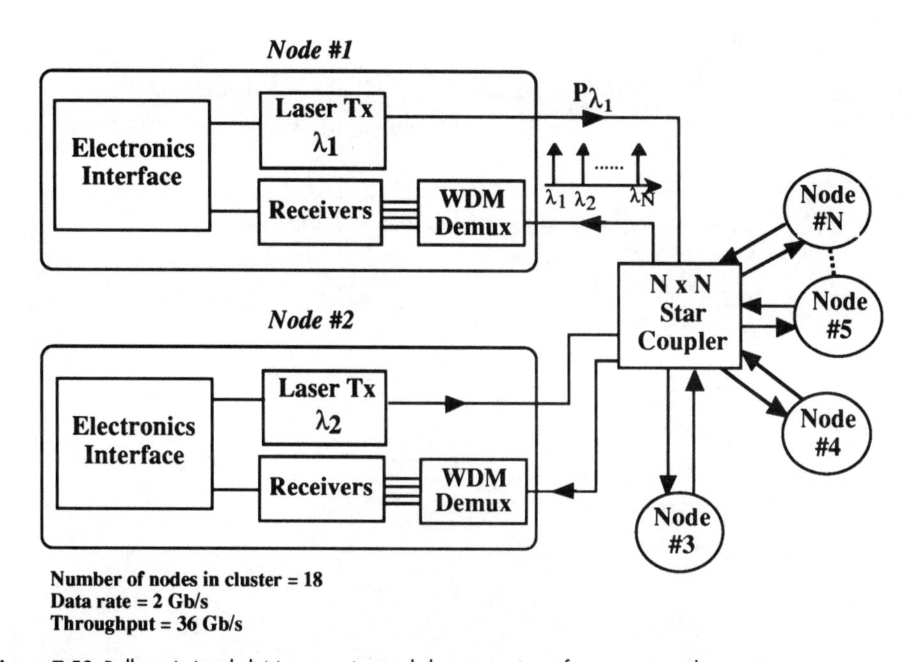

Figure 7.53 Bellcore's LambdaNet experimental demonstration of a star network.

emphasize that it is the optical amplifier that perhaps has been the most important recent key enabling technology for WDM systems, because it can simultaneously amplify signals over a very wide bandwidth (~25 nm). As shown in Figure 7.54, optical amplifiers are used in multichannel WDM systems to compensate for fiber attenuation losses in transmission, component excess losses, and optical network splitting losses [84]. These optical splitting losses can occur in a passive star, in which the optical power is divided by the number of users (N), or in a ring/bus, in which there possibly may be optical tapping losses at each node.

The nearly-ideal characteristics of the EDFA make it the amplifier of choice for systems requiring WDM transmission. However, one characteristic that makes the EDFA the clear choice over semiconductor optical amplifiers is the long carrier lifetime of erbium compared to the short carrier lifetime of conduction-band electrons in an SOA [85]. The long carrier lifetime of erbium guarantees that the gain will not change on the time scale of a single bit resulting in negligible cross-talk if the amplifier is operated near gain saturation and negligible four-wave mixing. Therefore, we will discuss exclusively EDFAs for WDM amplification unless otherwise stated. (We will not discuss any specifics of the PDFA [praseodymium-doped fiber amplifier] in this section, although much of the discussion will hold true for the PDFA as well. The PDFA can also provide simultaneous amplification over a wide bandwidth, but its high required pump power may limit its eventual usefulness in systems.)

7.3.5.1 Demultiplexing of Amplified Channels

Based on much of the groundwork laid in Chapter 5, we can determine the recovered SNR when demultiplexing many optically amplified WDM channels using an optical

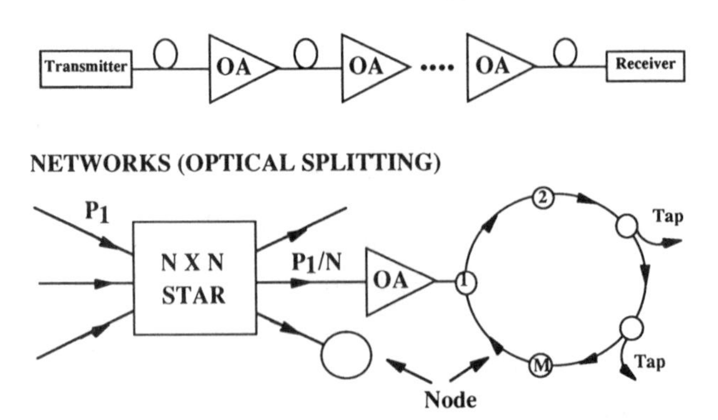

Figure 7.54 Uses of optical amplification in a WDM point-to-point system and in a network.

filter (Fig. 7.55). In addition to a filter passing a selected channel and blocking all other channels, there exists wideband amplified-spontaneous-emission noise that will leak through the filter passband. On absorption within the square-law detector, this ASE noise will produce signal-spontaneous beat noise ($N_{s\text{-}sp}$) between the selected channel and the ASE and spontaneous-spontaneous beat noise ($N_{sp\text{-}sp}$) between the ASE and itself. The general equations describing these two noise terms can be found in Chapter 5 when the ASE is assumed very broadband and can be assumed constant over the small electrical bandwidth of the detector. However, using a Lorentzian-lineshaped optical filter to demultiplex many amplified channels will alter the equations of the generated noise terms. In addition to the filter passing the selected channel and mostly blocking the rejected channels, the main differences of an amplified WDM system from either an unamplified WDM system or from a single-channel amplified system are the following [86]:

- The ASE is significantly attenuated as it passes through the filter.
- The ASE passing through the filter attains a similar spectral shape as the filter passband itself (i.e., Lorentzian if using an FP filter).
- The equations describing $N_{s\text{-}sp}$ and $N_{sp\text{-}sp}$ will be altered since the ASE itself is different after passing through the filter.
- Each of the rejected channels will itself beat with the ASE that passes through the filter and is spectrally colocated with any given rejected channel.

We can begin by following arguments similar to those given in Section 7.4.3.1.

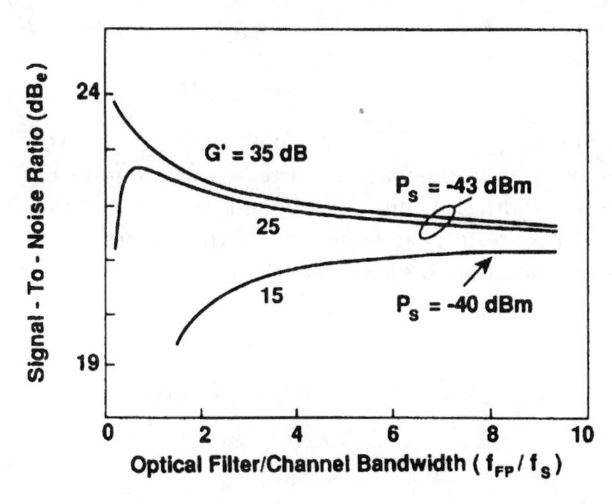

Figure 7.55 SNR for a single amplified channel as a function of optical filter/selected channel bandwidth for different gain and signal power values. B_e is 500 MHz, f_s is 1 GHz, n_{sp} is 1.4, and $N_{eir} = 5 \cdot 10^{-14}$ A^2. (*After:* [87]. © 1990 IEEE.)

The detected photocurrent of the selected channel power passing through the Lorentzian-lineshaped filter is:

$$S = \frac{(cP_S G f_{FP})}{(f_{FP} + f_S)} \tag{7.18}$$

and the sum of all the rejected channels' power leaking through the demultiplexing filter is [86]

$$R = \sum_i \frac{cP_R G f_{FP}}{(f_{FP} + f_R)\{1 + [2if_d/(f_{FP} + f_R)]^2\}}, \qquad i = 1,2, \ldots, N-1 \tag{7.19}$$

where P_s is the selected channel's power before the filter, G is the amplifier gain, f_{FP} is the 3-dB bandwidth of the filter passband, $c = (q/h\nu)$ and is a conversion factor between photons and electrons, f_s is the 3-dB bandwidth of the selected channel, P_R is the power of each of the rejected channels before the filter, f_d is the channel separation in frequency, f_R is the 3-dB bandwidth of each of the rejected channels, and i represents the channel number; in general, we can assume that $P_S = P_R$ and $f_S = f_R$. (Note that some nomenclature is slightly different in this section than in Section 7.3.4.1.)

The shape and power of the ASE (in one polarization) as it passes through the FP filter is Lorentzian:

$$P_{sp}(f - f_o) = \frac{2P_{sp}^o}{(\pi f_{FP})(1 + [2(f - f_o) / f_{FP}]^2)} \tag{7.20}$$

where f_o is the center frequency of the filter passband, $P_{sp}^o = n_{sp}(G - 1)h\upsilon f_{FP}$ and is the ASE optical power as it emanates from the optical amplifier (assuming a rectangular bandwidth of width f_{FP}), and n_{sp} is the spontaneous emission factor. This ASE power will now contribute to beat-noise terms. After appropriately convolving the signal and spontaneous terms, the beat noise terms are [86]

$$N_{s-sp} = \frac{4N_{sp}^2 G(G-1)qcP_s f_{FP}^2 B_e}{(f_{FP} + f_s)(f_{FP} + f_s + 2B_e)} \tag{7.21}$$

and

$$N_{sp-sp} = \frac{2N_{sp}^2 (G-1)^2 q^2 f_{FP}^2 B_e}{f_{FP} + 2B_e} \tag{7.22}$$

where B_e is the electrical bandwidth of the detector. Additionally, the rejected channels will each beat with some of the ASE that exists at the same spectral location as that channel and that has leaked through the demultiplexing filter. For this analysis, we can approximate the ASE to be relatively flat over the bandwidth of the rejected channel, since a Lorentzian acts as an almost uniform function when a channel is located away from the passband center. Therefore, the beat noise between the rejected channels and the ASE noise passing through the filter is

$$N_{r-sp} = \sum_i \frac{4G(G-1)N_{sp}qcP_R B_e}{[1+(2if_d/f_{FP})^2]^2} \tag{7.23}$$

When (7.21) to (7.23) are combined, the electrical SNR of the detected demultiplexed signal is [86]

$$SNR = 10 \log\left(\frac{(S-R)^2}{N_{s-sp}+N_{sp-sp}+N_{r-sp}+N_{cir}}\right) \tag{7.24}$$

where N_{cir} is the circuit noise and is typically dominated by the thermal noise of the receiver.

Figure 7.55 shows the SNR for a single recovered amplified channel as a function of the normalized optical filter bandwidth. Note that the SNR critically depends on the amplifier gain. Let's discuss the middle case, which considers a 25-dB-gain amplifier. As the optical filter width is increased from 0, more channel power can pass through the filter and the recovered SNR will increase. However, an SNR peak is quickly reached when the signal-spontaneous beat noise begins to dominate since (1) all the selected channel has already been completely transmitted through the filter, and (2) any increase in the filter bandwidth will only increase the noise terms. When the filter is increased further, $N_{s\text{-}sp}$ reaches a maximum, and then the only term to increase is the spontaneous-spontaneous beat noise. A good design guideline is that the optical filter should be more than three times the channel bandwidth. Recall from Chapter 5 that the optimum regime in which to operate an amplified system is when the detector is signal-spontaneous beat noise limited.

Figure 7.56 isolates the effects on SNR of the rejected channels by showing the system power penalty as a function of channel separation given 1, 2, and 100 rejected channels; also plotted, for comparison, is the case for 1 rejected channel in an unamplified system [86]. As is the case in an unamplified system, the vast majority of the SNR degradation is due to the most adjacent rejected channel that reduces the contrast ratio of the recovered selected channel signal power. The minor difference between 1 rejected channel in the amplified and in the unamplified cases is due to the noise term in which the rejected channel beats with the ASE. As a design guideline, the channel separation should be more than 1.5 times the FP filter bandwidth.

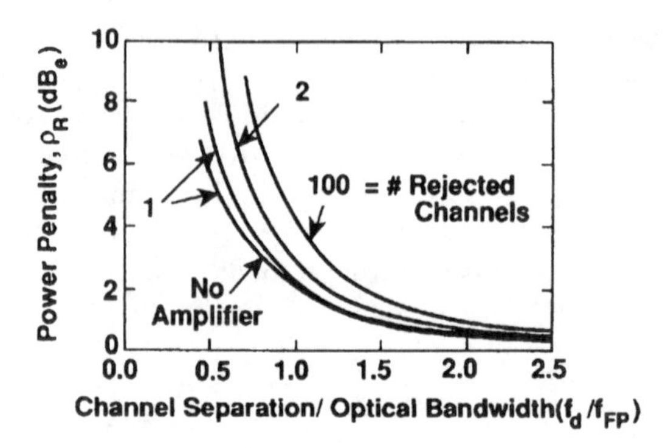

Figure 7.56 System power penalty generated from rejected channels as a function of the channel separation normalized to the filter bandwidth. For comparison, the penalty from 1 rejected unamplified channel is also plotted. (*After:* [87]. © 1990 IEEE.)

7.3.5.2 Amplification in Stars

In a multichannel WDM star network, an optical amplifier can be used to compensate for the optical splitting losses inherent in a passive star coupler. Each output port of the star would require an optical filter to simultaneously demultiplex the WDM channels and reduce the ASE noise. Figure 7.57 shows the BER (bit error rate) curve as a function of received optical power when a 25-dB-gain EDFA is placed at each output port. The system sensitivity has increased by 14 dB, thus accommodating 20 times the number of potential network users [84]. Additionally, Figure 7.58 shows a system in which 10 EDFAs are used to periodically compensate for star-based splitting losses in a 1–to–40 million–way–split 16-channel WDM system [87].

However, requiring each output port to have its own EDFA can be quite expensive. One method to significantly reduce the cost of implementing an EDFA is to share the optical power of one pump laser among many erbium-doped fibers [88]. This can be accomplished by reserving one input port of the star for an input pump source to equally distribute the pump power to each output port. Figure 7.59 shows this for a single star, in which one pump is used by many output-port erbium-doped fibers, and the need for separate pump signal combiners (i.e., wavelength multiplexers) is eliminated. An experiment was performed in which such a distributed pumping scheme enabled the full sensitivity compensation in an 8-by-8 star [88]. The cost savings is dramatic because 80–90% of the cost of an EDFA is due to the pump laser. The concept of using one pump to provide gain to many erbium-doped fibers can be extended to large multiple-stage stars. As is shown in Figure 7.60, a two-stage M-by-K large star can be constructed by utilizing K input star couplers, M output stage star couplers, erbium-doped fibers fully interconnecting the input stage with the output stage, and each input star reserving one input port for a pump laser [88].

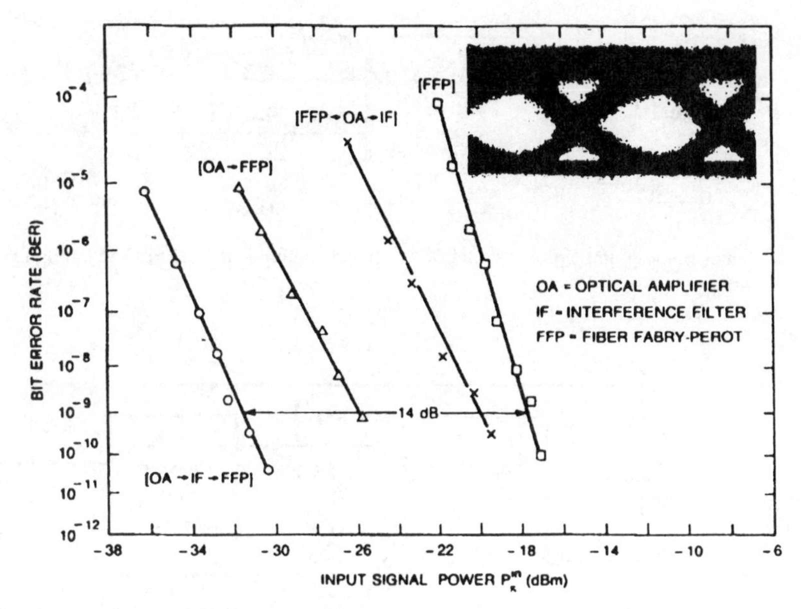

Figure 7.57 BER versus received optical power in a WDM star network. (*After:* [84]. © 1990 IEEE.)

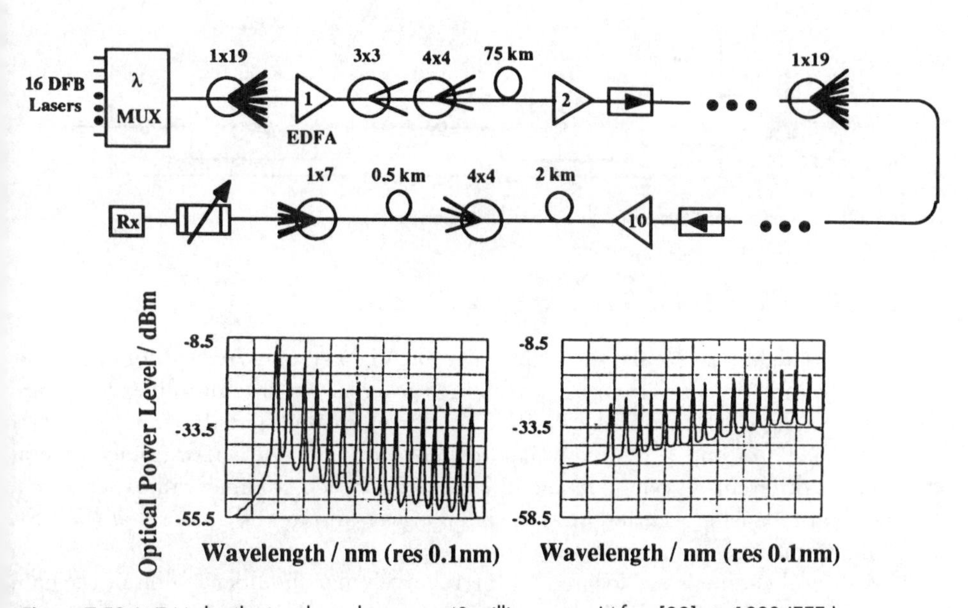

Figure 7.58 WDM distribution through stars to 40 million users. (*After:* [88]. © 1990 IEEE.)

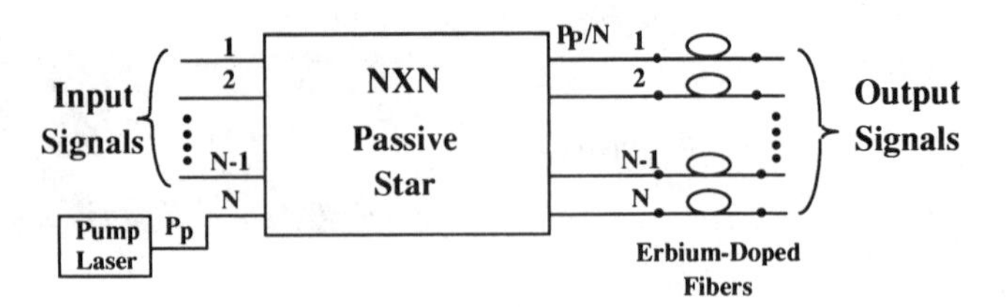

Figure 7.59 Simultaneous pumping of several EDFAs when the pump light is distributed through the central star.

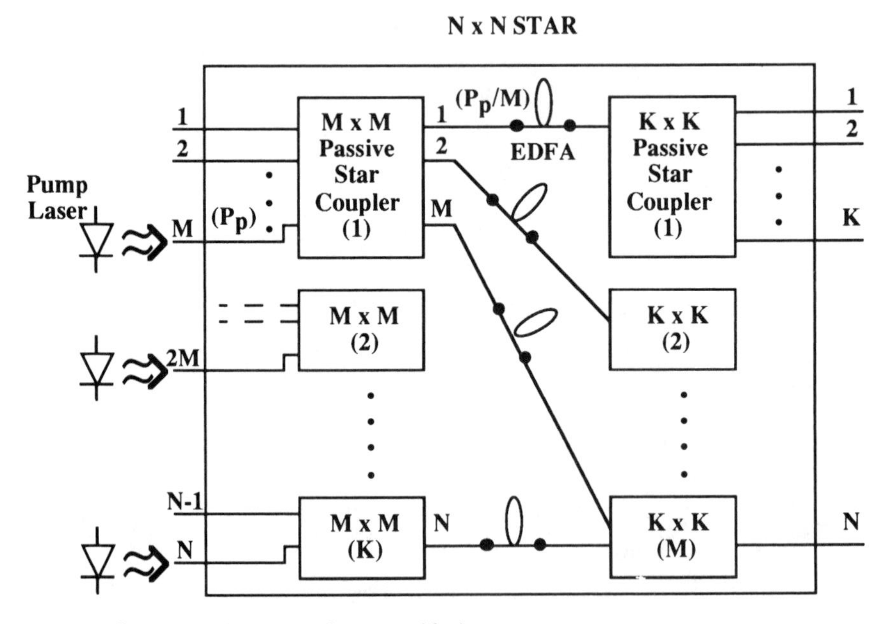

Figure 7.60 Fully connected two-stage large amplified star.

7.3.5.3 Nonuniform EDFA Gain and EDFA Cascades

The EDFA is an almost ideal optical amplifier for WDM systems, except for one major flaw: The gain is not uniform with wavelength whereas the interamplifier losses are nearly wavelength independent [89–92]. For a single amplifier, the gain exhibits a peak at 1.531 μm and a relatively flat region near 1.557 μm. If we place several channels at different wavelengths, as in Figure 7.61 each channel will experience a different gain, causing a differential in signal power and in SNR; note that the ASE and gain spectra are very nearly the same.

If several channels are located on the relatively flat shoulder region of the gain spectrum, the gain differential after a single amplifier will be within just a few dB.

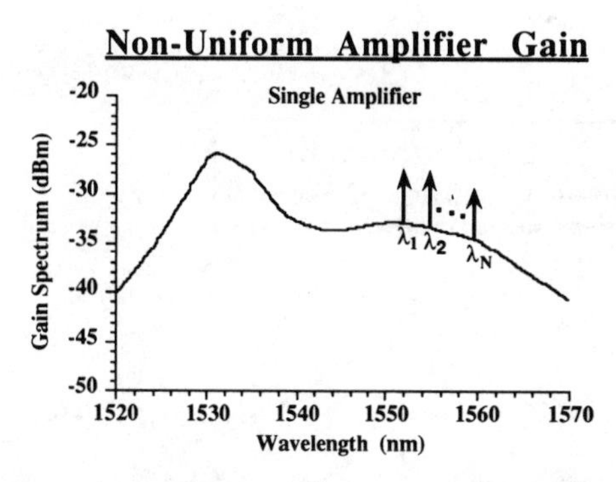

Figure 7.61 Several WDM channels are placed at different wavelengths within the nonuniform EDFA gain spectrum.

However, when a cascade of EDFAs is used to periodically compensate for losses, the differential in gain and resultant SNR can become severe. A large differential in SNR among many channels can be quite deleterious for proper system performance. Figure 7.62 shows the gain spectrum after a single amplifier and after 200 cascaded amplifiers, both in the absence of any input signals. The gain does not accumulate linearly from stage to stage, and the resultant wavelength-dependent gain dramatically changes shape in a cascade. The peak at 1.531 μm is attenuated and a newly generated peak appears near 1.560 μm [92]. This shift in the gain to longer wavelengths can be explained by the relative overlap of the emission (i.e., gain) and absorption (i.e., loss) cross-sections of erbium, as shown in Figure 7.63 [93]. There is more loss than gain at the shorter wavelengths, and there is more gain than loss at the longer wavelengths. Along the cascade, gain is gradually "pulled" away from the shorter wavelengths and is available at the longer wavelengths, with the relative overlap of the cross-sections determining that ~1.560 μm will experience a gain peak. Figure 7.64 shows the SNR for 20 WDM channels at the output of 20 and 100 EDFAs when the channels are all placed at the flattest portion of the EDFA gain curve. After 100 amplifiers, the SNR differential is ~16 dB; in this example, the EDFAs have 25-dB small-signal gain, the interamplifier loss is 10.6 dB, and the interamplifier distance is ~40 km. (Note that nonlinearities have not been included in this theoretical result, which considers only nonuniform EDFA gain. We will address nonlinearities in Section 7.3.6.)

There have been reported several methods used to equalize the nonuniform EDFA gain.

- *Pre-emphasis.* The powers of the input channels are selectively attenuated, with the higher attenuation values used for the wavelengths that receive the higher EDFA gain [94]. Figure 7.65 shows the results for four 2.5-Gbps channels

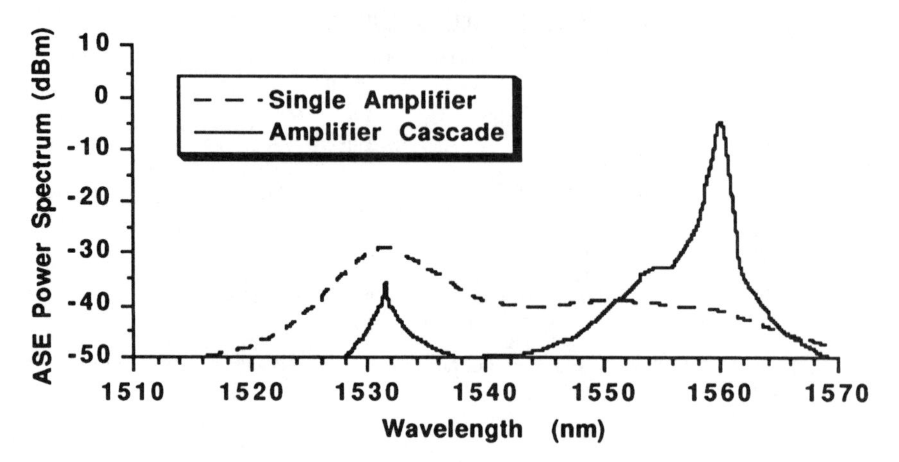

Figure 7.62 Nonuniform gain spectrum at the output of a single EDFA and after a cascade of 200 EDFAs. No input signals are present. (*After:* [92]. © 1995 IEEE.)

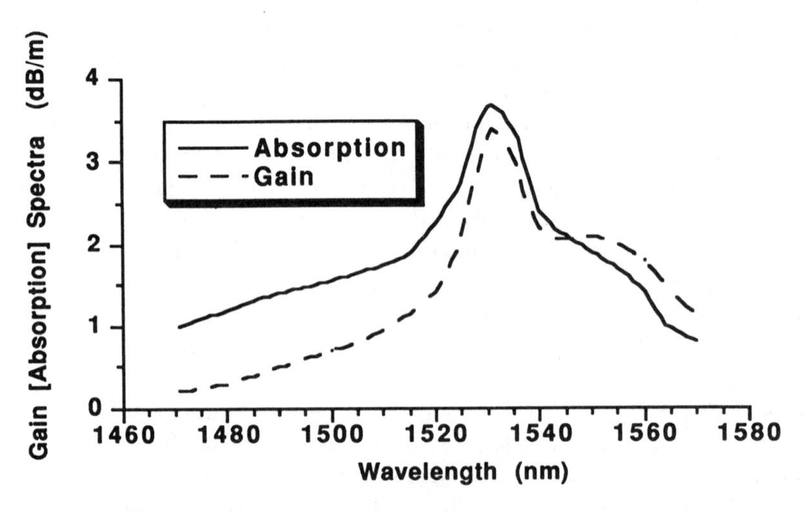

Figure 7.63 The absorption and emission cross-sections of erbium ions. (*After:* [92]. © 1995 IEEE.)

transmitted over 9,000 km using 294 cascaded EDFAs. This technique is viable, in a WDM point-to-point system.

- *Passive transmission filters.* A transmission filter can be used to pass the shoulder region and attenuate the 1.531-μm peak [95]. This is only useful for cascades of a few EDFAs, since this 1.531-μm gain peak will eventually become attenuated in a longer cascade.
- *Passive notch filters.* A notch, or attenuation, filter at 1.531 μm can be used in

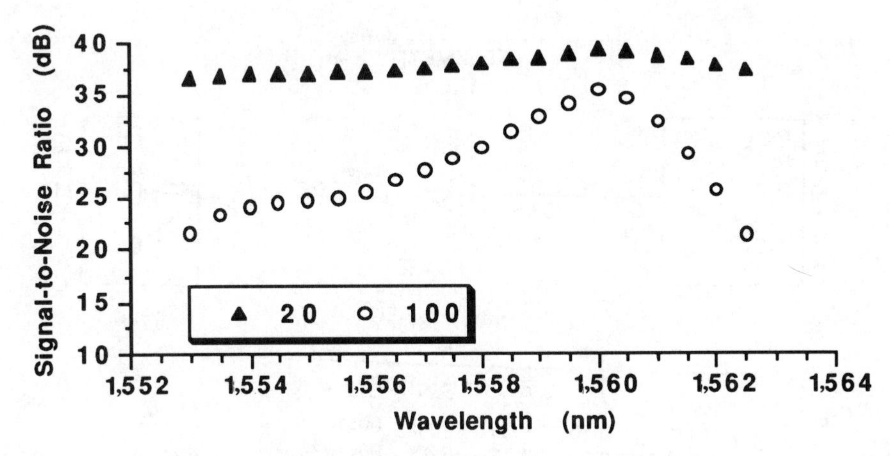

Figure 7.64 The theoretical SNR for 20 WDM channels at the output of 20 and 100 cascaded EDFAs. (*After:* [92]. © 1995 IEEE.)

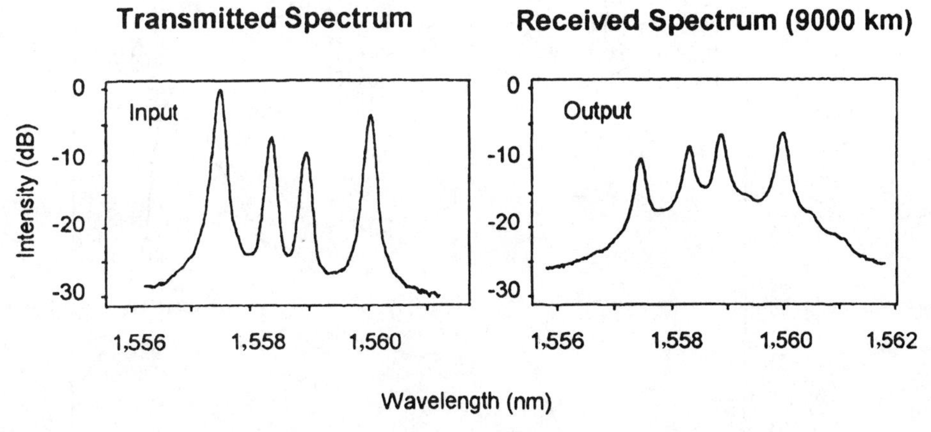

Figure 7.65 Four 2.5-Gbps channels transmitted over 9,000 km. (*After:* [95]. © 1994 OSA.)

short cascades to attenuate the shorter-wavelength gain peak; [96] and a notch filter at 1.560 μm can be used in longer cascades to attenuate the newly generated longer-wavelength 1.560-μm gain peak [97].

- *Telemetry.* A simple algorithm can be used in which information about the output signals is transmitted back to the input side. In just a few iterations, the SNRs can be made equal (Fig. 7.66) [98]. This method is limited to point-to-point WDM systems.
- *Inhomogeneous EDFAs.* The EDFAs can be cooled down to the temperature of liquid nitrogen [99]. The EDFA becomes inhomogeneous at this temperature,

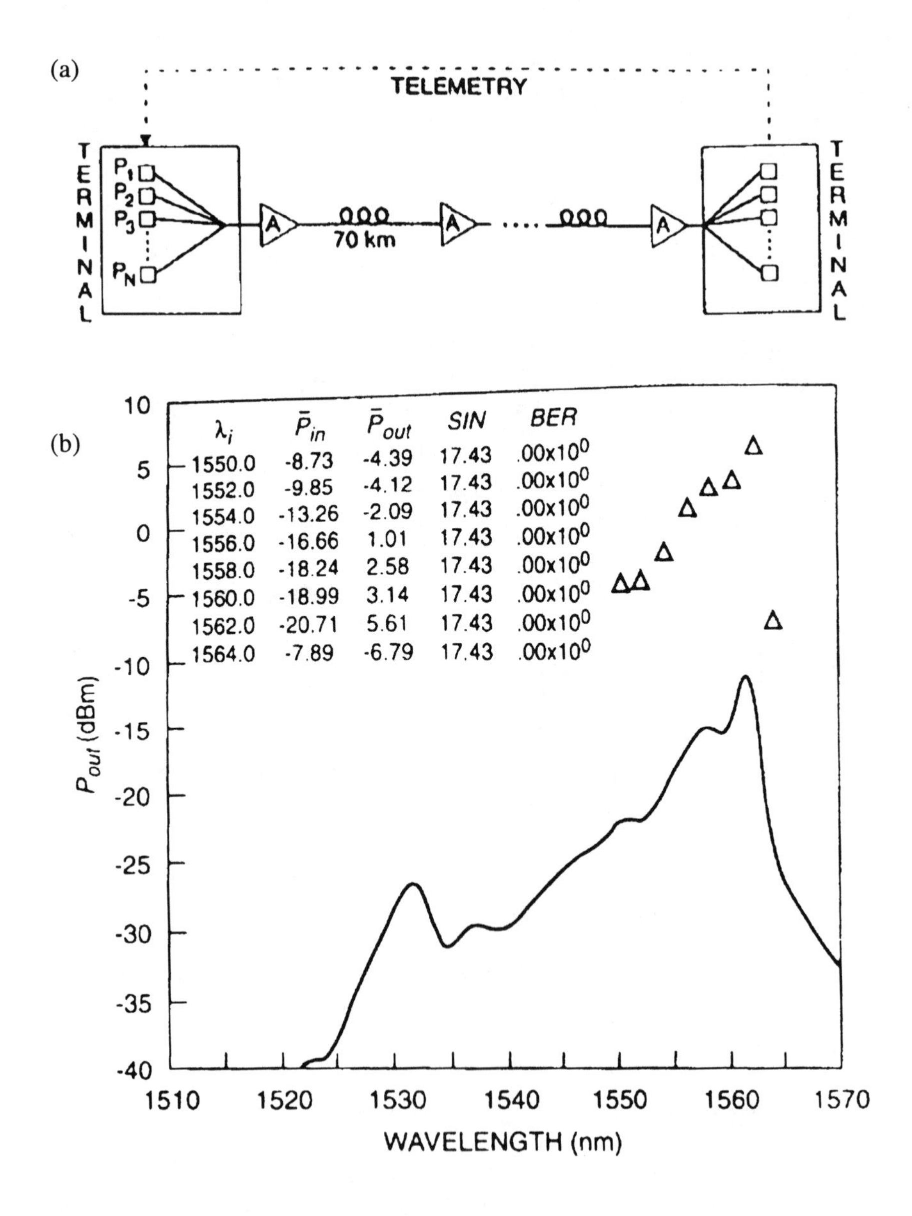

Figure 7.66 Telemetry used to equalize the output SNR of many WDM channels: (a) telemetry system and (b) equalized output SNR spectrum. (*After:* [98]. © 1993 IEEE.)

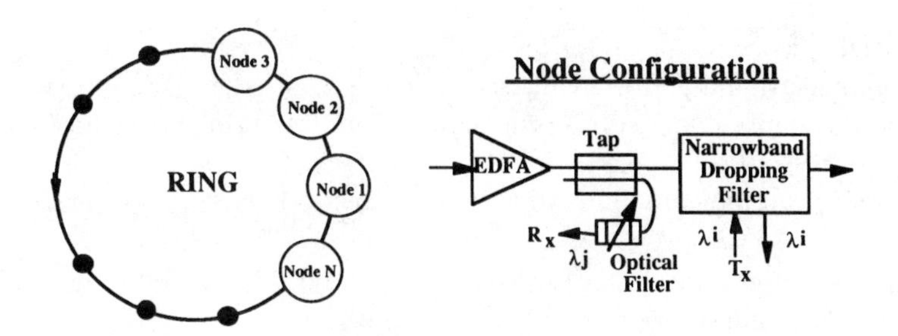

Figure 7.67 Node configuration in a WDM ring network.

and gain saturation at one wavelength does not affect any other wavelength. Gain accumulates nearly linearly from stage to stage.

- *Parallel EDFAs.* The WDM signal can be wavelength demultiplexed, with each different wavelength feeding into one of several parallel EDFAs [100]. Each EDFA is designed to provide the same gain value to the different wavelengths.
- *Active equalization.* Active equalization with feedback can be used for more accurate compensation. In theory, the nonuniform EDFA gain can be compensated by providing a nonuniform loss element that is the exact inverse of the gain. One possibility is to use an AOTF to act as the nonuniform loss element; the AOTF is faster than the EDFA gain dynamics and can have multiple independent loss resonances [101].

In addition to a point-to-point linear cascade of EDFAs, amplifiers can be used in closed-loop rings to compensate for the passive optical tapping losses at each node. A ring is similar to a linear cascade of EDFAs, because a closed-loop represents an infinite cascade of EDFAs. Several functions must be performed at each node in a WDM ring (assuming fixed transmitters and tunable receivers):

- *Amplification.* Amplification is necessary to compensate for fiber-attenuation losses, splitting losses, and excess losses,
- *New-signal injection.* A new signal on wavelength i must be injected onto the ring,
- *Old-signal removal.* Once the signal on wavelength i has propagated around the loop, it must be removed from the ring. Otherwise, it will interfere with any new signal being injected on wavelength i, causing intrawavelength cross-talk, which cannot be minimized by any filtering.
- *Signal recovery.* Data recovery can take place on any wavelength j.

A node configuration that holds promise for accommodating many nodes in a WDM ring is shown in Figure 7.67. Each node would include the following:

- An EDFA to compensate for losses;
- A channel-dropping filter to inject wavelength *i* onto the ring and to remove that same wavelength once it has propagated around the ring (to reduce intrawavelength crosstalk);
- A passive tap and tunable optical filter to recover data on wavelength *j*.

The channel-dropping filter is a four-port device and has the following "notching" transmission function in the straight-through path of the device [59]:

$$H_{CDF}(\lambda) = \frac{H_{CDF}(\lambda_C)}{1 + \left(\frac{\lambda - \lambda_C}{\lambda_{FWHM}/2}\right)^2} \tag{7.25}$$

The EDFA should be placed at the front end of the node to allow for relative insensitivity to slight changes in system variables, because the limited saturation output power of the EDFA will self-equalize the SNR and the signal powers [92,102]. Assuming that each node transmits at 2.5-Gbps and each wavelength is spaced 0.5 nm apart, then (1) 6 nodes can be accommodated when no EDFA and channel-dropping filter are used; (2) 14 nodes can be accommodated when an EDFA is added but no channel-dropping filter is used; and (3) 27 nodes can be accommodated and robust system operation enabled when a channel-dropping filter is added.

Another possible solution for the nonuniform EDFA gain is to dope the amplifier with fluoride, producing the *erbium-doped fluoride-fiber amplifier* (EFFA). The gain (ASE) profile of a single EFFA is shown in Fig. 7.68 and is relatively flat over a ~25-nm range with very little remnant of the 1.531-nm gain peak [103]. However, this uniformity due to doping occurs only in a single amplifier, with the gain profile still skewing toward longer wavelengths in amplifier cascades. Therefore, in long cascades, either equalization techniques will still be needed or only a few channels will be allowed to propagate.

7.3.6 Fiber Nonlinearities and Dispersion

Prior to the widespread use of EDFAs, long-distance transmission required periodic regeneration of a signal. Such detection and retransmission compensated for both attenuation and fiber chromatic dispersion effects (dispersion was discussed in detail in Chapter 2). However, the use of EDFAs compensates only for attenuation, allowing other effects such as dispersion and nonlinearities to accumulate unimpeded along a transmission link. Although dispersion-shifted fiber can be used in EDFA-based systems to minimize the effect of dispersion when channels are operated near the dispersion-zero wavelength, various nonlinear effects will also accumulate along the link and cause severe limitations in WDM transmission. These nonlinear effects include

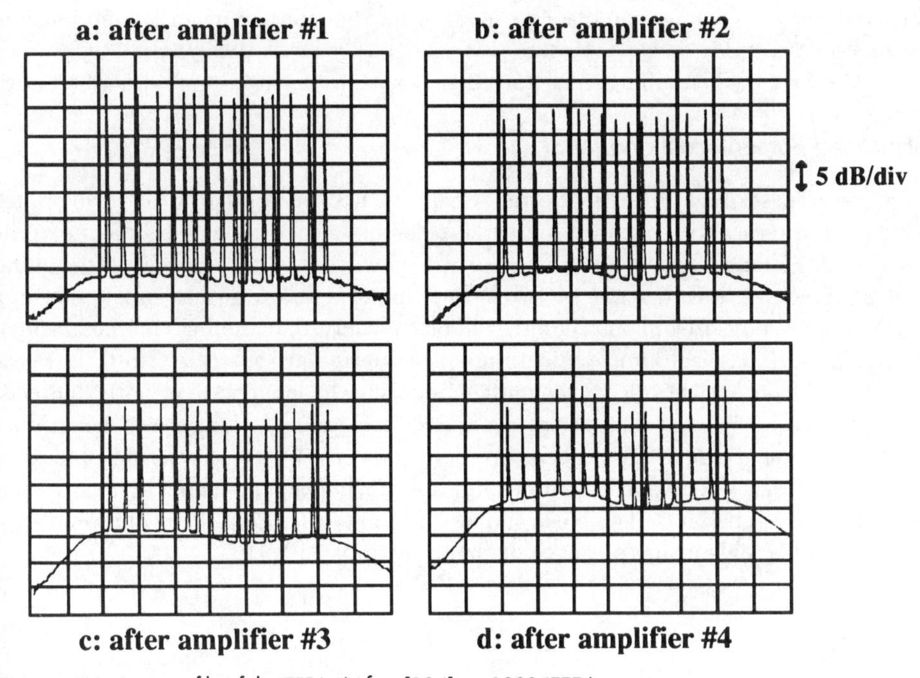

Figure 7.68 Gain profile of the EFFA. (*After:* [104]. © 1990 IEEE.)

stimulated scattering processes (i.e., Raman and Brillouin scattering) and material refractive-index interactions (i.e., self- and cross-phase modulation and four-wave mixing) [104,105]. We will describe these phenomena and discuss various techniques for compensating their deleterious effects.

(Another nonlinear effect is the nondispersive propagation of solitons. This important topic was discussed in detail in Chap. 6, including the propagation of two soliton streams, each at a different wavelength. WDM propagation of several soliton bit streams will incur soliton collisions due to the different group velocity of the different wavelengths, but these collisions do not tend to impair WDM transmission of solitons. In fact, a recent experiment that transmitted eight soliton channels across 9,000 km has been reported [106].)

7.3.6.1 Nonlinear Effects

The silica optical fiber has certain nonlinearities associated with it. The major nonlinear effects are each dependent on the power (i.e., intensity) of the propagating signals, with weak signals not incurring significant effects. Furthermore, these effects require a certain propagation distance to allow for the interactions to accumulate and become relevant.

The four basic nonlinearities and their basic characteristics are described next.

Stimulated Raman Scattering (SRS)

SRS is a process of nonlinear scattering within the fiber medium in which two optical waves will interact with each through a vibrational wave as they all propagate in the forward direction along the fiber. The two optical waves typically are called the pump and signal waves, but they can be two WDM channels in a transmission system. In a WDM system, power from one channel will be transferred to another channel through a broadband (i.e., ~200 nm) optical Raman scattering process (Fig. 7.69). Power is transferred from the shorter-wavelength channels to the longer-wavelength channels, causing the shorter wavelengths to experience loss and the longer wavelengths to experience gain (Figure 7.70). The threshold intensity for the scattering process to occur is ~500 mW for a single channel, but it is much less per channel in the presence of many WDM channels. The basic simple equation describing the evolution of two propagating signals in the presence of Raman gain is [107]

$$\frac{dI_1}{dz} = g_R I_1 I_2 \tag{7.26}$$

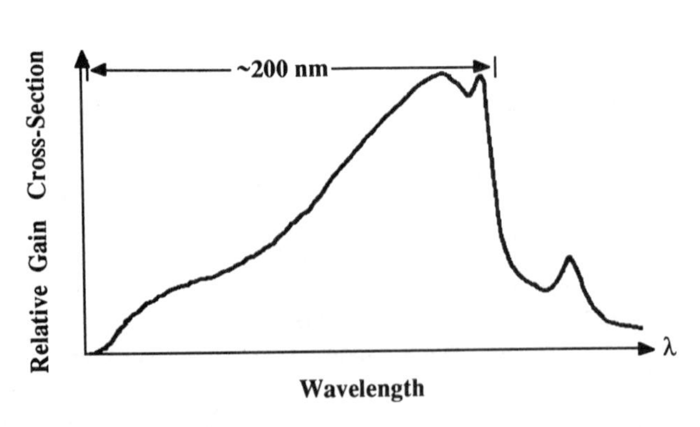

Figure 7.69 Wideband SRS gain profile.

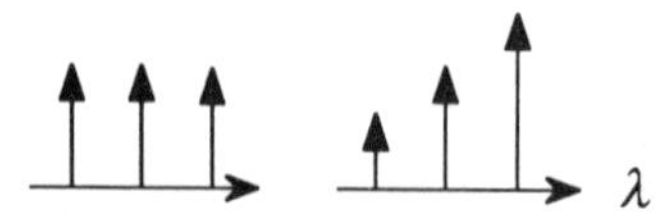

Figure 7.70 SRS providing gain to the longer-wavelength channels at the expense of the shorter-wavelength channels.

where I represents the intensity for the two different WDM channels, z is the direction of propagation, and g_R is the Raman gain coefficient.

Stimulated Brillouin Scattering (SBS)

Another scattering process in the fiber occurs when two optical waves interact through an acoustical wave, with one optical wave acting as a pump for the second optical wave. In SBS, the first signal wave propagates in the forward direction, but the scattered second optical wave propagates in the backward direction. As shown in Figure 7.71, the original forward propagating wave will lose power to the newly generated backward propagating wave. The power threshold for SBS is fairly low, being only ~2.4 mW for >20 km of propagation. This process is a very narrowband process for which the two waves must be within only ~10 MHz of each other. The simple equation describing the evolution between the forward and backward optical waves is [108]

$$\frac{dI_1}{dz} = -g_B I_1 I_2 \tag{7.27}$$

where g_B is the Brillouin gain coefficient.

Self- and Cross-Phase Modulation (SPM and CPM)

The remaining nonlinearities we will discuss owe their origin to the fact that the index of refraction of an optical fiber is nonlinear with signal power [105]

$$n = n_{linear} + \overline{n_2}\left(\frac{P}{A_{eff}}\right) \tag{7.28}$$

where $\overline{n_2}$ is the nonlinear index coefficient and is $(3.2 \cdot 10^{-16}\ \text{cm}^2/\text{W})$ for a silica fiber, n_{linear} is the linear (i.e., standard) index of refraction, P is the optical power, and A_{eff} is the mode cross-section. Although the nonlinear portion of the index is only 1 part in 10 billion, nonlinear refractive index effects can seriously affect high-speed long-distance optical systems.

Because the local refractive index is a function of the optical intensity of a propagating signal, a nonrectangular-shaped optical pulse will experience a varying

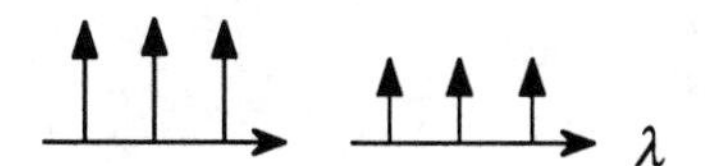

Figure 7.71 SBS causing a power loss to each channel when a backward traveling wave at each wavelength is generated.

refractive index depending on the optical power at that temporal location. As we know, the speed of an optical wave in a medium is dependent on the refractive index. Therefore, the varying intensity and varying refractive index pulse will cause the different intensities to propagate at different speeds along the fiber. Optical power fluctuations are converted to phase fluctuations. This group velocity phase shift causes the pulse to temporally disperse in the fiber and limits transmission distance and speed. This effect in a single optical pulse is known as *self-phase modulation* (SPM) and will distort a transmitted optical pulse, as shown in Figure 7.72. Both temporal and spectral broadening results from this nonlinearity.

This effect of intensity-dependent propagation speeds is quite important when we are considering WDM transmission of several channels. The optical power from data pulses on one channel will affect the refractive index, propagation speed, and dispersion/distortion of data pulses on another channel if they are temporally colocated. Since each wave is located at a different transmission wavelength, the different channels will propagate at different speeds and data pulses from one channel will propagate through the data pulses from another channel, causing overall smearing of the distortion and dispersion from one channel to the next. This cross-channel interference is known as *cross-phase modulation* (CPM). CPM is not symmetric as two pulses pass through each other the optical power is not constant along a fiber span given lumped amplification.

The general equation that describes several channels affecting channel 1 is [109]

$$\frac{dA_1}{dz} = i\gamma_1[|A_1|^2 + 2|A_2|^2 + \ldots + 2|A_n|^2]A_1 \tag{7.29}$$

where A_1 is the propagating optical field, the nonlinear coefficient γ_1 is defined by $2\pi n/\gamma_1 A_{eff}$ (where $n = 3.2 \cdot 10^{-16}$ cm$^2/W$ and is the nonlinear index coefficient, and A_{eff} is the effective fiber core area and equal to 50 μm.

Four-Wave Mixing (FWM)

We discussed in Chapter 5 how two waves will mix with each other in a nonlinear semiconductor medium, generating new waves at the sum and difference frequencies, known as four-wave mixing. Since the fiber is also a nonlinear medium, Figure 7.73 shows the situation in which two channels (or three, etc.) will mix with each other producing optical power at the sum and difference beat frequencies. An important

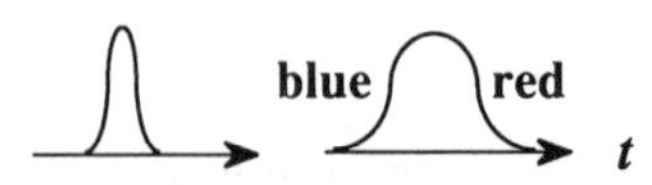

Figure 7.72 Dispersion-like temporal broadening of a single pulse due to SPM and CPM.

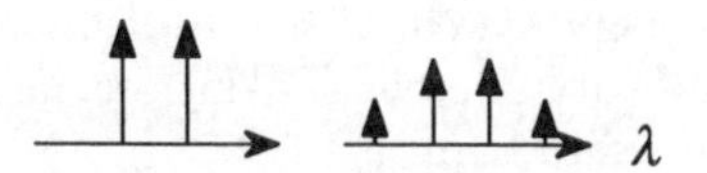

Figure 7.73 Four-wave mixing showing two original signals and the generated products.

feature is that FWM occurs only when the two mixing channels are located near the dispersion-zero wavelength of the fiber and the channel spacing is less than a few Å. These conditions exist since the two waves must maintain certain phase-matching requirements. The less critical effect of FWM is that the original channels suffer a power loss due to power transfer to the newly generated products. However, the more critical problem occurs when there are three or more original WDM signal channels producing many mixing products. It is quite probable that some of the products will appear at the same wavelength as an original signal. When this original signal is ultimately recovered by an optical filter, both the original signal and FWM product will be detected. The product has some modulation on it based on the combination of the channels that created it, and severe crosstalk will result. The equation describing the formation of FWM products between two waves A_1 and A_2 is: [110]

$$\frac{dA_1}{dz} = i\gamma_1 A_2^* A_3 A_4 e^{i\Delta kz}, \qquad \Delta k = k_3 + k_4 - k_1 - k_2 \tag{7.30}$$

where * denotes the complex conjugate of the wave, and k is the phase of the propagating wave.

Dispersion and the nonlinear effects of SRS, SPM, and CPM have been incorporated into one equation; FWM has been left out for simplicity. The propagation of the optical bits through the nonlinear fiber can be fully defined by solving the nonlinear Schrödinger equation for an individual channel given a total of four WDM channels and each subscript denoting channel number [105]:

$$\frac{\partial A_1}{\partial z} + \frac{1}{\nu_1}\frac{\partial A_1}{\partial t} + \frac{i}{2}\beta(\lambda_1)\frac{\partial A_1}{\partial t^2} + \frac{1}{2}\alpha A_1 = i\gamma_1(|A_1|^2 + 2|A_2|^2 + 2|A_3|^2 + 2|A_4|^2)A_1$$

$$+\left[\frac{g_{12}}{2}|A_2|^2 + \frac{g_{13}}{2}|A_3|^2 + \frac{g_{14}}{2}|A_4|^2\right]A_1 \tag{7.31}$$

where (for channel 1) λ_1 is the wavelength; A_1 is the propagating optical field; ν_1 is the group velocity of λ_1; $\beta(\lambda_1)$ is the linear group velocity dispersion given by $-\lambda_1^2 D/2\pi c$; α is the fiber loss coefficient, which we assume to be wavelength independent; the nonlinear coefficient λ_1 is the nonlinear coefficient; and g_{12}, g_3, g_{14} are the Raman gain coefficients between channel 1 and channels 2, 3, 4. It is instructive to mention the different contributions in (7.31). Starting from the left side of the equation, the six main terms represent:

1. The evolution of the optical field with distance of propagation in the fiber;
2. The change of the field with time;
3. The wavelength-dependent dispersion;
4. The wavelength-independent fiber loss;
5. The SPM and CPM of the optical field;
6. The Raman gain provided to a given channel.

7.3.6.2 Systems Limitations and Compensation Techniques

Once high-bandwidth optical amplification was adequately developed, dispersion and nonlinearities emerged as the major obstacles for implementing ultra-high-speed and ultra-long-distance WDM transmission systems. Let's first discuss the basic systems limitations due to nonlinear effects. These effects should be evaluated for higher power signals (>1 mW), longer fiber lengths (>10 km), and higher speeds (>5 Gbps). The theoretical limits are shown in Figure 7.74, and the actual system limits may be much more constraining. Figure assumes a 10-GHz channel spacing and operation near 1.55 μm using dispersion-shifted fiber. The reader is encouraged to see [104] for more detail.

We have painted a less than rosy picture when it comes to the effects of dispersion

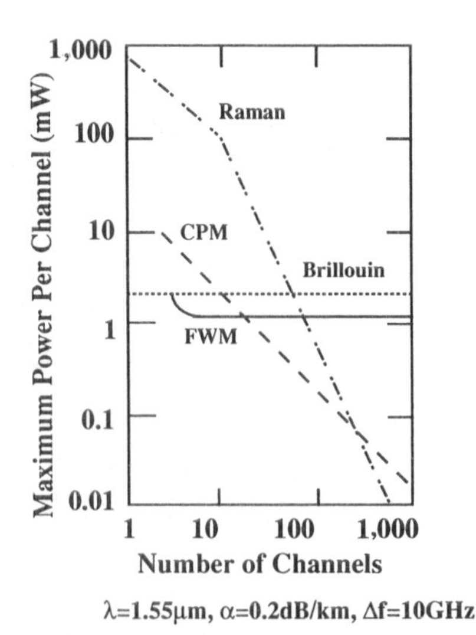

Figure 7.74 Limitations on optical power per channel due to nonlinearities in a WDM system. (*After:* [104]. © 1990 IEEE.)

and nonlinearities, especially when we consider that reducing dispersion in a system will increase the effects of CPM and FWM. However, there has been much recent research on minimizing all the effects we have discussed. Some clever compensation techniques are listed next. Some of these techniques are not mutually exclusive and may be used to complement each other.

Dispersion Management

Although fiber dispersion generally is considered a negative characteristic, it does have the desirable effect of inhibiting FWM and CPM. Therefore, a clever technique is to establish a long fiber link composed of alternating short lengths of two different types of fiber. One type of fiber has the opposite sign but a similar dispersion magnitude as that of the other type of fiber for a given wavelength; these two types of fiber are similar except that their dispersion-zero wavelength is shifted from each other. Each short length of fiber may have a dispersion value of, for instance, either +2 or −2 ps/(nm · km), depending on which type of fiber is used. After a given distance of propagation, the total effective dispersion is close to zero (Fig. 7.75), but there is an absolute value of dispersion at any given point along the fiber destroying the FWM and CPM. One astounding experimental result utilizing this method is the transmission of seventeen 20-Gbps channels along a ~175 km link [111]. This represents an aggregate system capacity of 340 Gbps, or 1/3 Tbps!

Mid-Span Spectral Inversion

Fiber dispersion will cause a temporal broadening based on the spectral content of a pulse, with the higher-frequency components of a pulse propagating faster than the lower-frequency components. To correct for dispersion effects, a phase conjugator can be placed at the mid-span point in a fiber link, forcing the faster higher-frequency components to now be in the trailing end of a pulse. The conjugator reverses the phase of a pulse, so that any additional dispersion will tend to temporally narrow the width of the pulse as it propagates. As shown in Figure 7.76, the sequence of events is as follows: (1) the pulse will disperse in the first half of the fiber link, (2) the phase

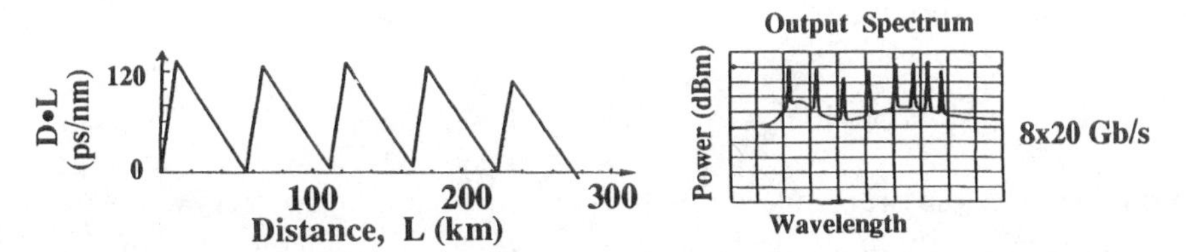

Figure 7.75 Dispersion management system in which different types of fiber are combined to ensure a zero total dispersion after a long link.

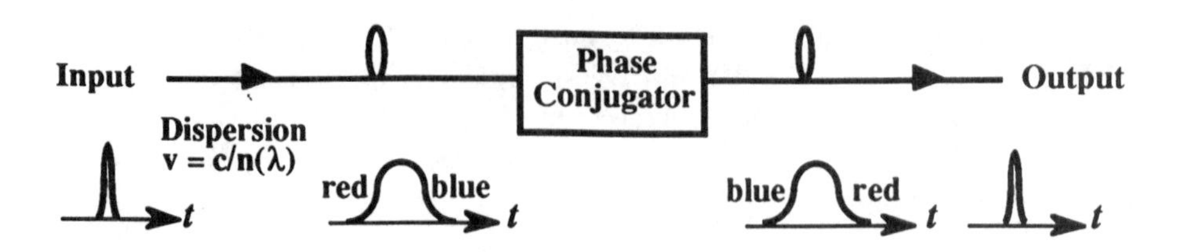

Figure 7.76 Mid-span spectral inversion using phase conjugation to compensate for dispersive broadening.

conjugator will reverse the spectrum of the pulse, and, finally, (3) the pulse will narrow in the second half of the link [112].

Unequal Channel Spacing

If the original signals are equally spaced in wavelength, many of the sum and difference mixing frequency products in FWM will appear at the wavelengths occupied by the original signals. However, if we use an algorithm to compute the wavelength location of the mixing products, the original channels can be spaced unequally so that all FWM products will not appear at any original signal's wavelength (Fig. 7.77) [113]. Power loss still occurs due to FWM, but interchannel cross-talk is now controllable. This technique cannot, however, accommodate WDM systems having many channels.

Pre-emphasis

Judiciously attenuating the longer-wavelength input channels will partially compensate for the Raman gain, which provides gain to the longer-wavelength channels at the expense of the shorter-wavelength channels.

Increased Channel Spacing

Increasing channel spacing will reduce the interchannel effects of FWM and SBS. Channel spacings on the order of 100 GHz would make FWM negligible, and channel spacings even as small as 1 GHz would inhibit interchannel SBS [114].

Decreased Channel Power

As was explicit from the results of Figure 7.74, reducing the channel power below some value will negate the nonlinear effects. Of course, a transmission link with optical amplifiers and many WDM channels may not be able to operate properly with very small channel powers.

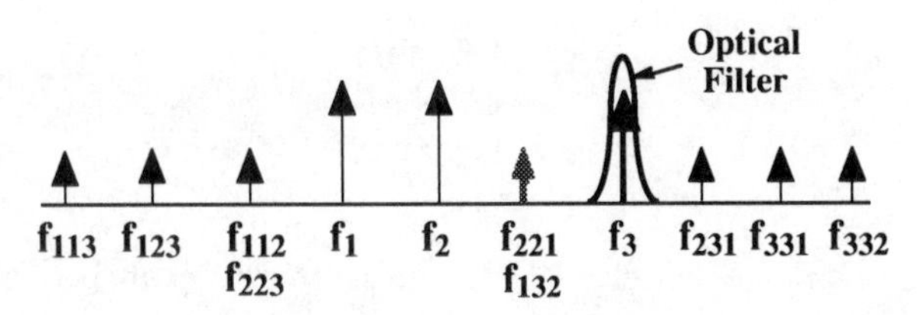

Figure 7.77 Original signals and FWM-generated products in a WDM system given unequal channel spacing.

Increased Channel Bandwidth

Since SBS is such a narrowband process, almost any method that broadens the channel bandwidth will significantly reduce the SBS efficiency within a single channel.

Example 7.3

A longer fiber link consists of alternate 25 km spans of two types of dispersion-shifted fiber, each type having a dispersion-zero wavelength of either 1550 or 1570 nm. The dispersion slope in both fibers is 0.075 ps/nm^2·km. A 10 Gbps externally-modulated in this fiber link. After what distance will this signal spread in time by 20%?

Solution

The signal wavelength of 1555 nm will experience D_1 and D_2 in two different DSF fibers where:

$$D_1 = 0.075 \times 5 = 0.375 \text{ ps/nm·km}$$
$$D_2 = -0.075 \times 15 = -1.125 \text{ ps/nm·km}$$

The average value of dispersion, D, which the signal experiences in 50 km of two links of dispersion-shifted fibers is given by:

$$D = \frac{D_1 + D_2}{2} = -0.375 \text{ ps/nm·km}$$

Without any additional laser-induced chirp, the pulse width of a 10 Gbps signal is 100 ps and the bandwidth (BW) is approximately 0.08 nm. Due to dispersion, the signall will spread 20% (i.e., spread $\Delta T = 20$ ps to a total of 120 ps) after propagating a fiber length L:

$$\Delta T = DL(BW)$$
$$\Rightarrow L = \frac{\Delta T}{D(BW)} = 666.67 \text{ km}$$

This corresponds to, approximately, 13 total links consisting of six of fiber type 1 and seven of fiber type 2.

7.3.7 WDM Transmission Capacity

We have detailed many of the specifics of transmitting and recovering WDM channels. It is extremely educational to show a figure that encapsulates much of the potency of using WDM. Figure 7.78 shows the 1984–1994 time-line transmission capacity records of various experimental systems, including electrical TDM, optical TDM, and WDM [115]. The highest demonstrated capacity in a given time frame has been achieved using WDM, with the most recent achievement being the 0.33-THz transmission mentioned in Section 7.3.6.2. TDM systems have heroically demonstrated 100-Gbps with one channel [116], but that was with using complex multiplexing and demultiplexing schemes, and significant challenges remain toward achieving much further progress using TDM. One interesting point predicted a decade ago by T. Li of AT&T Bell Labs should be mentioned. The trend, which has existed for more than a decade, is that the transmission capacity doubles every two years. WDM has kept the

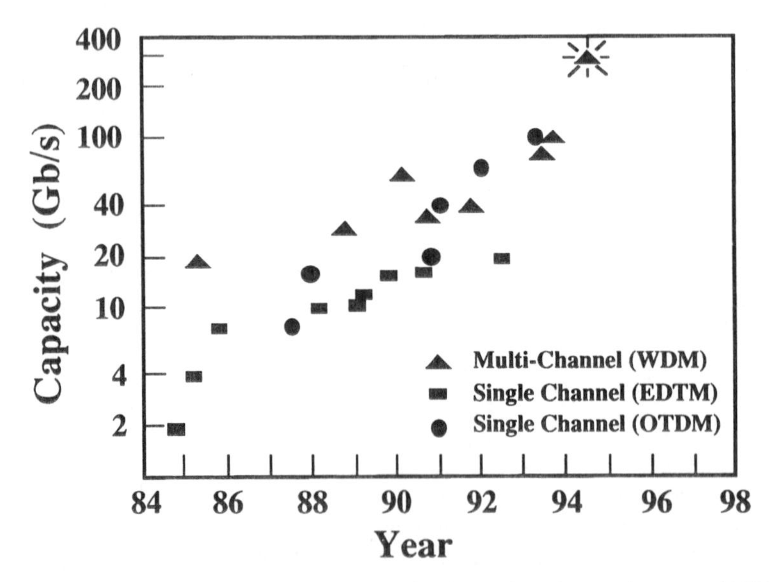

Figure 7.78 Transmission capacity records versus year for electrical TDM, optical TDM, and WDM systems.

capacity line still on this slope, with no reason to assume that WDM will not continue to produce dramatic progress.

7.3.8 Wavelength Shifting and Wavelength Reuse

In an ideal WDM network interconnecting many users, each user would have its own unique signature wavelength. Routing in such a network would be fairly straightforward. This situation may be possible in a small network, but it is extremely unlikely in a large network accommodating many users. For example, technology cannot provide 10,000 different wavelengths for a network of 10,000 users. In fact, 20 distinct wavelengths are the state of the art in relatively reproducible distinct wavelengths. Some technological limitations in providing a large number of wavelengths include the following:

- Channels must have a minimum wavelength spacing, due to channel-broadening effects and nonideal optical filtering. This minimum spacing is 1–5 Å.
- The EDFA, which will probably be required in a large network, has only a limited gain bandwidth of ~25 nm in which to place all the wavelengths.
- Wavelength range, accuracy, and stability are extremely difficult to control in either tunable lasers or multiple-wavelength arrays, thereby necessitating larger interchannel wavelength spacing over a finite wavelength range.

Therefore, it is quite possible that a given network may have more users than wavelengths, which will necessitate the reuse of a given set of wavelengths at different points in the network. We now will discuss the main issues involved with passive and active wavelength routing.

7.3.8.1. Passive Wavelength Routing

If a limited number of wavelengths are available, a network can take advantage of passive routing of a signal through the network based solely on its wavelength. The routing can be designed so the given set of wavelengths will have a different destination depending on the origin of the signal. Figure 7.79 shows this concept of passive wavelength routing enabling the reuse of a limited set of wavelengths in various portions of a large network [117]. In the figure, there are five users, three wavelengths, and two passive WDM cross-connects. User I can use wavelength λ_1 to establish a link with user II, while simultaneously user V can reuse the same wavelength, λ_1, to establish a connection with user III. This functionality is enabled by the proper arrangement of the two cross-connects that route an input signal to a wavelength-determined output.

A simple 2-by-2 example of the functionality of a passive WDM cross-connect is shown in Figure 7.80 for two wavelengths (λ_1 and λ_2) and two possible input ports

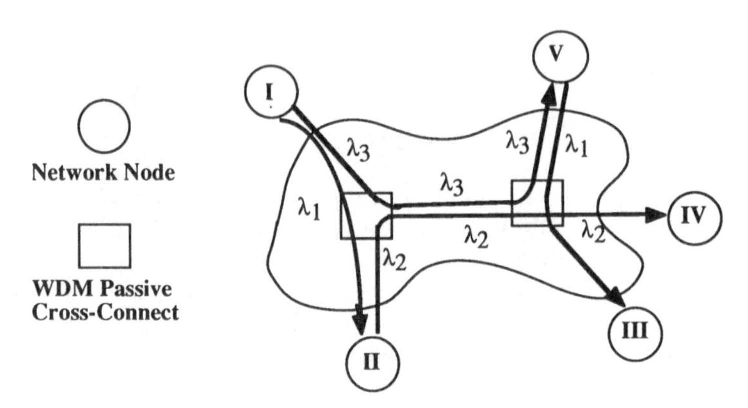

Figure 7.79 Passive wavelength routing in a network utilizing wavelength reuse.

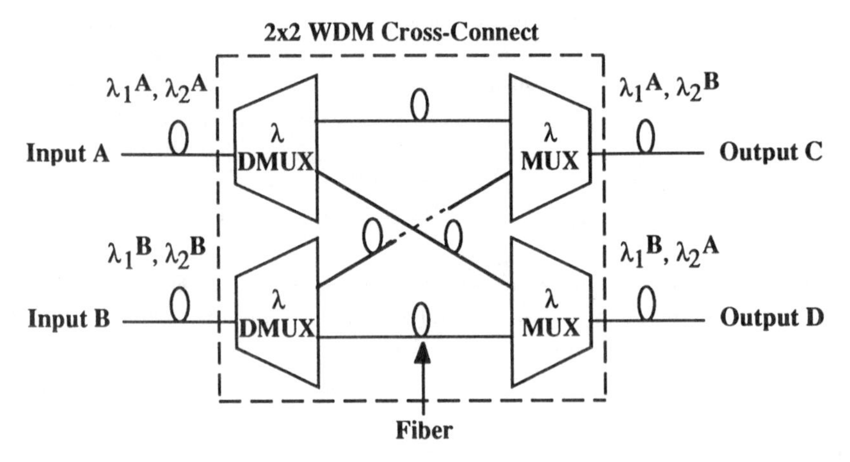

Figure 7.80 A 2-by-2 wavelength cross-connect in which output-port routing is determined both by the specific input wavelength and by the specific input port.

(A and B). The cross-connect is composed of an input stage of wavelength demultiplexers, an output stage of wavelength multiplexers, and fibers interconnecting the two stages. Instead of each wavelength having a specific output port associated with it, the wavelength output-port routing depends on both the signal wavelength and the specific input port. Although there are only two wavelengths, there are four possible noninterfering routing paths based on both wavelength and origin. Instead of N wavelengths and N possible connection paths, now there are N wavelengths and N^2 connections. The same wavelength could be reused by any of the input ports to access a completely different output port and establish an additional connection. This allows for increased capacity in a WDM network.

The functionality of a WDM cross-connect is the same as the functionality of the wavelength multiplexers discussed in Section 7.3.3.5. Simple wavelength multiplexers have a 1-by-N configuration, and the integrated frequency router has an N-by-N configuration, which enables wider flexibility in passive wavelength routing. The integrated frequency router serves the purpose of a passive N-by-N WDM cross-connect.

7.3.8.2. Active Wavelength Shifting

Passive routing is limited to static network conditions. However, active wavelength shifting must take place from wavelength i to wavelength j if a network of limited wavelengths is dynamic, that is, routing will change depending on the available links and wavelengths. This concept of a network requiring active wavelength shifting is illustrated in Figure 7.81, in which each of two small LANs is connected to a larger WAN, and each small network can transmit on only two available wavelengths (λ_a and λ_b). Node I in the left LAN wishes to communicate with node II in the right LAN. When node I wishes to transmit, the only wavelength available is λ_a. However, when the signal reaches the right LAN, it is revealed that λ_a is already being used by the right LAN. Therefore, the only way for the signal to reach node II is to be actively switched onto the available λ_b. (Appropriate header reading must be performed at the switching node to determine the intended routing.)

An additional scenario that would require active wavelength switching is where one set of wavelengths are reused exclusively by each LAN (i.e., locally), whereas another set of wavelengths is used exclusively for global communications between LANs (Fig. 7.82). Those wavelengths set aside for intra-LAN communication can be reused by each LAN since it will not interfere with another LAN. A signal originating

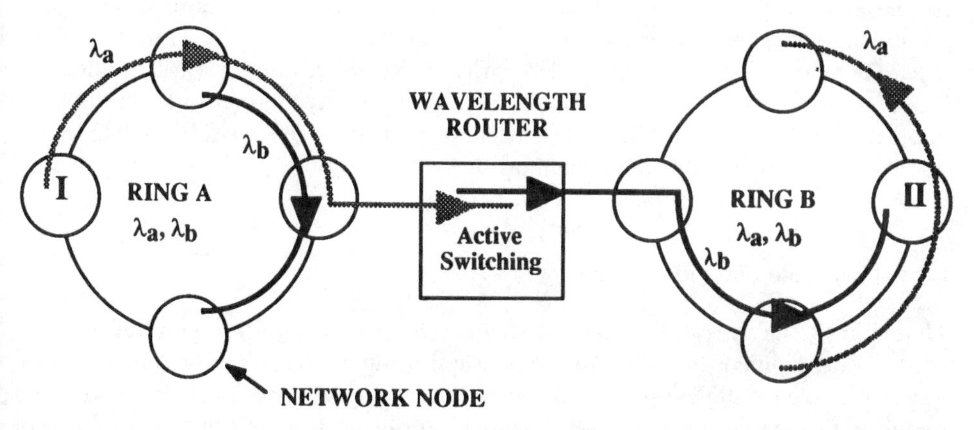

Figure 7.81 Active wavelength switching in a dynamic WAN in which two smaller LANs can transmit on only a limited set of wavelengths.

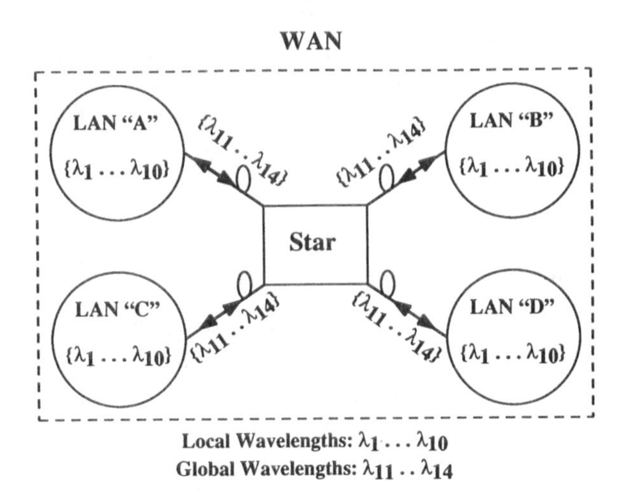

Figure 7.82 A set of local wavelengths (λ_{local}) are reused by each LAN, and a set of global wavelengths (λ_{global}) are used to interconnect the LANs.

at an individual node from one LAN would require (1) initial transmission on a λ_{local}, (2) active switching to a λ_{global}, and (3) active switching back to a λ_{local} to arrive at the appropriate destination node in another LAN [118].

One possible brute-force solution for facilitating active wavelength switching is to employ optoelectronic wavelength shifters, in which a signal on wavelength i is detected and retransmitted on wavelength j [117]. This method, shown in Figure 7.83, necessitates optoelectronic conversions and will cause an eventual optoelectronic speed bottleneck in much the same manner as a regenerator does in long-distance systems. A goal is to achieve all-optical active wavelength shifting to retain a high-speed all-optical transparent data path.

There are several methods for all-optical wavelength shifting. Each method has its advantages and disadvantages, and it is not clear if any method will eventually be implemented. There is certainly room for more research in this area. We will limit our discussion to four possible methods. For illustrative purposes, we will have a slightly longer discussion of the first method.

Cross-Gain Saturation in an SOA

The rapid (<1 ns) carrier lifetime of an SOA (semiconductor optical amplifier) can be used to shift a signal's wavelength. The basic shifting mechanism relies on an intense signal, causing significant SOA gain saturation. If a high-power intensity-modulated signal at λ_1 having power $P^{in}_{\lambda_1}(t)$ is coupled into the SOA, the SOA gain (G) will decrease rapidly when the input power is high (a 1 bit) and will increase rapidly to

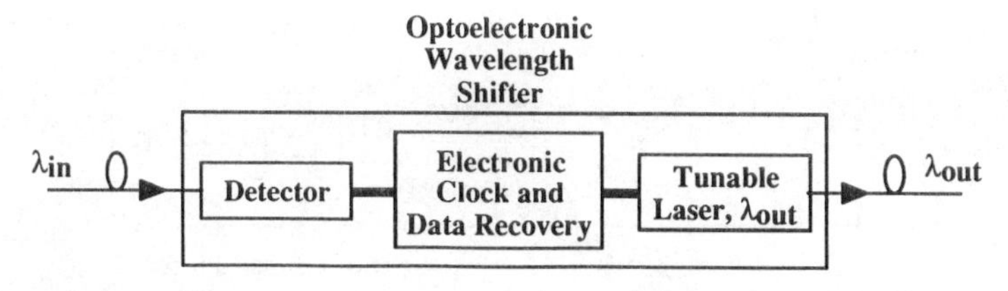

Figure 7.83 Active wavelength shifting by detection and retransmission.

its original value when the input power is low (a 0 bit) (Fig. 7.84). This relationship results in the SOA gain, $G(t)$, acquiring the inverse modulation of $P^{in}_{\lambda_1}(t)$:

$$G(t) = f[\overline{P^{in}_{\lambda_1}(t)}] \tag{7.32}$$

where f denotes some functional dependence. If this high-power intensity-modulated signal at λ_1 and a weak cw signal at λ_2 having power $P^{in}_{\lambda_2}(cw)$ are simultaneously coupled into the SOA, then the time-dependent output at λ_2 will be equal to $P^{in}_{\lambda_2}(cw)$ multiplied by the modulated gain, $G(t)$:

$$P^{out}_{\lambda_2}(t) = G(t) \times P^{in}_{\lambda_2}(cw) = f' [\overline{P^{in}_{\lambda_1}(t)}] \tag{7.33}$$

Therefore, $P^{out}_{\lambda_2}(t)$ also attains the inverse modulation of $P^{in}_{\lambda_1}(t)$. Figure 7.85 illustrates the mechanism through which an incoming signal modulation is effectively copied, inverted, and shifted from λ_1 to λ_2. This wavelength shifter has been operated at bit rates up to 20 Gbps and over a wavelength range up to 19 nm [119–121].

Figures of merit for the wavelength-shifted signal are a high contrast ratio and a high SNR. The contrast ratio is directly determined by the ratio between the gain for a 1 bit divided by the saturated gain for a 0 bit. Since the gain has a given profile, and the gain decreases and shifts to longer wavelengths under saturation, the optimal operational conditions can be analyzed. Following are some guidelines for optimal system operation [122]

- λ_1 should be located at longer wavelengths than the initial SOA gain peak. This design condition is because the gain peak shifts to longer wavelengths under saturation, and it is at the gain peak that λ_1 can achieve the highest gain and induce the most saturation.
- λ_2 should be located at shorter wavelengths than the initial SOA gain peak. This condition for using a shorter wavelength for λ_2 is because the gain peak is asymmetrical and shifts to longer wavelengths, thereby causing the contrast ratio to be higher at shorter wavelengths.

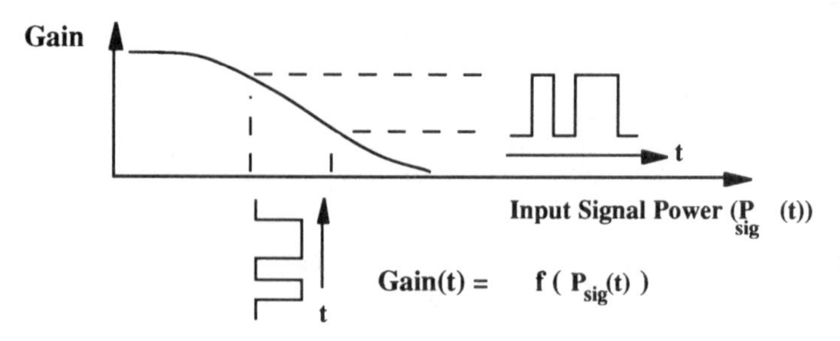

Figure 7.84 Gain saturation of an SOA versus input signal power.

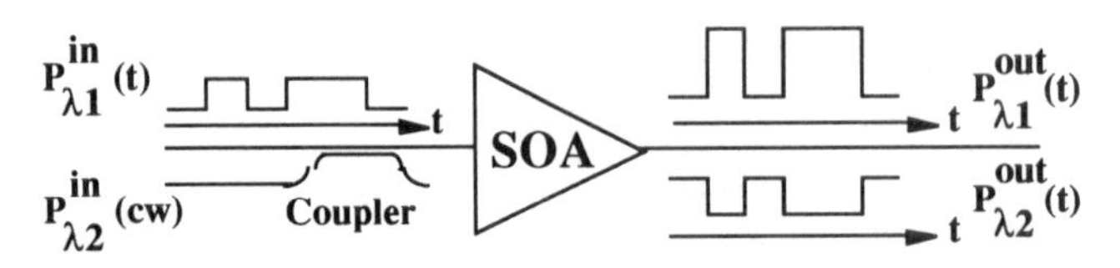

Figure 7.85 Block diagram of a wavelength shifter based on cross-gain compression in an SOA.

- The power in λ_1 should be high to induce more saturation and achieve a higher contrast ratio.
- The power in λ_2 should not be too high, because then the second wavelength would contribute (in an unmodulated way) to the gain saturation. Gain saturation would occur without contributing to wavelength shifting.

Michelson Interferometer

An interferometer operates on the principle of splitting an optical wave into two halves, having these two waves propagate inside separate and parallel interferometer arms, and then recombining them under either constructive or destructive interference conditions. Figure 7.86 shows a Michelson interferometer operated as an all-optical wavelength shifter [123]. An unmodulated signal at λ_1 is input into the interferometer and split into two halves, with the two waves denoted as A and B. Each wave propagates through a different SOA and is then reflected at the semiconductor facet interface. Those two backward-propagating waves will recombine at the same original splitter, producing an interference between A and B. If the two arms of the interferometer are equal, then constructive interference occurs, and the addition of (A + B) will appear at the lower left output as a maximum. However, if the effective propagation distance (i.e., phase delay) of each arm of the interferometer is different, then the output at the lower left output of the interferometer will decrease. In fact, destructive interference occurs, and a minimum of (A + B) is reached when the phase delay between the two arms is an odd multiple of half-wavelengths. This relative phase

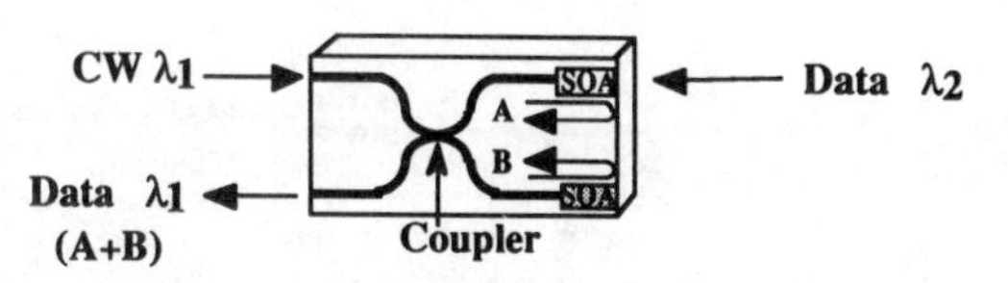

Figure 7.86 A Michelson interferometer operated as a wavelength shifter.

delay can be induced between the two interferometer arms by inputting from the right of the device a second input signal on λ_2. This second signal will be absorbed and will affect the properties of the SOA, changing its refractive index and therefore its effective length and phase delay. When this second signal on λ_2 is low, the lower left output on λ_1 is high, since there is not relative phase delay between the two arms. When λ_2 is high, the output on λ_1 is low based on destructive interference. Therefore, the modulated signal on λ_2 has effectively been copied onto λ_1 by changing the interference in an interferometer.

Distributed Bragg Reflector Laser With a Saturable Absorber

A multiple-electrode wavelength-tunable DBR laser also can be used as a wavelength-tunable wavelength shifter. Figure 7.87 shows a tunable DBR laser that has an intracavity saturable absorber [124]. Even when the DBR laser is biased with an injected current, it still will not lase, because the saturable absorber is a very lossy element preventing the gain from becoming higher than the loss. However, we can "bleach" the saturable absorber, making the absorber into a relatively neutral element. This bleaching can be produced by inputting into the cavity an external optical signal and having the incoming photons absorbed in this saturable region. When the input from this external signal is high (i.e., 1), the DBR output will also be high, since the DBR will begin to lase. When the external signal is low (i.e., 0), the DBR output is also low, since the cavity will not lase. This functionality can produce a wavelength shifter. If the input external signal is modulated and on λ_1 and the DBR cavity will naturally lase at λ_2, then the modulation signal on λ_1 is effectively copied onto λ_2. Furthermore, since the DBR is wavelength tunable, the wavelength shifting is also tunable over the tunability range of the DBR.

Four-Wave Mixing

We discussed in Chapter 5 the phenomenon of four-wave mixing in an SOA, in which two wavelengths input to an SOA will produce two additional mixing products at the sum and difference frequencies. Figure 7.88 shows two wavelengths input to an SOA, one input containing a modulated signal at $\lambda_1(t)$ and a second local unmodulated mixing signal at λ_2. These two signals will produce products at $\lambda_3(t) = [2\lambda_1(t) - \lambda_2]$ and at $\lambda_4(t) = [2\lambda_2 - \lambda_1(t)]$. As is evident, the signal modulation has been copied onto

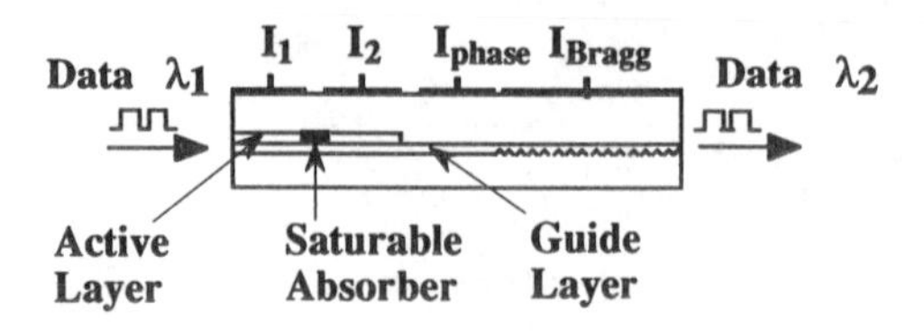

Figure 7.87 A wavelength-tunable DBR laser with an intracavity saturable absorber acting as a wavelength shifter.

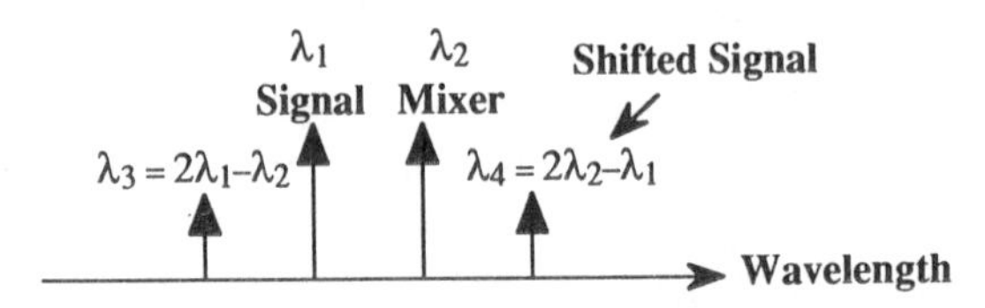

Figure 7.88 SOA four-wave mixing used for shifting a modulated signal from one wavelength to another.

the products, which appear at different wavelengths [125, 126]. The efficiency of the modulation copying is not high, because four-wave mixing is not itself an extremely efficient process.

Example 7.4

A given pump signal at λ_1 is shifted to two possible probe wavelengths of λ_2 or λ_3 using SOA cross gain compression. The "1" and "0" pump levels have the powers of 8 and 2 dBm, respectively. Gain compression for the λ_2 (1565 nm) probe is 3 and 10 dB at 2 and 8 dBm pump power levels, and gain compression for λ_3 (1530 nm) is 2 and 7 dB at 2 and 8 dBm pump power levels. The small signal gain (G) for signals at λ_2 and λ_3 is the same, and n_{sp} at λ_2 and λ_3 is 4 and 2 respectively. The optical bandwidth B_0 is 1 nm. Compare the optical SNR for both cases assuming that the output is signal-spontaneous-beat-noise limited.

Solution

Assuming that the input probe powers are both the same, the *relative* signal power in the two output probes is the gain in the "1"s minus the power in the "0"s:

$$P_{S(\lambda_2)} = 10 - 3 = 7\text{dB}$$

$$P_S(\lambda_3) = 7 - 2 = 5\text{d}$$

Since the output is s-sp noise limited, the ratio of SNR_2 and SNR_3 is given by:

$$\frac{SNR_2}{SNR_3} = \frac{P_s(\lambda_2)/P_n(\lambda_2)}{P_s(\lambda_3)/P_n(\lambda_3)}$$

where the noise power is $P_n = n_{sp}\ (G - 1)\ h\nu B_0$.

By inserting the values in the above equation we find that

$$\frac{SNR_2}{SNR_3} = \frac{1.75}{2.5} = 0.7$$

The contrast ratio for λ_2 shifting is more than for λ_3 shifting, but the $NF(=2n_{sp})$ is higher in for λ_2 shifting. Therefore, the resulting SNR is better for the case of λ_3 shifting. This result is, of course, very dependent on the specific system being considered.

7.4 SUBCARRIER MULTIPLEXING

This chapter has dealt entirely with digital optical communications due to the potential for very high-speed communication per channel. However, analog optical communications has some very appealing characteristics that make it a possible solution for some near-term (and perhaps long-term) general communications problems. These include CATV, video distribution, LANs of >>1 Gbps/channel and personal communication system interfaces. We will describe some of the basic elements of analog scm and its application to multichannel systems. (WDM is similar to SCM in that each channel is located at a different frequency.)

7.4.1 Modulation and Demodulation

The essence of SCM is to take all the (de)modulating, (de)multiplexing, and routing functions, which could be performed optically, and instead perform them electrically [127]. The only optical functions that remain are (1) optical generation using a laser, (2) optical transmission over an optical fiber, (3) optical detection using a photodetector, and (4) some passive optical coupling and splitting. The advantages of performing these functions electrically is that, under current circumstances, electrical components are cheaper and more reliable than optical components, and electrical filters can be of an efficient and near-ideal multipole design, whereas optical filters are only single pole.

Figure 7.89 shows a single-channel SCM system. In digital transmission, the optical beam is a carrier wave directly modulated by the data. We now remove the data one step further from the carrier wave by using a ~gigahertz subcarrier electrical

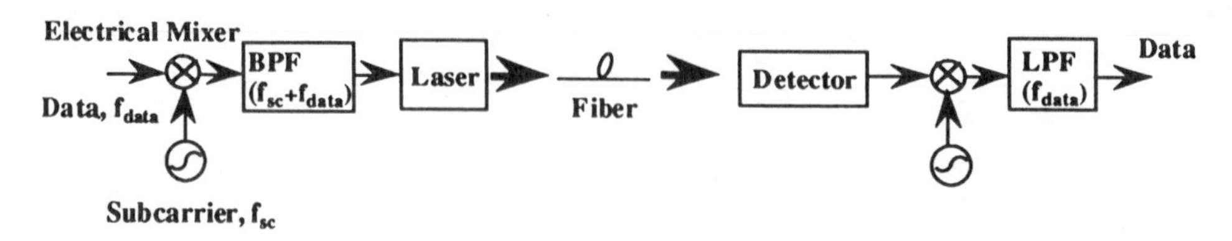

Figure 7.89 Basic diagram of SCM.

wave, which is modulated by the data, and which, in turn, modulates the ~terahertz optical carrier wave. The ~megahertz data are electrically mixed with the electrical subcarrier, producing sum and difference frequencies as results in standard heterodyning; an electrical bandpass filter is used to allow only one product, typically the sum frequency to pass and be transmitted. The resultant modulation directly modulates either the current on a laser or the optical transmission function of an external modulator. The electrical subcarrier is a simple sinusoidal wave that has an amplitude frequency (typically one hundreds of megahertz or gigahertz), and phase, in which the optical wave, P, can be roughly described by [128]

$$P = P_c[1 + Am_{data} \cos(2\pi f_{sc}t + \varphi_{SC})] \tag{7.34}$$

where f_{sc} is the frequency of the subcarrier, A is the amplitude, m_{data} is the data at a modulation rate of data, modulation index, and P_c is the unmodulated optical output. The subcarrier will appear as a modulated tone in the electrical frequency spectrum before optical transmission. After transmission, the optical wave is detected at the receiving end. The optical carrier wave is far above the detector bandwidth, but the electrical subcarrier will fall within the detector bandwidth and be detected. We employ the same principle as in the transmitter to demultiplex the data from the subcarrier. At the receiver, the same subcarrier wave frequency of f_{sc} is mixed with the recovered transmitted signal, again producing sum and difference frequencies. The sum frequency is $(2f_{sc} + f_{data})$ and the difference frequency is simply f_{data}. A lowpass filter is then used to allow only the difference frequency to pass, allowing the data to be unambiguously recovered. This method allows all the multiplexing and demultiplexing to occur in the electrical domain, and efficient electrical filters can be incorporated.

One of the key reasons for using analog optical transmission is that it is compatible with much of the analog transmission used today in video transmission of CATV signals [129]. The modulation format in CATV is *amplitude-modulation vestigial-sideband* (AM-VSB), in which the subcarrier is amplitude modulated, and the AM sidebands created by the modulation are manipulated so as to occupy a minimum bandwidth without sacrificing transmission quality [130]. Due to the compatibility issue, we will discuss only AM, although FM is also quite possible and requires a smaller signal power and larger bandwidth [131].

The key performance criterion is to achieve a high *carrier-to-noise ratio* (CNR), which is comparable to the SNR in digital communications. In a typical system, the receiver still contains random shot noise and thermal noise, which are generated in the photodetector. In addition, the laser itself produces RIN (relative intensity noise) over a wide bandwidth [132]. RIN is due to the random phase fluctuations in the laser light (i.e., stimulated emission), which are converted into amplitude fluctuations by reflections back into the laser or by multiple reflections along the transmission path; it is critically important to keep reflections low (<-65 dB) in an analog system by using high-quality connectors, splices, and isolators. Even fiber dispersion will have the effect of increasing the RIN. The RIN can be described by the statistical fluctuations in the photogenerated current [132];

$$RIN = \frac{\langle i_{ph}^2 \rangle}{\langle i_{ph} \rangle^2} \tag{7.35}$$

and the RIN-generated noise power, σ_{RIN}^2, is

$$\sigma_{RIN}^2 = (RIN)\overline{P}\, B_e \tag{7.36}$$

where $\overline{P}$ is the average received optical power, and B_e is the electrical bandwidth of the receiver. RIN is typically a very small value, but AM communications requires an extremely high CNR for good signal fidelity, requiring that the noise be kept to a minimum and that the optical power be kept high. Since analog transmission requires a very high average optical power and since the RIN noise power is proportional to $\overline{P}^4$, RIN usually becomes the dominating noise factor. The entire equation for the CNR for AM optical transmission is [133]:

$$CNR = \frac{\dfrac{(m\overline{P})^2}{2}}{\sigma_{sh}^2 + \sigma_{th}^2 + \sigma_{RIN}^2} \tag{7.37}$$

To achieve high fidelity and "error-free" signal recovery, the CNR typically must be >50 dB, much higher than the ~20 dB required in direct-detection ASK digital systems!

When an external modulator is not used, the laser acts to directly convert the electrical signal into an optical signal, with a modulation of the laser bias current producing a direct modulation of the laser light output. Any deviation in the linearity of the laser light as a function of the bias current will produce a decrease in the CNR, since the electrical modulation is not exactly replicated by the optical signal output [134]. Since the CNR must be extremely high in an analog system, we therefore require that the laser light output be extremely linear with bias current.

There are several limitations in SCM systems, and we will mention just the issue of "clipping." In general for a communications channel, it is desirable to produce as high a bias swing (i.e., light output swing) in the laser as possible to produce a high modulation index in the carrier power. Moreover, a large bias swing is needed to support many channels since each channel must individually contribute to a minimum modulation swing. However, there is a limit to the bias swing that can be supported by the laser since (1) the laser will produce no light when the bias current is below the laser threshold current, and the laser linearity will degrade above a certain bias current [135]. If the electrical modulation falls below the threshold current, then the SCM signal is "clipped," and the CNR will not be adequate for signal recovery (see Fig. 7.90). If the modulation is above a certain current, nonlinearities will destroy the signal fidelity and produce intermodulation products and distortions in the presence of other channels.

7.4.2. Multichannel Frequency-Division Multiplexing

We have explained SCM in terms of single-channel transmission. However, the powerful advantage of SCM is the transmittal of many channels simultaneously on one laser, with each channel transmitted on its own subcarrier frequency [136]. This is known as *frequency-division multiplexing* (FDM) and is, conceptually, analogous to WDM systems. Figure 7.91(a) shows a schematic of a multichannel SCM system, with each mixed with a different subcarrier frequency and collectively sharing the same laser transmitter. Each transmitted channel has its own electrical subcarrier generator, and the receiver can have a fixed or tunable oscillator to recover one of the

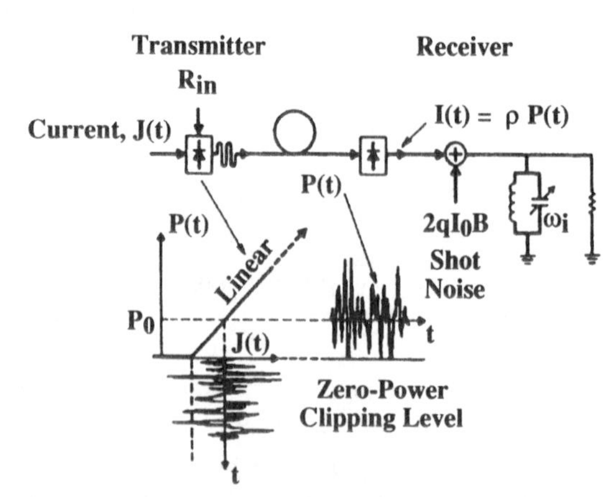

Figure 7.90 Clipping of an analog SCM signal due to the bias-current modulation appearing below the threshold current. (*After:* [136]. © 1989 IEEE.)

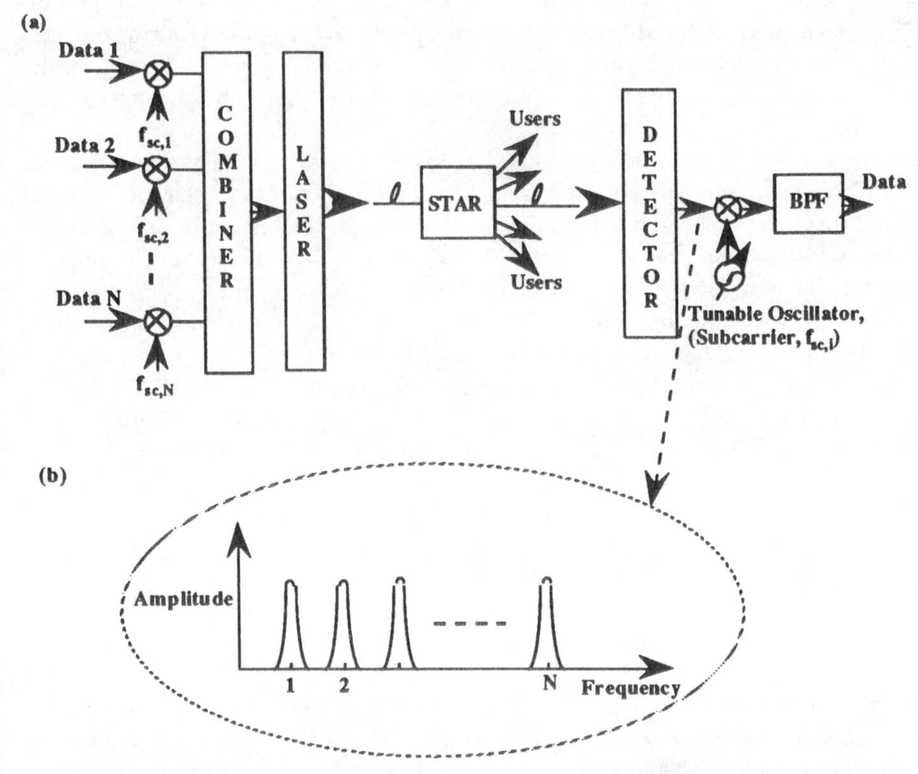

Figure 7.91 (a) FDM optical SCM system and (b) the electrical spectrum of the transmitted FDM system.

subcarrier frequencies, that is, to selectively demultiplex the many channels. Shown in Figure 7.91(b) is a rough representation of the electrical frequency spectrum of the multichannel transmission. One advantage of this SCM scheme is that many channels share the cost of the same expensive optical hardware. One potential disadvantage is that the channels sharing the laser must all be in close proximity.

Another major disadvantage of SCM is that there is a significant limitation on the transmission speed of the individual channels and on the number of channels due to the finite bandwidth of the laser transmitter and receiver. Reasonably priced optoelectronic devices do not, at present, exceed several gigahertz in bandwidth. If we assume that each channel occupies roughly its data rate and we separate the channels by twice the data rate to allow for filtering (and neglecting the intricacies of nonlinear distortions), we have the following rough approximation for the total number of channels, N:

$$N \approx \frac{B_e}{2B_{ch}} \tag{7.38}$$

where B_e is the electrical bandwidth of the system, and B_{ch} is the information bandwidth of each channel. It is difficult to have many 100-MHz channels sharing one laser, and gigahertz-per-channel transmission is not considered practical. However, transmitting many video channels (with bandwidths of a few megahertz each) is quite reasonable.

Since a real system is not perfectly linear in converting the electrical signal into an optical signal, another major limitations in the number of total channels is due to two and three channels mixing together nonlinearly, producing sum- and difference-frequency *intermodulation distortion* (IMD) products [137]. If the optical output as a function of bias current is not exactly linear, then these IMD products will be produced. If we assume three channels for simplicity, then the second-order nonlinear products appear at ($f_1 \pm f_2$, $f_1 \pm f_3$, and $f_2 \pm f_3$) and the third-order products appear at ($f_1 \pm f_2 \pm f_3$), where the frequencies represent the subcarrier frequencies of the individual channels. Given several original channels, these products may appear at the same frequency as an original channel, decreasing the CNR as follows [138].

$$CNR = \frac{(m\overline{P})^2}{\sigma_{sh}^2 + \sigma_{th}^2 + \sigma_{RIN}^2 + \sigma_{IMD}^2} \tag{7.39}$$

where σ_{IMD}^2 is the sum of all the IMD products contributing to system noise within the bandwidth of a single channel. The sum of all the second-order products and the sum of all the third-order products are called, respectively, the *composite second order* (CSO) and the *composite triple beat* (CTB). The effect of CSO and CTB is to reduce the number of channels that can be accommodated by an analog system.

It is quite likely that analog systems will incorporate EDFAs to compensate for losses. As discussed in Section 7.3.5.3, the EDFA gain spectrum is not uniform with wavelength. If a directly modulated laser produces a chirp of several gigahertz of the output optical spectrum, then the nonuniform EDFA gain will result in a nonlinear gain (i.e., transfer function) for the different frequency portions of the signal [139]. Higher CSO and CTB will result. There are several methods to reduce this problem, but the most straightforward solution is to use a low-chirp external modulator. Using an external modulator also helps to reduce the adverse affects on CNR due to simple fiber dispersion.

Probably the most obvious application of analog SCM and FDM is for distribution of CATV channels. Several channels can be transmitted on a single laser since the signals all are colocated at the CATV head end. Furthermore, analog optical transmission of AM-VSB modulation format is completely compatible with the present-day CATV format. Figure 7.92 shows a CATV system in which fiber is used as a backbone, transmitting the channels to a user's local area by optical means and then transmitting the channels for the final short distance along a conventional coaxial-cable electrical system. The optical fiber is high bandwidth and low loss,

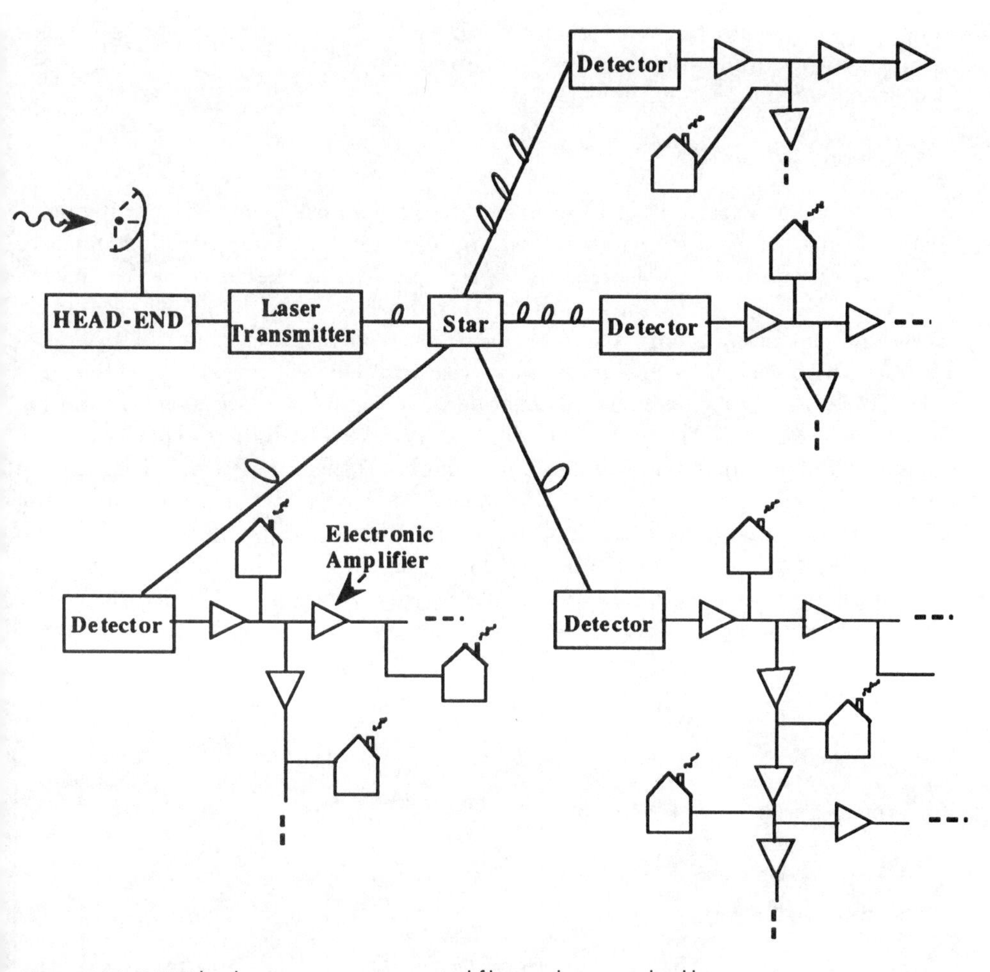

Figure 7.92 CATV distribution system using optical fiber as the system backbone.

significantly reducing the number of unreliable cascaded electrical amplifiers that must be traversed between head end and user. Today's CATV systems use optical transmission. Future ultra-high-capacity CATV systems may use more than one laser and also use EDFAs. One experiment demonstrated the potential distribution of 16 different wavelengths to 40 million users using EDFAs to compensate for the optical splitting losses [87].

Two important points should be mentioned. First, there is an anticipated explosion in wireless *personal communication systems* (PCS). Such a system is expected to be analog based, and it is possible that analog SCM optical transmission will be used to seamlessly connect the wireless system along a fiber-based system. Second, analog subcarriers can be used in conjuction with digital baseband transmission. In fact, one optical network application is to use subcarrier signals to help control the network traffic and communication paths [140].

7.4.3 Hybrid WDM/SCM

SCM can enable a single laser to transmit 20–50 channels, depending on the modulation format. If a system wants to transmit more than this number of channels or if the channels' origins are geographically separated by more than a few kilometers, it may be advantageous to transmit the SCM channels on more than one laser. Future ultra-high-capacity systems may use many lasers to distribute hundreds of SCM channels, with each laser at a different wavelength. One encouraging experiment is shown in Figure 7.93, in which four separate wavelength lasers are each transmitting 50 AM-modulated SCM channels for a total system distribution capacity of 200 channels [141]. Optical filters are used to demultiplex the different wavelengths, and electrical filters are used to demultiplex each single laser's 50 channels. One feature of this system is that it used 10 km of standard single-mode fiber, making this system compatible with the existing embedded fiber base in the United States. This was accomplished using 1.55-μm lasers and an EDFA power booster. This hybrid WDM/SCM system combines some of the best qualities of both types of multiplexing schemes,

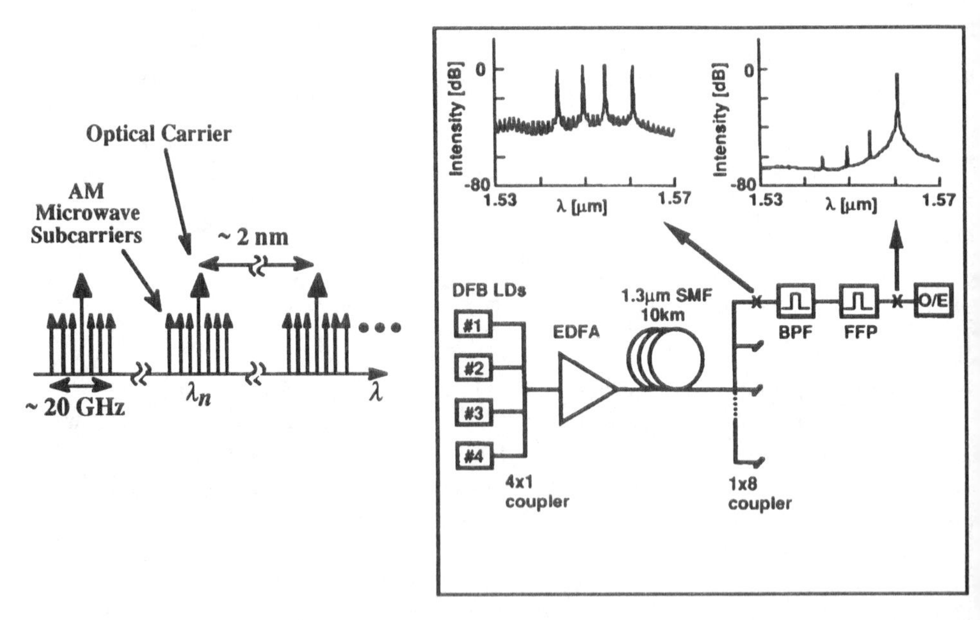

Figure 7.93 A hybrid WDM/SCM distribution system in which four lasers each transmit 50 AM-SCM channels.

allowing the modulation and demodulation to occur in the electrical domain, compatibility with existing CATV transmission, and the transmission of hundreds of channels, thereby utilizing a larger part of the fiber bandwidth.

7.5 CODE-DIVISION MULTIPLEXING (CDM)

CDM, a method of encoding that predates optical communications, establishes the unique communications link by encoding the destination inside of each bit time [142,143]. Figure 7.94(a) shows how a single bit time is divided into M chip slots. Each user is assigned a specific code of 1s and 0s within the chip times. N users would require N codes. Each transmitter must encode the correct sequence within each bit time for a given destination. Each destination has a receiver that will decode the intended data with the **inverse** of the transmitter coding hardware but will not decode data intended for another destination. (A fixed transmitter code and a tunable receiver code is another alternative.) If a decoder has the correct inverse hardware as the transmitter, the output will produce a peak in the optical signal called the autocorrelation. This autocorrelation signal will trigger a detected 1 bit in the receiver if it is above a certain threshold level. If the receiver does not have the correct decoder, the cross-correlation signal, which appears as a noisy temporally-broad background signal will contain all the energy and the autocorrelation will be small, producing only 0 bits. A straightforward method of encoding and decoding a chip sequence is to use a parallel array of different-length fiber delay lines; see Figure 7.94(b). Two desirable features of this method are that several different signals can be simultaneously transmitted on

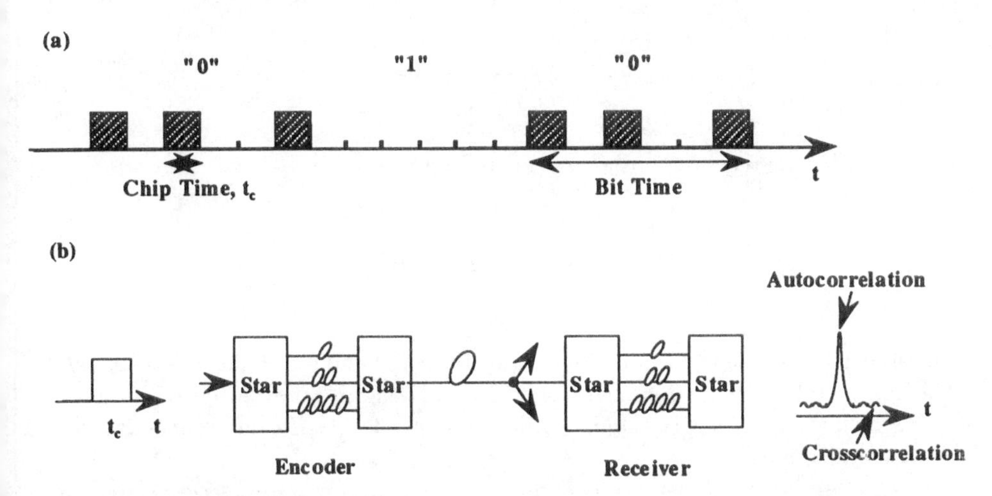

Figure 7.94 (a) An address-encoded bit composed of many chips and (b) encoding (and decoding) of an optical bit with variable-length optical delay lines.

the same fiber and individually decoded at a receiver, and that the decoding into 1 and 0 bits is performed asynchronously without requiring clock synchronization by the decoder. This method of encoding and decoding incurs high optical splitting losses, which must be compensated.

Figures of merit for operating code-division multiple access (CDMA) include (1) achieving a high autocorrelation signal when the receiver's decoder scheme inversely matches the encoder scheme and (2) achieving a low cross-correlation signal when the decoding scheme does **not** inversely match the encoding scheme; cross-correlation is cross-talk due to other channels sharing the same fiber medium. These two factors will determine the SNR of the recovered signal when several channels are simultaneously transmitted over the fiber. We have mentioned that N users require N different codes. However, this is misleading in the sense that N distinct possible codes can be achieved naively with ($\log_2 N$) number of chips per bit. For example, eight users could use only three chips per bit to uniquely identify each user. However, this means that one user would have a "111" code, and another user would have a "110" code. If both users are transmitting simultaneously, then two of the three coding chips will align in the decoder and produce a high cross-correlation cross-talk power, significantly reducing the recovered channel's SNR. This effect is increased if several channels are simultaneously transmitted. Therefore, many more than ($\log_2 N$) chips per bit are required to minimize the possible cross-correlation power from simultaneously transmitted channels. This increased number of chips per bit will minimize the possibility that chips for any nonselected channel will align in the decoder. One performance gauge for a CDMA system is the achievable SNR given a certain number of users (N) and a certain number of chips per bit (K) [144]:

$$SNR \cong \frac{1}{0.29}\left[\frac{K}{(N-1)}\right] \tag{7.40}$$

Furthermore, the receiver output, $r(t)$, can be given by [145]

$$r(t) = \int_{-\infty}^{+\infty} s(z) f(z-t)\, dz \tag{7.41}$$

where each bit has been encoded into a waveform $s(t)$ and the receiver address is $f(t)$. The autocorrelation signal appears when $s(t) = f(t)$, and the cross-correlation signal appears when $s(t) = f(t)$. To minimize the cross-correlation, the different codes should be (pseudo)orthogonal.

One fundamental result of encoding bits into short chip times is that the bandwidth, or spectrum, of the transmitted signal has increased severalfold. A 1-Gbps signal would normally occupy ~1 GHz information bandwidth, but this same signal will occupy 10 times that bandwidth if 10 chips/bit are used. This effect is known as spread-spectrum transmission [146]. If several channels are simultaneously transmitted on the same fiber, the channels will each have a different code, but their broad spectra will overlap. The decoding is temporal, not spectral, and has the effect of narrowing the transmitted spectrum on decoding.

We have already mentioned that the decoding can be performed asynchronously. Another advantage of CDM is the increased transmission security. This stems from the following.

- The spectrum of the signal is so broad so that narrowband noises or jamming signals do not significantly affect the data transmission and recovery.
- The data can only be recovered by a unique and private decoding sequence, or key, which can be kept private. Without a decoder, the spectra of all the channels overlap and cannot be demultiplexed.

The disadvantages of CDM include the following.

- A limitation in the number of users for which a fixed number of M chip slots will produce a high autocorrelation and a low cross-correlation signal when the decoder and encoder match;
- A limitation in the speed of the system, since very short pulses are required for the chip times within each bit time, thereby limiting the bit rate for a finite-pulse-width transmitter;
- High optical splitting losses at the encoder/decoder.

The second disadvantage listed can be thought of as the most critical drawback at present since ~gigabits transmission, which is a standard speed for optical transmission systems, can be achieved in CDM only with great difficulty. Ultra-short-pulse laser transmitters can produce ultra-short chip times, but the complexity and cost may not make this system achievable for practical use.

CDM received much attention in the mid to late 1980s, but it has not been popular recently in the optical communications community due to the fundamental limitations in modulation speed, number of users, and optical encoding/decoding losses. CDM may re-emerge in the future if these problems can be solved or if security is of paramount concern.

7.6 SPACE-DIVISION MULTIPLEXING

SDM is a fairly mature technology in which each switching element uniquely defines an optical path and connection [147]. An optical signal will enter an SDM switch and can be routed from any input port to any output port by appropriately enabling individual space-switching elements. Therefore, the spatial path is unique and defines the connection link. The most basic element in an SDM switch is a 2-by-2 crossbar switch whose functionality is described in Figure 7.95. This switching element can route an optical signal from input 1 to either output 1 or output 2, depending on whether the switch is biased in the bar or cross state, respectively. Many technologies are available for making optical crossbar switches, including electroabsorption in waveguides, lithium-niobate interferometers, and tunable directional couplers [148]. The concept of space multiplexing can be extended to large switches that incorporate these 2-by-2 switches as their basic building-block elements. For example, a 4-by-4 SDM switch can be composed of four 2-by-2 crossbar switches, with two switches at the input stage and two switches at the output stage. These switches must be properly set by an electronic controller to enable the routing of an incoming signal to the appropriate output port; optical control for such switches is still only a research topic. If two input ports request transmission to the same output port, only one signal will be allowed to reach its proper destination, and the other signal must be deflected to an alternate output port. Output-port-contention resolution, deflection routing fairness and priority, and local control are all network management issues that must be considered by any fully functioning switch.

As another illustrative example of an optical SDM switch, the two-dimensional optical crossbar switch of Figure 7.96 is composed of N input fibers, N output fibers, and an N-by-N array of optical switching elements, such as spatial light modulators. Each input fiber is split into N branches, with each branch connected to a different spatial element in a given **row.** Each output fiber is also split into N branches, with each branch connected to a different spatial element in a given **column.** Since each spatial switching element can be independently chosen to transmit or block the light, a unique path connecting any input fiber to any output fiber can be established by allowing the appropriate spatial element to transmit. Although SDM is a simple technique, it suffers from some disadvantages: (1) the number of elements scales as the square of the number of users, making this technology limited to a small number

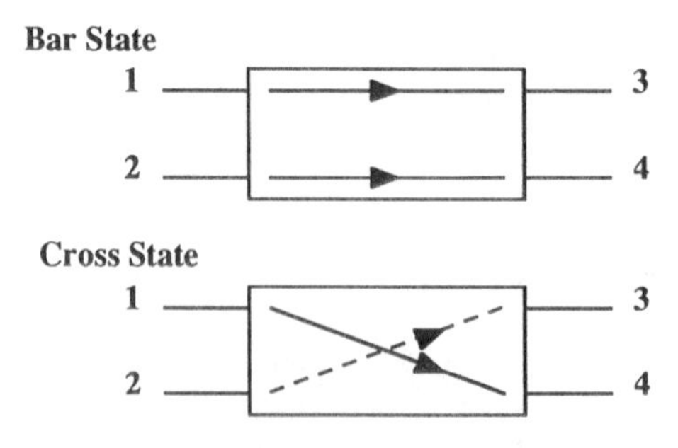

Figure 7.95 Functionality of a simple 2-by-2 crossbar switching element.

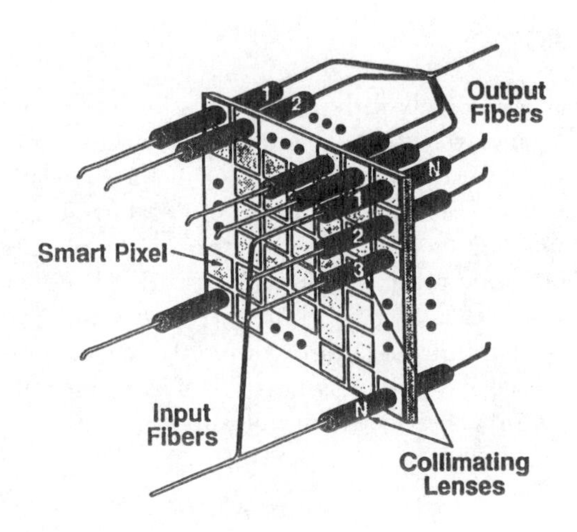

Figure 7.96 SDM optical crossbar switch (© 1989 SPIE).

of users due to fabrication limitations, and (2) optical splitting and recombining when using passive optical N by N star couplers will each incur a $1/N$ optical splitting loss, which limits operation. We can compensate for these losses by using external EDFAs or an array of SOAs as the switching elements themselves [149]. Since many switching technologies (that is, liquid crystal) offer slow millisecond switching speeds, fast switching SOAs can be employed to obviate the speed bottleneck. An alternate method for implementing SDM is to use an array of electro-optic lithium-niobate switches arranged in a multistage planar orientation [150]. The switching speeds are high (gigahertz), but there is a practical fabrication-limited size associated with such switches.

A desirable feature of SDM is that much of the technology is fairly mature and can be implemented in the short term, even though further technological advances are probable. It is quite possible that high-capacity multichannel systems will incorporate elements of SDM in conjunction with other multiplexing techniques.

SDM is different from the other multiplexing methods we have discussed in this chapter. All the other techniques simultaneously multiplex many signals onto a single optical fiber, thereby utilizing a larger fraction of the enormous fiber bandwidth. SDM however, allows for transmitting only a single signal on an individual fiber. The usefulness of SDM is that it enables the efficient routing of many high-speed signals through a switching matrix. It is the switching matrix that has an extremely high capacity, but the fiber-optic transmission system outside the space switch is still using only a small fraction of the fiber bandwidth.

7.7 NETWORK ISSUES

This chapter has dealt extensively with the transmission and recovery of multichannel signals. Additionally, an extremely important topic is the network switching issues as they relate to controlling a multichannel system. Some higher-level network issues include high-efficiency protocols, algorithms, and architectures. We will deal with these topics only in passing, instead concentrating on the physical-layer optical implementations of control and routing in optical networks. These topics incorporate an enormous amount of work, and we will limit our discussion to just a few examples. Furthermore, we have concentrated mostly on WDM systems in this chapter. Therefore, this section will discuss switching issues as they relate mostly to WDM, with some discussion applying to other forms of multiplexing.

7.7.1 Switching

We mentioned in Section 7.1.4 that networks establish communication links based on either circuit or packet switching. Circuit switching requires switching speeds of ~milliseconds, and packet switching requires switching speeds of microseconds or less. For high-speed optical transmission, packet switching holds the promise for more efficient data transfer in which no long-distance handshaking is required and high-bandwidth links are used more efficiently. Alternatively, many types of communication links and distribution systems may be interconnected satisfactorily by circuit switching, which is relatively simple to operate.

Network packet switching can be accomplished in a conceptually straightforward manner by requiring a node to optoelectronically detect and retransmit each and every incoming optical data packet. The control and routing information is contained in the newly detected electronic packet, and all the switching functions can occur in the electrical domain prior to optical retransmission of the signal. Unfortunately, this approach implies that an optoelectronic speed bottleneck will eventually occur in this system. Alternatively, much research is focused toward maintaining an all-optical data path and performing the switching functions all optically with only some electronic control of the optical components. The ultimate goal is focused toward a transparent all-optical network. The difficulties with optical switching include the following: (1) an optical path is not easily redirected since photons do not have as strong an interaction with their environment as electrons do, (2) the high speed of the incoming signal, and (3) switching nodes cannot easily tap a signal and acquire information about the channel. For many switching applications, a switching node must acquire information about the incoming signal before the node can determine the appropriate switching path. Figure 7.97 shows a generic solution that passively taps an incoming optical signal. Information about the signal is made known electrically to the node, but the signal itself remains in the optical domain. The routing information may be contained in the packet header or in some other form (i.e., wavelength, etc.). Note that

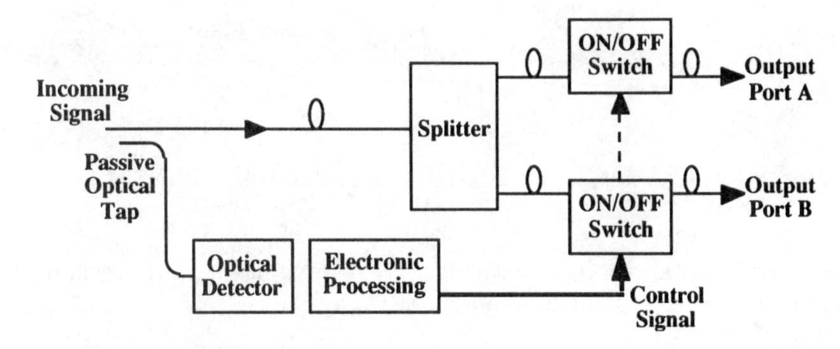

Figure 7.97 Passive optical tapping of an optical packet to determine routing information and allow a node to electronically control an optical switch.

an optical delay is required to allow sufficient time for the electronic circuitry to decide on the appropriate switching action.

Many challenging optical switching issues require solutions [151]. Most of these issues will relate to packet switching, although circuit switching will also require attention. We will be discussing packet switching issues mostly, with some discussion reserved for circuit switching. The specific issues that will be addressed include contention resolution, routing control, synchronization, header replacement, and data-format conversion. We will be giving some examples for each of these topics, but the reader should note that this whole area is extremely young and very dependent on the specific transmission systems that will eventually emerge. Our goal with this section is only to sensitize the reader to some of the many issues that must be addressed and the importance of an optical network becoming as functional as electronic networks have been.

7.7.1.1 Contention Resolution

Potentially the most limiting issue for a high-speed network is a situation in which two or more input ports request a communications path with the same output port, known as output-port contention. Since we are dealing with a high-speed system, we require rapid contention resolution in which one signal is allowed to reach its destination and the other signal is delayed or rerouted in some fashion; this resolution can be based on any type of network fairness algorithm. When considering our multiplexing schemes, the issue of contention exists when signals from two input ports would request routing to the same output port and contain identical wavelengths (or codes or time slots) as shown in Figure 7.98.

Several approaches exist for resolving contention. Following is a short and incomplete list, but it does give a flavor of possible approaches. The items listed are

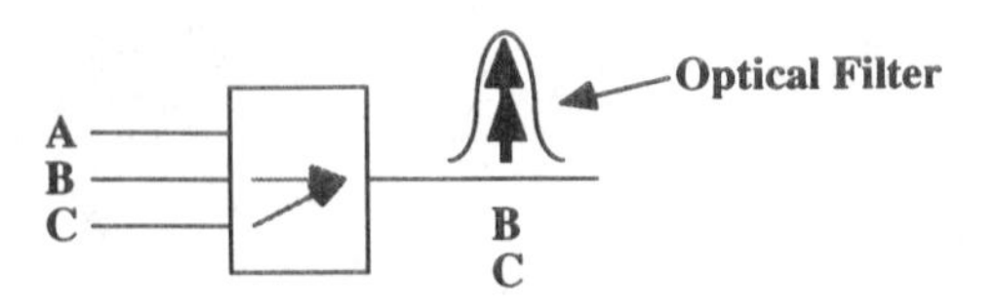

Figure 7.98 Output-port contention in a multichannel optical network.

not exclusive and can be complementary. Each track has its own advantages and disadvantages, with a complete solution not yet available:

- *Electronically determining contention and resolution.* If several signals are input into an N-by-N switch, one straightforward approach for recognizing contention is to have a central device read the packet headers of each input signal and determine which signals should reach their desired output port (Fig. 7.99). The decision making as to setting the switching matrix is performed in the electronic domain [152]. Reading of the routing information would preferably occur while the packets remain in optical form. Such packet output-port contention must be resolved rapidly to maintain high throughput and efficiency in our high-speed network. This approach is reasonable only given a small number of users since electronic circuitry cannot rapidly compare large numbers of parallel header streams. Optical approaches that determine if contention exists by comparing wavelength have also been examined [153–155].
- *Deflection routing.* Once it is determined that contention exists, one signal is routed to its desired output port and the other signal can be routed to another output port, known as deflection routing (Figure 7.100). The deflected packet is now traversing the network and eventually will be routed back to its desired destination. Increased delay will be incurred since the packet is being switched by more nodes than is the minimum number, adding extra hops to the deflected packet. One type of multi-hop network is called ShuffleNet [156].
- *Buffering.* An alternative to deflection routing is to retain the packet locally at the switching node and then switch it to the appropriate output port when that port is available. This requires local buffering of the packet, either in electrical or optical form. Electronic buffering is straightforward but requires undesirable optoelectronic conversions and may require very large buffers. On the other hand, optical buffering is difficult because many buffering schemes require updating a priority bit (it is difficult to change a priority bit of an optical data stream), and optical memory is not an advanced art, consisting mostly of using an optical delay line [157].
- *Time-slot interchanger.* In a TDM network, incoming signals at two different input ports may have data destined for the same time slot at a single output port. A time-slot interchanger (TSI), in which the time slot is changed for one of

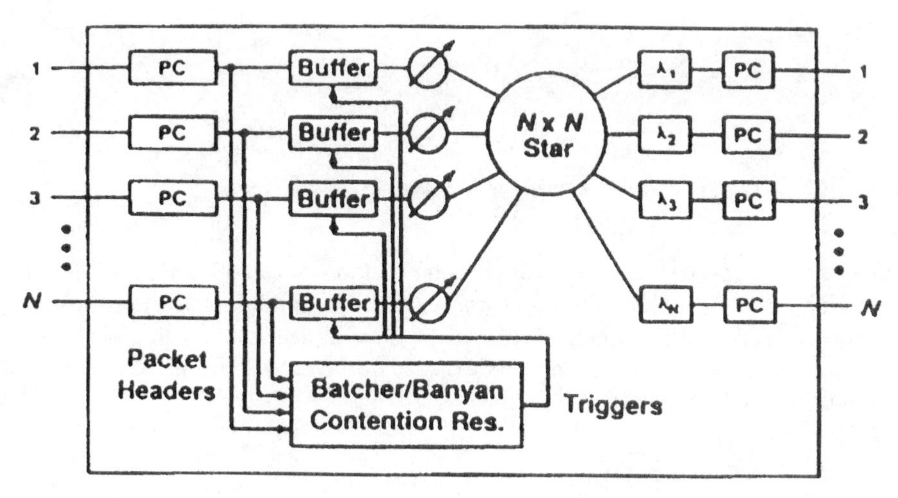

Figure 7.99 Electronic contention resolution of an N-by-N optical switch. (*After:* [153]. © 1990 IEEE.)

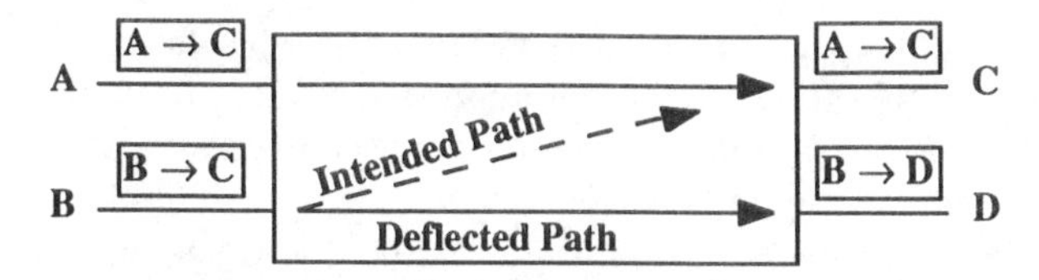

Figure 7.100 Deflection routing in a packet-switched network.

the incoming data streams, is quite common in electrical networks. A TSI can also be implemented in the optical domain by using a series of optical delay lines [158].

7.7.1.2 Routing Control

In a network, control and routing of signals can be centralized or decentralized [21]. A centralized system requires that a central controller have access to all signal information and can command the individual nodes to connect a given link. In a decentralized system, each node has its own autonomy, connecting to any available link it wishes. Each scenario depends critically on the specific network.

Signal routing is a critical network issue. As a signal enters a node or a switch, it is crucial that the node determines (1) if that signal it intended for it, (2) if the signal must be transmitted unaffected, or (3) if the signal requires active routing to a specific output port (e.g., see WDM ring in Fig. 7.101). In other words, how does a node know if the incoming packet is intended for it, and, if so, which wavelength should it tune its optical filter to given a WDM network? The straightforward approach of passive

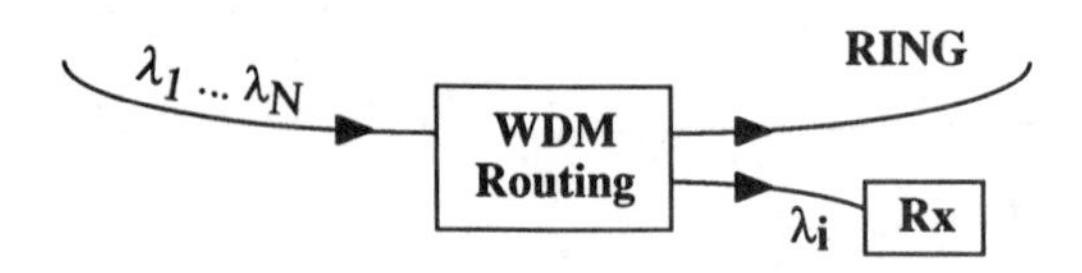

Figure 7.101 An example of routing concerns in a WDM ring network.

optical tapping and electronic circuit decision making is one approach, but it cannot easily accommodate a system that has many channels simultaneously transmitted on the same fiber. As with contention resolution, there are several approaches to attack this issue of multichannel routing, with no single favored solution. We will briefly mention three such approaches.

Common Wavelength Control

In a centralized WDM environment, each node can have its own signature wavelength, and a common central controller can have its own wavelength. The nodes and the controller also each transmit a subcarrier signal at a specific subcarrier frequency. Each individual node monitors the common wavelength from the controller and deciphers the low-speed routing information in the controller's subcarrier signal. This routing information commands the node to tune its optical filter to a specific wavelength. Each node can communicate back to the controller on its own individual subcarrier frequency. Each node uses a dual-passband liquid-crystal optical filter for which one fixed wavelength monitors the controller and the other tunable wavelength recovers data from another node [159].

Pilot-Tone Headers

Figure 7.102 shows a scenario in which a node can handle an incoming packet by tuning its optical filter on the fly in a self-routing scheme. Each node passively taps the signals on an incoming fiber. Each transmitter not only transmits data on its own wavelength, but it also transmits a subcarrier pilot tone on a specific frequency, f_k. The key factor in this method is that each receiver has a tunable acousto-optic filter. The acoustic wave and the wavelength selectivity of the filter are determined by f_{RF}, which will be f_k minus a specific local oscillator frequency at the receiver, f_{LO}. Therefore, a given transmitter can tune any receiver's passband by transmitting a tunable subcarrier frequency pilot tone [160]. Several different wavelength passbands (i.e., subcarriers) can be simultaneously accommodated by the AOTF.

Sequential Polling Using Optical Filters

A receiver with an optical filter can sequentially poll N distinct wavelengths in an N-user network [32]. As the filter is sequentially tuned to a given wavelength, the

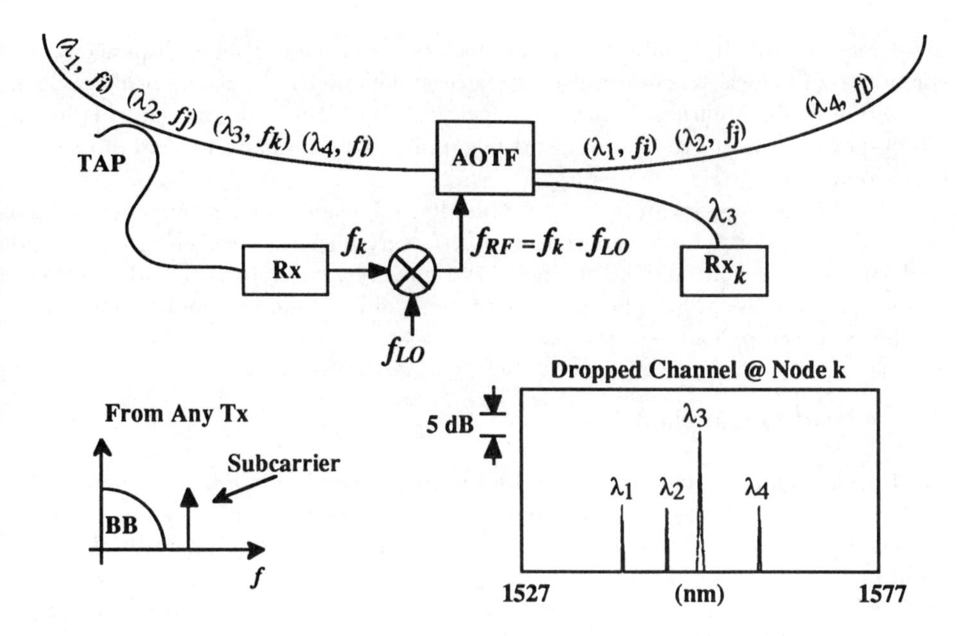

Figure 7.102 Self-routing control in a WDM ring network in which transmitted subcarrier pilot tones can tune the passband of a receiver's AOTF. (*After:* [160]. © 1992 IEEE.)

receiver listens momentarily to determine if that wavelength requests to speak with it. This method is useful in a circuit-switched environment, but is too slow (especially given large N) in a packet-switched network.

7.7.1.3 Synchronization

A high-speed network transmitting digital signals must have adequate time synchronization to unambiguously recover the data stream. This is especially true with packet switching, asynchronous packet arrival times, and long-distance transmission. A digital receiver can recover a clock from an incoming data signal by "synching" onto a series of bits at the beginning of a packet, and a figure of merit is to recover the clock accurately in as few bits as possible. Any ambiguity in the clock will cause a reduction in the sensitivity of the receiver and possibly errors in data recovery. These synching bits represent wasted overhead, since no useful data can be transmitted in that synching time. Faster data rates will cause more difficulty in clock recovery. Furthermore, there are factors in a long-distance fiber-optic network that cause difficulties in clock recovery: (1) any phase noise (i.e., in a transmitter) will cause timing jitter in the receiver and thus ambiguity in the clock recovery, and (2) fiber dispersion (i.e., spreading) of a clock-synching pulse will cause a reduction in the accuracy of the clock recovery. Possible solutions to network clock distribution include the transmission of a high-power ultra-narrow pulse at the beginning of a data packet to aid in receiver synchronization [161].

In a WDM network, it is also possible that wavelength synchronization will be required in addition to time synchronization. In such a scenario, a wavelength standard could be broadcast throughout the network. However, the hope is that the network wavelength stability and accuracy will be robust and will not require its own system overhead and complexity.

7.7.1.4 Header Replacement

Each packet typically has its routing and destination information contained in the packet header. In a large network, it is quite possible that the routing of a packet may change as it traverses the network due to dynamic traffic patterns and link availabilities. If the routing is changed, the packet header information itself must be changed. Moreover, if this rerouting is done in a WDM system, active wavelength shifting may also be required. This can be done optoelectronically by detection and retransmission. Alternatively, an all-optical approach is shown in Figure 7.103 in which the header is replaced and the packet is wavelength shifted, all in one simultaneous operation [162]. The experiment used the cross-gain compression wavelength-shifting method outlined earlier in this chapter. The incoming packet on λ_1 is passively tapped and its old header is read. The data is copied onto a second λ_2. Whereas inverse data copying is accomplished when the second wavelength is unmodulated, we can now modulate this second wavelength only during the header bits, thereby forcing the output on the second wavelength to be high or low, depending on the new header bits required for the wavelength-shifted packet. Five bits are changed in the header, the bit rate for the header and packet payload is 1 Gbps, no guard bits are required around the new or old packet header, and the wavelength is shifted by 19 nm.

7.7.1.5 Data-Format Conversion

In a large network, it is quite possible that a combination of data formats will be used in the local, regional, or global environments. This may occur if some links may more efficiently use TDM signaling, whereas other links may more effectively use WDM. There is the potential need for data-format conversion at network gateways, as illustrated in Figure 7.104. For example, one long-distance high-speed signal on λ_{TDM} may contain data for several different end users, with each user having its own signature wavelength. This scenario may exist if only a few wavelengths can be

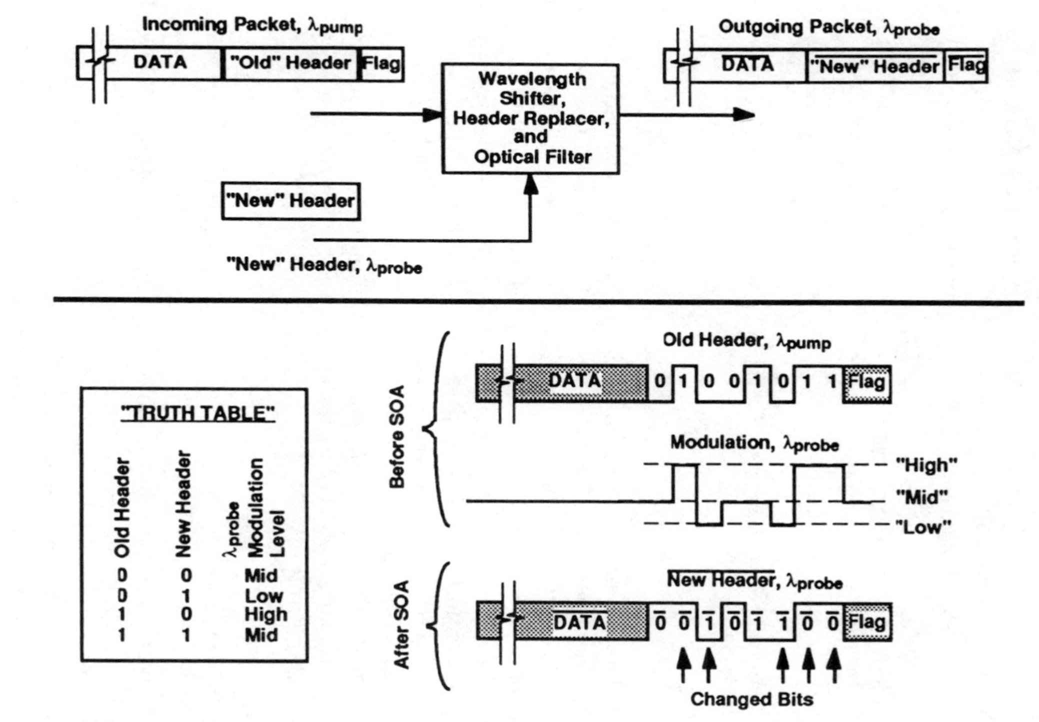

Figure 7.103 Simultaneous all-optical header replacement and wavelength shifting using SOA cross-gain saturation in a dynamically reconfigurable WDM network.

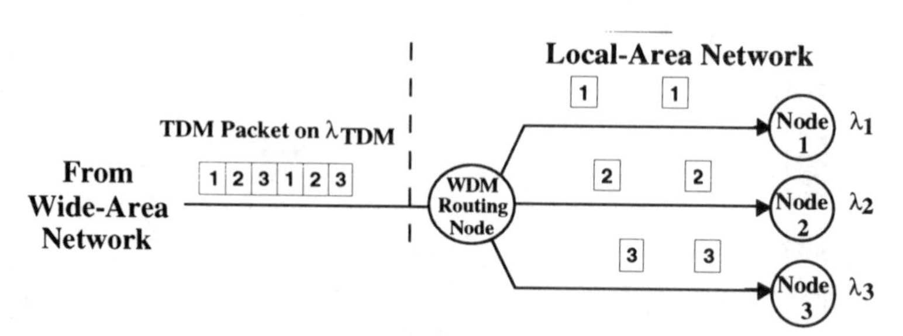

Figure 7.104 Data-format conversion from one TDM signal to several lower-speed WDM signals at a network gateway.

transmitted along a long distance of fiber due to limitations placed on the system by dispersion and nonlinearities. Therefore, one TDM signal would require time demultiplexing and wavelength routing. Figure 7.105 shows an experimental implementation of all-optical data-format conversion in which cross-gain saturation is again used in an SOA-based wavelength shifter [163]. An incoming high-speed "pump" signal on λ_{TDM} is simultaneously time demultiplexed and wavelength-shifted onto four different lower-speed "probe" wavelengths. The time demultiplexing and wavelength shifting is accomplished for each bit individually by sequentially biasing on only that probe laser whose wavelength corresponds to that particular bit time slot. WDM-to-TDM conversion can also be accomplished by using a wavelength shifter based on four-wave mixing [164].

7.7.2 Sample Networks

There has been much government and industrial funding for optical network projects to determine if high-speed optical networks are feasible and practical for general implementation. It is beyond the scope of this textbook to discuss in detail the specifics of the many systems being demonstrated for potential use. However, we will briefly mention some of the more well known testbed projects. An important feature is that each project relies on the use of multiple wavelengths for network routing.

- *All-Optical Network.* A team led by AT&T Bell Laboratories and MIT Lincoln Labs has developed a multilevel system in which a set of wavelengths is reused at each local network and a different set of wavelengths is used for inter-LAN traffic [165]. See Figure 7.106.
- *Optical Network Technology Consortium.* A seven-member team led by Bellcore has developed a high-speed LAN that uses AOTFs as WDM cross-connects [166]. See Figure 7.107.

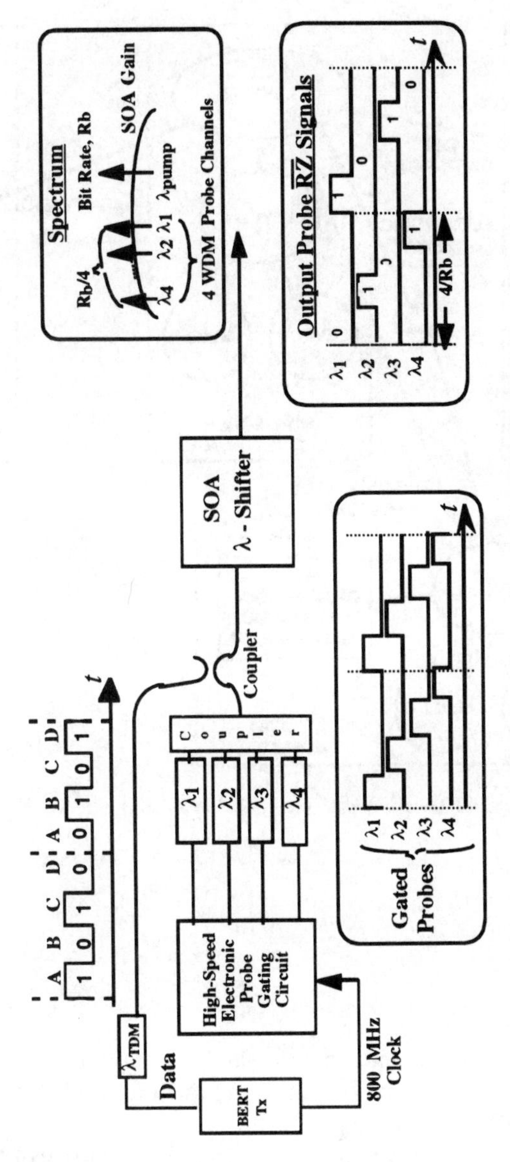

Figure 7.105 Experimental implementation of an all-optical TDM-to-WDM data-format converter.

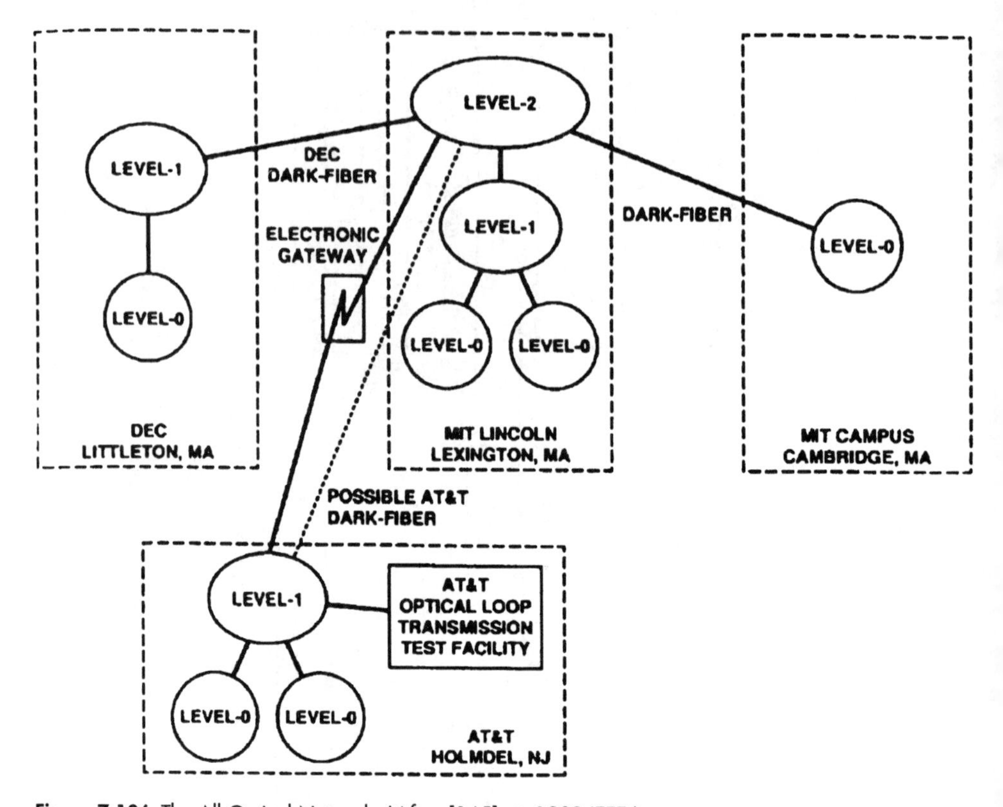

Figure 7.106 The All-Optical Network. (*After:* [165]. © 1993 IEEE.)

- *Rainbow.* An IBM-led team has worked on a multiuser high-speed LAN to interconnect workstations. The signaling includes both WDM and SCM [167]. See Figure 7.108.
- *STARNET.* Stanford University, GTE, and the University of Massachusetts have made progress on STARNET, an optical-network testbed that incorporates the possibility of using ASK, FSK, and PSK onto baseband and subcarrier signals [168]. See Figure 7.109.
- *RACE Program.* A consortium of industrial and academic groups in Europe has been pursuing a multifunctional WDM data switching structure [169]. See Figure 7.110.

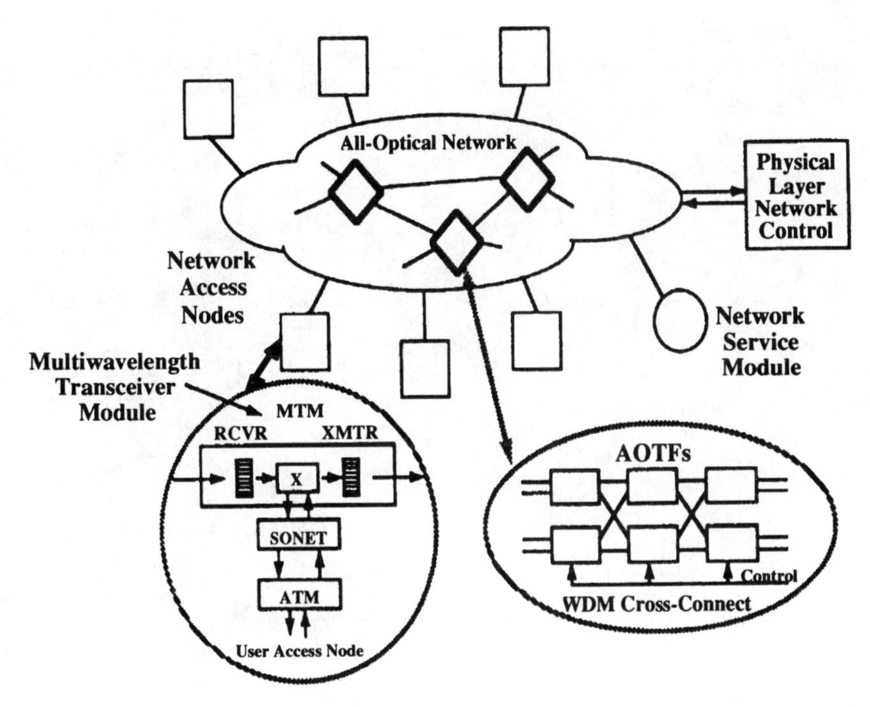

Figure 7.107 The Optical Network Technology Consortium system design. (*After:* [166]. © 1993 IEEE.)

7.7.3 Protocols

A standardized network protocol must be used to ensure that data packets are all formatted with recognizable routing information so that the packet can be switched through the network with full global compatibility. The two standards (or their foreign equivalents) that show the most promise of full adoption for a global optical network is *synchronous optical network* (SONET) [170] and *asynchronous transfer mode* (ATM) [171]; are shown in Figure 7.111 in very simple form. Data and header information are bunched into small 53-byte ATM packets. These packets arrive at a switching node at random times and are grouped together into a large 125-μs SONET frame, which makes its way in predetermined synchronous time slots through the network. The ATM packets are "unloaded" by the SONET frame when its direction is switched through the network and it can be placed into a different SONET frame. The analogy has been made that the ATM packets represent people randomly boarding a time-scheduled SONET train [172].

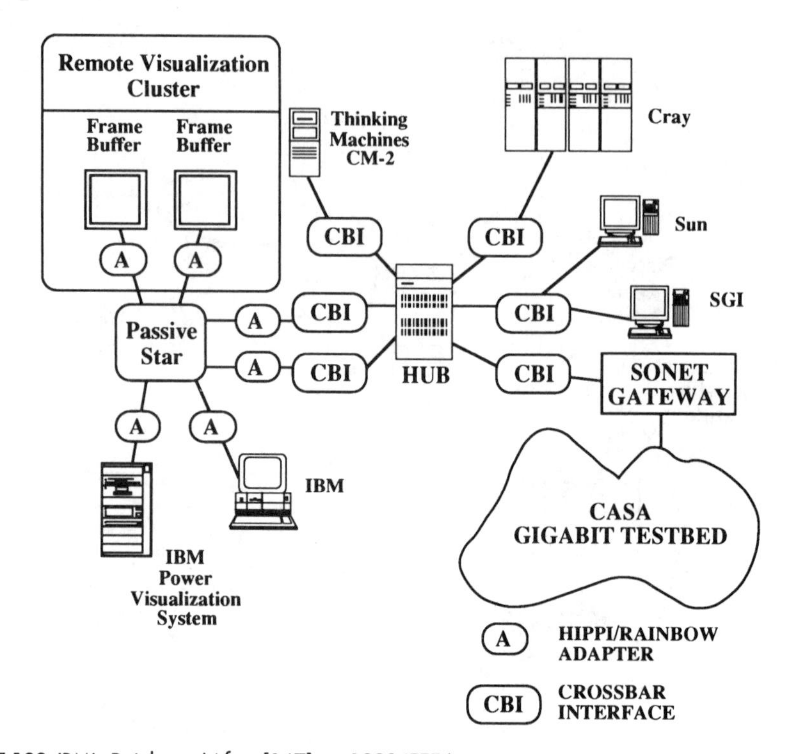

Figure 7.108 IBM's Rainbow. (*After:* [167]. © 1993 IEEE.)

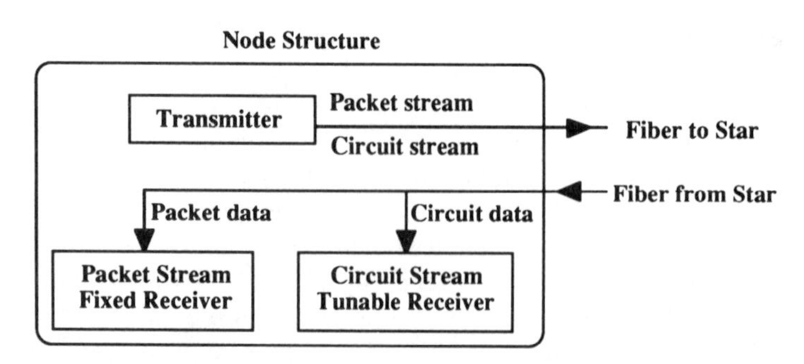

Figure 7.109 The STARNET optical network. (*After:* [168]. © 1993 IEEE.)

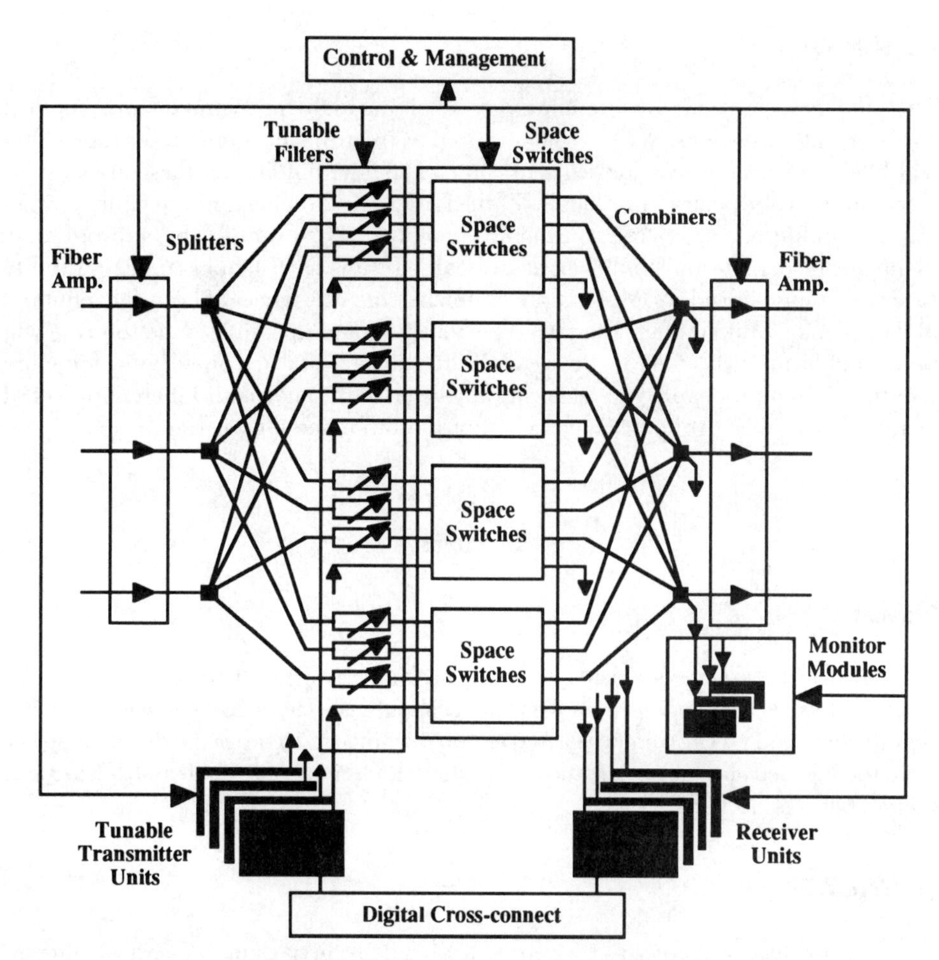

Figure 7.110 The RACE program's optical interconnection. (*After:* [169]. © 1993 IEEE.)

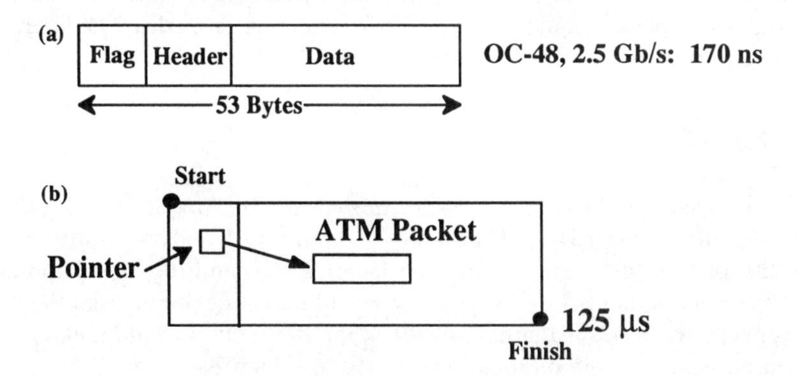

Figure 7.111 (a) An ATM packet and (b) a SONET frame containing many ATM packets and a header field.

7.8 SUMMARY

This chapter covered many different topics involving high-speed multichannel optical communication systems. We have endeavored to treat the most important topics that will likely affect networks for years to come. The glue binding all these areas is the recent and revolutionary introduction of the EDFA. Additionally, multichannel transmission limitations are being continually redefined. For example, by appropriately managing dispersion and nonlinearities, research results have demonstrated the ability to transmit high-speed WDM signals over much-longer distances of both conventional and dispersion-shifted fibers than was previously thought possible. What bit rate will be available to each of us in the year 2000? 2010? 2020? The striving for more bandwidth in multichannel systems continues due to the enormous inherent potential of the optical fiber. Although the sky is not the limit, the fiber certainly is!

Problems

Problem 7.1

A WDM transmission system is composed of four channels. If we assume that the channels have zero bandwidth, how far apart should the channels be placed to ensure that the detected power from a selected channel would be 10 times the detected power from the rejected channels? Assume a standard FP filter is used as a demultiplexer and its passband is 30 GHz.

Problem 7.2

We wish to directly detect a 1.5-μm FSK signal by first using an optical filter to demodulate the two transmitted tones. What filter transmission passband is required to limit the power penalty due to contrast-ratio reduction to <1 dB? The filter has the following hypothetical filter function: $T(f) = 1/(1 + (f/f_{FP})^{3/2})$.

Problem 7.3

Three 0.5-nm-separated channels are transmitted in a WDM system in which an FP demultiplexer follows an EDFA. The gain for channels 1, 2, and 3 is, respectively, 24, 22, and 20 dB. The spontaneous emission factor is 2.2, and the filter bandwidth is 0.2 Å. What attenuation before the receiver would make the power from the rejected channels below the receiver thermal noise? What attenuation would make the ASE-generated noises below the thermal noise?

Problem 7.4

A TDM system is simultaneously transmitting twenty 622-Mbps channels. What must the bandwidth of the demultiplexing switch be if the system is for bit interleaving? What must the bandwidth be if the system is for ATM-packet interleaving?

Problem 7.5

A receiver bandwidth is 10 GHz. How many SCM channels can be recovered if each channel is amplitude modulated at 125 MHz and the channel spacing must be twice the modulation rate?

Problem 7.6

To combat dispersion and nonlinearities, a fiber link can be established in which different lengths of fiber can be placed together in an alternating fashion in which one fiber has a dispersion zero wavelength of 1.50 μm and the second fiber has a dispersion zero wavelength of 1.59 μm. Three 10-Gbps channels are transmitted at 1.551, 1.552, and 1.553 μm. The dispersion slope for each fiber is 0.1 ps/(nm^2·km). What is the total dispersion in picoseconds for each channel for a 200-km link in which each subportion of fiber is 40 km long?

Problem 7.7

Design a four-channel WDM system that is robust to drifts in wavelength. The transmitter wavelength, filter passband, and wavelength multiplexer can each drift by ±1 Å.

Problem 7.8

Design a multihop WDM network in which 4 wavelengths are available at each of the 16 nodes. In an 8-node multihop system, what is the average number of hops if each node has either 2 or 4 wavelengths available?

Problem 7.9

A two-channel 1.55-μm WDM demultiplexer is made by cascading several MZ interferometers together. Design a filter demultiplexer made up of cascaded MZ filters in which the channel spacing is 30 GHz and the contrast ratio must be 20 dB. Concentrate on the number of cascaded interferometers and the relative lengths of the two interferometer arms for the first and last interferometers.

Problem 7.10

How many users can be accommodated in a CDMA system if there are 6 chip times within each bit time and we wish to have the autocorrelation power to be 20 times the crosscorrelation power. If the bit time is 5 ns, what is the spread-spectrum bandwidth of each channel? What is the total aggregate system bandwidth? In the encoder/decoder of each channel, design the delay lines that produce the specific code of a random node.

Problem 7.11

A hybrid WDM/TDM system is composed of only two channels. If two time-multiplexed channels are time demultiplexed and wavelength shifted, how long can each of the two separate signals propagate before a dispersion-induced power penalty of 2 dB is incurred on time multiplexing the channels? The original TDM signal is 10 Gbps, and the dispersion parameter is 5 ps/(nm · km).

References

[1] F.P. Kapron, "Fiber-Optic System Tradeoffs," *IEEE Spectrum Magazine*, March, pp. 68–75, 1985.
[2] N.S. Bergano, C.R. Davidson, G.M. Homsey, D.J. Kalmus, P.R. Trischitta, J. Aspell, D.A. Gray, R.L. Maybach, S. Yamamoto, H. Taga, N. Edagawa, Y. Yoshida, Y. Horiuchi, T. Kawazawa, Y. Namihira, and S. Akiba, "9000 km, 5 Gb/s NRZ Transmission Experiment Using 274 Erbium-Doped Fiber Amplifiers," *Topical Meeting on Optical Amplifiers and Their Applications* '92, Technical Digest, paper PD-11, Santa Fe, NM (Optical Society of America, Wash., D.C., 1992).
[3] S. Kawanishi, et al., "200 Gbit/s, 100 km TDM Transmission Using Supercontinuum Pulses With Prescaled PLL Timing Extraction and All-Optical Demultiplexing," *Electronics Letters*, vol. 31, p. 816, 1995.
[4] D.P. Berztsekas and R.G. Gallager, *Data Networks*, Englewood Cliffs, New Jersey: Prentice Hall, 1987.
[5] P. Kaiser and D.B. Keck, "Fiber Types and Their Status," in *Optical Fiber Telecommunications II*, chapter 2, p. 40, S.E. Miller and I.P. Kaminow, eds., New York: Academic Press, 1988.
[6] M. Schwartz, *Telecommunication Networks, Protocols, Modeling, and Analysis*, New York: Addison Wesley, 1987.
[7] A.R. Chraplyvy, "Limitations on Lightwave Communications Imposed by Optical Fiber Nonlinearities," *IEEE/OSA J. of Lightwave Tech.*, vol. 8, pp. 1548–1557, 1990.
[8] P.E. Green, Jr., *Fiber Optic Networks*, Englewood Cliffs: Prentice Hall, NJ, 1993.
[9] Special Issue on Wavelength Division Multiplexin, *IEEE J. on Selected Areas in Comm.*, vol. 8, 1990.
[10] C.A. Brackett, "Dense Wavelength Division Multiplexing: Principles and Applications," *ibid.*, p. 948, 1990.
[11] I.P. Kaminow, "FSK with Direct Detection in Optical Multiple-Access FDM Networks," *ibid*, p. 1005, 1990.
[12] N. Kashima, *Optical Transmission for the Subscriber Loop*, chapters 3 and 9, Norwood, MA: Artech House, 1993.
[13] M.S. Roden, *Analog and Digital Communication Systems*, 4th ed., Upper Saddle River, NJ: Prentice Hall, 1995.

[14] L.W. Couch II, *Digital and Analog Communication Systems,* New York: Macmillan, 1983.
[15] H.S. Hinton, *An Introduction to Photonic Switching Fabrics,* New York: Plenum Press, 1993.
[16] A.S. Acampora, *An Introduction to Broadband Networks,* New York: Plenum Press, 1994.
[17] J.E. Berthold, "Networking Fundamentals," *Conf. on Optical Fiber Communications,* Tutorial TuK, pp. 3–20, San Jose, CA, Feb., 1994 (Optical Society of America, Wash., D.C., 1994).
[18] P.S. Henry, R.A. Linke, and A.H. Gnauck, "Introduction to Lightwave Systems," in *Optical Fiber Telecommunications II,* chapter 21, S.E. Miller and I.P. Kaminow, eds., New York: Academic Press, 1988.
[19] A.B. Buckman, *Guided-Wave Photonics,* p. 201, New York: Saunders College Publishing, 1992.
[20] M. Schwartz, *Telecommunication Networks, Protocols, Modeling, and Analysis,* New York: Addison Wesley, 1987.
[21] J.Y. Hui, *Switching and Traffic Theory for Integrated Broadband Networks,* Boston: Kluwer Academic Publishers, 1990.
[22] J.A. McEachern, "Gigabit Networking on the Public Transmission Network," *IEEE Comm. Mag,* pp. 70–78, April, 1992.
[23] K. Oshima, T. Kitayama, M. Yamaki, T. Matsui, and K. Ito, "Fiber-Optic Local Area Passive Network Using Burst TDMA Scheme," J. of Lightwave Tech., vol. 3, pp. 502–510, 1985.
[24] P.R. Prucnal, M.A. Santoro, and S.K. Sehgal, "Ultrafast All-Optical Synchronous Multiple-Access Fiber Networks, IEEE J. on Selected Areas in Comm., vol. 4, pp. 1484–1493, 1986.
[25] R.S. Tucker, G. Eisenstein, S.K. Korotky, G. Raybon, J.J. Veselka, L.L. Buhl, B.L. Kasper, R.C. Alferness, "16Gbit/s Fibre Transmission Experiment Using Optical Time-Division Multiplexing," Electronics Lett, vol. 23, no. 24, 1987.
[26] R.C. Alferness, "Titanium-Diffused Lithium Niobate Waveguide Devices," chapter 4, in *Guided-Wave Optoelectronics,* T. Tamir, ed., Springer-Verlag, New York, 1989.
[27] N.J. Doran and D. Wood, "Nonlinear Optical Loop Mirror," *Optics Letters,* vol. 13, pp. 56–58, 1988.
[28] C.A. Brackett, "Dense Wavelength Division Multiplexing: Principles and Applications," *IEEE J. on Selected Areas in Comm.,* vol. 8, p. 948, 1990.
[29] I.P. Kaminow, "FSK with Direct Detection in Optical Multiple-Access FDM Networks," *ibid.,* p. 1005, 1990.
[30] G.P. Agrawal, *Fiber-Optic Communication Systems,* New York: Wiley-Interscience, 1992, chapter 7.
[31] A.E. Willner, I.P. Kaminow, M. Kuznetsov, J. Stone, and L.W. Stulz, "1.2 Gb/s Closely-Spaced FDMA-FSK Direct-Detection Star Network," Photon. Tech. Lett., vol. 2, pp. 223–226, 1990.
[32] N.R. Dono, P.E. Green, K. Liu, R. Ramaswami, and F.F. Tong, "A Wavelength Division Multiple Access Network for Computer Communication," *IEEE J. on Selected Areas in Comm.,* vol. 8, pp. 983–994, 1990.
[33] W.I. Way, D.A. Smith, J.J. Johnson, and H. Izadpanah, "A Self-Routing WDM High-Capacity SONET Ring Network, *IEEE Photonics Tech. Lett.,* vol. 4, pp. 402–405, 1992.
[34] A.S. Acampora, M.J. Karol, M.G. Hluchyj, "Terabit Lightwave Networks: The Multihop Approach," *AT&T Technical Journal,* vol. 66, no. 6, pp. 21–34, November/December 1987.
[35] K. Aiki, M. Nakamura, J. Umeda, A. Yariv, A. Katzir, and H.W. Yen, "GaAs-GaAlAs Distributed Feedback Laser With Separate Optical and Electrical Confinement," *Appl. Physics Lett.,* vol. 27, p. 145, 1975.
[36] K. Inoue, H. Toba, and K. Nosu, "Multichannel Amplification Utilizing an Er^{3+}-Doped Fiber Amplifier," *IEEE/OSA J. of Lightwave Tech.,* vol. 9, pp. 368–374, 1991.
[37] T.L. Koch, U. Koren, R.P. Gnall, C.A. Burrus, and B.I. Miller, "Continuously Tunable 1.5μm Multiple-Quantum-Well GaInAs/GaAsInP Distributed-Bragg-Reflector Lasers," *Electron. Lett,* vol. 21, pp. 283–285, 1988.
[38] M. Kuznetsov, A.E. Willner, and I.P. Kaminow, "Frequency Modulation Response of Two-Segment Distributed Feedback Laser," *Applied Physics Lett.,* vol. 55, pp. 1826–1828, 1989.

[39] B.E.A. Saleh and M.C. Teich, *Fundamentals of Photonics,* p. 801, New York: Wiley, 1991.
[40] M. Born and E. Wolf, *Principles of Optics,* 6th ed., chapter 10, New York: Pergamon Press, 1959, 1980.
[41] N.K. Shankaranarayanan, U. Koren, B. Glance, and G. Wright, "Two-Section DBR Laser Transmitters With Accurate Channel Spacing and Fast Arbitrary-Sequence Tuning for Optical FDMA Networks," *Conf. on Optical Fiber Communications,* paper TuI2, pp. 3–20, San Jose, CA, Feb., 1994 (Optical Society of America, Wash., D.C., 1994).
[42] B. Glance, U. Koren, C.A. Burrus, and J.D. Evankow, *Electronics Letters,* vol. 27, p. 15, 1991.
[43] S.K. Korotky, "Three-Space Representation of Phase-Mismatch Switching on Coupled Two-State Optical Systems," *IEEE J. of Quantum Electronics,* vol. QE-22, pp. 952–958, 1984.
[44] C.E. Zah, F.J. Favire, B. Pathak, R. Bhat, C. Caneau, P.S.D. Lin, A.S. Gozdz, N.C. Andreadakis, M.A. Koza, and T.P. Lee, "Monolithic Integration of Multiwavelength Compressive-Strained Multiquantum-Well Distributed-Feedback Laser Array with Star Coupler and Optical Amplifiers," *Electronics Lett.,* vol. 28, pp. 2361–2362, 1992.
[45] C.J. Chang-Hasnain, J.P. Harbison, C.E. Zah, M.W. Maeda, L.T. Florez, N.G. Stoffel, and T.P. Lee, "Multiple Wavelength Tunable Surface-Emitting Laser Arrays," IEEE J. Quant. Electron., vol. 27, p. 1368, 1991.
[46] T.L. Koch, and U. Koren, "Semiconductor Photonic Integrated Circuits," *IEEE J. of Quantum Electronics,* vol. 27, pp. 641–653, 1991.
[47] C. Fabry and A. Perot, "Theorie et Applications d'Une Nouvelle de Spectroscopie Interferentielle," *Ann. Chim. Phys.,* vol. 16, p. 115, 1899.
[48] A. Yariv, *Introduction to Optical Electronics,* 2nd ed., p. 64, New York: Holt, Rinehart, and Winston, 1976.
[49] F.S. Choa, and T.L. Koch, "Static and Dynamical Characteristics of Narrow-Band Tunable Resonant Amplifiers as Active Filters and Receivers," J. of Lightwave Tech., vol. 9, pp. 73–83, 1991.
[50] J. Stone and L.W. Stulz, "Pigtailed High-Finesse Tunable Fiber Fabry-Perot Interferometers with Large, Medium and Small Free Spectral Ranges," Electron. Lett., vol. 23, pp. 781–783, 1987.
[51] J.S. Patel, M.A. Saifi, D.W. Berreman, C. Lin, N.C. Andreadakis, and S.D. Lee, "Electrically Tunable Optical Filter for Infrared Wavelengths using Liquid Crystals in a Fabry-Perot Etalon," Appl. Phys. Lett., vol. 57, pp. 1718–1720, 1990.
[52] N. Takato, T. Kominato, A. Sujita, K. Jinguji, H. Toba, and M. Kawachi, "Silica-Based Integrated-Optic Mach-Zehnder Multi/Demultiplexer Family With Channel Spacing of 0.01–250 nm," *IEEE J. on Selected Areas in Comm.,* vol. 8, pp. 1120–1127, 1990.
[53] D.A. Smith, J.E. Baran, J.J. Johnson, and K.-W. Cheung, "Integrated-Optic Acoustically-Tunable Filters for WDM Networks," *ibid.,* pp. 1151–1159, 1990.
[54] I.P. Kaminow, "Photonic Local Networks," in *Optical Fiber Telecommunications II,* chapter 26, pp. 933–967, S.E. Miller and I.P. Kaminow, eds., Academic Press, New York, 1988.
[55] C. Dragone, "Efficient N X N Couplers Using Fourier Optics," J. of Lightwave Tech., vol. 3, pp. 467–471, 1989.
[56] K. Okamoto, H. Takahashi, S. Suzuki, A. Sujita, and Y. Ohmori, *Electronics Lett.,* vol. 27, 1991.
[57] H.M. Presby, S. Yang, A.E. Willner, and C.A. Edwards, "Connectorized Integrated Star Couplers on Silicon," *Optical Engineering,* vol. 31, pp. 1323–1327, 1992.
[58] J.B.D. Soole, et al., *Conf. on Optical Fiber Communications,* Technical Digest, paper ThB1, San Jose, CA, Feb., 1994 (Optical Society of America, Washington, D.C., 1994).
[59] H.A. Haus and Y. Lai, "Narrow-Band Optical Channel Dropping Filter," *IEEE/OSA J. Lightwave Tech.,* vol. 10, pp. 57–62, 1992.
[60] C. Dragone, "An NxN Optical Multiplexer Using a Planar Arrangement of Two Star Couplers," *IEEE Photonics Tech. Lett.,* vol. 3, pp. 812–815, 1991.
[61] G.-K. Chang, T.P. Liu, J.L. Gimlett, H. Shirokmann, M.Z. Iqbal, J.R. Hayes, and K.C. Wang, "A

Direct-Current Coupled, All-Differential Optical Receiver for High-Bit-Rate SONET Systems," *IEEE Photonics Technology Letters*, vol. 4, pp. 339–342, 1992.
[62] R. Ramaswami and K.N. Sivarajan, "Design of Logical Topologies for Wavelength-Routed Optical Networks," *IEEE J. on Selected Areas of Communications*, vol. 14, pp. 840–851, 1996.
[63] P.A. Humblet and W.M. Hamdy, "Crosstalk Analysis and Filter Optimization of Single- and Doulbe-Cavity Fabry-Perot Filters," *IEEE J. on Selected Areas of Communications*, vol. 8, pp. 1095–1107, 1990.
[64] M. Born and E. Wolf, *Principles of Optics, 6th Ed.*, chapter 1 (New York: Pergamon Press, 1980).
[65] I.P. Kaminow, P.P. Iannone, J. Stone, and L.W. Stulz, "FDMA-FSK Star Network with a Tunable Optical Fiber Demultiplexer," *J. Lightwave Tech.*, vol. 6, pp. 1406–1414, 1988.
[66] A.B. Carlson, *Communication Systems*, 3rd ed., New York: McGraw Hill, 1986.
[67] A.E. Siegman, *Lasers*, Mill Valley, CA: Univ. Science Books, 1986.
[68] T.L. Koch and J.E. Bowers, "Nature of Wavelength Chirping in Directly Modulated Semiconductor Lasers," *Electronics Lett.*, vol. 20, pp. 1038–1039, 1984.
[69] K.C. Reichman, P.D. Magill, U. Koren, B.I. Miller, M. Young, M. Newkirk, M.D. Chien, "2.5 Gbit/s Transmission over 674 km at Multiple Wavelengths Using a Tunable DBR Laser With an Integrated Electroabsorption Modulator," *IEEE Photonics Technology Letters*, vol. 5, pp. 1098–1100, 1993.
[70] A.E. Willner, "Simplified Model of a FSK-to-ASK Direct-Detection System Using a Fabry-Perot Demodulator," *IEEE Photonics Technology Lett.*, vol. 2, pp. 363–366, 1990.
[71] A.E. Willner, I.P. Kaminow, M. Kuznetsov, J. Stone, and L.W. Stulz, "FDMA-FSK Noncoherent Star Network Operated at 600 Mb/s Using Two-Electrode DFB Lasers and a Fiber Optical Filter Demultiplexer," *Electronics Lett.*, vol. 25, pp. 1600–1601, 1989.
[72] Y.C. Chung, K.J. Pollock, P.J. Fitzgerald, B. Glance, R.W. Tkach, and A.R. Chraplyvy, Electronics Lett., vol. 24, p. 1313, 1988.
[73] B. Glance, T.L. Koch, O. Scaramucci, K.C. Reichmann, U. Koren, and C.A. Burrus, "Densely Spaced FDM Coherent Optical Star Network Using Monolithic Widely Frequency-Tunable Lasers," *Electron. Lett.*, vol. 25, pp. 672–673, 1989.
[74] A.J. Keating and A.J. Lowery, "Wavelength Stabilization in WDM Packet-Switched Networks," *Conf. on Optical Fiber Communications*, paper WR2, San Diego, CA, Feb., 1995 (Optical Society of America, Wash., D.C., 1995).
[75] W.B. Sessa, R.E. Wagner, and P.C. Li, "Frequency Stability of DFB Lasers Used in FDM Multi-Location Networks," *Conf. on Optical Fiber Communications*, paper ThC3, San Jose, CA, Feb., 1992 (Optical Society of America, Wash., D.C., 1992).
[76] Y.C. Chung and J. Jeong, "Aging-Induced Wavelength Shifts in 1.5-μm DFB Lasers," *Conf. on Optical Fiber Communications*, paper WG6, San Jose, CA, Feb., 1994 (Optical Society of America, Wash., D.C., 1994).
[77] T.L. Koch, and U. Koren, "Semiconductor Photonic Integrated Circuits," *IEEE J. of Quantum Electronics*, vol. 27, pp. 641–653, 1991.
[78] A.E. Willner, M. Kuznetsov, I.P. Kaminow, U. Koren, T.L. Koch, C.A. Burrus, and G. Raybon, "Multi-Gigahertz Bandwidth FM Response of Frequency Tunable Two-Electrode DFB Lasers," *IEEE Photonics Technology Lett.*, vol. 1, pp. 360–363, 1989.
[79] A.R. Chraplyvy, R.W. Tkach, A.H. Gnauck, B.L. Kasper, and R.M. Derosier, "8 Gbit/s FSK Modulation of DFB Lasers With Optical Demodulation," *Electronics Letters*, vol. 25, pp. 319–321, 1989.
[80] B.S. Glance, K. Pollock, C.A. Burrus, B.L. Kasper, G. Eisenstein, and L.W. Stulz, "WDM Coherent Optical Star Network," *IEEE J. of Lightwave Technology*, vol. 6, pp. 67–72, 1988.
[81] L.G. Kazovsky and J.L. Gimlett, *IEEE/OSA J. of Lightwave Technology*, vol. 6, p. 1353, 1988.
[82] M.S. Goodman, H. Kobrinski, M.P. Vecchi, R.M. Bulley, and J.L. Gimlett, "The LAMBDANET Multiwavelength Network: Architecture, Applications, and Demonstrations," *IEEE J. on Selected Areas in Comm.*, vol. 8, pp. 995–1004, 1990.

[83] R. Ramaswami, "Multiwavelength Lightwave Networks for Computer Communication," *IEEE Communications Magazine*, Feb., pp. 78–88, 1993.

[84] A.E. Willner, E. Desurvire, H.M. Presby, C.A. Edwards, and J. Simpson, "LD-Pumped Erbium-Doped Fiber Preamplifiers with Optimal Noise Filtering in a FDMA-FSK 1Gb/s Star Network," *IEEE Photonics Technology Lett.*, vol. 2, pp. 669–672, 1990.

[85] E. Desurvire, *Erbium-Doped Fiber Amplifiers: Principles and Applications*, New York: Wiley-Interscience, 1994.

[86] A.E. Willner, "SNR Analysis of Crosstalk and Filtering Effects in an Amplified Multi-Channel Direct-Detection Dense-WDM System," *IEEE Photonics Technology Lett.*, vol. 4, pp. 186–189, 1992.

[87] A.M. Hill, R. Wyatt, J.F. Massicott, K.J. Blyth, D.S. Forrester, R.A. Lobbett, P.J. Smith, and D.B. Payne, "40-Million-Way WDM Broadcast Network Employing Two Stages of Erbium-Doped Fibre Amplifiers," Electronics Lett., vol. 26, pp. 1882–1884, 1990.

[88] A.E. Willner, A.A.M. Saleh, H.M. Presby, D.J. DiGiovanni, and C.A. Edwards, "Star Couplers with Gain Using Multiple Erbium-Doped Fibers Pumped with a Single Laser," *IEEE Photonics Technology Lett.*, vol. 3, pp. 250–252, 1991.

[89] E.L. Goldstein, A.F. Elrefaie, N. Jackman, and S. Zaidi, "Multiwavelength Fiber-Amplifier Cascades in Unidirectional Interoffice Ring Networks," *Conf. on Optical Fiber Communications* '93, Tech. Dig., paper TuJ3, San Jose, CA, Feb., 1993 (Optical Society of America, Wash., D.C., 1993).

[90] J.P. Blondel, A. Pitel, and J.F. Marcerou, "Gain-Filtering Stability in Ultralong-Distance Links," *Conference on Optical Fiber Communications '93*, Tech. Dig., paper TuI3, San Jose, CA, Feb. 1993 (Optical Society of America, Wash., D.C., 1993).

[91] H. Taga, N. Edagawa, Y. Yoshida, S. Yamamoto, and H. Wakabayashi, "IM-DD Four-Channel Transmission Experiment Over 1500 km Employing 22 Cascaded Optical Amplifiers," *Electron. Lett.*, vol. 29,. p. 485, 1993.

[92] A.E. Willner and S.-M. Hwang, "Transmission of Many WDM Channels Through a Cascade of EDFA's in Long-Distance Link and Ring Networks," *IEEE/OSA J. of Lightwave Technology, Special Issue on Optical Amplifiers and Their Applications*, vol. 13, pp. 802–816, 1995.

[93] C.R. Giles and E. Desurvire, "Modeling Erbium-Doped Fiber Amplifiers," J. Lightwave Tech., 9, p. 271, 1991.

[94] N.S. Bergano and C.R. Davidson, "Four-Channel WDM Transmission Experiment Over Transoceanic Distances," Topical Meeting on Optical Amplifiers and Their Applications, paper PD7, Breckenridge, CO (OSA, Washington, D.C., 1994).

[95] K. Inoue, T. Kominato, and H. Toba, "Tunable Gain Equalization Using a Mach-Zehnder Optical Filter in Multistage Fiber Amplifiers," *IEEE Photon. Tech. Lett.*, 3, p. 718, 1991.

[96] M. Tachibana, R.I. Laming, P.R. Morkel, and D.N. Payne, "Erbium-Doped Fiber Amplifier with Flattened Gain Spectrum," *IEEE Photon. Tech. Lett.*, 3, p. 118, 1991.

[97] A.E. Willner and S.-M. Hwang, "Passive Equalization of Non-Uniform EDFA Gain by Optical Filtering for Megameter Transmission of 20 WDM Channels through a Cascade of EDFA's," *IEEE Photonics Technology Lett.*, vol. 5, pp. 1023–1026, 1993.

[98] A.R. Chraplyvy, J.A. Nagel, and R.W. Tkach, "Equalization in Amplified WDM Lightwave Transmission Systems," *IEEE Photon. Tech. Lett.*, 4, p. 920, 1992.

[99] L. Eskildsen, E. Goldstein, V. Da Silva, M. Andrejco, and Y. Silberberg, "Optical Power Equalization for Multiwavelength Fiber-Amplifier Cascades Using Periodic Inhomogeneous Broadening," *IEEE Photonics Technology Letters*, vol. 5, pp. 1188–1190, 1993.

[100] E.L. Goldstein, "Multiwavelength Optical-Amplifier Cascades," *Conference on Optical Fiber Communications '95*, Tech. Dig., Tutorial, WK1, San Diego, CA, Feb. 1995 (Optical Society of America, Wash., D.C., 1995).

[101] S.H. Huang, X.Y. Zou, S.-M. Hwang, A.E. Willner, Z. Bao, and D.A. Smith, "Experimental Demonstration of Active Equalization and ASE Suppression of Three 2.5 Gbit/s WDM-Network Channels

over 2,500 km Using AOTFs as Transmission Filters," *Conference on Lasers and Electro-Optics,* paper CMA4, Anaheim, CA, June 1996 (Optical Society of America, Wash., D.C., 1996).

[102] C.R. Giles and E. Desurvire, "Propagation of Signal and Noise in Concatenated Erbium-Doped Fiber Optical Amplifiers," *IEEE/OSA J. of Lightwave Tech.,* vol. 9, pp. 147–154, 1991.

[103] B.B. Clesca, D. Bayart, C. Coeurjolly, L. Berthelon, L. Hamon, J.-L. Beylat, "Over 25-nm, 16 Wavelength-Multiplexed Signal Transmission Through Four Fluoride-Based Fiber-Amplifier Cascade and 440 km Standard Fiber," *Optical Fiber Communications '94,* Tech. Dig., paper PD20, San Jose, CA, Feb. 1994 (Optical Society of America, Wash., D.C., 1994).

[104] A.R. Chraplyvy, "Limitations on Lightwave Communications Imposed by Optical Fiber Nonlinearities," *IEEE/OSA J. of Lightwave Tech.,* vol. 8, pp. 1548–1557, 1990.

[105] G.P. Agrawal, *Nonlinear Fiber Optics,* New York: Academic Press, 1990.

[106] B.M. Nyman, S.G. Evangelides, G.T. Harvey, L.F. Mollenauer, P.V. Mamyshev, M. Saylors, S.K. Korotky, U. Koren, V. Mizrahi, T.A. Strasser, J.J. Veselka, J.D. Evankow, A.J. Lucero, J.A. Nagel, J.W. Sulhoff, J.L. Zyskind, P.C. Corbett, M.A. Mills, and G.A. Ferguson, "Soliton WDM Transmission of 8 × 2.5 Gb/s, Error Free over 10 Megameters," *Conf. on Optical Fiber Communications '95,* Technical Digest, paper PD21, San Diego, CA, March, 1995 (Optical Society of America, Wash., D.C., 1995).

[107] R.H. Stolen and E.R. Ippen, *Applied Physics Lett.,* vol. 22, p. 276, 1973.

[108] R.Y. Chiao, C.H. Townes, and B.P. Stoicheff, *Phys. Rev. Lett.,* vol. 12, p. 592, 1965.

[109] R.H. Stolen and C. Lin, *Phys. Rev. A,* vol. 17, p. 1448, 1978.

[110] K.D. Hill, D.C. Johnson, B.S. Kawasaki, and R.I. MacDonald, *J. Appl. Phys.,* vol. 49, p. 5098, 1978.

[111] A.R. Chraplyvy et al., "One-Third Terabit/s Transmission Through 150 km of Dispersion-Managed Fiber," *IEEE Photonics Tech. Lett.,* vol. 7, no. 1, 1995.

[112] A.H. Gnauck, R.M. Jopson, and R.M. Derosier, "10-Gb/s 360 km Transmission Over Dispersive Fiber using Midsystem Spectral Inversion," *IEEE Photonics Tech. Lett.,* vol. 5, pp. 663–666, 1993.

[113] F. Forghieri, R.W. Tkach, and A.R. Chraplyvy, "WDM Systems With Unequally Spaced Channels," *IEEE/OSA J. Lightwave Technology,* vol. 13, pp. 889–897, 1995.

[114] G.P. Agrawal, *Fiber-Optic Communication Systems,* p. 298, Wiley-Interscience, New York, 1992.

[115] T. Li, private communication; modified from P.S. Henry, R.A. Linke, and A.H. Gnauck, "Introduction to Lightwave Systems," in *Optical Fiber Telecommunications II,* chapter 21, p. 782, S.E. Miller and I.P. Kaminow, eds., Academic Press, New York, 1988.

[116] R.A. Barry, V.W.S. Chan, K.L. Hall, E.S. Kintzer, J.D. Moores, K.A. Rauschenbach, E.A. Swanson, L.E. Adams, C.R. Doerr, S.G. Finn, H.A. Haus, E.P. Ippen, W.S. Wong, and M. Haner, "All-Optical Network Consortium—Ultrafast TDM Networks," *IEEE J. on Selected Areas in Communications,* vol. 14, p. 999, 1996.

[117] C.A. Brackett, "Status and Early Results of the Optical Network Technology Consortium," *Conf. on Optical Fiber Communications,* Technical Digest, paper WD3, San Jose, CA, Feb., 1994.

[118] I.P. Kaminow, R.E. Thomas, and S.B. Alexander, "Early Results of the Research Consortium on Wideband All-Optical Networks," *Conf. on Optical Fiber Communications '94,* Technical Digest, paper WD1, San Jose, CA, Feb., 1994 (Optical Society of America, Wash., D.C., 1994).

[119] B. Glance, J.M. Wiesenfeld, U. Koren, A.H. Gnauck, H.M. Presby, and A. Jourdon, "High Performance Optical Wavelength Shifter," *Electronics Lett.,* vol. 28, pp. 1714–1715, 1992.

[120] J.M. Wiesenfeld, J.S. Perino, A.S. Gnauck and B. Glance, "Bit Error Rate Performance for Wavelength Conversion at 20 Gb/s," *Electron. Lett.,* vol. 30, pp. 720–721, 1994.

[121] W. Shieh and A.E. Willner, "Optimal Conditions for High-Speed All-Optical SOA-Based Wavelength Shifting," *IEEE Photonics Technology Letters,* vol. 7, pp. 1273–1275, 1995.

[122] A.E. Willner and W. Shieh, "Optimal Spectral and Power Parameters for All-Optical Wavelength Shifting: Single Stage, Fanout, and Cascadability," *IEEE/OSA J. of Lightwave Technology, Special Issue on Optical Amplifiers and Their Applications,* vol. 13, pp. 771–781, 1995.

[123] B. Mikkelson, T. Durhuus, C. Joergenson, R.J.S. Pedersen, C. Braagaard, and K.E. Stubkjaer, "Polarization Insensitive Wavelength Conversion of 10 Gbit/s Signals in a Michelson Interferometer," *Electronics Letters,* 1994, Vol. 30, p. 260–261.

[124] T. Durhuus, R.J.S. Pedersen, B. Mikkelsen, K.E. Stubkjaer, M. Oberg, and S. Nilsson, "Optical Wavelength Conversion Over 18 nm at 2.5 Gb/s by a DBR Laser," *IEEE Photon. Technol. Lett.,* vol. 5, no. 1, pp. 86–88, Jan. 1993.

[125] J. Zhou, N. Park, K.J. Vahala, M.A. Newkirk, and B.I. Miller, "Four-Wave Mixing Wavelength Conversion Efficiency in Semiconductor Traveling-Wave Amplifiers Measured to 65 nm of Wavelength Shift," *IEEE Photonics Tech. Lett.,* vol. 6, pp. 984, 1994.

[126] J.P.R. Lacey, M.V. Chan, R.S. Tucker, A.J. Lowery and M.A. Summerfield, "All-optical WDM to TDM Transmultiplexer," *Electron. Lett.,* vol. 30, pp. 1612–1613, 1994.

[127] D.J.G. Mestdagh, *Fundamentals of Multiaccess Optical Fiber Networks,* chapter 8, Norwood, MA: Artech House, 1995.

[128] N. Kashima, *Optical Transmission for the Subscriber Loop,* chapters 3 and 9, Norwood, MA: Artech House, 1993.

[129] W.I. Way, "Subcarrier Multiplexed Lightwave System Design Considerations for Subscriber Loop Applications," *IEEE J. of Lightwave Tech.,* vol. 7, p. 1806–1818, 1989.

[130] T.E. Darcie, "Subcarrier Multiplexing for Multiple-Access Lightwave Networks," *IEEE J. of Lightwave Tech.,* vol. 5, p. 1103–1110, 1987.

[131] W.I. Way, M.W. Maeda, A.Y. Yan, M.J. Andrejco, M.M. Choy, M. Saifi, and C. Lin, "160-Channel FM-Video Transmission Using FM/FDM and Subcarrier Multiplexing and an Erbium Doped Optical Fiber Amplifier," *Electronics Lett.,* vol. 26, p. 139, 1990.

[132] K. Sato, "Intensity Noise of Semiconductor Laser Diodes in Fiber Optic Analog Video Transmission," *IEEE J. of Quantum Electronics,* vol. QE-19, p. 1380, 1983.

[133] R. Olshansky, V.A. Lanzisera, and P.M. Hill, "Subcarrier Multiplexed Lightwave Systems for Broad-Band Distribution, *IEEE J. of Lightwave Tech.,* vol. 7, p. 1329–1342, 1989.

[134] K. Stubkjaer and M. Danielsen, "Nonlinearity of GaAlAs Lasers—Harmonic Distortion," *IEEE J. of Quantum Electronics,* vol. QE-16, p. 531, 1980.

[135] A.A.M. Saleh, "Fundamental Limit on Number of Channels in Subcarrier-Multiplexed Lightwave CATV System," *Electronics Lett.,* vol. 25, pp. 776–777, 1989.

[136] R.D. Gitlin, J.F. Hayes, and S.B. Weinstein, *Data Communications Principles,* New York: Plenum Press, 1992.

[137] T.E. Darcie, R.S. Tucker, and G.J. Sullivan, *Electronics Lett.,* vol. 21, p. 665, 1985.

[138] G.P. Agrawal, *Fiber-Optic Communication Systems,* p. 306, New York: Wiley-Interscience, 1992.

[139] C.Y. Kuo, "AM-VSB Transmission Using Erbium-Doped Fiber Amplifiers," *IEEE J. of Lightwave Tech.,* vol. 10, p. 235, 1992.

[140] W.I. Way, D.A. Smith, J.J. Johnson, and H. Izadpanah, "A Self-Routing WDM High-Capacity SONET Ring Network, *IEEE Photonics Tech. Lett.,* vol. 4, pp. 402–405, 1992.

[141] T. Uno, M. Mitsuda, J. Ohya, "Low Distortion Characteristics in Amplified 4 X 50-Channel WDM AM SCM Transmission," *Conf. on Optical Fiber Communications '94,* Technical Digest, paper WM2, San Jose, CA, Feb., 1994 (Optical Society of America, Wash., D.C., 1994).

[142] P.R. Prucnal, M.A. Santoro, and T.R. Fran, "Spread Spectrum Fiber-Optic Local Area Network Using Optical Processing," *IEEE/OSA J. of Lightwave Tech.,* vol. 4, p. 547, 1986.

[143] J.A. Salehi, A.M. Weiner, and J.P. Heritage, *IEEE/OSA J. of Lightwave Tech.,* vol. 8, p. 478, 1990.

[144] F.R.K. Chung, J.A. Salehi, V.K. Wei, "Optical Orthogonal Codes: Design, Analysis, and Applications," *IEEE Transactions on Information Theory,* vol. 35, pp. 595–604, 1989.

[145] R.S. Cheng, "Total Capacities of Maximum Numbers of Users in Strinctly Band Limited CDMA & FDMA Gaussian Channels," *MILCOM '92,* Proceedings, pp. 247–251, 1992.

[146] W.C. Lindsey and M.K. Simon, *Telecommunications Systems Engineering,* New York: Dover Publications, 1977.

[147] A.R. Diaz, R.F. Kalman, J.W. Goodman, and A.A. Sawchuk, "Fiber-Optic Crossbar Switch with Broadcast Capability," *Optical Engineering,* vol. 27, pp. 1087–1095.

[148] R.C. Alferness, "Titanium-Diffused Lithium Niobate Waveguide Devices," pp. 174–194, in *Guided-Wave Optoelectronics,* T. Tamir, ed., Springer-Verlag, New York, 1989.

[149] M. Fujiwara, H. Nishimoto, T. Kajitani, M. Itoh, and S. Suzuki, "Studies on Semiconductor Optical Amplifiers for Line Capacity Expansion in Photonic Space-Division Switching System," J. of Lightwave Tech., vol. 9, pp. 155–160, 1991.

[150] M. Kondo, N. Takado, K. Komatsu, and Y. Ohta, "32 Switch-Elements Integrated Low-Crosstalk Ti:LiNbO3 Optical Matrix Switch," IOOC-ECOC, Technical Digest, pp. 361–364, 1987.

[151] J.E. Berthold, "Networking Fundamentals," *Conf. on Optical Fiber Communications,* Tutorial TuK, pp. 3–20, San Jose, CA, Feb., 1994.

[152] M.S. Goodman, "Multiwavelength Networks and New Approaches to Packet Switching," *IEEE Comm. Mag.,* Oct., pp. 27–35, 1989.

[153] A. Cisneros, S.F. Habiby, and A.E. Willner, "Photonic Contention Resolution Devices, A Laboratory Demonstration," *Conf. on Optical Fiber Communications '93,* Technical Digest, pp. 94–95, San Jose, CA, Feb. 1993 (Optical Society of America, Washington, D.C., 1993).

[154] C.-Li Lu, D.J.M. Sabido IV, P. Poggiolini, R.T. Hofmeister, and L.G. Kazovsky, "CORD—A WDMA Optical Network: Subcarrier-based Signaling and Control Scheme," *IEEE Photonics Technology Letters,* vol. 7, pp. 555–557, 1995.

[155] D.J. Blumenthal, K.Y. Chen, J. Ma, R.J. Feuerstein, and J.R. Sauer, "Demonstration of a Deflection Routing 2X2 Photonic Switch for Computer Interconnects," *IEEE Photonics Technology Letters,* vol. 4, p. 169–173, 1992.

[156] A.S. Acampora, M.J. Karol, M.G. Hluchyj, "Terbit Lightwave Networks: The Multihop Approach," *AT&T Technical Journal,* vol. 66, no. 6, pp. 21–34, November/December 1987.

[157] K.K. Goel, "Nonrecirculating and Recirculating Delay Line Loop Topologies of Fiber-Optic Delay Line Filters," IEEE Photonics Technology Letters, vol. 5, pp. 1086–1088, 1993.

[158] H. Goto, K. Nagashima, and S. Suzuki, "Photonic Time-Division Switching Technology," *Photonic Switching: Proceedings of the First Topical Meeting,* pp. 151–157, Springer-Verlag, Berlin, 1987.

[159] M.W. Maeda, A.E. Willner, J.R. Wullert II, J. Patel, and M. Allersma, "Wavelength-Division Multiple-Access Network Based on Centralized Common-Wavelength Control, *IEEE Photonics Technology Lett.,* vol. 5, pp. 83–86, 1993.

[160] W.I. Way, D.A. Smith, J.J. Johnson, H. Izadpanah, and H. Johnson, "Self-Routing WDM High-Capacity SONET Ring Network," *Conf. on Optical Fiber Communications '91,* Technical Digest, paper TuO2, San Jose, CA, Feb. 1992 (Optical Society of America, Washington, D.C., 1992).

[161] D.H. Hartman, P.J. Delfyett, and S.Z. Ahmed, "Optical Clock Distribution Using a Mode-Locked Semiconductor Laser Diode System," *Conf. on Optical Fiber Communications '91,* Technical Digest, paper FC3, San Diego, CA, Feb. 1991 (Optical Society of America, Washington, D.C., 1991).

[162] E. Park, D. Norte, and A.E. Willner, "Simultaneous All-Optical Header Replacement and Wavelength Shifting for a Dynamically-Reconfigurable WDM," *IEEE Photonics Technology Letters,* vol. 7, pp. 810–812, 1995.

[163] D. Norte and A.E. Willner, "All-Optical Data Format Conversions and Reconversions Between the Wavelength and Time Domains for Dynamically Reconfigurable WDM Networks," *IEEE/OSA J. of Lightwave Technology and IEEE J. on Selected Areas in Communications, Special Issue on Multiple-Wavelength Technologies and Networks,* vol. 14, pp. 1170–1182, 1996.

[164] J.P.R. Lacey, M.V. Chan, R.S. Tucker, A.J. Lowery, and M.A. Summerfield, "All-Optical WDM-to-TDM Transmultiplexer," IEEE LEOS Summer Topical in Optical Networks and Their Enabling Technologies, paper PD4, Lake Tahoe, NV, 1994 (IEEE, Piscataway, NJ, 1994).

[165] S.B. Alexander, R.S. Bondurant, D. Byrne, V.W.S. Chan, S.G. Chan, R. Gallger, B.S. Glance, H.A. Haus, P. Humblet, R. Jain, I.P. Kaminow, M. Karol, R.S. Kennedy, A. Kirby, H.Q. Le, A.M. Saleh, B.A. Schofield, J.H. Shapiro, N.K. Shankaranarayanan, R.E. Thomas, R.C. Williamson, and

R.W. Wilson," A precompetitive consortium on wide-band all-optical networks," *IEEE/OSA J. of Lightwave Technology*, vol. 11, pp. 714–735, 1993.

[166] C.A. Brackett, A.S. Acampora, J. Sweitzer, G. Tangonan, M.T. Smith, W. Lennon, K. Wang, and R.H. Hobbs, "A scalable multiwavelength multihop optical network: a proposal for research on all-optical networks," *IEEE/OSA J. of Lightwave Technology*, vol. 11, pp. 736–753, 1993.

[167] P. Green, L.A. Cauldren, K.M. Johnson, J.G. Lewis, C.M. Miller, J.F. Morrison, R. Olshansky, R. Ramaswami, and E.H. Smith, Jr., "All-Optical Packet-Switched Metropolitan-Area Network Proposal," *IEEE/OSA J. of Lightwave Technology*, vol. 11, pp. 754–763, 1993.

[168] L.G. Kazovsky and P.T. Poggiolini, "STARNET: A Multi-Gigabit-per-Second Optical LAN Utilizing a Passive WDM Star," *IEEE/OSA J. of Lightwave Tech.*, vol. 11, pp. 1009–1027, 1993.

[169] G.R. Hill, P.J. Chidgey, F. Kaufhold, T. Lynch, O. Sahlen, M. Gustavsson, M. Janson, B. Lagerstrom, G. Grasso, F. Meli, S. Johansson, J. Ingers, L. Fernandez, S. Rotolo, A. Antonielli, S. Tebaldini, E. Vezzoni, R. Caddedu, N. Caponio, F. Testa, A. Scavennec, M.J. O'Mahony, J. Zhou, A. Yu, W. Sohler, U. Rust, and H. Herrmann, "A transport network layer based on otpical network elements," *IEEE/OSA J. of Lightwave Technology*, vol. 11, pp. 667–679, 1993.

[170] Y.-C. Ching, "SONET Implementation," IEEE Comm. Mag., pp. 34–40, Sept., 1993.

[171] J.A. McEachern, "Gigabit Networking on the Public Transmission Network," IEEE Comm. Mag, pp. 70–78, April, 1992.

[172] A. Cisneros, "Large Scale ATM Switching and Optical Technology," *Conf. on Optical Fiber Communications '93*, Technical Digest, paper TuJ3, San Jose, CA, Feb. 1992 (Optical Society of America, Washington, D.C., 1992).

Appendix A
Derivation of the General Form of the Signal-to-Noise Ratio, Λ_1

We will show here how to get from (3.101) to (3.105) for the SNR (signal-to-noise ratio), Λ_1. Let us start with (3.101), here reported for simplicity:

$$\Lambda_1 = \frac{(2GRR_t\overline{P}_R r_0)^2}{2G_2RR_t^2 q\overline{P}_R w_0 + G_2R_t^2 q i_d \int_{-\infty}^{\infty} h_L^2(t)dt + \sigma_{th}^2} \tag{A.1}$$

Substituting (3.104), the definition of the figure of merit, ρ_A, into (A.1), we obtain

$$\Lambda_1 = \frac{(2GRR_t\overline{P}_R r_0 / v_{ph})^2}{(2G_2RR_t^2 q\overline{P}_R w_0 + G_2R_t^2 q i_d \int_{-\infty}^{\infty} h_L^2(t)dt / v_{ph}^2 + \rho_A^2} \tag{A.2}$$

Recalling now the definition in (3.80), which becomes here

$$v_{ph} = \frac{qR_t r_0}{T} \tag{A.3}$$

substituting it and

$$\overline{P}_R = \frac{\overline{N}_R h\nu}{T} \tag{A.4}$$

into (A.2), we obtain

$$\Lambda_1 = \frac{(2\eta G\overline{N}_R)^2}{\frac{2G_2\eta\overline{N}_R w_0 T}{r_0^2} + \frac{G_2\lambda_0 T^2}{r_0^2}\int_{-\infty}^{\infty} h_L^2(t)dt + \rho_A^2} \tag{A.5}$$

Recalling now the sampling instant, t_0 the following expressions hold:

$$w_0 = \int_{-\infty}^{\infty} W(f)e^{-j2\pi f t_0}df \tag{A.6}$$

$$r_0 = \int_{-\infty}^{\infty} R(f)e^{-j2\pi f t_0}df \tag{A.7}$$

$$\int_{-\infty}^{\infty} h_L^2(t)dt = H_{L2}(0) = \int_{-\infty}^{\infty} |H_L(f)|^2 df \tag{A.8}$$

where we have defined $H_{L2}(f) \overset{\text{def}}{=} H_L(f) * H_L(f)$. Substituting (A.6) to (A.8) into (A.5) and then taking into account the definitions

$$N_0 \overset{\text{def}}{=} \lambda_0 T \tag{A.9}$$

$$B_L \overset{\text{def}}{=} \frac{\int_{-\infty}^{\infty} |H_L(f)|^2 df}{H_L^2(0)} \tag{A.10}$$

$$B_{RL} \overset{\text{def}}{=} \frac{\int_{-\infty}^{\infty} P(f)H_L(f)e^{-j2\pi f t_0}df}{P(0)H_L(0)} = \frac{\int_{-\infty}^{\infty} R(f)e^{-j2\pi f t_0}df}{R(0)} \tag{A.11}$$

$$B_{RL2} \overset{\text{def}}{=} \frac{\int_{-\infty}^{\infty} P(f)\, H_{L2}e^{-j2\pi f t_0}(f)df}{P(0)H_{L2}(0)} = \frac{\int_{-\infty}^{\infty} W(f)e^{-j2\pi f t_0}df}{W(0)} \tag{A.12}$$

and then

$$A_0 \overset{\text{def}}{=} \frac{B_{RL2}B_L}{B_{RL}^2} \tag{A.13}$$

$$A_1 \stackrel{\text{def}}{=} \frac{B_L}{B_{RL}^2 T} \tag{A.14}$$

yields finally

$$\Lambda_1 = \frac{(2\eta G\overline{N}_R)^2}{G_2(2\eta\overline{N}_R A_0 + N_0 A_1) + \rho_A^2} \tag{A.15}$$

Appendix B
Statistics of Phase Noise to Amplitude Conversion

From Section 4.8, we know that the phase noise of a semiconductor laser can be modeled as a Wiener-Lévy random process $\phi(t)$:

$$\phi(t) = 2\pi \int_0^t \mu(\tau) d\tau \tag{B.1}$$

where $\mu(t)$ is a zero-mean Gaussian noise with spectral density $G_\mu(f) = \Delta\nu/2\pi$, $\Delta\nu$ being the laser linewidth. This has been called Lorentzian model of the phase noise. The random process $\phi(t)$ is a nonstationary Markov process, with variance $2\pi\Delta\nu t$, whose pdf (probability density function) can be obtained as the solution of a partial differential Fokker-Planck equation [1].

In Section 4.10, we saw that the error probability performance of asynchronous coherent systems like ASK and FSK depend, under suitable hypotheses, on the statistics of RVs like

$$Z_i \stackrel{\text{def}}{=} \left| \frac{1}{T'} \int_{(i-1)T'}^{iT'} e^{j\phi(\tau)} d\tau \right|^2 \tag{B.2}$$

and

$$W = \sum_{i=1}^{N} Z_i \tag{B.3}$$

We have also proved that the RVs Z_i have the same statistics and are independent for distinct values of i, corresponding to disjoint integration intervals. Thus, all we need is the statistical characterization of the RV Z_i. From its pdf, we can easily derive the pdf of W, which turns out to be the sum of N RVs independent with the same pdf.

As a consequence, the main issue consists in the statistical characterization of the random process

$$z(t) = \int_0^t e^{j\phi(\tau)}\, d\tau \tag{B.4}$$

This objective can be achieved in several different ways.[1] We will sketch the most important ones and explain in detail the solution adopted to obtain the numerical results presented in Section 4.10, that is, the method of moments.

B.1 SIMULATION OF THE FILTERED ENVELOPE

The most direct way to obtain a statistical characterization of the random process $z(t)$ consists in its numerical simulation. For a given t, we can divide the interval $(0, t)$ into many disjoint segments and exploit the fact that the increments of $\phi(t)$ in disjoint intervals are independent Gaussian RVs. This makes it easy to generate a sufficient number of samples through standard techniques, like the Montecarlo method, and to get an estimate of the pdf of $z(t)$. We need, however, an accurate estimate of the pdf, since we are interested in events whose probability is of the order of 10^{-9}. This requires a huge number of simulation trials, so as to make prohibitive the direct approach.

A more efficient simulation approach [3] is based on the Radon-Nykodim theorem, a way that is known as *importance sampling* in the simulation of telecommunication systems, and consists of modifying the random process under study to make more likely the events with very low probability. In the problem at hand, this method can save order of magnitudes in the number of required simulations.

However, the simulation approach requires a simulation run for every value of the linewidth and leads to a numerical estimate of the pdf that is not suited to the further computations required. In the literature, it has been used only as a mean to validate approximate analytical solutions.

B.2 NUMERICAL SOLUTION OF THE FOKKER-PLANCK EQUATION

Let us represent $z(t)$ through its real and imaginary parts:

$$z(t) = z_R(t) + jz_I(t)$$

[1] In [2] a different approach with respect to the ones explained here is taken. The authors approximate the statistics of the decision RVs using a Gaussian pdf and evaluate the performance by characterizing the pdf in terms of mean and variance. This simplified approach leads to a fairly good approximation of the true performance of ASK and FSK systems in the presence of an optimized postdetection filter, while giving a poor approximation for nonoptimized IF and postdetection filters' bandwidth.

$$z_R(t) = \int_0^t \cos \phi(\tau) d\tau$$

$$z_I(t) = \int_0^t \sin \phi(\tau) d\tau \tag{B.5}$$

or through its magnitude and phase:

$$z(t) = \rho_z(t) e^{j\theta_z(t)}$$

$$\rho_z(t) = |z(t)| = \sqrt{z_R^2(t) + z_I^2(t)}$$

$$\theta_z(t) = \tan^{-1} \frac{z_I(t)}{z_R(t)} \tag{B.6}$$

Unlike $\phi(t)$, $z(t)$ (see Problem B.1) is not a Markov process, and neither are $z_R(t)$ and $z_T(t)$ or $\rho_z(t)$. However, the three-component vector process $[\phi(t), z_R(t), z_I(t)]$ or, equivalently, $[\phi(t), \rho_z(t), \theta_z(t)]$ is indeed a Markov vector process, satisfying the stochastic differential equations

$$\frac{\partial \phi(t)}{\partial t} = 2\pi\mu(t)$$

$$\frac{\partial z_R(t)}{\partial t} = \cos\phi(t)$$

$$\frac{\partial z_I(t)}{\partial t} = \sin\phi(t) \tag{B.7}$$

with initial conditions $\phi(0) = 0$, $z_R(0) = 0$, $z_I(0) = 0$. A similar system of equations can be obtained in terms of the magnitude and phase of $z(t)$.

The vector being a Markov process, its joint transition pdf, together with its initial conditions, gives a complete statistical specification of it. Now, owing to the theory of multidimensional Fokker-Planck equations [1], the joint transition pdf of $\phi(t)$, $z_R(t)$, $z_I(t)$ defined as

$$f(a, b, c; t) = \frac{\partial^3 F(a, b, c; t)}{\partial a \partial b \partial c}$$

$$F(a, b, c; t) = P[\phi(t) \le a, z_R(t) \le b, z_I(t) \le c] \tag{B.8}$$

satisfies the three-dimensional Fokker-Planck equation:

$$\frac{\partial f}{\partial t} = \pi \Delta \nu \frac{\partial^2 f}{\partial a^2} - \cos a \frac{\partial f}{\partial b} - \sin a \frac{\partial f}{\partial c} \tag{B.9}$$

with initial condition $f(a,b,c;0) = \delta(a)\delta(b)\delta(c)$, δ being the Dirac delta function. The partial differential equation (B.9) was first considered in [3], where approaches for numerical solutions were discussed.

Using a slight modification of the random process $z(t)$, namely,

$$z_m(t) = \int_0^t e^{j[\phi(\tau)-\phi(t)]} d\tau \tag{B.10}$$

we obtain (see Problem B.1) a Markov process with the same statistical characteristics as $z(t)$ that differs from it only in the details of how it evolves with time. Exploiting the fact that the new process is Markov, it is possible to remove the dependence on the process $\phi(t)$ and obtain the following two-dimensional Fokker-Planck equation for the joint pdf $f(r,\theta;t)$ of magnitude and phase of $z_m(t)$:

$$\frac{\partial f}{\partial t} = -\cos\theta \frac{\partial f}{\partial r} + \frac{\sin\theta}{r}\frac{\partial f}{\partial \theta} + \pi\Delta\nu \frac{\partial^2 f}{\partial \theta^2} \tag{B.11}$$

with initial condition $f(r,\theta;0) = \delta(r)\delta(\theta)$.

The Fokker-Planck equation (B.11) has been solved numerically in [4], obtaining two-dimensional plots of the joint pfd of phase and magnitude. Integrating the joint pdf over θ in the interval $(0,2\pi)$ allows one to get the marginal pdf of the magnitude $\rho_z = |z(t)|$. A plot of this pdf, that is, $f_{\rho_z}(r)$ is plotted in Figure B.1 as a function of the normalized amplitude r/t for various values of the parameter $\Delta\nu t$.

Remembering that in the ideal case of absence of phase noise we would have a magnitude $\rho_z(t)$ equal to t (its pdf being a Dirac delta function located at $r/t = 1$), we can see from the figure what are the effects of the phase noise to amplitude conversion. The pdf of the envelope departs from the ideal delta function more and more as long as $\Delta\nu t$ increases, causing increasing penalties in the system sensitivity.

Accurate numerical solutions of the Fokker-Planck equations permit us to obtain the true pdf of the filtered envelope. However, they require sophisticated numerical algorithms and long computer runs and give a solution in numerical form, which, owing to numerical error propagations, is not suited for use in further computations, like the ones required in Section 4.10 to get very low values of the error probability.

In the following sections, we will explain other methods leading to analytical estimates.

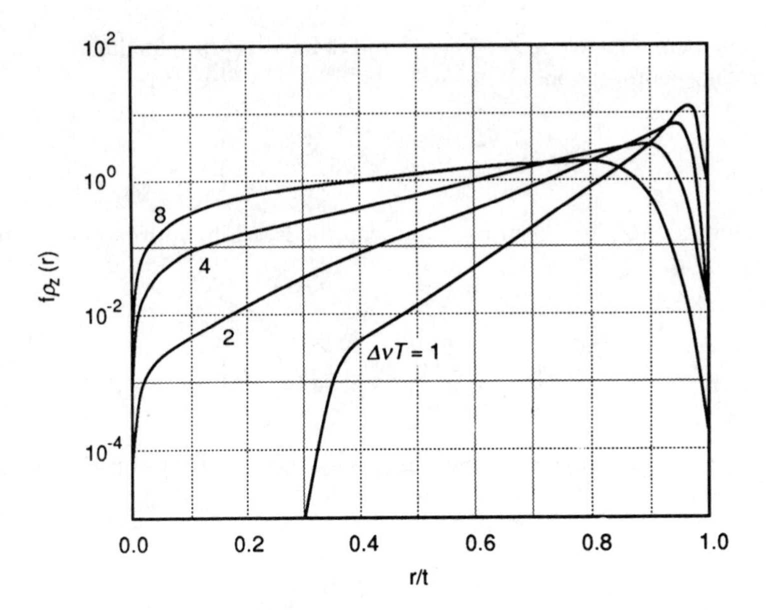

Figure B.1 Probability density function of the magnitude of a filtered carrier affected by phase noise.

B.3 ANALYTICAL APPROXIMATION USING TRUNCATED SERIES

Let us start again from (B.4) and define a new random process as

$$w(t) \stackrel{\text{def}}{=} \frac{1}{t}\, z(t)$$

Using a change of variable in the integral definition of $z(t)$, we can rewrite $w(t)$ in terms of a standard (i.e., one with variance t) Brownian motion, $\psi(t)$, as

$$w(\gamma) = \int_0^1 e^{j\sqrt{\gamma}\psi(\tau)}\, d\tau \tag{B.12}$$

where $\gamma = 2\pi\Delta\nu t$ and $\psi(t)$ is a standard Wiener-Lévy process with zero mean and variance equal to t.

In [5] the exponential on the right side of (B.12) is expressed as a power series:

$$\begin{aligned} w(\gamma) &= \int_0^1 e^{i\sqrt{\gamma}\,\psi(\tau)}\, d\tau \\ &= \int_0^1 \left(1 + j\sqrt{\gamma}\,\psi(\tau) - \frac{\gamma}{2}\psi^2(t) + \cdots + (j\sqrt{\gamma})^n \frac{\psi^n(\tau)}{n!} + \cdots\right) d\tau \end{aligned} \tag{B.13}$$

Taking the squared envelope $Z = |w(\gamma)|^2$ and then truncating the linear term in γ, we get, after easy computations,

$$Z_F = 1 - \gamma A \tag{B.14}$$

where the notation Z_F has been used to mean the Foschini approximation to the RV Z, and where

$$A = \int_0^1 \psi^2(\tau)d\tau - \left(\int_0^1 \psi(\tau)d\tau\right)^2 \tag{B.15}$$

Expanding now $\psi(t)$ in cosine Fourier series in the interval (0,1), substituting the series into (B.15), and integrating term by term yields

$$A = \sum_{i=1}^{\infty} \left(\frac{a_i}{i\pi}\right)^2 \tag{B.16}$$

where a_i are independent Gaussian RVs, with zero mean and unitary variance. Each term a_i^2 in the summation (B.16) is thus a χ-square RV with characteristic function

$$C_{a_i^2}(s) = (1 - 2s)^{-1/2}$$

and thus the characteristic function of A becomes

$$C_A(s) = \prod_{i=1}^{\infty} \left(1 - \frac{2s}{(i\pi)^2}\right)^{-\frac{1}{2}} = \sqrt{\frac{\sqrt{-2s}}{\sinh\sqrt{-2s}}}, \quad \mathrm{Re}(s) < \frac{\pi^2}{2} \tag{B.17}$$

Taking into account (B.14), we obtain

$$C_{Z_F}(s) = \mathrm{E}[\mathrm{e}^{sZ_F}] = \mathrm{E}[\mathrm{e}^{s(1-\gamma A)}] = \mathrm{e}^s \sqrt{\frac{\sqrt{2\gamma s}}{\sinh\sqrt{-2\gamma s}}}, \quad \mathrm{Re}(s) < \frac{\pi^2}{2} \tag{B.18}$$

The characteristic function of Z_F can be used to obtain the pdf in semiclosed form through inverse Laplace transform. Similarly, by taking the Nth power of the characteristic function and then the inverse transform, it is possible to obtain the pdf of the RV W, which is required when the system uses a postdetection filter.

The linear (in γ) approximation Z_F is a lower bound to the phase noisy envelope, and, as a consequence, it leads to an upper bound to the error probability. Comparisons with the true result have shown that the upper bound becomes quite loose for high values of the product SNR times the linewidth.

Using the approximation Z_F for the evaluation of the performance of FSK without postdetection filter (see Problem B.2) leads to

$$P_e = \begin{cases} \dfrac{1}{2} e^{-\rho/2} (\sqrt{\rho\gamma}/\sin\sqrt{\rho\gamma})^{1/2}, & \rho\gamma < \pi^2 \\ \infty, & \rho\gamma \geq \pi^2 \end{cases} \tag{B.19}$$

which fails to give sensible results for high values of $\rho\gamma$, ρ being the IF SNR defined in Section 4.10.

The failure of this approximation is due to the fact that, although Z can take values only in the interval $(0,1)$, Z_F can become negative. In fact, its pdf has a nonzero tail for negative arguments, which becomes increasingly dominant as γ increases.

To overcome this problem, the negative tail of the distribution is truncated in [5]; however, it is not clear how tight the results become with the truncation.

B.4 ANALYTICAL APPROXIMATION USING AN EXPONENTIAL PDF

In [5], observing that the tail of the pdf of the RV Z shows an exponential behavior, the authors also propose an exponential approximation with a fitting parameter, α. Taking into account that the range of Z is $(0,1)$ and normalizing the pdf, we can approximate the pdf of Z as

$$f_{Z_E}(x) = \frac{\alpha}{1 - e^{-\alpha}} e^{-\alpha(1-x)}, \qquad x \in (0, 1) \tag{B.20}$$

This approximation works for different kinds of IF filters. When using an integrate-and-dump filter, as we did in Section 4.10, the fitting value of α becomes

$$\alpha = 2.1306 \frac{2.507 + 0.5\sqrt{\gamma}}{\gamma}$$

The approximation f_{Z_E}, where E stands for "exponential," does not lead to a true upper bound in the evaluation of the error probability. Its main advantages lie in its simplicity and in the fact that it makes possible getting an analytical expression for the pdf of W, in the form (see Problem B.3)

$$f_{W_E}(x) = \left(\frac{\alpha e^{-\alpha}}{1 - e^{-\alpha}}\right)^N e^{\alpha x} \sum_{i=0}^{N} \binom{N}{i} \frac{(-1)^i}{(N-1)!} u(x - i)\, (x - i)^N - 1, \qquad x \in (0, N) \quad \text{(B.21)}$$

where $u(x)$ is the unit step function.

B.5 THE HUMBLET ANALYTICAL APPROXIMATION

Starting from a critical analysis of the diverging nontruncated approximation Z_F, the authors of [6] propose to use an approximate RV Z_H such as

- For all values of γ and all Brownian sample paths $\psi(t)$ Z_H lies in the interval (0,1);
- Z_H and Z match to the first order in γ for all $\psi(t)$, as it was for the approximation Z_F.

It can be proved that both requirements are satisfied by the choice

$$Z_H = e^{-\gamma A} \quad \text{(B.22)}$$

where A is the RV already defined in (B.16). In fact, for $A \in (0, \infty)$, $Z_H \in (0,1)$. Moreover, expressing the exponential (B.22) in series and truncating to the first term yields

$$Z_H = 1 - \gamma A$$

which coincides with Z_F.

Thus, Z_H can be considered a refinement of Z_F. A nice feature of the approximation Z_H is that its moments are easily obtained from the characteristics function of A, as

$$m_k^{(H)} \stackrel{\text{def}}{=} E[Z_H^k] = E[e^{-k\gamma A}] = C_A(-\gamma k) = \sqrt{\frac{\sqrt{2\gamma k}}{\sinh \sqrt{2\gamma k}}} \quad \text{(B.23)}$$

B.6 THE EXACT MOMENTS OF Z

In [7], analytical closed-form expressions for the moments of the RV Z are obtained as follows:

$$m_k \stackrel{\text{def}}{=} \mathrm{E}[Z^k] = (k!)^2 \left(\frac{2t}{\gamma}\right)^{2k+1} q_k^{2k} \left(\frac{\gamma}{2}\right) \tag{B.24}$$

where $q_m^n(t)$ is the inverse Laplace transform of the function $Q_m^n(s)$, obeying to the recursive equation

$$Q_m^n(s) = \frac{1}{s + (2m - n)^2} [Q_m^{n-1}(s) + Q_{m-1}^{n-1}(s)] \tag{B.25}$$

where the first of the two terms in the brackets vanishes if $m = n$, and the second vanishes if $m = 0$. The recursive equation (B.25), together with the initial condition

$$Q_0^0(s) = \frac{1}{s} \tag{B.26}$$

enable us to recursively evaluate the rational functions Q and, through easy inverse Laplace transform, the functions q in (B.24).

This method makes it possible to obtain a closed-form exact expression for all the moments of Z. However, it requires a high computational complexity and extremely accurate computer routines, because the numbers involved soon become very large.

B.7 THE MOMENTS OF THE RV *W*

The RV Z is the one involved in the averages that yield the error probability in the case of no postdetection filtering. When we insert the postdetection filter, which is often the case, the RV needed in the computation of averages is the RV W defined in (B.3), which is obtained as the sum of N independent RVs Z_i. To perform the averages with respect to W using the Gauss quadrature rules approach, we need a certain number of moments $m_k^{(W)}$ of W, which can be obtained in a simple and fast recursive way from the moments of Z_i.

The procedure involves the following steps:

1. Compute the moments of the individual RVs Z_i:

$$m_k^{(i)} \stackrel{\text{def}}{=} \mathrm{E}[Z_i^k]$$

2. Define the partial sums:

$$W_m \stackrel{\text{def}}{=} \sum_{i=1}^{m} Z_i$$

with $W_N \equiv W$, and compute recursively, for each k, the moments

$$\lambda_k^{(m)} \stackrel{\text{def}}{=} \mathrm{E}[W_m^k]$$

through the recursion

$$\lambda_k^{(m+1)} = \mathrm{E}[W_{m+1}^k] = \mathrm{E}[(W_m + Z_{m+1})^k]$$

$$= \sum_{j=0}^{k} \binom{k}{j} \mathrm{E}[W_m^j]\mathrm{E}[Z_{m+1}^{k-j}] = \sum_{j=0}^{k} \binom{k}{j} \lambda_j^{(m)} m_{k-j}^{(m+1)} \tag{B.27}$$

where the independence between the RVs W_m and Z_{m+1} has been exploited.

3. When $m = N$, stop the procedure, because

$$\lambda_k^{(N)} = \mathrm{E}[W_N^k] = \mathrm{E}[W^k]$$

B.8 COMPARISON BETWEEN DIFFERENT ANALYTICAL METHODS

In computing the performance of asynchronous modulation schemes with phase noise, we need to evaluate averages of known functions with respect to the pdf of the RV Z or W. These averages can be performed directly through an analytical expression of the pdf's or, using Gauss quadrature rules, through knowing a certain number of moments of the RVs.

In the results of Section 4.10, we used the method of moments, described in Appendix C. We know two ways of computing the moments of the RVs Z and W. The first one, based on the approximation Z_H, is simpler than the second, which, however, allows us to obtain the exact moments.

To check the accuracy in the simpler moments' evaluation using the approximations Z_H^2, we have computed the exact expressions of the first 10 moments of Z using the exact procedure and the approximated moments m_H. The relative error

[2]The approximation Z_F will not be considered here since it has been proved to be quite inaccurate in its nontruncated versions. Moreover, dealing with Z_H is almost as simple as with Z_F, and always leads to more accurate results.

$\epsilon_k = (m_k^{(H)} - m_k)/m_k$ is reported in Figure B.2 for a value of the normalized linewidth $\Delta\nu t$ in the range (0,0.1) and in Figure B.3 for a value of the normalized linewidth $\Delta\nu t$ up to 4. The relative error is less than $4 \cdot 10^{-3}$ for $\Delta\nu t$ less than 0.1 and less than $2 \cdot 10^{-2}$ for $\Delta\nu t$ less than 1.5.

Through knowledge of the moments m_k or $m_k^{(H)}$, we have applied the Gauss quadrature rules method (described in Appendix C) to perform the averages involved in the error probability computations. The results are reported in Figure B.4, in terms of power penalty in decibels for the case of the matched IF filter, and in Figure B.5 for the case of an optimized IF filter and postdetection filter. In that case, we also report the results obtained through the truncated exponential approximation Z_E.

It can be observed that the case of matched filter is the most critical one, because a value of the normalized linewidth of 0.09 already leads us to underestimate the power penalty of 0.5 dB.

On the other hand, when the postdetection filter with optimized IF bandwidth is used, the approximation Z_H gives very accurate results for values of $\Delta\nu T$ larger than 1, which encompass all practical applications. The reason is that, for large values of $\Delta\nu T$, the optimum IF bandwidth becomes much larger than $1/T$, so that the integration time T' of the IF filter is much smaller than T, and, consequently, the phase noise acts in a reduced time span. As an example, we saw in Figure 4.46 of Section 4.10 that the optimum value of T' for $\Delta\nu T = 1$ is $T/7$. Moreover, in the case of postdetection

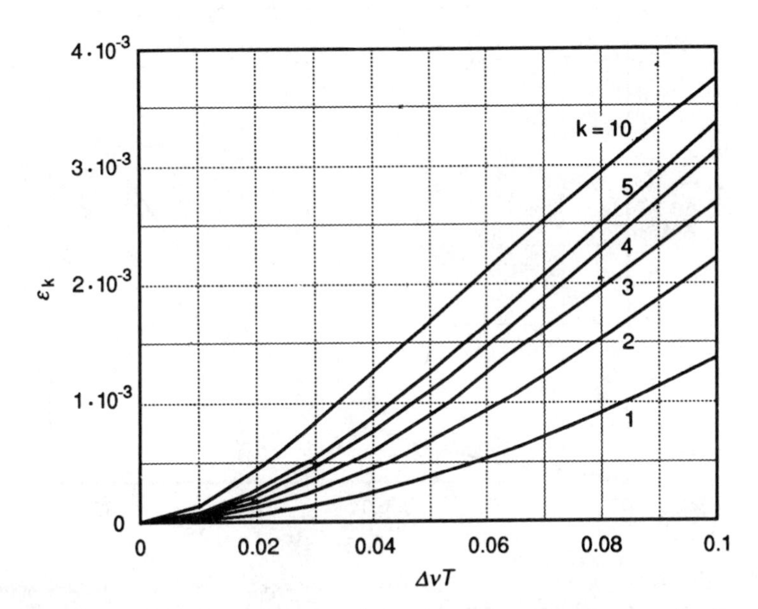

Figure B.2 Relative error in the moments for the Humblet approximation versus the normalized laser linewidth in the range (0,0.1). The parameter k is the order of the moments.

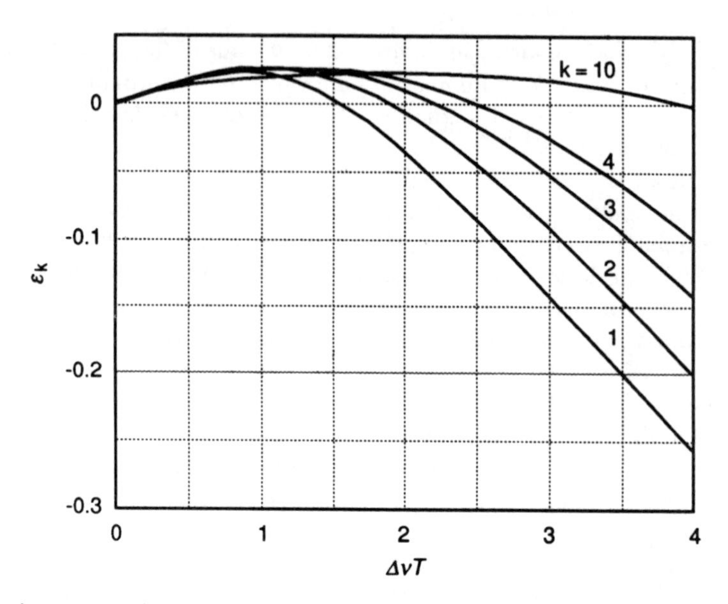

Figure B.3 Relative error in the moments for the Humblet approximation versus the normalized laser linewidth in the range (0,4). The parameter k is the order of the moments.

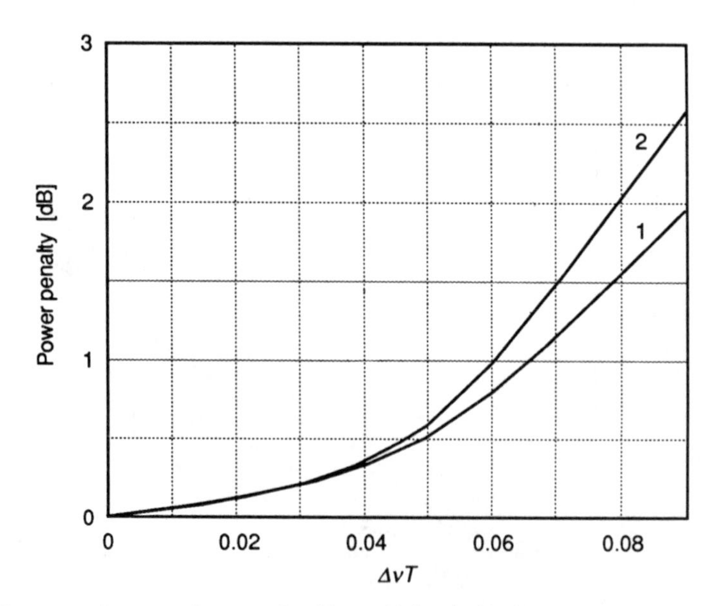

Figure B.4 Power penalty versus the normalized linewidth for double-filter FSK with phase noise, matched IF filter, and no postdetection filter. The two curves refer to the use of exact moments (2) and Humblet approximation (1).

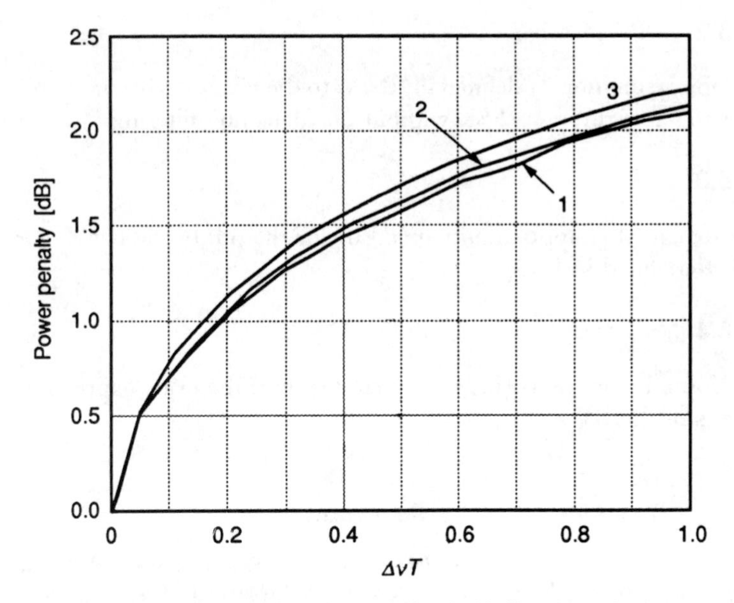

Figure B.5 Power penalty versus the normalized linewidth for double-filter FSK with phase noise, optimized IF filter, and postdetection filter. The three curves refer to the use of exact moments (2), Humblet approximation (1), and exponential fitting (3).

filtering, we must use in the computations the moments of the RV W, which is obtained summing seven independent RVs Z. That fact also contributes to increase the accuracy of the approximation Z_H.

B.9 METHOD USED IN SECTION 4.10

To obtain the results presented in Section 4.10, we used the method of the Gauss quadrature rules (described in Appendix C) in connection with the approximate moments $m_k^{(H)}$ of (B.23) to obtain the moments of the RV W.

Problems

Problem B.1

1. Prove that the random process $\phi(t)$ in (B.1) describing the phase noise is a Markov process.
2. Prove that the random process $z(t)$ defined in (B.4) is *not* a Markov process.
3. Prove that the random process $z_m(t)$ defined in (B.10) is a Markov process.

Problem B.2

Using the approximation Z_F defined in (B.14) to the RV Z, derive the error probability (B.19) for the asynchronous FSK without postdetection filtering.

Problem B.3

Using the exponential approximation (B.20) for the pdf of the RV Z, derive the pdf of the RV W defined in (B.3).

Problem B.4

Using the formal definitions (B.24) and (B.25), find the exact expression of the first 10 moments of the RV Z.

References

[1] Papoulis, A., *Probability, Random Variables, and Stochastic Processes,* New York: McGraw-Hill, 1984.

[2] Kazovsky, L. G., and O. K. Tonguz, "ASK and FSK Coherent Lightwave Systems: A Simplified Approximate Analysis," *Journal of Lightwave Technology,* 8(3):338–352, 1990.

[3] Foschini, G. J., and G. Vannucci, "Characterizing Filtered Light Waves Corrupted by Phase Noise," *IEEE Transactions on Information Theory,* 34:1437–1448, 1988.

[4] Garrett, I., et al., "Impact of Phase Noise in Weakly Coherent Systems: A New and Accurate Approach," *Journal of Lightwave Technology,* 8(3):329–337, 1990.

[5] Foschini, G. J., G. Vannucci, and L. J. Greenstein, "Envelope Statistics for Filtered Optical Signals Corrupted by Phase Noise," *IEEE Transactions on Communications,* 37(12):1293–1302, 1989.

[6] Azizoglu, M., and P. A. Humblet, "Envelope Detection of Orthogonal Signals With Phase Noise," *Journal of Lightwave Technology,* 9(10):1398–1410, 1991.

[7] Pierobon, G. L., and L. Tomba, "Moment Characterization of Phase Noise in Coherent Optical Systems," *Journal of Lightwave Technology,* 9(8):996–1005, 1991.

Appendix C Evaluation of Averages by Quadrature Rules

This appendix describes a technique to approximate the values of averages of functions of an RV Z in the form

$$\mathrm{E}[g(Z)] = \int_{\mathfrak{X}} g(z) f_Z(z)\, dz \simeq \sum_{i=1}^{N} w_i\, g(z_i) \tag{C.1}$$

where $\mathfrak{X}$ is the range of the RV Z, $f_Z(z)$ is the pdf of Z, and $\{w_i\}_{i=1}^N$, $\{z_i\}_{i=1}^N$ represent the *weights* and *abscissas* of the quadrature rule.

By choosing $g(z)$ to be a polynomial, it is said that the quadrature rule (C.1) has *degree of precision m* if it is exact whenever $g(\cdot)$ is a polynomial of degree $\leq m$.

Given an RV Z with range $\mathfrak{X}$ and all of whose moments exist, it is possible to define a sequence of polynomials $P_0(z), P_1(z), \ldots,$ with deg $P_i(z) = i$, orthonormal with respect to Z, that is satisfying

$$\mathrm{E}[P_i(z)P_j(z)] = \delta_{ij}, \qquad i, j = 0, 1, 2, \ldots$$

Denote by $z_1 < z_2 < \ldots < z_N$ the N roots of the polynomial $P_N(z)$ and by k_N the coefficient of z^N in the polynomial $P_N(z)$. By defining

$$w_i \stackrel{\text{def}}{=} -\frac{k_{N+1}}{k_N} \cdot \frac{1}{P_{N+1}(z_i) P'_N(z_i)}, \qquad i = 1, 2, \cdots, N$$

the set $\{z_i, w_i\}_{i=1}^N$ forms a quadrature rule with degree of precision $2N - 1$. This is the highest degree of precision that can be attained by any quadrature rule with N weights and abscissas.[1]

[1] A systematic introduction to the theory of quadrature rules (often called *Gauss* quadrature rules because they were first studied by Gauss) is given in [1].

For a number of pdf's $f_Z(\cdot)$, the weights and abscissas are available in tabular form. That, unfortunately, is not the case for the pdf of the phase noisy envelope. In the general case, however, it is possible to compute the N weights and abscissas of a Gauss quadrature rule with respect to an RV Z from the knowledge of its first $2N$ moments m_k. In fact, since

$$m_k \stackrel{\text{def}}{=} \mathrm{E}[Z^k] = \int_{\mathfrak{X}} z^k f_Z(z) dz$$

and since for any $0 \le k \le 2N - 1$ the quadrature rule is exact, we have

$$m_k = \sum_{i=1}^{N} w_i z_i^k, \qquad k = 0, 1, \cdots, 2N - 1$$

This system of $2N$ nonlinear equations has the weights and abscissas as unknowns; by solving it, the Gauss quadrature rule can be found.

In general, a direct solution of the nonlinear system is not convenient. A computationally effective technique to determine Gauss quadrature rules based on the moments of an RV Z has been proposed by Golub and Welsch [2]. Their technique consists essentially of three steps:

1. Evaluate the coefficients $(\alpha_n)_{n=1}^N$, $(\beta_n)_{n=1}^N$ of the three-term recurrence relationship satisfied by the polynomials $P_0(z), \ldots, P_N(z)$:

   $$\beta_n P_n(z) = (z - \alpha_n) P_{n-1}(z) - \beta_{n-1} P_{n-2}(z), \qquad n = 1, \ldots, N$$

   with $P_{-1}(z) \equiv 0$, $P_0(z) \equiv 1$. This step is based on the knowledge of the moments of the RV Z.
2. Generate a symmetric tridiagonal matrix whose entries depend on the coefficients α, β.
3. Evaluate the weights as the first components of the eigenvectors of the tridiagonal matrix and of the abscissas as the corresponding eigenvalues.

This computational technique has been used successfully in dealing with the problem of evaluating the performance of digital communication schemes in the presence of intersymbol interference (see, for example, [3]).

C.1 CONVERGENCE AND ACCURACY OF GAUSS QUADRATURE RULES

As $N \to \infty$ the right side of (C.1) converges to the left side for almost any conceivable functions we may meet in practice.

As for the truncation error $R_N[g]$, that is, the error involved in approximating the integral of $g(\cdot)$ in the left side of (C.1) using a truncated series with N terms, as in the right side of (C.1), it is equal to

$$R_N[g] = \frac{1}{(2N)!k_N^2} g^{(2N)}(\eta) \tag{C.2}$$

where η is a point in $\mathfrak{X}$ and provided that $g(\cdot)$ has continuous derivatives of order $2N$.

Computing the truncation error through (C.2) is often impractical, because it requires knowledge of the $2N$th derivative of $g(\cdot)$ and an upper bound for η ranging in $\mathfrak{X}$. A most practical rule of thumb for stopping the summation, that is, for choosing N, consists in keeping N increasing until the value of the right side of (C.1) remains stable.

As an example, let us consider the evaluation of the error probability of an FSK asynchronous receiver, as described in Section 4.10. Using the moments of the RVs Z and W evaluated as described in Appendix B, we have computed the error probability as a function of the number of points N in the quadrature rule. The results are plotted in Figures C.1 and C.2 for the case of matched IF filter without postdetection filtering and for the case of optimized IF filter with postdetection filtering, respectively. For

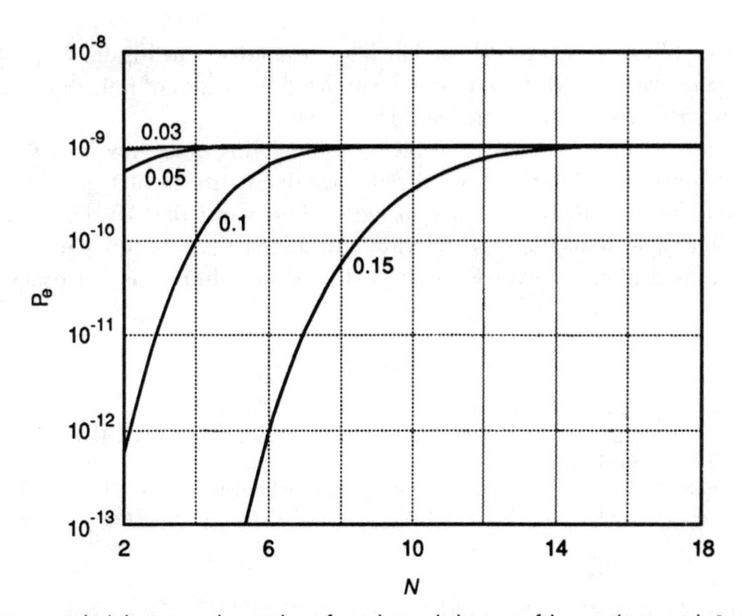

Figure C.1 Error probability versus the number of weights and abscissa of the quadrature rule for four values of the normalized laser linewidth for double-filter FSK, matched IF filter, and no postdetection filter.

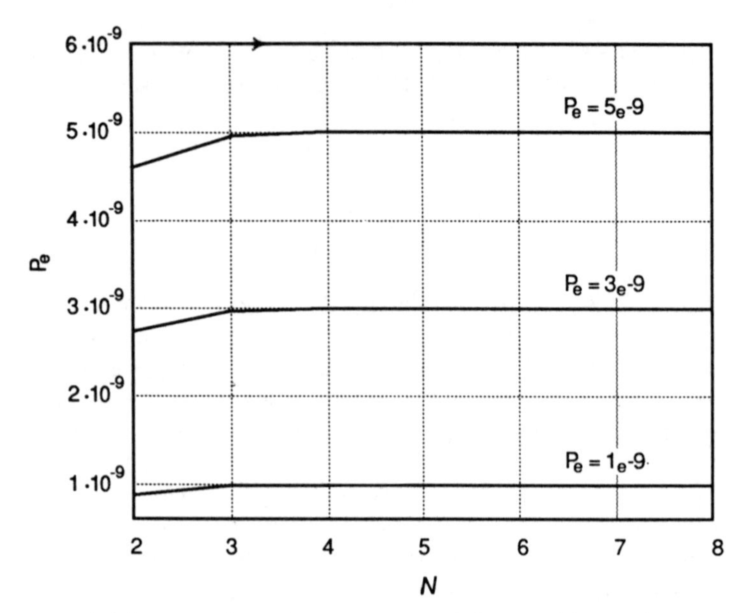

Figure C.2 Error probability versus the number of weights and abscissa of the quadrature rule for a value of the normalized laser linewidth of 0.4 for double-filter FSK, optimized IF filter, and no post-detection filter.

optimized IF filtering and postdetection filter, the curves in Figure C.2, derived for different values of the SNR ρ and for a normalized linewidth of 0.4, show that a value of $N = 4$ is sufficient to guarantee stable results.

The case of matched filter receiver is more critical, as shown by the curves in Figure C.1, parametric in the normalized laser linewidth. In fact, a value of $N = 14$ is required to have stable results for a normalized linewidth of 0.15. That does not pose problems in real systems, because for values of the normalized linewidth greater than 0.1 the matched filter receiver without postdetection filtering is not practical at all.

References

[1] Krylor, V.J. *Approximate Calculation of Integrals*, MacMillan, 1962.

[2] Golub, G. H., and J. H. Welsch, "Calculation of Gauss Quadrature Rules," *Mathematics of Computation*, 23:221–230, 1969.

[3] Benedetto, S., G. De Vincentiis, and A. Luvison, "Error Probability in the Presence of Intersymbol Interference and Additive for Multilevel Digital Signals," *IEEE Transactions on Communications*, 21:181–188, 1973.

List of Symbols

α	Fiber attenuation coefficient (dB/km)
β	Propagation constant (rad/m)
C	Strength of stimulated emission ($m^3/J \cdot s^2$)
c	Speed of light (2.998 · 108m/s)
D	Fiber dispersion coefficient (ps/nm · km)
E	Electric field (V/m)
ε	Dielectric constant (F/m)
F	Noise figure
ϕ	Photon density ($1/m^3$)
f,ν	Frequency of the optical carrier (Hz)
G	Gain of an optical amplifier (dB)
$h\nu$	Photon energy (J)
H	Magnetic field (A/m)
k	Propagation constant (1/m)
L	Length of the optical fiber (km)
L_{pert}	distance between adjacent amplifiers (km)
L_{coll}	collision length (km)
λ	Wavelength of the optical carrier (μm)
λ_0	Zero-dispersion wavelength (μm)
$\Delta\lambda$	Channel spacing (nm)
μ	Magnetic constant (H/m)
NA	Numerical aperture
n	Refractive index
n	Electron density ($1/m^3$)
n_{sp}	Excess spontaneous emission factor
η	Quantum efficiency
ω	Angular frequency (rad/sec)
P	Optical power (mW)
q	Electron charge (C)
R	Responsivity (A/W)

R_b	Information transmission rate (bits/sec)
θ_C	Critical angle (degree)
T	Absolute temperature (degree); bit duration (sec)
τ	Pulse width of a soliton (ps)
t_{coll}	collision time (ps)
τ_{ph}	Photon lifetime in a laser cavity (sec)
τ_r	Recombination time (sec)
τ_{sp}	Spontaneous emission lifetime (sec)
V	Normalized cut-off frequency
v_g	Group velocity (m/s)
W_e	Soliton energy
z_0	Spatial length of a soliton (km)

About the Authors

Dr. Leonid G. Kazovsky (M-80, SM-83, F-90), a professor of electrical engineering at Stanford University, was born in Leningrad, USSR, in 1947. He received his M.Sc. (summa cum laude) and Ph.D degrees from the Leningrad Electrotechnical Institute of Communications, Leningrad, U.S.S.R., in 1969 and 1972, respectively, both in electrical engineering.

He moved to Israel in 1973. From 1974 to 1984 (with a one-year interruption for active military service), Dr. Kazovsky taught and carried out research at Israeli and U.S. universities. From 1984 to 1990, he was with Bellcore in Red Bank, New Jersey, doing research on coherent and high-speed WDM optical fiber communication systems. In 1990, Dr. Kazovsky joined Stanford University as a professor of electrical engineering.

Dr. Kazovsky has published in the areas of optical communications, high-speed networks, applied optics, and signal processing. His current research interests are in the area of WDM networks, fiber nonlinearities, and analog optical systems. Professor Kazovsky is the author or co-author of roughly 120 journal technical papers, numerous conference papers, and two books published by Wiley. Dr. Kazovsky has acted as a reviewer for various IEEE and IEE transactions, proceedings, and journals, as well as for funding agencies (e.g., the National Science Foundation and the Energy Research Council) and publishers (e.g., John Wiley & Sons and Macmillan). He serves or has served on technical program committees of OFC, CLEO, SPIE, and GLOBECOM and is an associate editor of *IEEE Transactions on Communications, IEEE Photonics Technology Letters, and Wireless Networks.* Dr. Kazovsky is a fellow of the IEEE and a fellow of the OSA.

Sergio Benedetto received the "Laurea in Ingegneria Elettronica" (summa cum laude) from Politecnico di Torino, Italy, in 1969. From 1970 to 1979 he was with the Istituto di Elettronica e Telecomunicazioni, first as a research engineer, then as an associate professor. In 1980, he was made a professor in radio communications at the Università di Bari. In 1981 he rejoined Politecnico di Torino as professor of data transmission theory in the Dipartimento di Elettronica. He spent nine months in

1980–1981 at the System Science Department of the University of California, Los Angeles, as a visiting professor and three months at the University of Canterbury, New Zealand, as an Erskine Fellow.

Mr. Benedetto has coauthored two books in Italian on signal theory and probability and random variables and the book *Digital Transmission Theory* (Prentice-Hall, 1987), as well as over 200 papers for leading engineering journals and conferences. He is area editor for *Signal Design, Modulation and Detection* for the *IEEE Transactions on Communications.*

Active in the field of digital transmission systems since 1970, his current interests are in the field of performance evaluation and simulation of digital communication systems, trellis-coded modulation and concatenated coding schemes, and optical fiber communications systems.

Mr. Benedetto is a senior member of the IEEE.

Alan E. Willner received his B.A. (1982) in physics from Yeshiva University and his M.S. (1984) and Ph.D. (1988) degrees in electrical engineering from Columbia University. He was a postdoctoral member of the technical staff at AT&T Bell Laboratories (Crawford Hill Laboratory) and a member of technical staff at Bellcore. He is currently an associate professor in the Dept. of Electrical Engineering-Systems at the University of Southern California. He is also the associate director of the USC Center for Photonic Technology and an associate director for student affairs of the USC NSF Engineering Research Center in Integrated Media Systems.

Professor Willner has received the NSF-Sponsored Presidential Faculty Fellows Award from the White House, the David and Lucile Packard Foundation Fellowship in Science and Engineering, the NSF Young Investigator Award, the USC/Northrop Outstanding Junior Engineering Faculty Research Award, the USC School of Engineering Outstanding Teacher Award, and the Armstrong Foundation Memorial Prize for the highest-ranked EE graduate student at Columbia University. He was also an NSF Alan T. Waterman Award Finalist. He is a fellow of the Semiconductor Research Corporation and an IEEE Senior Member.

Alan Willner is the IEEE Lasers and Electro-Optics Society (LEOS) vice-president for Technical Affairs and chair of the LEOS Optical Networks Technical Committee and was chair of the LEOS Optical Communications Technical Committee. He is also vice-chair of the Optical Communications Group in the Technical Council of the Optical Society of America (OSA). He was general co-chair of the IEEE LEOS '95 Summer Topical Meeting on Technologies for a Global Information Infrastructure. He is a program committee member for: the '96 and '97 Conference on Optical Fiber Communications, the '96 Topical Meeting on Optical Amplifiers, the '97 Conference on Lasers and Electro-Optics, and the LEOS '96 Summer Topical on Broadband Optical Networks. He is an awards committee member for the IEEE William Streifer Achievement Award. He is an associate editor of the *IEEE/OSA Journal of Lightwave Technology* (JLT) and a guest editor for a *JLT/JSAC Special Issue on*

Multiple-Wavelength Technologies and Networks. He was the chair of Information and Communications at the NSF Forum on Optical Science & Engineering ('94; '95). He is listed in Marquis' *Who's Who in America.*

Prof. Willner's research is in high-capacity optical communications. He is working on wavelength-division multiplexing, optical amplification, all-optical WDM networks, and WDM optical interconnections. He has more than 150 publications, including 32 invited talks, three patents, one book, and two book chapters.

Index

The Artech House Optoelectronics Library

Brian Culshaw and Alan Rogers, Series Editors

Amorphous and Microcrystalline Semiconductor Devices, Volume II: Materials and Device Physics, Jerzy Kanicki, editor

Bistabilities and Nonlinearities in Laser Diodes, Hitoshi Kawaguchi

Chemical and Biochemical Sensing With Optical Fibers and Waveguides, Gilbert Boisdé and Alan Harmer

Coherent and Nonlinear Lightwave Communications, Milorad Cvijetic

Coherent Lightwave Communication Systems, Shiro Ryu

Elliptical Fiber Waveguides, R. B. Dyott

Field Theory of Acousto-Optic Signal Processing Devices, Craig Scott

Frequency Stabilization of Semiconductor Laser Diodes, Tetsuhiko Ikegami, Shoichi Sudo, Yoshihisa Sakai

Fundamentals of Multiaccess Optical Fiber Networks, Denis J. G. Mestdagh

Germanate Glasses: Structure, Spectroscopy, and Properties, Alfred Margaryan and Michael A. Piliavin

Handbook of Distributed Feedback Laser Diodes, Geert Morthier and Patrick Vankwikelberge

High-Power Optically Activated Solid-State Switches, Arye Rosen and Fred Zutavern, editors

Highly Coherent Semiconductor Lasers, Motoichi Ohtsu

Iddq Testing for CMOS VLSI, Rochit Rajsuman

Integrated Optics: Design and Modeling, Reinhard März

Introduction to Lightwave Communication Systems, Rajappa Papannareddy

Introduction to Glass Integrated Optics, S. Iraj Najafi

Introduction to Radiometry and Photometry, William Ross McCluney

Introduction to Semiconductor Integrated Optics, Hans P. Zappe

Laser Communications in Space, Stephen G. Lambert and William L. Casey

Optical Document Security, Second Edition, Rudolf L. van Renesse, editor

Optical FDM Network Technologies, Kiyoshi Nosu

Optical Fiber Amplifiers: Design and System Applications, Anders Bjarklev

Optical Fiber Amplifiers: Materials, Devices, and Applications, Shoichi Sudo, editor

Optical Fiber Communication Systems, Leonid Kazovsky, Sergio Benedetto, Alan Willner

Optical Fiber Sensors, Volume Two: Systems and Applicatons, John Dakin and Brian Culshaw, editors

Optical Fiber Sensors, Volume Three: Components and Subsystems, John Dakin and Brian Culshaw, editors

Optical Fiber Sensors, Volume Four: Applications, Analysis, and Future Trends, John Dakin and Brian Culshaw, editors

Optical Interconnection: Foundations and Applications, Christopher Tocci and H. John Caulfield

Optical Measurement Techniques and Applications, Pramod Rastogi

Optical Network Theory, Yitzhak Weissman

Optoelectronic Techniques for Microwave and Millimeter-Wave Engineering, William M. Robertson

Reliability and Degradation of LEDs and Semiconductor Lasers, Mitsuo Fukuda

Reliability and Degradation of III-V Optical Devices, Osamu Ueda

Semiconductor Raman Laser, Ken Suto and Jun-ichi Nishizawa

Semiconductors for Solar Cells, Hans Joachim Möller

Smart Structures and Materials, Brian Culshaw

Ultrafast Diode Lasers: Fundamentals and Applications, Peter Vasil'ev

For further information on these and other Artech House titles, including previously considered out-of-print books now available through our In-Print-Forever™ (IPF™) program, contact:

Artech House
685 Canton Street
Norwood, MA 02062
781-769-9750
Fax: 781-769-6334
Telex: 951-659
email: artech@artech-house.com

Artech House
Portland House, Stag Place
London SW1E 5XA England
+44 (0) 171-973-8077
Fax: +44 (0) 171-630-0166
Telex: 951-659
email: artech-uk@artech-house.com

Find us on the World Wide Web at:
www.artech-house.com